Computational Statistics in Data Science

Computational Statistics in Data Science

Edited by

Walter W. Piegorsch
University of Arizona

Richard A. Levine
San Diego State University

Hao Helen Zhang
University of Arizona

Thomas C. M. Lee
University of California–Davis

WILEY

This edition first published 2022

Registered Office(s)
John Wiley & Sons, Inc., 111 River Street, Hoboken, NJ 07030, USA
John Wiley & Sons Ltd, The Atrium, Southern Gate, Chichester, West Sussex, PO19 8SQ, UK

Editorial Office
9600 Garsington Road, Oxford, OX4 2DQ, UK

For details of our global editorial offices, customer services, and more information about Wiley products visit us at www.wiley.com.

Wiley also publishes its books in a variety of electronic formats and by print-on-demand. Some content that appears in standard print versions of this book may not be available in other formats.

Library of Congress Cataloging-in-Publication Data

ISBN 9781119561071 (hardback)

Cover Design: Wiley
Cover Image: © goja1/Shutterstock

Set in 9.5/12.5pt STIXTwoText by Straive, Chennai, India

10 9 8 7 6 5 4 3 2 1

Contents

List of Contributors

Ayodele Adebiyi
Landmark University
Omu-Aran, Kwara
Nigeria

Anirban Bhattacharya
Texas A&M University
College Station, TX
USA

Peter Calhoun
Jaeb Center for Health Research
Tampa, FL
USA

Wu Changye
Université Paris Dauphine PSL
Paris
France

Xueying Chen
Novartis Pharmaceuticals Corp.
East Hanover, NJ
USA

Jerry Q. Cheng
New York Institute of Technology
New York, NY
USA

Hugh Chipman
Acadia University
Wolfville, Nova Scotia
Canada

Olawande Daramola
Cape Peninsula University of Technology
Cape Town
South Africa

Katarina Domijan
Maynooth University
Maynooth
Ireland

Víctor Elvira
School of Mathematics
University of Edinburgh, Edinburgh
UK

Juanjuan Fan
Department of Mathematics and Statistics
San Diego State University
San Diego, CA
USA

James M. Flegal
University of California
Riverside, CA
USA

Marc G. Genton
King Abdullah University of Science and Technology
Thuwal
Saudi Arabia

Edward George
The Wharton School
University of Pennsylvania
Philadelphia, PA
USA

Robert B. Gramacy
Virginia Polytechnic Institute and State University
Blacksburg, VA
USA

Richard Hahn
The School of Mathematical and Statistical Sciences
Arizona State University
Tempe, AZ
USA

Botao Hao
DeepMind
London
UK

Andrew J. Holbrook
University of California
Los Angeles, CA
USA

Mingyi Hong
University of Minnesota
Minneapolis, MN
USA

Cho-Jui Hsieh
University of California
Los Angeles, CA
USA

Jessica Hullman
Northwestern University
Evanston, IL
USA

David R. Hunter
Penn State University
State College, PA
USA

Catherine B. Hurley
Maynooth University
Maynooth
Ireland

Xiang Ji
Tulane University
New Orleans, LA
USA

Adam M. Johansen
University of Warwick
Coventry
UK

James E. Johndrow
University of Pennsylvania
Philadelphia, PA
USA

Galin L. Jones
University of Minnesota
Twin-Cities Minneapolis, MN
USA

Seung Jun Shin
Korea University
Seoul
South Korea

Matthew Kay
Northwestern University
Evanston, IL
USA

Alexander D. Knudson
The University of Nevada
Reno, NV
USA

Taiwo Kolajo
Federal University Lokoja
Lokoja
Nigeria

and

Covenant University
Ota
Nigeria

Alfonso Landeros
University of California
Los Angeles, CA
USA

Kenneth Lange
University of California
Los Angeles, CA
USA

Thomas C.M. Lee
University of California at Davis
Davis, CA
USA

Richard A. Levine
Department of Mathematics and Statistics
San Diego State University
San Diego, CA
USA

Hongzhe Li
University of Pennsylvania
Philadelphia, PA
USA

Jia Li
The Pennsylvania State University
University Park, PA
USA

Lexin Li
University of California
Berkeley, CA
USA

Yao Li
University of North Carolina at Chapel Hill
Chapel Hill, NC
USA

Yufeng Liu
University of North Carolina at Chapel Hill
Chapel Hill, NC
USA

Rong Ma
University of Pennsylvania
Philadelphia, PA
USA

Shiqian Ma
University of California
Davis, CA
USA

Luca Martino
Universidad Rey Juan Carlos de Madrid
Madrid
Spain

Robert McCulloch
The School of Mathematical
and Statistical Sciences
Arizona State University
Tempe, AZ
USA

Weibin Mo
University of North Carolina at Chapel Hill
Chapel Hill, NC
USA

Edward Mulrow
NORC at the University of Chicago
Chicago, IL
USA

Akihiko Nishimura
Johns Hopkins University
Baltimore, MD
USA

Lace Padilla
University of California
Merced, CA
USA

Vincent A. Pisztora
The Pennsylvania State University
University Park, PA
USA

Matthew Pratola
The Ohio State University
Columbus, OH
USA

Christian P. Robert
Université Paris Dauphine PSL
Paris
France

and

University of Warwick
Coventry
UK

Alfred G. Schissler
The University of Nevada
Reno, NV
USA

Rodney Sparapani
Institute for Health and Equity
Medical College of Wisconsin
Milwaukee, WI
USA

Kelly M. Spoon
Computational Science Research Center
San Diego State University
San Diego, CA
USA

Xiaogang Su
Department of Mathematical Sciences
University of Texas
El Paso, TX
USA

Marc A. Suchard
University of California
Los Angeles, CA
USA

Ying Sun
King Abdullah University of Science and Technology
Thuwal
Saudi Arabia

Nola du Toit
NORC at the University of Chicago
Chicago, IL
USA

Dootika Vats
Indian Institute of Technology Kanpur
Kanpur
India

Matti Vihola
University of Jyväskylä
Jyväskylä
Finland

Justin Wang
University of California at Davis
Davis, CA
USA

Will Wei Sun
Purdue University
West Lafayette, IN
USA

Leland Wilkinson
H_2O.ai, Mountain View
California
USA

and

University of Illinois at Chicago
Chicago, IL
USA

Joong-Ho Won
Seoul National University
Seoul
South Korea

Yichao Wu
University of Illinois at Chicago
Chicago, IL
USA

Min-ge Xie
Rutgers University
Piscataway, NJ
USA

Ming Yan
Michigan State University
East Lansing, MI
USA

Yuling Yao
Columbia University
New York, NY
USA

and

Center for Computational Mathematics
Flatiron Institute
New York, NY
USA

Chun Yip Yau
Chinese University of Hong Kong
Shatin
Hong Kong

Hao H. Zhang
University of Arizona
Tucson, AZ
USA

Hua Zhou
University of California
Los Angeles, CA
USA

Preface

Computational statistics is a core area of modern statistical science and its connections to data science represent an ever-growing area of study. One of its important features is that the underlying technology changes quite rapidly, riding on the back of advances in computer hardware and statistical software. In this compendium we present a series of expositions that explore the intermediate and advanced concepts, theories, techniques, and practices that act to expand this rapidly evolving field. We hope that scholars and investigators will use the presentations to inform themselves on how modern computational and statistical technologies are applied, and also to build springboards that can develop their further research. Readers will require knowledge of fundamental statistical methods and, depending on the topic of interest they peruse, any advanced statistical aspects necessary to understand and conduct the technical computing procedures.

The presentation begins with a thoughtful introduction on how we should view **Computational Statistics & Data Science in the 21st Century** (Holbrook, et al.), followed by a careful tour of contemporary **Statistical Software** (Schissler, et al.). Topics that follow address a variety of issues, collected into broad topic areas such as Simulation-based Methods, Statistical Learning, Quantitative Visualization, High-performance Computing, High-dimensional Data Analysis, and Numerical Approximations & Optimization.

Internet access to all of the articles presented here is available via the online collection *Wiley StatsRef: Statistics Reference Online* (Davidian, et al., 2014–2021); see https://onlinelibrary.wiley.com/doi/book/10.1002/9781118445112.

From **Deep Learning** (Li, et al.) to **Asynchronous Parallel Computing** (Yan), this collection provides a glimpse into how computational statistics may progress in this age of big data and transdisciplinary data science. It is our fervent hope that readers will benefit from it.

We wish to thank the fine efforts of the Wiley editorial staff, including Kimberly Monroe-Hill, Paul Sayer, Michael New, Vignesh Lakshmikanthan, Aruna Pragasam, Viktoria Hartl-Vida, Alison Oliver, and Layla Harden in helping bring this project to fruition.

Tucson, Arizona	*Walter W. Piegorsch*
San Diego, California	*Richard A. Levine*
Tucson, Arizona	*Hao Helen Zhang*
Davis, California	*Thomas C. M. Lee*

Reference

Davidian, M., Kenett, R.S., Longford, N.T., Molenberghs, G., Piegorsch, W.W., and Ruggeri, F., eds. (2014–2021). *Wiley StatsRef: Statistics Reference Online*. Chichester: John Wiley & Sons. doi:10.1002/9781118445112.

Part I

Computational Statistics and Data Science

1

Computational Statistics and Data Science in the Twenty-First Century

Andrew J. Holbrook[1], Akihiko Nishimura[2], Xiang Ji[3], and Marc A. Suchard[1]

[1] *University of California, Los Angeles, CA, USA*
[2] *Johns Hopkins University, Baltimore, MD, USA*
[3] *Tulane University, New Orleans, LA, USA*

1 Introduction

We are in the midst of the data science revolution. In October 2012, the Harvard Business Review famously declared data scientist the sexiest job of the twenty-first century [1]. By September 2019, Google searches for the term "data science" had multiplied over sevenfold [2], one multiplicative increase for each intervening year. In the United States between the years 2000 and 2018, the number of bachelor's degrees awarded in either statistics or biostatistics increased over 10-fold (382–3964), and the number of doctoral degrees almost tripled (249–688) [3]. In 2020, seemingly every major university has established or is establishing its own data science institute, center, or initiative.

Data science [4, 5] combines multiple preexisting disciplines (e.g., statistics, machine learning, and computer science) with a redirected focus on creating, understanding, and systematizing workflows that turn real-world data into actionable conclusions. The ubiquity of data in all economic sectors and scientific disciplines makes data science eminently relevant to cohorts of researchers for whom the discipline of statistics was previously closed off and esoteric. Data science's emphasis on practical application only enhances the importance of *computational statistics*, the interface between statistics and computer science primarily concerned with the development of algorithms producing either statistical inference[1] or predictions. Since both of these products comprise essential tasks in any data scientific workflow, we believe that the pan-disciplinary nature of data science only increases the number of opportunities for computational statistics to evolve by taking on new applications[2] and serving the needs of new groups of researchers.

This is the natural role for a discipline that has increased the breadth of statistical application from the beginning. First put forward by R.A. Fisher in 1936 [6, 7], the permutation test allows the scientist (who owns a computer) to test hypotheses about a broader swath of functionals of a target population while making fewer statistical assumptions [8].

Computational Statistics in Data Science.
Edited by Walter W. Piegorsch, Richard A. Levine, Hao Helen Zhang and Thomas C. M. Lee.
 ISBN 978-1-11956107-1

With a computer, the scientist uses the bootstrap [9, 10] to obtain confidence intervals for population functionals and parameters of models too complex for analytic methods. Newton–Raphson optimization and the Fisher scoring algorithm facilitate linear regression for binary, count, and categorical outcomes [11, 12]. More recently, Markov chain Monte Carlo (MCMC) [13, 14] has made Bayesian inference practical for massive, hierarchical, and highly structured models that are useful for the analysis of a significantly wider range of scientific phenomena.

While computational statistics increases the diversity of statistical applications historically, certain central difficulties exist and will continue to remain for the rest of the twenty-first century. In Section 2, we present the first class of Core Challenges, or challenges that are easily quantifiable for generic tasks. Core Challenge 1 is Big N, or statistical inference when the number "N" of observations or data points is large; Core Challenge 2 is Big P, or statistical inference when the model parameter count "P" is large; and Core Challenge 3 is Big M, or statistical inference when the model's objective or density function is multimodal (having many modes "M")[3]. When large, each of these quantities brings its own unique computational difficulty. Since well over 2.5 exabytes (or 2.5×10^{18} bytes) of data come into existence each day [15], we are confident that Core Challenge 1 will survive well into the twenty-second century.

But Core Challenges 2 and 3 will also endure: data complexity often increases with size, and researchers strive to understand increasingly complex phenomena. Because many examples of big data become "big" by combining heterogeneous sources, big data often necessitate big models. With the help of two recent examples, Section 3 illustrates how computational statisticians make headway at the intersection of big data and big models with model-specific advances. In Section 3.1, we present recent work in Bayesian inference for big N and big P regression. Beyond the simplified regression setting, data often come with structures (e.g., spatial, temporal, and network), and correct inference must take these structures into account. For this reason, we present novel computational methods for a highly structured and hierarchical model for the analysis of multistructured and epidemiological data in Section 3.2.

The growth of model complexity leads to new inferential challenges. While we define Core Challenges 1–3 in terms of generic target distributions or objective functions, Core Challenge 4 arises from inherent difficulties in treating complex models generically. Core Challenge 4 (Section 4.1) describes the difficulties and trade-offs that must be overcome to create fast, flexible, and friendly "algo-ware". This Core Challenge requires the development of statistical algorithms that maintain efficiency despite model structure and, thus, apply to a wider swath of target distributions or objective functions "out of the box". Such generic algorithms typically require little cleverness or creativity to implement, limiting the amount of time data scientists must spend worrying about computational details. Moreover, they aid the development of flexible statistical software that adapts to complex model structure in a way that users easily understand. But it is not enough that software be flexible and easy to use: mapping computations to computer hardware for optimal implementations remains difficult. In Section 4.2, we argue that Core Challenge 5, effective use of computational resources such as central processing units (CPU), graphics processing units (GPU), and quantum computers, will become increasingly central to the work of the computational statistician as data grow in magnitude.

2 Core Challenges 1–3

Before providing two recent examples of twenty-first century computational statistics (Section 3), we present three easily quantified Core Challenges within computational statistics that we believe will always exist: big N, or inference from many observations; big P, or inference with high-dimensional models; and big M, or inference with nonconvex objective – or multimodal density – functions. In twenty-first century computational statistics, these challenges often co-occur, but we consider them separately in this section.

2.1 Big *N*

Having a large number of observations makes different computational methods difficult in different ways. A worst case scenario, the *exact* permutation test requires the production of $N!$ datasets. Cheaper alternatives, resampling methods such as the Monte Carlo permutation test or the bootstrap, may require anywhere from thousands to hundreds of thousands of randomly produced datasets [8, 10]. When, say, population means are of interest, each Monte Carlo iteration requires summations involving N expensive memory accesses. Another example of a computationally intensive model is Gaussian process regression [16, 17]; it is a popular nonparametric approach, but the exact method for fitting the model and predicting future values requires matrix inversions that scale $\mathcal{O}(N^3)$. As the rest of the calculations require relatively negligible computational effort, we say that matrix inversions represent the *computational bottleneck* for Gaussian process regression.

To speed up a computationally intensive method, one only needs to speed up the method's computational bottleneck. We are interested in performing Bayesian inference [18] based on a large vector of observations $\mathbf{x} = (x_1, \ldots, x_N)$. We specify our model for the data with a likelihood function $\pi(\mathbf{x}|\boldsymbol{\theta}) = \prod_{n=1}^{N} \pi(x_n|\boldsymbol{\theta})$ and use a prior distribution with density function $\pi(\boldsymbol{\theta})$ to characterize our belief about the value of the P-dimensional parameter vector $\boldsymbol{\theta}$ *a priori*. The target of Bayesian inference is the posterior distribution of $\boldsymbol{\theta}$ conditioned on $\mathbf{x}$

$$\pi(\boldsymbol{\theta}|\mathbf{x}) = \pi(\mathbf{x}|\boldsymbol{\theta})\pi(\boldsymbol{\theta}) \bigg/ \int \pi(\mathbf{x}|\boldsymbol{\theta})\pi(\boldsymbol{\theta})\, d\boldsymbol{\theta} \tag{1}$$

The denominator's multidimensional integral quickly becomes impractical as P grows large, so we choose to use the MetropolisHastings (M–H) algorithm to generate a Markov chain with stationary distribution $\pi(\boldsymbol{\theta}|\mathbf{x})$ [13, 19, 20]. We begin at an arbitrary position $\boldsymbol{\theta}^{(0)}$ and, for each iteration $s = 0, \ldots, S$, randomly generate the proposal state $\boldsymbol{\theta}^*$ from the transition distribution with density $q(\boldsymbol{\theta}^*|\boldsymbol{\theta}^{(s)})$. We then accept proposal state $\boldsymbol{\theta}^*$ with probability

$$a = \min\left(1, \frac{\pi(\boldsymbol{\theta}^*|\mathbf{x})q(\boldsymbol{\theta}^{(s)}|\boldsymbol{\theta}^*)}{\pi(\boldsymbol{\theta}^{(s)}|\mathbf{x})q(\boldsymbol{\theta}^*|\boldsymbol{\theta}^{(s)})}\right) \tag{2}$$

The ratio on the right no longer depends on the denominator in Equation (1), but one must still compute the likelihood and its N terms $\pi(x_n|\boldsymbol{\theta}^*)$.

It is for this reason that likelihood evaluations are often the computational bottleneck for Bayesian inference. In the best case, these evaluations are $\mathcal{O}(N)$, but there are many situations in which they scale $\mathcal{O}(N^2)$ [21, 22] or worse. Indeed, when P is large, it is often advantageous to use more advanced MCMC algorithms that use the gradient of the log-posterior to generate better proposals. In this situation, the log-likelihood gradient may also become a computational bottleneck [21].

2.2 Big *P*

One of the simplest models for big P problems is ridge regression [23], but computing can become expensive even in this classical setting. Ridge regression estimates the coefficient θ by minimizing the distance between the observed and predicted values $\mathbf{y}$ and $\mathbf{X}\theta$ along with a weighted square norm of θ:

$$\hat{\theta} = \text{argmin}\{\,||\mathbf{y} - \mathbf{X}\theta||^2 + ||\mathbf{\Phi}^{1/2}\theta||^2\} = (\mathbf{X}^\intercal\mathbf{X} + \mathbf{\Phi})^{-1}\mathbf{X}^\intercal\mathbf{y}$$

For illustrative purposes, we consider the following direct method for computing $\hat{\theta}$.[4] We can first multiply the $N \times P$ design matrix $\mathbf{X}$ by its transpose at the cost of $\mathcal{O}(N^2P)$ and subsequently invert the $P \times P$ matrix $(\mathbf{X}^\intercal\mathbf{X} + \mathbf{\Phi})$ at the cost of $\mathcal{O}(P^3)$. The total $\mathcal{O}(N^2P + P^3)$ complexity shows that (i) a large number of parameters is often sufficient for making even the simplest of tasks infeasible and (ii) a moderate number of parameters can render a task impractical when there are a large number of observations. These two insights extend to more complicated models: the same complexity analysis holds for the fitting of generalized linear models (GLMs) as described in McCullagh and Nelder [12].

In the context of Bayesian inference, the length P of the vector $\boldsymbol{\theta}$ dictates the dimension of the MCMC state space. For the M-H algorithm (Section 2.1) with P-dimensional Gaussian target and proposal, Gelman *et al.* [25] show that the proposal distribution's covariance should be scaled by a factor inversely proportional to P. Hence, as the dimension of the state space grows, it behooves one to propose states $\boldsymbol{\theta}^*$ that are closer to the current state of the Markov chain, and one must greatly increase the number S of MCMC iterations. At the same time, an increasing P often slows down rate-limiting likelihood calculations (Section 2.1). Taken together, one must generate many more, much slower MCMC iterations. The wide applicability of latent variable models [26] (Sections 3.1 and 3.2) for which each observation has its own parameter set (e.g., $P \propto N$) means M-H simply does not work for a huge class of models popular with practitioners.

For these reasons, Hamiltonian Monte Carlo (HMC) [27] has become a popular algorithm for fitting Bayesian models with large numbers of parameters. Like M-H, HMC uses an accept step (Equation 2). Unlike M-H, HMC takes advantage of additional information about the target distribution in the form of the log-posterior gradient. HMC works by doubling the state space dimension with an auxiliary Gaussian "momentum" variable $\mathbf{p} \sim \text{Normal}_P(\mathbf{0}, \mathbf{M})$ independent to the "position" variable $\boldsymbol{\theta}$. The constructed Hamiltonian system has energy function given by the negative logarithm of the joint distribution

$$H(\theta, \mathbf{p}) \propto -\log(\pi(\theta|\mathbf{X}) \times \exp(-\mathbf{p}^T\mathbf{M}^{-1}\mathbf{p}/2)) \propto -\log \pi(\theta|\mathbf{X}) + \mathbf{p}^T\mathbf{M}^{-1}\mathbf{p}/2$$

and we produce proposals by simulating the system according to Hamilton's equations

$$\dot{\theta} = \frac{\partial}{\partial \mathbf{p}} H(\theta, \mathbf{p}) = M^{-1}\mathbf{p}/2$$
$$\dot{\mathbf{p}} = -\frac{\partial}{\partial \theta} H(\theta, \mathbf{p}) = \nabla \log \pi(\theta|\mathbf{X})$$

Thus, the momentum of the system moves in the direction of the steepest ascent for the log-posterior, forming an analogy with first-order optimization. The cost is repeated gradient evaluations that may comprise a new computational bottleneck, but the result is effective MCMC for tens of thousands of parameters [21, 28]. The success of HMC has inspired research into other methods leveraging gradient information to generate better MCMC proposals when P is large [29].

2.3 Big *M*

Global optimization, or the problem of finding the minimum of a function with arbitrarily many local minima, is NP-complete in general [30], meaning – in layman's terms – it is impossibly hard. In the absence of a tractable theory, by which one might prove one's global optimization procedure works, brute-force grid and random searches and heuristic methods such as particle swarm optimization [31] and genetic algorithms [32] have been popular. Due to the overwhelming difficulty of global optimization, a large portion of the optimization literature has focused on the particularly well-behaved class of *convex* functions [33, 34], which do not admit multiple local minima. Since Fisher introduced his "maximum likelihood" in 1922 [35], statisticians have thought in terms of maximization, but convexity theory still applies by a trivial negation of the objective function. Nonetheless, most statisticians safely ignored *concavity* during the twentieth century: exponential family log-likelihoods are log-concave, so Newton–Raphson and Fisher scoring are guaranteed optimality in the context of GLMs [12, 34].

Nearing the end of the twentieth century, multimodality and nonconvexity became more important for statisticians considering high-dimensional regression, that is, regression with many covariates (big P). Here, for purposes of interpretability and variance reduction, one would like to induce *sparsity* on the weights vector $\hat{\boldsymbol{\theta}}$ by performing best subset selection [36, 37]:

$$\hat{\theta} = \underset{\theta \in \mathbb{R}^P}{\operatorname{argmin}} ||\mathbf{y} - \mathbf{X}\theta||_2^2 \quad \text{subject to} \quad ||\theta||_0 \leq k \tag{3}$$

where $0 < k \leq P$, and $|| \cdot ||_0$ denotes the ℓ_0-norm, that is, the number of nonzero elements. Because best subset selection requires an immensely difficult nonconvex optimization, Tibshirani [38] famously replaces the ℓ_0-norm with the ℓ_1-norm, thereby providing sparsity, while nonetheless maintaining convexity.

Historically, Bayesians have paid much less attention to convexity than have optimization researchers. This is most likely because the basic theory [13] of MCMC does not require such restrictions: even if a target distribution has one million modes, the well-constructed Markov chain explores them all in the limit. Despite these theoretical guarantees, a small literature has developed to tackle multimodal Bayesian inference [39–42] because multimodal target distributions *do* present a challenge in practice. In analogy with Equation (3), Bayesians seek to induce sparsity by specifiying priors such as the spike-and-slab [43–45], for example,

$$\mathbf{y} \sim \text{Normal}_N(\mathbf{X}\boldsymbol{\Gamma}\theta, \sigma^2 \mathbf{I}_N) \text{ for } [\boldsymbol{\Gamma}]_{pp'} = \begin{cases} \gamma_p \sim \text{Bernoulli}(\pi) & p = p' \\ 0 & p \neq p' \end{cases} \quad \text{and } \pi \in (0, 1)$$

As with the best subset selection objective function, the spike-and-slab target distribution becomes heavily multimodal as P grows and the support of $\mathbf{\Gamma}$'s discrete distribution grows to 2^P potential configurations.

In the following section, we present an alternative Bayesian sparse regression approach that mitigates the combinatorial problem along with a state-of-the-art computational technique that scales well both in N and P.

3 Model-Specific Advances

These challenges will remain throughout the twenty-first century, but it is possible to make significant advances for specific statistical tasks or classes of models. Section 3.1 considers Bayesian sparse regression based on continuous shrinkage priors, designed to alleviate the heavy multimodality (big M) of the more traditional spike-and-slab approach. This model presents a major computational challenge as N and P grow, but a recent computational advance makes the posterior inference feasible for many modern large-scale applications.

And because of the rise of data science, there are increasing opportunities for computational statistics to grow by enabling and extending statistical inference for scientific applications previously outside of mainstream statistics. Here, the science may dictate the development of structured models with complexity possibly growing in N and P. Section 3.2 presents a method for fast phylogenetic inference, where the primary structure of interest is a "family tree" describing a biological evolutionary history.

3.1 Bayesian Sparse Regression in the Age of Big *N* and Big *P*

With the goal of identifying a small subset of relevant features among a large number of potential candidates, sparse regression techniques have long featured in a range of statistical and data science applications [46]. Traditionally, such techniques were commonly applied in the "$N \leq P$" setting, and correspondingly computational algorithms focused on this situation [47], especially within the Bayesian literature [48].

Due to a growing number of initiatives for large-scale data collections and new types of scientific inquiries made possible by emerging technologies, however, increasingly common are datasets that are "big N" and "big P" at the same time. For example, modern observational studies using health-care databases routinely involve $N \approx 10^5 \sim 10^6$ patients and $P \approx 10^4 \sim 10^5$ clinical covariates [49]. The UK Biobank provides brain imaging data on $N = 100\,000$ patients, with $P = 100 \sim 200\,000$, depending on the scientific question of interests [50]. Single-cell RNA sequencing can generate datasets with N (the number of cells) in millions and P (the number of genes) in tens of thousands, with the trend indicating further growths in data size to come [51].

3.1.1 Continuous shrinkage: alleviating big *M*

Bayesian sparse regression, despite its desirable theoretical properties and flexibility to serve as a building block for richer statistical models, has always been relatively computationally intensive even before the advent of "big N and big P" data [45, 52, 53]. A major source of its computational burden is severe posterior multimodality (big M) induced by the discrete binary nature of spike-and-slab priors (Section 2.3). The class of *global–local* continuous shrinkage priors is a more recent alternative to shrink θ_ps in a more continuous manner, thereby alleviating (if not eliminating) the multimodality issue [54, 55]. This class of prior is represented as a scale mixture of Gaussians:

$$\theta_p \mid \lambda_p, \tau \sim \text{Normal}_N(0, \tau^2\lambda_p^2), \;\; \lambda_p \sim \pi_{\text{local}}(\cdot), \;\; \tau \sim \pi_{\text{global}}(\cdot)$$

The idea is that the *global scale* parameter $\tau \leq 1$ would shrink most θ_ps toward zero, while the *local scale* λ_ps, with its heavy-tailed prior $\pi_{\text{local}}(\cdot)$, allow a small number of $\tau\lambda_p$ and hence θ_ps to be estimated away from zero. While motivated by two different conceptual frameworks, the spike-and-slab can be viewed as a subset of global–local priors in which $\pi_{\text{local}}(\cdot)$ is chosen as a mixture of delta masses placed at $\lambda_p = 0$ and $\lambda_p = \sigma/\tau$. Continuous shrinkage mitigates the multimodality of spike-and-slab by smoothly bridging small and large values of λ_p.

On the other hand, the use of continuous shrinkage priors does not address the increasing computational burden from growing N and P in modern applications. Sparse regression posteriors under global–local priors are amenable to an effective Gibbs sampler, a popular class of MCMC we describe further in Section 4.1. Under the linear and logistic models, the computational bottleneck of this Gibbs sampler stems from the need for repeated updates of θ from its conditional distribution

$$\theta \mid \tau, \lambda, \mathbf{\Omega}, \mathbf{y}, \mathbf{X} \sim \text{Normal}_P(\mathbf{\Phi}^{-1}\mathbf{X}^\intercal\mathbf{\Omega}\mathbf{y}, \mathbf{\Phi}^{-1})) \text{ for } \mathbf{\Phi} = \mathbf{X}^\intercal\mathbf{\Omega}\mathbf{X} + \tau^{-2}\mathbf{\Lambda}^{-2} \tag{4}$$

where $\mathbf{\Omega}$ is an additional parameter of diagonal matrix and $\mathbf{\Lambda} = \text{diag}(\lambda)$.[5] Sampling from this high-dimensional Gaussian distribution requires $\mathcal{O}(NP^2 + P^3)$ operations with the standard approach [58]: $\mathcal{O}(NP^2)$ for computing the term $\mathbf{X}^\intercal\mathbf{\Omega}\mathbf{X}$ and $\mathcal{O}(P^3)$ for Cholesky factorization of $\mathbf{\Phi}$. While an alternative approach by Bhattacharya *et al.* [48] provides the complexity of $\mathcal{O}(N^2P + N^3)$, the computational cost remains problematic in the big N and big P regime at $\mathcal{O}(\min\{N^2P, NP^2\})$ after choosing the faster of the two.

3.1.2 Conjugate gradient sampler for structured high-dimensional Gaussians

The *conjugate gradient* (CG) sampler of Nishimura and Suchard [57] combined with their *prior preconditioning* technique overcomes this seemingly inevitable $\mathcal{O}(\min\{N^2P, NP^2\})$ growth of the computational cost. Their algorithm is based on a novel application of the CG method [59, 60], which belongs to a family of *iterative methods* in numerical linear algebra. Despite its first appearance in 1952, CG received little attention for the next few decades, only making its way into major software packages such as MATLAB in the 1990s [61]. With its ability to solve a large and structured linear system $\mathbf{\Phi}\theta = \boldsymbol{b}$ via a small number of matrix–vector multiplications $\boldsymbol{v} \to \mathbf{\Phi}\boldsymbol{v}$ without ever explicitly inverting $\mathbf{\Phi}$, however, CG has since emerged as an essential and prototypical algorithm for modern scientific computing [62, 63].

Despite its earlier rise to prominence in other fields, CG has not found practical applications in Bayesian computation until rather recently [57, 64]. We can offer at least two explanations for this. First, being an algorithm for solving a deterministic linear system, it is not obvious how CG would be relevant to Monte Carlo simulation, such as sampling from $\text{Normal}_P(\mu, \mathbf{\Phi}^{-1})$; ostensively, such a task requires computing a "square root" $\boldsymbol{L}$ of the precision matrix so that $Var(\boldsymbol{L}^{-1}\boldsymbol{z}) = \boldsymbol{L}^{-1}\boldsymbol{L}^{-\intercal} = \mathbf{\Phi}^{-1}$ for $\boldsymbol{z} \sim \text{Normal}_P(\mathbf{0}, \boldsymbol{I}_P)$. Secondly, unlike direct linear algebra methods, iterative methods such as CG have a variable computational cost that depends critically on the user's choice of a preconditioner and thus *cannot* be used as a "black-box" algorithm.[6] In particular, this novel application of CG to Bayesian computation is a reminder that other powerful ideas in other computationally intensive fields may remain untapped by the statistical computing community; knowledge transfers will likely be facilitated by having more researchers working at intersections of different fields.

Nishimura and Suchard [57] turns CG into a viable algorithm for Bayesian sparse regression problems by realizing that (i) we can obtain a Gaussian vector $\boldsymbol{b} \sim \text{Normal}_P(\mathbf{X}^\intercal\boldsymbol{\Omega}\mathbf{y}, \boldsymbol{\Phi})$ by first generating $\boldsymbol{z} \sim \text{Normal}_P(\mathbf{0}, \boldsymbol{I}_P)$ and $\zeta \sim \text{Normal}_N(\mathbf{0}, \boldsymbol{I}_N)$ and then setting $\boldsymbol{b} = \mathbf{X}^\intercal\boldsymbol{\Omega}\mathbf{y} + \mathbf{X}^\intercal\boldsymbol{\Omega}^{1/2}\zeta + \tau^{-1}\boldsymbol{\Lambda}^{-1}\boldsymbol{z}$ and (ii) subsequently solving $\boldsymbol{\Phi\theta} = \boldsymbol{b}$ yields a sample $\boldsymbol{\theta}$ from the distribution (4). The authors then observe that the mechanism through which a shrinkage prior induces sparsity of θ_ps also induces a tight clustering of eigenvalues in the prior-preconditioned matrix $\tau^2\boldsymbol{\Lambda\Phi\Lambda}$. This fact makes it possible for prior-preconditioned CG to solve the system $\boldsymbol{\Phi\theta} = \boldsymbol{b}$ in K matrix–vector operations of form $\boldsymbol{v} \rightarrow \boldsymbol{\Phi v}$, where K roughly represents the number of significant θ_ps that are distinguishable from zeros under the posterior. For $\boldsymbol{\Phi}$ having a structure as in (4), $\boldsymbol{\Phi v}$ can be computed via matrix–vector multiplications of form $\boldsymbol{v} \rightarrow \mathbf{X}\boldsymbol{v}$ and $\boldsymbol{w} \rightarrow \mathbf{X}^\intercal\boldsymbol{w}$, so each $\boldsymbol{v} \rightarrow \boldsymbol{\Phi v}$ operation requires a fraction of the computational cost of directly computing $\boldsymbol{\Phi}$ and then factorizing it.

Prior-preconditioned CG demonstrates an order of magnitude speedup in posterior computation when applied to a comparative effectiveness study of atrial fibrillation treatment involving $N = 72\,489$ patients and $P = 22\,175$ covariates [57]. Though unexplored in their work, the algorithm's heavy use of matrix–vector multiplications provides avenues for further acceleration. Technically, the algorithm's complexity may be characterized as $\mathcal{O}(NPK)$, for the K matrix–vector multiplications by $\mathbf{X}$ and $\mathbf{X}^\intercal$, but the theoretical complexity is only a part of the story. Matrix–vector multiplications are amenable to a variety of hardware optimizations, which in practice can make orders of magnitude difference in speed (Section 4.2). In fact, given how arduous manually optimizing computational bottlenecks can be, designing algorithms so as to take advantage of common routines (as those in Level 3 BLAS) and their ready-optimized implementations has been recognized as an effective principle in algorithm design [65].

3.2 Phylogenetic Reconstruction

While big N and big P regression adapts a classical statistical task to contemporary needs, the twenty-first century is witnessing the application of computational statistics to the entirety of applied science. One such example is the tracking and reconstruction of deadly global viral pandemics. Molecular phylogenetics has become an essential analytical tool for understanding the complex patterns in which rapidly evolving pathogens propagate throughout and between countries, owing to the complex travel and transportation patterns evinced by modern economies [66], along with other factors such as increased global population and urbanization [67]. The advance in sequencing technology is generating pathogen genomic data at an ever-increasing pace, with a trend to real time that requires the development of computational statistical methods that are able to process the sequences in a timely manner and produce interpretable results to inform national/global public health organizations.

The previous three Core Challenges are usually interwound such that the increase in the sample size (big N) and the number of traits (big P) for each sample usually happen simultaneously and lead to increased heterogeneity that requires more complex models (big M). For example, recent studies in viral evolution have seen a continuing increase in the sample size that the West Nile virus, Dengue, HIV, and Ebola virus studies involve 104, 352, 465, and 1610 sequences [68–71], and the GISAID database has collected 92 000 COVID-19 genomic sequences by the end of August 2020 [72].

To accommodate the increasing size and heterogeneity in the data and be able to apply the aforementioned efficient gradient-based algorithms, Ji *et al.* [73] propose a linear-time algorithm for calculating an $O(N)$-dimensional gradient on a tree w.r.t. the sequence evolution. The linear-time gradient algorithm calculates each branch-specific derivative through a preorder traversal that complements the postorder traversal from the likelihood calculation of the observed sequence data at the tip of the phylogeny by marginalizing over all possible hidden states on the internal nodes. The pre- and postorder traversals complete the Baum's forward–backward algorithm in a phylogenetic framework [74]. The authors then apply the gradient algorithm with HMC (Section 2.2) samplers to learn the branch-specific viral evolutionary rates.

Thanks to these advanced computational methods, one can employ more flexible models that lend themselves to more realistic reconstructions and uncertainty quantification. Following a random-effects relaxed clock model, they model the evolutionary rate r_p of branch p on a phylogeny as the product of a global treewise mean parameter μ and a branch-specific random effect ϵ_p. They model the random-effect ϵ_ps as independent and identically distributed from a lognormal distribution such that ϵ_p has mean 1 and variance ψ^2 under a hierarchical model where ψ is the scale parameter. To accommodate the difference in scales of the variability in the parameter space for the HMC sampler, the authors adopt preconditioning with adaptive mass matrix informed by the diagonal entries of the Hessian matrix. More precisely, the nonzero diagonal elements of the mass matrix truncate the values from the first s HMC iterations of $H_{pp}^{(s)} = \frac{1}{\lfloor s/k \rfloor} \sum_{s:s/k \in \mathbb{Z}^+} \left[-\frac{\partial^2}{\partial^2 \theta_p} \log \pi(\theta) \Big|_{\theta=\theta^{(s)}} \right] \approx \mathbb{E}_{\pi(\theta)} \left[-\frac{\partial^2}{\partial^2 \theta_i} \log \pi(\theta) \right]$ so that the matrix remains positive-definite and numerically stable. They estimate the treewise (fixed-effect) mean rate μ with posterior mean 4.75 (95% Bayesian credible interval: 4.05, 5.33) $\times 10^{-4}$ substitutions per site per year with rate variability characterized by scale parameter with posterior mean $\psi = 1.26[1.06, 1.45]$ for serotype 3 of Dengue virus with a sample size of 352 [69]. Figure 1 illustrates the estimated maximum clade credible evolutionary tree of the Dengue virus dataset.

The authors report relative speedup in terms of the effective sample size per second (ESS/s) of the HMC samplers compared to a univariate transition kernel. The "vanilla" HMC sampler with an identity mass matrix gains 2.2× speedup for the minimum ESS/s and 2.5× speedup for the median ESS/s, whereas the "preconditioned" HMC sampler gains 16.4× and 7.4× speedups, respectively. Critically, the authors make these performance gains available to scientists everywhere through the popular, open-source software package for viral phylogenetic inference *Bayesian evolutionary analysis by sampling trees* (BEAST) [75].

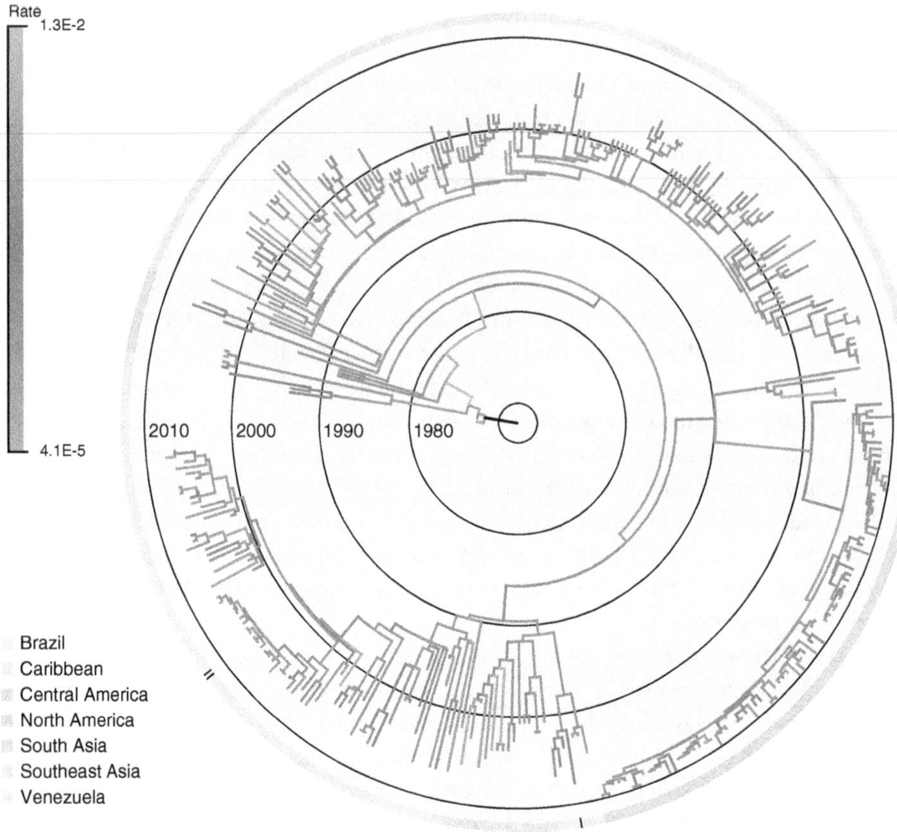

Figure 1 A nontraditional and critically important application in computational statistics is the reconstruction of evolutionary histories in the form of phylogenetic trees. Here is a maximum clade credible tree of the Dengue virus example. The dataset consists of 352 sequences of the serotype 3 of the Dengue virus. Branches are coded by the posterior means of the branch-specific evolutionary rates according to the gradient bar on the top left. The concentric circles indicate the timescale with the year numbers. The outer ring indicates the geographic locations of the samples by the color code on the bottom left. '**I**' and '**II**' indicate the two Brazilian lineages as in the original study.

In Section 4.1, we discuss how software package such as BEAST addresses Core Challenge 4, the creation of fast, flexible, and friendly statistical algo-ware.

4 Core Challenges 4 and 5

Section 3 provides examples of how computational statisticians might address Core Challenges 1–3 (big N, big P, and big M) for individual models. Such advances in computational methods must be accompanied by easy-to-use software to make them accessible to end users. As Gentle *et al.* [76] put it, "While referees and editors of scholarly journals determine what statistical theory and methods are published, the developers of the major statistical software packages determine what statistical methods are used." We would like statistical software to be widely applicable yet computationally efficient at the same time. Trade-offs invariably arise between these two desiderata, but one should nonetheless strive to design algorithms that are general enough to solve an important class of problems and as efficiently as possible in doing so.

Section 4.1 presents Core Challenge 4, achieving "algo-ware" (a neologism suggesting an equal emphasis on the statistical algorithm and its implementation) that is sufficiently efficient, broad, and user-friendly to empower everyday statisticians and data scientists. Core Challenge 5 (Section 4.2) explores the mapping of these algorithms to computational hardware for optimal performance. Hardware-optimized implementations often exploit model-specific structures, but good, general-purpose software should also optimize common routines.

4.1 Fast, Flexible, and Friendly Statistical Algo-Ware

To accommodate the greatest range of models while remaining simple enough to encourage easy implementation, inference methods should rely solely on the quantities that can be computed algorithmically for any given model. The log-likelihood (or log-density in the Bayesian setting) is one such quantity, and one can employ the computational graph framework [77, 78] to evaluate conditional log-likelihoods for any subset of model parameters as well as their gradients via backpropagation [79]. Beyond being efficient in terms of the first three Core Challenges, an algorithm should demonstrate robust performance on a reasonably wide range of problems without extensive tuning if it is to lend itself to successful software deployment.

HMC (Section 2.2) is a prominent example of a general-purpose algorithm for Bayesian inference, only requiring the log-density and its gradient. The generic nature of HMC has opened up possibilities for complex Bayesian modeling as early as Neal [80], but its performance is highly sensitive to model parameterization and its three tuning parameters, commonly referred to as trajectory length, step size, and mass matrix [27]. Tuning issues constitute a major obstacle to the wider adoption of the algorithm, as evidenced by the development history of the popular HMC-based probabilistic programming software Stan [81], which employs the *No-U-Turn* sampler (NUTS) of Hoffman and Gelman [82] to make HMC user-friendly by obviating the need to tune its trajectory length. Bayesian software packages such as Stan empirically adapt the remaining step size and mass matrix [83]; this approach helps make the use of HMC automatic though is not without issues [84] and comes at the cost of significant computational overhead.

Although HMC is a powerful algorithm that has played a critical role in the emergence of general-purpose Bayesian inference software, the challenges involved in its practical deployment also demonstrate how an algorithm – no matter how versatile and efficient at its best – is not necessarily useful unless it can be made easy for practitioners to use. It is also unlikely that one algorithm works well in all situations. In fact, there are many distributions on which HMC performs poorly [83, 85, 86]. Additionally, HMC is incapable of handling discrete distributions in a fully general manner despite the progresses made in extending HMC to such situations [87, 88].

But broader applicability comes with its own challenges. Among sampling-based approaches to Bayesian inference, the Gibbs sampler [89, 90] is, arguably, the most versatile of the MCMC methods. The algorithm simplifies the task of dealing with a complex multidimensional posterior distribution by factorizing the posterior into simpler conditional distributions for blocks of parameters and iteratively updating parameters from their conditionals. Unfortunately, the efficiency of an individual Gibbs sampler depends on its specific factorization and the degree of dependence between its blocks of parameters. Without a careful design or in the absence of effective factorization, therefore, Gibbs samplers' performance may lag behind alternatives such as HMC [91].

On the other hand, Gibbs samplers often require little tuning and can take advantage of highly optimized algorithms for each conditional update, as done in the examples of Section 3. A clear advantage of the Gibbs sampler is that it tends to make software implementation quite modular; for example, each conditional update can be replaced with the latest state-of-the-art samplers as they appear [92], and adding a new feature may amount to no more than adding a single conditional update [75]. In this way, an algorithm may not work in a completely model-agnostic manner but with a broad enough scope can serve as a valuable recipe or meta-algorithm for building model-specific algorithms and software. The same is true for optimization methods. Even though its "E"-step requires a derivation (by hand) for each new model, the EM algorithm [93] enables maximum-likelihood estimation for a wide range of models. Similarly, variational inference (VI) for approximate Bayes requires manual derivations but provides a general framework to turn posterior computation into an optimization problem [94]. As meta-algorithms, both EM and VI expand their breadth of use by replacing analytical derivations with Monte Carlo estimators but suffer losses in statistical and computational efficiency [95, 96]. Indeed, such trade-offs will continue to haunt the creation of fast, flexible, and friendly statistical algo-ware well into the twenty-first century.

4.2 Hardware-Optimized Inference

But successful statistical inference software must also interact with computational hardware in an optimal manner. Growing datasets require the computational statistician to give more and more thought to how the computer implements any statistical algorithm. To effectively leverage computational resources, the statistician must (i) identify the routine's computational bottleneck (Section 2.1) and (ii) algorithmically map this rate-limiting step to available hardware such as a multicore or vectorized CPU, a many-core GPU, or – in the future – a quantum computer. Sometimes, the first step is clear theoretically: a naive implementation of the high-dimensional regression example of Section 3.1 requires an order $\mathcal{O}(N^2P)$ matrix multiplication followed by an order $\mathcal{O}(P^3)$ Cholesky decomposition. Other times, one can use an instruction-level program profiler, such as INTEL VTUNE (Windows, Linux) or INSTRUMENTS (OSX), to identify a performance bottleneck. Once the bottleneck is identified, one must choose between computational resources, or some combination thereof, based on relative strengths and weaknesses as well as natural parallelism of the target task.

Multicore CPU processing is effective for parallel completion of multiple, mostly independent tasks that do not require intercommunication. One might generate 2 to, say, 72 independent Markov chains on a desktop computer or shared cluster. A positive aspect is that the tasks do not have to involve the same instruction sets at all; a negative is *latency*, that is, that the slowest process dictates overall runtime. It is possible to further speed up CPU computing with single instruction, multiple data (SIMD) or vector processing. A small number of vector processing units (VPUs) in each CPU core can carry out a single set of instructions on data stored within an extended-length register. Intel's streaming SIMD extensions (SSE), advance vector extensions (AVX), and AVX-512 allow operations on 128-, 256-, and 512-bit registers. In the context of 64-bit double precision, theoretical speedups for SSE, AVX, and AVX-512 are two-, four-, and eightfold. For example, if a computational bottleneck exists within a for-loop, one can unroll the loop and perform operations on, say, four consecutive loop bodies at once using AVX [21, 22]. Conveniently, languages such as OpenMP [97] make SIMD loop optimization transparent to the user [98]. Importantly, SIMD and multicore optimization play well together, providing multiplicative speedups.

While a CPU may have tens of cores, GPUs accomplish fine-grained parallelization with thousands of cores that apply a single instruction set to distinct data within smaller workgroups of tens or hundreds of cores. Quick communication and shared cache memory within each workgroup balance full parallelization across groups, and dynamic on- and off-loading of the many tasks hide the latency that is so problematic for multicore computing. Originally designed for efficiently parallelized matrix math calculations arising from image rendering and transformation, GPUs easily speed up tasks that are tensor multiplication intensive such as deep learning [99] but general-purpose GPU applications abound. Holbrook *et al.* [21] provide a larger review of parallel computing within computational statistics. The same paper reports a GPU providing 200-fold speedups over single-core processing and 10-fold speedups over 12-core AVX processing for likelihood and gradient calculations while sampling from a Bayesian multidimensional scaling posterior using HMC at scale. Holbrook *et al.* [22] report similar speedups for inference based on spatiotemporal Hawkes processes. Neither application involves matrix or tensor manipulations.

A quantum computer acts on complex data vectors of magnitude 1 called qubits with gates that are mathematically equivalent to unitary operators [100]. Assuming that engineers overcome the tremendous difficulties involved in building a practical quantum computer (where practicality entails simultaneous use of many quantum gates with little additional noise), twenty-first century statisticians might have access to quadratic or even exponential speedups for extremely specific statistical tasks. We are particularly interested in the following four quantum algorithms: quantum search [101], or finding a single 1 amid a collection of 0s, only requires $\mathcal{O}(\sqrt{N})$ queries, delivering a quadratic speedup over classical search; quantum counting [102], or finding the number of 1s amid a collection of 0s, only requires $\mathcal{O}(\sqrt{N/M})$ (where M is the number of 1s) and could be useful for generating p-values within Monte Carlo simulation from a null distribution (Section 2.1); to obtain the gradient of a function (e.g., the log-likelihood for Fisher scoring or HMC) with a quantum computer, one only needs to evaluate the function once [103] as opposed to $\mathcal{O}(P)$ times for numerical differentiation, and there is nothing stopping the statistician from using, say, a GPU for this single function call; and finally, the HHL algorithm [104] obtains the scalar value $\mathbf{q}^T\mathbf{M}\mathbf{q}$ for the P-vector $\mathbf{q}$ satisfying $\mathbf{A}\mathbf{q} = \mathbf{b}$ and

$\mathbf{M}$ and $P \times P$ matrix in time $\mathcal{O}(\log(P\kappa^2))$, delivering an exponential speedup over classical methods. Technical caveats exist [105], but HHL may find use within high-dimensional hypothesis testing (big P). Under the null hypothesis, one can rewrite the score test statistic

$$\mathbf{u}^T(\hat{\theta}_0)\, \mathcal{I}^{-1}(\hat{\theta}_0)\, \mathbf{u}(\hat{\theta}_0) \quad \text{as} \quad \mathbf{u}^T(\hat{\theta}_0)\, \mathcal{I}^{-1}(\hat{\theta}_0)\, \mathcal{I}(\hat{\theta}_0)\, \mathcal{I}^{-1}(\hat{\theta}_0)\, \mathbf{u}(\hat{\theta}_0)$$

for $\mathcal{I}(\hat{\theta}_0)$ and $\mathbf{u}(\hat{\theta}_0)$, the Fisher information and log-likelihood gradient evaluated at the maximum-likelihood solution under the null hypothesis. Letting $\mathbf{A} = \mathcal{I}(\hat{\theta}_0) = \mathbf{M}$ and $\mathbf{b} = \mathbf{u}(\hat{\theta}_0)$, one may write the test statistic as $\mathbf{q}^T\mathbf{M}\mathbf{q}$ and obtain it in time logarithmic in P. When the model design matrix $\mathbf{X}$ is sufficiently sparse – a common enough occurrence in large-scale regression – to render $\mathcal{I}(\hat{\theta}_0)$ itself sparse, the last criterion for the application of the HHL algorithm is met.

5 Rise of Data Science

Core Challenges 4 and 5 – fast, flexible, and user-friendly algo-ware and hardware-optimized inference – embody an increasing emphasis on application and implementation in the age of data science. Previously undervalued contributions in statistical computing, for example, hardware utilization, database methodology, computer graphics, statistical software engineering, and the human–computer interface [76], are slowly taking on greater importance within the (rather conservative) discipline of statistics. There is perhaps no better illustration of this trend than Dr. Hadley Wickham's winning the prestigious COPSS Presidents' Award for 2019

> [for] influential work in statistical computing, visualization, graphics, and data analysis; for developing and implementing an impressively comprehensive computational infrastructure for data analysis through R software; for making statistical thinking and computing accessible to large audience; and for enhancing an appreciation for the important role of statistics among data scientists [106].

This success is all the more impressive because Presidents' Awardees have historically been contributors to statistical theory and methodology, not Dr. Wickham's scientific software development for data manipulation [107–109] and visualization [110, 111].

All of this might lead one to ask: *does the success of data science portend the declining significance of computational statistics and its Core Challenges?* Not at all! At the most basic level, data science's emphasis on application and implementation underscores the need for computational thinking in statistics. Moreover, the scientific breadth of data science brings new applications and models to the attention of statisticians, and these models may require or inspire novel algorithmic techniques. Indeed, we look forward to a golden age of computational statistics, in which statisticians labor within the intersections of mathematics, parallel computing, database methodologies, and software engineering with impact on the entirety of the applied sciences. After all, significant progress toward conquering the Core Challenges of computational statistics requires that we use every tool at our collective disposal.

Acknowledgments

AJH is supported by NIH grant K25AI153816. MAS is supported by NIH grant U19AI135995 and NSF grant DMS1264153.

Notes

1 *Statistical inference* is an umbrella term for hypothesis testing, point estimation, and the generation of (confidence or credible) intervals for population functionals (mean, median, correlations, etc.) or model parameters.
2 We present the problem of phylogenetic reconstruction in Section 3.2 as one such example arising from the field of molecular epidemiology.
3 The use of "N" and "P" to denote observation and parameter count is common. We have taken liberties in coining the use of "M" to denote mode count.
4 A more numerically stable approach has the same complexity [24].
5 The matrix parameter $\mathbf{\Omega}$ coincides with $\mathbf{\Omega} = \sigma^{-2}\mathbf{I}_N$ for linear regression and $\mathbf{\Omega} = \text{diag}(\omega)$ for auxiliary Pólya-Gamma parameter ω for logistic regression [56, 57].
6 See Nishimura and Suchard [57] and references therein for the role and design of a preconditioner.

References

1 Davenport, T.H. and Patil, D. (2012) Data scientist. *Harvard Bus. Rev.*, 90, 70–76.
2 Google Trends (2020) Data source: Google trends. https://trends.google.com/trends (accessed 12 July 2020).
3 American Statistical Association (2020) *Statistics Degrees Total and By Gender*, https://ww2.amstat.org/misc/StatTable1987-Current.pdf (accessed 01 June 2020).
4 Cleveland, W.S. (2001) Data science: an action plan for expanding the technical areas of the field of statistics. *Int. Stat. Rev.*, 69, 21–26.
5 Donoho, D. (2017) 50 Years of data science. *J. Comput. Graph. Stat.*, 26, 745–766.
6 Fisher, R.A. (1936) Design of experiments. *Br Med J 1.3923*, 554–554.
7 Fisher, R.A. (1992) Statistical methods for research workers, in Kotz S., Johnson N.L. (eds) *Breakthroughs in Statistics*, Springer Series in Statistics (Perspectives in Statistics). Springer, New York, NY. (Especially Section 21.02). doi: 10.1007/978-1-4612-4380-9_6.
8 Wald, A. and Wolfowitz, J. (1944) Statistical tests based on permutations of the observations. *Ann. Math. Stat.*, 15, 358–372.
9 Efron B. (1992) Bootstrap methods: another look at the jackknife, in *Breakthroughs in Statistics*. Springer Series in Statistics (Perspectives in Statistics) (eds S. Kotz and N.L. Johnson), Springer, New York, NY, pp. 569–593. doi: 10.1007/978-1-4612-4380-9_41.
10 Efron, B. and Tibshirani, R.J. (1994) *An Introduction to the Bootstrap*, CRC press.
11 Bliss, C.I. (1935) The comparison of dosage-mortality data. *Ann. Appl. Biol.*, 22, 307–333 (Fisher introduces his scoring method in appendix).
12 McCullagh, P. and Nelder, J. (1989) *Generalized Linear Models*, 2nd edn, Chapman and Hall, London. Standard book on generalized linear models.

13 Tierney, L. (1994) Markov chains for exploring posterior distributions. *Ann. Stat.*, 22, 1701–1728.

14 Brooks, S., Gelman, A., Jones, G., and Meng, X.-L. (2011) *Handbook of Markov Chain Monte Carlo*, CRC press.

15 Chavan, V. and Phursule, R.N. (2014) Survey paper on big data. *Int. J. Comput. Sci. Inf. Technol.*, 5, 7932–7939.

16 Williams, C.K. and Rasmussen, C.E. (1996) Gaussian processes for regression. Advances in Neural Information Processing Systems, pp. 514–520.

17 Williams, C.K. and Rasmussen, C.E. (2006) *Gaussian Processes for Machine Learning*, vol. 2, MIT press, Cambridge, MA.

18 Gelman, A., Carlin, J.B., Stern, H.S. *et al.* (2013) *Bayesian Data Analysis*, CRC press.

19 Metropolis, N., Rosenbluth, A.W., Rosenbluth, M.N. *et al.* (1953) Equation of state calculations by fast computing machines. *J. Chem. Phys.*, 21, 1087–1092.

20 Hastings, W.K. (1970) Monte Carlo sampling methods using Markov chains and their applications, *Biometrika*, 57 (1), 97–109. doi: 10.1093/biomet/57.1.97

21 Holbrook, A.J., Lemey, P., Baele, G. *et al.* (2020) Massive parallelization boosts big Bayesian multidimensional scaling. *J. Comput. Graph. Stat.*, 1–34.

22 Holbrook, A.J., Loeffler, C.E., Flaxman, S.R. *et al.* (2021) Scalable Bayesian inference for self-excitatory stochastic processes applied to big American gunfire data, *Stat. Comput.* 31, 4.

23 Seber, G.A. and Lee, A.J. (2012) *Linear Regression Analysis*, vol. 329, John Wiley & Sons.

24 Trefethen, L.N. and Bau, D. (1997) Numerical linear algebra. *Soc. Ind. Appl. Math.*

25 Gelman, A., Roberts, G.O., and Gilks, W.R. (1996) Efficient metropolis jumping rules. *Bayesian Stat.*, 5, 42.

26 Van Dyk, D.A. and Meng, X.-L. (2001) The art of data augmentation. *J. Comput. Graph. Stat.*, 10, 1–50.

27 Neal, R.M. (2011) MCMC using Hamiltonian dynamics, in *Handbook of Markov Chain Monte Carlo* (eds S. Brooks, A. Gelman, G. Jones and X.L. Meng), Chapman and Hall/CRC Press, 113–162.

28 Holbrook, A., Vandenberg-Rodes, A., Fortin, N., and Shahbaba, B. (2017) A Bayesian supervised dual-dimensionality reduction model for simultaneous decoding of LFP and spike train signals. *Stat*, 6, 53–67.

29 Bouchard-Côté, A., Vollmer, S.J., and Doucet, A. (2018) The bouncy particle sampler: a nonreversible rejection-free Markov chain Monte Carlo method. *J. Am. Stat. Assoc.*, 113, 855–867.

30 Murty, K.G. and Kabadi, S.N. (1985) Some NP-Complete Problems in Quadratic and Nonlinear Programming. *Tech. Rep.*

31 Kennedy, J. and Eberhart, R. (1995) *Particle Swarm Optimization*. Proceedings of ICNN'95-International Conference on Neural Networks, vol. 4, pp. 1942–1948. IEEE.

32 Davis, L. (1991) *Handbook of Genetic Algorithms*, Van Nostrand Reinhold, New York.

33 Hunter, D.R. and Lange, K. (2004) A tutorial on MM algorithms. *Am. Stat.*, 58, 30–37.

34 Boyd, S., Boyd, S.P., and Vandenberghe, L. (2004) *Convex Optimization*, Cambridge University Press.

35 Fisher, R.A. (1922) On the mathematical foundations of theoretical statistics. *Philos. Trans. R. Soc. London, Ser. A*,222,309–368.

36 Beale, E., Kendall, M., and Mann, D. (1967) The discarding of variables in multivariate analysis. *Biometrika*, 54, 357–366.

37 Hocking, R.R. and Leslie, R. (1967) Selection of the best subset in regression analysis. *Technometrics*, 9, 531–540.

38 Tibshirani, R. (1996) Regression shrinkage and selection via the lasso. *J. R. Stat. Soc., Ser. B*, 58,267–288.

39 Geyer, C. (1991) *Markov Chain Monte Carlo Maximum Likelihood*. Computing Science and Statistics: Proceedings of 23rd Symposium on the Interface Interface Foundation, Fairfax Station, 156–163.

40 Tjelmeland, H. and Hegstad, B.K. (2001) Mode jumping proposals in MCMC. *Scand. J. Stat.*, 28, 205–223.

41 Lan, S., Streets, J., and Shahbaba, B. (2014) *Wormhole Hamiltonian Monte Carlo*. Twenty-Eighth AAAI Conference on Artificial Intelligence.

42 Nishimura, A. and Dunson, D. (2016) Geometrically tempered Hamiltonian Monte Carlo. *arXiv preprint arXiv:1604.00872*.

43 Mitchell, T.J. and Beauchamp, J.J. (1988) Bayesian variable selection in linear regression. *J. Am. Stat. Assoc.*, 83, 1023–1032.

44 Madigan, D. and Raftery, A.E. (1994) Model selection and accounting for model uncertainty in graphical models using Occam's window. *J. Am. Stat. Assoc.*, 89, 1535–1546.

45 George, E.I. and McCulloch, R.E. (1997) Approaches for Bayesian variable selection. *Statistica Sinica*, 7, 339–373.

46 Hastie, T., Tibshirani, R., and Wainwright, M. (2015) *Statistical Learning with Sparsity: The Lasso and Generalizations*, CRC Press.

47 Friedman, J., Hastie, T., and Tibshirani, R. (2010) Regularization paths for generalized linear models via coordinate descent. *J. Stat. Softw.*, 33, 1.

48 Bhattacharya, A., Chakraborty, A., and Mallick, B.K. (2016) Fast sampling with Gaussian scale mixture priors in high-dimensional regression. *Biometrika*, 103, 985–991.

49 Suchard, M.A., Schuemie, M.J., Krumholz, H.M. *et al.* (2019) Comprehensive comparative effectiveness and safety of first-line antihypertensive drug classes: a systematic, multinational, large-scale analysis. *The Lancet*, 394, 1816–1826.

50 Passos, I.C., Mwangi, B., and Kapczinski, F. (2019) *Personalized Psychiatry: Big Data Analytics in Mental Health*, Springer.

51 Svensson, V., da Veiga Beltrame, E., and Pachter, L. (2019) A curated database reveals trends in single-cell transcriptomics. *bioRxiv 742304*.

52 Nott, D.J. and Kohn, R. (2005) Adaptive sampling for Bayesian variable selection. *Biometrika*, 92, 747–763.

53 Ghosh, J. and Clyde, M.A. (2011) Rao–Blackwellization for Bayesian variable selection and model averaging in linear and binary regression: a novel data augmentation approach. *J. Am. Stat. Assoc.*, 106,1041–1052.

54 Carvalho, C.M., Polson, N.G., and Scott, J.G. (2010) The horseshoe estimator for sparse signals. *Biometrika*, 97,465–480.

55 Polson, N.G. and Scott, J.G. (2010) Shrink globally, act locally: sparse Bayesian regularization and prediction. *Bayesian Stat.*, 9, 501–538.

56 Polson, N.G., Scott, J.G., and Windle, J. (2013) Bayesian inference for logistic models using Pólya–Gamma latent variables. *J. Am. Stat. Assoc.*, 108, 1339–1349.

57 Nishimura, A. and Suchard, M.A. (2018) Prior-preconditioned conjugate gradient for accelerated gibbs sampling in "large n & large p" sparse Bayesian logistic regression models. *arXiv:1810.12437*.

58 Rue, H. and Held, L. (2005) *Gaussian Markov Random Fields: Theory and Applications*, CRC Press.

59 Hestenes, M.R. and Stiefel, E. (1952) Methods of conjugate gradients for solving linear systems. *J. Res. Nat. Bur. Stand.*, 49, 409–436.

60 Lanczos, C. (1952) Solution of systems of linear equations by minimized iterations. *J. Res. Nat. Bur. Stand.*, 49, 33–53.

61 Van der Vorst, H.A. (2003) *Iterative Krylov Methods for Large Linear Systems*, vol. 13, Cambridge University Press.

62 Cipra, B.A. (2000) The best of the 20th century: editors name top 10 algorithms. *SIAM News*, 33, 1–2.

63 Dongarra, J., Heroux, M.A., and Luszczek, P. (2016) High-performance conjugate-gradient benchmark: a new metric for ranking high-performance computing systems. *Int. J. High Perform. Comput. Appl.*, 30, 3–10.

64 Zhang, L., Zhang, L., Datta, A., and Banerjee, S. (2019) Practical Bayesian modeling and inference for massive spatial data sets on modest computing environments. *Stat. Anal. Data Min.*, 12, 197–209.

65 Golub, G.H. and Van Loan, C.F. (2012) *Matrix Computations*, vol. 3, Johns Hopkins University Press.

66 Pybus, O.G., Tatem, A.J., and Lemey, P. (2015) Virus evolution and transmission in an ever more connected world. *Proc. R. Soc. B: Biol. Sci.*, 282, 20142878.

67 Bloom, D.E., Black, S., and Rappuoli, R. (2017) Emerging infectious diseases: a proactive approach. *Proc. Natl. Acad. Sci. U.S.A.*, 114, 4055–4059.

68 Pybus, O.G., Suchard, M.A., Lemey, P. *et al.* (2012) Unifying the spatial epidemiology and molecular evolution of emerging epidemics. *Proc. Natl. Acad. Sci. U.S.A.*, 109, 15066–15071.

69 Nunes, M.R., Palacios, G., Faria, N.R. *et al.* (2014) Air travel is associated with intracontinental spread of dengue virus serotypes 1–3 in Brazil. *PLoS Negl. Trop. Dis.*, 8, e2769.

70 Bletsa, M., Suchard, M.A., Ji, X. *et al.* (2019) Divergence dating using mixed effects clock modelling: an application to HIV-1. *Virus Evol.*, 5, vez036.

71 Dudas, G., Carvalho, L.M., Bedford, T. *et al.* (2017) Virus genomes reveal factors that spread and sustained the Ebola epidemic. *Nature*, 544, 309–315.

72 Elbe, S. and Buckland-Merrett, G. (2017) Data, disease and diplomacy: GISAID's innovative contribution to global health. *Glob. Chall.*, 1, 33–46.

73 Ji, X., Zhang, Z., Holbrook, A. *et al.* (2020) Gradients do grow on trees: a linear-time O(N)-dimensional gradient for statistical phylogenetics. *Mol. Biol. Evol.*, 37, 3047–3060.

74 Baum, L. (1972) An inequality and associated maximization technique in statistical estimation of probabilistic functions of a Markov process. *Inequalities*, 3, 1–8.

75 Suchard, M.A., Lemey, P., Baele, G. *et al.* (2018) Bayesian phylogenetic and phylodynamic data integration using BEAST 1.10. *Virus Evol.*, 4, vey016.

76 Gentle, J.E., Härdle, W.K., and Mori, Y. (eds) (2012) How computational statistics became the backbone of modern data science, in *Handbook of Computational Statistics*, Springer, pp. 3–16.

77 Lunn, D., Spiegelhalter, D., Thomas, A., and Best, N. (2009) The BUGS project: evolution, critique and future directions. *Stat. Med.*, 28, 3049–3067.

78 Bergstra, J., Breuleux, O., Bastien, F. *et al.* (2010) *Theano: A CPU and GPU Math Expression Compiler*. Proceedings of the Python for Scientific Computing Conference (SciPy) Oral Presentation.

79 Rumelhart, D.E., Hinton, G.E., and Williams, R.J. (1986) Learning representations by back-propagating errors. *Nature*, 323, 533–536.

80 Neal, R.M. (1996) *Bayesian Learning for Neural Networks*, Springer-Verlag.

81 Gelman, A. (2014) Petascale Hierarchical Modeling Via Parallel Execution. U.S. Department of Energy. Report No: DE-SC0002099.

82 Hoffman, M.D. and Gelman, A. (2014) The no-U-turn sampler: adaptively setting path lengths in Hamiltonian Monte Carlo. *J. Mach. Learn. Res.*, 15, 1593–1623.

83 Stan Development Team (2018) Stan Modeling Language Users Guide and Reference Manual. Version 2.18.0.

84 Livingstone, S. and Zanella, G. (2019) On the robustness of gradient-based MCMC algorithms. *arXiv:1908.11812*.

85 Mangoubi, O., Pillai, N.S., and Smith, A. (2018) Does Hamiltonian Monte Carlo mix faster than a random walk on multimodal densities? *arXiv:1808.03230*.

86 Livingstone, S., Faulkner, M.F., and Roberts, G.O. (2019) Kinetic energy choice in Hamiltonian/hybrid Monte Carlo. *Biometrika*, 106, 303–319.

87 Dinh, V., Bilge, A., Zhang, C., and Matsen IV, F.A. (2017) *Probabilistic Path Hamiltonian Monte Carlo*. Proceedings of the 34th International Conference on Machine Learning, vol. 70, pp. 1009–1018.

88 Nishimura, A., Dunson, D.B., and Lu, J. (2020) Discontinuous Hamiltonian Monte Carlo for discrete parameters and discontinuous likelihoods. *Biometrika*, 107, 365–380.

89 Geman, S. and Geman, D. (1984) Stochastic relaxation, Gibbs distributions, and the Bayesian restoration of images. *IEEE Trans. Pattern Anal. Mach. Intell.*, PAMI-6, 721–741.

90 Gelfand, A.E. and Smith, A.F. (1990) Sampling-based approaches to calculating marginal densities. *J. Am. Stat. Assoc.*, 85, 398–409.

91 Monnahan, C.C., Thorson, J.T., and Branch, T.A. (2017) Faster estimation of Bayesian models in ecology using Hamiltonian Monte Carlo. *Methods Ecol. Evol.*, 8, 339–348.

92 Zhang, Z., Zhang, Z., Nishimura, A. *et al.* (2020) Large-scale inference of correlation among mixed-type biological traits with phylogenetic multivariate probit models. *Ann. Appl. Stat.*

93 Dempster, A.P., Laird, N.M., and Rubin, D.B. (1977) Maximum likelihood from incomplete data via the EM algorithm. *J. R. Stat. Soc., Ser. B*, 39, 1–22.

94 Jordan, M.I., Ghahramani, Z., Jaakkola, T.S., and Saul, L.K. (1999) An introduction to variational methods for graphical models. *Mach. Learn.*, 37, 183–233.

95 Wei, G.C. and Tanner, M.A. (1990) A Monte Carlo implementation of the EM algorithm and the poor man's data augmentation algorithms. *J. Am. Stat. Assoc.*, 85, 699–704.

96 Ranganath, R., Gerrish, S., and Blei, D.M. (2014) *Black Box Variational Inference.* Proceedings of the Seventeenth International Conference on Artificial Intelligence and Statistics.

97 Dagum, L. and Menon, R. (1998) OpenMP: an industry standard API for shared-memory programming. *IEEE Comput. Sci. Eng.*, 5, 46–55.

98 Warne, D.J., Sisson, S.A., and Drovandi, C. (2019) Acceleration of expensive computations in Bayesian statistics using vector operations. *arXiv preprint arXiv:1902.09046.*

99 Bergstra, J., Bastien, F., Breuleux, O. *et al.* (2011) *Theano: Deep Learning on GPUS with Python.* NIPS 2011, BigLearning Workshop, Granada, Spain vol. 3, pp. 1–48. Citeseer.

100 Nielsen, M.A. and Chuang, I. (2002) *Quantum computation and quantum information*, Cambridge University Press.

101 Grover, L.K. (1996) *A Fast Quantum Mechanical Algorithm for Database Search.* Proceedings of the Twenty-Eighth Annual ACM Symposium on Theory of Computing, pp. 212–219.

102 Boyer, M., Brassard, G., Høyer, P., and Tapp, A. (1998) Tight bounds on quantum searching. *Fortschritte der Physik: Progress of Physics*, 46, 493–505.

103 Jordan, S.P. (2005) Fast quantum algorithm for numerical gradient estimation. *Phys. Rev. Lett.*,95, 050501.

104 Harrow, A.W., Hassidim, A., and Lloyd, S. (2009) Quantum algorithm for linear systems of equations. *Phys. Rev. Lett.*, 103, 150502.

105 Aaronson, S. (2015) Read the fine print. *Nat. Phys.*, 11, 291–293.

106 COPSS (2020) Committee of Presidents of Statistical Societies, https://community.amstat.org/copss/awards/winners (accessed 31 August 2020).

107 Wickham, H. (2007) Reshaping data with the reshape package. *J. Stat. Soft.*, 21, 1–20.

108 Wickham, H. (2011) The split-apply-combine strategy for data analysis. *J. Stat. Soft.*, 40, 1–29.

109 Wickham, H. (2014) Tidy data. *J. Stat. Soft.*, 59, 1–23.

110 Kahle, D. and Wickham, H. (2013) ggmap: spatial visualization with ggplot2. *R J.*, 5, 144–161.

111 Wickham, H. (2016) *ggplot2: Elegant Graphics for Data Analysis*, Springer.

2

Statistical Software

Alfred G. Schissler and Alexander D. Knudson

The University of Nevada, Reno, NV, USA

This chapter discusses selected statistical software in a format that will inform users transitioning from basic applications to more advanced applications, including elaborate statistical modeling and machine learning (ML), simulation design, and big data situations. We begin with discussions on the most popular statistical software. In the course of these expositions, we provide some historical context for the computing environment, discuss the foundational principles for the development of the language (purpose), discuss user environments/workflows, and analyze strengths and shortcomings for the language (compared to other popular/notable statistical software), language support, among other software features.

Next, we briefly mention an array of software used for statistical applications. We discuss the specific purpose of each software and how the tool fills a need for data scientists. The aim here is to be fairly complete to provide a comprehensive viewpoint of the statistical software ecosystem and to leave readers with some familiarity with the most prevalent languages and software.

After the presentation of noteworthy software, we transition to describing a handful of emerging and promising statistical computing technologies. Our goal in these sections is to guide users who wish to be early adopters for a software application or readers facing a scale-limiting aspect to their current statistical programming language. Some of the latest tools for big data statistical applications are discussed in these sections.

To orientate the reader to the discussion below, two tables are provided. Table 1 includes a list of the software described in the chapter. Throughout, we discuss user environments and workflow considerations to provide practical guidance, aiming to increase efficiency and describe typical use cases. Table 2 summarizes these environments included in the sections that follow.

1 User Development Environments

We begin by discussing user environments rather than focusing on specific statistical programming languages. The subsections below contain descriptions of some selected user development environments and related tools. This introductory material may be omitted if

Computational Statistics in Data Science.
Edited by Walter W. Piegorsch, Richard A. Levine, Hao Helen Zhang and Thomas C. M. Lee.
 ISBN 978-1-11956107-1

Table 1 Summary of selected statistical software.

Software	Open source	Classification	Style	Notes
Python	Y	Popular	Programming	Versatile, popular
R	Y	Popular	Programming	Academia/Industry, active community
SAS	N	Popular	Programming	Strong historical following
SPSS	N	Popular	GUI: menu, dialogs	Popular in scholarly work
C++	Y	Notable	Programming	Fast, low-level
Excel	N	Notable	GUI: menu, dialogs	Simple, works well for rectangular data
GNU Octave	Y	Notable	Mixed	Open source counterpart to MATLAB
Java	Y	Notable	Programming	Cross-platform, portable
JavaScript, Typescript	Y	Notable	Programming	Popular, cross-platform
Maple	N	Notable	Mixed	Academia, algebraic manipulation
MATLAB	N	Notable	Mixed	Speedy, popular among engineers
Minitab	N	Notable	GUI: menu, dialogs	Suitable for teaching and simple analysis
SQL	Y	Notable	Programming	Necessary tool for databases
Stata	N	Notable	GUI: menu, dialogs	Popular in scholary works
Tableau	N	Notable	GUI: menu, dialogs	Popular for business analytics
Julia	Y	Promising	Programming	Speedy, underdeveloped
Scala	Y	Promising	Programming	Typed version of Java, less boilerplate code

Table 2 Summary of selected user environments/workflows.

Software	Virtual environment	Multiple languages	Remote integration	Notes
Emacs, Vim	N	Y	Y	Extensible, steep learning curve
Jupyter project	Y	Y	Y	Open source, interactive data science
RStudio	Y	Y	Y	Excellent at creating reproducible reports/docs

desired, and one can safely proceed to Section 2 for descriptions of the most popular statistical software.

1.1 Extensible Text Editors: Emacs and Vim

GNU's text-editor Emacs (https://www.gnu.org/software/emacs/) is completely free software and offers a powerful solution to working with statistical software. Emacs (or EMACS) is an extensible and customizable text editor that could be used to complete the majority of all computer-based tasks. Once a user learns the keyboard-centric user interface through muscle memory, editing text for reports or coding becomes rapid and outpaces point-and-click style approaches. Emacs works on all major operating systems and gives near-seamless interaction on Linux-based computing clusters. The extensibility ensures that while the latest tools develop and change, your interface will remain constant. This quality will provide confidence to adopt new tools and adapt to new trends in software.

Using Emacs for specifically statistical computing, we note the excellent add-on package called *Emacs Speaks Statistics* (ESS) that offers a unified user interface for R, S-Plus, SAS, Stata, and OpenBUGS/JAGS, among other popular statistical packages. An easy-to-use package manager provides quick ESS installation. Once installed, a basic workflow would be to open an associated file type (.R,.Rmarkdown, etc.) to trigger ESS mode. In ESS mode, code is highlighted, tab completion enabled for rapid code generation and editing, and help documentation integrated. Code can be interactively evaluated in separate processes (e.g., a single or even multiple R sessions), or code can be run noninteractively through Emacs-displayed shell processes. Statistical visualizations are displayed in separate windows for easy plot development. As mentioned above, one can work seamlessly on remote servers (using TRAMP mode). This greatly reduces the inefficiencies inherent to switching between local and remote machines.

We also mention another popular extensible text editor Vim (https://www.vim.org/). Vim offers many of the same benefits as Emacs. There is a constant debate over the superiority of either Vim or Emacs. We avoid this discussion here and simply admit that the first author is an Emacs user, leading to the discussion above. This is not a vote of confidence toward Emacs over Vim but simply a reflection of familiarity.

1.2 Jupyter Notebooks

The Jupyter Project is an effort to develop open-source software and services for interactive computing across a variety of popular programming languages such as Python, R, Julia, and C++. The interactive environment is based on *notebooks* which contain text cells and code cells. Text cells can utilize a mix of plain text, markdown, and render LaTeX through the Mathjax engine. Code cells can be run, modified, and rerun in any order. This functionality makes it easy to perform data analyses and document your work as you go.

The Jupyter IDE (integrated development environment) is run locally in a web browser and can be configured for remote and multiuser workflows. Since reproducible data science is a core feature of the Jupyter Project, they have made it so that notebooks can be exported and shared online as an interactive document or as a static HTML or PDF document. Services such as *mybinder.org* let a user upload and run notebooks online so that an analysis is instantly reproducible by anyone.

1.3 RStudio and Rmarkdown

RStudio is an organization that develops free and enterprise-ready tools for working with the R language. Their IDE (also called *RStudio*) integrates the R console, file browser, script editor, and more in one unified user interface. Through the use of project-associated directories/files, the entire projects are nearly self-contained and easily shared among different systems.

Similar to Jupyter Notebooks, RStudio supports a file format called *Rmarkdown* that allows for code to be embedded and executed in a markdown-style document. The basic setup is a YAML (https://yaml.org/) header, markdown text, and *code chunks*. This simple structure can be built upon through the use of the *knitr* package that can build PDF, HTML, or XML (MS Word) documents and – via the R package *rticles* – build journal-style documents from the same basic file format. *Knitr* can also create slideshows just by changing a parameter in the YAML header. This kind of flexibility for document creation is a huge (and unique) advantage to using Rmarkdown, and it is easily done using the RStudio IDE. Notably, Rmarkdown supports many other programming engines besides R, such as Python, C++, and Julia.

2 Popular Statistical Software

With introductory matters behind, we now transition to discussions of the most popular statistical computing languages. We begin with R, our preferred statistical programming language. This leads to an unbalanced discussion compared to the other most popular statistical software (Python, SAS, and SPSS); yet we hope to provide objective recommendations despite the unequal coverage.

2.1 R

R [1] began at the University of Auckland, New Zealand, in the early 1990s. Ross Ihaka and Robert Gentleman needed a statistical environment to use in their teaching lab. At the time, their computer labs featured only Macintosh computers that lacked suitable software. Ihaka and Gentleman decided to implement a language based on an S-like syntax [2]. R's initial versions were provided to *Statlib* at Carnegie Mellon University, and the user feedback indicated a positive reception.

R's success encouraged its release under the Open Source Initiative (https://opensource.org/). Developers released the first version in June 1995. A software system under the open-source paradigm benefits from having "many pairs of eyes to develop the software." R developed a huge following, and it soon became difficult for the developers to maintain. As a response, a 10-member core group was formed in 1997. The core team handles any changes to the R source code. The massive R community provides support via online mailing lists (https://www.r-project.org/mail.html) and statistical computing forums – such as Talk Stats (http://www.talkstats.com/), Cross Validated (https://stats.stackexchange.com/), and Stack Overflow (https://stackoverflow.com/). Often users receive responses within a matter of minutes.

Since humble beginnings, R has developed into a popular, complete, and flexible statistical computing environment that is appreciated by academia, industry, and government. R's main benefits include support on all major operating systems and comprehensive package archives. Further, R integrates well with document formats (such as LaTeX (https://www.latex-project.org/), HTML, and Microsoft Word) through R Markdown (https://rmarkdown.rstudio.com/) and other file formats to enhance literate programming and reproducible data analysis.

R provides extensive statistical capacity. Nearly any method is available as an R package – the trick is locating the software. The *base* package and default included packages perform most standard analyses and computation. If the included packages are insufficient, one can use CRAN (the comprehensive R archive network) that houses nearly 13 000 packages (visit https://cran.r-project.org/ for more information). To help navigate CRAN, "CRAN Task Views" organizes packages into convenient topics (https://cran.r-project.org/web/views/). For bioinformatics, over 1500 packages reside on Bioconductor [3]. Developers also distribute their packages via git repositories, such as github (https://github.com/). For easy retrieval from github, the *devtools* package allows direct installation.

2.1.1 Why use R over Python or Minitab?

R is tailored to working with data and performing statistical analysis in a way that is more consistent and extensible than Python. The syntax for accessing data in lists and data frames is convenient with tab completion showing what elements are in an object. Creating documents, reports, notebooks, presentations, and web pages is possible through Rmarkdown/RStudio.

Through the use of the metapackage *tidyverse* or the library *data.table*, working with tabular data is direct, efficient, and intuitive. Because R is a scripted language, reproducible workflows are possible, and steps in the process of extracting and transforming data are easy to go back and modify without disrupting the analysis. While this is a virtue shared among all scripting languages, the nature of reproducible results and modular code saves time compared to a point-and-click interface like that of Excel or Minitab.

2.1.2 Where can users find R support?

R has a large community for support online and even built-in documentation within the software. Most libraries provide documentation and examples for their functions and objects that can be accessed via the ? in the command line (e.g., type `?glm` for help about creating a generalized linear model). These help documents are displayed directly in the console, or if using RStudio, they are displayed in the *help* panel with extra links to related functions. For more in-depth documentation, some developers provide *vignettes* for their packages. Vignettes are long-form documentation that demonstrates how to use the functionality in the package and tie it together with a working example.

The online R community is lively, and the people are often helpful. Searching for any question about R or its packages will often lead you to a post on Stack Overflow (https://stackoverflow.com/) or Reddit (either r/rstats or r/RStudio). There is also the RStudio Community (https://community.rstudio.com/) where you can go to ask questions about features specific to the IDE. It is rare to encounter an R programming challenge that has not been addressed somewhere online and, in that case, a well-posed question posted on such forums is quickly answered. Twitter also has an active community of developers that can sometimes respond directly (such as # RSTUDIO or HADLEYWICKHAM).

2.1.3 How easy is R to develop?

R is becoming easier and easier to develop packages and analyses with. This is largely due to the efforts of RStudio, bringing slick new tools and support software on a regular basis. Their software "combine robust and reproducible data analysis with tools to effectively share data products." One package that integrates well with RStudio is *devtools* written by Dr Hadley

Wickham, the chief scientist at RStudio. *devtools* provides a plethora of tools to create, test, and export R packages. *devtools* has grown so comprehensive that developers have split the project into several smaller packages such as *testthat* (for writing tests), *roxygen2* (for writing R documentation), *usethis* (for automating package setup, data, imports, etc.), and a few others that provide convenient tools for building and testing packages.

2.1.4 What is the downside of R?

R is slow. Or at least that is the perception and sometimes the case. This is because R is not a compiled language, so methods of flow control such as for-loops are not optimized. This shortcoming is easily circumvented by taking advantage of the vectorization offered through other built-in functions like those from the *apply* family in R, but these faster techniques often go unused through lack of proficiency or because it is easier to write a for-loop. Intrinsically slow functions can be written in C++ and run via *Rcpp*, but then that negates the simplicity of writing R. This is a special case where Python easily surpasses R. Python is also a scripted language, but through the use of *NumPy* and *numba* it can gain fast vectorized operations, loops, and utilize a just-in-time (JIT) compiler. Ergo, any performance shortcoming of Python can be taken care of through a decorator.

Packages are not written by programmers, or at least not programmers by trade or education. A great deal of libraries for R are written by researchers and analysts who needed a tool and created the tool. Because of this, there is often fragmentation in the syntax or incompatibility between packages, or generally a lack of best practices that leads to poorly performing code, or, in the most drastic setting, code that simply gives erroneous results.

2.1.5 Summary of R

R is firmly entrenched as a premier statistical software package. Its open-source, community-based approach has taken the statistical software scene by storm. R's interactive and scripting programming style makes it an attractive and flexible analytic tool. R does lack the speed/flexibility of other languages; yet, for a specialist in statistics, R provides a near-complete solution. RStudio's efforts further solidify R as a key player moving forward in the modern statistical software ecosystem. We see the popularity of R continuing – however, big data's demands could force R programmers to adapt other tools in conjunction with R, if companies/developers fail to keep up with tomorrow's challenges.

2.2 Python

Created by Guido van Rossum and released in 1991, Python is a hugely popular programming language [4]. Python features readable code, an interactive workflow, and an object-oriented design. Python's architecture affords rapid application development from prototyping to production. Additionally, many tools integrate nicely with Python, facilitating complex workflows. Python also possesses speed, as most of its high-performance libraries are implemented in C/C++.

Python's core distribution lacks statistical features, prompting developers to create supplementary libraries. Below, we detail four well-supported statistical and mathematical libraries: *NumPy* [5], *SciPy* [6], *Pandas* [7], and *Statsmodels* [8].

NumPy is a general and fundamental package for scientific computing [5]. *NumPy* provides functions for operations on large arrays and matrices, optimized for speed via a C implementation. The package features a dense, homogeneous array called *ndarray*. ndarray provides computational efficiency and flexibility. Developers consider *NumPy* a low-level tool as only foundational functions are available. To enhance capabilities, other statistical libraries and packages use *NumPy* to provide richer features.

One widely used higher level package, *SciPy*, employs *NumPy* to enable engineering and data science [6]. *SciPy* contains modules addressing standard problems in scientific computing, such as mathematical integration, linear algebra, optimization, statistics, clustering, image, and signal processing.

Another higher level Python package built upon *NumPy*, *Pandas*, is designed particularly for data analysis, providing standard models and cohesive frameworks [7]. *Pandas* implements a data type named *DataFrame* – a concept similar to the *data.frame* object in R. DataFrame's structure features efficient methods for data sorting, splicing, merging, grouping, and indexing. *Pandas* implements robust input/output tools – supporting flat files, Excel files, databases, and HDF files. Additionally, *Pandas* provides visualization methods via *Matplotlib* [9].

Lastly, the package *Statsmodels* facilitates data exploration, estimation, and statistical testing [8]. Built at even a higher level than the other packages discussed, *Statsmodels* employs *NumPy*, *SciPy*, *Pandas*, and *Matplotlib*. Many statistical models exist, such as linear regression, generalized linear models, probability distributions, and time series. See http://www.statsmodels.org/stable/index.html for the full feature list.

In addition to the four libraries discussed above, Python features numerous other bespoke packages for a particular task. For ML, the *TensorFlow* and *PyTorch* packages are widely used, and for Bayesian inference, *Pyro* and *NumPyro* are becoming popular (see more on these packages in Section 4). For big data computations, *PySpark* provides scalable tools to handle memory and computation time issues. For advanced data visualization, *pyplot*, *seaborn*, and *plotnine* may be worth adopting for a Python-inclined data scientist.

Python's easy-to-learn syntax, speed, and versatility make it a favorite among programmers. Moreover, the packages listed above transform Python into a well-developed vehicle for data science. We see Python's popularity only increasing in the future. Some believe that Python will eventually eliminate the need for R. However, we feel that the immediate future lies in a Python + R paradigm. Thus, R users may well consider exploring what Python offers as the languages have complementary features.

2.3 SAS®

SAS was born during the late 1960s, within the Department of Experimental Statistics at North Carolina State University. As the software developed, the SAS Institute was formed in 1976. Since its infancy, SAS has evolved into an integrated system for data analysis and exploration. The SAS system has been used in numerous business areas and academic institutions worldwide.

SAS provides packages to support various data analytic tasks. The SAS/STAT component contains capabilities one normally associates with data analysis. SAS/STAT supports analysis of variance (ANOVA), regression, categorical data analysis, multivariate analysis, survival analysis, psychometric analysis, cluster analysis, and nonparametric analysis. The SAS/INSIGHT package implements visualization strategies. Visualizations can be linked across multiple windows to uncover trends, spot outliers, and readily discern subtle patterns. Finally, SAS provides the user with a matrix-programming language via the SAS/IML system. The matrix-based language allows custom statistical algorithm development.

Recently, SAS's popularity has diminished [4]; yet, it remains widely used. Open-source competitors threaten SAS's previous overall market dominance. Rather than complete removal, we see SAS becoming a niche product in the future. Now, however, SAS expertise remains desired in certain roles and industries.

2.4 SPSS®

Norman H. Nie, C. Hadlai (Tex) Hul, and Dale Brent developed SPSS in the late 1960s. The trio were Stanford University graduate students at the time. SPSS was founded in 1968 and incorporated in 1975. SPSS became publicly traded in 1993. Now, IBM owns the rights to SPSS. Originally, developers designed SPSS for mainframe use. In 1984, SPSS introduced SPSS/PC+ for computers running MS-DOS, followed by a UNIX release in 1988 and a Macintosh version in 1990. SPSS features an intuitive point-and-click interface. This design empowers a broad user base to conduct standard analyses.

SPSS features a wide variety of analytic capabilities including one for regression, classification trees, table creation, exact tests, categorical analysis, trend analysis, conjoint analysis, missing value analysis, map-based analysis, and complex samples analysis. In addition, SPSS supports numerous stand-alone products including Amos™ (a structural equation modeling package), SPSS Text Analysis for Surveys™ (a survey analysis package utilizing natural language processing (NLP) methodology), SPSS Data Entry™ (a web-based data entry package; *see* Web Based Data Management in Clinical Trials), AnswerTree® (a market segment targeting package), SmartViewer® Web Server™ (a report-generation and dissemination package), SamplePower® (sample size calculation package), DecisionTime® and What if?™ (a scenario analysis package for the nonspecialist), SmartViewer® for Windows (a graph/report sharing utility), SPSS WebApp Framework (web-based analytics package), and the Dimensions Development Library (a data capture library).

SPSS remains popular, especially in scholarly work [4]. For many researchers whom apply standard models, SPSS gets the job done. We see SPSS remaining a useful tool for practitioners across many fields.

3 Noteworthy Statistical Software and Related Tools

Next, we discuss noteworthy statistical software, aiming to provide essential details for a fairly complete survey of the most commonly used statistical software and related tools.

3.1 BUGS/JAGS

The BUGS (Bayesian inference using Gibbs sampling) project led to some of the most popular general-purpose Bayesian posterior sampling programs – WinBUGS [10] and, later, OpenBUGS, the open-source equivalent. BUGS begin in 1989 in the MRC Biostatistics Unit,

Cambridge University. The project in part led to a rapid expansion of applied Bayesian statistics due its pioneering timing, relative ease of use, and broad range of applicable models.

JAGS (Just Another Gibbs Sampler) [11] was developed as a cross-platform engine for the BUGS modeling language. A secondary goal was to provide extensibility, allowing user-specific functions, distributions, and sampling algorithms. The BUGS/JAGS approach to specifying probabilistic models has become standard in other related software (e.g., NIMBLE). Both BUGS and JAGS are still widely used and are well suited for tasks of small-to-medium complexity. However, for highly complex models and big data problems there are similar, more-powerful Bayesian inference engines emerging, for example, STAN and Pyro (see Section 4 for more details).

3.2 C++

C++ is a general-purpose, high-performance programming language. Unlike other scripting languages for statistics such as R and Python, C++ is a compiled language – adding complexity (such as memory management) and strict syntax requirements. As such, C's design may complicate prototyping. Thus, data scientists typically turn to C++ to optimize/scale a developed algorithm at the production level.

C++'s standard libraries lack many mathematical and statistical operations. However, since C++ can be compiled cross-platform, developers often interface C++ functions from different languages (e.g., R and Python). Thus, C++ can be used to develop libraries across languages, offering impressive computing performance.

To enable analysis, developers created mathematical and statistical libraries in C++. The packages often employ of BLAS (basic linear algebra subprograms) libraries, written in C/Fortran and offer numerous low-level, high-performance linear algebra operations on numbers, vectors, and matrices. Some popular BLAS-compatible libraries include Intel Math Kernel Library (MKL) [12], automatically tuned linear algebra software (ATLAS) [13], OpenBLAS [14], and linear algebra package (LAPACK) [15].

Among the C++ libraries for mathematics and statistics built on top BLAS, we detail three popular, well-maintained libraries: Eigen [16], Armandillo [17], and Blaze [18] below:

> *Eigen* is a high-level, header-only library developed by Guennebaud *et al.* [16]. *Eigen* provides classes dealing with vector types, arrays, and dense/sparse/large matrices. It also supports matrix decomposition and geometric features. *Eigen* uses single instruction multiple data vectorization to avoid dynamic memory allocation. *Eigen* also implements extra features to optimize the computing performance, including unrolling techniques and processor-cache utilization. *Eigen* itself does not take much advantage from parallel hardware, currently supporting parallel processing only for general matrix–matrix products. However, since *Eigen* uses BLAS-compatible libraries, users can utilize external BLAS libraries in conjunction with *Eigen* for parallel computing. Python and R users can call *Eigen* functions using the *minieigen* and *RcppEigen* packages.

The National ICT Australia (NICTA) developed the open-source library *Armadillo* to facilitate science and engineering [17]. *Armadillo* provides a fast, easy-to-use matrix library with MATLAB-like syntax. *Armadillo* employs template meta-programming techniques to avoid unnecessary operations and increase library performance. Further, *Armadillo* supports 3D objects and provides numerous utilities for matrices manipulation and decomposition. *Armadillo* automatically utilizes open multiprocessing (OpenMP) [19] to increase speed. Developers designed *Armadillo* to provide a balance between speed and ease of use. *Armadillo* is widely used for many applications in ML, pattern recognition, signal processing, and bioinformatics. R users may call *Armadillo* functions through the *RcppArmadillo* package.

Blaze is a high-performance math library for dense/sparse arithmetic developed by Iglberger *et al.* [18]. *Blaze* extensively uses LAPACK functions for various computing tasks, such as matrix decomposition and inversion, providing high-performance computing. *Blaze* supports high-performance parallex (HPX) [20] and OpenMP to enable parallel computing.

The difficulty to develop C++ programs limits its use as a primary statistical software package. Yet, C++ appeals when a fast, production-quality program is desired. Therefore, R and Python developers may find C++ knowledge beneficial to optimize their code prior to distribution. We see C/C++ as the standard for speed and, as such, an attractive tool for big data problems.

3.3 Microsoft Excel/Spreadsheets

Much of statistical work today involves the use of Microsoft Excel and other spreadsheet-style applications (Google Sheets, Apple Numbers, etc.). A spreadsheet application provides a simple and interactive way to collect data. This has an appeal for any manual data entry process. The sheets are easy to share, both through traditional file sharing (e.g., e-mail attachments) and cloud-based solutions (Google Drive, Dropbox, etc.). Simple numeric summaries and plots are easy to construct. More advanced macros/scripts are possible, yet most data scientists would prefer to switch to a more full-featured environment (such as R or Python). Yet, as nearly all computer workers have some level of familiarity with spreadsheets, spreadsheets remain hugely popular and ubiquitous in organizations. Thus, we wager that spreadsheet applications will likely always be involved in statistical software and posit they can be quite efficient for appropriate tasks.

3.4 Git

Very briefly, we mention Git, a free and open-source distributed version control system (https://git-scm.com/). As the complexities of modern data science workflows increase, statistical programmers are increasingly reliant on some type of version control system, with Git being the most popular. Git allows for a branching scheme to foster experimentation in projects and to converge to a final product. By compiling a complete history of a project, Git provides transparent data analyses for reproducible research. Further, projects and software can be shared easily via web-based repositories, such as GitHub (https://github.com/).

3.5 Java

Java is one of the most popular programming languages (according to the TIOBE index, www.tiobe.com/tiobe-index/), partially due to its extensive library ecosystem. Java's design seduces programmers – it is simple, object oriented, and portable. Java applications run on any machine, from personal laptops to high-performance supercomputers, even game consoles and internet of things (IoT) devices. Notably, Android (based on Java) development has driven recent Java innovations. Java's "write once, run anywhere" adage provides versatility, triggering interest even at the research level.

Developers may prefer Java for intensive calculations performing slowly within scripted languages (e.g., R). For speed-up purposes, Java's cross-platform design could even be preferred to C/C++ in certain cases. Alternatively, Java code can be wrapped nicely in an R package for faster processing. For example, the *rJava* package allows one to call java code in an R script and also reversely (calling *R functions* in Java). On the other hand, Java can be used independently for statistical analysis, thanks to a nice set of statistical libraries.

Popular sources of native Java statistical and mathematical functionalities are JSC (Java Statistical Classes) and Apache Commons Math application programming interfaces (APIs) (http://commons.apache.org/proper/commons-math/). JSC and Apache Commons Math libraries perform many methods including univariate statistics, parametric and nonparametric tests (*t*-test, chi-square test, and Wilcoxon test), random number generation, random sampling/resampling, regression, correlation, linear or stochastic optimization, and clustering.

Additionally, Java boasts an extensive number of machine-learning packages and big data capabilities. For example, Java enables the WEKA [21] tool, the JSAT library [22], and the TensorFlow framework [23]. Moreover, Java provides one of the most desired and useful big data analysis tools – Apache Spark [24]. Spark provides ML support through modules in the Spark MLlib library [25].

As with other discussed software, Java APIs often require importing other packages/libraries. For example, developers commonly use external matrix-operation libraries, such as JAMA (Java matrix package, https://math.nist.gov/javanumerics/jama/) or EJML (efficient Java matrix library, http://ejml.org/wiki/). Such packages allow for routine computation – for example, matrix decomposition and dense/sparse matrix calculation. *JFreeCHart* enables data visualization by generating scatter plots, histograms, barplots, and so on. Recently, these Java libraries are being replaced by more popular JavaScript libraries such as *Plot.ly* (https://plot.ly/), *Bokeh* (bokeh.pydata.org), *D3* [26], or *Highcharts* (www.highcharts.com).

As outlined above, Java could serve as a useful statistical software solution, especially for developers familiar with it or who have interest in cross-platform development. We would then recommend its use for seasoned programmers looking to add some statistical punch to their desktop, web, and mobile apps. For the analysis of big data, Java offers some of the best ML tools available.

3.6 JavaScript, Typescript

JavaScript is one of the most popular programming languages, outpacing even Java and Python. It is fully featured, flexible, and fast, leading to its broad appeal. JavaScript excels at visualization through *D3.js*. JavaScript even features interactive, browser-based ML via *TensorFlow.js*. For real-time data collection and analysis, JavaScript provides streaming tools through *MongoDB*. JavaScript's unsurpassed popularity alone makes it worth a look, especially if tasked with a complex real-time data analytic challenge across heterogeneous architectures.

3.7 Maple

Maple is a "math software that combines the world's most powerful math engine with an interface that makes it extremely easy to analyze, explore, visualize, and solve mathematical problems." (https://www.maplesoft.com/products/Maple/). While not specifically a statistical software package, Maple's computer algebra system is a handy supplement to an analyst's toolkit. Often in statistical computing, a user may employ Maple to check a hand calculation or reduce the workload/error rate in lengthy derivations. Moreover, Maple offers add-on packages for statistics, calculus, analysis, linear algebra, and more. One can even create interactive plots and animations. In sum, Maple is a solid choice for a computer algebra system to aid in statistical computing.

3.8 MATLAB, GNU Octave

MATLAB began as FORTRAN subroutines for solving linear (LINPACK) and eigenvalue (EISPACK) problems. Cleve Moler developed most of the subroutines in the 1970s for use in the classroom. MATLAB quickly gained popularity, primarily through word of mouth. Developers rewrote MATLAB in C during the 1980s, adding speed and functionality. The parent company of MATLAB, the Mathworks, Inc., was created in 1984, and MATLAB has since become a fully featured tool that is often used in engineering and developer fields where integration with sensors and controls is a primary concern.

MATLAB has a substantial user base in government, academia, and the private sector. The MATLAB base distribution allows reading/writing data in ASCII, binary, and MATLAB proprietary formats. The data are presented to the user as an array, the MATLAB generic term for a matrix. The base distribution comes with a standard set of mathematical functions including trigonometric, inverse trigonometric, hyperbolic, inverse hyperbolic, exponential, and logarithmic. In addition, MATLAB provides the user with access to *cell arrays*, allowing for heterogeneous data across the cells and creation analogous to a C/C++. MATLAB provides the user with numerical methods, including optimization and quadrature functions.

A highly similar yet free and open-sourced programming language is GNU Octave. Octave offers many if not all features of the core MATLAB distribution, although MATLAB has many add-on packages for which Octave has no equivalent, and that may prompt a user to choose MATLAB over Octave. We caution analysts against using MATLAB/Octave as their primary statistical computing solution as MATLAB's popularity is diminishing [4] – likely due to open-source, more fully featured competitors such as R and Python.

3.9 Minitab®

Barbara F. Ryan, Thomas A. Ryan, Jr., and Brian L. Joiner created Minitab in 1972 at the Pennsylvania State University to teach statistics. Now, Minitab Inc. owns the proprietary software. Academia and industry widely employ Minitab [4]. The intuitive point-and-click design and spreadsheet-like interface allow users to analyze data with little learning curve. Minitab feels like Excel, but with many more advanced features. This greatly reduces the Minitab learning curve compared to more flexible programming environments.

Minitab offers import tools and a comprehensive set of statistical capabilities. Minitab's features include basic statistics, ANOVA, fixed and mixed models, regression analyses, measurement systems analysis, and graphics including contour and rotating 3D plots. A full feature list resides at http://www.minitab.com/en-us/products/minitab/features-list/. For advanced users, a command-line editor exists. Within the editor, users may customize macros (functions).

Minitab serves its user base well and will continue to be viable in the future. For teaching academics, Minitab provides near immediate access to many statistical methods and graphics. For industry, Minitab offers tools to produce standardized analyses and reports with little training. However, Minitab's flexibility and big data capabilities are limited.

3.10 Workload Managers: SLURM/LSF

Working on shared computing clusters has become commonplace in contemporary data science applications. Some working knowledge of workload managing programs (aka *schedulers*) is essential to running statistical software in these environments. Two popular workload managers are SLURM (https://slurm.schedmd.com/documentation.html) and IBM's platform load sharing facility (LSF), another popular workload management platform for distributed high-performance computing. These schedulers can be used to execute batch jobs on networked Unix and Windows systems on many different architectures. A user would typically interface with a scheduling program via a command line tool or through a scripting language. The user specifies the hardware resources and program inputs. The scheduler then distributes the work across resources, and jobs are run based on system-prioritization schemes. In such a way, hundreds or even thousands of programs can be run in parallel, increasing the scale of statistical computations possible within a reasonable time frame. For example, simulations for a novel statistical method could require many thousands of runs at various configurations, and this could be done in days rather than months.

3.11 SQL

Structured Query Language (SQL) is the standard language for relationship database management systems. While not strictly a statistical computing environment, the ability to query databases through SQL is an essential skill for data scientists. Nearly all companies seeking a data scientist require SQL knowledge as much of an analyst's job is extracting, transforming, and loading data from an established relational database.

3.12 Stata®

Stata is commercial statistical software, developed by William Gould in 1985. StatCorp currently owns/develops Stata and markets the product as "fast, accurate, and easy to use with both a point-and-click interface and a powerful, intuitive command syntax" (https://www.stata.com/). However, most Stata users maintain the point-and-click workflow. Stata strives to provide user confidence through regulatory certification.

Stata provides hundreds of tools across broad applications and methods. Even Bayesian modeling and maximum-likelihood estimation are available. With its breadth, Stata targets all sectors – academia, industry, and government.

Overall, Stata impresses through active support and development while possessing some unique characteristics. Interestingly, in scholarly work over the past decade, only SPSS, R, and SAS have overshadowed Stata [4]. Taken together, we anticipate Stata to remain popular. However, Stata's big data capabilities are limited, and we have reservations whether industry will adopt Stata over competitors.

3.13 Tableau®

Tableau stemmed from visualization research by Stanford University's computer science department in 1999. The Seattle-based company was founded in 2003. Tableau advertises itself as a data exploration and visualization tool, not a statistical software *per se*. Tableau targets the business intelligence market primarily. However, Tableau provides a free, less powerful version for instruction.

Tableau is versatile and user-friendly: providing MacOS and Windows versions while supporting web-based apps on iOS and Android. Tableau connects seamlessly to SQL databases, spreadsheets, cloud apps, and flat files. The software appeals to nontechnical "business" users via its intuitive user interface but also allows "power users" to develop analytical solutions by connecting to an R server or installing TabPy to integrate Python scripts.

Tableau could corner the data visualization market with its easy-to-learn interface, yet intricate features. We contend that big data demands visualization as many traditional methods are not well suited for high-dimensional, observational data. Based on its unique characteristics, Tableau will appeal broadly and could even emerge as a useful tool to supplement an R or Python user's toolkit.

4 Promising and Emerging Statistical Software

With a forward-thinking mindset, our final section describes a few emerging and promising statistical software languages/packages that have the ability to meet tomorrow's complex modeling demands. If a reader encounters scalability challenges in their current statistical programming language, one of the following options may turn a computationally infeasible model into a useful one.

4.1 Edward, Pyro, NumPyro, and PyMC3

Recently, there have been several important probabilistic programming libraries released for Python, namely, *Edward*, *Pyro*, *NumPyro*, and *PyMC3*. These packages are characterized by the capacity to fit broad classes of models, with massive number of parameters, using advanced particle simulators (such as Hamiltonian Monte Carlo (HMC)).

These packages differ in implementation, but all provide world-class computational solutions to probabilistic inference and Monte Carlo techniques. These packages provide the latest and optimized algorithms for many classes of models: directed graphs, neural networks, implicit generative models, Bayesian nonparametrics, Markov Chains, variational inference, Bayesian multilevel regression, Gaussian processes, mixture modeling, and survival analysis. Edward is built on a *TensorFlow* backend, while *Pyro* is built using *PyTorch* (and *NumPyro* is based on *NumPy*). *Pyro* uses the universal probabilistic programming language (PPL) to specify models. *NumPy* complies code to either central processing unit (CPU) or Graphical Processing Unit (GPU), greatly increasing computation speed in many statistical/linear algebra computations. *PyMC3* is built on a *Theno* backend and uses an intuitive syntax to specify models.

4.2 Julia

Julia is a new language designed by Bezanson *et al.* and was released in 2012 [27]. Julia's first stable version (1.0) was released in August 2018. The developers describe themselves as "greedy" – they want a software application that does it all. Users no longer would create prototypes in scripting languages than port to C or Java for speed. Below, we quote from Julia's public announcement (https://julialang.org/blog/2012/02/why-we-created-julia):

> We want a language that's open source, with a liberal license. We want the speed of C with the dynamism of Ruby. We want a language that's homoiconic, with true macros like Lisp, but with obvious, familiar mathematical notation like MATLAB. We want something as usable for general programming as Python, as easy for statistics as R, as natural for string processing as Perl, as powerful for linear algebra as MATLAB, as good at gluing programs together as the shell. Something that is dirt simple to learn, yet keeps the most serious hackers happy. We want it interactive and we want it compiled.

Despite the stated goals, we classify Julia as an analysis software at this early stage. Indeed, Julia's syntax exhibits elegance and friendliness to mathematics. The language natively implements an extensive mathematical library. Julia's core distribution includes multidimensional arrays, sparse vectors/matrices, linear algebra, random number generation, statistical computation, and signal processing.

Julia's design affords speeds comparable to C due to it being an interpreted, embeddable language with a JIT compiler. The software also implements concurrent threading, enabling parallel computing natively. Julia integrates nicely with other languages including calling C directly, Python via *PyCall*, and R via *RCall*.

Julia exhibits great promise but remains nascent. We are intrigued by a language that does it all and is easy to use. Yet, Julia's underdevelopment limits its statistical analysis capability. On the other hand, Julia is growing fast with active support and positive community outlook. Coupling Julia's advantages and MATLAB's diminishing appeal, we anticipate Julia to contribute in the area for years to come.

4.3 NIMBLE

NIMBLE (https://r-nimble.org/) provides a framework for building and sharing computationally intensive statistical models. The software has gained instant recognition due to the adoption of the familiar BUGS modeling language. This feature appeals to a broad base of Bayesian statisticians who have limited time to invest in learning new computing skills. NIMBLE is implemented as an R package, but all the under-the-hood work is completed in compiled C++ code, providing near-optimal speed. Even if a user does not desire the BUGS language, NIMBLE accelerates R for general-purpose numerical work via *nimbleFunctions* without the burden of writing native C++ source code.

4.4 Scala

An emerging data science tool, Scala (https://www.scala-lang.org/), combines object-oriented and functional paradigms in a high-level programming language. Scala is built for complex applications and workflows. To meet such applications, *static* object typing keeps the code bug-free, even during numerous parallelized computations or asynchronous programming (dependent jobs). Scala is designed for interoperability with Java/JavaScript as it runs on Java Virtual Machine. This provides access to the entire Java ecosystem. Scala interfaces with Apache Spark (as does Python and R) for scalable, accurate, and numeric operations. In short, Scala scales Java for high-performance computing.

4.5 Stan

Stan [28] is a PPL for specifying models, most often Bayesian. Stan samples posterior distributions using HMC – a variant of Markov Chain Monte Carlo (MCMC). HMC boasts a more robust and efficient approach over Gibbs or Metropolis-Hastings sampling for complex models, while providing insightful diagnostics to assess convergence and mixing. This may explain why Stan is gaining popularity over other Bayesian samplers (such as BUGS [10] and JAGS [11]).

Stan provides a flexible and principled model specification framework. In addition to fully Bayesian inference, Stan computes log densities and Hessians, variational Bayes, expectation propagation, and approximate integration. Stan is available as a command line tool or R/Python interface (RStan and PyStan, respectively).

Stan has the ability to become the *de facto* Bayesian modeling software. Designed by thought leader Andrew Gelman and a growing, enthusiastic community, Stan possesses much promise. The language architecture promotes cross-compatibility and extensibility, and the general-purpose posterior sampler with innovative diagnostics appeals to novice and advanced modelers alike. Further, to our knowledge, Stan is the only general-purpose Bayesian modeler that scales to thousands of parameters – a boon for big data analytics.

5 The Future of Statistical Computing

Two key drivers will dictate statistical software moving forward: (i) Increased model complexity and (ii) increased data collection speed and sheer size (big data). These two factors will require software to be highly flexible – the languages must be easy to work with for small-to-medium data sets/models, while easily scaling to massive data sets/models. The software must give easy access to the latest computer hardware (including GPUs) and provide hassle-free parallel distribution of tasks. To this end, successful statistical software must feature compiled/optimized code of the latest algorithms, parallelization, and cloud/cluster computing support. Likely, one tool will not meet all the demands, and therefore cross-compatibility standards must be developed. Moreover, data visualization will become increasingly important (including virtual reality) for large, complex data sets where conventional inferential tools are suspect or without use.

The advantages of open-source, community-based development have been emphasized throughout – especially in the scholarly arena and with smaller businesses. The open-source paradigm enables rapid software development with limited resources. However, commercial software with dedicated support services will appeal to certain markets, including medium-to-large businesses.

6 Concluding Remarks

We attempted to evaluate the current statistical software landscape. Admittedly, our treatment has been focused by our experience. We have, however, sought to be fair in our appraisal and provide the burgeoning statistical programmer the information required to make strong tool selection choices and increase their performance. We begin by in-depth discussions of the most-popular statistical software, followed by brief descriptions of many other noteworthy tools, and then finally highlighted a handful of emerging statistical software. We hope that this organization is useful, but note that it is solely based on our experiences and informal popularity studies [4]. We also provided a limited prognostication with regard to the statistical software future by identifying issues and applications likely to shape software development. We realize, of course, that the future is usually full of surprises and only time will tell what actually occurs.

Acknowledgments

The work of the two authors, AG Schissler and A Knudson, was partially supported by the NIH grant (1U54GM104944) through the National Institute of General Medical Sciences (NIGMS) under the Institutional Development Award (IDeA) program. The authors thank the Wiley staff and editor of this chapter, Dr Walter W. Piegorsch, for their expertise and support.

References

1 R Core Team (2018) *R: A Language and Environment for Statistical Computing*. R Foundation for Statistical Computing, Vienna, Austria.

2 Venables, W. and Ripley, B.D. (2013) *S Programming*, Springer Science & Business Media, New York, NY, USA.

3 Gentleman, R.C., Carey, V.J., Bates, D.M., *et al.* (2004) Bioconductor: open software development for computational biology and bioinformatics. *Genome Biol.*, 5 (10), R80.

4 Muenchen, R.A. (2019) The Popularity of Data Science Software, r4stats.com/articles/popularity.

5 Oliphant, T.E. (2006) *A Guide to NumPy*, vol. 1, Trelgol Publishing, Provo, UT, USA, p. 85.

6 Jones, E., Oliphant, T., and Peterson, P. (2001) SciPy: open source scientific tools for Python.

7 McKinney, W. (2011) pandas: a foundational Python library for data analysis and statistics. *Python High Performance Sci. Comput.*, 14 (9), 1–9.

8 Seabold, S. and Perktold, J. (2010) *Econometric and Statistical Modeling with Python Skipper Seabold 1 1*. Proceedings of the 9th Python in Science Conference, vol. 57, p. 61.

9 Hunter, J.D. (2007) Matplotlib: a 2D graphics environment. *Comput. Sci. Eng.*, 9 (3), 90–95.

10 Thomas, A., Spiegelhalter, D.J., and Gilks, W.R. (1992) BUGS: a program to perform Bayesian inference using Gibbs sampling. *Bayesian Stat.*, 4 (9), 837–842.

11 Plummer, M. (2005) JAGS: just another Gibbs sampler. Proceedings of the 3rd International Workshop on Distributed Statistical Computing (DSC 2003), Vienna, Austria.

12 Intel (2007) Intel® Math Kernel Library Reference Manual, https://software.intel.com/en-us/mkl.

13 Whaley, R.C. and Petitet, A. (2005) Minimizing development and maintenance costs in supporting persistently optimized BLAS. *Softw. Pract. Exp.*, 35 (2), 101–121.

14 Xianyi, Z., Qian, W., and Chothia, Z. (2012) OpenBLAS, p. 88, http://xianyi.github.io/OpenBLAS.

15 Anderson, E., Bischof, C., Demmel, J., *et al.* (1990) Prospectus for an Extension to LAPACK. Working Note ANL-90-118, Argonne National Laboratory.

16 Guennebaud, G., *et al.* (2010) Eigen v3.

17 Sanderson, C., and Curtin, R. (2016) Armadillo: a template-based C++ library for linear algebra. *J. Open Source Softw.*, 1 (2), 26.

18 Iglberger, K., Hager, G., Treibig, J., and Rüde, U. (2012) *High Performance Smart Expression Template Math Libraries*. 2012 International Conference on High Performance Computing and Simulation (HPCS) (pp. 367–373) IEEE.

19 Dagum, L., and Menon, R. (1998) OpenMP: an industry standard API for shared-memory programming. *IEEE Comput. Sci. Eng.*, 5 (1), 46–55.

20 Heller, T., Diehl, P., Byerly, Z., *et al.* (2017) *Hpx-An Open Source C++ Standard Library for Parallelism and Concurrency*. Proceedings of OpenSuCo, p. 5.

21 Frank, E., Hall, M.A., and Witten, I.H. (2016) *The WEKA Workbench*, Morgan Kaufmann, Burlington, MA.

22 Raff, E. (2017) JSAT: Java statistical analysis tool, a library for machine learning. *J. Mach. Learn. Res.*, 18 (1), 792–796.

23 Abadi, M., Agarwal, A., Barham, P., *et al.* (2015) TensorFlow: large-scale machine learning on heterogeneous systems.

24 Zaharia, M., Xin, R.S., Wendell, P., *et al.* (2016) Apache spark: a unified engine for big data processing. *Commun. ACM*, 59 (11), 56–65.

25 Meng, X., Bradley, J., Yavuz, B., *et al.* (2016) Mllib: machine learning in Apache Spark. *J. Mach. Learn. Res.*, 17 (1), 1235–1241.
26 Bostock, M., Ogievetsky, V., and Heer, J. (2011) D3 data-driven documents. *IEEE Trans. Vis. Comput. Graph.*, 17 (12), 2301–2309.
27 Bezanson, J., Karpinski, S., Shah, V.B., and Edelman, A. (2012) Julia: a fast dynamic language for technical computing. *arXiv preprint arXiv:1209.5145.*
28 Carpenter, B., Gelman, A., Hoffman, M.D., *et al.* (2017) Stan: a probabilistic programming language. *J. Stat. Softw.*, 76 (1), 1–32.

Further Reading

de Leeuw, J. (2009) Journal of Statistical Software, *Wiley Interdiscip. Rev. Comput. Stat.*, 1 (1), 128–129.

3

An Introduction to Deep Learning Methods

Yao Li[1], *Justin Wang*[2], *and Thomas C. M. Lee*[2]

[1] *University of North Carolina at Chapel Hill, Chapel Hill, NC, USA*
[2] *University of California at Davis, Davis, CA, USA*

1 Introduction

Many models in the field of machine learning, such as deep neural networks (DNNs) and graphical models, are naturally represented in a layered network structure. The more layers we use in such models, the more complex the functions that are able to be represented. However, models with many layers are difficult to estimate optimally, and thus those in the machine learning field have generally opted to restrict their model to fewer layers, trading model expressivity for simplicity [1]. Deep learning explores ways to effectively train models with many hidden layers in order to retain the model's expressive powers. One of the most effective approaches to deep learning has been proposed by Hinton and Salakhutdinov [2]. Traditionally, estimating the parameters of network-based models involves an iterative algorithm with the initial parameters being randomly chosen. Hinton's proposed method involves *pretraining*, or deliberately presetting in an effective manner, the parameters of the model as opposed to randomly initializing them. In this chapter, we review the architectures and properties of DNNs and discuss their applications.

We first briefly discuss the general machine learning framework and basic machine learning methodology in Section 2. We then discuss feedforward neural networks and backpropagation in Section 3. In Section 4, we explore convolutional neural networks (CNNs), the type of architectures that are usually used in computer vision. In Section 5, we discuss autoencoders, the unsupervised learning models that learn latent features without labels. In Section 6, we discuss recurrent neural networks (RNNs), which can handle sequence data.

2 Machine Learning: An Overview

2.1 Introduction

Machine learning is a field focusing on the design and analysis of algorithms that can learn from data [3]. The field originated from artificial intelligence research in the late 1950s, developing independently from statistics. However, by the early 1990s, machine learning

Computational Statistics in Data Science.
Edited by Walter W. Piegorsch, Richard A. Levine, Hao Helen Zhang and Thomas C. M. Lee.
 ISBN 978-1-11956107-1

researchers realized that a lot of statistical methods could be applied to the problems they were trying to solve. Modern machine learning is an interdisciplinary field that encompasses theory and methodology from both statistics and computer science.

Machine learning methods are grouped into two main categories, based on what they aim to achieve. The first category is known as *supervised learning*. In supervised learning, each observation in a dataset comes attached with a *label*. The label, similar to a response variable, may represent a particular class the observation belongs to (categorical response) or an output value (real-valued response). In either case, the ultimate goal is to make inferences on possibly unlabeled observations outside of the given dataset. Prediction and classification are both problems that fall into the supervised learning category. The second category is known as *unsupervised learning*. In unsupervised learning, the data come without labels, and the goal is to find a pattern within the data at hand. Unsupervised learning encompasses the problems of clustering, density estimation, and dimension reduction.

2.2 Supervised Learning

Here, we state the problem of supervised learning explicitly. We have a set of training data $\boldsymbol{X} = (\boldsymbol{x}_1, ..., \boldsymbol{x}_n)$, where $\boldsymbol{x}_i \in \mathbb{R}^p$ for all i, and a corresponding set of labels $\boldsymbol{y} = (y_1, ..., y_n)$, which can represent either a category membership or a real-valued response. We aim to construct a function $\delta : \mathbb{R}^p \to \mathbb{R}$ that maps each input $\boldsymbol{x}_i$ to a predicted label $\hat{y}_i$. A given supervised learning method $\mathcal{M}$ chooses a particular form $\delta = \delta(\boldsymbol{X}, \theta_{\mathcal{M}})$, where $\theta_{\mathcal{M}}$ is a vector of parameters based on $\mathcal{M}$.

We wish to choose $\delta(\boldsymbol{X}, \theta_{\mathcal{M}})$ to minimize an *error function* $E(\delta, \boldsymbol{y})$. The error function is most commonly taken to be the sum of square errors in which case the goal is to choose an optimal $\delta^*(\boldsymbol{X}, \theta_{\mathcal{M}})$ such that

$$\delta^*(\boldsymbol{X}, \theta_{\mathcal{M}}) = \arg\min_{\delta} E(\delta, \boldsymbol{y}) = \arg\min_{\delta} \sum_{i=1}^{n} \ell(\delta(\boldsymbol{x}_i, \theta_{\mathcal{M}}), y_i)$$

where ℓ can be any loss function that evaluates the distance between $\delta(\boldsymbol{x}_i, \theta_{\mathcal{M}})$ and y_i, such as cross-entropy loss and square loss.

2.3 Gradient Descent

The form of the function δ will usually be fairly complex, so attempting to find $\delta^*(\boldsymbol{X}, \theta_{\mathcal{M}})$ via direct differentiation will not be feasible. Instead, we use *gradient descent* to minimize the error function.

Gradient descent is a general optimization algorithm that can be used to find the minimizer of any given function. We pick an arbitrary starting point, and then at each time point, we take a small step in the direction of the greatest decrease, which is given by the gradient. The idea is that if we repeatedly do this, we will eventually arrive at a minimum. The algorithm guarantees a local minimum, but not necessarily a global one [4]; see Algorithm 1.

Gradient descent is often very slow in machine learning applications, as finding the true gradient of the error criterion usually involves iterating through the entire dataset. Since we need to calculate the gradient at each time step of the algorithm, this leads to having

Algorithm 1. Gradient Descent

Input: a function $g(\theta)$ to be minimized
Output: a minimizer, $\hat{\theta}$
initialization: θ_i = random value for $i = 1, \dots, p$; $t = 0$; η = value in $(0, 1)$ (η is known as the *learning rate*);
while *not converged* **do**
  calculate $\nabla g(\theta_t)$, the gradient of $g(\theta)$ evaluated at θ_t;
  $\theta_{t+1} \leftarrow \theta_t + \eta \cdot \nabla g(\theta_t)$;
  $t \leftarrow t + 1$;
end

to iterate through the entire dataset a very large number of times. To speed up the process, we instead use a variation on gradient descent known as *stochastic gradient descent*. Stochastic gradient descent involves approximating the gradient at each time step with the gradient at a single observation, which significantly speeds up the process [5]; see Algorithm 2.

Algorithm 2. Stochastic Gradient Descent

Input: a function $g(\theta) = \sum_{i=1}^{n} g_i(\theta)$ to be minimized
Output: a minimizer, $\hat{\theta}$
initialization: θ_i = random value for $i = 1, \dots, p$; $t = 0$; η = value in $(0, 1)$ (η is known as the *learning rate*); random.obs = random permutation of $[1, \dots, n]$;
while *not converged* **do**
  for *i in random.obs* **do**
    calculate $\nabla g_i(\theta_t)$, the gradient of $g(\theta)$ evaluated at θ_t with the ith observation;
    $\theta_{t+1} \leftarrow \theta_t + \eta \cdot \nabla g_i(\theta_t)$;
    $t \leftarrow t + 1$;
  end
end

3 Feedforward Neural Networks

3.1 Introduction

A feedforward neural network, also known as a multilayer perceptron (MLP), is a popular supervised learning method that provides a parameterized form for the nonlinear map δ from an input to a predicted label [6]. The form of δ here can be depicted graphically as a directed layered network, where the directed edges go upward from nodes in one layer to nodes in the next layer. The neural network has been seen to be a very powerful model, as they are able to approximate any Borel measurable function to an arbitrary degree, provided that parameters are chosen correctly.

3.2 Model Description

We start by describing a simple MLP with three layers, as depicted in Figure 1.

The bottom layer of a three-layer MLP is called the *input layer*, with each node representing the respective elements of an input vector. The top layer is known as the *output layer* and represents the final output of the model, a predicted vector. Again, each node in the output layer represents the respective predicted score of different classes. The middle layer is called the *hidden layer* and captures the unobserved latent features of the input. This is the only layer where the number of nodes is determined by the user of the model, rather than the problem itself.

The directed edges in the network represent *weights* from a node in one layer to another node in the next layer. We denote the weight from a node x^i in the input layer to a node h^j in the hidden layer as w_{ij}. The weight from a node h^j in the hidden layer to a node $\hat{y}^k$ in the output layer will be denoted v_{jk}. In each of the input and hidden layers, we introduce *intercept nodes*, denoted x^0 and h^0, respectively. Weights from them to any other node are called *biases*. Each node in a given layer is connected by a weight to every node in the layer above except the intercept node.

The value of each node in the hidden and output layers is determined as a nonlinear transformation of the linear combination of the values of the nodes in the previous layers and the weights from each of those nodes to the node of interest. That is, the value of h^j, $j = 1, \ldots, m$, is given by $\gamma(\boldsymbol{w}_j^T \boldsymbol{x})$, where $\boldsymbol{w}_j = (w_{0j}, \ldots, w_{pj})^T$, $\boldsymbol{x} = (1, x^1, \ldots, x^p)^T$, and $\gamma(\cdot)$ is a nonlinear transformation with range in the interval $(0, 1)$. Similarly, the value of $\hat{y}^k$, $k = 1, \ldots, c$, is given by $\tau(\boldsymbol{v}_k^T \boldsymbol{h})$, where $\boldsymbol{v}_k = (v_{0k}, \ldots, v_{mk})^T$, $\boldsymbol{h} = (1, h^1, \ldots, h^m)^T$, and $\tau(\cdot)$ is also a nonlinear transformation with a range in the interval $(0, 1)$.

More formally, the map δ provided by an MLP from a sample $\boldsymbol{x}_i$ to $\hat{y}_i$ can be written as follows:

$$\delta(\boldsymbol{x}_i, \theta_{\mathcal{M}}) = \hat{y}_i = \tau(\boldsymbol{V}^T \gamma(\boldsymbol{W}^T \boldsymbol{x}_i))$$

where $\boldsymbol{V} = (\boldsymbol{v}_0, \ldots, \boldsymbol{v}_m)$, $\boldsymbol{W} = (\boldsymbol{w}_0, \ldots, \boldsymbol{w}_m)$, $\boldsymbol{x}_i = (x_i^0, x_i^1, \ldots, x_i^p)$, and $\tau(\cdot)$ and $\gamma(\cdot)$ are nonlinear functions.

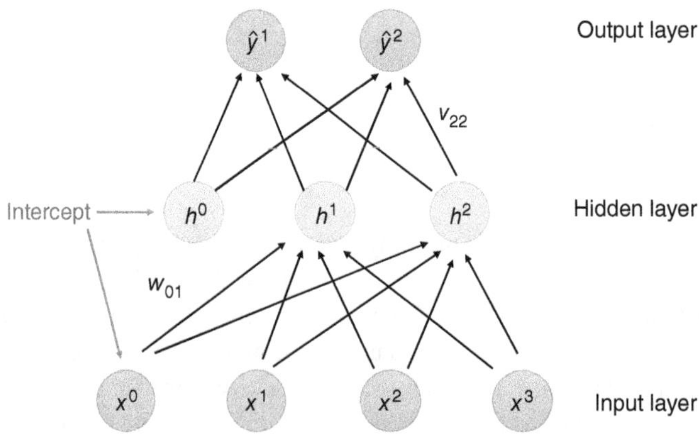

Figure 1 An MLP with three layers.

Most often, $\tau(\cdot)$ and $\gamma(\cdot)$ are chosen to be the logistic function $\sigma(z) = \frac{1}{1+e^{-z}}$. This function is often chosen for the following desirable properties: (i) it is highly nonlinear, (ii) it is monotonically increasing, (iii) it is asymptotically bounded at some finite value in both the negative and positive directions, and (iv) its output lies in the interval $(0, 1)$, so that it stays relatively close to 0. However, Yann LeCun recommends that a different function be used: $1.7159 \tanh(\frac{2}{3}x)$. This function retains all of the desirable properties of the logistic function and has the additional advantage of being symmetric about the origin, which results in outputs closer to 0 than the logistic function.

3.3 Training an MLP

We want to choose the weights and biases in such a way that they minimize the sum of squared errors within a given dataset. Similar to the general supervised learning approach, we want to find an optimal prediction $\delta^*(\boldsymbol{X}, \boldsymbol{W}, \boldsymbol{V})$ such that

$$\delta^*(\boldsymbol{X}, \boldsymbol{W}, \boldsymbol{V}) = \arg\min_{\boldsymbol{W},\boldsymbol{V}} \sum_{i=1}^{n} \ell(\hat{y}_i, y_i) \tag{1}$$

where $\boldsymbol{X} = (\boldsymbol{x}_1, \boldsymbol{x}_2, ..., \boldsymbol{x}_n)$, and $\ell(\cdot, \cdot)$ is cross-entropy loss.

$$\ell(\hat{y}_i, y_i) = -\sum_{c=1}^{m} y_{i,c} \log \hat{y}_{i,c} \tag{2}$$

where m is the total number of classes; $y_{i,c} = 1$ if the ith sample belongs to class c, otherwise it is equal to 0; and $\hat{y}_{i,c}$ is the predicted score of the ith sample belonging to class c.

Function (1) cannot be minimized through differentiation, so we must use gradient descent. The application of gradient descent to MLPs leads to an algorithm known as backpropagation. Most often, we use stochastic gradient descent as that is far faster. Note that backpropagation can be used to train different types of neural networks, not just MLP.

We would like to address the issue of possibly being trapped in local minima, as backpropagation is a direct application of gradient descent to neural networks, and gradient descent is prone to finding local minima, especially in high-dimensional spaces. It has been observed in practice that backpropagation actually does not typically get stuck in local minima and generally reaches the global minimum. There do, however, exist pathological data examples in which backpropagation will not converge to the global minimum, so convergence to the global minimum is certainly not an absolute guarantee. It remains a theoretical mystery why backpropagation does in fact generally converge to the global minimum, and under what conditions it will do so. However, some theoretical results have been developed to address this question. In particular, Gori and Tesi [7] established that for linearly separable data, backpropagation will always converge to the global solution.

So far, we have discussed a simple MLP with three layers aimed at classification problems. However, there are many extensions to the simple case. In general, an MLP can have any number of hidden layers. The more hidden layers there are, the more complex the model, and therefore the more difficult it is to train/optimize the weights. The model remains almost exactly the same, except for the insertion of multiple hidden layers between the first hidden layer and the output layer. Values for each node in a given layer are determined in

Algorithm 3. Backpropagation for a three-layer MLP

Input: dataset of input $\{(\boldsymbol{x}_i, y_i)\}_{i=1}^n$
Output: optimal weights $\boldsymbol{W}$ and $\boldsymbol{V}$
initialization: randomly initial weight matrices $\boldsymbol{W}$ and $\boldsymbol{V}$; η = value in (0, 1) (η is known as the *learning rate*);

```
while not converged do
    random.obs ← random permutation of [1, ..., n];
    for i in random.obs do
        h_i ← hidden.values(W, V, x_i);
        ŷ_i ← output.values(W, V, x_i);
        ℓ ← − Σ_{k=1}^{2} y_i^k log(ŷ_i^k), y_i^k = 1 if x_i belongs to class k;
        for each hidden node j do
            ε_j^(1) ← δℓ/δŷ_i × δŷ_i/δv_j;
            ε_j^(2) ← δℓ/δŷ_i × δŷ_i/δh_i × δh_i/δw_j;
            v_j ← v_j + η * ε_j^(1), for each column j in V;
            w_j ← w_j + η * ε_j^(2), for each column j in W;
        end
    end
end
```

the same way as before, that is, as a nonlinear transformation of the values of the nodes in the previous layer and the associated weights. Training the network via backpropagation is almost exactly the same.

4 Convolutional Neural Networks

4.1 Introduction

A CNN is a modified DNN that is particularly well equipped to handling image data. CNN usually contains not only fully connected layers but also *convolutional layers* and *pooling layers*, which make a difference. Image is a matrix of pixel values, which should be flattened to vectors before feeding into DNN as DNN takes a vector as input. However, spatial information might be lost in this process. The convolutional layer can take a matrix or tensor as input and is able to capture the spatial and temporal dependencies in an image.

In the convolutional layer, the weight matrix (kernel) scans over the input image to produce a feature matrix. This process is called convolution operation. The pooling layer operates similar to the convolutional layer and has two types: Max Pooling and Average Pooling. The Max Pooling layer returns the maximum value from the portion of the image covered by the kernel matrix. The Average Pooling layer returns the average of all values covered by the kernel matrix. The convolution and pooling process can be repeated by adding additional convolutional and pooling layers. Deep convolutional networks have been successfully trained and used in image classification problems.

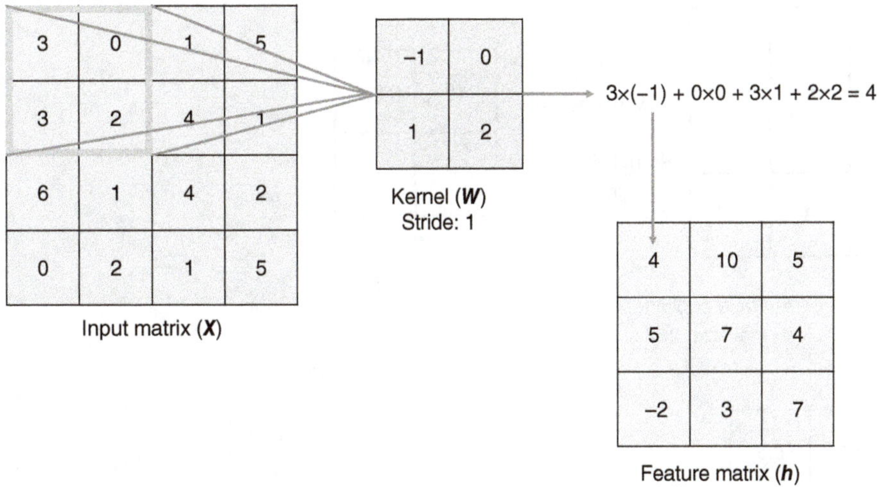

Figure 2 Convolution operation with stride size 1.

4.2 Convolutional Layer

The convolution operation is illustrated in Figure 2. The weight matrix of the convolutional layer is usually called the kernel matrix. The kernel matrix ($\boldsymbol{W} \in \mathbb{R}^{d\times d}$) shifts over the input matrix and performs elementwise multiplication between the kernel matrix ($\boldsymbol{W}$) and the covered portion of the input matrix ($\boldsymbol{X} \in \mathbb{R}^{n\times m}$), resulting in a feature matrix ($\boldsymbol{h} \in \mathbb{R}^{(n-d+1)\times(m-d+1)}$). The stride of the kernel matrix determines the amount of movement in each step. In the example in Figure 2, the stride size is 1, so the kernel matrix moves one unit in each step. In total, the kernel matrix shifts 9 times, resulting in a 3×3 feature matrix. The stride size does not have to be 1, and a larger stride size means fewer shifts.

Another commonly used structure in a CNN is the pooling layer, which is good at extracting dominant features from the input. Two main types of pooling operation are illustrated in Figure 3. Similar to a convolution operation, the kernel shifts over the input matrix with a specified stride size. If Max Pooling is applied to the input, the maximum of the covered portion will be taken as the result. If Average Pooling is applied, the mean of the covered portion will be calculated and taken as the result. The example in Figure 3 shows the result of pooling with kernel size that equals 2×2 and stride that equals 1 on a 3×3 input matrix.

4.3 LeNet-5

LeNet-5 is a CNN introduced by LeCun *et al.* [8]. This is one of the earliest structures of CNNs and was initially introduced to do handwritten digit recognition on the MNIST dataset [9]. The structure is straightforward and simple to understand, and details are shown in Figure 4.

The LeNet-5 architecture consists of seven layers, where three are convolutional layers, two are pooling layers, and two are fully connected layers. LeNet-5 takes images of

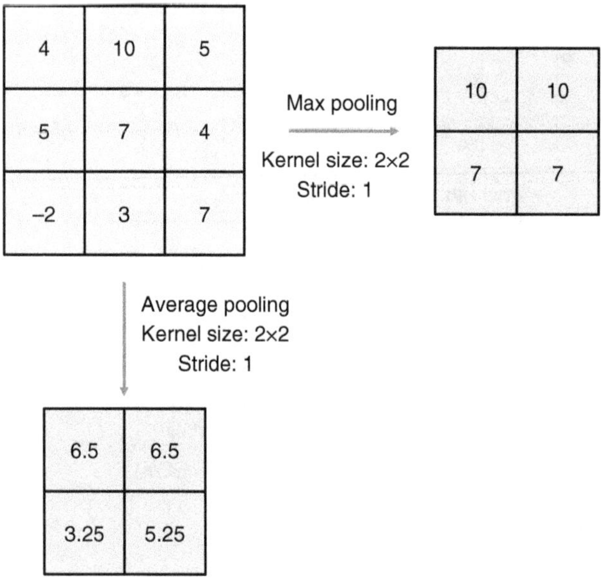

Figure 3 Pooling operation with stride size 1.

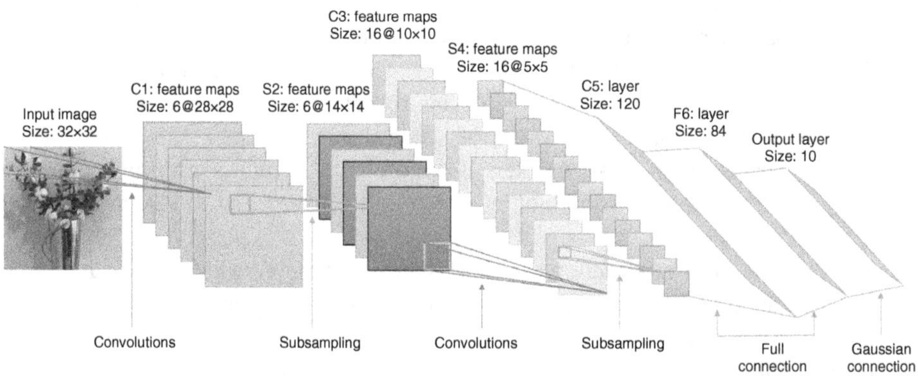

Figure 4 LeNet-5 of LeCun *et al.* [8]. *Source*: Modified from LeCun *et al.* [8].

size 32×32 as input and outputs a 10-dimensional vector as the predict scores for each class.

The first layer (C1) is a convolutional layer, which consists of six kernel matrices of size 5×5 and stride 1. Each of the kernel matrices will scan over the input image and produce a feature matrix of size 28×28. Therefore, six different kernel matrices will produce six different feature matrices. The second layer (S2) is a Max Pooling layer, which takes the 28×28 matrices as input. The kernel size of this pooling layer is 2×2, and the stride size is 2. Therefore, the outputs of this layer are six 14×14 feature matrices.

Table 1 Connection between input and output matrices in the third layer of LeNet-5 [8].

	Indices of output matrices									
1	1	5	6	7	10	11	12	13	15	16
2	1	2	6	7	8	11	12	13	14	16
3	1	2	3	7	8	9	12	14	15	16
4	2	3	4	7	8	9	10	13	15	16
5	3	4	5	8	9	10	11	13	14	16
6	4	5	6	9	10	11	12	14	15	16

The row names are indices of input matrices, and the second column shows indices of output matrices that are connected to the corresponding input matrix. There are 60 connections in total, meaning 60 different kernel matrices.
Source: LeCun *et al.* [8].

The third layer (C3) is the second convolutional layer in LeNet-5. It consists of 60 kernel matrices of size 5×5, and the stride size 1. Therefore, the output feature matrices are of size 10×10. Note that the relationship between input matrices and output matrices in this layer is not fully connected. Each of the input matrices is connected to a part of the output matrices. Details of the connection can be found in Table 1. Input matrices connected to the same output matrix will be used to produce the output matrix. Take the first output matrix, which is connected to the first three input matrices, as an example. The first three input matrices will be filtered by three different kernel matrices and result in three 10×10 feature matrices. The three feature matrices will first be added together, and then a bias is added elementwise, resulting in the first output matrix. There are 16 feature matrices of size 10×10 produced by layer C3.

The fourth layer (S4) is a Max Pooling layer that produces 16 feature matrices with size 5×5. The kernel size of this layer is 2×2, and the stride is 2. Therefore, each of the input matrices is reduced to 5×5. The fifth layer (C5) is the last convolutional layer in LeNet-5. The 16 input matrices are fully connected to 120 output matrices. Since both the input matrices and kernel matrices are of size 5×5, the output matrices are of size 1×1. Therefore, the output is actually a 120-dimensional vector. Each number in the vector is computed by applying 16 different kernel matrices on the 16 different input matrices and then combining the results and bias.

The sixth and seventh layers are fully connected layers, which are introduced in the previous section. In the sixth layer (S6), 120 input neurons are fully connected to 84 output neurons. In the last layer, 84 neurons are fully connected to 10 output neurons, where the 10-dimensional output vector contains predict scores of each class. For the classification task, cross-entropy loss between the model output and the label is usually used to train the model.

There are many other architectures of CNNs, such as AlexNet [10], VGG [11], and ResNet [12]. These neural networks demonstrated state-of-the-art performances on many machine learning tasks, such as image classification, object detection, and speech processing.

5 Autoencoders

5.1 Introduction

An autoencoder is a special type of DNN where the target of each input is the input itself [13]. The architecture of an autoencoder is shown in Figure 5, where the encoder and decoder together form the autoencoder. In the example, the autoencoder takes a horse image as input and produces an image similar to the input image as output. When the embedding dimension is greater than or equal to the input dimension, there is a risk of overfitting, and the model may learn an identity function. One common solution is to make the embedding dimension smaller than the input dimension. Many studies showed that the intrinsic dimension of many high-dimensional data, such as image data, is actually not truly high-dimensional; thus, they can be summarized by low-dimensional representations. Autoencoder summarizes the high-dimensional data information with low-dimensional embedding by training the framework to produce output that is similar to the input. The learned representation can be used in various downstream tasks, such as regression, clustering, and classification. Even if the embedding dimension is as small as 1, overfitting is still possible if the number of parameters in the model is large enough to encode each sample to an index. Therefore, regularization [15] is required to train an autoencoder that reconstructs the input well and learns a meaningful embedding.

5.2 Objective Function

Autoencoder is first introduced in Rumelhart *et al.* [16] as a model with the main goal of learning a compressed representation of the input in an unsupervised way. We are essentially creating a network that attempts to reconstruct inputs by learning the identity function. To do so, an autoencoder can be divided into two parts, $\boldsymbol{E} : \mathbb{R}^n \to \mathbb{R}^p$ (encoder) and $\boldsymbol{D} : \mathbb{R}^p \to \mathbb{R}^n$ (decoder), that minimize the following loss function w.r.t. the input $\boldsymbol{x}$:

$$||\boldsymbol{x} - \boldsymbol{D}(\boldsymbol{E}(\boldsymbol{x}))||^2$$

The encoder ($\boldsymbol{E}$) and decoder ($\boldsymbol{D}$) can be any mappings with the required input and output dimensions, but for image analysis, they are usually CNNs. The norm of the distance can be different, and regularization can be incorporated. Therefore, a more general form of the loss function is

$$L(\boldsymbol{x}, \hat{\boldsymbol{x}}) + \text{regularizer} \tag{3}$$

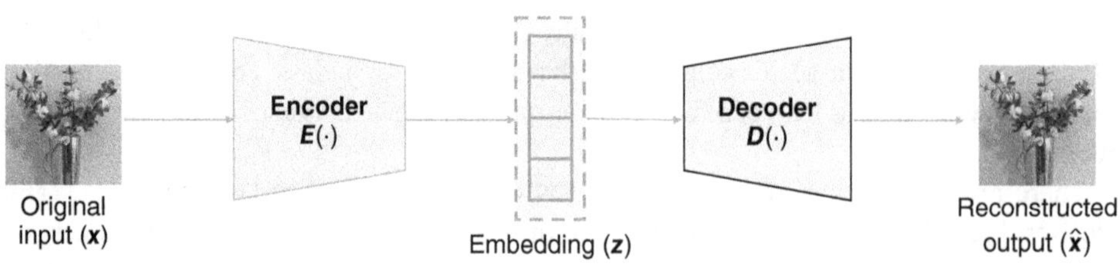

Figure 5 Architecture of an autoencoder. *Source*: Krizhevsky [14]

where $\hat{\boldsymbol{x}}$ is the output of an autoencoder, and $L(\cdot,\cdot)$ represents the loss function that captures the distance between an input and its corresponding output.

The output of the encoder part is known as the embedding, which is the compressed representation of input learned by an autoencoder. Autoencoders are useful for dimension reduction, since the dimension of an embedding vector can be set to be much smaller than the dimension of input. The embedding space is called the latent space, the space where the autoencoder manipulates the distances of data. An advantage of the autoencoder is that it can perform unsupervised learning tasks that do not require any label from the input. Therefore, autoencoder is sometimes used in pretraining stage to get a good initial point for downstream tasks.

5.3 Variational Autoencoder

Many different variants of the autoencoder have been developed in the past years, but the variational autoencoder (VAE) is the one that achieved a major improvement in this field. VAE is one of the frameworks, which attempts to describe an observation in latent space in a probabilistic manner. Instead of using a single value to describe each dimension of the latent space, the encoder part of VAE uses a probability distribution to describe each latent dimension [17].

Figure 6 shows the structure of the VAE. The assumption is that each input data $\boldsymbol{x}_i$ is generated by some random process conditioned on an unobserved random latent variable $\boldsymbol{z}_i$. The random process consists of two steps, where $\boldsymbol{z}_i$ is first generated from some prior distribution $p_\theta(\boldsymbol{z})$, and then $\boldsymbol{x}_i$ is generated from a conditional distribution $p_\theta(\boldsymbol{x}|\boldsymbol{z})$. The probabilistic decoder part of VAE performs the random generation process. We are interested in the posterior over the latent variable $p_\theta(\boldsymbol{z}|\boldsymbol{x}) = p_\theta(\boldsymbol{x}|\boldsymbol{z})p_\theta(z)/p_\theta(x)$, but it is intractable since the marginal likelihood $p_\theta(\boldsymbol{x})$ is intractable. To approximate the true posterior, the posterior distribution over the latent variable $\boldsymbol{z}$ is assumed to be a distribution $q_\phi(\boldsymbol{z}|\boldsymbol{x})$ parameterized by ϕ.

Given an observed dataset $\{\boldsymbol{x}_i\}_{i=1}^n$, the marginal log-likelihood is composed of a sum over the marginal log-likelihoods of all individual data points: $\log p_\theta(\boldsymbol{x}_1, \boldsymbol{x}_2, \ldots, \boldsymbol{x}_n) = \sum_{i=1}^n \log p_\theta(\boldsymbol{x}_i)$, where each marginal log-likelihood can be written as

$$\log p_\theta(\boldsymbol{x}_i) = \mathrm{KL}(q_\phi(\boldsymbol{z}|\boldsymbol{x}_i)||p_\theta(\boldsymbol{z}|\boldsymbol{x}_i)) + \ell(\theta,\phi;\boldsymbol{x}_i) \tag{4}$$

where the first term is the KL divergence [18] between the approximate and the true posterior, and the second term is called the variational lower bound. Since KL divergence is nonnegative, the variational lower bound is defined as

$$\begin{aligned}\log p_\theta(\boldsymbol{x}_i) \geq \ell(\theta,\phi;\boldsymbol{x}_i) &= \mathbb{E}_{q_\phi(\boldsymbol{z}|\boldsymbol{x}_i)}[-\log q_\phi(\boldsymbol{z}|\boldsymbol{x}) + \log p_\theta(\boldsymbol{x},\boldsymbol{z})] \\ &= \mathbb{E}_{q_\phi(\boldsymbol{z}|\boldsymbol{x}_i)}[\log p_\theta(\boldsymbol{x}_i|\boldsymbol{z})] - \mathrm{KL}(q_\phi(\boldsymbol{z}|\boldsymbol{x}_i)||p_\theta(\boldsymbol{z}))\end{aligned} \tag{5}$$

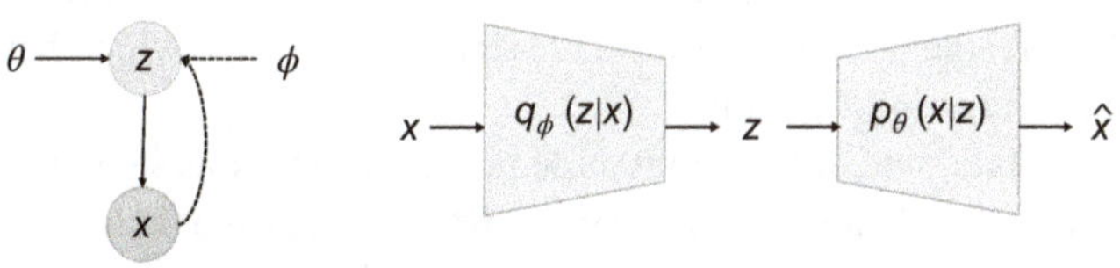

Figure 6 Architecture of variational autoencoder (VAE).

Therefore, the loss function of training a VAE can be simplified as

$$L(\boldsymbol{x}, \hat{\boldsymbol{x}}) + \mathrm{KL}(q_\phi(\boldsymbol{z}|\boldsymbol{x})||p_\theta(\boldsymbol{z})) \tag{6}$$

where the first term captures the reconstruction loss, and the second term is regularization on the embedding. To optimize the loss function (6), a reparameterization trick is used. For a chosen approximate posterior $q_\phi(\boldsymbol{z}|\boldsymbol{x})$, the latent variable $\tilde{\boldsymbol{z}} \sim q_\phi(\boldsymbol{z}|\boldsymbol{x})$ is approximated by

$$\tilde{\boldsymbol{z}} = g_\phi(\epsilon, \boldsymbol{x}), \epsilon \sim p(\epsilon) \tag{7}$$

where ϵ is an auxiliary variable with independent marginal $p(\epsilon)$, and $g_\phi(\cdot)$ is some vector-valued function parameterized by ϕ. With this reparameterization trick, the variational lower bound can be estimated by sampling a batch of ϵ from $p(\epsilon)$:

$$\ell(\theta, \phi; \boldsymbol{x}_i) \approx \sum_{j=1}^{B}[-\log q_\phi(\boldsymbol{z}_{(i,j)}|\boldsymbol{x}_i) + \log p_\theta(\boldsymbol{x}_i, \boldsymbol{z}_{(i,j)})] \tag{8}$$

where $\boldsymbol{z}_{(i,j)} = g_\phi(\epsilon_{(i,j)}, \boldsymbol{x}_i)$ and $\epsilon_{(i,j)} \sim p(\epsilon)$. The selections of $p(\epsilon)$ and $g_\phi(\cdot)$ are discussed in detail in Kingma and Welling [17].

6 Recurrent Neural Networks

6.1 Introduction

The previously introduced models have the same assumptions on the data, that is, the independence among the samples and the fixed input size. However, these assumptions may not be true in some cases, thus limiting the application of these models. For example, videos can have different lengths, and frames of the same video are not independent, and sentences of an chapter can have different lengths and are not independent.

A RNN is another modified DNN that is used primarily to handle sequential and time series data. In a RNN, the hidden layer of each input is a function of not just the input layer but also the previous hidden layers of the inputs before it. Therefore, it addresses the issues of dependence among samples and does not have any restriction on the input size. RNNs are used primarily in natural language processing applications, such as document modeling and speech recognition.

6.2 Architecture

As illustrated in Figure 7, a general neural network N takes in input $\boldsymbol{x}$ and outputs $\boldsymbol{h}$. The output of one sample will not influence the output of another sample. To capture the dependence between inputs, RNN adds a loop to connect the previous information with the current state. The graph on the left side of Figure 8 shows the structure of RNN, which has a loop connection to leverage previous information.

RNN can work with sequence data, which has input as sequence or target as sequence or both. An input sequence data can be denoted as $(\boldsymbol{x}^{(1)}, \boldsymbol{x}^{(2)}, ..., \boldsymbol{x}^{(T)})$, where each data point $\boldsymbol{x}^{(t)}$ is a real-valued vector. Similarly, the target sequence can be denoted as $(\boldsymbol{y}^{(1)}, \boldsymbol{y}^{(2)}, ..., \boldsymbol{y}^{(T)})$. A sample from the sequence dataset is typically a pair of one input sequence and one target

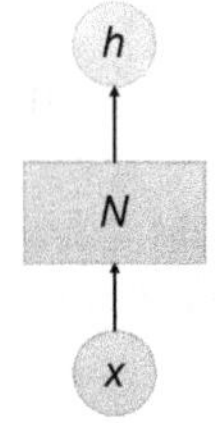

Figure 7 Feedforward network.

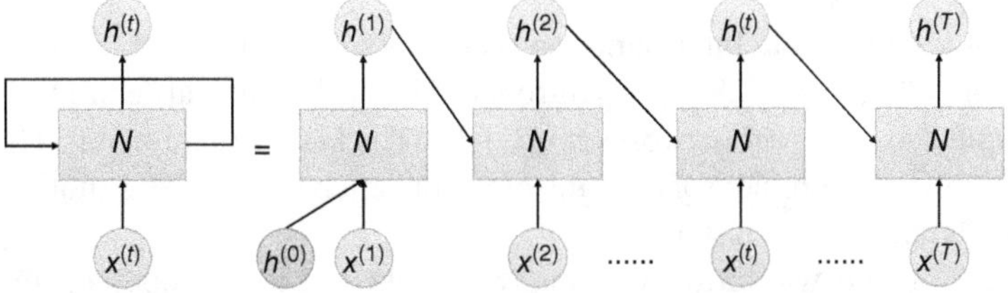

Figure 8 Architecture of recurrent neural network (RNN).

sequence. The right side of Figure 8 shows the information passing process. At $t = 1$, network N takes in a random initialed vector $\boldsymbol{h}^{(0)}$ together with $\boldsymbol{x}^{(1)}$ and outputs $\boldsymbol{h}^{(1)}$, and then at $t = 2$, N takes in both $\boldsymbol{x}^{(2)}$ and $\boldsymbol{h}^{(1)}$ and outputs $\boldsymbol{h}^{(2)}$. This process is repeated over all data points in the input sequence.

Though multiple network blocks are shown on the right side of Figure 8, they share the same structure and weights. A simple example of the process can be written as

$$\boldsymbol{h}^{(t)} = \sigma(\boldsymbol{W}_1\boldsymbol{x}^{(t)} + \boldsymbol{W}_2\boldsymbol{h}^{(t-1)} + \boldsymbol{b}) \tag{9}$$

where $\boldsymbol{W}_1$ and $\boldsymbol{W}_2$ are weight matrices of network N, $\sigma(\cdot)$ is an activation function, and $\boldsymbol{b}$ is the bias vector. Depending on the task, the loss function is evaluated, and the gradient is backpropagated through the network to update its weights. For the classification task, the final output $\boldsymbol{h}^{(T)}$ can be passed into another network to make prediction. For a sequence-to-sequence model, $\hat{\boldsymbol{y}}^{(t)}$ can be generated based on $\boldsymbol{h}^{(t)}$ and then compared with $\boldsymbol{y}^{(t)}$.

However, a drawback of RNN is that it has problem "remembering" remote information. In RNN, long-term memory is reflected in the weights of the network, which memorizes remote information via shared weights. Short-term memory is in the form of information flow, where the output from the previous state is passed into the current state. However, when the sequence length T is large, the optimization of RNN suffers from vanishing gradient problem. For example, if the loss $\ell^{(T)}$ is evaluated at $t = T$, the gradient w.r.t. $\boldsymbol{W}_1$ calculated via backpropagation can be written as

$$\frac{\delta \ell^{(T)}}{\delta \boldsymbol{W}_1} = \sum_{t=0}^{T} \frac{\delta \ell^{(T)}}{\delta \boldsymbol{h}^{(T)}} \left(\prod_{j=t+1}^{T} \frac{\delta \boldsymbol{h}^{(j)}}{\delta \boldsymbol{h}^{(j-1)}} \right) \frac{\delta \boldsymbol{h}^{(t)}}{\delta \boldsymbol{W}_1} \tag{10}$$

where $\prod_{j=t+1}^{T} \frac{\delta \boldsymbol{h}^{(j)}}{\delta \boldsymbol{h}^{(j-1)}}$ is the reason for the vanishing gradient. In RNN, the tanh function is commonly used as the activation function, so

$$\boldsymbol{h}^{(j)} = \tanh(\boldsymbol{W}_1\boldsymbol{x}^{(j)} + \boldsymbol{W}_2\boldsymbol{h}^{(j-1)} + \boldsymbol{b}) \tag{11}$$

Therefore, $\prod_{j=t+1}^{T} \frac{\delta \boldsymbol{h}^{(j)}}{\delta \boldsymbol{h}^{(j-1)}} = \prod_{j=t+1}^{T} \tanh' \boldsymbol{W}_1$, and $\tanh'$ is always smaller than 1. When T becomes larger, the gradient will get closer to zero, making it hard to train the network and update the weights with remote information. However, it is possible that relevant information is far apart in the sequence, so how to leverage remote information of a long sequence is important.

6.3 Long Short-Term Memory Networks

To solve the problem of losing remote information, researchers proposed long short-term memory (LSTM) networks. The idea of LSTM was introduced in Hochreiter and Schmidhuber [19], but it was applied to recurrent networks much later. The basic structure of LSTM is shown in Figure 9. It solves the problem of the vanishing gradient by introducing another hidden state $\mathbf{c}^{(t)}$, which is called the cell state.

Since the original LSTM model was introduced, many variants have been proposed. Forget gate was introduced in Gers *et al.* [20]. It has been proven effective and is standard in most LSTM architectures. The forwarding process of LSTM with a forget gate can be divided into two steps. In the first step, the following values are calculated:

$$\begin{aligned} \boldsymbol{z}^{(t)} &= \tanh(\boldsymbol{W}_{1z}\boldsymbol{x}^{(t)} + \boldsymbol{W}_{2z}\boldsymbol{h}^{(t-1)} + \boldsymbol{b}_z) \\ \boldsymbol{i}^{(t)} &= \sigma_g(\boldsymbol{W}_{1i}\boldsymbol{x}^{(t)} + \boldsymbol{W}_{2i}\boldsymbol{h}^{(t-1)} + \boldsymbol{b}_i) \\ \boldsymbol{f}^{(t)} &= \sigma_g(\boldsymbol{W}_{1f}\boldsymbol{x}^{(t)} + \boldsymbol{W}_{2f}\boldsymbol{h}^{(t-1)} + \boldsymbol{b}_f) \\ \mathbf{o}^{(t)} &= \sigma_g(\boldsymbol{W}_{1o}\boldsymbol{x}^{(t)} + \boldsymbol{W}_{2o}\boldsymbol{h}^{(t-1)} + \boldsymbol{b}_o) \end{aligned} \tag{12}$$

where W and b are weight matrix and bias, and $\sigma_g(z) = \frac{1}{1+\exp(z)}$ is the sigmoid function.

The two hidden states $\boldsymbol{h}^{(t)}$ and $\mathbf{c}^{(t)}$ are calculated by

$$\mathbf{c}^{(t)} = \boldsymbol{f}^{(t)} \circ \mathbf{c}^{(t-1)} + \boldsymbol{i}^{(t)} \circ \boldsymbol{z}^{(t)} \tag{13}$$

$$\boldsymbol{h}^{(t)} = \mathbf{o}^{(t)} \circ \tanh(\mathbf{c}^{(t)}) \tag{14}$$

where $\circ$ represents elementwise product between matrices. In Equation (13), the first term multiplies $\boldsymbol{f}^{(t)}$ with $\mathbf{c}^{(t-1)}$, controlling what information in the previous cell state can be passed to the current cell state. As for the second term, $\boldsymbol{z}^{(t)}$ stores the information passed from $\boldsymbol{x}^{(t)}$ and $\boldsymbol{h}^{(t-1)}$, and $\boldsymbol{i}^{(t)}$ controls how much information from the current state is preserved in the cell state. The hidden state $\boldsymbol{h}^{(t)}$ depends on the current cell state and $\mathbf{o}^{(t)}$, which decides how much information from the current cell state will be passed to the hidden state $\boldsymbol{h}^{(t)}$.

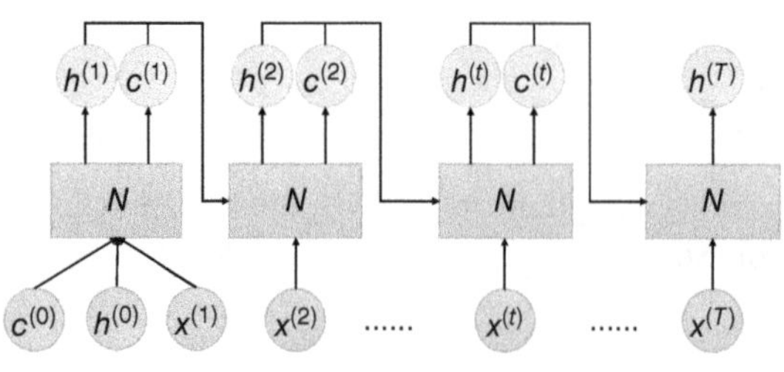

Figure 9 Architecture of long short-term memory network (LSTM).

In LSTM, if the loss $\ell^{(T)}$ is evaluated at $t = T$, the gradient w.r.t. $\boldsymbol{W}_{1f}$ calculated via backpropagation can be written as

$$\begin{aligned}\frac{\delta \ell^{(T)}}{\delta \boldsymbol{W}_{1f}} &= \sum_{t=0}^{T} \frac{\delta \ell^{(T)}}{\delta \boldsymbol{h}^{(T)}} \frac{\delta \boldsymbol{h}^{(T)}}{\delta \boldsymbol{c}^{(T)}} \left(\prod_{j=t+1}^{T} \frac{\delta \boldsymbol{c}^{(j)}}{\delta \boldsymbol{c}^{(j-1)}} \right) \frac{\delta \boldsymbol{c}^{(t)}}{\delta \boldsymbol{W}_{1f}} \\ &= \sum_{t=0}^{T} \frac{\delta \ell^{(T)}}{\delta \boldsymbol{h}^{(T)}} \frac{\delta \boldsymbol{h}^{(T)}}{\delta \boldsymbol{c}^{(T)}} \left(\prod_{j=t+1}^{T} \boldsymbol{f}^{(t)} + A^{(t)} \right) \frac{\delta \boldsymbol{c}^{(t)}}{\delta \boldsymbol{W}_{1f}}\end{aligned} \tag{15}$$

where $A^{(t)}$ represents other terms in the partial derivative calculation. Since the sigmoid function is used when calculating the values of $\boldsymbol{i}^{(t)}, \boldsymbol{f}^{(t)}, \boldsymbol{o}^{(t)}$, this implies that they will be close to either 0 or 1. When $\boldsymbol{f}^{(t)}$ is close to 1, the gradient does not vanish, and when it is close to 0, it means that the previous information is not useful for the current state and should be forgotten.

7 Conclusion

We discussed the architectures of four types of neural networks and their extensions in this chapter. There have been many other neural networks proposed in the past years, but the ones discussed in this chapter are the classical ones that served as foundations for many other works. Though DNNs have achieved breakthroughs in many fields, the performances in many fields are far from perfect. Developing new architectures that can improve the performances on various tasks or solve new problems is an important research direction. Analyzing the properties and problems of existing architectures is also of great interest to the community.

References

1 Larochelle, H., Bengio, Y., Louradour, J., and Lamblin, P. (2009) Exploring strategies for training deep neural networks. *J. Mach. Learn. Res.*, 1, 1–40.

2 Hinton, G.E. and Salakhutdinov, R.R. (2006) Reducing the dimensionality of data with neural networks. *Science*, 313, 504–507.

3 Hastie, T., Tibshirani, R., and Friedman, J. (2002) *The Elements of Statistical Learning*, Springer, New York.

4 Boyd, S., Boyd, S.P., and Vandenberghe, L. (2004) *Convex Optimization*, Cambridge university press.

5 Nocedal, J. and Wright, S. (2006) *Numerical Optimization*, Springer Science & Business Media.

6 Izenman, A.J. (2008) Modern multivariate statistical techniques. *Regression Classif. Manifold Learn.*, 10, 978–980.

7 Gori, M. and Tesi, A. (1992) On the problem of local minima in backpropagation. *IEEE Trans. Pattern Anal. Mach. Intell.*, 14, 76–86.

8 LeCun, Y., Bottou, L., Bengio, Y., and Haffner, P. (1998) Gradient-based learning applied to document recognition. *Proc. IEEE*, 86, 2278–2324.

9 LeCun, Y. (1998) *The MNIST Database of Handwritten Digits*, http://yann.lecun.com/exdb/mnist/ (accessed 20 April 2021).

10 Krizhevsky, A., Sutskever, I., and Hinton, G.E. (2012) Imagenet classification with deep convolutional neural networks. *Adv. Neural Inf. Process. Syst.*, 25, 1097–1105.

11 Simonyan, K. and Zisserman, A. (2014) Very deep convolutional networks for large-scale image recognition. *arXiv preprint arXiv:1409.1556.*

12 He, K., Zhang, X., Ren, S., and Sun, J. (2016) *Deep Residual Learning for Image Recognition.* Proceedings of the IEEE Conference on Computer Vision and Pattern Recognition, pp. 770–778.

13 Goodfellow, I., Bengio, Y., and Courville, A. (2016) *Deep Learning*, MIT Press.

14 Krizhevsky, A. (2009) Learning multiple layers of features from tiny images.

15 Bickel, P.J., Li, B., Tsybakov, A.B. *et al.* (2006) Regularization in statistics. *Test*, 15, 271–344.

16 Rumelhart, D.E., Hinton, G.E., and Williams, R.J. (1986) Learning Internal Representations by Error Propagation. Tech. report. California Univ San Diego La Jolla Inst for Cognitive Science.

17 Kingma, D.P. and Welling, M. (2014) *Auto-Encoding Variational Bayes.* International Conference on Learning Representations.

18 Kullback, S. and Leibler, R.A. (1951) On information and sufficiency. *Ann. Math. Stat.*, 22, 79–86.

19 Hochreiter, S. and Schmidhuber, J. (1997) Long short-term memory. *Neural Comput.*, 9, 1735–1780.

20 Gers, F., Schmidhuber, J., and Cummins, F. (1999) *Learning to Forget: Continual Prediction with LSTM.* 1999 Ninth International Conference on Artificial Neural Networks ICANN 99. (Conf. Publ. No. 470), vol. 2, pp. 850–855.

4

Streaming Data and Data Streams

Taiwo Kolajo[1,2], Olawande Daramola[3], and Ayodele Adebiyi[4]

[1]*Federal University Lokoja, Lokoja, Nigeria*
[2]*Covenant University, Ota, Nigeria*
[3]*Cape Peninsula University of Technology, Cape Town, South Africa*
[4]*Landmark University, Omu-Aran, Kwara, Nigeria*

1 Introduction

As at the dawn of 2020, the amount of the world data generated was estimated to be 44 zettabytes (i.e., 40 times more than the number of stars in the observable universe). The amount of data generated daily is projected to be 463 exabytes globally by 2025 [1]. Not only that, data are growing in volume but also in structure, in complexity, and geometrically [2]. These high-volume data, generated at a high-velocity, lead to what is called *streaming data*. Data streams can originate from IoT devices and sensors, spreadsheets, text files, images, audio and video recordings, chat and instant messaging, email, blogs and social networking sites, web traffic, financial transactions, telephone usage records, customer service records, satellite data, smart devices, GPS data, and network traffic and messages.

There are different schools of thought when it comes to defining streaming data and data stream, and it is difficult to situate a position between these two concepts. One school of thought defined streaming data as the act of sending data bit by bit instead of a whole package while data stream is the actual source of data. That is, streaming data is the act, the verb, the action while data stream is the product. In the field of Engineering, streaming data is the process or art of collecting the streamed data. It is the main activity or operation, while data stream is the pipeline through which streaming is performed. It is the engineering architecture, that is the line-up of tools that will perform the streaming. In the context of data science, streaming data and data streams are used interchangeably. To better understand the concepts, let us first define what a stream is. A stream S is a possibly infinite bag of elements (x, t) where x is a tuple belonging to the schema S and $t \in T$ is the timestamp of the elements [3]. Data stream refers to an unbounded and ordered sequence of instances of data arriving over time [4]. Data stream can be formally defined as an infinite sequence of tuples $S = (x_1, t_i), (x_2, t_2), \ldots, (x_n, t_n), \ldots$ where x_i is a tuple and t_i is a timestamp [5]. Streaming data can be defined as frequently changing, and potentially infinite data flow generated from disparate sources [6]. Formally, streaming data $X = (x_t^1, \ldots, x_t^m)^T$ is a set of count values of a variable x of an event that happened at timestamp $t(0 < t \leq T)$, where T is the lifetime

Computational Statistics in Data Science.
Edited by Walter W. Piegorsch, Richard A. Levine, Hao Helen Zhang and Thomas C. M. Lee.
 ISBN 978-1-11956107-1

Table 1 Streaming data versus static data [9, 10]

Dimension	Streaming data	Static data
Hardware	Typical single constrained measure of memory	Multiple CPUs
Input	Data streams or updates	Data chunks
Time	A few moments or even milliseconds	Much longer
Data size	Infinite or unknown in advance	Known and finite
Processing	A single or few pass over data	Processes in multiple rounds
Storage	Not store or store a significant portion in memory	Store
Applications	Web mining, traffic monitoring, sensor networks	Widely adopted in many domains

Source: Tozi, C. (2017). Dummy's guide to batch vs streaming. Retrieved from Trillium Software, Retrieved from http://blog.syncs ort.com/2017/07/bigdata/; Kolajo, T., Daramola, O. & Adebiyi, A. (2019). Big data stream analysis: A systematic literature review, Journal of Big Data 6(47).

of the streaming data [7]. Looking at the definitions of both data stream and streaming data in the context of data science, the two concepts are trickily similar. All the different schools of thought slightly agree with these slightly confusing and closely related concepts except for the Engineering school of thought that refers to data stream as an architecture. Although this is still left open for further exploration, we will use them interchangeably in this chapter.

The ocean of streaming data continuously generated through various mediums such as sensors, ATM transactions, and the web is tremendously increasing, and recognizing patterns in these mediums is equally challenging [8]. Most methods used for data stream mining are adapted from techniques designed for a finite or static dataset. Data stream mining imposes a high number of constraints on canonical algorithms. To quickly appreciate these constraints, the differences between static and streaming scenarios are presented in Table 1.

In the big data era, data stream mining serves as one of the vital fields. Since streaming data is continuous, unlimited, and with nonuniform distribution, there is the need for efficient data structures and algorithms to mine patterns from this high volume, high traffic, often imbalanced data stream that is also plagued with concept drift [11].

This chapter intends to broaden the existing knowledge in the domain of data science, streaming data, and data streams. To do this, relevant themes including data stream mining issues, streaming data tools and technologies, streaming data pre-processing, streaming data algorithms, strategies for processing data streams, best practices for managing data streams, and suggestions for the way forward are discussed in this chapter. The structure of the rest of this chapter is as follows. Section 2 presents a brief background on data stream computing; Section 3 discusses issues in data stream mining, tools, and technologies for data streaming are presented in Sections 4 while streaming data pre-processing is discussed in Section 5. Sections 6 and 7 present streaming data algorithms and data stream processing strategies, respectively. This is followed by a discussion on best practices for managing data streams in Section 8, while the conclusion and some ideas on the way forward are presented in Section 9.

2 Data Stream Computing

Data stream computing alludes to the real-time processing of vast measures of data produced at high speed from numerous sources, with different schemas, and different temporal resolutions [12]. It is another required worldview given the new wellsprings of data-generation situations, which incorporates the cell phones, ubiquity of location services, and sensor universality [13].

The principal presumption of stream computing is that the likelihood estimation of data lies in its newness. Thus, the analysis of data is done the moment they arrive in a stream instead of what obtains in batch processing where data are first stored before they are analyzed. This is a serious requirement for suitable platforms for scalable computing with parallel architectures [14]. With stream computing, it is feasible for organizations to analyze and respond to speedily changing data in real-time [15]. Integrating streaming data into the decision-making process brings about a programming concept called *stream computing*. Stream processing solutions ought to have the option to deal with the high volume of data from different sources in real-time by giving due consideration to accessibility, versatility, and adaptation to noncritical failure. Datastream analysis includes the ingestion of data as a boundless tuple, analysis, and creation of significant outcomes in a stream [16].

In a stream processor, the representation of an application is done with the data flow graph, which is comprised of operations and interconnected streams. A stream processing workflow consists of programs. Formally, a *composition* $C = (\mathcal{P}, <_p)$,where $\mathcal{P} = \{P_1, P_2, \dots, P_n\}$ is a set of *transaction* programs and $<_p$ is the *program* order, also called *partial order*. The partial order contains the dataflow and control order of the data stream. The *composition graph* $\mathcal{GC}(C)$ is the acyclic graph representing the partial order. Input streams to the *composition* are called *source* streams, while the output streams are called *derived* streams [17]. In a streaming analytics system, the application comes as continuous queries, data are continuously ingested, analyzed, and interrelated to produce results in streaming fashion. Streaming analytics frameworks must be able to recognize new data, build models incrementally, and detect deviation from model predictions [18].

3 Issues in Data Stream Mining

One of the challenges of data stream mining is concept drift. Concept drift is a phenomenon that bothers on how data stream evolves [19]. The presence of concept drift affects the fundamental characteristics that the learning system seeks to uncover, thus leading to degraded results by the classifier as the change progresses [20].

Concept drift in data stream can be broadly classified into two main categories, which are concept drift based on classification boundaries and concept drift concerning types of change. The former influences the classification boundaries and can be further subdivided into virtual concept drift and real concept drift. Virtual concept drift affects the conditional probability density functions, though the influence on the decision boundary is insignificant on the currently used learning models. On the other hand, real concept drift often impacts the unconditional probability density functions, leading to degraded results of the learning models. Concept drift concerning change is subdivided into sudden, gradual, and

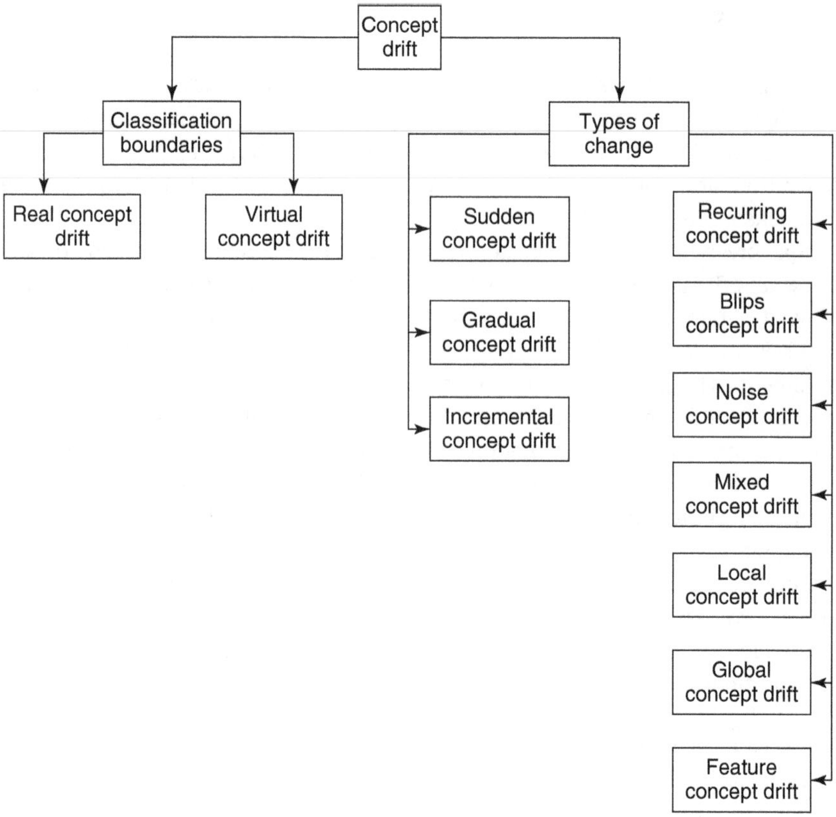

Figure 1 Taxonomy of concept drift in data stream.

incremental concept drift. Other categories based on types of change include blip, noise, mixed, local, global, feature, and adversarial concept drifts [21]. The taxonomy of concept drift is presented in Figure 1.

Three standard solutions to address concept drift are (i) to detect changes and retrain classifiers when the degree of changes is significantly high, (ii) retraining of the classification model at the arrival of a new chunk or instance, and (iii) the use of adaptive learning methods. However, option number 2 is practically not feasible due to computational cost. The four main approaches for addressing concept drift are (i) concept drift detectors [22], (ii) sliding windows [23], (iii) online learners [24], (iv) and ensemble learners [25]. Other challenges for data stream are briefly highlighted below.

3.1 Scalability

Another fundamental challenge in streaming data analysis is the issue of scalability. The rate at which data stream is growing is much faster than the resources available to the computer. While the processors keep Moore's law, the data size is experiencing exponential growth. Subsequently, research endeavors must be equipped toward creating scalable frameworks and machine learning algorithms that will adjust data stream computing mode, manage resource allocation strategy effectively, and react to parallelization issues to adapt to the high-volume and complexity in data. While data streams provide the opportunity for machine learning algorithms to uncover useful and interesting patterns, traditional machine learning algorithms face the challenge of scalability to truly uncover the hidden value in the data stream [26].

3.2 Integration

Building a distributed framework with every node having a data stream flow view implies that each node is liable for performing analysis with few sources. Aggregating these views to build a complete view is inconsequential. This calls for the development of an integration technique that can perform efficient operations through disparate datasets [27].

3.3 Fault-Tolerance

For life-critical systems, high fault-tolerance is required. In streaming computing environments, where unbounded data are generated in real-time, an amazing high adaptation to noncritical failure procedure and scalable system, is required to allow an application to keep working without interruption despite component failure. The most widely recognized adaptation to internal failure is checkpointing, where the framework state is intermittently persisted to recapture the computational state after system failures. However, the overhead incurred with checkpointing can negatively affect system performance. An improved checkpointing to minimize the overhead cost was proposed by [28, 29].

3.4 Timeliness

Time is essential for time-sensitive processes, which incorporate foiling fraud, mitigating security threats, or responding to a natural disaster. Such architectures or platforms must be scalable to enable consistent handling of data streams [30]. The fundamental challenge bothers on implementing a distributed architecture for data aggregation with insignificant latency between the communicating nodes.

3.5 Consistency

Achieving high consistency or stability in the data stream computing environments is nontrivial as it is hard to figure out which data are required and which nodes ought to be consistent [31, 32]. Thus, a good framework is required.

3.6 Heterogeneity and Incompleteness

Data streams are heterogeneous in structure, semantics, organizations, granularity, and accessibility. Different data in disparate sources, different formats, combined with the volume of data, make the integration, retrieval, and reasoning over the data stream a challenging task [33]. The challenge here is how to deal with ever-growing data, extract, aggregate, and correlate data streams from numerous sources in real-time. There is need to design a competent data presentation to mirror the structure, hierarchy, and diversity of data streams.

3.7 Load Balancing

In an ideal situation, a data stream framework should be self-adaptive and avoid load shedding. Be that as it may, this is challenging because the possibility of dedicated resources to cover peak loads 24/7 is slim, and load shedding is not realistic, most especially when the variance between the average load and the peak load is high [34]. Consequently, a distributing environment that can stream, analyze, and aggregate partial data streams to a global center when local resources become deficient is required.

3.8 High Throughput

Decision concerning the identification of the data stream portion that needs replication, number of these replicas that is required, and which of the data stream to assign to each replica is an issue in data stream computing environment. Proper multiple instances replication is required if high throughput is to be achieved [35].

3.9 Privacy

Data stream analytics open doors for real-time analysis of massive amount of data but also made a colossal danger to individual privacy [36]. As indicated by the International Data Cooperation (IDC), half of the aggregate data that needs protection is not adequately protected. Relevant and efficient privacy-preserving solutions for interpretation, observation, evaluation, and decision for data stream mining should be designed [37]. The sensitive nature of data necessitates privacy-preserving techniques. One of the leading privacy-preserving techniques is perturbation [38].

3.10 Accuracy

Developing efficient methods that can precisely predict future observations is one of the leading goals of data stream analysis. Yet, the intrinsic features of data stream, which include noisy characteristics, velocity, volume, variety, value, veracity, variability, and volatility, data stream analysis strongly constrain processing algorithms spatiotemporally. Hence, to guarantee high accuracy, mitigation of these challenges must be taken into consideration as they can negatively influence the accuracy of data stream analysis [39].

4 Streaming Data Tools and Technologies

The demand for stream processing is on the increase, and data have to be processed fast to make decisions in real-time. Because of the developing interest in streaming data analysis, a huge number of enormous streaming data solutions have been created both by the open-source community and enterprise technology vendors [10]. As indicated by Millman [40], there are a few elements to consider while choosing data stream tools and technologies in request to settle on viable data management decisions. Those elements include the shape of the data, data accessibility, availability, and consistency requirement, and workload. Some prominent open-source tools and technologies for data stream analytics include NoSQL [41], Apache Spark [42–44], Apache Storm [45], Apache Samza [46, 47], Yahoo! S4 [48], Photon [49], Apache Aurora [50], EsperTech [51], SAMOA [52], C-SPARQL [53], CQELS [54], ETALIS [55], SpagoWorld [56]. Some proprietary tools and technologies for streaming data are Cloudet [57], Sentiment Brand Monitoring [58], Elastic Streaming Processing Engine [59], IBM InfoSphere Streams [16, 60, 61], Google MillWheel [46], Infochimps Cloud [56], Azure Stream [62], Microsoft Stream Insight [63], TIBCO StreamBase [64], Lambda Architecture [6], IoTSim-Stream [65], and Apama Stream [62].

5 Streaming Data Pre-Processing: Concept and Implementation

Data stream pre-processing, which aims at reducing the inherent complexity associated with streaming data for a faster, more understandable, and interpretable, and more precise learning process is an essential technique in knowledge discovery. However, despite the recorded growth in online learning, data stream pre-processing methods still have a long way to go due to the high level of noise [66]. These noisy terms incorporate a short length of messages, slangs, abbreviations, acronyms, blended dialects, linguistic and spelling mistakes, sporadic, casual, shortened words, and ill-advised sentence structure, which make it hard for learning algorithms to perform productively and adequately [67]. Additionally, error from sensor reading due to low battery, damage, incorrect calibrations, among others, can render data delivered from such sensors unsuitable for analysis [68].

Data quality is a fundamental determinant in the knowledge discovery pipeline as low-quality data yields low-quality models and choices [69]. There is need to strengthen data stream pre-processing stage in the face of multi-label [70], imbalance [71], and multi-instance [72] problems associated data stream [66]. Also, data stream pre-processing techniques with low computational requirement [73] needs to be evolved as this is still open for research. Moreover, the representation of social media posts must be in a way that the semantics of social media content is preserved [74, 75]. To improve the result of analysis in the data stream, there is need to develop frameworks that will cope with the noisy characteristics, redundancy, heterogeneity, data imbalance, transformation, feature representation, or selection issues in data streams [26]. Some of the new frameworks developed for pre-processing and enriching data stream for better results are SlangSD [76], N-gram and Hidden Markov Model [77], SLANGZY [78], and SMFP [67].

6 Streaming Data Algorithms

Data stream poses a significant number of challenges to mining algorithms and research community due to the high-traffic, high-velocity, and brief life span of streaming data [79]. Many algorithms that are suitable for mining data at rest are not suited to streaming data due to the inherent characteristics of streaming data [80]. Some of the constraints that are naturally imposed on mining algorithms by streaming data include (i) the concept of a single pass, (ii) the probability distribution of data chunk is not known in advance, (iii) no limitation on the amount of generated data, (iv) the size of incoming data may vary, (v) the incoming data may belong to various sub-clusters, and (vi) access to correct class labels is limited due to overhead incurred by label query for each arriving instance [81]. The constraints further generate other problems, which include: (i) capturing sub-cluster data within the bounded learning time complexity, (ii) the minimum number of epochs required to achieve the learning time complexity, and (iii) making algorithm robust in the face of dynamically evolving and irregular streaming data.

Different streaming data mining tasks include clustering, similarity search, prediction, classification, and object detection, among others [82, 83]. Algorithms used for streaming data analysis can be grouped into four: Unsupervised learning, semi-supervised learning, supervised learning, and ontology-based techniques. These are subsequently described.

6.1 Unsupervised Learning

Unsupervised learning is a type of learning that draws inductions from the unlabeled dataset [84]. Data stream source is nonstationary, and for clustering algorithms, there is no information about the data distribution in advance [85]. Due to several iterations required to compute similarity or dissimilarity in the observed dataset, the entirety of the datasets ought to be accessible in memory before running the algorithm in most cases. However, with data stream clustering, the challenge is searching for a new structure in data as it evolves, which involves characterizing the streaming data in the form of clusters to leverage them to report useful and interesting patterns in the data stream [86]. Unsupervised learning algorithms are suitable for analyzing data stream as it does not require a predefined label [87]. Clusters are ordered dependent on scoring function, for example, catchphrase or keyword, hashtags, the semantic relationship of terms, and segment extraction [88].

Data stream clustering can be grouped into five categories, which are partitioning methods, hierarchical methods, model-based methods, density-based methods, and grid-based methods.

Partition-based techniques try to find out k-partitions based on some measurement. Partitioning clustering methods are not suitable for streaming scenarios since they require earlier information on cluster number. Examples of partition-based methods include Incremental K-Mean, STREAMKM++, Stream LSearch, HPStream, SWClustering, and CluStream.

Hierarchical methods can be further subdivided into divisive and agglomerative. With divisive hierarchical clustering, a cluster is divided into small clusters until it cannot be split further. In contrast, agglomerative hierarchical clustering merges up separate clusters until the distance between two clusters reaches a required threshold. Balanced iterative reducing and clustering using hierarchies (BIRCH), open distributed application construction (ODAC), E-Stream, clustering using representatives (CURE), and HUE- are some hierarchical algorithms for data stream analysis.

In model-based methods, a hypothesized model is run for each cluster to check which data properly fits a cluster. Some of the algorithms that fit into this category are CluDistream, Similarity Histogram-based Incremental Clustering, sliding window with expectation maximization (SWEM), COBWEB, and Evolving Fractal-Based Clustering of Data Streams.

Density-based methods separate data into density regions (i.e., nonoverlapping cells) of different shapes and sizes. Density-based algorithms require a single pass and can handle noise. Stating the number of clusters in advance is not also required. Some density-based algorithms include DGStream, MicroTEDAclus, clustering of evolving data-streams into arbitrary shapes (CEDAS), Incremental DBSCAN (Density-Based Spatial Clustering with Noise), DenStream, r-DenStream, DStream, DBstream, data stream clustring (DSCLU), MR-Stream, Ordering Points to Identify Clustering Structure (OPTICS), OPClueStream, and MBG-Stream.

6.2 Semi-Supervised Learning

Semi-supervised learning belongs to a class of AI frameworks that trains on the combination of both the unlabeled and labeled data [89]. Semi-supervised learning in data stream context is challenging because data are being generated at real-time and the labels may be missing due to different factors, which include communication errors, network delays, expensive labeling processes, among others [90]. According to Zhu and Li [91], a semi-supervised learning problem in a data stream context is defined as follows. Let $S = \{(x_t, y_t)\}_{t=1}^{T_0}$ as the data in the first T_0 time period and S denote streaming data. Let $Y = \{1, 2, \ldots, K\}$ be the known label set. The arriving data stream has an instance x_t and $y_t \in Y' = \{-1, 1, 2, \ldots, K\}$. If $y_t = -1$, x_t is an unlabelled instance, but the true label is in set Y. As time goes on, evolution happens, a data stream $S' = \{(x_t, y_t)\}_{t=T_0+1}^{\infty}$, which contains novel classes. That is, $\exists\{x_{t'}, y_{t'}\} \in S'$ where $y_{t'} = -1$, but the true label of $x_{t'}$ is not in set Y. Note that if $y_{t'} \neq -1, y_{t'} \in Y$ holds forever.

Semi-supervised learning on streaming data may return similar results to that of the supervised approach. However, there are observations with semi-supervised learning on streaming data, which include (i) to balance out classifiers, considerably more objects ought to be labeled, and (ii) more significant threshold adversely impacts the strength of classifiers with the increase in standard deviation and a bigger threshold [19]. Some of the semi-supervised learning techniques for data streams include ensemble techniques, graph-based methods, deep learning, active learning, linear neighborhood propagation.

6.3 Supervised Learning

Supervised learning is the type of machine learning that infers function from trained labeled data. The training examples contain a couple of input (vector) and output (supervisory signal). Let data stream $S = \{\ldots, d_{t-1}, d_t, d_{t+1}, \ldots\}$, where $d_t = \{x_i, y_i\}$, x_i is the value set of the ith datum in each attribute and y_i is the class of the instance. Data stream classification aims to train a classifier $f : x \rightarrow y$ that establishes a mapping relationship between feature vectors and class labels [92].

Supervised learning approaches can be subdivided into two major categories, which are regression and classification. When the class attribute is continuous, it is called *regression*, but when the class attribute is discrete, it is referred to as classification. Manual labeling is difficult, time-consuming, and could be very costly [93]. In a streaming scenario with high velocity and volume, label data are very scarce, thus leading to poorly trained classifiers as a result of the constrained measure of labeled data accessible for building the models [94].

Some of the supervised learning algorithms for streaming scenario are grouped as presented in [95] (i) Tree-based algorithms: OLIN, Ultra-Fast Forest Tree system (UFFT), Very Fast Decision Tree learner (VFDT), VFDTc, Random Forest, and Vertical Hoeffding Tree, Concept-adapting Evolutionary Algorithm for Decision Tree (CEVOT); (ii) Rule-based algorithms: On-demand classifier, Fuzzy Passive-aggressive classification, Similarity-based data stream classification (SimC), Prequential area under curve (AUC) based classifier, one-class classifier with incremental learning and forgetting, and Classifying recurring concept using fuzzy similarity function; (iii) Ensemble-based algorithms: Streaming ensemble algorithm, Weighted classifier ensemble, Distance-based ensemble online classifier with kernel clustering; (iv) Nearest-neighbor: Adaptive nearest neighbor classification algorithm, anytime nearest neighbor algorithm; (v) Statistical: Evolving Naïve Bayes; (vi) Deep learning: Activity recognition [96].

6.4 Ontology-Based Methods

Performing streaming data analysis over ontologies and linked open data are a challenging and emerging research area. Semantic web technology, an extension of the World Wide Web, is used to improve the interoperability of heterogeneous sources with a data model called *Resource Description Framework* (RDF) and ontological languages such as Web Ontology Language (OWL). Some of the works done using ontology or linked open data on data stream include [97–99]. Due to the dynamic nature of data stream, current solutions for reasoning over the data model and ontological languages are not suited to streaming data context. This gap brought about what is referred to as stream reasoning. Stream reasoning is the set of inference approaches and deduction mechanisms concerned with the provision of continuous inference over a data stream, leading to a better decision support system [100]. Stream reasoning has been applied in remote health monitoring [101], smart cities [102], semantic analysis of social media [103], maritime safety, and securities [104], amongst others. Another attempt to improve semantic web ontology is to lift the existing streams to RDF streams using intuitive configuration mechanisms. Some of the techniques for RDF stream modeling include Semantic Sensor Network (SSN) ontology [105], Stream Annotation Ontology (SOA) [106], smart appliances reference (SAREF) ontology [107], and Linked Stream Annotation Engine (LSane) [108].

7 Strategies for Processing Data Streams

Data stream processing includes techniques, models, and systems for processing data as soon as they arrive to detect trends and patterns in a low latency [109]. Data stream processing requires two factors which include storage capability and computational power in the face of an unbounded generation of data with high velocity and brief life span. To cope with these requirements, approximate computing, which aims at low latency at the expense of acceptable quality loss, has been a practical solution [110]. The ideology behind approximate computing is based on returning approximate answer instead of the exact answer for user queries. This is done by choosing a representative sample of data instead of the whole data [111]. The two main techniques for approximate computing includes

(i) sampling [4], which constructs data stream summaries by probability selection, and (ii) sketches [112], which compress data using data structure (such as histogram or hash tables), prediction-based method (such as Bayesian Inference), and transformation-based method (such as wavelet).

Fixed window and sliding window are two computation models for the partitioning of the data stream. Fixed window partitions data stream into nonoverlapping time segments, and the current data are removed after processing, resetting the window size back to zero. The sliding window contains a historical snapshot of the data stream at any point in time. When the arriving data are at variance with the current window elements, tuples are updated by discarding the oldest data [5]. The sliding window can be further sub-divided into a count-based window and time-based window. In the count-based window, the progressive step is expressed in tuple counts, while items with the oldest timestamp are replaced with items with the latest timestamp in the time-based window [113].

8 Best Practices for Managing Data Streams

A data stream is so dynamic that dealing with data in motion is not just limited to design-time but also a run-time problem that requires an operation that must be managed in real-time. Stream computing has emerged as a capability of real-time applications in smart cities, monitoring systems, manufacturing, and financial markets [15]. Data stream management systems should be able to update the answers to continuous queries as new data arrives. Choosing the right processing model for streaming data is challenging, given the growing number of frameworks with various and similar services [114]. When a high volume of data from disparate sources is needed to be processed at a short time interval, Storm and Flink may be considered. For purely stream processing, Storm is recommended for high stream-oriented applications as it can process millions of events per second. When it comes to durability, scalability, high-throughput, and low-latency capabilities, Apache Kafka is a good option [115]. Yahoo! S4 has capabilities for real-time response, fault-tolerance, and scalability [116]. Spark framework may be suitable for periodic processing tasks such as fraud detection, web usage mining, and so on. For a task that combines both batch and streaming programming models such as IoT and healthcare, Spark and Flink may be good candidates [117]. Some of the frameworks that support iterative processing or machine learning tasks are Flink (FlinkML) Spark (Spark MLlib), GraphX with Spark, and Flinkgelly with Flink. Other graph processing frameworks include Bladgy, Graphlab, and Trinity.

IBM InfoSphere Streams can handle millions of messages or events in a second with high throughput rates, making it one of the leading proprietary solutions for real-time applications [61]. Apama Stream Analytics is suitable for real-time and high-volume business operations [62]. Azure Stream is another proprietary solution for driving streaming analytics and IoT goals [62]. Other reasonable proprietary solutions include Kinesis, PieSync, TIBCO Spotfire, Google Cloud Pub/Sub, Azure Event Hubs, Kibana, Amazon Elastic Search Service, and Kibana.

In an ideal case, choosing a single streaming data technology that supports all the system requirements such as the state of data, use case, and kind of results seems the best as this alleviates the problems of interoperability constraints.

9 Conclusion and the Way Forward

In this chapter, we have considered cutting-edge issues concerning data stream or streaming data. The interest in stream processing is on the increase, and data must be handled quickly to make decisions in real-time. The key presumption of stream computing is that the likelihood estimation of data lies in its newness. Thus, data analysis is done the moment they arrive in a stream instead of what is obtained in batch processing where data are first stored before they are explored. Challenges for data stream analysis include concept drift, scalability, integration, fault tolerance, timeliness, consistency, heterogeneity and incompleteness, load balancing, privacy issues, and accuracy [27, 28, 30–32, 34, 35], which emerges from the nature of data streams.

Streaming is an active research area. However, there are still some aspects of streaming that have received little attention. One of them is transactional guarantees. Current stream processing can provide basic guarantees such as processing each data point in the stream exactly once or at least once but cannot provide guarantees that span multiple operations or stream elements. Another area to intensify research effort is data stream pre-processing. Data quality is a vital determinant in the knowledge discovery pipeline as low-quality data yields low-quality models and choices [69]. There is need to reinforce data stream pre-processing stage [67] in the face of multi-label [70], imbalance [71], and multi-instance [72] problems associated data stream [66]. Also, the representation of social media posts must be such that the semantics of social media content is preserved [74, 75]. Moreover, data stream pre-processing techniques with low computational requirement [73] need to be evolved as this is still open for research.

Data stream processing requires two factors which include storage capability and computational power in the face of an unbounded generation of data with high velocity and brief life span. To cope with these requirements, approximate computing, which aims at low latency at the expense of acceptable quality loss, has been a practical solution [110]. Even though approximate computing has been extensively used for the processing of data stream, combining it with distributed processing models brings new research directions. Such research directions include approximation with heterogeneous resources, pricing models with approximation, intelligent data processing, and energy-aware approximation.

References

1 World Economic Forum (2019) *How Much Data is Generated Each Day? Visual Capitalist*, https://www.visualcapitalist.com/how-much-data-is-generated-each-day.

2 Huynh, V. and Phung, D. (2017) Streaming clustering with Bayesian nonparametric models. *Neurocomputing*, 258, 52–62. doi: 10.1016/j.neucom.2017.02.078.

3 Ray, I., Adaikkalavan, R., Xie, X., and Gamble, R. (2015) *Stream Processing with Secure Information Flow Constraints*. 29th IFIP Annual Conference on Data and Applications Security and Privacy. Fairfax, USA, pp. 311–329. doi: 10.1007/978-3-319-20810-7_22.

4 Sibai, R.E., Chabchoub, Y., Demerjian, J. *et al.* (2016) *Sampling Algorithms in Data Stream Environment*. 2016 International Conference on Digital Economy Carthage. IEEE, Tunisia, pp. 29–36. doi: 10.1109/ICDEC.2016.7563142.

5 Youn, J., Shim, J., and Lee, S.G. (2018) Efficient data stream clustering with sliding windows based on locality sensitive hashing. *IEEE Access*, 6, 63757–63776. doi: 10.1109/ACCESS.2018.2877138.

6 Das, S., Beheraa, R.K., Kumar, M., and Rath, S.K. (2018) Real-time sentiment analysis of twitter streaming data for stock prediction. *Procedia Comput. Sci.*, 132, 956–964.

7 Wang, J., Zhu, R., and Liu, S. (2018) A differentially private unscented Kalman filter for streaming data in IoT. *IEEE Access*, 6 (1), 6487–6495. doi: 10.1109/ACCESS.2018.2797159.

8 Kolchinsky, I. and Schuster, A. (2019) *Real-Time Multi-Pattern Detection Over Event Streams*. Proceedings of the 2019 International Conference on Management of Data, Amsterdam Netherlands: New York, NY, USA: ACM, pp. 589–606. doi: 10.1145/3299869.3319869.

9 Tozi, C. (2017) *Dummy's Guide to Batch vs Streaming. Retrieved from Trillium Software*, https://www.precisely.com/blog/big-data/big-data-101-batch-stream-processing.

10 Kolajo, T., Daramola, O., and Adebiyi, A. (2019) Big data stream analysis: a systematic literature review. *J. Big Data*, 6, 47.

11 Kusumakumari, V., Sherigar, D., Chandran, R., and Patil, N. (2017) Frequent pattern mining on stream data using Hadoop CanTree-GTree. *Procedia Comput. Sci.*, 115, 266–273.

12 Giustozzia, F., Sauniera, J., and Zanni-Merk, C. (2019) Abnormal situations interpretation in industry 4.0 using stream reasoning. *Procedia Comput. Sci.*, 159, 620–629.

13 Liu, R., Li, Q., Li, F. *et al.* (2014) *Big Data Architecture for IT Incident Management*. Proceedings of IEEE international conference on service operations and logistics, and informatics. Qingdao, China, pp. 424–429.

14 Sakr, S. (2013) *An Introduction to Infosphere Streams: A Platform for Analyzing Big Data in Motion, IBM*, https://www.ibm.com/developerworks/library/bd-streamsintro/index.html.

15 Inoubli, W., Aridhi, S., Mezni, H. *et al.* (2018) An experimental survey on big data frameworks. *Future Gener. Comp. System*, 86, 546–564. doi: 10.1016/j.future.2018.04.032.

16 International Business Machine (2019) *Stream Computing Platforms, Applications and Analytics*, https://researcher.watson.ibm.com/researcher/view_group.php?id=2531.

17 Vidyasankar, K. (2017) On continuous queries in stream processing. *Procedia Comput. Sci.*, 109C, 640–647.

18 Joseph, S., Jasmin, E.A., and Chandran, S. (2015) Stream computing: opportunities and challenges in smart grid. *Procedia Tech.*, 21, 49–53.

19 Wozniak, M., Ksieniewicz, P., Cyganek, B. *et al.* (2016) Active learning classification of drifted streaming data. *Procedia Comput. Sci.*, 80, 1724–1733.

20 Kim, T. and Park, C.H. (2020) Anomaly pattern detection for streaming data. *Expert Syst. Appl.*, 149, 113252. doi: 10.1016/j.eswa.2020.113252.

21 Sethi, T.S. and Kantardzic, M. (2018) Handling adversarial concept drift in streaming data. *Expert Syst. Appl.*, 97, 18–40.

22 Toor, A.A., Usman, M., Younas, F. *et al.* (2020) Mining massive e-health data streams for IoMT enabled healthcare systems. *Sensors*, 20 (7), 2131. doi: 10.3390/s20072131.

23 Shan, J., Luo, J., Ni, G. *et al.* (2016) CVS: fast cardinality estimation for large-scale data streams over sliding windows. *Neurocomputing*, 194, 107–116.

24 Liu, W., Wang, Z., Liu, X. *et al.* (2017) A survey of deep neural network architectures and their applications. *Neurocomputing*, 234, 11–26.

25 Priya, S. and Uthra, R.A. (2020) Comprehensive analysis for class imbalance data with concept drift using ensemble based classification. *J. Ambient Intell. Humaniz. Comput.* doi: 10.1007/s12652-020-01934-y.

26 Zhou, L., Pan, S., Wang, J., and Vasilakos, A.V. (2017) Machine learning on big data: opportunities and challenges. *Neurocomputing*, 237, 350–361. doi: 10.1016/j.neucom.2017.01.026.

27 O'Donovan, P., Leahy, K., Bruton, K., and O'Sullivan, D.T.J. (2015) An industrial big data pipeline for data-driven analytics maintenance applications in large-scale smart manufacturing facilities. *J. Big Data*, 2, 25. doi: 10.1186s40537-015-0034-z.

28 Zaharia, M., Das, T., Li, H. *et al.* (2013) *Discretized Streams: Fault-Tolerant Streaming Computation at Scale*. Proceedings of the 24th ACM Symposium on Operating System Principles (SOSP 2013), Farmington: ACM Press, pp. 423–438.

29 Jayasekara, S., Harwood, A., and Karunasekera, S. (2020) A utilization model for optimization of checkpoint intervals in distributed stream processing systems. *Futur. Gener. Comput. Syst.*, 110, 68–79. doi: 10.1016/j.future.2020.04.019.

30 Chong, D. and Shi, H. (2015) Big data analytics: a literature review. *J. Manag. Anal.*, 2 (3), 175–201.

31 Qian, Z., He, Y., Su, C. *et al.* (2013) *TimeStream: Reliable Stream Computation in the Cloud*. Proceedings of the 8th ACM European Conference on Computer Systems. ACM, Prague, pp. 1–14. doi: 10.1145/2465351.2465353.

32 Shi, P., Cui, Y., Xu, K. *et al.* (2019) Data consistency theory and case study for scientific big data. *Information*, 10, 137. doi: 10.3390/info10040137.

33 Santipantakis, G., Kotis, K., and Vouros, G.A. (2017) OBDAIR: ontology-based distributed framework for accessing, integrating and reasoning with data in disparate data sources. *Expert Syst. Appl.*, 90, 464–483.

34 Cortes, R., Bonnaire, X., Marin, O., and Sens, P. (2015) Stream processing of healthcare sensor data: studying user traces to identify challenges from a big data perspective. *Procedia Comput. Sci.*, 52, 1004–1009.

35 D'Argenio, V. (2018) The high-throughput analyses era: are we ready for the data struggle. *High Throughput*, 7 (1), 8. doi: 10.3390/ht7010008.

36 Qiu, Y. and Ma, M. (2018) Secure group mobility support for 6LoWPAN networks. *IEEE Internet Things J.*, 5 (2), 1131–1141.

37 Wanga, J., Luo, J., Liu, X. *et al.* (2019) Improved kalman filter based differentially private streaming data release in cognitive computing. *Futur. Gener. Comput. Syst.*, 98, 541–549.

38 Denham, B., Pears, R., and Naeem, A.M. (2020) Enhancing random projection with independent and cumulative additive noise for privacy-preserving data stream mining. *Expert Syst. Appl.*, 152, 113380. doi: 10.1016/j.eswa.2020.113380.

39 Hariri, R.H., Fredericks, E.M., and Bowers, K.M. (2019) Uncertainty in big data analytics: survey, opportunities, and challenges. *J. Big Data*, 6, 44. doi: 10.1186/s40537-019-0206-3.

40 Millman, N. (2014) *Analytics for Business. Computerworld*, https://www.computerworld.com/article/24758 40/bigdata/8-considerations-when-selecting-big-data-technology.html.

41 Brook, C. (2014) *Enterprise NoSQL for Dummies*, John Wiley & Sons, Hoboken.

42 Shanahan, J.G. and Dai, L. (2015) *Large Scale Distributed Data Science using Apache Spark*. Proceedings of the 21st ACM SIGKDD International Conference on Knowledge Discovery and Data Mining. ACM, New York, pp. 2323–2324. doi: 10.1145/2783258.2789993.

43 Sharma, S. (2016) Expanded cloud plumes hiding big data ecosystem. *Futur. Gener. Comput. Syst.*, 59, 63–92.

44 Meng, X., Bradley, J., Yavuz, B. *et al.* (2016) Mllib: machine learning in apache spark. *J. Mach. Learn. Res.*, 17 (1), 1235–1241.

45 Mazumder, S. (2016) Big data application in engineering and science, in *Big Data Concepts, Theories, and Applications* (eds S. Yu and S. Guo), Springer, Cham, pp. 29–128. doi: 10.1007/978-3-319-27763-9_2.

46 Liao, X., Gao, Z., Ji, W., and Wang, Y. (2016) *An Enforcement of Real-Time Scheduling in Spark Streaming*. Sixth IEEE International Green Computing Conference and Sustainable Computing Conference (IGSC). IEEE, Las Vegas, pp. 1–6. doi: 10.1109/IGCC.2015.7393730.

47 Jayanthi, D. and Sumathi, G. (2016) *A Framework for Real-Time Streaming Analytics Using Machine Learning Approach*. Proceedings of the National Conference on Communication and Informatics, Sriperumbudur, India, pp. 85–90.

48 Agha, G. (1986) *Actors: A Model of Concurrent Computation in Distributed Systems*, MIT Press, Cambridge.

49 Ananthanarayanan, R., Basker, V., Das, S. *et al.* (2013). *Photon: Fault-Tolerant and Scalable Joining of Continuous Data Streams*. Proceedings of 2013 ACM SIGMOD International Conference on Management of Data. ACM, New York, pp. 577–588. doi: 10.1145/2463676.2465272.

50 Apache Software Foundation (2017) *Apache Aurora: System Overview*, http://aurora.apache.org/documentation/latest/getting-started/overview.

51 Yang, W., DaSilva, A., and Picard, M.L. (2015) Computing data quality indicators on big data streams using a CEP, in *2015 IEEE International Workshop on Computational Intelligence for Multimedia Understanding*, IEEE, Prague, pp. 1–5.

52 Morales, F.G. (2013) *SAMOA: A Platform for Mining Big Data Streams*. Proceedings of the 22nd International Conference on World Wide Web. ACM, Rio de Janeiro, pp. 777–778.

53 Ren, X., Khrouf, H., Kazi-Aoul, Z. *et al.* (2018) *On Measuring Performances of C-SPARQL and CQELS*, Kobe, Japan https://hal-upec-upem.archives-ouvertes.fr/hal-01740520.

54 Keeney, J., Fallon, L., Tai, W., and O'Sullivan, D. (2015) *Towards Composite Semantic Reasoning for Real-Time Network Management Data Enrichment*. Proceedings of the 2015 IEEE 11th International Conference on Network and Service Management (CNSM), Barcelona. pp. 246–250. doi: 10.1109/CNSM.2015.7367365.

55 Gao, F., Ali, M.I., Cury, E., and Mileo, A. (2017) Automated discovery and integration of semantic urban data streams: the ACEIS middleware. *Futur. Gener. Comput. Syst.*, 76, 561–581.

56 Toll, W. (2014) *Top 45 Big Data Tool for Developers*, https://blog.profitbricks.com/top-45-big-data-tools-for-developers.

57 Baciu, G., Li, C., Wang, Y., and Zhang, X. (2015) Cloudet: a cloud-driven visual cognition for large streaming data. *Int. J. Cognitive Inform. Nat. Intel.*, 10 (1), 12–31. doi: 10.4018/IJCINI.2016010102.

58 Chen, X.J. and Ke, J. (2015) *Fast Processing of Conversion Time Data Flow in Cloud Computing via Weighted FP-Tree Mining Algorithms*. Ubiquitous Intelligence and Computing and 2015 IEEE 12th Intl Conference on Autonomic and Trusted Computing and 2015 IEEE 15th Intl Conference on Scalable Computing and Communications and Its Associated Workshops (UIC-ATC-ScalCom), Beijing, China, pp. 386–391.

59 Chen, X., Chen, H., Zhang, N. *et al.* (2015) Large-scale real-time semantic processing framework for internet of things. *Int. J. Distrib. Sens. Net.*, 11 (10), 365–372. doi: 10.1155/2015/365372.

60 Kropivnitskaya, Y., Qin, J., Tiampo, K.F., and Bauer, M.A. (2015) A pipelining implementation for high resolution seismic hazard maps production. *Procedia Comput. Sci.*, 51, 1473–1482.

61 Birjali, M., Beni-Hssane, A., and Erritali, M. (2017) Analyzing social media through big data using infosphere biginsights and apache flume. *Procedia Comput. Sci.*, 113, 280–285. doi: 10.1016/j.procs.2017.08.299.

62 Warner, J (2019) *5 Streaming Analytics Platforms for All Real-Time Applications*, https://www.google.com/amp/s/datafloq.com/read/amp/streaming-analytics-platforms-real-time-apps/4658.

63 Yang, H., Lee, Y., Lee, H. *et al.* (2015) A study on word vector models for representing Korean semantic information. *Phone. Speech Sci.*, 7, 41–47. doi: 10.13064/KSSS.2015.7.4.041.

64 Joseph, S. and Jasmin, E.A. (2016) *Stream Computing Framework for Outage Detection in Smart Grid*. Proceedings of 2015 IEEE International Conference on Power Instrumentation, Control and Computing, Thrissur, India, pp. 1–5. doi: 10.1109/PICC.2015.7455744.

65 Barika, M., Garg, S., Chan, A. *et al.* (2019) IoTSim-stream: modelling stream graph application in cloud simulation. *Futur. Gener. Comput. Syst.*, 99, 86–105.

66 Ramírez-Gallego, S., Krawczyk, B., García, S., and Woniak, M. (2017) A survey on data preprocessing for data stream mining: current status and future directions. *Neurocomputing*, 239, 39–57. doi: 10.1016/j.neucom.2017.01.078.

67 Kolajo, T., Daramola, O., Adebiyi, A., and Seth, A. (2020) A framework for pre-processing of social media feeds based on local knowledge base. *Inf. Process. Manag.*, 57 (6), 102348.

68 Gill, S. and Lee, B. (2015) A framework for distributed cleaning of data streams. *Procedia Comput. Sci.*, 52, 1186–1191.

69 Ramírez-Gallego, S., García, S., and Herrera, F. (2018) Online entropy-based discretization for data streaming classification. *Future Gener. Comp. Syst.*, 86, 59–70. doi: 10.1016/j.future.2018.03.008.

70 Herrera, F., Charte, F., Rivera, A.J., and del Jesús, M.J. (2016) *Multi-Label Classification – Problem Analysis, Metrics and Techniques*, 1st edn, Springer, Cham.

71 Krawczyk, B. (2016) GPU-accelerated extreme learning machines for imbalanced data streams with concept drift. *Procedia Comput. Sci.*, 80, 1692–1701.

72 Herrera, F., Ventura, S., Bello, R. *et al.* (2016) *Multiple Instance Learning – Foundations and Algorithms*, Cham, Switzerland Springer.

73 García, S., Ramírez-Gallego, S., Luengo, J. *et al.* (2016) Big data preprocessing: methods and prospects. *Big Data Anal.*, 1, 9. doi: 10.1186/s41044-016-0014-0.

74 Hasan, M., Orgun, M.A., and Schwitter, R. (2019) Real-time event detection from the twitter data stream using the twitterNews + framework. *Inf. Process. Manag.*, 56 (3), 1146–1165.

75 Pagliardini, M., Gupta, P., and Jaggi, M. (2018) *Unsupervised Learning of Sentence Embeddings using Compositional n-Gram Features*. Proceedings of NAACL-HLT. ACM, New Orleans, LA, USA, pp. 528–540.

76 Wu, L., Morstatter, F., and Liu, H. (2018) SlangSD: building, expanding and using a sentiment dictionary of slang words for short-text sentiment classification. *Lang Res. Eval.*, 52 (3), 839–852. doi: 10.1007/s10579-018-9416-0.

77 Wankhede, S., Patil, R., Sonawane, S., and Save, A. (2018) *Data Pre-Processing for Efficient Sentimental Analysis*. 2018 Second International Conference on Inventive Communication and Computational Technologies (ICICCT), Coimbatore, India, pp. 723–726.

78 Gupta, A., Taneja, S.B., Malik, G. *et al.* (2019) SLANGZY: a fuzzy logic-based algorithm for english slang meaning selection. *Prog. Artif. Intell.*, 8, 111–121. doi: 10.1007/s13748-018-0159-3.

79 Mehta, J.S. (2017) Concept drift in streaming data classification: algorithms, platforms and issues. *Procedia Comput. Sci.*, 122, 804–811.

80 BakshiRohit, P. and Agarwal, S. (2016) Stream data mining: platforms, algorithms, performance evaluators and research trends. *Int. J. Database Theory App.*, 9 (9), 201–218.

81 Wei, X., Liu, Y., and Wanga, X. (2019) A survey on quality-assurance approximate stream processing and applications. *Futur. Gener. Comput. Syst.*, 101, 1062–1080.

82 Hu, Y., Jiang, Z., Zhan, P. *et al.* (2018) A novel multi-resolution representation for streaming time series. *Procedia Comput. Sci.*, 129, 178–184. doi: 10.1016/j.procs.2018.03.069.

83 Yaseen, M.U., Anjum, A., Rana, O., and Hill, R. (2018) Cloud-based scalable object detection and classification in video streams. *Futur. Gener. Comput. Syst.*, 80, 286–298. doi: 10.1016/j.future.2017.02.003.

84 Boushaki, S.I., Kamel, N., and Bendjeghaba, O. (2018) High-dimensional text datasets clustering algorithm based on cuckoo search and latent semantic indexing. *J. Inf. Knowl. Manag.*, 17 (3), 1–24.

85 Neto, J.M., Severiano Junior, C.A., Guimarães, F.G. *et al.* (2020) Evolving clustering algorithm based on mixture of typicalities for stream. *Futur. Gener. Comput. Syst.*, 106, 672–684.

86 Ibrahim, O.A., Du, Y., and Keller, J.M. (2018) Extended robust online streaming clustering (EROLSC), in *Information Processing and Management of Uncertainty in Knowledge-Based Systems: Theory and Foundations* (eds J. Medina *et al.*), Springer, Cadiz.

87 Sharma, N., Masih, S., and Makhija, P. (2018) A survey on clustering algorithms for data streams. *Int. J. Comput. Appl.*, 182 (22), 18–24.

88 Panagiotou, N., Katakis, I., and Gunopulos, D. (2016) Detecting events in online social networks: definitions, trends and challenges, in *Solving Large Scale Learning Tasks: Challenges and Algorithms* (ed. S. Michaelis), Springer, Cham, pp. 42–84.

89 Li, Y., Guo, L., and Zhou, Z. (2019) Towards safe weakly supervised learning. *IEEE Trans. Pattern Anal. Mach. Intell.*, 43 (1), 334–346. doi: 10.1109/TPAMI.2019.2922396.

90 Le Nguyen, M.H., Gomes, H.M., and Bifet, A. (2019). *Semi-Supervised Learning Eover Streaming Data Using MOA*. 2019 IEEE International Conference on Big Data (Big Data). IEEE, Los Angeles, CA, USA, pp. 553–562. doi: 10.1109/BigData47090.2019.9006217.

91 Zhu, Y. and Li, Y.-F. (2020) Semi-supervised streaming learning with emerging new labels. *Proc. Thirty-Fourth AAAI Conf. Artif. Intel.*, 34, 7015–7022. doi: 10.1609/aaai.v34i04.6186.

92 Li, P., Wu, X., Hu, X., and Wang, H. (2015) Learning concept-drifting data streams with random ensemble decision trees. *Neurocomputing*, 166, 68–83.

93 Sethi, T.S. and Kantardzic, M. (2017) On the reliable detection of concept drift from streaming unlabeled data. *Expert Syst. Appl.*, 82, 77–99. doi: 10.1016/j.eswa.2017.04.008.

94 Masud, M.M., Gao, J., Khan, L. *et al.* (2008) *A Practical Approach to Classify Evolving Data Streams: Training with Limited Amount of Labeled Data*. 2008 Eighth IEEE International Conference on Data Mining. IEEE, Pisa, pp. 929–934. doi: 10.1109/ICDM.2008.152.

95 BakshiRohit, P. and Agarwal, S. (2017) Critical parameter analysis of vertical hoeffding tree for optimized performance using SAMOA. *Int. J. Mach. Learn. Cybern.*, 8, 1389–1402.

96 Ullah, A., Muhammad, K., Haq, I.U., and Baik, S.W. (2019) Action recognition using optimized deep autoencoder and CNN for surveillance data streams of non-stationary environments. *Futur. Gener. Comput. Syst.*, 96, 386–397. doi: 10.1016/j.future.2019.01.029.

97 Elsaleh, T., Enshaeifar, S., Rezvani, R. *et al.* (2020) IoT-stream: a lightweight ontology for internet of things data streams and its use with data analytics and event detection services. *Sensors (Basel)*, 20 (4), 953. doi: 10.3390/s20040953.

98 Janowicz, K., Haller, A., Cox, S.J. *et al.* (2019) SOSA: a lightweight ontology for sensors, observations, samples, and actuators. *J. Web Semant.*, 56, 1–10. doi: 10.2139/ssrn.3248499.

99 Gonzalez-Gil, P., Skarmeta, A.F., and Martinez, J.A. (2019) *Towards an Ontology for IoT Context-Based Security Evaluation*. Proceedings of the 2019 Global IoT Summit (GIoTS), Aarhus, Denmark, pp. 1–6.

100 Bazoobandi, H.R., Beck, H., and Urbani, J. (2017) Towards expressive stream reasoning with laser, in *The Semantic Web*, vol. 10587 (ed. C.E. d'Amato), *LNCS*, pp. 87–103.

101 Albahri, O.S., Albahri, A.S., Mohammed, K.I. *et al.* (2018) Systematic review of real-time remote health monitoring system in triage and priority-based sensor technology: Taxonomy, open challenges, motivation and recommendations. *J. Med. Syst.*, 42, 80. doi: 10.1007/s10916-018-0943-4.

102 D'Aniello, G., Gaeta, M., and Orciuoli, F. (2018) An approach based on semantic stream reasoning to support decision processes in smart cities. *Telemat. Inform.*, 35 (1), 68–81. doi: 10.1016/j.tele.2017.09.019.

103 Mondal, J. and Deshpande, A. (2018) Stream querying and reasoning on social data, in *Encyclopedia of Social Network Analysis and Mining* (eds R. Alhajj and J. Rokne), Springer, New York. doi: 10.1007/978-1-4939-7131-2_391.

104 Wen, Y., Zhang, Y., Huang, L. *et al.* (2019) Semantic modelling of ship behavior in harbor based on ontology and dynamic bayesian network. *Int. J. Geogr. Inf. Sci.*, 8 (3), 107. doi: 10.3390/ijgi8030107.

105 Compton, M., Barnaghi, P., Bermudez, R.G. *et al.* (2012) The SSN ontology of the W3C semantic sensor network incubator group. *J. Web Semant.*, 17, 25–32.

106 Daniele, L., den Hartog, F., and Roes, J. (2015) Created in close einteraction with the industry: the smart appliances reference (saref) ontology, in *Formal Ontologies Meet Industries*, vol. 225 (eds R. Cuel and R. Young), LNBIP, pp. 100–112. doi: 10.1007/978-3-319-21545-7_9.

107 Franka, M.T., Baderb, S., Simko, V., and Zander, S. (2018) LSane: collaborative validation and enrichment of heterogeneous observation streams. *Procedia Comput. Sci.*, 137, 235–241. doi: 10.1016/j.procs.2018.09.022.

108 Kolozali, S., Bermudez-Edo, M., Puschmann, D. *et al.* (2014) *A knowledge-Based Approach for Real-Time IoT Data Stream Annotation and Processing*. 2014 IEEE International Conference on Internet of Things (iThings), and IEEE Green Computing and Communications (GreenCom) and IEEE Cyber, Physical and Social Computing (CPSCom). IEEE, Taipei, pp. 215–222. doi: 10.1109/iThings.2014.39.

109 Cardellini, V., Mencagli, G., Talia, D., and Torquati, M. (2019) New landscapes of the data stream processing in the era of fog computing. *Futur. Gener. Comput. Syst.*, 99, 646–650. doi: 10.1016/j.future.2019.03.027.

110 Wei, X., Liu, Y., Wanga, X. *et al.* (2019) A survey on quality-assurance approximate stream processing and applications. *Futur. Gener. Comput. Syst.*, 101, 1062–1080.

111 Quoc, D.L., Krishnan, D.R., Bhatotia, P. *et al.* (2018) Incremental approximate computing, in *Encyclopedia of Big Data Technologies* (eds S. Sakr and A. Zomaya), Springer, Cham.

112 Sigurleifsson, B., Anbarasu, A., and Kangur, K. (2019) An overview of count-min sketch and its application. *EasyChair*, 879, 1–7.

113 Garofalakis, M., Gehrke, J., and Rastogi, R. (eds) (2016) *Data Stream Management: Processing High-Speed Data Streams*, Springer, Berlin, Heidelberg.

114 Sakr, S. (2016) *Big Data 2.0 Processing Systems: A Survey*, Springer, Switzerland. doi: 10.1007/978-3-319-38776-5.

115 Yates, J. (2020) *Stream Processing with IoT Data: Challenges, Best Practices, and Techniques*, https://www.confluent.io/blog/stream-processing-iot-data-best-practices-and-techniques.

116 Zhao, X., Garg, S., Queiroz, C., and Buyya, R. (2017) A taxonomy and survey of stream processing systems, in *Software Architecture for Big Data and the Cloud* (eds I. Mistrik, R. Bahsoon, N. Ali, *et al.*), Elsevier, pp. 183–206. doi: 10.1016/B978-0-12-805467-3.00011-9.

117 Landset, S., Khoshgoftaar, T.M., Richter, A.N., and Hasanin, T. (2015) A survey of open source tools for machine learning with big data in the Hadoop ecosystem. *J. Big Data*, 2 (1), 1–36.

Part II

Simulation-Based Methods

5

Monte Carlo Simulation: Are We There Yet?

Dootika Vats[1], James M. Flegal[2], and Galin L. Jones[3]

[1] *Indian Institute of Technology Kanpur, Kanpur, India*
[2] *University of California, Riverside, CA, USA*
[3] *University of Minnesota, Twin-Cities Minneapolis, MN, USA*

1 Introduction

Monte Carlo simulation methods generate observations from a chosen distribution in an effort to estimate unknowns of that distribution. A rich variety of methods fall under this characterization, including classical Monte Carlo simulation, Markov chain Monte Carlo (MCMC), importance sampling, and quasi-Monte Carlo.

Consider a distribution F defined on a d-dimensional space $\mathcal{X}$, and suppose that $\theta \in \mathbb{R}^p$ are features of interest of F. Specifically, θ may be a combination of quantiles, means, and variances associated with F. Samples $X_1, \dots, X_n$ are obtained via simulation either approximately or exactly from F, and a consistent estimator of θ, $\hat{\theta}$, is constructed so that, as $n \to \infty$,

$$\hat{\theta}(X_1, \dots, X_n) \stackrel{a.s.}{\to} \theta \tag{1}$$

Thus, even when F is a complicated distribution, Monte Carlo simulation allows for estimation of features of F. Throughout, we assume that either independent and identically distributed (IID) samples or MCMC samples from F can be obtained efficiently; see Refs 1–5 for various techniques.

The foundation of Monte Carlo simulation methods rests on asymptotic convergence as indicated by (1). When enough samples are obtained, $\hat{\theta} \approx \theta$, and simulation can be terminated with reasonable confidence. For many estimators, an asymptotic sampling distribution is available in order to ascertain the variability in estimation via a central limit theorem (CLT) or application of the delta method on a CLT. Section 2 introduces estimators of θ, while Section 3 discusses sampling distributions of these estimators for IID and MCMC sampling.

Although Monte Carlo simulation relies on large-sample frequentist statistics, it is fundamentally different in two ways. First, data is generated by a computer, and so often there is little cost to obtaining further samples. Thus, the reliance on asymptotics is reasonable. Second, data is obtained sequentially, so determining when to terminate the simulation

Computational Statistics in Data Science.
Edited by Walter W. Piegorsch, Richard A. Levine, Hao Helen Zhang and Thomas C. M. Lee.
 ISBN 978-1-11956107-1

can be based on the samples already obtained. As this implies a random simulation time, additional safeguards are necessary to ensure asymptotic validity. This has led to the study of sequential stopping rules, which we present in Section 5.

Sequential stopping rules rely on estimating the limiting Monte Carlo variance–covariance matrix (when $p = 1$, this is the standard error of $\hat{\theta}$). This is a particularly challenging problem in MCMC due to serial correlation in the samples. We discuss these challenges in Section 4 and present estimators appropriate for large simulation sizes.

Over a variety of examples in Section 7, we conclude that the simulation size required for a reliable estimation is often higher than what is commonly used by practitioners (see also Refs 6, 7. Given modern computational power, the recommended strategies can easily be adopted in most estimation problems. We conclude the introduction with an example illustrating the need for careful sample size calculations.

Example 1. Consider IID draws $X_1, \dots, X_m \sim N(\theta, \sigma^2)$. An estimate of θ is $\overline{X} = m^{-1}\sum_{i=1}^{m} X_i$, and σ^2 is estimated with the sample variance, s^2. Let z_u be the uth quantile of a standard normal distribution, for $0 < u < 1$. A large-sample $(1-\alpha)100\%$ confidence interval for θ is

$$\overline{X} \pm z_{1-\alpha/2}\frac{s}{\sqrt{m}}$$

Confidence intervals are notoriously difficult to understand at a first instance, and thus a standard Monte Carlo experiment in an introductory statistics course is that of repeating the above experiment multiple times and illustrating that on average about $(1-\alpha)$ proportion of such confidence intervals will contain the true mean. That is, for $t = 1, \dots, n$, we generate $X_{t1}, \dots, X_{tm} \sim N(\theta, \sigma^2)$, calculate the mean $\overline{X}_t$ and the sample variance s_t^2, and define p_t to be

$$p_t = I\left\{\overline{X}_t - z_{1-\alpha/2}\frac{s_t}{\sqrt{m}} < \theta < \overline{X}_t + z_{1-\alpha/2}\frac{s_t}{\sqrt{m}}\right\}$$

where $I\{\cdot\}$ is the indicator function. By the law of large numbers, $\overline{p} = n^{-1}\sum_{t=1}^{n} p_t \to 1-\alpha$ with probability 1, as $n \to \infty$, and the following CLT holds:

$$\sqrt{n}(\overline{p} - (1-\alpha)) \xrightarrow{d} N\left(0, \frac{\alpha(1-\alpha)}{n}\right)$$

In conducting this experiment, we must choose the Monte Carlo sample size n. A reasonable argument here is that our estimator $\overline{p}$ must be accurate up to the second significant digit with roundoff. That is, we may allow a margin of error of 0.005. This implies that n must be chosen so that

$$n > \frac{\alpha(1-\alpha)}{(0.005^2)}$$

That is, to construct, say a 95% confidence interval, an accurate Monte Carlo study in this simple example requires at least 1900 Monte Carlo samples. A higher precision would require an even larger simulation size! This is an example of an absolute precision stopping rule (Section 5) and is unique since the limiting variance is known. For further discussion of this example, see Frey [8].

2 Estimation

Recall that F is a d-dimensional target distribution, and interest is in estimating different features of F. In Monte Carlo simulation, we generate $X_1, \dots, X_n \sim F$ either via IID sampling or via a Markov chain that has F as its limiting distribution. For MCMC samples, we assume throughout that a Harris ergodic Markov chain is employed ensuring convergence of sample statistics to (finite) population quantities (see Roberts and Rosenthal [9], for definitions).

2.1 Expectations

The most common quantity of interest in Monte Carlo simulations is the expectation of a function of the target distribution. Let $\|\cdot\|$ denote the Euclidean norm, and let $h : \mathcal{X} \to \mathbb{R}^p$, so that interest is in estimating

$$\theta_h = \int h(x)F(\mathrm{d}x)$$

where we assume $\mathrm{E}_F\|h(X)\| < \infty$. If h is identity, then the mean of the target is of interest. Alternatively, h can be chosen so that moments or other quantities are of interest. A Monte Carlo estimator of θ_h is

$$\hat{\theta}_h = \frac{1}{n}\sum_{t=1}^{n} h(X_t)$$

For IID and MCMC sampling, the ergodic theorem implies that $\hat{\theta}_h \overset{a.s.}{\to} \theta_h$ as $n \to \infty$. The Monte Carlo average $\hat{\theta}_h$ is naturally unbiased as long as the samples are either IID or the Markov chain is stationary.

2.2 Quantiles

Quantiles are particularly of interest when making credible intervals in Bayesian posterior distributions or making boxplots from Monte Carlo simulations. In this section, we assume that h is one-dimensional (i.e., $p = 1$). Extensions to $p > 1$ are straightforward but notationally involved [10]. For $V = h(X)$, interest may be in estimating a quantile of V. Let $F_h(v)$ be the distribution function of $h(X)$, assumed to be absolutely continuous with a continuous density $f_h(v)$. The q-quantile associated with F_h is

$$\phi_q = F_h^{-1}(q) = \inf\{v : F_h(v) \geq q\}$$

Sample statistics are used to estimate ϕ_q. That is, let $\hat{\phi}_q = h(X)_{\lceil nq \rceil :n}$ be the $\lceil nq \rceil$th order statistic of V. Then, standard arguments for IID sampling and MCMC [11] show that $\hat{\phi}_q \overset{a.s.}{\to} \phi_q$ as $n \to \infty$.

2.3 Other Estimators

Other quantities of interest that cannot naturally be presented as expectations (i.e., coefficient of variation) can be estimated by standard plug-in estimation techniques. We focus on estimating the $p \times p$ variance–covariance matrix of h under F

$$\Lambda = \mathrm{Var}_F[h(X)] = \mathrm{E}_F[(h(X) - \theta_h)(h(X) - \theta_h)^T]$$

A natural estimator is the sample covariance matrix

$$\hat{\Lambda}_n = \frac{1}{n-1} \sum_{t=1}^{n} (h(X_t) - \hat{\theta}_h)(h(X_t) - \hat{\theta}_h)^T$$

The strong law of large numbers and the continuous mapping theorem imply that $\hat{\Lambda}_n \xrightarrow{a.s.} \Lambda$ as $n \to \infty$. For IID samples, $\hat{\Lambda}_n$ is unbiased, but for MCMC samples under stationarity, $\hat{\Lambda}_n$ is typically biased from below [12]

$$\mathrm{E}_F[\hat{\Lambda}_n] = \frac{n}{n-1}(\Lambda - \mathrm{Var}_F(\hat{\theta}_h))$$

For MCMC samples, $\mathrm{Var}_F(\hat{\theta}_h)$ is typically larger than Λ/n, yielding biased-from-below estimation. If obtaining an unbiased estimator of Λ is desirable, a bias correction should be done by estimating $\mathrm{Var}(\hat{\theta}_h)$ using methods described in Section 4.

3 Sampling Distribution

An asymptotic sampling distribution for estimators in the previous section can be used to summarize the Monte Carlo variability, provided it is available and the limiting variance is estimable. For IID sampling, moment conditions for the function of interest, h, with respect to the target distribution, F, suffice. For MCMC sampling, more care needs to be taken to ensure that a limiting distribution holds. We present a subset of the conditions under which the estimators exhibit a normal limiting distribution [9, 13]. The main Markov chain assumption is that of *polynomial ergodicity*. Let $\|\cdot\|_{TV}$ denote the total-variation distance. Let P^t be the t-step Markov chain transition kernel, and let $M : \mathcal{X} \to \mathbb{R}^+$ such that $\mathrm{E}M < \infty$ and for $\xi > 0$,

$$\|P^t(x, \cdot) - F(\cdot)\|_{TV} \leq M(x)t^{-\xi}$$

for all $x \in \mathcal{X}$. The constant ξ dictates the rate of convergence of the Markov chain. Ergodic Markov chains on finite state spaces are polynomially ergodic. On general state spaces, demonstrating at least polynomial ergodicity usually requires a separate study of the sampler, and we provide some references in Section 6.

3.1 Means

Recall that $\Lambda = \mathrm{Var}_F(h(X))$. For MCMC sampling, a key quantity of interest will be

$$\begin{aligned}\Sigma &= \sum_{k=-\infty}^{\infty} \mathrm{Cov}_F(h(X_1), h(X_{1+k})) \\ &= \Lambda + \sum_{k=1}^{\infty} [\mathrm{Cov}_F(h(X_1), h(X_{1+k})) + \mathrm{Cov}_F(h(X_1), h(X_{1+k}))^T]\end{aligned}$$

which we assume is positive-definite. A CLT for a Monte Carlo average, $\hat{\theta}_h$, is available under both IID and MCMC sampling.

Theorem 1.

1. ***IID.*** *Let* $X_1, X_2, \dots, X_n \overset{iid}{\sim} F$. *If* $\mathrm{E}_F\|h(X_1)\|^2 < \infty$, *then, as* $n \to \infty$,

$$\sqrt{n}(\hat{\theta}_h - \theta_h) \xrightarrow{d} N_p(0, \Lambda)$$

2. ***MCMC.*** *Let* $\{X_t\}$ *be polynomially ergodic of order* $\xi > (2+\delta)/\delta$ *where* $\delta > 0$ *such that* $\mathrm{E}_F\|h(X_1)\|^{2+\delta} < \infty$, *then if* Σ *is positive-definite, as* $n \to \infty$,

$$\sqrt{n}(\hat{\theta}_h - \theta_h) \xrightarrow{d} N_p(0, \Sigma)$$

Typically, MCMC algorithms exhibit positive correlation implying that Σ is larger Λ. This naturally implies that MCMC simulations require more samples than IID simulations. Using Theorem 1 to assess the simulation reliability requires estimation of Λ and Σ, which we describe in Section 4.

3.2 Quantiles

Let

$$\begin{aligned}\sigma^2(\phi_q) &= \sum_{k=-\infty}^{\infty} \mathrm{Cov}(I(V_1 \le \phi_q), I(V_{1+k} \le \phi_q)) \\ &= \mathrm{Var}(I(V_1 \le \phi_q)) + 2\sum_{k=1}^{\infty} \mathrm{Cov}(I(V_1 \le \phi_q), I(V_{1+k} \le \phi_q))\end{aligned}$$

An asymptotic distribution for sample quantiles is available under both IID Monte Carlo and MCMC.

Theorem 2. *Let* F_h *be absolutely continuous, twice differentiable with density* f_h, *and let* f_h' *be bounded within some neighborhood of* $\hat{\phi}_q$.

1. ***IID.*** *Let* $X_1, \dots, X_n \overset{iid}{\sim} F$, *then*

$$\sqrt{n}(\hat{\phi}_q - \phi_q) \xrightarrow{d} N\left(0, \frac{q(1-q)}{f_v(\phi_q)^2}\right)$$

2. ***MCMC.*** *[11] If the Markov chain is polynomially ergodic of order* $m > 1$ *and* $\sigma^2(\phi_q) > 0$, *then*

$$\sqrt{n}(\hat{\phi}_q - \phi_q) \xrightarrow{d} N\left(0, \frac{\sigma^2(\phi_q)}{f_v(\phi_q)^2}\right)$$

The density value, $f_v(\phi_q)$, can be estimated using a Gaussian kernel density estimator. In addition, $\sigma^2(\phi_q)$ is replaced with $\sigma^2(\hat{\phi}_q)$, the univariate version of Σ for $h(V_t) = I(V_t \le \hat{\phi}_q)$. We present methods of estimating $\sigma^2(\hat{\phi}_q)$ in Section 4.

3.3 Other Estimators

For many estimators, a delta method argument can yield a limiting normal distribution. For example, a CLT for $\hat{\theta}_h$ and a delta method argument yields an elementwise asymptotic distribution of Λ. Let Λ_{ij} denote the (i,j)th element of Λ. If h_i and $\hat{\theta}_{i,h}$ denote the components of h and $\hat{\theta}_h$, respectively, then the ith diagonal of $\hat{\Lambda}_n$ is

$$\hat{\Lambda}_{ii,n} = \frac{1}{n}\sum_{t=1}^{n}(h_i(X_t) - \hat{\theta}_{i,h})^2 = \frac{1}{n}\sum_{t=1}^{n}[h_i(X_t)]^2 - [\hat{\theta}_{i,h}]^2$$

We obtain the asymptotic distribution of $\hat{\Lambda}_{ii,n}$. A similar argument can be made for the off-diagonals of Λ. Under the conditions of Theorem 1,

$$\sqrt{n}\left(\begin{pmatrix} n^{-1}\sum h_i(X_t)^2 \\ \hat{\theta}_{i,h} \end{pmatrix} - \begin{pmatrix} \mathrm{E}_F[h_i^2] \\ \theta_{i,h} \end{pmatrix}\right) \xrightarrow{d} N_2\left(0, \sum_{\Lambda_{ii}}\right)$$

where $\Sigma_{\Lambda_{ii}}$ is

$$\Sigma_{\Lambda_{ii}} = \sum_{k=-\infty}^{\infty}\begin{pmatrix} \mathrm{Cov}_F(h_i(X_1)^2, h_i(X_{1+k})^2) & \mathrm{Cov}_F(h_i(X_1)^2, h_i(X_{1+k})) \\ [\mathrm{Cov}_F(h_i(X_1)^2, h_i(X_{1+k}))]^T & \mathrm{Cov}_F(h_i(X_1), h_i(X_{1+k})) \end{pmatrix}$$

Under IID sampling, the infinite sum above reduces to

$$\Sigma_{\Lambda_{ii}}^{\mathrm{IID}} = \begin{pmatrix} \mathrm{Var}_F(h_i(X_1)^2) & \mathrm{Cov}_F(h_i(X_1)^2, h_i(X_1)) \\ [\mathrm{Cov}_F(h_i(X_1)^2, h_i(X_1))]^T & \mathrm{Var}_F(h_i(X_1)) \end{pmatrix}$$

Applying the delta method for function $\phi(x,y) = x - y^2$, we obtain

$$\sqrt{n}(\hat{\Lambda}_{ii,n} - \Lambda_{ii}) \xrightarrow{d} N\left(0, \begin{bmatrix} 1 & -2\mu_h \end{bmatrix} \Sigma_{\Lambda_{ii}} \begin{bmatrix} 1 \\ -2\mu_h \end{bmatrix}\right)$$

3.4 Confidence Regions for Means

Suppose that A_n is an estimate of the limiting Monte Carlo variance–covariance matrix, Λ for IID sampling, and Σ for MCMC sampling. Let $\chi^2_{1-\alpha,p}$ be the $(1-\alpha)$-quantile of a χ^2_p distribution. The CLT yields a large-sample confidence region around $\hat{\theta}_h$ as

$$C_\alpha^E(\hat{\theta}_h) = \{\theta \in \mathbb{R}^p : n(\theta_n - \theta)^T A_n^{-1}(\theta_n - \theta) < \chi^2_{1-\alpha,p}\}$$

Let $|\cdot|$ denote the determinant. The volume of this ellipsoidal confidence region, which depends on p, α, and $|A_n|$, is given by

$$V_n^E = \mathrm{Volume}(C_\alpha(\hat{\theta}_h)) = \frac{2\pi^{p/2}}{p\Gamma(p/2)}\left(\frac{\chi^2_{1-\alpha,p}}{n}\right)^{p/2} |A_n|^{1/2} \tag{2}$$

Sometimes a joint sampling distribution may be difficult to obtain, or the limiting variance–covariance matrix may be too complicated to estimate. In such cases, one can consider hyperrectangular confidence regions. Let $z(\alpha)$ be a quantile of a standard normal distribution possibly chosen to correct for simultaneous inference. Recall that

$\theta = (\theta_1, \dots, \theta_p)$, let $\hat{\theta}_{h_i}$ denote the ith component of $\hat{\theta}_h$. Further, let $A_{ii,n}$ denote the ith diagonal of A_n. Then

$$C_\alpha^R(\hat{\theta}_h) = \prod_{i=1}^{p} \left\{ \theta_i : \hat{\theta}_{h_i} - z(\alpha)\frac{A_{ii,n}}{\sqrt{n}} < \theta_i < \hat{\theta}_{h_i} + z(\alpha)\frac{A_{ii,n}}{\sqrt{n}} \right\}$$

The volume of this hyperrectangular confidence region is

$$V_n^R = \text{Volume}(C_\alpha^R(\hat{\theta}_h)) = \prod_{i=1}^{p} \left[2z(\alpha)\frac{A_{ii,n}}{\sqrt{n}} \right] \tag{3}$$

As more samples are obtained, V_n^E and V_n^R converge to 0 so that the variability in the estimator $\hat{\theta}_h$ disappears. Sequential stopping rules in Section 5 will utilize this feature to terminate simulation.

4 Estimating Σ

To construct confidence regions, the asymptotic variance requires estimation. For IID sampling, Λ is estimated by the sample covariance matrix, as discussed in Section 2.3. For MCMC sampling, a rich literature of estimators of Σ is available including spectral variance [14, 15], regeneration-based [16, 17], and initial sequence estimators [5, 18–20]. Considering the size of modern simulation output, we recommend the computationally efficient batch means estimators.

The multivariate batch means estimator considers nonoverlapping batches and constructs a sample covariance matrix from the sample mean vectors of each batch. More formally, let $n = ab$, where a is the number of batches, and b is the batch sizes. For $k = 0, \dots, a-1$, define $\overline{Y}_k = b^{-1}\sum_{t=1}^{b} h(X_{kb+t})$. The batch means estimator of Σ is

$$\hat{\Sigma}_b = \frac{b}{a-1}\sum_{k=0}^{a-1}(\overline{Y}_k - \hat{\theta}_h)(\overline{Y}_k - \hat{\theta}_h)^T$$

Univariate and multivariate batch means estimators have been studied in MCMC and operations research literature [21–26]. Although the batch means estimator has desirable asymptotic properties, it suffers from underestimation in finite samples, particularly for slowly mixing Markov chains. Specifically, let

$$\Gamma = -\sum_{k=-\infty}^{\infty} |k|\text{Cov}_F(X_1, X_{1+k})$$

Then, Vats and Flegal [27] show (ignoring smaller order terms)

$$\text{E}[\hat{\Sigma}_b] = \Sigma + \frac{\Gamma}{b}$$

When the autocorrelation in the Markov chain is large, or b is small, there is significant underestimation in Σ. To combat this issue, Vats and Flegal [27] propose lugsail batch means estimators formed by a linear combination of two batch means estimators with different batch sizes. For $r \geq 1$ and $0 \leq c < 1$, the lugsail batch means estimator is

$$\hat{\Sigma}_L = \frac{1}{1-c}\hat{\Sigma}_b - \frac{c}{1-c}\hat{\Sigma}_{\lfloor b/r \rfloor} \tag{4}$$

It is then easy to see

$$\mathrm{E}[\hat{\Sigma}_L] = \Sigma + \left(\frac{1-rc}{1-c}\right)\frac{\Gamma}{b}$$

When $r > 1/c$, the finite-sample bias is positive. Vats and Flegal [27] recommend $r = 3$ and $c = 1/2$, which induces a positive bias of $-\Gamma/b$ offsetting the original bias in the opposite direction. For $r = 1/c$, this estimator corresponds to the flat-top batch means estimator of Liu and Flegal [28]. Under polynomial ergodicity and additional conditions on the batch size b, the lugsail batch means estimators are strongly consistent [26].

5 Stopping Rules

Monte Carlo simulations are often terminated according to a fixed-time regime. That is, before the start of the simulation, it is decided that some n^* steps of the process will be generated. The fixed-time termination rule may be formally written as

$$T_f = \inf\{n \geq 0 : I(n < n^*) < 1\} \tag{5}$$

By construction, $T_f = n^*$, and simulation terminates when the criterion is met. The representation in Equation (5) allows further adjustments to our termination rule with an ϵ-fixed-time approach, where for some $0 < \epsilon < 1$, the simulation terminates at

$$T_f(\epsilon) = \inf\{n \geq 0 : \epsilon I(n < n^*) + n^{-1} \leq \epsilon\} \tag{6}$$

The termination time is deterministically dependent on ϵ. Specifically, $T_f(\epsilon) = \max\{n^*, \lceil \epsilon^{-1} \rceil\}$. Glynn and Whitt [29] show that $T_f(\epsilon) \to \infty$ as $\epsilon \to 0$. However, since the structure of the underlying distribution F and the quantity of interest θ_h are unknown, there is often little intuition as to what n^* and ϵ should be for any given problem.

Alternatively, one could terminate according to a random-time regime such as when the volume of a confidence region (possibly relative to some quantity) is below a prespecified threshold. These confidence region volumes, V_n, could be either V_n^E at Equation (2) or V_n^R at Equation (3). Glynn and Whitt [29] and Vats *et al.* [26] show that the resulting confidence regions created at termination have the correct coverage, asymptotically. Since the simulation ends at a random time, the estimate of the limiting Monte Carlo variance–covariance matrix, A_n used in construction of V_n, is required to be strongly consistent. Glynn and Whitt [29] further show that weak consistency is not sufficient. We discuss stopping rules of this type for IID and MCMC sampling in the following sections.

5.1 IID Monte Carlo

The absolute precision sequential stopping rule terminates simulation when the variability in the simulation is smaller than a prespecified tolerance, ϵ. Specifically, simulation is terminated at time T_1 where

$$T_a(\epsilon) = \inf\{n \geq 0 : V_n^{1/p} + \epsilon I(n < n^*) + n^{-1} \leq \epsilon\}$$

Here, n^* ensures a minimal simulation effort. By definition, $T_a(\epsilon) \geq T_f(\epsilon) \to \infty$ as $\epsilon \to 0$. Thus, as the tolerance decreases, the required simulation size increases. The stopping rule explained in the motivating example in the introduction is a one-dimensional absolute precision sequential stopping rule. This rule works best in small dimensions when each component is on the same scale and an informed choice of ϵ can be made (as in the motivating example).

In situations where the components of θ_h are in different units, stopping simulation when the variability in the estimator is small compared to the size of the estimate is natural. For a choice of norm $\|\cdot\|_a$, a relative-magnitude sequential stopping rule terminates simulation at

$$T_m(\epsilon) = \inf\{n \geq 0 : V_n^{1/p} + \epsilon\|\hat{\theta}_h\|_a I(n < n^*) + n^{-1} \leq \epsilon\|\hat{\theta}_h\|_a\}$$

This termination rule essentially controls the coefficient of variation for $\hat{\theta}_h$. An advantage here is that problem-free choices of ϵ can be used since problems where $\|\theta_h\|_a$ is small will automatically require smaller cutoff. A clear disadvantage is that this rule is ineffective when $\theta_h = 0$.

5.2 MCMC

Although both $T_a(\epsilon)$ and $T_m(\epsilon)$ may be used in MCMC, a third alternative arises due to the correlation in the Markov chain. A relative-standard deviation sequential stopping rule terminates the simulation when the Monte Carlo variability (as measured by the volume of the confidence region) is small compared to the underlying variability inherent to the problem (Λ). That is,

$$T_s(\epsilon) = \inf\{n \geq 0 : V_n^{1/p} + \epsilon|\hat{\Lambda}_n|^{1/2p} I(n < n^*) + n^{-1} \leq \epsilon|\hat{\Lambda}_n|^{1/2p}\}$$

If this rule is used for IID Monte Carlo, then A_n in Equation (2) is $\hat{\Lambda}_n$, and $T_s(\epsilon) \approx T_a(\epsilon')$ for some other (deterministic) ϵ'. For MCMC, this sequential stopping rule connects directly to the concept of effective sample size [26]. That is, stopping at $T_s(\epsilon)$ is equivalent to stopping when

$$\mathrm{ESS}_n = n\left(\frac{|\hat{\Lambda}_n|}{|\hat{\Sigma}_n|}\right)^{1/p} > \frac{2^{2/p}\pi}{(p\Gamma(p/2))^{2/p}} \frac{\chi^2_{1-\alpha,p}}{\epsilon^2} \quad (7)$$

Thus, simulation is terminated when the number of effective samples is larger than the lower bound in Equation (7). Effective sample size measures the number of equivalent IID samples that would produce equivalent variability in $\hat{\theta}_h$. Terminating simulation using Equation (7) is intuitive and easy to implement in MCMC sampling once appropriate estimators of Λ and Σ have been obtained.

6 Workflow

We have presented tools for determining when to stop a Monte Carlo simulation. The workflow starts by identifying F and θ and then running a chosen sampler for some small n^* iterations. Preliminary estimates of θ and Λ or Σ are obtained along with visualizations determining quality of the sampler. The simulation continues until a chosen stopping rule indicates termination using a prespecified ϵ. In the following section, we present three examples where we demonstrate this workflow.

In our examples, we assume that a CLT (or asymptotic distribution) for Monte Carlo estimators exists. However, extra care must be taken when working with a generic Monte Carlo procedure. Particularly, importance sampling can often yield estimators with infinite variances, where a CLT cannot hold. See Refs 3, 4 for more details. A CLT is particularly difficult to establish for MCMC due to serial correlation in the Markov chain. However, many individual Markov chains have been shown to be at least polynomially ergodic, for examples, see Jarner and Hansen [30], Roberts and Tweedie [31], Vats [32], Khare and Hobert [33], Tan *et al.* [34], Hobert and Geyer [35], Jones and Hobert [36].

A similar workflow can be adopted for embarrassingly parallel implementations of Monte Carlo samplers. Given the power of the modern personal computer, most Monte Carlo samplers can run on multiple cores simultaneously, producing more samples in the same clock time. For IID Monte Carlo, averaging estimators across all independent runs is reasonable. However, for estimating Σ in MCMC, estimation quality can be improved by sharing information across multiple runs at the end of the simulation, see Gupta and Vats [37] for more details.

Sequential stopping rules, particularly in MCMC, should not be implemented as a black-box procedure. Each implementation of the stopping rule must be accompanied with visualizations that give qualitative insights about the quality of the samplers. A better quality sampler can significantly improve estimation and lead to smaller run times. We illustrate this point by comparing samplers in our examples.

7 Examples

7.1 Action Figure Collector Problem

Consider the general coupon collector problem [38] where the goal is to collect N distinct objects (e.g., coupons, trading cards, and action figures). Specifically, independent draws of size n are made from N with replacement, and interest is in the number of draws necessary, say W, to draw all N objects at least once. The classical case where $n = 1$ and all N objects are equally likely yields a closed-form solution (related to random sampling of digits). We consider a variation where $n = 1$ and $N = 15$ action figures appear in cereal boxes with probabilities in Table 1.

We estimate the expected number of boxes needed to collect all 15 action figures and the probability we needed to buy more than 100 and 200 total boxes. Denote these as $\mathrm{E}[W]$, $P(W > 100)$, and $P(W > 200)$, respectively. Additionally, we implement an absolute precision sequential stopping rule to simulate until 95% confidence interval lengths for the three quantities of interest are below 1, 0.01, and 0.01, respectively. Specifically, we set $n^* = 100$ and simulate an additional 100 Monte Carlo sample between checking the stopping rule. The sequential stopping rule terminates at $n = 51\,500$ with estimates of (116.4, 0.527, 0.085). We note that stopping is based on $\mathrm{E}[W]$ since its 95% confidence interval criteria is the most restrictive. The left panel of Figure 1 provides a histogram of the Monte Carlo samples along with vertical bold lines corresponding to 100 and 200 boxes.

Table 1 Probabilities for each action figure

Figures	A	B	C	D	E	F	G	H	I	J	K	L	M	N	O
Probability	0.2	0.1	0.1	0.1	0.1	0.1	0.05	0.05	0.05	0.05	0.02	0.02	0.02	0.02	0.02

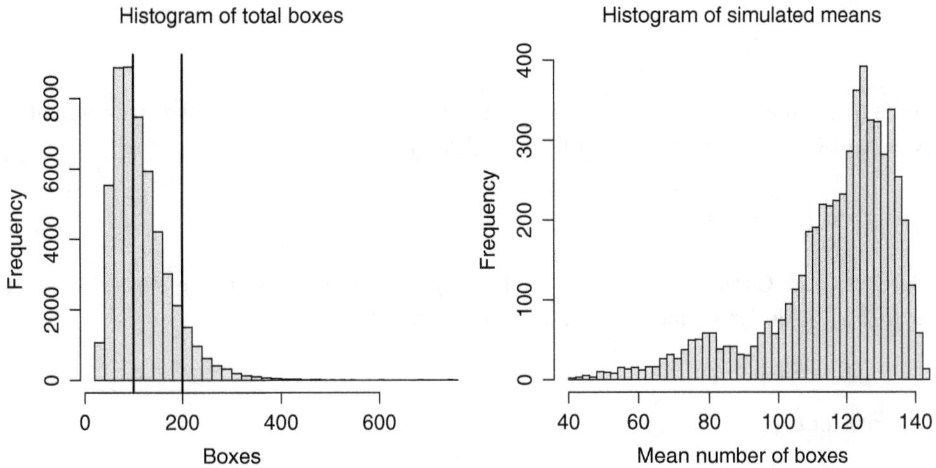

Figure 1 Histograms of simulated boxes and mean number of boxes for two Monte Carlo sampling strategies in the collector problem.

A more efficient Monte Carlo experiment is available if we only wish to estimate $\mathrm{E}[W]$. Suppose that Z is the set of all permutations of the set $\{A, B, \ldots, O\}$ representing the order in which the action figures were collected. Then, for any $z \in Z$, we can calculate $\mathrm{E}[W|Z = z]$ and notice

$$\mathrm{E}[W] = \sum_{z \in Z} \mathrm{E}[W|Z = z]P[Z = z]$$

This calculation is unavailable since there are over 3 trillion partitions in Z. However, we can simulate $Z_1, \ldots, Z_n$ equally likely permutations from Z and estimate $\mathrm{E}[W]$ with

$$\frac{1}{n} \sum_{t=1}^{n} \mathrm{E}[W|Z = Z_t]$$

Using this sampler, we simulate until the 95% confidence interval length for $\mathrm{E}[W]$ is below 1. Again, we set $n^* = 100$ and simulate an additional 100 Monte Carlo sample between checking the stopping rule. Now the sequential stopping rule terminates at $n = 5500$ with an estimate of 116.1, which is approximately 10 times more efficient than the naive Monte Carlo sampling. The right panel of Figure 1 provides a histogram of the Monte Carlo simulated means.

7.2 Estimating Risk for Empirical Bayes

Risk of empirical Bayes estimators is often not available in closed form, and Monte Carlo simulation is used to estimate it. Consider Example 3.3 from Robert and Casella [4] where for a fixed λ,

$$X|\theta \sim N_p(\theta, I_p) \quad \text{and} \quad \theta \sim N_p(0, \lambda I_p)$$

The posterior distribution of θ (given λ) is

$$\theta \mid x, \lambda \sim N\left(\frac{\lambda x}{\lambda+1}, \frac{\lambda I_p}{\lambda+1}\right)$$

If the true value of λ is unknown, it is often estimated from the marginal distribution of X, $X \sim N_p(0, (\lambda+1)I_p)$ via maximum-likelihood estimation as

$$\hat{\lambda} = \left(\frac{\|x\|^2}{p} - 1\right)^+$$

Robert and Casella [4] consider estimating $h(\theta) = \|\theta\|^2$ using the posterior mean $\mathrm{E}[\|\theta\|^2|x, \hat{\lambda}]$. Under a quadratic loss, the Bayes estimator is

$$\hat{h}_{eb} = (\|x\|^2 - p)^+$$

The risk for $\hat{h}_{eb}$

$$\eta_{eb}(\|\theta\|) = \mathrm{E}[(\|\theta\|^2 - (\|x\|^2 - p)^+)^2 \mid \theta]$$

is difficult to obtain analytically (although not impossible, see Robert and Casella [4]). Instead, we can estimate the risk over a grid of $\|\theta\|$ values using Monte Carlo. To do this, we fix m choices $\theta_1, \dots, \theta_m$ over a grid, and for each $k = 1, \dots, m$, generate n Monte Carlo samples from $X|\theta_k \sim N(\theta_k, 1)$ yielding estimates

$$\hat{\eta}_{eb}(\|\theta_k\|) = \frac{1}{n}\sum_{t=1}^{n}(\|\theta_k\| - (\|X_t\|^2 - p)^+)^2$$

The resulting estimate of the risk is an m-dimensional vector of means, for which we can utilize the sampling distribution in Theorem 1 to construct large-sample confidence regions. An appropriate choice of a sequential stopping rule here is the relative-magnitude sequential stopping rule, which stops simulation when the Monte Carlo variance is small relative to the average risk over all values of θ considered. It is important to note that the risk at a particular θ could be zero, but it is unlikely.

For illustration, we set $p = 5$ and simulate a data point from the true model with $\lambda = 1$. To evaluate risk we choose a grid of θ values with $m = 50$. In order to assess the appropriate Monte Carlo sample size n, we set $n^* = 10^3$ so that at least 10^3 Monte Carlo samples are obtained. With $\epsilon = 0.05$, and Λ estimated using the sample covariance matrix, the sequential stopping rule terminates simulation at 21 100 steps. Figure 2 demonstrates the estimated risk at $n^* = 10^3$ iterations and the estimated risk at termination. Pointwise Bonferroni corrected confidence intervals are presented as an indication of variability for each component[1].

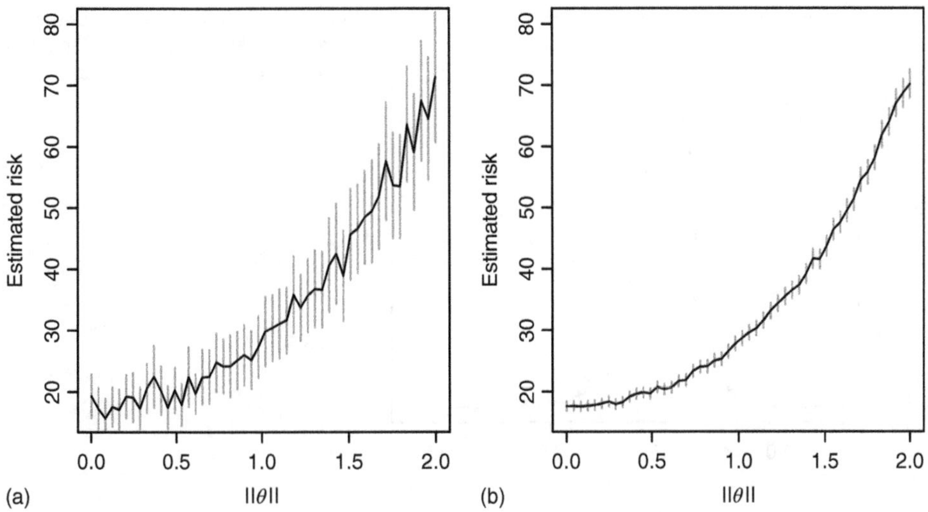

Figure 2 Estimated risk at $n^* = 10^3$ (a) and at $n = 21\,100$ (b) with pointwise Bonferroni corrected confidence intervals.

7.3 Bayesian Nonlinear Regression

Consider the biomedical oxygen demand (BOD) data collected by Marske [39] where BOD levels were measured periodically from cultured bottles of stream water. Bates and Watts [40] and Newton and Raftery [41] study a Bayesian nonlinear model with a fixed rate constant and an exponential decay as a function of time. The data is available in Bates and Watts [40, Section A4.1]. Let x_i, $i = 1, \dots, 8$ be the time points, and let Y_i be the BOD at time x_i. Assume for $\epsilon_i \overset{\text{iid}}{\sim} N(0, \sigma^2)$ and $Y_i | \beta_1, \beta_2, \sigma^2 = \beta_1(1 - e^{-\beta_2 x_i}) + \epsilon_i$. Newton and Raftery [41] assume a default prior on σ^2, $\pi(\sigma^2) \propto \sigma^{-2}$, and a transformation invariant design-dependent prior for $\beta = (\beta_1, \beta_2)$ such that $\pi(\beta) \propto |V^T V|^{1/2}$, where V is an 8×2 where the (i,j)th element of $V = \partial \mathrm{E}[y_i | x_i] / \partial \beta_j$. The resulting posterior distribution of (β, σ^2) is intractable and up to normalization and can be written as

$$\pi(\beta, \sigma^2 \mid y_1, \dots, y_8) \propto \left(\frac{1}{\sigma^2}\right)^{n/2+1} |V^T V|^{1/2} \exp\left\{ -\frac{1}{2\sigma^2} \sum_{i=1}^{8} (y_i - \beta_1(1 - e^{-\beta_2 x_i}))^2 \right\}$$

The goal is to estimate the posterior mean of (β, σ^2). We implement an MCMC algorithm to estimate the posterior mean and implement the relative-standard deviation sequential stopping rule via effective sample size.

We sample from the posterior distribution via a componentwise random walk Metropolis–Hastings algorithm updating β first and then σ^2, with step size for both components chosen so that the acceptance probability is around 30%. Since the posterior distribution is three-dimensional, the minimum ESS required for $\epsilon = 0.05$ and $\alpha = 0.05$ in Equation (7) is 8123. Thus, we first run the sampler for $n^* = 8123$ and obtain early estimates of $\mathrm{E}[\beta, \sigma^2 | y]$ and the corresponding effective sample size. We then proceed to run the sampler until ESS_n using $\hat{\Lambda}_n$ and $\hat{\Sigma}_L$ with $b = \lfloor n^{1/2} \rfloor$ in Equation (4) is more than 8123.

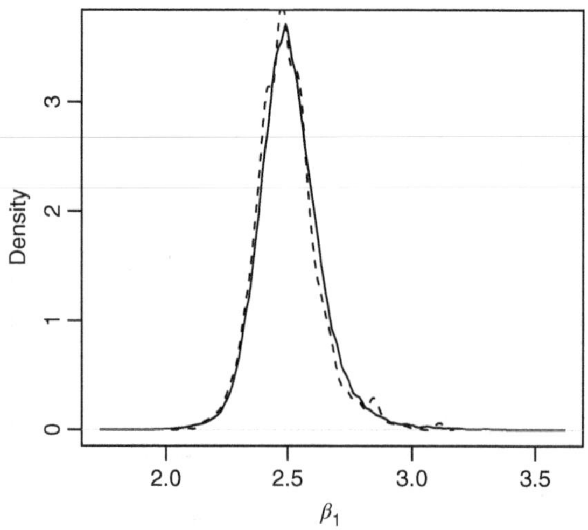

Figure 3 Estimated density of the marginal posterior for β from an initial run of $n^* = 8123$ (dashed) and at termination (solid).

At $n^* = 8123$, ESS_n was 237, and the estimated density plot is presented in Figure 3 by the dashed line. We verify the termination criteria in Equation (7) incrementally, and simulation terminates at $n = 276\ 053$ iterations. The final estimated density is presented in Figure 3 by the solid line.

At termination, the estimated posterior mean is $(2.5074, 0.2034, 0.00\ 654)$, and 80% credible intervals are $(2.357, 2.665)$, $(0.178, 0.229)$, and $(0.00\,246, 0.01\,200)$ for β_1, β_2, and σ^2, respectively.

It is possible to run a more efficient linchpin sampler [42] by integrating out σ^2 from the posterior. That is, $\pi(\beta, \sigma^2|y) = \pi(\sigma^2|\beta, y)\pi(\beta|y)$, where

$$\sigma^2|\beta, y \sim \text{Inverse Gamma}\left(\frac{n}{2}, \frac{1}{2}\sum_{i=1}^{8}(y_i - \beta_1(1 - e^{-\beta_2 x_i}))^2\right)$$

and

$$\pi(\beta|y) \propto |V^T V|^{1/2}\left(\sum_{i-1}^{8}(y_i - \beta_1(1 - e^{-\beta_2 x_i}))^2\right)^{-n/2}$$

The sampler then proceeds to implement a random walk Metropolis–Hastings step to update β, and a draw from $\sigma^2|\beta, y$ yields a joint MCMC draw from the posterior. We empirically note that this linchpin variable sampler yields lower marginal autocorrelation in σ^2 as illustrated by Figure 4.

Repeating the previous procedure with the linchpin sampler, we have an estimated ESS at $n^* = 8123$ of 652, and the sequential stopping rule terminates at $n = 183\,122$. The resulting estimates of posterior mean and quantiles are similar. Thus, using a more efficient sampler requires substantially fewer iterations to obtain estimates of similar quality.

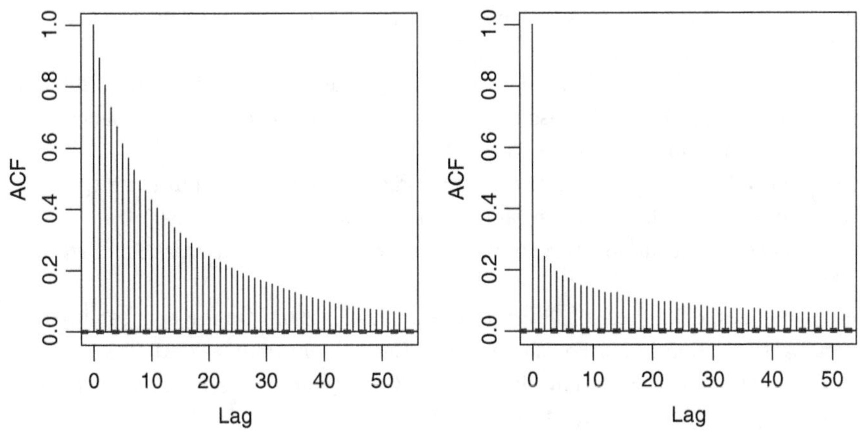

Figure 4 Estimated autocorrelations for nonlinchpin sampler (a) and linchpin sampler (b).

Note

1 For constructing simultaneous confidence intervals with approximately correct coverage, see Robertson *et al.* [10].

References

1 Caffo, B.S., Booth, J.G., and Davison, A.C. (2002) Empirical sup rejection sampling. *Biometrika*, 89, 745–754.

2 Chib, S. and Greenberg, E. (1995) Understanding the Metropolis-Hastings algorithm. *Am. Stat.*, 49, 327–335.

3 Fishman, G.S. (1996) *Monte Carlo: Concepts, Algorithms, and Applications*, Springer, New York.

4 Robert, C.P. and Casella, G. (2013) *Monte Carlo Statistical Methods*, Springer, New York.

5 Robert, C.P., Elvira, V., Tawn, N., and Wu, C. (2018) Accelerating MCMC algorithms. *Wiley Interdiscip. Rev. Comput. Stat.*, 10, e1435.

6 Flegal, J.M., Haran, M., and Jones, G.L. (2008) Markov chain Monte Carlo: can we trust the third significant figure? *Stat. Sci.*, 23, 250–260.

7 Koehler, E., Brown, E., and Haneuse, S.J.-P. (2009) On the assessment of Monte Carlo error in simulation-based statistical analyses. *Am. Stat.*, 63, 155–162.

8 Frey, J. (2010) Fixed-width sequential confidence intervals for a proportion. *Am. Stat.*, 64, 242–249.

9 Roberts, G.O. and Rosenthal, J.S. (2004) General state space Markov chains and MCMC algorithms. *Probab. Surv.*, 1, 20–71.

10 Robertson, N., Flegal, J.M., Vats, D., and Jones, G.L. (2019) Assessing and visualizing simultaneous simulation error. *arXiv preprint arXiv:1904.11912.*

11 Doss, C.R., Flegal, J.M., Jones, G.L., and Neath, R.C. (2014) Markov chain Monte Carlo estimation of quantiles. *Electron. J. Stat.*, 8, 2448–2478.

12 Brooks, S.P. and Gelman, A. (1998) General methods for monitoring convergence of iterative simulations. *J. Comput. Graph. Stat.*, 7, 434–455.

13 Jones, G.L. (2004) On the Markov chain central limit theorem. *Probab. Surv.*, 1, 299–320.

14 Andrews, D.W. (1991) Heteroskedasticity and autocorrelation consistent covariance matrix estimation. *Econometrica*, 59, 817–858.

15 Vats, D., Flegal, J.M., and Jones, G.L. (2018) Strong consistency of multivariate spectral variance estimators in Markov chain Monte Carlo. *Bernoulli*, 24, 1860–1909.

16 Seila, A.F. (1982) Multivariate estimation in regenerative simulation. *Oper. Res. Lett.*, 1, 153–156.

17 Hobert, J.P., Jones, G.L., Presnell, B., and Rosenthal, J.S. (2002) On the applicability of regenerative simulation in Markov chain Monte Carlo. *Biometrika*, 89, 731–743.

18 Geyer, C.J. (1992) Practical Markov chain Monte Carlo (with discussion). *Stat. Sci.*, 7, 473–511.

19 Dai, N. and Jones, G.L. (2017) Multivariate initial sequence estimators in Markov chain Monte Carlo. *J. Multivar. Anal.*, 159, 184–199.

20 Kosorok, M.R. (2000) Monte Carlo error estimation for multivariate Markov chains. *Stat. Probab. Lett.*, 46, 85–93.

21 Chen, D.-F.R. and Seila, A.F. (1987) *Multivariate Inference in Stationary Simulation Using Batch Means*. Proceedings of the 19th Conference on Winter simulation, pp. 302–304. ACM.

22 Jones, G.L., Haran, M., Caffo, B.S., and Neath, R. (2006) Fixed-width output analysis for Markov chain Monte Carlo. *J. Am. Stat. Assoc.*, 101, 1537–1547.

23 Chien, C.-H. (1988) Small sample theory for steady state confidence intervals, in *Proceedings of the Winter Simulation Conference* (eds M. Abrams, P. Haigh, and J. Comfort), Association for Computing Machinery, New York, NY, USA, pp. 408–413, doi: https://doi.org/10.1145/318123.318225.

24 Chien, C.-H., Goldsman, D., and Melamed, B. (1997) Large-sample results for batch means. *Manage. Sci.*, 43, 1288–1295.

25 Flegal, J.M. and Jones, G.L. (2010) Batch means and spectral variance estimators in Markov chain Monte Carlo. *Ann. Stat.*, 38, 1034–1070.

26 Vats, D., Flegal, J.M., and Jones, G.L. (2019) Multivariate output analysis for Markov chain Monte Carlo. *Biometrika*, 106, 321–337.

27 Vats, D. and Flegal, J.M. (2020) Lugsail lag windows for estimating time-average covariance matrices. *arXiv preprint arXiv:1809.04541*.

28 Liu, Y. and Flegal, J.M. (2018) Weighted batch means estimators in Markov chain Monte Carlo. *Electron. J. Stat.*, 12, 3397–3442.

29 Glynn, P.W. and Whitt, W. (1992) The asymptotic validity of sequential stopping rules for stochastic simulations. *Ann. Appl. Probab.*, 2, 180–198.

30 Jarner, S.F. and Hansen, E. (2000) Geometric ergodicity of Metropolis algorithms. *Stoch. Proc. Appl.*, 85, 341–361.

31 Roberts, G.O. and Tweedie, R.L. (1996) Geometric convergence and central limit theorems for multidimensional Hastings and Metropolis algorithms. *Biometrika*, 83, 95–110.

32 Vats, D. (2017) Geometric ergodicity of Gibbs samplers in Bayesian penalized regression models. *Electron. J. Stat.*, 11, 4033–4064.

33 Khare, K. and Hobert, J.P. (2013) Geometric ergodicity of the Bayesian lasso. *Electron. J. Stat.*, 7, 2150–2163.

34 Tan, A., Jones, G.L., and Hobert, J.P. (2013) On the geometric ergodicity of two-variable Gibbs samplers, in *Advances in Modern Statistical Theory and Applications: A Festschrift in Honor of Morris L. Eaton* (eds G. L. Jones and X. Shen), Institute of Mathematical Statistics, Beachwood, Ohio, pp. 25–42.

35 Hobert, J.P. and Geyer, C.J. (1998) Geometric ergodicity of Gibbs and block Gibbs samplers for a hierarchical random effects model. *J. Multivar. Anal.*, 67, 414–430.

36 Jones, G.L. and Hobert, J.P. (2004) Sufficient burn-in for Gibbs samplers for a hierarchical random effects model. *Ann. Stat.*, 32, 784–817.

37 Gupta, K. and Vats, D. (2020) Estimating Monte Carlo variance from multiple Markov chains. *arXiv preprint arXiv:2007.04229.*

38 Dawkins, B. (1991) Siobhan's problem: the coupon collector revisited. *Am. Stat.*, 45 (1), 76–82.

39 Marske, D.M. (1967) *BOD Data Interpretation Using the Sum of Squares Surface*, University of Wisconsin, Madison.

40 Bates, D.M. and Watts, D.G. (1988) *Nonlinear Regression Analysis and Its Applications*, vol. 2, Wiley, New York.

41 Newton, M.A. and Raftery, A.E. (1994) Approximate Bayesian inference with the weighted likelihood bootstrap. *J. R. Stat. Soc., Ser. B*, 56, 3–26.

42 Archila, F.H.A. (2016) Markov chain Monte Carlo for linear mixed models. PhD thesis. University of Minnesota.

6

Sequential Monte Carlo: Particle Filters and Beyond

Adam M. Johansen

University of Warwick, Coventry, UK

1 Introduction

Sequential Monte Carlo (SMC) methods are a broad class of algorithms for approximating distributions of interest, integrals with respect to those distributions, and their normalizing constants. They employ an ensemble of weighted samples which targets each of a sequence of distributions in turn. In some settings this sequence arises naturally from the problem being addressed and in others it is specified as a design choice. This chapter presents a generic framework in which these methods can be described, arguing that the vast majority of SMC approaches admit an interpretation directly within this framework and that the remainder require only small extensions of it, before dedicating some space to a number of major statistical applications of these methods. A high-level view is taken, with many details left to references so that a broad overview of this area can be provided. This allows us to showcase a number of the areas in which SMC finds natural applications, not just the particle filtering setting in which it has particular prominence but also in many other contexts including Bayesian inference, approximate Bayesian computation (ABC), and rare event estimation.

2 Sequential Importance Sampling and Resampling

We will be interested in providing weighted sample approximations to each of a sequence of related distributions in turn. In some settings, each of these distributions will be interesting in its own right and might arise naturally from a problem at hand; this is the case in the filtering context, explored in Section 3.1, for example. In other cases, the sequence of distributions is a computational device with only the final distribution in the sequence being of independent interest but with the others used to allow it to be approximated efficiently, typically by constructing a sequence which moves from a tractable distribution to that of interest. We will consider throughout probability distributions defined on Euclidean spaces which admit (Lebesgue) densities; the generalization to arbitrary Polish spaces is essentially direct but significantly complicates the required notation.

Computational Statistics in Data Science.
Edited by Walter W. Piegorsch, Richard A. Levine, Hao Helen Zhang and Thomas C. M. Lee.
 ISBN 978-1-11956107-1

Consider a sequence of probability distributions, $\{\pi_n\}_{n\in\mathbb{N}}$, defined on an increasing sequence of state spaces, $\tilde{E}_n = \otimes_{p=1}^{n} E_p$, with $E_p = \mathbb{R}^{d_p}$ so that, for each n, π_n is a density over $\mathbb{R}^{d_1+\ldots+d_n}$. Assume that this sequence of densities may be evaluated up to a possibly unknown normalizing constant, that is, for each n, $\pi_n = \gamma_n/Z_n$, where $\gamma_n : \tilde{E}_n \to (0,\infty)$ is an unnormalized probability density, and $Z_n := \int_{\tilde{E}_n} \gamma_n(x_{1:n})dx_{1:n}$ may not be available.

A simple importance sampling solution to the problem of approximating both π_n and Z_n would be to draw some number, N, of independent samples from a proposal distribution Q_n with respect to which π_n was absolutely continuous and to use it to approximate both of these quantities via the standard importance sampling identities:

$$\hat{\pi}_n^N(\varphi) := \frac{1}{N \cdot \hat{Z}_n^N} \sum_{i=1}^{N} \frac{\gamma_n(X_{1:n}^i)}{Q_n(X_{1:n}^i)} \varphi(X_{1:n}^i) \qquad \hat{Z}_n^N := \frac{1}{N} \sum_{i=1}^{N} \frac{\gamma_n(X_{1:n}^i)}{Q_n(X_{1:n}^i)}$$

where $\varphi : \tilde{E}_n \to \mathbb{R}$ is any suitably integrable test function, and $\hat{\pi}_n^N(\varphi)$ denotes the N-particle approximation of the expectation of $\varphi(X_{1:n})$ with $X_{1:n}$ distributed according to π_n.

However, if one seeks to approximate each distribution in turn, such an approach seems wasteful. It is natural in this context to consider $Q_n(x_{1:n})$, which can be decomposed as

$$Q_n(x_{1:n}) = q_1(x_1)q_2(x_2|x_1)\ldots q_n(x_n|x_{1:n-1})$$

In this case, given an importance weighted ensemble of samples, $\{W_{n-1}^i, X_{1:n-1}^i\}_{i=1}^N$, which targets π_{n-1} (and which were drawn from Q_{n-1}), one can extend the sample to approximate π_n by drawing $X_n^i \sim q_n(\cdot|X_{1:n-1}^i)$ independently for $i = 1, \ldots, N$ and updating the weights accordingly, setting

$$W_n^i = \frac{W_{n-1}^i G_n(X_{1:n}^i)}{\sum_{j=1}^N W_{n-1}^j G_n(X_{1:n}^j)}$$

where $G_n(x_{1:n}) := \gamma_n(x_{1:n})/\gamma_{n-1}(x_{1:n-1})q_n(x_n|x_{1:n-1})$ is termed the incremental weight function. (In most settings in which SMC methods find application, further simplification arises via a Markovian decomposition of the unnormalized target, $\gamma_n(x_{1:n}) = \gamma_{n-1}(x_{1:n-1})\gamma_n(x_n|x_{n-1})$, and proposal, $q_n(x_n|x_{1:n-1}) = q_n(x_n|x_{n-1})$, distributions, which means that $G_n(x_{1:n}) = G_n(x_{n-1:n}) = \gamma_n(x_n|x_{n-1})/q_n(x_n|x_{n-1})$.) This gives a mechanism by which each distribution can be approximated in turn, at a computational cost per iteration which does not increase with n. However, this *sequential importance sampling* (SIS) strategy is of limited direct usefulness because the variance of the importance weights and associated estimators will grow with the length of the sequence [1], typically exponentially [2], and only sequences of modest length can be approximated adequately using a reasonable sample size.

In order to make further progress, it is necessary to constrain the class of problems which we hope to solve a little. In particular, approximating $\pi_n(x_{1:n})$ is a task which becomes harder and harder as n increases because the dimension of the space on which these distributions are defined is growing. If we instead settle for approximating only the final time marginal of these distributions $\pi_n(x_n) = \int \pi_n(x_{1:n})dx_{1:n-1}$, then we arrive at a sequence of problems which are of comparable difficulty. Within this regime, an approach known as resampling can be combined with the SIS strategy described above. Resampling is a process by which a weighted sample $\{W_n^i, X_{1:n}^i\}$ is replaced with the equally weighted population

$\{1/N, \tilde{X}^i_{1:n}\}$ in such a way that the expected number of copies of each member of the original ensemble is proportional to its weight. A simple algorithmic description of this generic *sequential importance resampling* (SIR) scheme is provided in Algorithm 1. In practice, in the sequential setting described above, in which $G_n(x_{1:n})$ is dependent on only $x_{n-1:n}$, it is not necessary to store the entire history of the surviving particles as a direct implementation of this algorithm would suggest, a feature which is important, for example, in the filtering context described in Section 3.1.1; when one does need to store the entire history of every surviving particle, space-efficient methods for doing so exist [3].

Algorithm 1. The generic sequential importance resampling algorithm.

Initialization: $n = 1$

Sample $X^1_1, \dots, X^N_1 \sim q_1$.

Compute $W^i_1 = G_1(X^i_1)/\sum_{j=1}^N G_1(X^j_1)$ for $i = 1, \dots, N$.

Iteration: $n \leftarrow n + 1$

Resample $(W^i_{n-1}, X^i_{1:n-1})_{i=1}^N$ to obtain $(1/N, \tilde{X}^i_{1:n-1})_{i=1}^N$.

Sample $X^i_n \sim q_n(\cdot|\tilde{X}^i_{1:n-1})$ for $i = 1, \dots, N$.

Concatenate $X^i_{1:n} \leftarrow (\tilde{X}^i_{1:n-1}, X^i_n)$ for $i = 1, \dots, N$.

Compute $W^i_n = G_n(X^i_{1:n})/\sum_{j=1}^N G_n(X^j_{1:n})$ for $i = 1, \dots, N$.

Where, to keep notation light, we slightly abusively allow $X^i_{1:n-1}$ *to be overwritten with new values in the concatenation step.*

The simplest approach to resampling is known as multinomial resampling – because one can view the number of copies made of each particle in the present generation under this scheme as a multinomial random variable with N trials and categorical probabilities given by the vector of particle weights. Multinomial resampling often features in theoretical work on SMC. In practical implementations, there are often advantages to employing lower variance resampling schemes. One comparative review of common resampling schemes [4] compares the properties of a number of simple schemes; more recently, the properties of a broad class of algorithms have been studied in detail [5] – in both cases, there is evidence that better performance can be obtained using more sophisticated schemes than the simple multinomial one. In the particular case of finite state spaces, a specialized resampling scheme can be shown to outperform generic techniques [6].

Resampling is often viewed as a *selection* mechanism in which the "fittest" particles survive and are replicated and the least fit produce no offspring. The act of resampling clearly introduces additional variance into the estimators associated with SMC algorithms in the sense that one obtains better lower variance estimates immediately before a resampling operation than immediately after it; however, the immediate increase in variance is justified by the stability that it provides to the system in the future. Consequently, it may be desirable to avoid resampling more often than is necessary – particularly if a simple scheme, such as the multinomial one is being used. Occasional resampling, for example when the effective sample size [1] falls below a threshold, is one way to limit the number of resampling events. This approach is widespread and intuitive but was only shown to inherit many favorable convergence properties from standard SIR schemes rather more recently [7]. Resampling only at some iterations makes no fundamental change to the algorithm, but doing so at iterations which are selected based on the properties of the collection of particles introduces some additional considerations which require additional steps to justify theoretically (one successful strategy [7] being, essentially, to demonstrate that for large enough sample sizes the times at which resampling occurs converge, almost surely, to a deterministic limit).

A number of estimators can be associated with this algorithm, one of the normalizing constant, Z_n, of γ_n:

$$Z_n^N := \prod_{p=1}^{n} \frac{1}{N} \sum_{i=1}^{N} G_p(X_p^i) \tag{1}$$

and one of expectations with respect to each of the target distributions in turn:

$$\pi_n^N(\varphi) := \sum_{i=1}^{N} W_n^i \varphi(X_n^i)$$

There is now a considerable literature on the theoretical properties of these algorithms, the rigorous analysis of which dates back to the mid-1990s [8]. Methods within this broad general class can be profitably interpreted as mean field approximations of Feynman–Kac formulae [9, 10], which provide an elegant framework within which central limit theorem and law of large number results among many others have been obtained. Direct analysis of these methods is, of course, also possible [11–15].

Although a detailed theoretical survey is beyond the scope of this chapter, it is convenient to sketch some of the most prominent results as these provide a formal justification for the use of SMC methods. We present below three results taken from a single monograph [9] by way of illustration; in each case, variants whose proofs hold under different (and often weaker) assumptions can also be found in the literature. The unbiasedness of normalizing constant estimates is a consequence of Theorem 7.4.2 of the monograph and holds under minimal conditions, although some care is required if the potential functions can take the value zero. Corollary 7.4.2 of the monograph provides a strong law of large numbers for particle approximations. The development in Chapter 9 of the monograph provides a central limit theorem.

Proposition 1 (Unbiasedness). *If the potential functions are uniformly bounded above, $\sup_{p\leq n,x} G_p(x) < \infty$, and we set $Z_n^N = 0$ if the system becomes extinct (i.e., all the associated weights have never simultaneously taken the value zero), then $\mathbb{E}[Z_n^N] = Z_n$.*

Proposition 2 (Strong Law of Large Numbers). *Provided G_p is bounded above and away from zero for all $p \leq n$, or other technical conditions are met, for bounded measurable $\varphi : E_n \to \mathbb{R}$, $\pi_n^N(\varphi) \xrightarrow{a.s.} \pi_n(\varphi)$.*

Proposition 3 (Central Limit Theorem). *If the potential functions are uniformly bounded above and away from zero, $\sup_{p\leq n,x,y} G_p(x)/G_p(y) < \infty$, then, for bounded measurable $\varphi : E_n \to \mathbb{R}$,*

$$\sqrt{N}[\pi_n^N(\varphi) - \pi_n(\varphi)] \rightharpoonup \mathcal{N}(0, \sigma_n^2(\varphi))$$

where $\rightharpoonup$ denotes convergence in distribution, and the asymptotic variance $\sigma_n^2(\varphi)$ can be written either recursively or as a sum of integral expressions. Explicit forms for the asymptotic variance can be found for particle filters [12], auxiliary particle filters [16], and SMC samplers [17], for example.

2.1 Extended State Spaces and SMC Samplers

It is often the case that we are interested in a sequence of distributions over a common space (rather than distributions defined on spaces of increasing dimension) or even a single distribution, π. In order to use SMC in the first of these cases it is necessary to define a sequence of distributions of the correct form to allow the SIR paradigm to be deployed while retaining the distributions of interest as marginals; the second case can be handled by constructing an artificial sequence of distributions which lead from a tractable distribution to that of interest. Examples of both cases in the context of Bayesian inference are provided in Section 3.2.

An explicit technique for doing exactly this in some degree of generality was introduced by Del Moral *et al.* in 2006 [17]. Given a sequence of target distributions over some space, E, $\pi_1, \dots, \pi_T$, one can define a sequence of distributions over $\tilde{E}_n = E^n$, say, $\tilde{\pi}_1, \dots, \tilde{\pi}_T$ such that $\tilde{\pi}_n$ is a distribution over vectors of n elements of E, in such a way that they satisfy the requirements for the deployment of SMC and such that they admit the distributions of interest as marginal (and, in particular, the final time marginal which SMC algorithms are best able to approximate). In order to do this, it is convenient to introduce a sequence of Markov kernels $L_1. \dots, L_{T-1}$ which operate *backward* in time so that we can define $\tilde{\pi}_n(x_{1:n}) = \pi_n(x_n) \prod_{p=n-1}^{1} L_p(x_{p+1}, x_p)$. If one denotes the proposal distribution at iteration n of such an algorithm q_n, then one arrives at importance weights:

$$G_n(x_{1:n}) = \frac{\tilde{\pi}_n(x_{1:n})}{\tilde{\pi}_{n-1}(x_{1:n-1})q_n(x_n|x_{n-1})}$$

$$= \frac{\pi_n(x_n) \prod_{p=n-1}^{1} L_p(x_{p+1}, x_p)}{\pi_{n-1}(x_{n-1}) \prod_{p=n-2}^{1} L_p(x_{p+1}, x_p) q_n(x_n|x_{n-1})} = \frac{\pi_n(x_n) L_{n-1}(x_n, x_{n-1})}{\pi_{n-1}(x_{n-1}) q_n(x_n|x_{n-1})}$$

The simple form of these weights and the lack of dependence on any but the final two coordinates is a consequence of the Markovian approach to extending these distributions. The remaining question of how to choose the backward kernels, L_n, can be (partially) answered by considering the variance of the resulting importance weights [17]. The optimal choice for finite sample sizes is intractable as it depends on the actual marginal sampling distribution of the particles at iteration n which is hard to characterize as a consequence of the resampling mechanism, but asymptotic arguments suggest that neglecting the departure of the approximation from the target at time n is a reasonable way to proceed. This suggests that a near optimal strategy would be to choose

$$L_p(x_p, x_{p-1}) = \frac{\pi_{p-1}(x_{p-1}) q_p(x_p|x_{p-1})}{\int \pi_{p-1}(x'_{p-1}) q_p(x_p|x'_{p-1}) dx'_{p-1}}$$

but in general this will lead to intractable importance weights (loosely speaking, it can be seen as an attempt to integrate out the history of the particle system). In the case in which q_p is a π_p-invariant Markov kernel[1], a small departure from the optimal expression gives rise to the time reversal of q_p with respect to its invariant distribution:

$$L_p(x_p, x_{p-1}) = \frac{\pi_p(x_{p-1})q_p(x_p|x_{p-1})}{\int \pi_p(x'_{p-1})q_p(x_p|x'_{p-1})dx'_{p-1}} = \frac{\pi_p(x_{p-1})q_p(x_p|x_{p-1})}{\pi_p(x_p)}$$

which can be more readily used. This is a rather natural choice when one uses π_p-invariant Markov kernels in the proposal mechanism; indeed, this auxiliary kernel appears in the proof of a central limit theorem for the resample-move algorithm [18]. In this particular setting one can also arrive at the same importance weights using more direct arguments [19].

The ease with which adaptation can be incorporated within SMC methods is one of their great strengths, and several strategies have been proposed [20–23] and theoretical support provided [24]. There are two areas in which adaptation is most commonly employed: first within the sequence of target distributions (in settings in which a single distribution is of ultimate interest) and in the parameters of the proposal distribution employed at each step. Appropriate adaptive methods naturally vary between contexts, but at least in contexts in which one expects consecutive distributions within the sequence of targets to be broadly "similar" and Metropolis–Hastings kernels are used, there are two common approaches to tuning the proposal distribution: using the particle population at time $n - 1$ to estimate the moments of the target at that time and to employ a proposal at time n which would be optimal, for example, for a Gaussian target with those moments or to adjust the proposal scale whenever the acceptance rate falls outside some target region (motivated by optimal scaling considerations). In settings in which one builds an artificial sequence of distributions in order to reach a single distribution of interest, it is a common practice to specify a sequence of distributions which differ from one another by approximately equal amounts; strategies that control the effective sample size (or variants in the case of occasional resampling [25]) aim, essentially, to control the χ-squared discrepancy between consecutive distributions, which is intuitively appealing if not, in general, optimal.

2.2 Particle MCMC and Related Methods

The particle Markov chain Monte Carlo (MCMC) [26] approach essentially employs SMC algorithms *within* MCMC algorithms – in some sense the counterpart of the use of MCMC moves within SMC algorithms – in order to provide good proposal distributions. It is intuitive that as SMC provides good approximations to its target distributions that it could provide good approximations to, for example, block-Gibbs sampler proposals and intractable marginal distributions, considerable care is required to justify this: one cannot simply use approximations naively within an MCMC algorithm and expect to obtain the correct invariant distribution.

In order to justify this type of algorithm, it is necessary to characterize the distribution of all of the random variables generated in the running of an SMC algorithm, and to do this it is convenient to reinterpret the resampling slightly as sampling an *ancestor* for each member of the resulting population. Having done this, the joint distribution over the variables simulated in the proposal step and in the selection of ancestors can be characterized straightforwardly, allowing for a variety of MCMC algorithms which make use of SMC as a constituent part to be justified by an extended state-space construction in which the distribution of interest is admitted as a marginal variable, and the additional variables involved in the SMC algorithm can be viewed as auxiliary variables.

More precisely, let $\mathbf{a}_p = (a_p^1, \ldots, a_p^N)$ denote the vector of time p ancestors of the particles at time $p+1$ so that, for example, x_{p+1}^i is an offspring of $x_p^{a_p^i}$. Similarly, let $\mathbf{x}_p = (x_p^1, \ldots, x_p^N)$. For simplicity, consider the case in which $q_p(x_p|x_{1:p-1}) = q_p(x_p|x_{p-1})$ and $G_p(x_{1:p}) = G_p(x_{p-1}, x_p)$; the general case follows by identical arguments but with somewhat more cumbersome notational requirements. The random variables simulated in the course of running Algorithm 1 up to time n are the states $\mathbf{x}_{1:n} \in \prod_{p=1}^n E_p^N$ and $\mathbf{a}_{1:n-1} \in \{1, \ldots, N\}^{Nn}$ and have the joint distribution

$$\psi_n(\mathbf{x}_{1:n}, \mathbf{a}_{1:n-1}) := \prod_{i=1}^N q_1(x_1^i) \cdot \prod_{p=2}^n \left[r(\mathbf{a}_{p-1}|\mathbf{w}_{p-1}) \prod_{i=1}^N q_p(x_p^i | x_{p-1}^{a_p^i}) \right]$$

where $r(\cdot|\mathbf{w})$ denotes the conditional distribution of ancestors arising from a resampling operation with weight vector $\mathbf{w}$, and the weight vectors are included to simplify the notation but are formally redundant as $\mathbf{w}_p = (w_p^1, \ldots, w_p^N)$ is a deterministic function of $\mathbf{x}_{1:p}$ and $\mathbf{a}_{1:p-1}$ with $w_p^i = G_p(x_{p-1}^{a_{p-1}^i}, x_p^i) / \sum_{j=1}^N G_p(x_{p-1}^{a_{p-1}^j}, x_p^j)$ in the context described. For a concrete example of such a construction, consider the case in which multinomial resampling is used, in which case

$$r(\mathbf{a}_{p-1}|\mathbf{w}_{p-1}) = \prod_{i=1}^N w_{p-1}^{a_{p-1}^i}$$

Two broad categories of algorithms arise from the use of this construction within an MCMC context. Algorithms within the first category mimic a marginal form of Metropolis–Hastings algorithm in settings in which a completed likelihood is tractable but the marginal one is not; such *particle marginal Metropolis–Hastings* (PMMH) algorithms can be justified directly as pseudomarginal algorithms [27], noting that the normalizing constant estimates provided by SMC algorithms are unbiased, or via the type of auxiliary variable construction described above. Algorithms in the second category mimic an idealized block-Gibbs sampler in which the full vector of random variables being updated are drawn from their joint conditional distribution; these algorithms are a little more complex requiring the introduction of the so-called conditional SMC (cSMC) algorithms and admitting a justification as partially collapsed Gibbs samplers [28]. The cSMC algorithm corresponds essentially to an SMC algorithm which is modified, so one particular particle trajectory is fixed in advance and guaranteed to survive through resampling steps; although notationally awkward to describe in full generality, such algorithms are simple to implement and enjoy good mixing properties, potentially justifying a little additional implementation effort.

cSMC algorithms warrant a little discussion in their own right; it can be shown that running a cSMC algorithm and drawing a single-particle trajectory from the resulting weighted ensemble correspond to a Markov kernel which is invariant with respect to a particular distribution (the smoothing distribution in the context of hidden Markov models (HMMs) as described in Section 3.1.2) and can enjoy uniform ergodicity [29].

The basic algorithm can be further improved in many cases by sampling not from the population of particle trajectories obtained naturally by the resampling mechanism but employing a backward simulation approach [30] reminiscent of the backward simulation smoother described in Section 3.1.2 or a forward-only representation of the same known as ancestor sampling [31] – with these modifications it can be possible to employ very modest population sizes.

The SMC^2 algorithm [32] embeds the particle MCMC approach within a SMC sampler and, to some extent, allows for online parameter estimation within state-space models. Roughly speaking, a data-tempered SMC sampler is used to approximate the distribution over the parameter space with the importance weights associated with this algorithm being obtained from an ensemble of particle filters approximating the distribution in the latent variable space although, of course, some care is needed in dealing with the details.

3 SMC in Statistical Contexts

3.1 SMC for Hidden Markov Models

Perhaps the most widely known application of SMC methods is to Bayesian inference for general state-space HMMs or state-space models (SSMs) as they are sometimes known. This approach dates back at least to the early 1990s [33, 34] along with the terms bootstrap filter [34] and interacting particle filter [8]. One fairly recent survey of SMC [2] in the HMM context demonstrates that almost all particle filtering methods can be viewed within the simple SIR framework described above, which is also the perspective which we take here.

Consider an $\mathbb{R}^{d_x}$–valued discrete-time Markov process $\{X_n\}_{n\geq 1}$ such that

$$X_1 \sim \mu(x_1) \text{ and } X_n|(X_{n-1} = x_{n-1}) \sim f(x_n|x_{n-1}) \tag{2}$$

where "$\sim$" means distributed according to, $\mu(x)$ is a probability density function, and $f(x|x')$ denotes the probability density associated with moving from x' to x. We are interested in estimating $\{X_n\}_{n\geq 1}$ but only have access to the $\mathbb{R}^{d_y}$–valued process $\{Y_n\}_{n\geq 1}$. We assume that, given $\{X_n\}_{n\geq 1}$, the observations $\{Y_n\}_{n\geq 1}$ are statistically independent, and their marginal densities are given by

$$Y_n|(X_n = x_n) \sim g(y_n|x_n) \tag{3}$$

For the sake of simplicity, we have considered only the homogeneous case here; that is, the transition and observation densities are independent of the time index n. The extension to the inhomogeneous case is straightforward.

There are several inferential problems associated with this class of models: *filtering* corresponds to the sequential characterization of the law of the latent state, X_n at time n given observations $y_{1:n}$ for each n as observations become available; *smoothing* to the characterization of the law of the entire vector of latent states $X_{1:n}$ up until time n given observations $y_{1:n}$ again often sequentially as observations become available; *prediction* to the characterization of the law of X_{n+p} for $p \geq 1$ given observations $y_{1:n}$ for each n and can often be treated as a straightforward extension of filtering; and parameter estimation corresponds to the estimation of static model parameters which do not evolve over time. Until Section 3.1.3 it will be assumed that any model parameters are known.

3.1.1 Filtering

Perhaps the most natural approach to filtering within the SMC framework described above is to simply set

$$\gamma_n(x_{1:n}) = p(x_{1:n}, y_{1:n}) = \mu(x_1)g(y_1|x_1)\prod_{p=2}^{n} f(x_n|x_{n-1})g(y_n|x_n)$$

where p denotes the joint density of the latent and observation processes over the time horizon indicated by its arguments as well as associated conditional and marginal distributions as is common in this literature. In this case, $Z_n = p(y_{1:n})$ and $\pi_n(x_{1:n}) = p(x_{1:n}|y_{1:n})$. If one also sets $q_n(x_n|x_{1:n-1}) \equiv f(x_n|x_{n-1})$, one arrives at a particularly simple algorithm known as the bootstrap particle filter.

There are numerous strategies to improve the performance of SMC in the context of filtering problems. A number of the more prominent strategies are summarized below; for more details and a demonstration that all of these methods can be viewed as SIR algorithms (sometimes on suitably extended state spaces), see Ref. 2.

Alternative proposals can improve the performance of the algorithm; the *locally optimal* proposal $q_n(x_n|x_{1:n-1}) = p(x_n|x_{n-1}, y_n)$ minimizes the conditional variance of the importance weights within the class of algorithms being considered here [35].

Auxiliary particle filters [16, 36, 37] attempt to further improve performance by deferring resampling until after the influence of the next observation has been (partially) incorporated into the importance weights.

Lookahead methods [38] / block-sampling [39] techniques extend these ideas further into the future, albeit at the expense of immediacy. They do this either by modifying the target distribution to further into the future approximately incorporate the influence of several subsequent observations or by sampling new values for the most recently estimated states (using an extended state-space construction similar to that employed within SMC samplers) during each iteration to allow for the influence of the most recent observations to be incorporated. Some recent work [40] attempts to address the difficulty of designing good high-dimensional proposals via an iterative scheme appropriate only *outside* the online filtering framework; this idea was recently explored more extensively outside the HMM context [41].

MCMC moves can be included within particle filters. There are two broad approaches to the inclusion of MCMC-based innovations within SMC algorithms. The so-called resample-move [18]-based approaches add a Markov kernel with respect to which the target distribution is invariant to each iteration of the algorithm; this provides a mechanism to improve sample diversity but does not fundamentally change the structure of the algorithm. Another approach, often termed sequential MCMC, replaces the simulation of a collection of conditionally independent samples during each iteration with the simulation of a Markov chain with an appropriate invariant distribution; such approaches have been present in the literature for some time [42], and good empirical performance has been observed [43, 44], although convergence results appear to have become available only recently [45]. The ensemble HMM method [46], in which a grid of points is obtained at each time via the simulation of a Markov chain of an appropriate invariant distribution, prior to the performance of inference using that grid as a discrete state space can be shown to be closely connected with sequential MCMC methods [47] combined with particle MCMC.

There is also considerable work on the use of SMC for filtering in the continuous time setting; good recent surveys [48, 49] and references therein provide a good overview, but a detailed survey falls outside the scope of this chapter.

3.1.2 Smoothing

In principle, Algorithm 1 applied to a sequence of target distributions coinciding with $p(x_{1:n}|y_{1:n})$ provides an approximation of each smoothing distribution in turn. However, this naive approach sometimes known as the "smoother mode" of the particle filter is doomed to fail eventually as it corresponds to an importance sampling-like approach on a space of ever-increasing dimension. In fact, the situation is a little worse as every resampling step reduces the number of unique paths at earlier times, and eventually $p(x_1|y_{1:n})$ is approximated by only a single surviving path. There has been considerable attention in the literature to the problem of better approximating smoothing distributions.

Fixed-lag methods provide one simple approximate scheme [50] which allows for smoothing in an online fashion as observations become available. Rather than attempting to approximate the distribution of x_p given all of the observations received, one settles for an approximation given all of the observations obtained up until a time, some fixed time after p, that is, making the approximation $p(x_p|y_{1:n}) \approx p(x_p|y_{1:\min(p+L,n)})$ which is intuitively reasonable for sufficiently large L provided that the process under study is sufficiently ergodic. The resulting approximation error can be controlled under mixing assumptions, at least for the estimation of additive functionals [51].

Several more sophisticated methods are possible; see Ref. 52. In particular, the forward-filtering backward-simulation (FFBSi) approach which revolves around the decomposition of the smoothing distribution as

$$p(x_{1:n}|y_{1:n}) = p(x_n|y_{1:n}) \prod_{p=1}^{n-1} p(x_p|y_{1:p}, x_{p+1})$$

with

$$p(x_p|x_{p+1}, y_{1:p}) = \frac{p(x_p|y_{1:p})f(x_{p+1}|x_p)}{p(x_{p+1}|y_{1:p})}$$

This allows us to write

$$p(x_{1:n}|y_{1:n}) = p(x_n|y_{1:n}) \prod_{p=1}^{n-1} \frac{p(x_p|y_{1:p})f(x_{p+1}|x_p)}{p(x_{p+1}|y_{1:p})}$$

and within the SMC framework, one can obtain a sample approximation of the smoothing distribution by first running a standard particle filter forward to the final time computing and storing all of the marginal filtering distributions along the way and then to run a backward pass using the resulting particle approximation of $p(x_p|x_{p+1}, y_{1:p})$. A theoretical analysis of this and related approaches is provided by Douc *et al.* [53].

The FFBSi approach has a computational cost of $O(Nn)$ per backward sample path (where N is the number of particles used in the forward filtering phase, and n is the length of the time series) and hence a cost of $O(N^2n)$ if one wishes to obtain an N-particle approximation. Some work has been done to mitigate this in the literature, including a slightly different approximation of the distribution which can reduce the cost to something linear in the sample size if one is interested in only marginal smoothing distributions [54] and methods which allow efficient estimation of smoothing expectations of additive functionals [53, 55, 56].

Offline approaches to smoothing via particle MCMC or iterated cSMC have recently been developed [57] and are closely related to the problem of static parameter estimation, which is discussed in the following section.

3.1.3 Parameter estimation

Estimating static parameters, that is, those parameters which take a single value which is common to all time points, is a challenging problem in the HMM context – particularly in online contexts. Online, here, means providing an estimate each time a new observation is obtained which incorporates the influence of all observations received to date at an iterative cost which is bounded in time. The particular difficulties arise from the nontrivial dependence structure in which the static parameter and the entire latent state vector have complex dependencies; the path degeneracy problem of the particle filter makes dealing with the full joint distribution challenging.

Broadly speaking, methods can be characterized as online or offline and make use of either maximum-likelihood or Bayesian approaches to parameter estimation. Offline inference, a competitor to MCMC for the same problem, is generally easier, and likelihood-based methods are less computationally demanding than fully Bayesian ones, especially in the online setting in which it is possible to leverage ideas based around Fisher scoring on stochastic expectation maximization algorithms. Good approaches to these problems are somewhat specialized, but an excellent recent survey exists [58].

3.2 SMC for Bayesian Inference

There are many ways in which SMC finds application in the context of Bayesian inference; there is a good recent review of methods applicable in the context of graphical models [59]. One common application of SMC in the statistical literature is in the approximation of the Bayesian posterior distribution for some parameter θ for which one has prior distribution $p(\theta)$ and a likelihood $p(y|\theta)$, where y denotes the full set of data available.

Approaches to this problem date back in the statistics literature have existed for approximately two decades [19], with related ideas to be found in the earlier literature [60, 61], and a framework incorporating this and many other algorithms is provided by Del Moral *et al.* [17]. Two common approaches, widely identified as data tempering and likelihood tempering, to the specification of a suitable sequence of distributions are widespread. In the data-tempering setting, one defines a sequence of distributions by adding additional observations at each step, arriving at a sequence of partial posteriors of the form

$$\pi_n(\theta) \propto \gamma_n(\theta) = p(\theta)p(y_{1:m_n}|\theta)$$

for some sequence (m_n) of data sizes increasing from zero to the actual size of the data set, whereas in likelihood tempering

$$\pi_n(\theta) \propto \gamma_n(\theta) = p(\theta)p(y|\theta)^{\alpha_n}$$

for some monotonically increasing real-valued sequence, (α_n), which increases from zero to one. Both m_n and α_n can be specified adaptively.

In the context of Bayesian inference for static parameters with either of these sequences of target distributions, it is natural to employ π_n-invariant Markov kernels as the mutation elements of the SMC algorithm giving rise to incremental importance weights at time n of the form $p(y_{m_{n-1}+1:m_n}|\theta)$ and $p(y|\theta)^{\alpha_n-\alpha_{n-1}}$, respectively, if one operates within the SMC sampler framework using the time reversal of these invariant Markov kernels as the associated auxiliary kernels.

Of course, the SMC framework provides very considerable flexibility and we need not be constrained to sequences of distributions which temper from prior to posterior. In the context of generalized linear mixed models, for example, it has been found that starting with a distribution motivated by quasi-likelihood arguments and moving from that to the posterior leads to somewhat better performance [62].

3.2.1 SMC for model comparison

Similar to parameter estimation in HMMs, Bayesian model comparison centers around some computation of the marginal likelihoods, that is, the marginal probability under a given model of observing the data actually observed, with unknown model parameters marginalized out. In the context of any sequence of distributions which begins with a properly normalized distribution over the space of unknown parameters and finishes with the posterior characterized as the product of the complete likelihood and parameter priors divided by an unknown normalizing constant, that normalizing constant corresponds exactly with the marginal likelihood and is estimated unbiasedly by the associated SMC scheme via Equation (1) (i.e., $Z_n = \int \gamma_n(\theta)d\theta = \int p(\theta)p(y|\theta)d\theta$ when n is the final distribution within either the data- or likelihood-tempering schemes described in Section 3.2), so that either m_n corresponds to the size of the data set or $\alpha_n = 1$.

As the estimation of normalizing constants and marginal likelihoods is somewhat natural in the SMC setting, these algorithms lend themselves to this problem. A number of different approaches to this problem have been explored and found to perform well in many settings [25, 63]. These approaches include simultaneously addressing model and parameter inference in a similar manner to reversible jump MCMC methods [64], explicitly approximating the marginal likelihoods of each of a family of competing models and directly computing the ratio of marginal likelihoods of pairs of competing models, the so-called Bayes factor.

3.2.2 SMC for ABC

ABC (introduced in [65]; recent survey [66]) is another area in which SMC has been widely applied [21, 67]. ABC is a technique for performing computational inference in settings in which the likelihood cannot be evaluated, but it is possible to simulate from the associated data-generating model for given parameter values. A detailed survey of ABC methods is outside the scope of this chapter, but in essence, the fit of a parameter value to a given data set is assessed by simulating a data set from the generative model for that parameter value and comparing it with the actually observed data, typically by determining the distance between summary statistics computed using the real and simulated data sets. For example, by considering a target distribution of the form

$$\pi_\epsilon(\theta, y) \propto p(\theta)f(y|\theta)\mathbb{I}_{[0,\epsilon]}(d(S(y), S(y_{\text{obs}})))$$

where ϵ denotes a tolerance, θ the unknown parameters of interest, $p(\theta)$ a prior distribution, y the auxiliary simulated data, $f(y|\theta)$ the modeled generative relationship between parameters and data, S a mapping from the data space to a low-dimensional summary statistic space, d some appropriate distance, and y_{obs} the actually observed data. In the SMC context it is natural to make use of a sequence of distributions which require an increasing degree of fidelity between the observed and simulated data, that is, considering a (possibly adaptive) decreasing sequence of values of ϵ.

In an ABC context, the need to resimulate synthetic data whenever a new parameter value is proposed limits the ability for SMC to benefit from local exploration as it does in standard Bayesian inferential settings; one remedy to this is to adopt an appropriate non-centered parameterization when this is possible [68].

It is also possible to compute estimates of model evidence within the ABC framework using SMC [69], although considerable caution is required in doing so, particularly in the selection of summary statistics, and interpreting the conclusions [69, 70].

3.3 SMC for Maximum-Likelihood Estimation

It is worthwhile noting that, although SMC like many Monte Carlo methods is widely used within the Bayesian domain, it also finds application in other statistical paradigms.

Maximum-likelihood estimation (MLE) is, at heart, an optimization problem, and it is no surprise that simulated-annealing-like methods can be used in this context; within the marginal MLE setting, SMC samplers and data cloning provide one natural approach to this problem [71]. A more direct use of a simulated annealing strategy was explored by Rubenthaler [72], and a pseudomarginal [27] variant also shows promise [73]. All of these approaches essentially involve the construction of a sequence of distributions which become progressively more concentrated on the set of maximizers of the likelihood function and targetting this sequence using SMC sampler algorithms.

3.4 SMC for Rare Event Estimation

Estimating the probabilities of rare events (i.e., those with small probability of occurrence) is a natural application of SMC methods; in this context, one can begin from the law of some underlying random variable and move via a sequence of intermediate distributions to the restriction of that law to the rare event of interest obtaining both an approximation of the probability of this event (via the normalizing constant of this restriction) and also an approximation of the law of the random variable restricted to that set (via the final particle set).

SMC provides natural approaches to the so-called dynamic rare event problem in which one is interested in establishing the probability that a Markov process hits a specified rare set of interest before its next entrance into some recurrent set [74] and the static rare event problem in which the question is whether a random variable/process takes a value within some set which has small probability under its law [75–77]. In the dynamic case it is common to employ a sequence of intermediate distributions in order to characterize the probability of hitting each of a sequence of increasingly rare sets before the recurrent set; the latter one simply needs to construct a sequence of distributions which begins with the law of the random quantity of interest and becomes increasingly concentrated on the rare set of interest.

4 Selected Recent Developments

This section concludes with a brief summary of some exciting emerging topics within the field of SMC.

One, perhaps surprising, recent development is the emergence of a methodology which permits the consistent estimation of the variance and asymptotic variance of SMC algorithms using the output from a single realization of the particle system [78, 79]. This has recently been extended to the case of a class of adaptive algorithms [80]. In the context of online inference in state-space models, a "fixed lag" approach was explored by Olsson and Douc [81]. These methods provide an avenue to the characterization of the quality of estimates obtained from SMC algorithms *without* recourse to multiple costly runs of those algorithms.

Considering the "genealogical properties" of SMC algorithms (i.e., the trees which one obtains by tracing back particles surviving until the current generation and producing a tree containing all particles in previous generations which are ancestors to surviving particles) has provided another avenue to understanding the behavior of these algorithms. Both bounds on the properties of these trees [3] and a characterization of the limiting tree [82] have been obtained and provide information about storage costs of algorithms as well as efficient data structures for storing the entire history of the currently surviving particles.

Efficient distributed implementation via modifications of the fundamentally synchronous resampling operation [83–85] or via more fundamental changes to the methodology suitable for offline inference [86] has been the subject of substantial recent research, and further developments in this direction are to be expected in the future.

Quasi-Monte Carlo (QMC) methods eschew the use of random numbers in favor of low discrepancy sequences which seek, in a suitable sense, to fill space as regularly as possible. Leveraging these techniques in an SMC setting is challenging, in part because of the increasing-state-space justification of most SMC methods and in part due to complications arising from resampling, but substantial progress in this direction was made in the form of sequential QMC [87], which employs QMC within a marginal framework, at iteration n sampling (a_{n-1}^i, x_n^i) jointly according to $r(a_{n-1}^i|\mathbf{w}_{n-1})q_n(x_n^i|x_{n-1}^{a_{n-1}^i})$, in the notation of Section 2.2, via QMC methods. It shows particularly substantial performance gains in relatively low-dimensional filtering-type problems.

Acknowledgments

The author's research is partially supported by the Alan Turing Institute–Lloyd's Register Foundation programme on Data-Centric Engineering and the Engineering and Physical Sciences Council Grants EP/R034710/1 and EP/T004134.

Note

1 Including those, like those arising from Metropolis-like accept–reject mechanisms, which do not admit Lebesgue densities; a more careful treatment allows it to be established that absolute continuity of $\pi_p(x_p)L_{p-1}(x_p, x_{p-1})$ with respect to $\pi_{p-1}(x_{p-1})q_p(x_p|x_{p-1})$ is all that is really required, and the time reversal kernel described here readily satisfies that requirement.

References

1 Kong, A., Liu, J.S., and Wong, W.H. (1994) Sequential imputations and Bayesian missing data problems. *J. Am. Stat. Assoc.*, 89 (425), 278–288.

2 Doucet, A. and Johansen, A.M. 2011 A tutorial on particle filtering and smoothing: fifteen years later, in *The Oxford Handbook of Nonlinear Filtering* (eds D. Crisaned and B. Rozovskiui), Oxford University Press, pp. 656–704.

3 Jacob, P.E., Murray, L., and Rubenthaler, S. (2015) Path storage in the particle filter. *Stat. Comput.*, 25 (2), 487–496.

4 Douc, R., Cappé, O., and Moulines, E. (2005) *Comparison of Resampling Schemes for Particle Filters*. Proceedings of the 4th International Symposium on Image and Signal Processing and Analysis, vol. I, IEEE, pp. 64–69.

5 Gerber, M., Chopin, N., and Whiteley, N. (2019) Negative association, ordering and convergence of resampling methods. *Ann. Stat.*, 47 (4), 2236–2260.

6 Fearnhead, P. and Clifford, P. (2003) On-line inference for hidden Markov models via particle filters. *J. Royal Stat. Soc. B*, 65 (4), 887–899.

7 Del Moral, P., Doucet, A., and Jasra, A. (2012) On adaptive resampling procedures for sequential Monte Carlo methods. *Bernoulli*, 18 (1), 252–278.

8 Del Moral, P. (1995) Nonlinear filtering using random particles. *Theory Probab. Appl.*, 40 (4), 690–701.

9 Del Moral, P. (2004) *Feynman-Kac Formulae: Genealogical and Interacting Particle Systems with Applications, Probability and Its Applications*, Springer Verlag, New York.

10 Del Moral, P. (2013) *Mean Field Integration*, Chapman Hall.

11 Crisan, D. and Doucet, A. (2002) A survey of convergence results on particle filtering methods for practitioners. *IEEE Trans. Signal Process*, 50 (3), 736–746.

12 Chopin, N. (2004) Central limit theorem for sequential Monte Carlo methods and its applications to Bayesian inference. *Ann. Stat.*, 32 (6), 2385–2411.

13 Künsch, H.R. (2005) Recursive Monte Carlo filters: algorithms and theoretical analysis. *Ann. Stat.*, 33 (5), 1983–2021.

14 Cappé, O., Moulines, E., and Ryden, T. (2005) *Inference in Hidden Markov Models*, Springer Verlag, New York.

15 Douc, R. and Moulines, E. (2008) Limit theorems for weighted samples with applications to sequential Monte Carlo methods. *Ann. Stat.*, 36 (5), 2344–2376.

16 Johansen, A.M. and Doucet, A. (2008) A note on the auxiliary particle filter. *Stat Probab Lett.*, 78 (12), 1498–1504.

17 Del Moral, P., Doucet, A., and Jasra, A. (2006) Sequential Monte Carlo samplers. *J. Royal Stat. Soc. B*, 63 (3), 411–436.

18 Gilks, W.R. and Berzuini, C. (2001) Following a moving target – Monte Carlo inference for dynamic Bayesian models. *J. Royal Stat. Soc. B*, 63 (1), 127–146.

19 Chopin, N. (2002) A sequential particle filter method for static models. *Biometrika*, 89 (3), 539–551.

20 Jasra, A., Stephens, D.A., Doucet, A., and Tsagaris, T. (2010) Inference for Lévy-driven stochastic volatility models via adaptive sequential Monte Carlo. *Scand. J. Stat.*, 38 (1), 1–22.

21 Del Moral, P., Doucet, A., and Jasra, A. (2012) An adaptive sequential Monte Carlo method for approximate Bayesian computation. *Stat. Comput.*, 22 (5), 1009–1020.

22 Schäfer, C. and Chopin, N. (2013) Sequential Monte Carlo on large binary sampling spaces. *Stat. Comput.*, 23 (2), 163–184.

23 Fearnhead, P. and Taylor, B. (2013) An adaptive sequential Monte Carlo sampler. *Bayesian Anal.*, 8 (2), 411–438.

24 Beskos, A., Jasra, A., Kantas, N., and Thiéry, A.H. (2016) On the convergence of adaptive sequential Monte Carlo methods. *Ann. Appl. Probab.*, 26 (2), 1111–1146.

25 Zhou, Y., Johansen, A.M., and Aston, J.A.D. (2016) Towards automatic model comparison: an adaptive sequential Monte Carlo approach. *J. Comput. Graph. Stat.*, 25 (3), 701–726. doi: 10.1080/10618600.2015.1060885.

26 Andrieu, C., Doucet, A., and Holenstein, R. (2010) Particle Markov chain Monte Carlo. *J. Royal Stat. Soc. B*, 72 (3), 269–342.

27 Andrieu, C. and Roberts, G.O. (2009) The pseudo-marginal approach for efficient Monte Carlo computations. *Ann. Stat.*, 37 (2), 697–725.

28 Van Dyk, D.A. and Park, T. (2008) Partially collapsed Gibbs samplers: theory and methods. *J. Am. Stat. Assoc.*, 103 (482), 790–796.

29 Andrieu, C., Lee, A., and Vihola, M. (2018) Uniform ergodicity of the iterated conditional SMC and geometric ergodicity of particle Gibbs samplers. *Bernoulli*, 24 (2), 842–872.

30 Whiteley, N. (2010) Contribution to the discussion on 'Particle Markov chain Monte Carlo methods' by Andrieu, C., Doucet, A., and Holenstein, R. *J. Royal Stat. Soc. B*, 72 (3), 306–307.

31 Lindsten, F., Jordan, M.I., and Schön, T.B. (2012) Ancestor sampling for particle Gibbs, in *Proceedings of the 2012 Conference on Neural Information Processing Systems (NIPS)* (eds F. Pereira, C.J.C. Burges, L. Bottou, and K.Q. Weinberger), Curran Associates, Inc., Lake Tahoe, NV, pp. 2591–2599

32 Chopin, N., Jacob, P., and Papaspiliopoulos, O. (2013) SMC^2: an efficient algorithm for sequential analysis of state space models. *J. Royal Stat. Soc. B*, 75 (3), 397–426.

33 Stewart, L. and McCarty, P. (1992) *The Use of Bayesian Belief Networks to Fuse Continuous and Discrete Information for Target Recognition, Tracking and Situation Assessment*. Proceedings of SPIE Signal Processing, Sensor Fusion and Target Recognition, vol. 1699, pp. 177–185.

34 Gordon, N.J., Salmond, S.J., and Smith, A.F.M. (1993) Novel approach to nonlinear/non-Gaussian Bayesian state estimation. *IEE Proc.-F*, 140 (2), 107–113. April

35 Doucet, A., Godsill, S., and Andrieu, C. (2000) On sequential simulation-based methods for Bayesian filtering. *Stat. Comput.*, 10 (3), 197–208.

36 Pitt, M.K. and Shephard, N. (1999) Filtering via simulation: auxiliary particle filters. *J. Am. Stat. Assoc.*, 94 (446), 590–599.

37 Douc, R., Moulines, E., and Olsson, J. (2009) Optimality of the auxiliary particle filter. *Probab. Math. Stat.*, 29 (1), 1–28.

38 Lin, M., Chen, R., and Liu, J.S. (2013) Lookahead strategies for sequential Monte Carlo. *Stat. Sci.*, 28 (1), 69–94.

39 Doucet, A., Briers, M., and Sénécal, S. (2006) Efficient block sampling strategies for sequential Monte Carlo methods. *J. Comput. Graph. Stat.*, 15 (3), 693–711.

40 Guarniero, P., Johansen, A.M., and Lee, A. (2017) The iterated auxiliary particle filter. *J. Am. Stat. Assoc.*, 112 (520), 1636–1647.

41 Heng, J., Bishop, A.N., Deligiannidis, G., and Doucet, A. (2020) Controlled sequential Monte Carlo. *Ann. Stat.* (In press).

42 Berzuini, C., Best, N.G., Gilks, W.R., and Larizza, C. (1997) Dynamic conditional independence models and Markov chain Monte Carlo. *J. Am. Stat. Assoc.*, 92 (440), 1403–1412.

43 Septier, F., Kim Pang, S., Carmi, A., and Godsill, S. (2009) *On MCMC-Based Particle methods for Bayesian Filtering: Application to Multitarget Tracking*. 3rd IEEE International Workshop on Computational Advances in Multi-Sensor Adaptive Processing (CAMSAP), IEEE, pp. 360–363.

44 Septier, F. and Peters, G.W. (2016) Langevin and Hamiltonian based sequential MCMC for efficient Bayesian filtering in high-dimensional spaces. *IEEE J. Sel. Topics Signal Process.*, 10 (2), 312–327

45 Finke, A., Doucet, A., and Johansen, A.M. (2020) Limit theorems for sequential MCMC methods. *Adv. Appl. Probab.*, 52 (2) (In press).

46 Shestopaloff, A.Y. and Neal, R.M. (2013) MCMC for non-linear state space models using ensembles of latent sequences. arXiv:1305.0320.

47 Finke, A., Doucet, A., and Johansen, A.M. (2016) On embedded hidden Markov models and particle Markov chain Monte Carlo methods. arXiv:1610.08962.

48 Bain, A. and Crisan, D. (2009) *Fundamentals of Stochastic Filtering*, Stochastic Modelling and Applied Probability, Springer.

49 Crisan, D. and Rozovskiui, B. (eds) (2011) *The Oxford Handbook of Nonlinear Filtering*, Oxford University Press, Oxford.

50 G. Kitigawa and S. Sato (2001) Monte carlo smoothing and self-organising state-space model, in *Sequential Monte Carlo Methods in Practice* (eds A. Doucet, N. de.Freitas, and N. Gordon), Statistics for Engineering and Information Science. Springer Verlag, New York, pp. 177–195.

51 Olsson, J., Cappé, O., Douc, R., and Moulines, E. (2008) Sequential Monte Carlo smoothing with application to parameter estimation in non-linear state space models. *Bernoulli*, 14 (1), 155–179.

52 Briers, M., Doucet, A., and Maskell, S. (2010) Smoothing algorithms for state space models. *Ann. Inst. Stat. Math.*, 62 (1), 61–89.

53 Douc, R., Garivier, A., Moulines, E., and Olsson, J. (2011) Sequential Monte Carlo smoothing for general state space hidden Markov models. *Ann. Appl. Probab.*, 21 (6), 2109–2145.

54 Fearnhead, P., Wyncoll, D., and Tawn, J. (2010) A sequential smoothing algorithm with linear computational cost. *Biometrika*, 97 (2), 447–464.

55 Del Moral, P., Doucet, A., and Singh, S.S. (2010) Forward smoothing using sequential Monte Carlo. arXiv:1012.5390.

56 Olsson, J. and Westerborn, J. (2017) Efficient particle-based online smoothing in general hidden markov models: the PaRIS algorithm. *Bernoulli*, 23, 1951–1996.

57 Jacob, P., Lindsten, F., and Schön, T. (2019) Smoothing with couplings of conditional particle filters. *J. Am. Stat. Assoc.* doi: 10.1080/01621459.2018.1548856.

58 Kantas, N., Doucet, A., Singh, S.S., *et al.* (2015) On particle methods for parameter estimation in general state-space models. *Stat. Sci.*, 30 (3), 328–351.

59 Doucet, A. and Lee, A. (2018) Sequential Monte Carlo methods, in *Handbook of Graphical Models* (eds M. Maathuis, M. Drton, S. L. Lauritzen, and M. Wainwright), CRC Press, pp. 165–189.

60 Neal, R.M. (1998) Annealed importance sampling. *Technical Report 9805*. University of Toronto, Department of Statistics.

61 MacEachern, S.N., Clyde, M., and Liu, J.S. (1999) Sequential importance sampling for nonparametric Bayes models: the next generation. *Can. J. Stat.*, 27 (2), 251–267.

62 Fan, Y., Leslie, D., and Wand, M.P. (2008) Generalized linear mixed model analysis via sequential Monte Carlo sampling. *Electron. J. Stat.*, 2, 916–938.

63 Jasra, A., Doucet, A., Stephens, D.A., and Holmes, C.C. (2008) Interacting sequential Monte Carlo samplers for trans-dimensional simulation. *Comput. Stat. Data. An*, 52 (4), 1765–1791.

64 Green, P.J. (1995) Reversible jump Markov Chain Monte Carlo computation and Bayesian model determination. *Biometrika*, 82, 711–732.

65 Tavaré, S., Balding, D.J., Griffiths, R.C., and Donnelly, P. (1997) Inferring coalescence times from dna sequence data. *Genetics*, 145 (2), 505–518.

66 Sisson, S.A., Fan, Y., and Beaumont, M. (2018) *Handbook of Approximate Bayesian Computation*, Chapman and Hall/CRC.

67 Sisson, S.A., Fan, Y., and Tanaka, M.M. (2007) Sequential Monte Carlo without likelihoods. *Proc. Natl. Acad. Sci. USA*, 104 (4), 1760–1765.

68 Andrieu, C., Doucet, A., and Lee, A. (2012) Discussion of "constructing summary statistics for approximate Bayesian computation: semi-automatic approximate Bayesian computation" by Fearnhead and Prangle. *J. Royal Stat. Soc. B*, 74 (3), 451–452.

69 Didelot, X., Everitt, R.G., Johansen, A.M., and Lawson, D.J. (2011) Likelihood-free estimation of model evidence. *Bayesian Anal.*, 6 (1), 49–76.

70 Marin, J.-M., Pillai, N., Robert, C.P., and Rousseau, J. (2014) Relevant statistics for Bayesian model choice. *J. Royal Stat. Soc. B*, 76 (5), 833–859.

71 Johansen, A.M., Doucet, A., and Davy, M. (2008) Particle methods for maximum likelihood parameter estimation in latent variable models. *Stat. Comput.*, 18 (1), 47–57.

72 Rubenthaler, S., Rydén, T., and Wiktorsson, M. (2009) Fast simulated annealing in $\mathbb{R}^d$ with an application to maximum likelihood estimation in state-space models. *Stoch. Proc. Appl.*, 119 (6), 1912–1931.

73 Finke, A. (2015) On extended state-space constructions for Monte Carlo methods. Ph.D. thesis. University of Warwick.

74 Cérou, F., Del Moral, P., Le Gland, F., and Lezaud, P. (2006) Genetic genealogical models in rare event analysis. *ALEA: Lat. Am. J. Probab. Math. Stat.*, 1, 181–203.

75 Del Moral, P. and Garnier, J. (2005) Genealogical particle analysis of rare events. *Ann. Appl. Probab.*, 15 (4), 2496–2534.

76 Johansen, A.M., Del Moral, P., and Doucet, A. (2006) *Sequential Monte Carlo Samplers for Rare Events.* Proceedings of the 6th International Workshop on Rare Event Simulation, Bamberg, Germany, pp. 256–267.

77 Cérou, F., Del Moral, P., Furon, T., and Guyader, A. (2012) Sequential Monte Carlo for rare event estimation. *Stat. Comput.*, 22 (3), 795–808.

78 Chan, H.P. and Lai, T.L. (2013) A general theory of particle filters in hidden Markov models and some applications. *Ann. Stat.*, 41 (6), 2877–2904.

79 Lee, A. and Whiteley, N. (2018) Variance estimation in the particle filter. *Biometrika*, 105 (1), 609–625.

80 Du, Q. and Guyader, A. (2019) Variance estimation in adaptive sequential Monte Carlo. arXiv:1909.13602.

81 Olsson, J. and Douc, R. (2019) Numerically stable online estimation of variance in particle filters. *Bernoulli*, 25 (2), 1504–1535.

82 Koskela, J., Jenkins, P., Johansen, A.M., and Spanò, D. (2020) Asymptotic genealogies of interacting particle systems with an application to sequential Monte Carlo. *Ann. Stat.*, 48 (1), 560–583.

83 Murray, L., Lee, A., and Jacob, P. (2016) Parallel resampling in the particle filter. *J. Comput. Graph. Stat.*, 25 (3), 789–805.

84 Lee, A. and Whiteley, N. (2016) Forest resampling for distributed sequential Monte Carlo. *Stat. Anal. Data Min.*, 9 (4), 230–248.

85 Whiteley, N., Lee, A., and Heine, K. (2016) On the role of interaction in sequential Monte Carlo algorithms. *Bernoulli*, 22 (1), 494–429.

86 Lindsten, F., Johansen, A.M., Naesseth, C.A., *et al.* (2017) Divide and conquer with sequential Monte Carlo samplers. *J. Comput. Graph. Stat.*, 26 (2), 445–458.

87 Gerber, M. and Chopin, N. (2015) Sequential quasi Monte Carlo. *J. Royal Stat. Soc. B*, 77 (3), 509–579.

7

Markov Chain Monte Carlo Methods, A Survey with Some Frequent Misunderstandings

Christian P. Robert[1,2] *and Wu Changye*[1*]

[1] *Université Paris Dauphine PSL, Paris, France*
[2] *University of Warwick, Coventry, UK*

1 Introduction

When analyzing a complex probability distribution or facing an unsolvable integration problem, as in most of Bayesian inference, Monte Carlo methods on a large variety of solutions are mostly based on the ability to simulate a sequence of random variables and subsequently call for the law of large numbers (LLN). Techniques based on the simulation of Markov chains are a special case of these methods, in which the current simulation value (and its probability) is used to switch to a different simulation value (hence the Markovian nature of such techniques). While the working principle of Markov chain Monte Carlo (MCMC) methods was proposed almost as early as the original Monte Carlo algorithms, the variety and efficiency of these methods has grown significantly since Gelfand and Smith [1] (re)introduced them to the statistical community and in particular to its Bayesian component [2].

Given a likelihood function defined as a function of the parameter associated with the probability mass function or density function of the observations (x^{obs}), $L(\theta|x^{\text{obs}})$, a Bayesian approach means relying on a so-called prior distribution on the parameters, from which the resulting posterior distribution defined by

$$\pi(\theta|x^{\text{obs}}) = \frac{L(\theta|x^{\text{obs}})\pi(\theta)}{\int_{\Theta} L(\theta|x^{\text{obs}})\pi(\theta')\mathrm{d}\theta'} \tag{1}$$

is derived. The denominator is sometimes called the marginal likelihood and is denoted by $m_\pi(x^{\text{obs}})$. While most Bayesian procedures are by nature uniquely defined, the practice of this theory exposes various computational problems.

*This chapter is partly based on material found in the PhD thesis of the second author, which he successfully defended in 2018 at Université Paris Dauphine, under the supervision of the first author. Another related book chapter by the same authors is scheduled to appear in Mengersen, Pudlo and Robert (2020). The first author is grateful to Antonietta Mira for her comments. Ce travail a bénéficié d'une aide de l'Etat gérée par l'Agence Nationale de la Recherche au titre du programme d'Investissements d'Avenir portant la référence ANR-19-P3IA-0001.

Computational Statistics in Data Science.
Edited by Walter W. Piegorsch, Richard A. Levine, Hao Helen Zhang and Thomas C. M. Lee.
 ISBN 978-1-11956107-1

"Why is it necessary to sample from the posterior distribution if we already know the posterior distribution?" [cross-validated:307882]

When one states that we "know the posterior distribution", the meaning of "we know" is unclear. "Knowing" a function of θ to be proportional to the posterior density, namely

$$\pi(\theta)f(x^{\text{obs}}|\theta) \tag{2}$$

as, for instance, the completely artificial following target

$$\pi(\theta|x) \propto \exp\{-||\theta - x||^2 - ||\theta + x||^4 - ||\theta - 2x||^6\},\ x, \theta \in \mathbb{R}^{18}$$

does not mean a quick resolution for approximating the following entities:

- the posterior expectation of a function of θ, for example, $\mathbb{E}[\mathfrak{h}(\theta)|x]$, posterior mean that operates as a Bayesian estimator under standard losses;
- the optimal decision under an arbitrary utility function, decision that minimizes the expected posterior loss;
- a 90% or 95% range of uncertainty on the parameter(s), a subvector of the parameter(s), or a function of the parameter(s), aka higher posterior density region (HPD) region $\{h = \mathfrak{h}(\theta);\ \pi^{\mathfrak{h}}(h) \geq \underline{h}\}$, where $\pi^{\mathfrak{h}}(\cdot)$ denotes the marginal posterior distribution of $\mathfrak{h}$;

The above quantities are only examples of the infinity of usages made of a posterior distribution. In all cases but the most simple ones, the answers are mathematically derived from the posterior but cannot be found without analytical or numerical steps, like Monte Carlo and MCMC methods.

The existing solutions to this computing challenge are roughly divisible into deterministic and stochastic approaches. The former includes Laplace's approximation, expectation propagation [3], and Bayesian variational methods [4]. The resulting approximation error then is usually unknown and cannot be corrected from additional calculations. The alternative of Monte Carlo methods leads to approximations that converge when the computational effort becomes infinite. We focus on the latter.

"Why is variational Bayesian mixture model an alternative to MCMC? What are the similarities?" [cross-validated:386093]

Variational Bayes inference is a weak form of empirical Bayesian inference [2], in the sense that it estimates some parameters of the prior from the data for a simplified version of the true posterior, most often of a conjugate form. The variational Bayes approach to a Bayesian latent variable model [4] is producing a pseudoposterior distribution on the parameters of the model, including the latent variables $\mathbf{Z}$, by imposing a certain dependence structure (or graphical model) and estimating its hyperparameters of this model by a maximizing algorithm akin to the expectation-maximisation (EM) algorithm [5].

There is thus no clear direct connection with MCMC, since the variational Bayes posterior is made of standard distributions, thus does not require simulation but has hyperparameters that must be derived by an optimization program, hence the call to an EM-like algorithm.

2 Monte Carlo Methods

Monte Carlo approximations [6] are based on the LLN in the sense that an integral like

$$I_h := \mathbb{E}_P(h(X))$$

is the limiting value of an empirical average

$$\frac{1}{N}\sum_{i=1}^{N} h(x_i) \xrightarrow[N\to\infty]{P} I_h$$

when $x_1, x_2, \cdots$, are i.i.d. random variables with probability distribution P. In practice, the sample $x_1, x_2, \cdots$, is produced by a pseudorandom generator [7].

"How can you draw samples from the posterior distribution without first knowing the properties of said distribution?" [cross-validated:307882]

In Bayesian settings, Monte Carlo methods are based on the assumption that the product (2) can be numerically computed (hence is known) for a given (θ, x^{obs}), where x^{obs} denotes the observation, $\pi(\cdot)$ the prior, and $f(x^{\text{obs}}|\theta)$ the likelihood. This does not imply an in-depth knowledge about this function of θ. Still, from a mathematical perspective the posterior density is completely and entirely determined by Bayes' formula, hence derived from the product (2). Thus, it is not particularly surprising that simulation methods can be found using solely the input of the product (2). The most amazing feature of Monte Carlo methods is that some methods such as MCMC algorithms do not formally require anything further than this computation of the product, when compared with accept–reject algorithms, for instance, which call for an upper bound. A related software such as Stan [8] operates on this input and still delivers high-end performances with tools such as no-U-turn sampler (NUTS) [9] and Hamiltonian Monte Carlo (HMC), including numerical differentiation.

The normalizing constant of the posterior (1) is not particularly useful for conducting Bayesian inference in that, where one to "know" its exact numerical value in addition to the product (2), $\mathfrak{Z} = 3.17\ 232\ 10^{-23}$ say, one would not have made any progress toward finding Bayes estimates or credible regions. (The only exception when this constant matters is in conducting Bayesian model comparison.)

"If we do not know the normalising constant for a posterior distribution, why does it imply we can only sample dependent draws?" [cross-validated:182525]

This is mostly unrelated: missing normalizing constant and dependence have no logical connection. That is to say, one may have a completely defined density and yet be unable to produce i.i.d. samples, or one may have a density with a missing constant and nonetheless be able to produce i.i.d. samples.

If one knows a density $f(\cdot)$ up to a normalizing constant, $f(x) \propto p(x)$, there are instances when one can draw independent samples, using, for instance [accept–reject algorithms][1]: if one manages to find another density g such that

1. one can simulate from g
2. there exists a known constant M such that

$$p(x) \leq Mg(x)$$

then the algorithm

```
Repeat
  simulate y~g(y)
  simulate u~U(0,1)
until u<p(y)/Mg(y)
```

produces i.i.d. simulations from f, even though one only knows p.

For instance, if one wants to generate a Beta $\mathcal{B}e(a+1, b+1)$ distribution from scratch (with $a, b \geq 1$), the density up to a normalizing constant is

$$p(x) = x^a(1-x)^b \mathbb{I}_{(0,1)}(x)$$

which is bounded by 1. Thus, we can use $M = 1$ and $g(x) = 1$, the density of the uniform distribution in an accept–reject algorithm that produces a sample (with random size) that is i.i.d. from the Beta $\mathcal{B}e(3.3, 4.4)$ distribution. In practice, finding such a g may prove a formidable task, and an easier approach is to produce simulations (asymptotically) from f by MCMC algorithms.

When direct simulation from P, for instance a posterior distribution, is impossible, alternative stochastic solutions must be sought. A wide collection of such methods goes under the name of importance sampling, relying on a convenient if somewhat arbitrary auxiliary distribution.

"What is importance sampling?" [cross-validated:254114]

The intuition behind importance sampling is that a well-defined integral, such as

$$\mathfrak{I} = \int_{\mathfrak{X}} h(x)\,\mathrm{d}x$$

can be expressed as an expectation for a wide range of probability distributions with density f:

$$\mathfrak{I} = \mathbb{E}_f[H(X)] = \int_{\mathfrak{X}} H(x)f(x)\,\mathrm{d}x$$

where H is determined by h and f. (Note that $H(\cdot)$ is usually different from $h(\cdot)$.) The choice

$$H(x) = h(x)/f(x)$$

leads to the equalities $H(x)f(x) = h(x)$ and $\mathfrak{I} = \mathbb{E}_f[H(X)]$ – under some restrictions on the support of f, meaning $f(x) > 0$ when $h(x) \neq 0$. Hence, there is no unicity in the representation of an integral as an expectation, but on the opposite an infinite array of such representations, some of which are better than others once a criterion to compare them is adopted. For instance, it may mean choosing f toward reducing the variance of the estimator.

Once this elementary property is understood, the implementation means simulating – via a pseudorandom generator – an i.i.d. sample $(x_1, \ldots, x_n)$ distributed from f and using the average of the $H(x_i)$ as an unbiased approximation, $\hat{\mathfrak{I}}$. Depending on the choice of the distribution f, this estimator $\hat{\mathfrak{I}}$ may or may not have a finite variance. However, there always exist choices of f that allow for a finite variance and even for an arbitrarily small variance (albeit those choices may be unavailable in practice). And there also exist choices of f that make the importance sampling estimator $\hat{\mathfrak{I}}$ a very poor approximation of $\mathfrak{I}$. This includes all the choices where the variance gets infinite, even though Chatterjee and Diaconis [10] compare importance samplers with infinite variance. Figure 1 is taken from the first author's blog discussion of the paper and illustrates the poor convergence of infinite variance estimators.

A decisive appeal in using importance sampling is that the weight function w can be known up to a multiplicative constant, which most often occurs when sampling from a given posterior in Bayesian inference. Indeed, the multiplicative constant can be estimated by $\frac{1}{N}\sum_{i=1}^{N} w(X_i)$, and it is straightforward to deduce that the normalized (if biased) estimator

$$\sum_{i=1}^{N} h(X_i)w(X_i) / \sum_{i=1}^{N} w(X_i)$$

consistently approximates the integral of interest.

The importance distribution Q selected for the associated approximation significantly impacts the quality of the method. The sequence of pseudorandom variables that stands at the core of the method remains at this stage i.i.d., but the following section describes a new class of sampling algorithms, based on Markov chains, which produce correlated samples to approximate the target distribution or the integrals of interest.

The term "sampling" is somewhat confusing in that it does not intend to provide samples from a given distribution.

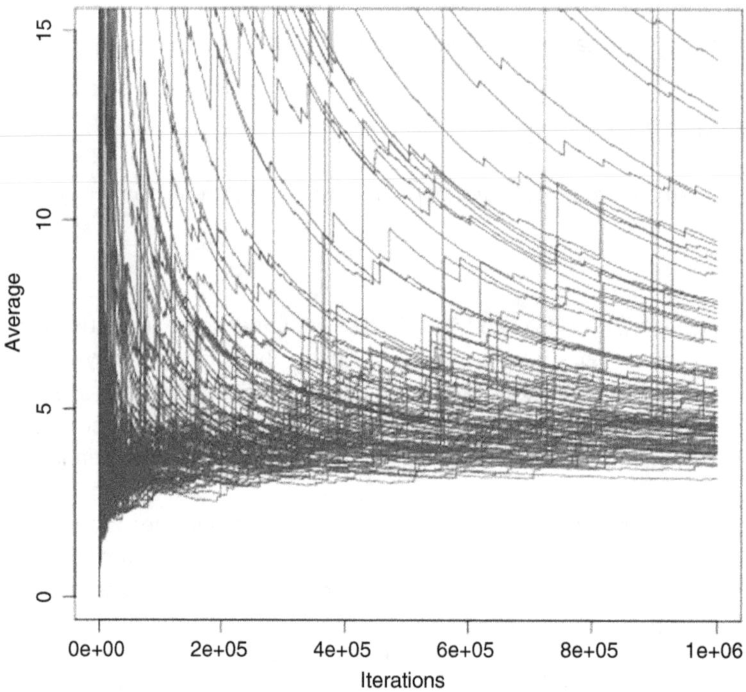

Figure 1 Importance sampling with importance distribution of an exponential $\mathcal{E}(1)$ distribution, target distribution, a $\mathcal{E}(1/10)$ distribution, and function of interest $h(x) = x$. The true value of the expectation is equal to 10. The 100 curves produced on this graph correspond to repeated simulations experiments, with each curve describing the evolution of the empirical average of the $h(X_i)$'s with the number of iterations. In this particular case, the importance sampling estimators have infinite variance.

"Can importance sampling be used as an actual sampling mechanism?" [cross-validated:436453]

The difficulty is that the resulting (re)sample is not marginally distributed from p. While

$$\mathbb{E}_q[h(Y)p(Y)/q(Y)] = \mathbb{E}_p[h(Y)]$$

for any integrable function $h(\cdot)$, weighting and resampling an i.i.d. sample $(Y_1, \dots, Y_n)$ from q does not produce a sample distributed from p, even marginally. The reason for the discrepancy is that the weighting-resampling step implies dividing the $p(Y_i)/q(Y_i)$ by the random sum of the weights, that is, the index i is selected with probability

$$p(Y_i)/q(Y_i)/\sum_j p(Y_j)/q(Y_j)$$

which modifies the marginal distribution of the resampled rvs, especially when the sum has an infinite variance.

Figure 2 provides an illustration when p is the density of a Student's t_5 distribution with mean 3, and q is the density of a standard Normal distribution. The original Normal sample fails to cover the rhs of the tail of the Student's t and hence that the weighted-resampled sample cannot recover with a manageable number of simulations. Obviously, as shown in Figure 3, when the target q has fatter tails than p, the method converges reasonably fast.

"What is the difference between Metropolis Hastings, Gibbs, Importance, and Rejection sampling?" [cross-validated:185921]

These methods all produce samples from a given distribution, with density f say, either to get an idea about this distribution or to solve an integration or optimization problem related with f. Instances include finding the value of

$$\int_{\mathcal{X}} h(x)f(x)\mathrm{d}x \qquad h(\mathcal{X}) \subset \mathbb{R}$$

or the mode of the distribution of $h(\mathcal{X})$ when $\mathcal{X} \sim f(x)$ or a quantile of this distribution.

Here are a few generic points that do not cover the complexity of the issue:

1. *Accept–reject methods* are intended to provide an i.i.d. sample from f, as explained above. The *pros* are that there is no approximation in the method: the outcome is

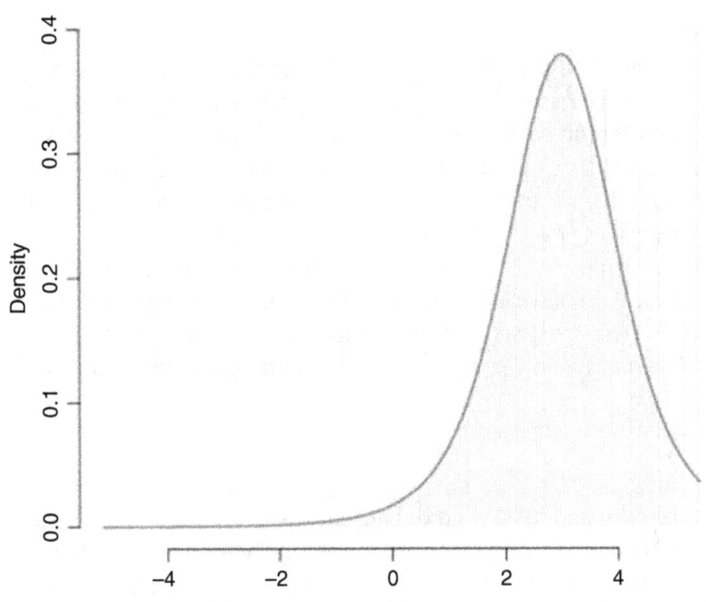

Figure 2 Failed simulation of a Student's t_5 distribution with mean 3 when simulating 10^7 realizations from a standard Normal importance distribution (with thinner tails).

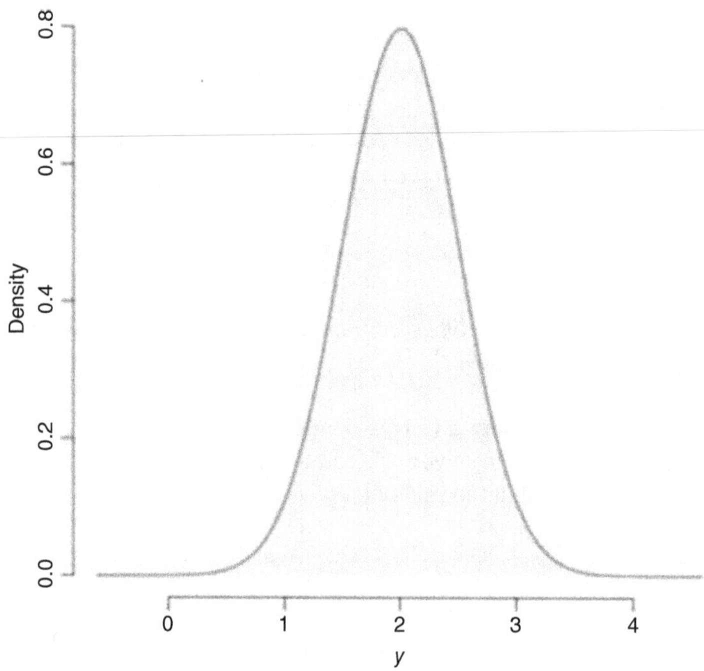

Figure 3 Recovery of a Normal $\mathcal{N}(2, 1/\sqrt{2})$ distribution when simulating 10^7 realizations from a standard Normal importance distribution (with fatter tails).

truly an i.i.d. sample from f. The *cons* are many: (i) designing the algorithm by finding an envelope of f that can be generated may be very costly in human time; (ii) the algorithm may be inefficient in computing time, that is, requires many uniforms to produce a single x; and (iii) those performances are decreasing with the dimension of X. In short, such methods cannot be used for simulating one or a few simulations from f unless they are already available in a computer language such as R.

2. *MCMC methods* are extensions of i.i.d. simulation methods when i.i.d. simulation is too costly. They produce a sequence of simulations $(x_t)_t$ whose limiting distribution is the distribution f. The *pros* are that (i) less information about f is needed to implement the method; (ii) f may be only known up to a normalizing constant or even as an integral

$$f(x) \propto \int_{Z} \tilde{f}(x, z)\mathrm{d}z$$

and still be associated with an MCMC method; (iii) there exist generic MCMC algorithms to produce simulations $(x_t)_t$ that require very little calibration; and (iv) dimension is less of an issue as large dimension targets can be broken into conditionals of smaller dimension (as in Gibbs sampling). The *cons* are that (i) the simulations $(x_t)_t$ are correlated, hence less informative than i.i.d. simulations; (ii) the validation of the method is only asymptotic, hence there is an approximation in considering x_t for a fixed t as a realization of f; (iii) convergence to f (in t) may be so slow that for all practical purposes the algorithm does not converge; and (iv) the universal validation of the method means there is an infinite number of potential implementations, with an equally infinite range of efficiencies.

3. *Importance sampling methods* are originally designed for integral approximations, namely generating from the wrong target $g(x)$ and compensating by an importance weight $f(x)/g(x)$. The resulting sample is thus weighted, which makes the comparison with the above awkward. Importance sampling can be turned into importance sampling resampling using an additional resampling step based on the weights, as shown in Figure 4 for a simulation based on a Beta(3,4) importance function still failing to produce an exact simulation from the target as discussed above. The *pros* of importance sampling are that (i) generation from an importance target g can be cheap and recycled for different targets f; (ii) the "right" choice of g can lead to huge improvements compared with regular or MCMC sampling; (iii) importance sampling is more amenable to numerical integration improvement, like for instance quasi-Monte Carlo (qMC) integration; and (iv) it can be turned into adaptive versions such as population Monte Carlo and sequential Monte Carlo. The *cons* are that (i) resampling induces inefficiency (which can be partly corrected by reducing the

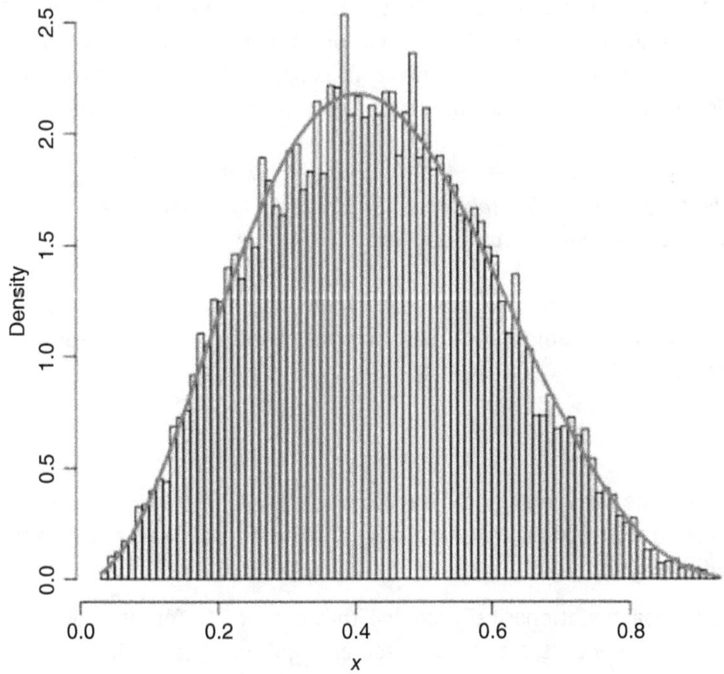

Figure 4 Histogram of 9781 simulations of a $\mathcal{B}e(3.3, 4.4)$ distribution with the target density in superposition. The sample size 9781 is a random realization, due to the underlying resampling mechanism.

noise as in systematic resampling or qMC); (ii) the "wrong" choice of g can lead to huge losses in efficiency and even to infinite variance; (iii) importance has trouble facing large dimensions, and its efficiency diminishes quickly with the dimension; and (iv) the method may be as myopic as local MCMC methods in missing important regions of the support of f.

A final warning is that *there is no such thing as an optimal simulation method.* Even in a specific setting like approximating an integral $\mathcal{I}$, costs of designing and running different methods intrude as to make a global comparison very delicate, if at all possible, while, from a formal point of view, *they can never beat the zero variance answer of returning the constant "estimate."* For instance, simulating from f is very rarely if ever the best option. This does not mean that methods cannot be compared, but that there always is a possibility for an improvement, which usually comes with additional costs.

3 Markov Chain Monte Carlo Methods

MCMC algorithms are now standard computing tools for analyzing Bayesian complex models [1] even though practitioners may still face difficulties with its implementations. The concept behind MCMC is quite simple in that it creates a sequence of dependent variables that converge (in distribution) to the distribution of interest (also called target). In that sense, MCMC algorithms are robust or universal, as opposed to the most standard Monte Carlo methods which require direct simulations from the target distribution.

"Is Markov chain-based sampling the "best" for Monte Carlo sampling? Are there alternative schemes available?" [cross-validated:131455]

There is no reason that MCMC sampling is the "best" Monte Carlo method! Usually, it is on the opposite *worse* than i.i.d. sampling, at least in terms of variance of the resulting Monte Carlo estimators

$$\frac{1}{T}\sum_{t=1}^{T} h(X_t).$$

Indeed, while this average converges to the expectation $\mathbb{E}_\pi[h(X)]$ when π is the stationary and limiting distribution of the Markov chain $(X_t)_t$, there are at least two drawbacks in using MCMC methods:

1. The chain needs to "reach stationarity," meaning that it needs to forget about its starting value X_0. In other words, t must be "large enough" for X_t to be distributed from π. Sometimes "large enough" may exceed by several orders of magnitude the computing budget available for the experiment.

2. The values X_t are correlated, leading to an asymptotic variance that involves

$$var_\pi(X) + 2\sum_{t=1}^{\infty} \text{cov}_\pi(X_0, X_t)$$

which generally exceeds $var_\pi(X)$ and hence requires longer simulations than for an i.i.d. sample, as well as more involved evaluation techniques.

This being said, MCMC is very useful for handling settings where regular i.i.d. sampling is impossible or too costly and where importance sampling is quite difficult to calibrate, in particular because of the dimension of the random variable to be simulated. However, sequential Monte Carlo methods [11] like particle filters may be more appropriate in dynamical models, where the data comes by bursts that need immediate attention and may even vanish (i.e., cannot be stored) after a short while.

From the early 1950s, MCMC methods [12–14] have been utilized to handle complex target distributions by simulation, where the meaning of complexity depends on the target density, the size of the associated data, the dimension of the object to be simulated, or the allocated budget. For instance, the density $p(x)$ is only expressed as a multidimensional integral that is analytically intractable,

$$p(x) = \int \omega(x, \xi)\mathrm{d}\xi$$

An evaluation of this density requires the simulation of the whole vector (x, ξ).

In cases when ξ has its dimension at least as large as the dimension of the data, such a simulation involves a significant increase in the dimension of the simulated object and hence leads to more severe computational difficulties, starting with manipulating the extended target $\omega(x, \xi)$. An MCMC algorithm provides an alternative solution to this computational issue through a simulated Markov chain evolving in the augmented space without requiring further information on the density p.

"What is the connection between Markov chain and Markov chain Monte Carlo?" [cross-validated:169518]

The connection between both concepts is that MCMC methods rely on Markov chain theory to produce simulations and Monte Carlo approximations from a complex target distribution π.

In practice, these simulation methods output a sequence $X_1, \ldots, X_N$ that is a Markov chain, that is, such that the distribution of X_i given the whole past $\{X_{i-1}, \ldots, X_1\}$ only depends on X_{i-1}. In other words,

$$X_i = f(X_{i-1}, \epsilon_i)$$

where f is a function specified by the algorithm, and the target distribution π and the ϵ_i's are i.i.d. The (ergodic) theory guarantees that X_i converges (in distribution) to π as i gets to ∞.

The easiest example of an MCMC algorithm is the slice sampler: at iteration i of this algorithm, do

1. simulate $\epsilon_i^1 \sim \mathcal{U}(0,1)$
2. simulate $X_i \sim \mathcal{U}(\{x; \pi(x) \geq \epsilon_i^1 \pi(X_{i-1})\})$ (which amounts to generating a second independent ϵ_i^2)

For instance, if the target is a Normal $\mathcal{N}(0,1)$ distribution[1] the above translates as

1. simulate $\epsilon_i^1 \sim \mathcal{U}(0,1)$
2. simulate $X_i \sim \mathcal{U}(\{x; x^2 \leq -2\log(\sqrt{2\pi}\epsilon_i^1\})$, that is,

$$X_i = \pm\epsilon_i^2\{-2\log(\sqrt{2\pi}\epsilon_i^1)\varphi(X_{i-1})\}^{1/2}$$

with $\epsilon_i^2 \sim \mathcal{U}(0,1)$

Figure 5 is a representation of the output, showing the right fit to the $\mathcal{N}(0,1)$ target and the evolution of the Markov chain (X_i). And Figure 6 zooms on the evolution of the Markov chain $(X_i, \epsilon_i^1 \pi(X_i))$ over the last 100 iterations, which follows vertical and horizontal moves of the Markov chain under the target density curve.

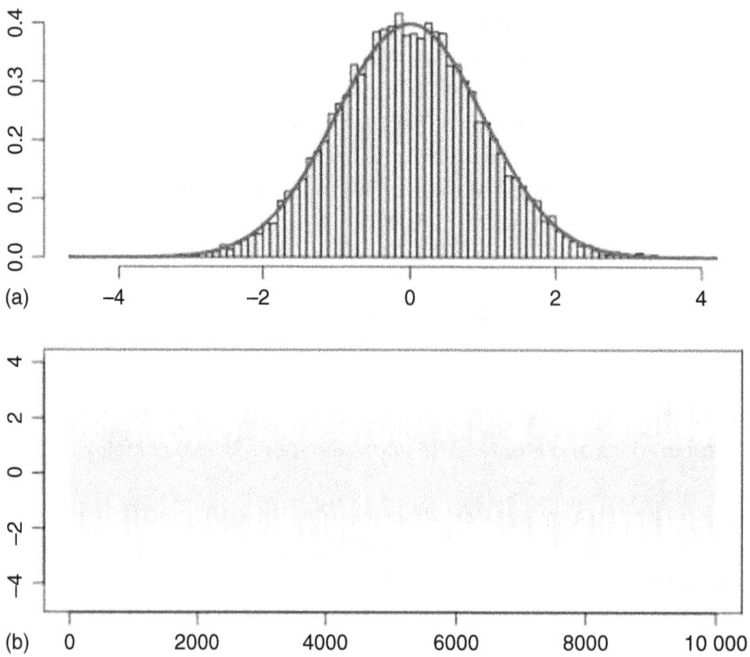

Figure 5 (a) Histogram of 10^4 iterations of a slice sampler with a Normal $\mathcal{N}(0,1)$ target; (b) sequence (X_i).

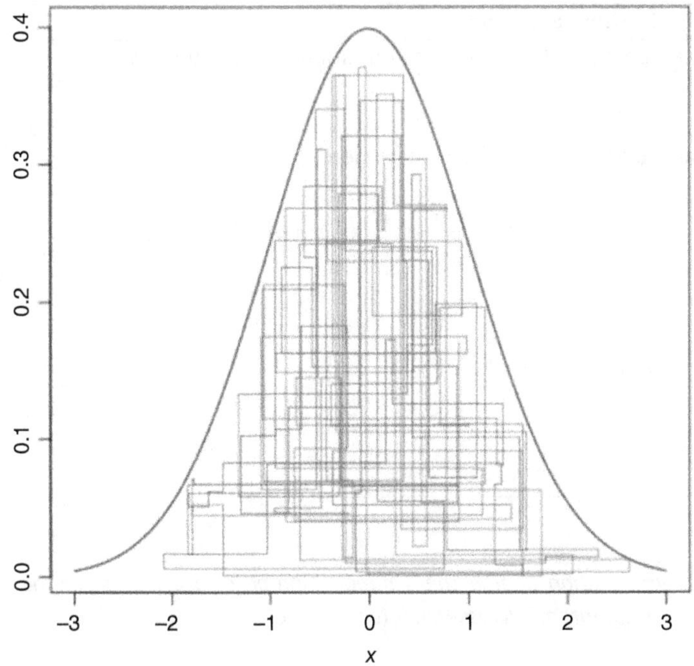

Figure 6 100 last moves of the above slice sampler.

The validation of the method [6] proceeds by establishing that the resulting Markov chain is ergodic [15], meaning that it converges to the distribution corresponding to π, making the starting value of the chain irrelevant. Akin to basic Monte Carlo methods, MCMC samples (usually) enjoy standard limit theorems.

3.1 Metropolis–Hastings Algorithms

The Metropolis–Hastings[2] algorithm is the "Swiss knife" of MCMC methods in that it offers a form of universal solution to the construction of an appropriate Markov chain. The algorithm requires a proposal distribution, with density $q(x'|x)$ and proceeds one step at a time based on simulations proposed from this distribution and accepted or rejected by a Metropolis–Hastings ratio, as described in Algorithm 1.

The accept–reject step in this algorithm is fundamental in that it turns p into its stationary distribution, assuming that the resulting Markov kernel is irreducible, provided the chain $(X_n)_n$ is irreducible, meaning it has a positive probability of hitting any part of the support of p on a finite number of steps. Stationarity follows from the transition satisfying the detailed balance condition, corresponding to the chain being reversible in time [6]. A special case when q is symmetric, that is, $q(x|y) = q(y|x)$, is called random walk MCMC, and the acceptance probability only involves the targeted p.

Algorithm 1. Metropolis–Hastings algorithm

Input: starting point X_0, proposal distribution q and number of iterations N.
for $n = 1, 2, \cdots, N$ **do**
Sample $X' \sim q(\cdot|X_{n-1})$
Compute the acceptance probability $\alpha(X_{n-1}, X')$, where

$$\alpha(X_{n-1}, X') = \min\left\{1, p(X')q(X_{n-1}|X')\Big/p(X_{n-1})q(X'|X_{n-1})\right\}$$

Sample $U \sim \mathcal{U}[0, 1]$;
if $U < \alpha(X_{n-1}, X')$
$X_n \rightarrow X'$
else
$X_n \rightarrow X_{n-1}$
endif
end for

"What is the deeper intuition behind the symmetric proposal distribution in the Metropolis–Hastings algorithm?" [cross-validated:262216]

1. the Normal and Uniform are symmetric probability density functions themselves; is this notion of "symmetry" the same as the "symmetry" above?
2. is there an intuitive way of seeing the deeper meaning behind the symmetry formula above?

Both Normal and Uniform distributions are symmetric around their mean. But the symmetry in Metropolis–Hastings signifies that $q(x|y) = q(y|x)$, which makes the ratio cancel in the Metropolis–Hastings acceptance probability. If one uses a Normal distribution not centered at the previous value in the Metropolis–Hastings proposal (as, e.g., in the Langevin version), the Normal distribution remains symmetric *as a distribution*, but the proposal distribution is no longer symmetric, and hence it must appear in the Metropolis–Hastings acceptance probability.

There is no particular depth in this special symmetric case, it simply makes life easier by avoiding the ratio of the proposals. It may save time or it may avoid computing complex or intractable densities. Note also that the symmetry depends on the parameterization of the model: if one changes the parameterization, a Jacobian appears and kills the symmetry.

"The independent Metropolis algorithm using the proposal $X' \sim f_V(x)$ should have $\alpha(X_0, X'_0) = 1$ and hence the chain always equal to X'_0." [cross-validated:396704]

The confusion stems from a misunderstanding of the notation $X' \sim f_V$, which means both (i) X' is a random variable with density f_V and (ii) X' is created by a pseudorandom generation algorithm that reproduces a generation of a random variable with density f_V. Each time a generation $X_i' \sim f_V$ occurs in the algorithm, a new realization of a random variable with density f_V occurs, which is independent of all previous realizations, hence different from these previous realizations. Equivalently, stating that the X_i' are all identically distributed from the same distribution f_V does not mean that their realizations all are numerically identical.

The starting point of the Metropolis–Hastings algorithm is arbitrary, either fixed $X_0 = 0$, for instance, or random, for instance $X_0 \sim f_V$ (a notation meaning that X_0 is distributed from f_V). This starting value *is always accepted*. For $i = 1$, one generates $X_1' \sim f_V$ (meaning that X_1' is distributed from f_V, independently and thus different from X_0)

$$X_1 = \begin{cases} X_1' & \text{if } U_1 \le \alpha_1 = \min\left(\frac{f_Y(X_1')}{f_V(X_1')}\frac{f_V(X_0)}{f_Y(X_0)}, 1\right) \\ X_0 & \text{if } U_1 > \alpha_1 \end{cases}$$

and $\alpha_1 \neq 1$ in general. Hence, sometimes X_1 is accepted and sometimes not. The same applies to the following steps. To make a toy illustration on how the algorithm applies, take f_V to be the density of a $\mathcal{N}(0, 1)$ distribution and f_Y to be the density of a $\mathcal{N}(1, 1)$ distribution. A sequence of i.i.d. generations from f_V is, for instance (by a call to R rnorm),

$$0.45735433, -0.99178415, -1.08312586, -0.85762451, 0.92186197, \\ -0.50442298, \ldots$$

(note that they are all different) and a sequence of generations from $\mathcal{U}$ is, for instance (by a call to R runif),

$$0.441328, 0.987837, 0.386258, 0.316593, 0.195910, 0.2772669, \ldots$$

(note that they are all different). Applying the algorithm with starting value $X_0 = 0$ means considering

$$\frac{f_Y(X_1')}{f_V(X_1')}\frac{f_V(X_0)}{f_Y(X_0)} = 0.9582509/0.6065307 = 1.579889 > 1$$

which implies that $X_1 = X_1' = 0.45735433$. Then,

$$\frac{f_Y(X_2')}{f_V(X_2')}\frac{f_V(X_1)}{f_Y(X_1)} = 0.2249709/0.9582509 = 0.2347724 < U_2 = 0.987837$$

which implies that $X_2 = X_1$. The algorithm can be applied step by step to the sequences provided above, which leads to

$$\frac{f_Y(X_3')}{f_V(X_3')}\frac{f_V(X_2)}{f_Y(X_2)} = 0.2053581/0.9582509 = 0.2143051 < U_3 \qquad Z_3 = Z_1$$

$$\frac{f_Y(X_4')}{f_V(X_4')}\frac{f_V(X_3)}{f_Y(X_3)} = 0.2572712/0.9582509 = 0.2684800 < U_4 \qquad Z_4 = Z_1$$

$$\frac{f_Y(X_5')}{f_V(X_5')}\frac{f_V(X_4)}{f_Y(X_4)} = 1.5247980/0.9582509 = 1.591230 > 1 \qquad Z_5 = V_5$$

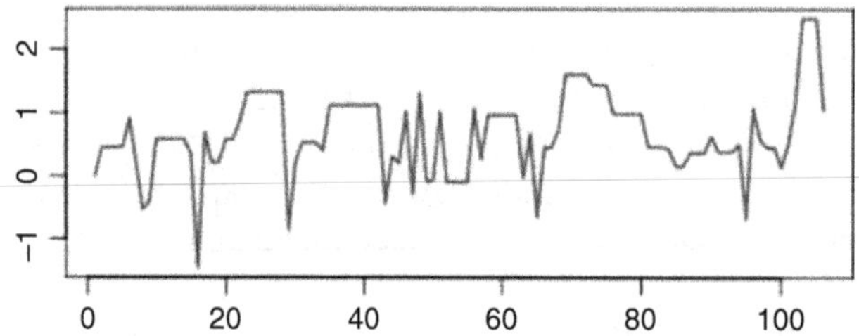

Figure 7 Independent Metropolis sequence with a proposal f_V equal to the density of a $\mathcal{N}(0, 1)$ distribution and a target f_Y being the density of a $\mathcal{N}(1, 1)$ distribution.

producing a sequence as in Figure 7 (notice the flat episodes in the graph, which correspond to a sequence of rejections).

As a final remark, the only potentially confusing part in the description in Casella and Berger (1990) is the very first sentence where the random variables Y and V are not needed. It could have been clearer to state "Let f_Y and f_V be two densities with common support."

Since the purpose of MCMC methods such as the Metropolis–Hastings algorithm is to simulate realizations from p, their performances are highly variable. These obviously depend on the connection between p and q. For instance, the Metropolis–Hastings algorithm is an i.i.d. sampler when $q(\cdot|X_n) = p(\cdot)$, a choice that is rarely available. Although it may happen that the Markov chain (X_n) achieves negative correlations between successive and further terms of the series, making it *more efficient* than i.i.d. sampling [18], it is more common that there exists a positive covariance between simulations (sometimes for all transforms [19]). This feature means a lesser efficiency of the algorithm which thus requires a greater number of simulations to achieve the same accuracy as the i.i.d. approach (regardless of the deficiencies in computing time). In general, the MCMC algorithm may require a large number of iterations to escape the attraction of the starting point X_0 and to converge. There is a real danger that some versions of these algorithms do not converge within the allotted time (in practice if not in theory).

"What is the Metropolis–Hastings acceptance ratio for a truncated proposal?" [cross-validated:345291]

If a Metropolis–Hastings algorithm uses a truncated Normal as proposal, for example, the positive Normal

$$\mathcal{N}^+(\mu_{t-1}, \sigma^2)$$

the associated Metropolis–Hastings acceptance ratio is

$$\frac{\pi(\mu')}{\pi(\mu_{t-1})} \times \frac{\varphi(\{\mu_{t-1} - \mu'\}/\sigma)}{\varphi(\{\mu' - \mu_{t-1}\}/\sigma)} \times \frac{\Phi(\mu_{t-1}/\sigma)}{\Phi(\mu'/\sigma)}$$

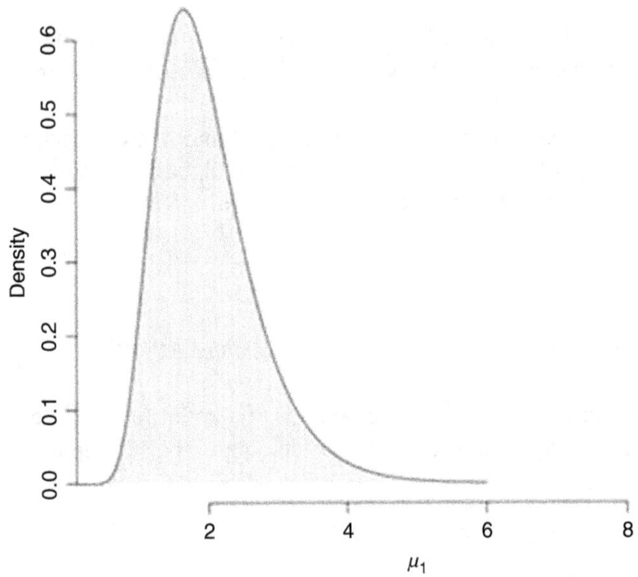

Figure 8 Fit of a Metropolis sample of size 10^4 to a target when using a truncated Normal proposal.

where $\mu' \sim \mathcal{N}^+(\mu_{t-1}, \sigma^2)$ is the proposed value, and π denotes the target of the simulation (e.g., the posterior distribution). This ratio simplifies into

$$\frac{\pi(\mu')}{\pi(\mu_{t-1})} \times \frac{\Phi(\mu_{t-1}/\sigma)}{\Phi(\mu'/\sigma)}$$

hence the truncation impacts the Metropolis–Hastings acceptance ratio.

Figure 8 provides an illustration for the target density

$$\pi(\mu) \propto \exp\{-(\log \mu - 1)^2\} \; \exp\{-(\log \mu - 3)^4/4\}$$

when using $\sigma = 0.1$ as the scale in the truncated Normal.

"What to do when rejecting a proposed point in MCMC?" [cross-validated:123113]

The validation of the Metropolis–Hastings algorithm relies on repeating the current value in the Markov chain if the proposed value is rejected. One should not consider the list of *accepted* points as one's sample but instead the Markov chain with transition

$$\begin{aligned} X_{t+1} &= Y_{t+1} && \text{if } U_{t+1} \le \pi(Y_{t+1})/\pi(X_t) \\ &= X_t && \text{otherwise} \end{aligned}$$

(assuming a symmetric proposal distribution). The repetition of the current value in the event of a rejection is what makes the algorithm valid, that is, why π is the stationary distribution.

It is always possible to study the distribution of the accepted and of the rejected values, with some recycling possible by Rao–Blackwellization [20], but this study is more advanced and far from necessary to understand the algorithm.

"How to account for impossible proposed values?" [cross-validated:51808]

It is indeed a popular belief that something needs to be done to account for restricted supports. However, there is no mathematical reason for doing so. The Metropolis–Hastings acceptance probability

$$\rho(x_t, y_{t+1}) = \min(1, \pi(y_{t+1})q(x_t|y_{t+1})/\pi(x_t)q(y_{t+1}|x_t))$$

with $y_t \sim q(y_{t+1}|x_t)$ can handle cases when y_t is outside the support of π by extending this support, defining $\pi(y) = 0$ outside the original support. Hence, if $\pi(y_{t+1}) = 0$, then $\rho(x_t, y_{t+1}) = 0$, which means the proposed value is automatically rejected, and $x_{t+1} = x_t$.

Consider the following illustration.

```
target=function(x)  (x>0)*(x<1)*dnorm(x,mean=4)
 mcmc=rep(0.5,10^5)
 for (t in 2:10^5){
 prop=mcmc[t-1]+rnorm(1,.1)
 if (runif(1)<target(prop)/target(mcmc[t-1]))
 mcmc[t]=prop
 else
 mcmc[t]=mcmc[t-1]
 }
 hist(mcmc,prob=TRUE)
 curve(dnorm(x-4)/(pnorm(-3)-pnorm(-4)),add=TRUE)
```

that is targeting a truncated Normal distribution using a Gaussian random walk proposal with support the entire real line. Then, the algorithm is properly converging as shown by the fit in Figure 9.

Unbiased MCMC [xianblog:25/08/2017]

Jacob *et al.* [21] propose an unbiased MCMC technique based on coupling. Associating MCMC with unbiasedness is rather challenging since MCMC algorithms are rarely

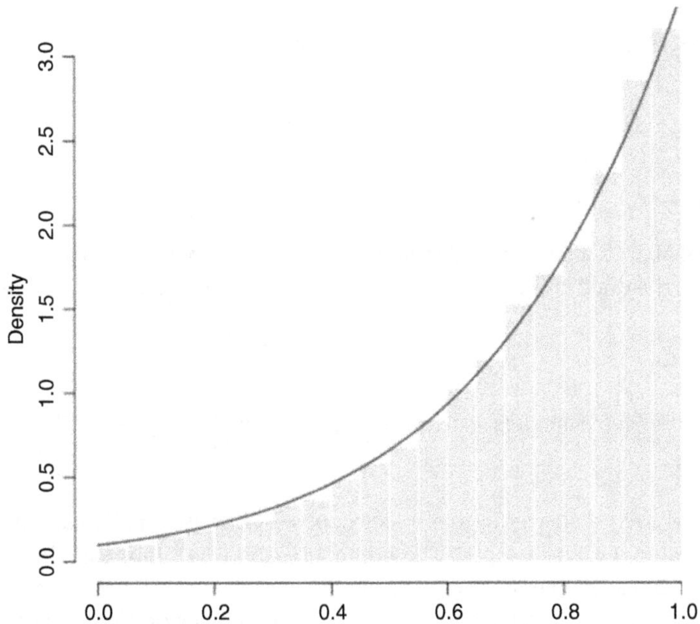

Figure 9 Graph of a truncated Normal density and fit by the histogram of an MCMC sample using a Gaussian random walk.

producing simulations from the exact target, unless specific tools like renewal can be produced in an efficient manner.

The central idea is coupling of two (MCMC) chains, associated with the debiasing formula used by Glynn and Rhee [22]. Having the coupled chains meet at some time with probability one implies that the debiasing formula does not need a (random) stopping time. The coupling time is sufficient. Furthermore, several estimators can be derived from the same coupled Markov chain simulations, obtained by starting the averaging at a later time than the first iteration. The average of these (unbiased) averages results into a weighted estimate that weights more the later differences. Although coupling is also at the basis of perfect simulation methods, the analogy between this debiasing technique and perfect sampling is hard to fathom, since the coupling of two chains is not a perfect sampling instant. (Something obvious in retrospect is that the variance of the resulting unbiased estimator is at best the variance of the original MCMC estimator.)

When discussing the implementation of coupling in Metropolis and Gibbs settings, the authors produce a simple optimal coupling algorithm, a form of accept–reject also found in perfect sampling. While I did not fully understood the way two random walk Metropolis steps are coupled, in that the Normal proposals seem at odds with the boundedness constraints, coupling is clearly working in this setting, while renewal does not. In toy examples, such as the Efron and [23] baseball data and the [1] pump failure data, the parameters of the algorithm can be optimized against the variance of the averaged averages. And this approach proves highly useful in the case of the cut distribution.

3.2 Gibbs Sampling

Historically, this form of MCMC algorithm is distinguished from the other types of MCMC methods for being both justified by other arguments and used for a specific class of models [6].

> *"Why would one use Gibbs sampling instead of Metropolis–Hastings?" [cross-validated: 244573]*

The question does not have an answer in that a Metropolis–Hastings sampler can be almost anything, including a Gibbs sampler. The primary reason why Gibbs sampling was introduced was to break the curse of dimensionality (which impacts both rejection and importance sampling) by producing a sequence of low-dimension simulations that still converge to the right target even though the dimension of the target impacts the speed of convergence. Metropolis–Hastings samplers are designed to create a Markov chain (like Gibbs sampling) based on a proposal (like importance and rejection sampling) by correcting for the wrong density through an acceptance–rejection step. But an important point is that they are not opposed, namely, Gibbs sampling may require Metropolis–Hastings steps when facing complex if low-dimension conditional targets, while Metropolis–Hastings proposals may be built on approximations to (Gibbs) full conditionals. In a formal definition, Gibbs sampling is a special case of Metropolis–Hastings algorithm with a probability of acceptance of one.

Usually, Gibbs sampling – understood as running a sequence of low-dimensional conditional simulations – is favored in settings where the decomposition into such conditionals is easy to implement and fast to run. In settings where such decompositions induce multimodality and hence a difficulty to move between modes (latent variable models like mixture models come to mind), using a more global proposal in a Metropolis–Hastings algorithm may produce a higher efficiency. But the drawback stands with choosing the proposal distribution in the Metropolis–Hastings algorithm.

3.3 Hamiltonian Monte Carlo

A more advanced (and still popular) form of MCMC algorithm is HMC [24–26]. While a special case of continuous time samplers, it can be implemented in discrete time and is actually behind the successful Stan package [8]. The construction of the process relies on an auxiliary variable v that augments the target into

$$\rho(x, v) = p(x)\varphi(v|x) \propto \exp\{-H(x, v)\}$$

where $\varphi(v|x)$ is the conditional density of v given x. This density obviously enjoys $p(v)$ as its marginal, and while it could be anything, the so-called momentum v is usually chosen of the same dimension as v, with $\varphi(v|x)$ often taken as a Normal density. The associated *Hamiltonian equations*

$$\frac{\mathrm{d}x_t}{\mathrm{d}t} = \frac{\partial H}{\partial v}(x_t, v_t) \qquad \frac{\mathrm{d}v_t}{\mathrm{d}t} = -\frac{\partial H}{\partial x}(x_t, v_t)$$

which keep the *Hamiltonian* target $H(\cdot)$ constant over time, as

$$\frac{\mathrm{d}H(x_t, v_t)}{\mathrm{d}t} = \frac{\partial H}{\partial v}(x_t, v_t)\,\frac{\mathrm{d}v_t}{\mathrm{d}t} + \frac{\partial H}{\partial x}(x_t, v_t)\,\frac{\mathrm{d}x_t}{\mathrm{d}t} = 0$$

Since there is no randomness in the above process, the HMC algorithm is completed with random changes of the momentum according to the correct conditional distribution, $v_t \sim \varphi(v|x_t)$, at times driven by a Poisson process $\{\tau_n\}_n$.

As noted above, the choice of the conditional density $\varphi(v|x_t)$ often is a Gaussian density with either a constant covariance matrix M calibrated from the target covariance or as a local curvature, depending on x in the version of Girolami and Calderhead [27] called Riemannian HMC. See, for example, Livingstone *et al.* [28] for an analysis of the impact of different types of kinetic energy on HMC performances.

When the fixed covariance matrix is equal to M, the Hamilton equations write as

$$\frac{\mathrm{d}x_t}{\mathrm{d}t} = M^{-1}v_t \qquad \frac{\mathrm{d}v_t}{\mathrm{d}t} = \nabla \log p(x_t)$$

where the last term is the score function. The velocity of the HMC process is thus connected to the gradient of the log-target.

In practice, implementing this rather simple remark proves rather formidable in that there is no direct approach for methodology for simulating this continuous time process, since the above equations are intractable. A natural resolution associates a numerical solver like Euler's method, usually unstable, with a numerical solver naturally suited to these equations.

This general method is called a symplectic integrator [29] with implementation in the constant covariance case resorting to time-discretization leapfrog steps

$$\begin{aligned}
v_{t+\epsilon/2} &= v_t + \epsilon\nabla \log p(x_t)/2 \\
x_{t+\epsilon} &= x_t + \epsilon M^{-1} v_{t+\epsilon/2} \\
v_{t+\epsilon} &= v_{t+\epsilon/2} + \epsilon\nabla \log p(x_{t+\epsilon})/2
\end{aligned}$$

which symmetrizes the two-step move, with ϵ standing for the time-discretization step. The proposed value of v_0 is generated from the true Gaussian target. The correction to the discretization approximation involves a Metropolis–Hastings step over the pair $(x_{t+\epsilon} v_{t+\epsilon})$ which reintroduces some reversibility into the picture.

Time-discretizing the Hamiltonian dynamics in the leapfrog integrator involves two quantities, ϵ and T, the trajectory length. One empirically sound calibration of these parameters is found in the NUTS of Hoffman and Gelman [9] which selects the value of N by primal–dual averaging and produces the trajectory length T as the length of the chain is taken for the path to fold back.

Algorithm 2. Leapfrog(x_0, v_0, ϵ, L)

Input: starting position x_0, starting momentum v_0, step-size ϵ, steps L
for $\ell = 0, 1, \cdots, L-1$ **do**
  $v_{\ell+1/2} = v_\ell + \epsilon\nabla \log p(x_\ell)$
  $x_{\ell+1} = x_\ell + \epsilon M^{-1} v_{\ell+1/2}$
  $v_{\ell+1} = v_{\ell+1/2} + \epsilon\nabla \log p(x_{\ell+1})$
end for
Output: (x_L, v_L)

Algorithm 3. Hamiltonian Monte Carlo algorithm

Input: step-size ϵ, steps of leapfrog integrator L, starting position x_0, desired number of iterations N.
for $n = 1, \cdots, N$ **do**
  Sample $v_{n-1} \sim \varphi(v)$;
  Compute $(x^*, v^*) \leftarrow$ Leapfrog($x_{n-1}, v_{n-1}, \epsilon, L$);
  Compute the acceptance ratio α, where

$$\alpha = \min\left\{1, \exp(-H(x^*, -v^*))\big/\exp(-H(x_{n-1}, v_{n-1}))\right\};$$

  Sample $u \sim \mathcal{U}[0, 1]$;
  if $u < \alpha$ **then**
    $x_n \leftarrow x^*$
  else
    $x_n \leftarrow x_{n-1}$
  end if
end for

In practice, it is important to note that discretizing Hamiltonian dynamics introduces two free parameters, the step size ϵ and the trajectory length T, both to be calibrated. As an empirically successful and popular variant of HMC, the NUTS of Hoffman and Gelman [9] adapts the value of ϵ based on primal–dual averaging. It also eliminates the need to choose the trajectory length T via a recursive algorithm that builds a set of candidate proposals for a number of forward and backward leapfrog steps and stops automatically when the simulated path retraces.

Unbiased HMC [xianblog:25/09/2017]

Heng and Jacob [30] propose to achieve unbiased HMC by coupling, following Jacob *et al.* [21] discussed earlier. The coupling within the HMC amounts to running two HMC chains with common random numbers, plus subtleties.

"As with any other MCMC method, HMC estimators are justified in the limit of the number of iterations. Algorithms which rely on such asymptotics face the risk of

becoming obsolete if computational power keeps increasing through the number of available processors and not through clock speed." Heng and Jacob (2019)

The main difficulty here is to have both chains meet (exactly) with large probability, since coupled HMC can only bring these chain close to one another. The trick stands in using both coupled HMC and coupled Hastings–Metropolis kernels, since the coupled MH kernel allows for exact meetings when the chains are already close, after which they remain forever identical. The algorithm is implemented by choosing at random between the kernels at each iteration. (Unbiasedness follows by the Glynn–Rhee trick, which is eminently well suited for coupling.) As pointed out from the start of the chapter, the appeal of this unbiased version is that the algorithm can be (embarrassingly) parallelized since all processors in use return estimators that are i.i.d. copies of one another, hence easily merged into a better estimator.

4 Approximate Bayesian Computation

The methods surveyed above share the common feature of exploiting the shape of the target density, $p(\cdot)$, namely that it is known exactly or known up to a normalizing constant $p(x) \propto \tilde{p}(x)$ or yet known as the marginal of another density

$$p(x) = \int_{\mathcal{Y}} q(x,y)\, \mathrm{d}y$$

It may, however, occur that the density of the target is not numerically available, in the sense that computing $p(x)$ or $\tilde{p}(x)$ is not feasible in a reasonable time or that completing $p(\cdot)$ into $q(\cdot)$ involves a massive increase in the dimension of the problem. This obviously causes difficulties in applying, for example, MCMC methods. A particularly common case occurs in the Bayesian analysis of intractable likelihoods.

"What would be a good example of a really simple model that has an intractable likelihood?" [cross-validated:127180]

Given an original Normal dataset

$$x_1, \dots, x_n \overset{\text{i.i.d.}}{\sim} \mathrm{N}(\theta, \sigma^2)$$

the reported data is made of the two-dimensional summary

$$S(x_1, \dots, x_n) = (\mathsf{med}(x_1, \dots, x_n), \mathsf{mad}(x_1, \dots, x_n))$$

where $\mathsf{mad}(x_1, \dots, x_n)$ is the median average deviation of the sample, which is not sufficient and which does not have a closed-form joint density.

Besides this simple example, there are numerous occurrences of well-defined likelihoods that cannot be computed, from latent variable models, including hidden Markov models, to likelihoods with a missing normalizing term depending on the parameter, including Ising models [31] and other nonstandard exponential families, to densities defined as solutions of differential equations, via their characteristic function, such as α-stable distributions [32], or via their quantile function such as Tukey's g-and-k distributions [33].

A different kind of algorithm is required for handling such situations. They are called "likelihood-free" or approximate Bayesian computation (ABC) methods, as they do not require the likelihood function and provide an approximation of the original posterior distribution.

"What does it mean for an inference or optimization method to be 'likelihood-free'?" [cross-validated:383731]

Specifically, likelihood-free methods are a rewording of the ABC algorithms, where ABC stands for approximate Bayesian computation. This intends to cover inference methods that do not require the use of a closed-form likelihood function but still intend to study a specific statistical model. They are free from the computational difficulty attached with the likelihood but not from the model that produces this likelihood. See, for instance, the recent handbook by Sisson *et al.* [34].

The basic ABC algorithm is based on the following principle: given a target posterior proportional to $\pi(\theta)f(x^{\text{obs}}|\theta)$, when the likelihood function $f(x^{\text{obs}}|\theta)$ is not available in closed form,[3] jointly simulating

$$\theta' \sim \pi(\theta), z \sim f(z|\theta')$$

until the auxiliary variable z is equal to the observed value, $z = x^{\text{obs}}$ does produce a realization from the posterior distribution without ever computing a numerical value of the likelihood function. It only requires that the model associated with this likelihood can be simulated, which often leads to the model being called a *generative model.*

"How can we prove that when accepting for $x = x^{\text{obs}}$ in this algorithm, we sample from the true posterior?" [cross-validated:380076]

This case is the original version of the algorithm, as in Refs 35, 36. Assuming that

$$\mathbb{P}_\theta(Z = x^{\text{obs}}) > 0$$

the values of θ that come out of the algorithm are distributed from a distribution with density proportional to

$$\pi(\theta) \times \mathbb{P}_\theta(Z = x^{\text{obs}})$$

since the algorithm generates the pair $(\theta, \mathbb{I}_{Z=x^{\text{obs}}})$ with joint distribution

$$\pi(\theta) \times \mathbb{P}_\theta(Z = x^{\text{obs}})^{\mathbb{I}_{Z=x^{\text{obs}}}} \times \mathbb{P}_\theta(Z \neq x^{\text{obs}})^{\mathbb{I}_{Z\neq x^{\text{obs}}}}$$

Conditioning on $\mathbb{I}_{Z=x^{\text{obs}}} = 1$ leads to

$$\theta | \mathbb{I}_{Z=x^{\text{obs}}} = 1 \sim \pi(\theta) \times \mathbb{P}_\theta(Z = x^{\text{obs}}) / \int \pi(\theta) \times \mathbb{P}_\theta(X = x^{\text{obs}})\, d\theta$$

which is the posterior distribution.

As noted in the above vignette, the principle can only be implemented when $\mathbb{P}_\theta(Z = x^{\text{obs}}) > 0$ and more accurately when the event $Z = x^{\text{obs}}$ has a nonnegligible chance to occur. This is, however, rarely the case in realistic settings, especially when Z is a continuous variable, and the first implementations [37] of the ABC algorithm replaced the constraint of equality $z = x^{\text{obs}}$ of a relaxed version,

$$\varrho(z, x^{\text{obs}}) \leq \epsilon$$

where ϱ is a distance, and $\epsilon > 0$ is called the tolerance. This approximation step makes the concept applicable in a wider range of settings with an intractable distribution, but it also implies that the simulated distribution is modified from the true posterior into

$$\pi(\theta | \varrho(Z, x^{\text{obs}}) < \epsilon) \propto \pi(\theta)\ \mathbb{P}_\theta\{\varrho(Z, x^{\text{obs}}) < \epsilon\}$$

It helps to visualize this alternative posterior distribution as truly conditioning on the event $\varrho(Z, x^{\text{obs}}) < \epsilon$ rather than x^{obs} as it gives a specific meaning to this distribution and explains the loss in information brought by the approximation.

In many settings, especially with large datasets, looking at a distance between the raw observed data and the raw simulated data is very inefficient. It is much more efficient [38, 39] to compare informative summaries of the data as the decrease in dimension allows for smaller tolerance, a higher signal-to-noise ratio, and outweighs the potential loss in information. A more common implementation of the algorithm is thus

Algorithm 4. Likelihood-free (ABC) rejection sampler

for $i = 1$ to N **do**
  repeat
    generate θ' from the prior distribution $\pi(\cdot)$
    generate z from the likelihood $f(\cdot|\theta')$
  until $\rho\{\eta(z), \eta(x^{\text{obs}})\} \leq \epsilon$
  set $\theta_i = \theta'$
end for

where $\eta(\cdot)$ denotes a (not necessarily sufficient) statistic, usually (needlessly) called a summary statistic. While there is a huge literature [34, 40–43] on the choice of the summary statistic, compelling arguments [38, 41] lead to opt for summaries of the same dimension as the parameter θ.

While the motivation for simulating from the prior distribution is clear from a theoretical perspective, given that the probability of accepting in Algorithm 4 is approximately the intractable likelihood, it is also often poor in efficiency since the posterior is much more concentrated. Subsequent versions of ABC have thus aimed at alternative approaches to increase the efficiency of the method. For instance, the proposal distribution on θ can be modified toward increase in the frequency of x's within the vicinity of x^{obs} [38, 44, 45]. Others have replaced the indicator function in Algorithm 4 with less rudimentary estimators of the likelihood [46–48], interpreting the tolerance ϵ as a bandwidth [38, 39] or a new component in the inferential framework [49].

Computational advances have seen MCMC, sequential Monte Carlo (SMC) [50], and Gibbs [51] versions of ABC. For instance, ABC–MCMC [44] is based on the property that the Markov chain $(\theta^{(t)})$ created via the transition function

$$\theta^{(t+1)} = \begin{cases} \theta' \sim K_\omega(\theta'|\theta^{(t)}) & \text{if } z \sim f(z|\theta') \text{ is such that } z = x^{\text{obs}} \\ & \text{and } u \sim \mathcal{U}(0,1) \le \frac{\pi(\theta')K_\omega(\theta^{(t)}|\theta')}{\pi(\theta^{(t)})K_\omega(\theta'|\theta^{(t)})} \\ \theta^{(t)} & \text{otherwise} \end{cases}$$

enjoys the posterior $\pi(\theta|x^{\text{obs}})$ as its stationary distribution. The corresponding algorithm is then

Algorithm 5. Likelihood-free MCMC sampler

Use Algorithm 4 to get $(\theta^{(0)}, z^{(0)})$
for $t = 1$ to N **do**
  Generate θ' from $K_\omega\left(\cdot|\theta^{(t-1)}\right)$,
  Generate z' from the likelihood $f(\cdot|\theta')$,
  Generate u from $\mathcal{U}_{[0,1]}$,
  if $u \le \frac{\pi(\theta')K_\omega(\theta^{(t-1)}|\theta')}{\pi(\theta^{(t-1)})K_\omega(\theta'|\theta^{(t-1)})} \mathbb{I}_{\varrho(\eta(z'),\eta(x^{\text{obs}}))\le\epsilon}$ **then**
   set $(\theta^{(t)}, z^{(t)}) = (\theta', z')$
  else
   $(\theta^{(t)}, z^{(t)})) = (\theta^{(t-1)}, z^{(t-1)})$,
  end if
end for

The choice of summary statistics in ABC method is paramount for the efficiency of the approximation and nowhere more than for model choice. Since the Bayes factor is given by

$$B_{12}(x^{\text{obs}}) = \frac{\Pr(M_1|x^{\text{obs}})}{\Pr(M_2|x^{\text{obs}})} \Bigg/ \frac{\Pr(M_1)}{\Pr(M_2)}$$

the ratio of frequencies of simulations from M_1 and M_2 that are accepted need to be divided by the prior probabilities of M_1 and M_2 if these reflect the number of times each model is simulated. Apart from this, the approximation is valid. Using inappropriate summary statistics in this setting has been pointed out in Refs 52–54.

A special instance of (almost) intractable is the setting of "Big Data" problems where the size of the data makes computing the likelihood quite expensive. In such cases, ABC can be seen as a convenient approach to scalable Monte Carlo.

> *"What difference does it make working with a big or small dataset in ABC?" [cross-validated:424712]*

It all depends on the structure of the dataset and the complexity of the model behind. In some settings the size of the data may be the reason for conducting an ABC inference as the likelihood takes too much time to compute. But there is no generic answer to the question since in the ultimate case when there exists a sufficient statistic of fixed dimension, size does not matter (and of course ABC is unlikely to be needed).

> *"Do we get any computational benefits by reducing a very big dataset when doing inference using ABC methods?"*

In most settings, ABC proceeds through a set of summary statistics that are of a much smaller dimension than the data. In that sense they are independent of the size of the data, except that to simulate values of the summaries, most models require simulations of the entire dataset first unless a proxy model is used as in synthetic likelihood.

> *"...the rejection criterion in ABC is related to how well we approximate the full likelihood of the dataset which is typically captured in some low-dimensional summary statistics vector."*

You have to realize that the rejection is relative to the distribution of the distances between the observed and the simulated summaries [simulated under the prior predictive], rather than absolute. In other words, there is no predetermined value for the tolerance. This comes in addition to the assessment being based on an insufficient statistics rather than the full data. This means that, for a given computing budget, the true likelihood of an accepted parameter may be quite low.

5 Further Reading

There are many reviews and retrospectives on the MCMC methods, not only in statistics but also in physics, econometrics, and several other fields, most of which provide different perspectives on the topic. For instance, Dunson and Johndrow [55] recently wrote a celebration of Hastings' 1970 paper in Biometrika, where they cover adaptive Metropolis [56, 57], the importance of gradient-based versions toward universal algorithms [58, 59], discussing the advantages of HMC over Langevin versions. They also recall the significant step represented by Green's [60] reversible jump algorithm for multimodal and multidimensional targets as well as tempering [61, 62]. They further cover intractable likelihood cases within MCMC (rather than ABC), with the use of auxiliary variables [63, 64] and pseudomarginal MCMC [65, 66]. They naturally insist upon the need to handle huge datasets, high-dimension parameter spaces, and other scalability issues, with links to unadjusted Langevin schemes [67–69]. Similarly, Dunson and Johndrow [55] discuss recent developments toward parallel MCMC and see nonreversible schemes such as partly deterministic Markov process (PDMP) as highly promising, with a concluding section on the challenges of automating and robustifying much further the said procedures, if only to reach a wider range of applications. Other directions that are clearly still relevant after decades of development include convergence assessment, for example, the comparison and aggregation of various approximation schemes, since this is a fairly common request from users, recycling schemes, such as Rao–Blackwellization [1, 20] and other postprocessing improvements that address the massive waste of simulation in most methods, the potential for mutual gains between machine-learning tools and MCMC refinements, and the theoretical difficulties presented by approximations such as synthetic likelihood [70], indirect inference [71], and incompatible conditionals [51, 72, 73].

Abbreviations and Acronyms

ABC	approximate Bayesian computation
EM	expectation-maximisation
HMC	Hamiltonian Monte Carlo
MCMC	Markov chain Monte Carlo
NUTS	no-U-turn sampler
PDMP	partly deterministic Markov process
PMC	population Monte Carlo
QMC	quasi-Monte Carlo
SMC	sequential Monte Carlo

Notes

1 For which one obviously does not need MCMC in practice: this is a toy example.
2 In reference to N. Metropolis, with whom the algorithm originated [16], although his contribution to the paper is somewhat disputed, and K. Hastings, for his generalization [17].
3 The notation x^{obs} is intended to distinguish the observed sample from simulated versions of this sample.

References

1 Gelfand, A. and Smith, A. (1990) Sampling based approaches to calculating marginal densities. *J. Am. Stat. Assoc.*, 85, 398–409.

2 Berger, J. (1985) *Statistical Decision Theory and Bayesian Analysis*, 2nd, Springer-Verlag, New York.

3 Gelman, A., Vehtari, A., Jylänki, P. *et al.* (2014) Expectation propagation as a way of life. *arXiv*.

4 Jaakkola, T. and Jordan, M. (2000) Bayesian parameter estimation via variational methods. *Stat. Comput.*, 10, 25–37.

5 Dempster, A., Laird, N., and Rubin, D. (1977) Maximum likelihood from incomplete data via the EM algorithm (with discussion). *J. R. Stat. Soc. Ser. B*, 39, 1–38.

6 Robert, C. and Casella, G. (2004) *Monte Carlo Statistical Methods*, 2nd, Springer-Verlag, New York.

7 Rubinstein, R.Y. (1981) *Simulation and the Monte Carlo Method*, J. Wiley, New York.

8 Carpenter, B., Gelman, A., Hoffman, M. *et al.* (2017) Stan: a probabilistic programming language. *J. Stat. Soft., Articles*, 76(1), 1–29.

9 Hoffman, M.D. and Gelman, A. (2014) The No-U-turn sampler: adaptively setting path lengths in Hamiltonian Monte Carlo. *Ann. Appl. Probab.*, 27(4), 2159–2194.

10 Chatterjee, S. and Diaconis, P. (2018) The sample size required in importance sampling. *Ann. Appl. Probab.*, 28(2), 1099–1135.

11 Liu, J., Chen, R., and Logvinenko, T. (2001) A theoretical framework for sequential importance sampling and resampling, in *Sequential Monte Carlo Methods in Practice* (eds A. Doucet, N. De. Freitas, and N. Gordon), Springer-Verlag, New York, pp. 225–246.

12 Cappé, O. and Robert, C. (2000) Ten years and still running! *J. Am. Stat. Assoc.*, 95(4), 1282–1286.

13 Robert, C. and Casella, G. (2010) A history of Markov chain Monte Carlo–Subjective recollections from incomplete data, in *Handbook of Markov Chain Monte Carlo: Methods and Applications* (eds S. Brooks, A. Gelman, X. Meng, and G. Jones), Chapman and Hall, New York. arXiv0808.2902

14 Green, P.J., Łatuszyński, K., Pereyra, M., and Robert, C.P. (2015) Bayesian computation: a summary of the current state, and samples backwards and forwards. *Stat. Comput.*, 25 (4), 835–862.

15 Meyn, S. and Tweedie, R. (1993) *Markov Chains and Stochastic Stability*, Springer-Verlag, New York.

16 Metropolis, N., Rosenbluth, A.W., Rosenbluth, M.N. *et al.* (1953) Equations of state calculations by fast computing machines. *J. Chem. Phys.*, 21, 1087–1092.

17 Hastings, W.K. (1970) Monte Carlo sampling methods using Markov chains and their applications. *Biometrika*, 57(1), 97–109.

18 Liu, J., Wong, W., and Kong, A. (1995) Covariance structure and convergence rates of the Gibbs sampler with various scans. *J. R. Stat. Soc. Ser. B*, 57, 157–169.

19 Liu, J., Wong, W., and Kong, A. (1994) Covariance structure of the Gibbs sampler with application to the comparison of estimators and augmentation schemes. *Biometrika*, 81, 27–40.

20 Casella, G. and Robert, C. (1996) Rao-Blackwellisation of sampling schemes. *Biometrika*, 83 (1), 81–94.

21 Jacob, P., Leary, J., and Atchadé, Y. (2020) Unbiased Markov chain monte carlo methods with couplings. *J. R. Stat. Soc. Ser. B*, 82, 1–32.

22 Glynn, P.W. and Rhee, C.-H. (2014) Exact estimation for Markov chain equilibrium expectations. *J. Appl. Probab.*, 51, 377–389.

23 Efron, B., and Morris, C. (1973) Stein's estimation rule and its competitors–An empirical Bayes approach. *J. Am. Stat. Assoc.*, 68(341), 117–130.

24 Duane, S., Kennedy, A.D., Pendleton, B.J., and Roweth, D. (1987) Hybrid Monte Carlo. *Phys. Lett. B*, 195, 216–222.

25 Neal, R. (1999) *Bayesian Learning for Neural Networks*, 118, Springer–Verlag, New York. Lecture Notes.

26 Neal, R. (2011) MCMC using Hamiltonian dynamics, in *In Handbook of Markov Chain Monte Carlo* (eds S. Brooks, A. Gelman, G. L. Jones, and X.-L. Meng), CRC Press, New York. pp. 113–162.

27 Girolami, M. and Calderhead, B. (2011) Riemann manifold Langevin and Hamiltonian Monte Carlo methods. *J. R. Stat. Soc.: Ser. B Stat. Methodol.*, 73, 123–214.

28 Livingstone, S., Faulkner, M.F., and Roberts, G.O. (2017) Kinetic energy choice in Hamiltonian/hybrid Monte Carlo. *arXiv preprint arXiv:1706.02649*.

29 Betancourt, M. (2017) A conceptual introduction to Hamiltonian Monte Carlo. *arXiv preprint arXiv:1701.02434*.

30 Heng, J. and Jacob, P.E. (2019) Unbiased Hamiltonian Monte Carlo with couplings. *Biometrika*, 106 (2), 287–302.

31 Potts, R.B. (1952) Some generalized order-disorder transitions. *Proc. Camb. Philos. Soc.*, 48, 106–109.

32 Peters, G., Sisson, S., and Fan, Y. (2012) Likelihood-free Bayesian inference for α-stable models. *Comput. Stat. Data Anal.*, 56 (11), 3743–3756.

33 Haynes, M.A., MacGillivray, H.L., and Mengersen, K.L. (1997) Robustness of ranking and selection rules using generalised g-and-k distributions. *J. Stat. Plan. Inference*, 65 (1), 45–66.

34 Sisson, S., Fan, Y., and Beaumont, M. (2019) *Handbook of Approximate Bayesian Computation*, CRC Press, Taylor & Francis Group, Boca Raton.

35 Rubin, D. (1984) Bayesianly justifiable and relevant frequency calculations for the applied statistician. *Ann. Stat.*, 12, 1151–1172.

36 Tavaré, S., Balding, D., Griffith, R., and Donnelly, P. (1997) Inferring coalescence times from DNA sequence data. *Genetics*, 145, 505–518.

37 Pritchard, J., Seielstad, M., Perez-Lezaun, A., and Feldman, M. (1999) Population growth of human Y chromosomes: a study of Y chromosome microsatellites. *Mol. Biol. Evol.*, 16, 1791–1798.

38 Li, W. and Fearnhead, P. (2018) On the asymptotic efficiency of approximate Bayesian computation estimators. *Biometrika*, 105 (2), 285–299.

39 Frazier, D.T., Martin, G.M., Robert, C.P., and Rousseau, J. (2018) Asymptotic properties of approximate Bayesian computation. *Biometrika*, 105 (3), 593–607.

40 Aeschbacher, S., Beaumont, M.A., and Futschik, A. (2012) A novel approach for choosing summary statistics in Approximate Bayesian Computation. *Genetics*, 192 (3), 1027–1047.

41 Fearnhead, P. and Prangle, D. (2012) Constructing summary statistics for Approximate Bayesian Computation: semi-automatic Approximate Bayesian Computation. *J. R. Stat. Soc. Ser. B Stat. Methodol.*, 74 (3), 419–474.

42 Estoup, A., Lombaert, E., Marin, J.-M. *et al.* (2012) Estimation of demo-genetic model probabilities with Approximate Bayesian Computation using linear discriminant analysis on summary statistics. *Mol. Ecol. Resour.*, 12 (5), 846–855.

43 Blum, M.G.B., Nunes, M.A., Prangle, D., and Sisson, S.A. (2013) A comparative review of dimension reduction methods in Approximate Bayesian computation. *Stat. Sci.*, 28 (2), 189–208.

44 Marjoram, P., Molitor, J., Plagnol, V., and Tavaré, S. (2003) Markov chain Monte Carlo without likelihoods. *Proc. Natl. Acad. Sci. USA*, 100 (26), 15324–15328.

45 Bortot, P., Coles, S., and Sisson, S. (2007) Inference for stereological extremes. *J. Am. Stat. Assoc.*, 102, 84–92.

46 Beaumont, M., Zhang, W., and Balding, D. (2002) Approximate Bayesian computation in population genetics. *Genetics*, 162, 2025–2035.

47 Blum, M. (2010) Approximate Bayesian Computation: a non-parametric perspective. *J. Am. Stat. Assoc.*, 105 (491), 1178–1187.

48 Mengersen, K., Pudlo, P., and Robert, C. (2013) Bayesian computation via empirical likelihood. *Proc. Nat. Acad. Sci.*, 110 (4), 1321–1326.

49 Ratmann, O., Andrieu, C., Wiuf, C., and Richardson, S. (2009) Model criticism based on likelihood-free inference, with an application to protein network evolution. *PNAS*, 106, 1–6.

50 Beaumont, M., Cornuet, J.-M., Marin, J.-M., and Robert, C. (2009) Adaptive approximate Bayesian computation. *Biometrika*, 96 (4), 983–990.

51 Clarté, G., Robert, C.P., Ryder, R., and Stoehr, J. (2019) Component-wise approximate Bayesian computation via Gibbs-like steps. *arXiv e-prints*, arXiv:1905.13599.

52 Didelot, X., Everitt, R., Johansen, A., and Lawson, D. (2011) Likelihood-free estimation of model evidence. *Bayesian Anal.*, 6, 48–76.

53 Robert, C., Cornuet, J.-M., Marin, J.-M., and Pillai, N. (2011) Lack of confidence in ABC model choice. *Proc. Nat. Acad. Sci.*, 108 (37), 15112–15117.

54 Marin, J., Pillai, N., Robert, C., and Rousseau, J. (2014) Relevant statistics for Bayesian model choice. *J. R. Stat. Soc. Ser. B*, 76 (5), 833–859.

55 Dunson, D. and Johndrow, J. (2020) The Hastings algorithm at fifty. *Biometrika*, 107, 1–23.

56 Haario, H., Saksman, E., and Tamminen, J. (1999) Adaptive proposal distribution for random walk Metropolis algorithm. *Comput. Stat.*, 14 (3), 375–395.

57 Roberts, G. and Rosenthal, J. (2005) Coupling and ergodicity of adaptive MCMC. *J. Appl. Probab.*, 44, 458–475.

58 Roberts, G. and Tweedie, R. (1995) Exponential Convergence for Langevin Diffusions and their Discrete Approximations. Technical report. Statistics Laboratory, University of Cambridge.

59 Neal, R. (2003) Slice sampling (with discussion). *Ann. Stat.*, 31, 705–767.

60 Green, P. (1995) Reversible jump MCMC computation and Bayesian model determination. *Biometrika*, 82 (4), 711–732.

61 Woodard, D.B., Schmidler, S.C., and Huber, M. (2009) Sufficient conditions for torpid mixing of parallel and simulated tempering. *Electron. J. Probab.*, 14, 780–804.

62 Miasojedow, B., Moulines, E., and Vihola, M. (2013) An adaptive parallel tempering algorithm. *J. Comput. Graph. Stat.*, 22 (3), 649–664.

63 Møller, J., Pettitt, A., Reeves, R., and Berthelsen, K. (2006) An efficient Markov chain Monte Carlo method for distributions with intractable normalising constants. *Biometrika*, 93, 451–458.

64 Friel, N. and Pettitt, A. (2008) Marginal likelihood estimation via power posteriors. *J. R. Stat. Soc. Ser. B*, 70 (3), 589–607.

65 Andrieu, C. and Roberts, G. (2009) The pseudo-marginal approach for efficient Monte Carlo computations. *Ann. Stat.*, 37, 697–725.

66 Andrieu, C. and Vihola, M. (2016) Establishing some order amongst exact approximations of MCMCs. *Ann. Appl. Probab.*, 26 (5), 2661–2696.

67 Welling, M. and Teh, Y. (2011) Bayesian Learning Via Stochastic Gradient Langevin Dynamics. *Proceedings of the 28th International Conference on Machine Learning (ICML-11)*, pp. 681–688.

68 Bardenet, R., Doucet, A., and Holmes, C. (2014) Towards Scaling Up Markov Chain Monte Carlo: An Adaptive Subsampling Approach. *Proc. 31st Intern. Conf. Machine Learning (ICML)*, pp. 405–413.

69 Durmus, A. and Moulines, E. (2017) Nonasymptotic convergence analysis for the unadjusted Langevin algorithm. *Ann. Appl. Probab.*, 27 (3), 1551–1587.

70 Wood, S. (2010) Statistical inference for noisy nonlinear ecological dynamic systems. *Nature*, 466, 1102–1104.

71 Drovandi, C., Pettitt, A., and Fddy, M. (2011) Approximate Bayesian computation using indirect inference. *J. R. Stat. Soc. Ser. A*, 60 (3), 503–524.

72 Plummer, M. (2015) Cuts in Bayesian graphical models. *Stat. Comput.*, 25 (1), 37–43.

73 Jacob, P.E., Murray, L.M., Holmes, C.C., and Robert, C.P. (2017) Better together? Statistical learning in models made of modules. *arXiv e-prints*, arXiv:1708.08719.

8

Bayesian Inference with Adaptive Markov Chain Monte Carlo

Matti Vihola

University of Jyväskylä, Jyväskylä, Finland

1 Introduction

The Markov chain Monte Carlo (MCMC) revolution in the 1990s and the following widespread popularity of the Bayesian methods was largely fuelled by the introduction of the BUGS software [1]. With BUGS, the user could focus on the statistically important part and let the software take care of the MCMC inference automatically. Unfortunately, the Gibbs sampling approach used by (variants of) BUGS has certain limitations, such as imposing some modeling constraints due to conjugacy and suffering poor mixing with high correlations.

This section provides a self-contained review of selected simple, robust, and general-purpose adaptive MCMC methods, which can deliver (nearly) automatic inference like BUGS but can overcome some of its limitations. We focus on methods based on random-walk Metropolis (RWM) [2] and parallel tempering (PT; also known as replica exchange) [3]. We also discuss guidelines on how the methods can be used with particle MCMC [4], in order to do inference for a wide class of Bayesian hidden Markov models.

Instead of rigorous theory, the aim is to give an intuitive understanding of why the methods work, what methods are suitable for certain problem classes, and how they can be combined with some other methods. The methods are explained algorithmically, and guidelines are given for parameter values. For more in-depth insight to the theory and methods of adaptive MCMC, the reader is advised to consult the review [5] and references therein and the articles about rigorous theoretical foundations [6–9]. The section is complemented by open-source Julia [10] packages[1,2] which implement the methods and illustrate them on examples.

2 Random-Walk Metropolis Algorithm

Suppose for now that π is a probability density of interest on $\mathbb{R}^d$. Let ℓ stand for the unnormalized log-target, that is, $\ell(x) = \log \pi(x) + c$, where $c \in \mathbb{R}$ is a constant whose

Computational Statistics in Data Science.
Edited by Walter W. Piegorsch, Richard A. Levine, Hao Helen Zhang and Thomas C. M. Lee.
 ISBN 978-1-11956107-1

value need not be known. In the case of Bayesian inference, ℓ will typically be the sum of the log-likelihood and the log-prior density. Algorithm 1 presents the pseudocode for a random-walk Metropolis algorithm [2] targetting π, with initial state $x_0 \in \mathbb{R}^d$, number of iterations n, a symmetric proposal distribution q on $\mathbb{R}^d$, which we will take as the standard Gaussian, and a (nonsingular) proposal shape $S \in \mathbb{R}^{d\times d}$.

Algorithm 1. $X_{1:n} \leftarrow \text{RWM}(\ell, x_0, n, S)$

Set $X_0 \leftarrow x_0$ and $P_0 \leftarrow \ell(x_0)$.
for $k = 1, \dots, n$ **do**:
  $(X_k, P_k; \alpha_k, Z_k) \leftarrow \text{RWMStep}(X_{k-1}, P_{k-1}, \ell, S)$

function RWMStep(X, P, ℓ, S):
  Draw $Z \sim q$ and set $X' \leftarrow X + SZ$.
  Calculate $P' \leftarrow \ell(X')$, let $\alpha \leftarrow \min\{1, \exp(P' - P)\}$and draw $U \sim U(0, 1)$.
  if $U \leq A$ **then return** $(X', P'; \alpha, Z)$; **else return** $(X, P; \alpha, Z)$.

The samples $X_{b:n} = (X_b, \dots, X_n)$ produced by Algorithm 1, for some sufficiently large "burn-in" length $1 \leq b \leq n$, say $b = 0.1n$, are approximately distributed as π. The samples are not independent, but if the chain is well behaved and n sufficiently large, they provide a reliable empirical approximation of π.

It is sufficient to choose any initial state x_0 such that $\ell(x_0) > -\infty$, but it is generally advisable to choose x_0 near the maximum of ℓ. In order to make the method efficient, the proposal increment shape S needs to be tuned based on the properties of the target π. There are two general "rules of thumb" for choosing S, originating from several theoretical results, starting from the seminal work [11]:

(R1) The proposal covariance $SS^T \approx 2.38^2 d^{-1}\Sigma_\pi$, where $\Sigma_\pi = \text{cov}(\pi)$.
(R2) Choose S such that $\text{avg}(\alpha_1, \dots, \alpha_n) \approx 0.234$ (or perhaps 0.44 if $d = 1$).

The random-walk adaptations discussed below implement automatic adjustment of S based on these rules.

3 Adaptation of Random-Walk Metropolis

All of the adaptive RWMs that we discuss may be written in a common form as summarized in Algorithm 2, where we use the RWM step of Algorithm 1. Table 1 summarizes the ingredients of the four commonly used instances of Algorithm 2, which are discussed below.

Algorithm 2. $X_{1:n} \leftarrow \text{ARWM}(\ell, x_0, n)$

Initialize ξ_0, set $X_0 \leftarrow x_0$ and $P_0 \leftarrow \ell(x_0)$.
for $k = 1, \dots, n$ **do**:
  $(X_k, P_k; \alpha_k, Z_k) \leftarrow \text{RWMStep}\left(X_{k-1}, P_{k-1}, \ell, \text{Shape}(\xi_k)\right)$
  $\xi_k \leftarrow \text{Adapt}(k, \xi_{k-1}, X_k, Z_k, \alpha_k)$.

Table 1 Summary of ingredients of Algorithm 2 for the four adaptive MCMC methods.

Method	Initialization ξ_0	State ξ_k	Domain of ξ_k	Adapt(·)	Shape(ξ_k)
AM	(x_0, I_d)	(μ_k, C_k)	$\mathbb{R}^d \times \mathbb{L}_d$	(1)	$2.38d^{-1/2}C_k$
ASM	1	η_k	$\mathbb{R}$	(2)	e^{η_k}
ASM + AM	$(x_0, I_d, \log(2.38d^{-1/2}))$	(μ_k, C_k, η_k)	$\mathbb{R}^d \times \mathbb{L}_d \times \mathbb{R}$	(1) & (2)	$e^{\eta_k}C_k$
RAM	I_d	S_k	$\mathbb{L}_d$	(3)	S_k

I_d stands for the identity matrix in $\mathbb{R}^d$, and $\mathbb{L}_d \subset \mathbb{R}^{d\times d}$ is the set of lower triangular matrices.

3.1 Adaptive Metropolis (AM)

The seminal adaptive Metropolis (AM) algorithm [6] is a direct implementation of the rule 1. The adaptation defines Shape(ξ_k) = Chol($2.38^2 d^{-1}\Sigma_k$), where Chol(S) stands for the lower triangular Cholesky factor L such that $LL^T = S$, and Σ_k is an estimator of cov(π). In the original work [6], the regularized empirical covariance $\Sigma_k = \text{Cov}(X_1, \ldots, X_k) + \epsilon I_d$ was used, where $\epsilon > 0$ was a user-defined parameter.

The follow-up work [8] suggested a slightly modified AM adaptation rule, where Σ_k is a recursively defined covariance estimator defined as follows:

$$\begin{aligned} \mu_k &= \mu_{k-1} + \gamma_k(X_k - \mu_{k-1}) \\ \Sigma_k &= \Sigma_{k-1} + \gamma_k[(X_k - \mu_{k-1})(X_k - \mu_{k-1})^T - \Sigma_{k-1}] \end{aligned} \tag{1}$$

where γ_k is a step size sequence decaying to zero, typically $\gamma_k = (k+1)^{-\beta}$, where $\beta \in (1/2, 1]$, and initial values may be set as $\mu_0 = x_0$ and $\Sigma_0 = I_d$, the identity matrix on $\mathbb{R}^d$.

We suggest to use (1) with the common choice $\gamma_k = (k+1)^{-1}$, which behaves asymptotically similar to the original rule [6], with $\epsilon = 0$. The update (1) is appealing because it avoids the need to choose the regularization factor ϵ and allows for calculation of $C_k =$ Chol(Σ_k) using rank-1 Cholesky updates $C_{k-1} \to C_k$ [12], which cost $O(d^2)$ in contrast with $O(d^3)$ cost of direct calculation of the Cholesky factor. We define the state of adaptation $\xi_k = (\mu_k, C_k)$.

In higher dimensions, the AM adaptation may sometimes suffer from poor initial behavior [13], which may be resolved by adding a fixed (nonadaptive) component in the proposal distribution [13, 14], or using a regularization factor $\epsilon > 0$ as in the original work. Stability may also be improved by adding a delayed rejection stage to the algorithm [15] or using a modified update with X_{k-1} and Y_k weighted by rejection and acceptance probabilities, respectively, which corresponds to one-step Rao–Blackwellization [5].

3.2 Adaptive Scaling Metropolis (ASM)

Automatic selection of the parameter S of the RWM based on rule 2 has been suggested at regeneration times [16] and attempting to directly optimize a loss function [17]. We consider the following simpler adaptation rule [8, 18], which is called here adaptive scaling Metropolis (ASM): set Shape(ξ_k) = e^{η_k}, where $\xi_k = \eta_k$ is adapted with

$$\eta_k = \eta_{k-1} + \gamma_k(\alpha_k - \alpha_*) \tag{2}$$

where $\alpha_* = 0.234$ (or 0.44 if $d = 1$) and with (recommended) step size $\gamma_k = k^{-2/3}$. This adaptation is simpler than the AM adaptation, and even more robust, in the sense that no specific initialization strategies or stabilizing mechanisms are necessary [19]. But because ASM is essentially univariate, it cannot (automatically) capture correlation structures, which may lead to inefficient sampling.

It is quite natural to also use covariance information in the ASM. If no prior information about cov_π is available, we may directly use the AM adaptation together with ASM [5, 7, 18], by setting $\text{Shape}(\xi_k) = e^{\eta_k} C_k$, where $\xi_k = (\mu_k, C_k, \eta_k)$ and (μ_k, C_k) is adapted with AM (1). In this approach, hereafter ASM + AM, it is recommended that a common step size, for instance $\gamma_k = (k+1)^{-2/3}$, is used for both the AM and ASM adaptations.

3.3 Robust Adaptive Metropolis (RAM)

There is an alternative to the combination of AM and ASM, which implements the rule 2 using directional information. The robust adaptive Metropolis (RAM) [20] uses the following direct update on $\text{Shape}(\xi_k) = S_k$:

$$S_k S_k^T = S_{k-1} S_{k-1}^T + \gamma_k(\alpha_k - \alpha_*) V_k V_k^T, \quad \text{where} \quad V_k = S_{k-1} Z_k / \|Z_k\| \tag{3}$$

which may also be implemented as $O(d^2)$ cost rank-1 Cholesky update/downdate [12].

In the univariate case, the RAM update shares similar behavior with the ASM (2), in the sense that then $S_k^2 \approx e^{\eta_k}$. This is because

$$2 \log S_k = 2 \log S_{k-1} + \log(1 + \gamma_k(\alpha_k - \alpha_*)) \approx 2 \log S_{k-1} + \gamma_k(\alpha_k - \alpha_*) \tag{4}$$

for small γ_k. This suggests that RAM can be seen as a multivariate extension of the ASM adaptation. The recommended step size of RAM is $\min\{1, d \cdot k^{-2/3}\}$, where the dimension d inflates the step size because of the directional adaptation [20].

Similar to the ASM, the RAM adaptation has been found stable empirically, typically not requiring specific initialization strategies. However, the ASM + AM adaptation has been suggested to be used initially, before starting the RAM adaptation [21].

3.4 Rationale behind the Adaptations

When looking at the adaptation formulae (1)–(3), it is evident that they all are similar: the previous value of the state is updated by an increment weighted by a decreasing positive step size γ_k. The fact that the changes in the adaptation get smaller and smaller is the key point for the validity of the methods and is called "diminishing" or "vanishing" adaptation [8, 9]. Roughly speaking, this combined with suitable uniform-in-S mixing assumption of the RWM ensures the validity of the algorithms.

The specific forms of adaptation considered here can all be viewed as stochastic gradient-type methods [22, 23] as pointed out in Refs 8 and 17. Their limiting behavior is intuitively characterized by replacing the increments with their stationary expectations, regarding ξ_{k-1} as constant. For instance, such an "averaged" version of the AM update (1) would be

$$\begin{aligned} \mu_k &= \mu_{k-1} + \gamma_k(\mu_\pi - \mu_{k-1}) \\ \Sigma_k &= \Sigma_{k-1} + \gamma_k[\Sigma_\pi - \Sigma_{k-1} - (\mu_\pi - \mu_{k-1})(\mu_\pi - \mu_{k-1})^T] \end{aligned} \tag{5}$$

where μ_π is the mean of π. If the averaged update has a limit, then the adaptation tends to the same limit, under technical assumptions [8]; see also Ref. 5 for further intuitive discussion about the behavior of this type of adaptation.

It is not hard to see that (5) has a unique fixed point (μ_π, Σ_π), so AM adaptation $C_k \to \mathrm{Chol}(\Sigma_\pi)$ under general conditions. Empirically, the convergence appears to happen always (as long as Σ_π is finite). Similarly, in case of the ASM, it is relatively easy to see [24] that the mean acceptance rate $\mathbb{E}[\alpha_k] \to 0$ as the proposal increments get smaller $\eta_{k-1} \to -\infty$, and vice versa, $\mathbb{E}[\alpha_k] \to 1$ as $\eta_{k-1} \to \infty$, suggesting that a limit always exists but might not be unique [25]. In case π is elliptically symmetric, the limit point of RAM coincides with the shape of π, up to a constant [20], as does the ASM + AM.

3.5 Summary and Discussion on the Methods

The adaptive RWM algorithms are simple and generally well behaved when the corresponding non-adaptive RWM are. This requires essentially the following:

- Moderate dimension d.
- Essentially unimodal target π, that is, π does not have well-separated nodes.
- Target π has bounded support or sufficiently regular tails that are fast decaying (superexponentially, such as Gaussian [26]).

The tail decay rate may be enforced by a suitably chosen prior, for instance a Gaussian. There are some theoretical results about the stability of the algorithms under further technical conditions [13, 19, 27]. If the algorithms are modified to include auxiliary stabilizing mechanisms, typically enforcing the values of ξ_k to a compact set, they may be guaranteed to be valid even more generally [8, 9, 18].

The recommended step sizes γ_k differ between the algorithms, due to their different characteristics. The step sizes must ensure that the adaptations remain "effective," in the sense that $\sum_k \gamma_k = \infty$. If this condition was not met, the algorithms could converge prematurely to a spurious limit. The limiting behavior of the methods may be guaranteed to satisfy a central limit theorem if $\sum_k \gamma_k^2 < \infty$ [8]. If we focus on sequences with polynomially decaying tails $O(n^{-\beta})$, then the above are satisfied with $\beta \in (1/2, 1]$. As commented earlier, the given step size for the AM makes the algorithm behave similarly in the limit to the original algorithm, where Σ_k were sample covariances. However, with bounded increments, such as with the ASM, the choice $\gamma_k = O(k^{-1})$ would lead to η_k that can deviate from η_0 at most of order $\log k$, rendering the adaptation ineffective. With ASM + AM, there is potential interaction between the covariance and scale adaptations, and using different step sizes might amplify this. Because RAM is similar to ASM, the suggested step size decay rate is similar, but because of directional adaptation, the step size is inflated with dimension.

In a univariate case, ASM is the recommended method because of its simplicity. In a general multivariate case, using AM, ASM + AM, or RAM is recommended, because these methods can adapt to different scaling of variables and correlations. In simple scenarios, they work equally well, but in some cases, differences may arise [20]. All of the adaptive RWM methods have good theoretical backing, but the results are not complete. If the user is in doubt, adaptation may also be stopped (typically after burn-in), to ensure theoretical validity with minimal conditions (irreducibility).

4 Multimodal Targets with Parallel Tempering

RWM is based on small increments of X_k, which are accepted or rejected individually. This makes RWM behave poorly with multimodal distributions, where reaching one mode from another would require several steps that are each accepted with small probability. The higher the dimension, the more easily this problem arises, because the steps made by the RWM need to be smaller in higher dimension, of order $O(d^{-1/2})$ [11].

If further information about the π, such as location of modes, is available, tailored transitions may be designed. We focus on the case where little is known about π a priori. Then, a general "tempering" procedure may be applied, where the target density $\pi(x)$ is modified to one proportional to $\pi^{\beta}(x)$, where $\beta \in (0,1)$ is an "inverse temperature" parameter; equivalently, the unnormalized log-density of the modified target is $\beta\ell(x)$. The lower the value of β, the more π is "flattened" by making the modes less pronounced and the unlikely states more likely.

The PT or replica exchange algorithm [3] uses a number $L \geq 2$ of levels, with inverse temperatures $1 = \beta^{(1)} > \beta^{(2)} > \cdots > \beta^{(L)} > 0$ and corresponding unnormalized log-targets $\tilde{\ell}_{\beta^{(i)}}(x) := \beta^{(i)}\ell(x)$. The algorithm updates a joint state $X_{k-1}^{(1:L)} \to X_k^{(1:L)}$ in two stages. The first step consists of independent updates $X_{k-1}^{(1)} \to X_k^{(1)}, \ldots, X_{k-1}^{(L)} \to X_k^{(L)}$ with MCMCs targetting $\tilde{\ell}_{\beta^{(1)}}, \ldots, \tilde{\ell}_{\beta^{(L)}}$, respectively. The second step involves an attempt to swap the states of two random adjacent levels, $X_k^{(I)} \longleftrightarrow X_k^{(I-1)}$, where $I \sim U\{2, \ldots, L\}$, which is accepted with probability

$$\min\left\{1, \frac{\pi^{\beta^{(I)}}(X^{(I-1)})\pi^{\beta^{(I-1)}}(X^{(I)})}{\pi^{\beta^{(I-1)}}(X^{(I-1)})\pi^{\beta^{(I)}}(X^{(I)})}\right\} \tag{6}$$

which ensures that $X_b^{(1)}, \ldots, X_n^{(1)}$ approximates the target distribution of interest π.

An adaptive version of this algorithm, the adaptive parallel tempering (APT) [28] which uses adaptive RWM together with inverse temperature adaptation, is summarized in Algorithm 3.

The temperature adaptation in Algorithm 3 implements the ASM adaptation (2) to $\rho^{(i)}$, which parameterize the log differences of the consecutive temperatures, via $1/\beta^{(i+1)} = 1/\beta^{(i)} + e^{\rho^{(i)}}$. The mean acceptance rate of the swaps between levels $\{i-1, i\}$ was shown in Ref. 28 to be monotonically decreasing with respect to $\rho^{(i)}$, and therefore the algorithm converges to $\beta_*^{(1:L)}$, which ensures constant $\alpha_* = 0.234$ acceptance rate of the swaps. This rule of thumb, which is equivalent with RWM rule 2, is loosely justified in the APT context [29] and appears to work well.

In a multimodal case, the lower level RWM moves act "locally," exploring one mode at a time. The AM often works well under unimodality, but in the multimodal case, the AM proposal may become too wide, leading to poor acceptance rate. Therefore, we suggest to use either ASM + AM or RAM within APT. We use the step size $\gamma_k = (L-1)(k+1)^{-2/3}$ for the temperature adaptation, which is similar to the one suggested with ASM, with an additional factor accounting for random update to one of $L-1$ temperature difference adaptations.

Algorithm 3. $X_{1:n}^{(1)} \leftarrow \mathrm{APT}(\ell, x_0, n, L)$

Initialize $\xi_0^{(i)}$, set $\rho_0^{(1:L-1)} \leftarrow 0$, $\beta_0^{(i)} = i^{-1}$ for $i \in \{1:L\}$, $X_0^{(i)} \leftarrow x_0$, and $P_0^{(i)} \leftarrow \ell_{\beta_0^{(i)}}(x_0)$.
for $k = 1, \ldots, n$ **do**:
  for $i = 1, \ldots, L$ **do**:
    $(\tilde{X}_k^{(i)}, \tilde{P}_k^{(i)}; \tilde{A}_k^{(i)}, \tilde{Z}_k^{(i)}) \leftarrow \mathrm{RWMStep}\left(X_{k-1}^{(i)}, \beta_{k-1}^{(i)} P_{k-1}^{(i)}, \tilde{\ell}_{\beta_{k-1}^{(i)}}, \mathrm{Shape}(\xi_k^{(i)})\right)$
    $\xi_k^{(i)} \leftarrow \mathrm{Adapt}(k, \xi_{k-1}^{(i)}, \tilde{X}_k^{(i)}, \tilde{Z}_k^{(i)}, \tilde{A}_k^{(i)})$
  $\tilde{L}_k^{(i)} \leftarrow \tilde{P}_k^{(i)} / \beta_{k-1}^{(i)}$ for $i = 1, \ldots, L$.
  $(X_k^{(1:L)}, L_k^{(1:L)}, A_k, I_k) \leftarrow \mathrm{SwapStep}(\tilde{X}_k^{(1:L)}, \tilde{L}_k^{(1:L)}, \beta_{k-1}^{(1:L)})$
  $(\rho_k^{(1:L-1)}, \beta_k^{(1:L)}) \leftarrow \mathrm{AdaptTemp}(k, \rho_{k-1}^{(1:L-1)}, A_k, I_k)$
  $P_k^{(i)} \leftarrow \beta_k^{(i)} L_k^{(i)}$ for $i = 1, \ldots, L$.

function $\mathrm{SwapStep}(X^{(1:L)}, L^{(1:L)}, \beta^{(1:L)})$:
  $I \sim U\{1, \ldots, L-1\}$, $A \leftarrow \min\left\{1, \exp\left((\beta^{(I)} - \beta^{(I+1)})(L^{(I+1)} - L^{(I)})\right)\right\}$ and
  $U \sim U(0,1)$
  if $U \le A$ **then** swap $(X^{(I+1)}, X^{(I)}) \leftarrow (X^{(I)}, X^{(I+1)})$ and $(L^{(I+1)}, L^{(I)}) \leftarrow (L^{(I)}, L^{(I+1)})$
  return $(X^{(1:L)}, L^{(1:L)}, A, I)$

function $\mathrm{AdaptTemp}(k, \rho^{(1:L)}, A, I)$:
  $\tilde{\rho}^{(I)} \leftarrow \rho^{(I)} + \gamma_k(A - \alpha^*)$, and $\tilde{\rho}^{(i)} \leftarrow \rho^{(i)}$ for $i \neq I$.
  $T^{(1)} \leftarrow 1$ and $T^{(i+1)} = T^{(i)} + \exp(\tilde{\rho}^{(i)})$ for $i = 2, \ldots, L$.
  return $(\tilde{\rho}^{(1:L-1)}, \tilde{\beta}^{(1:L)})$ where $\tilde{\beta}^{(i)} = 1/T^{(i)}$.

In Bayesian statistics, the target distribution $\pi(x) \propto \mathrm{pr}(x)\mathrm{lik}(x)$, product of the prior density and the likelihood, respectively. Equivalently, the log-target factorizes to $\ell(x) = \ell_{\mathrm{pr}}(x) + \ell_{\mathrm{lik}}(x)$. Often, the prior distribution is regular and unimodal, and the multimodality is caused by the likelihood term only. In this case, it is advisable to "temper" only the log-likelihood part, so that $\tilde{\ell}_{\beta^{(i)}}(x) := \ell_{\mathrm{pr}}(x) + \beta^{(i)}\ell_{\mathrm{lik}}(x)$ [30]. This leads to slight modification of Algorithm 3, so that $\tilde{L}_k^{(i)} \leftarrow (\tilde{P}_k^{(i)} - \ell_{\mathrm{pr}}(\tilde{X}_k^{(i)}))/\beta_{k-1}^{(i)}$ and $P_k^{(i)} \leftarrow \ell_{\mathrm{pr}}(X_k^{(i)}) + \beta_k^{(i)} L_k^{(i)}$.

It is possible to further refine the APT algorithm using different swap strategies, for instance by alternating between odd and even swaps with large L [31], or to reduce the number of levels L adaptively [32]. Multimodal distributions are also considered in the framework presented in Ref. 33, which consists of an "exploratory" phase aiming to find the modes and a consequent sampling phase. The APT could be used in the former phase. It is possible to extend the PT by adding a transformation to the swap step, based on the information of the modes [34].

5 Dynamic Models with Particle Filters

Hidden Markov models (HMMs, also known as state-space models) are a flexible class of models often used in modern time-series analysis [35, 36]. The data $y^{(1:T)} = (y^{(1)}, \ldots, y^{(T)})$ are modeled conditionally independent given the latent Markov process $x^{(1:T)}$, with initial distribution $f_\theta^{(1)}(x^{(1)})$ and transitions $f_\theta^{(k)}(x^{(k)} \mid x^{(k-1)})$ and with observation densities $g_\theta^{(k)}(y^{(k)} \mid x^{(k)})$, all parameterized by (hyper)parameters θ with prior pr(θ). The full joint posterior of the parameters and the latent state satisfies $\pi(\theta, x^{(1:T)}) \propto \mathrm{pr}(\theta) p_\theta(x^{(1:T)}, y^{(1:T)})$, where

$$p_\theta(x^{(1:T)}, y^{(1:T)}) = f_\theta^{(1)}(x^{(1)}) g_\theta^{(1)}(y^{(1)} \mid x^{(1)}) \prod_{k=2}^{T} f_\theta^{(k)}(x^{(k)} \mid x^{(k-1)}) g_\theta^{(k)}(y^{(k)} \mid x^{(k)}) \tag{7}$$

In the context of HMMs, the parameters $\theta \in \mathbb{R}^d$ are often of moderate dimension, but the dimension of the latent process $x^{(1:T)}$ is proportional to the data record length T, making direct MCMC for $(\theta, x^{(1:T)})$ inefficient. The pioneering work [4] introduced "particle MCMC" methods for sampling from π. They combine MCMC with particle filters, a generic class of Monte Carlo method tailored for HMMs. Adaptive MCMC has been suggested to automatically design proposals for the hyperparameters θ within particle MCMC [4, 37, 38], and we discuss some guidelines on how this may be done in practice.

Algorithms 4 and 5 summarize the two distinct particle MCMC methods, the particle marginal Metropolis–Hastings (PMMH) and the particle Gibbs (PG) [4], with adaptation. The algorithms are written with generic particle filter parameters: the "proposals" $M_\theta^{(k)}$ and the "potentials" $G_\theta^{(k)}$. The simplest valid choice is $M_\theta^{(k)} \equiv f_\theta^{(k)}$ and $G_\theta^{(k)}(x^{(k)}) = g_\theta^{(k)}(y^{(k)} \mid x^{(k)})$, which is known as the bootstrap filter [39], but any other choice is valid as long as

$$M_\theta^{(1)}(x^{(1)}) G_\theta^{(1)}(x^{(1)}) \prod_{k=2}^{T} M_\theta^{(k)}(x^{(k)} \mid x^{(k-1)}) G_\theta^{(k)}(x^{(k)}) \equiv p_\theta(x^{(1:T)}, y^{(1:T)}) \tag{8}$$

as a function of $(\theta, x^{(1:T)})$. (Note that both $M_\theta^{(k)}$ and $G_\theta^{(k)}$ may depend on $y^{(1:T)}$, but this dependence is suppressed from the notation.)

Algorithm 4. $(\Theta_{1:n}, X_{1:n}^{(1:T)}) \leftarrow \mathrm{AdaptivePMMH}(\ell_{\mathrm{pr}}, \theta_0, n, N, M_\theta^{(1:T)}, G_\theta^{(1:T)})$

Initialize ξ_0, $\Theta_0 \leftarrow \theta_0$, $P_0 \leftarrow \ell_{\mathrm{pr}}(\Theta_0)$ and $(V_0, X_0^{(1:T)}) \leftarrow \mathrm{PF}(M_{\Theta_0}^{(1:T)}, G_{\Theta_0}^{(1:T)}, N)$
for $k = 1, \ldots, n$ **do**:
  $\tilde{\Theta}_k \leftarrow \Theta_{k-1} + \mathrm{Shape}(\xi_{k-1}) Z_k$ where $Z_k \sim q$
  $\tilde{P}_k \leftarrow \ell_{\mathrm{pr}}(\tilde{\Theta}_k)$, $(\tilde{V}_k, \tilde{X}_k^{(1:T)}) \leftarrow \mathrm{PF}(M_{\tilde{\Theta}_k}^{(1:T)}, G_{\tilde{\Theta}_0}^{(1:T)}, N)$ and $U_k \sim U(0,1)$
  if $U_k \le \alpha_k := \min\{1, \exp(\tilde{P}_k + \tilde{V}_k - P_{k-1} - V_{k-1})\}$ **then**:
    $(\Theta_k, P_k, V_k, X_k^{(1:T)}) \leftarrow (\tilde{\Theta}_k, \tilde{P}_k, \tilde{V}_k, \tilde{X}_k^{(1:T)})$
  else:
    $(\Theta_k, P_k, V_k, X_k^{(1:T)}) \leftarrow (\Theta_{k-1}, P_{k-1}, V_{k-1}, X_{k-1}^{(1:T)})$
  $\xi_k \leftarrow \mathrm{Adapt}(k, \xi_{k-1}, \Theta_k, Z_k, \alpha_k)$.

The functions PF($\cdot$) and CPF($\cdot$) are abstractions of the "particle filter" and the "conditional particle filter," respectively [4]. More specifically, PF($\cdot$, N) refers to the particle filter run with N particles and the given parameters, and the output consists of the logarithm of the marginal likelihood estimate, and one trajectory picked from the generated particle system. PF only requires that $M_\theta^{(k)}(\,\cdot \mid x)$ can be sampled from, and that (logarithm of) $G_\theta^{(k)}$ can be calculated. The call of CPF($\cdot$) is similar, with the third argument being the previous (reference) trajectory. We refer the reader to consult the original paper [4] for

Algorithm 5. $(\Theta_{1:n}, X_{1:n}^{(1:T)}) \leftarrow \text{AdaptivePG}(\ell_{\text{pr}}, \theta_0, n, N, M_\theta^{(1:T)}, G_\theta^{(1:T)})$

Initialize ξ_0, $\Theta_0 \leftarrow \theta_0$, $P_0 \leftarrow \ell_{\text{pr}}(\Theta_0)$ and $(-, X_0^{(1:T)}) \leftarrow \text{PF}(M_{\Theta_0}^{(1:T)}, G_{\Theta_0}^{(1:T)}, N)$
for $k = 1, \dots, n$ **do**:
  $\tilde{\Theta}_k \leftarrow \Theta_{k-1} + \text{Shape}(\xi_{k-1})Z_k$ where $Z_k \sim q$, and $\tilde{P}_k \leftarrow \ell_{\text{pr}}(\tilde{\Theta}_k)$
  $V_{k-1} \leftarrow \log p_{\Theta_{k-1}}(X_{k-1}^{(1:T)}, y^{(1:T)})$, $\tilde{V}_k \leftarrow \log p_{\tilde{\Theta}_k}(X_{k-1}^{(1:T)}, y^{(1:T)})$ and $U_k \sim U(0,1)$
  if $U_k \le \alpha_k := \min\{1, \exp(\tilde{P}_k + \tilde{V}_k - P_{k-1} - V_{k-1})\}$ **then**:
    $(\Theta_k, P_k) \leftarrow (\tilde{\Theta}_k, \tilde{P}_k)$
  else:
    $(\Theta_k, P_k) \leftarrow (\Theta_{k-1}, P_{k-1})$
  $\xi_k \leftarrow \text{Adapt}(k, \xi_{k-1}, \Theta_k, Z_k, \alpha_k)$.
  $X_k^{(1:T)} \leftarrow \text{CPF}(M_{\Theta_k}^{(1:T)}, G_{\Theta_k}^{(1:T)}, X_{k-1}^{(1:T)}, N)$

details, but remark that the backward sampling variant of the CPF [40, 41] may be used if the (logarithmic) density values of $M_\theta^{(k)}(x' \mid x)$ can be calculated. It is recommended if applicable, because it can improve the performance dramatically and is provably stable with large T [42].

In principle, it is possible to apply any simple RWM adaptation of Section 3 within both Algorithms 4 and 5. However, in case of PMMH (Algorithm 4), the mean acceptance rate depends both on Shape(ξ_k) and on the number of particles N, making it difficult to know what desired acceptance rate value α_* should be used. Therefore, it is simpler to employ the AM adaptation, which does not rely on acceptance rate, but only on the posterior covariance, which is independent of N. The number of particles N needs to be chosen per application; some guidelines are given with related theoretical developments [43, 44]. When using adaptation within PMMH, the number of particles may be best chosen slightly higher than the guidelines suggest (yielding at least 10% acceptance rate, say), in order to avoid potential instability of the adaptation.

In the case of PG, the update of θ is a Metropolis-within-Gibbs update targetting the posterior conditional $\theta \mid x^{(1:T)}$. This step is independent of N, and the acceptance rate remains an effective proxy for adaptation. Therefore, we suggest to use either AM + ASM or the RAM adaptation with PG. The "global" nature of AM adaptation, as discussed in Section 4, makes it inappropriate for sampling the conditional distributions, which are typically more concentrated than the posterior marginal.

It may be possible to design more efficient independent proposals for the PMMH, by fitting a mixture distribution to the posterior marginal of θ [37, 45]. This may be achieved by first running Algorithm 4 or 5 and then using the simulated samples for mixture fitting.

6 Discussion

We reviewed a set of adaptive MCMC methods applicable for some general model classes. Our focus was on relatively simple methods, which require minimal user specification. More refined methods may improve the efficiency of the methods but often come with a cost of further user specification, in the form of more careful choice of algorithm or their parameters.

Table 2 Summary of recommended algorithms for specific problems and their step sizes.

Method	PMMH	PG	MwG-1	MwG-d	PT	γ_k
AM	✓	×	×	×	×	$(k+1)^{-1}$
ASM	×	×	✓	×	×	$k^{-2/3}$
ASM + AM	×	✓	×	✓	✓	$(k+1)^{-2/3}$
RAM	×	✓	×	✓	✓	$\min\{1, d \cdot (k+1)^{-2/3}\}$

Adaptation may be applied in a straightforward manner with hierarchical models, using multiple independent adaptations for individual Metropolis-within-Gibbs updates of either single parameters or blocks of parameters [46–48]. This avoids conjugacy constraints, and using block updates for tightly correlated variables may lead to improved mixing. Some variables could also be updated by pure Gibbs moves (if perfect sampling of the conditional is possible). However, to the knowledge of the author, there is no general-purpose software that would allow for this, even though such an extension of a BUGS-type implementation would be technically straightforward.

Table 2 summarizes the recommendations which RWM adaptations are appropriate in different contexts: dynamic models (PMMH and PG methods), hierarchical models (Metropolis-within-Gibbs, univariate and multivariate update), and with multimodal targets (PT). The recommended step size sequence is also shown.

Unfortunately, all MCMC methods come with their strengths and weaknesses, and therefore the "end user" may need to make certain choices. Hamiltonian Monte Carlo (HMC)-type methods, such as those implemented in STAN software [49], have recently become very popular. They have shown great promise for challenging inference problems but also come with limitations. For instance, HMC cannot be used to sample discrete variables, and the model may need to be rescaled and/or reparameterized before inference. The more domain-specific methods, such as particle MCMC in the time-series context, also tend to outperform general-purpose methods, such as HMC. Inference software that would allow for flexibly using all successful samplers to date, including the HMC-type methods, Gibbs sampling, particle MCMC, and adaptive methods, could provide a way forward and push the boundaries of ergonomic practical Bayesian inference.

Acknowledgments

The author was supported by Academy of Finland grants 274740, 312605, and 315619.

Notes

1 https://github.com/mvihola/AdaptiveMCMC.jl

2 https://github.com/mvihola/AdaptiveParticleMCMC.jl

References

1 Lunn, D.J., Thomas, A., Best, N., and Spiegelhalter, D. (2000) WinBUGS – a Bayesian modelling framework: concepts, structure, and extensibility. *Stat. Comput.*, 10 (4), 325–337.

2 Metropolis, N., Rosenbluth, A.W., Rosenbluth, M.N., *et al.* (1953) Equations of state calculations by fast computing machines. *J. Chem. Phys.*, 21 (6), 1087–1092.

3 Swendsen, R.H. and Wang, J.-S. (1986) Replica Monte Carlo simulation of spin-glasses. *Phys. Rev. Lett.*, 57 (21), 2607–2609.

4 Andrieu, C., Doucet, A., and Holenstein, R. (2010) Particle Markov chain Monte Carlo methods. *J. R. Stat. Soc. Ser. B Stat. Methodol.*, 72 (3), 269–342.

5 Andrieu, C. and Thoms, J. (2008) A tutorial on adaptive MCMC. *Stat. Comput.*, 18 (4), 343–373.

6 Haario, H., Saksman, E., and Tamminen, J. (2001) An adaptive Metropolis algorithm. *Bernoulli*, 7 (2), 223–242.

7 Atchadé, Y.F. and Rosenthal, J.S. (2005) On adaptive Markov chain Monte Carlo algorithms. *Bernoulli*, 11 (5), 815–828.

8 Andrieu, C. and Moulines, É. (2006) On the ergodicity properties of some adaptive MCMC algorithms. *Ann. Appl. Probab.*, 16 (3), 1462–1505.

9 Roberts, G.O. and Rosenthal, J.S. (2007) Coupling and ergodicity of adaptive Markov chain Monte Carlo algorithms. *J. Appl. Probab.*, 44 (2), 458–475.

10 Bezanson, J., Edelman, A., Karpinski, S., and Shah, V.B. (2017) Julia: a fresh approach to numerical computing. *SIAM Rev.*, 59 (1), 65–98.

11 Roberts, G.O., Gelman, A., and Gilks, W.R. (1997) Weak convergence and optimal scaling of random walk Metropolis algorithms. *Ann. Appl. Probab.*, 7 (1), 110–120.

12 Dongarra, J.J., Bunch, J.R., Moler, C.B., and Stewart, G.W. (1979) *LINPACK Users' Guide*, Society for Industrial and Applied Mathematics, Philadelphia, PA.

13 Vihola, M. (2011) Can the adaptive Metropolis algorithm collapse without the covariance lower bound? *Electron. J. Probab.*, 16, 45–75.

14 Bai, Y., Roberts, G.O., and Rosenthal, J.S. (2011) On the containment condition for adaptive Markov chain Monte Carlo algorithms. *Adv. Appl. Stat.*, 21 (1), 1–54.

15 Haario, H., Laine, M., Mira, A., and Saksman, E. (2006) DRAM: efficient adaptive MCMC. *Stat. Comput.*, 16 (4), 339–354.

16 Gilks, W.R., Roberts, G.O., and Sahu, S.K. (1998) Adaptive Markov chain Monte Carlo through regeneration. *J. Am. Stat. Assoc.*, 93 (443), 1045–1054.

17 Andrieu, C. and Robert, C.P. (2001) Controlled MCMC for Optimal Sampling. Technical Report Ceremade 0125, Université Paris Dauphine.

18 Atchadé, Y. and Fort, G. (2010) Limit theorems for some adaptive MCMC algorithms with subgeometric kernels. *Bernoulli*, 16 (1), 116–154.

19 Vihola, M. (2011) On the stability and ergodicity of adaptive scaling Metropolis algorithms. *Stochastic Process. Appl.*, 121 (12), 2839–2860.

20 Vihola, M. (2012) Robust adaptive Metropolis algorithm with coerced acceptance rate. *Stat. Comput.*, 22 (5), 997–1008.

21 Siltala, L. and Granvik, M. (2020) Asteroid mass estimation with the robust adaptive Metropolis algorithm. *Astron. Astrophys.*, 633, A46.

22 Robbins, H. and Monro, S. (1951) A stochastic approximation method. *Ann. Math. Stat.*, 22, 400–407.

23 Benveniste, A., Métivier, M., and Priouret, P. (1990) *Adaptive Algorithms and Stochastic Approximations*, Number 22 in Applications of Mathematics, Springer-Verlag, Berlin.

24 Vihola, M. (2010) On the convergence of unconstrained adaptive Markov chain Monte Carlo algorithms. PhD thesis. University of Jyväskylä.

25 Hastie, D. (2005) Toward automatic reversible jump Markov chain Monte Carlo. PhD thesis. University of Bristol.

26 Jarner, S.F. and Hansen, E. (2000) Geometric ergodicity of Metropolis algorithms. *Stochastic Process. Appl.*, 85 (2), 341–361.

27 Saksman, E. and Vihola, M. (2010) On the ergodicity of the adaptive Metropolis algorithm on unbounded domains. *Ann. Appl. Probab.*, 20 (6), 2178–2203.

28 Miasojedow, B., Moulines, E., and Vihola, M. (2013) An adaptive parallel tempering algorithm. *J. Comput. Graph. Stat.*, 22 (3), 643–664.

29 Roberts, G.O. and Rosenthal, J.S. (2014) Minimising MCMC variance via diffusion limits, with an application to simulated tempering. *Ann. Appl. Probab.*, 24 (1), 131–149.

30 Gwiazda, P., Miasojedow, B., and Rosińska, M. (2016) Bayesian inference for age-structured population model of infectious disease with application to varicella in Poland. *J. Theor. Biol.*, 407, 38–50.

31 Syed, S., Bouchard-Côté, A., Deligiannidis, G., and Doucet, A. (2019) Non-reversible parallel tempering: an embarassingly parallel MCMC scheme. Preprint arXiv:1905.02939.

32 Łacki, M.K. and Miasojedow, B. (2016) State-dependent swap strategies and automatic reduction of number of temperatures in adaptive parallel tempering algorithm. *Stat. Comput.*, 26 (5), 951–964.

33 Pompe, E., Holmes, C., and Łatuszyński, K. (2018) A framework for adaptive MCMC targeting multimodal distributions. Preprint arXiv:1812.02609.

34 Tawn, N.G. and Roberts, G.O. (2019) Accelerating parallel tempering: quantile tempering algorithm (QuanTA). *Adv. Appl. Probab.*, 51 (3), 802–834.

35 Durbin, J. and Koopman, S.J. (2012) *Time Series Analysis by State Space Methods*, 2nd edn, Oxford University Press, New York.

36 Cappé, O., Moulines, E., and Rydén, T. (2005) *Inference in Hidden Markov Models*, Springer, New York.

37 Silva, R., Giordani, P., Kohn, R., and Pitt, M. (2009) Particle filtering within adaptive Metropolis Hastings sampling. Preprint arXiv:0911.0230.

38 Peters, G.W., Hosack, G.R., and Hayes, K.R. (2010) Ecological non-linear state space model selection via adaptive particle Markov chain Monte Carlo (AdPMCMC). Preprint arXiv:1005.2238.

39 Gordon, N.J., Salmond, D.J., and Smith, A.F.M. (1993) Novel approach to nonlinear/non-Gaussian Bayesian state estimation. *IEE Proc.-F*, 140 (2), 107–113.

40 Whiteley, N. (2010) Discussion on particle Markov chain Monte Carlo methods. *J. R. Stat. Soc. Ser. B Stat. Methodol.*, 72 (3), 306–307.

41 Lindsten, F., Jordan, M.I., and Schön, T.B. (2014) Particle Gibbs with ancestor sampling. *J. Mach. Learn. Res.*, 15 (1), 2145–2184.

42 Lee, A., Singh, S.S., and Vihola, M. Coupled conditional backward sampling particle filter. *Ann. Stat.*, to appear.

43 Doucet, A., Pitt, M.K., Deligiannidis, G., and Kohn, R. (2015) Efficient implementation of Markov chain Monte Carlo when using an unbiased likelihood estimator. *Biometrika*, 102 (2), 295–313.

44 Sherlock, C., Thiery, A.H., Roberts, G.O., and Rosenthal, J.S. (2015) On the efficiency of pseudo-marginal random walk Metropolis algorithms. *Ann. Stat.*, 43 (1), 238–275.

45 Knape, J. and De Valpine, P. (2012) Fitting complex population models by combining particle filters with Markov chain Monte Carlo. *Ecology*, 93 (2), 256–263.

46 Haario, H., Saksman, E., and Tamminen, J. (2005) Componentwise adaptation for high dimensional MCMC. *Comput. Stat.*, 20 (2), 265–274.

47 Roberts, G.O. and Rosenthal, J.S. (2009) Examples of adaptive MCMC. *J. Comput. Graph. Stat.*, 18 (2), 349–367.

48 Vihola, M. (2010) Grapham: graphical models with adaptive random walk Metropolis algorithms. *Comput. Stat. Data Anal.*, 54 (1), 49–54.

49 Gelman, A., Lee, D., and Guo, J. (2015) Stan: a probabilistic programming language for Bayesian inference and optimization. *J. Educ. Behav. Stat.*, 40 (5), 530–543.

9

Advances in Importance Sampling

Víctor Elvira[1] and Luca Martino[2]

[1] *School of Mathematics, University of Edinburgh, Edinburgh, UK*
[2] *Universidad Rey Juan Carlos de Madrid, Madrid, Spain*

1 Introduction and Problem Statement

In many problems of science and engineering, intractable integrals must be approximated. Let us denote an integral of interest

$$I(f) = E_{\widetilde{\pi}}[f(\mathbf{x})] = \int f(\mathbf{x})\widetilde{\pi}(\mathbf{x})d\mathbf{x} \tag{1}$$

where $f : \mathbb{R}^{d_x} \rightarrow \mathbb{R}$, and $\widetilde{\pi}(\mathbf{x})$ is a distribution of the r.v. $\mathbf{X} \in \mathbb{R}^{d_x}$.[1] Note that although Equation (1) involves a distribution, more generic integrals could be targeted with the techniques described below.

The integrals of this form appear often in the Bayesian framework, where a set of observations are available in $\mathbf{y} \in \mathbb{R}^{d_y}$, and the goal is in inferring some hidden parameters and/or latent variables $\mathbf{x} \in \mathbb{R}^{d_y}$ that are connected to the observations through a probabilistic model. The information provided by the observations is compacted in the likelihood function $\ell(\mathbf{y}|\mathbf{x})$, and the prior knowledge on $\mathbf{x}$ is encoded in the prior distribution $p_0(\mathbf{x})$. Both sources of information are fused to create through the simple Bayes' rule the posterior probability density function (pdf), also called *target distribution*, given by

$$\widetilde{\pi}(\mathbf{x}) = p(\mathbf{x}|\mathbf{y}) = \frac{\ell(\mathbf{y}|\mathbf{x})p_0(\mathbf{x})}{Z(\mathbf{y})} \tag{2}$$

where $Z(\mathbf{y}) = \int \ell(\mathbf{y}|\mathbf{x})p_0(\mathbf{x})d\mathbf{x}$ is the marginal likelihood (a.k.a., partition function, Bayesian evidence, model evidence, or normalizing constant) [1, 2].

In most models of interest $Z(\mathbf{y})$ is unknown, and in many applications it must be approximated [1–3]. But even when its approximation is not needed, the unavailability of $Z(\mathbf{y})$ implies that the posterior can be evaluated only up to that (unknown) constant, that is, we can only evaluate

$$\pi(\mathbf{x}) = \ell(\mathbf{y}|\mathbf{x})p_0(\mathbf{x}) \tag{3}$$

that we denote as unnormalized target distribution.[2] Table 1 summarizes the notation of this chapter.

Computational Statistics in Data Science.
Edited by Walter W. Piegorsch, Richard A. Levine, Hao Helen Zhang and Thomas C. M. Lee.
 ISBN 978-1-11956107-1

Table 1 Summary of thc notation.

d_x	Dimension of the inference problem, $\mathbf{x} \in \mathbb{R}^d_x$
d_y	Dimension of the observed data, $\mathbf{y} \in \mathbb{R}^d_y$
$\mathbf{x}$	r.v. of interest; parameter to be inferred
$\mathbf{y}$	Observed data
$\ell(\mathbf{y}\|\mathbf{x})$	Likelihood function
$p_0(\mathbf{x})$	Prior pdf
$\tilde{\pi}(\mathbf{x})$	Posterior pdf (target), $\tilde{\pi}(\mathbf{x}) \equiv p(\mathbf{x}\|\mathbf{y})$
$\pi(\mathbf{x})$	Posterior density function (unnormalized target), $\pi(\mathbf{x}) \equiv \ell(\mathbf{y}\|\mathbf{x})g(\mathbf{x}) \propto \overline{\pi}(\mathbf{x})$
$q(\mathbf{x})$	Proposal density
Z	Normalizing constant or marginal likelihood, $Z \equiv Z(\mathbf{y})$
$I(f)$	Integral to be approximated, $I(f) \equiv E_{\tilde{\pi}}[f(\mathbf{x})]$.

The integral $I(f)$ cannot be computed in a closed form in many practical scenarios and hence must be approximated. The approximation methods can be divided into either deterministic or stochastic. While many deterministic numerical methods are available in the literature [4–7], it is in general accepted that they tend to become less efficient than stochastic approximations when the problem dimension d_x grows.

1.1 Standard Monte Carlo Integration

The Monte Carlo approach consists in approximating the integral $I(f)$ in Equation (1) with random samples [8–13]. In the standard Monte Carlo solution (often called instinctively vanilla/raw/classical/direct Monte Carlo), N samples $\mathbf{x}_n$ are independently simulated from $\widetilde{\pi}(\mathbf{x})$. The standard Monte Carlo estimator is built as

$$\overline{I}^N(f) = \frac{1}{N} \sum_{t=1}^{N} f(\mathbf{x}_n) \tag{4}$$

First, note that $\overline{I}^N(f)$ is unbiased since $E_{\tilde{\pi}}[\overline{I}^N(f)] = I(f)$. Moreover, due to the weak law of large numbers, it can be shown that $\overline{I}_N$ is consistent and then converges in probability to the true value I, that is, $\overline{I}^N(f) \xrightarrow{p} I(f)$, which is equivalent to stating that, for any positive number $\epsilon > 0$, we have $\lim_{N\to\infty} \Pr(|\overline{I}^N(f) - I(f)| > \epsilon) = 0$. The variance of $\overline{I}^N(f)$ is simply $\overline{\sigma}^2 = \frac{1}{N}(I(f^2) - I(f)^2)$. If the second moment is finite, $I(f^2) < \infty$, then the central limit theorem (CLT) applies, and the estimator converges in distribution to a well-defined Gaussian when N grows to infinity, that is,

$$\sqrt{N}\left(\overline{I}^N(f) - I(f)\right) \xrightarrow{d} \mathcal{N}(0, \overline{\sigma}^2) \tag{5}$$

There exist multiple families of Monte Carlo Methods [13]. We address the interested reader to the articles in Markov chain Monte Carlo (including Metropolis–Hastings) and Gibbs sampling and previous articles in importance sampling.

2 Importance Sampling

2.1 Origins

The first use of the importance sampling (IS) methodology dates from 1950 for rare event estimation in statistical physics, in particular for the approximation of the probability of nuclear particles penetrating shields [14]. IS was later used as a variance reduction technique when standard Monte Carlo integration was not possible and/or not efficient [15]. The renewed interest in IS has run in parallel with the hectic activity in the community of Bayesian analysis and its ever-increasing computational demands. In most cases, the posterior in (2) is not available due to the intractability of the normalizing constant. See Ref. 16 for a previous review in IS.

2.2 Basics

Let us start defining the proposal pdf, $q(\mathbf{x})$, used to simulate the samples. It is widely accepted that the proposal is supposed to have heavier tails than the target, that is, the target $\tilde{\pi}(\mathbf{x})$ decays faster than $q(\mathbf{x})$ when $\mathbf{x}$ is *far* from the region where most of the probability mass is concentrated. However, this usual restriction is too vague, and it will be clarified below. Here, we simply stick to the restriction that $q(\mathbf{x}) > 0$ for all $\mathbf{x}$ where $\tilde{\pi}(\mathbf{x})f(\mathbf{x}) \neq 0$. IS is constituted of two simple steps:

1. *Sampling*: N samples are simulated as

$$\mathbf{x}_n \sim q(x), \qquad n = 1, \dots, N \tag{6}$$

2. *Weighting*: Each sample receives an associated importance weight given by

$$w_n = \frac{\pi(\mathbf{x}_n)}{q(\mathbf{x}_n)}, \qquad n = 1, \dots, N \tag{7}$$

The importance weights describe how representative the samples simulated from $q(\mathbf{x})$ are when one is interested in computing integrals w.r.t. $\tilde{\pi}(\mathbf{x})$. The set of N weighted samples can be used to approximate the generic integral $I(f)$ of Equation (1) by the two following IS estimators:

- Unnormalized (or nonnormalized) IS (UIS) estimator:

$$\hat{I}^N(f) = \frac{1}{NZ}\sum_{n=1}^{N} w_n f(\mathbf{x}_n) \tag{8}$$

  Note that the UIS estimator can be used only when Z is known.
- Self-normalized IS (SNIS) estimator:

$$\tilde{I}^N(f) = \sum_{n=1}^{N} \overline{w}_n f(\mathbf{x}_n) \tag{9}$$

  where

$$\overline{w}_n = \frac{w_n}{\sum_{j=1}^{N} w_j} \tag{10}$$

  are the normalized weights.

The derivation of the SNIS estimator departs from the UIS estimator of Equation (8), substituting Z by its unbiased estimator [13]

$$\widehat{Z} = \frac{1}{N}\sum_{n=1}^{N} w_n \tag{11}$$

After a few manipulations, one recovers Equation (10). The normalized weights also allow to approximate the target distribution by

$$\widetilde{\pi}^N(\mathbf{x}) = \sum_{n=1}^{N} \overline{w}_n \delta(\mathbf{x} - \mathbf{x}_n) \tag{12}$$

where δ represents the Dirac measure.

2.3 Theoretical Analysis

The UIS estimator is unbiased since it can be easily proven that $\mathrm{E}_q[\widehat{I}^N(f)] = I(f)$. Its variance $\mathrm{E}_q[\widehat{I}^N(f)] = \frac{\sigma_q^2}{N}$ is given by

$$\sigma_q^2 = \int \frac{(f(\mathbf{x})\widetilde{\pi}(\mathbf{x}) - I(f)q(\mathbf{x}))^2}{q(\mathbf{x})} d\mathbf{x} \tag{13}$$

if $q(\mathbf{x}) > 0$ for all $\mathbf{x}$ where $\widetilde{\pi}(\mathbf{x})f(\mathbf{x}) \neq 0$, as we have stated above [17]. We remark that it is not strictly necessary to have a proposal with heavier tails than the target distribution as long as $\sigma_q^2 < \infty$. One counterexample is a case where $f(\mathbf{x})$ decays fast enough to compensate the heavier tails of the target distribution. Another counterexample is a case where $f(\mathbf{x})$ takes nonzero and finite values only in a bounded set.

Note that $q(\mathbf{x})$ is chosen by the practitioner, and a good choice is critical for the efficiency of IS. Let us first suppose that $\mathrm{sign}(f(\mathbf{x}))$ is constant for all $\mathbf{x}$ and $I(f) \neq 0$. Let us also suppose that it is possible to simulate from

$$q^*(\mathbf{x}) = \frac{f(\mathbf{x})\widetilde{\pi}(\mathbf{x})}{\int f(\mathbf{z})\widetilde{\pi}(\mathbf{z})d\mathbf{z}} \tag{14}$$

Then, the UIS estimator, for any $N \geq 1$ number of samples, yields a zero-variance unbiased estimator, since the numerator in (13) is zero, and hence $\sigma_q^2 = 0$. However, it is very unlikely to have access to the proposal of (14). The main reason is that its normalizing constant is exactly the intractable integral we are trying to approximate, $I(f)$. However, $q^*(\mathbf{x})$ gives the useful intuition that the proposal should have mass proportional to the targeted integrand in Equation (1). More precisely, inspecting (13), we see that the efficiency is penalized with the mismatch of $f(\mathbf{x})\widetilde{\pi}(\mathbf{x})$ and $q(\mathbf{x})$, with this penalization amplified inversely proportional to the density $q(\mathbf{x})$. This explains the usual safe practice of overspreading the proposal. The case where sign $(f(x))$ alternates can be easily modified by splitting the function as $f(x) = f_+(x) + f_-(x)$, where $f_+(x)$ is nonnegative, and $f_-(x)$ is nonpositive. It is easy to show that with the use of two proposals and $N = 2$, a zero-variance estimator is possible [17, Section 9.13]. In summary, the UIS estimator, $\widehat{I}^N(f)$, is unbiased, while the $\widetilde{I}^N(f)$ is only asymptotically unbiased, that is, with a bias that goes to 0 when N grows to infinity. Both UIS and SNIS are consistent estimators of I with a variance that depends on the discrepancy between $\pi(\mathbf{x})|f(\mathbf{x})|$ and $q(\mathbf{x})$, although the variance of the SNIS is more difficult to evaluate, and its bias also plays a central role when N is not large enough [17].

When several different moments f of the target must be estimated, a common strategy in IS is to decrease the mismatch between the proposal $q(\mathbf{x})$ and the target $\tilde{\pi}(\mathbf{x})$ [18]. This is equivalent to minimizing the variance of the weights and consequently the variance of the estimator $\widehat{Z}$, and it is closely linked to the diagnostics of Section 2.4.

2.4 Diagnostics

It is a legitimate question to wonder about the efficiency of the set of simulated weighted samples in the task of approximating the target distribution and/or moments of it. Usual metrics of efficiency involve the computation of the variance of the IS estimators. However, the computation of those variances is intractable, and even more, their approximation is usually a harder problem than computing Equation (1) (see Ref. 17, Chapter 9.3 for a discussion). A classic diagnostic metric in the IS literature [19] is

$$\widehat{\text{ESS}} = \frac{1}{\sum_{n=1}^{N} \bar{w}_n^2} \tag{15}$$

Note that $1 \leq \widehat{\text{ESS}} \leq N$, taking the value $\widehat{\text{ESS}} = 1$, when one $w_j = 1$ and hence $w_i = 0$, for all $i \neq j$. Therefore, $\widehat{\text{ESS}} = N$ only when $w_j = 1/N$ for all $j = 1, \ldots, N$. Hence, $\widehat{\text{ESS}}$ measures the discrepancy among normalized weights. This diagnostic is commonly called *effective sample size*, although it is an approximation of the more reasonable but intractable diagnostic given by [20]

$$\text{ESS}^* = N \frac{\text{Var}[\bar{I}^N]}{\text{MSE}[\tilde{I}^N]} \tag{16}$$

Then, ESS^* can be interpreted as the number of standard Monte Carlo that are necessary to obtain the same performance (in terms of MSE) as with the SNIS estimator with N samples. The interested reader can find the derivation from ESS^* to $\widehat{\text{ESS}}$ through a series of approximations and assumptions that rarely hold (see Ref. 20 for a thorough analysis). In practice, a low $\widehat{\text{ESS}}$ is a symptom of malfunctioning, but a high $\widehat{\text{ESS}}$ does not necessarily imply good behavior of the IS method.

New ESS-like methods have been proposed in the past years. In Refs 21 and 22, novel discrepancy measures with similar properties to $\widehat{\text{ESS}}$ are proposed and discussed, mitigating some of the deficiencies of the original diagnostic. For instance, an alternative to ESS^* is using $1/\max(\bar{w}_n)$, which preserves some of those properties (e.g., it takes values between 1 and N, being 1 if all the normalized weights are zero except one, and N if all weights are the same). Another metric in the same spirit has been recently proposed in Ref. 23. Finally, the use of the importance trick within quadrature schemes has been recently proposed [24, 25]. Note that these *importance quadrature* schemes are not stochastic but strongly inspired in IS and its variants.

2.5 Other IS Schemes

The research in IS methods has been very active in the past decade not only in the development of novel methodology but also for increasing the understanding and the theoretical behavior of IS-based methods. For instance, Agapiou *et al.* [26] unifies different perspectives about how many samples are necessary in IS for a given proposal and target densities, a problem that is usually related to some notion of distance (more precisely divergence) between the two densities. With a similar aim, in Ref. 27 it is shown that in a fairly general setting, IS requires a number of samples proportional to the exponential of the KL divergence between the target and the proposal densities. The notion of divergences between both densities is also explored in Ref. 28 through the Rényi generalized divergence, and in Ref. 29 in terms of the Pearson χ^2 divergence. Both divergences are connected with the variance of the $\widehat{Z}$ estimator in Equation (11).

2.5.1 Transformation of the importance weights

As described in Section 2.4, a large variability in the importance weights is usually responsible for a large variance in the IS estimators. One alternative is adapting the proposals in order to diminish the mismatch with the target, as we describe in Section 4. However, this usually means throwing away past weighted samples (or stick to large variance estimators from the early iterations). Another alternative is the nonlinear transformation of the IS weights. The first work in this line is the *truncated importance sampling* [30] where the standard unnormalized weights w_n are truncated as $w_n' = \min(w_n, \tau)$, where τ is a maximum value allowed for the transformed/truncated weights. The consistency of the method and a central limit theorem of the modified estimator are proved. This transformation of the weights was also proposed in Ref. 31 and called *nonlinear importance sampling* within an adaptive IS scheme (N-PMC algorithm). The convergence of this method is analyzed in Refs 29, 31, 32. The underlying problem that those methods fight is the right heavy tail in the distribution of the importance weights when the proposal is not well fit. In Ref. 33, the authors go a step beyond by characterizing the distribution of the importance weights with generalized Pareto distribution that fits the upper tail. Based on this fitting, a method is proposed for the stabilization of the importance weights. The authors provide proofs for consistency, finite variance, and asymptotic normality. See Ref. 34 for a review of the clipping methodologies.

2.5.2 Particle filtering (sequential Monte Carlo)

Particle filtering (also known as sequential Monte Carlo) is an IS-based methodology for performing approximate Bayesian inference on a hidden state that evolves over the time in state-space models, a class of probabilistic Markovian models. Due to the structure of the Bayesian network, it is possible to process sequentially and efficiently the observations related to the hidden state for building the sequence of filtering distributions (i.e., the posterior distribution of a given hidden state conditioned to all available observations). Particle filters (PFs) are based on IS, incorporating in most cases a resampling step that helps to increase the diversity of the particle approximation [35, 36]. Since the publication of the seminal paper [37] where the bootstrap PF is developed (BPF), a plethora of PFs have been proposed in the literature [38–42]. Advanced MIS and AIS techniques are often implicit in those algorithms, but they are rarely explicit. In Ref. 43, a novel perspective of BPF and auxiliary PF (APF) based on MIS is introduced. In these state-space models, the ESS and its approximations are also used as diagnostics metrics for PF (Section 2.4). Moreover, other metrics have been recently developed for these models (where the observations are dependent). These new metrics are based on the one-step-ahead predictive distribution of the observations [44–47].

3 Multiple Importance Sampling (MIS)

The IS methodology can be easily extended when the samples are simulated from M proposals, $\{q_m(\mathbf{x})\}_{m=1}^{M}$, instead of only one. In a generic setting, one can consider that n_m samples are simulated from each proposal ($\sum_{j=1}^{m} n_j = 1$) and weighted appropriately. This extension is usually called *multiple importance sampling* (MIS), and it has strong connections with the case of standard IS with a single mixture proposal with components that are distributions, which is sometimes called *mixture IS*. Here, we consider mixture IS as a subset of MIS methods when n_m are not deterministic number of samples but r.v.s instead.

3.1 Generalized MIS

A unifying framework of MIS has been recently proposed in Ref. 48. The framework encompasses most of the existing IS methods with multiple proposals, proposes new schemes, and compares them in terms of variance. For the sake of clarity, the framework is described in the case where (a) no prior information about the adequateness of the proposals is available and (b) $M = N$ proposals are available (i.e., exactly the same number of proposals than samples to be simulated). However, straightforward extensions are possible to more generic settings. According to this framework, a MIS is proper if it fulfills two conditions related to the sampling and weighting processes. A valid sampling scheme for the simulation of N samples, $\{\mathbf{x}_n\}_{n=1}^{N}$, can be agnostic to the dependence of those samples but must fulfill the following statistically property: a sample $\mathbf{x}$ randomly picked from the whole set of N simulated samples must be distributed as the mixture of proposals $\psi(\mathbf{x}) = \frac{1}{N}\sum_{n=1}^{N} q_n(\mathbf{x})$. A valid weighting scheme must yield an unbiased and consistent UIS estimator, $\widehat{I}^N$. These properness conditions extend the standard properness in IS established by Liu [12] and have also been used to assign proper importance weights to resampled particles [49]. The paper analyzes and ranks several resulting MIS schemes (different combinations of valid sampling and weighting procedures) in terms of variance. Due to space restrictions, here we show only two MIS schemes commonly used in the literature. Let us simulate exactly one sample per proposal (sampling scheme S_3 in Ref. 48) as

$$\mathbf{x}_n \sim q_n(\mathbf{x}), \qquad n = 1, \ldots, N \tag{17}$$

The next two weighting schemes are possible (among many others):

- *Option 1*: Standard MIS (s-MIS, also called `N1` scheme):

$$w_n = \frac{\pi(\mathbf{x}_n)}{q_n(\mathbf{x}_n)}, \qquad n = 1, \ldots, N \tag{18}$$

- *Option 2*: Deterministic mixture MIS (DM-MIS, also called `N3` scheme):

$$w_n = \frac{\pi(\mathbf{x}_n)}{\psi(\mathbf{x}_n)} = \frac{\pi(\mathbf{x}_n)}{\frac{1}{N}\sum_{j=1}^{N} q_j(\mathbf{x}_n)}, \quad n = 1, \dots, N$$

In both cases, it is possible to build the UIS and SNIS estimators. In Ref. 48, it is shown that

$$\text{Var}[\widetilde{I}_{\text{N3}}^{N}] \leq \text{Var}[\widetilde{I}_{\text{N1}}^{N}]$$

that is, that using the second weighting option with the whole mixture in the denominator is always better than using just the proposal that simulated the sample (the equality in the variance relation happens only when all the proposals are the same). The result is relevant since `N1` is widely used in the literature, but it should be avoided whenever possible. Note that both `N1` and `N3` require just one target evaluation per sample. However, `N3` requires N proposal evaluations per sample, while `N1` just one. For a small number of proposals, or when the target evaluation is very expensive (and hence the bottleneck), this extra complexity in `N3` may not be relevant, but it can become cumbersome otherwise. Several MIS strategies have been proposed in the literature to alleviate this problem. In Ref. 50, a partition of the proposals is done *a priori*, and then the `N3` scheme is applied within each cluster (i.e., small mixtures appear in the denominator of the weights). This method is called *partial deterministic mixture* and in some examples is able a similar variance reduction as in the `N3` method, while reducing drastically the number of proposal evaluations [[50], Figure 1]. The *overlapped partial deterministic mixture* method [51] extends the framework to the case where the proposals can belong to more than one cluster. However, the way the proposals are clustered remains an open problem, and few attempts have been done for optimizing the clustering (see Ref. 52 where the clusters are done after the sampling, using the information of the samples, and hence biasing the estimators).

When the selection of the proposals is also random, unlike in the sampling in (17), there exist options to evaluate only the proposals that have been used for sampling (scheme `R2` in Ref. 48) instead of using all of them in the numerator (scheme `R3` in Ref. 48). A recent paper explores the `R2` scheme and some of its statistical properties [53].

3.1.1 MIS with different number of samples per proposal

Since the seminal works of Hesterberg [15] and Veach and Guibas [54] in the computer graphics community, several works have addressed the case where the number of samples (also called counts) per proposal (also called techniques) can be different (see also Ref. 55 where the authors introduce control variates in MIS). In particular, the so-called balance heuristic estimator, proposed in Ref. 54 and very related to the scheme `N3` in Section 3.1, has attracted attention due to its high performance. The UIS balance heuristic estimator is given by

$$\widehat{I}^N(f) = \sum_{j=1}^{M}\sum_{i=1}^{n_j} \frac{f(\mathbf{x}_{j,i})\overline{\pi}(\mathbf{x}_{j,i})}{\sum_{k=1}^{M} n_k q_k(\mathbf{x}_{j,i})} \tag{19}$$

where again $\{q_m(\mathbf{x})\}_{m=1}^{M}$ is the set of available proposals, $\{n_m\}_{m=1}^{M}$ is the number of samples associated to each proposal, $N = \sum_{k=1}^{M} n_k$ is the total number of samples, and $\mathbf{x}_{j,i} \sim q_j(\mathbf{x})$, for $i = 1, \dots, n_j$, and for $j = 1, \dots, M$. Regarding the denominator in (19), it can be interpreted that the N samples are simulated from the mixture $\sum_{k=1}^{M} n_k q_k(\mathbf{x})$ via stratified sampling (a similar interpretation can be done in the aforementioned N3 scheme). In Ref. 56, this estimator is revisited, and novel bounds are obtained. In Ref. 57, the balance heuristic estimator of Equation (19) is generalized, introducing more degrees of freedom that detach the sampling and the denominator of the importance weights, being able to obtain unbiased estimators that reduce the variance with respect to the standard balance heuristic. In Ref. 58, control variates are introduced in an IS scheme with a mixture proposal (similarly to Ref. 55), and all parameters (including the mixture weights) are optimized to minimize the variance of the UIS estimator (which is jointly convex w.r.t. the mixture probabilities and the control variate regression coefficients). More works with a variable number of samples per proposal (either fixed or optimized) include Refs 59–61.

3.2 Rare Event Estimation

IS is often considered as a variance reduction technique, not only in the case when sampling from $\widetilde{\pi}$ is not possible, but also when it is possible but not efficient. A classical example is the case of Equation (1) when $f(\mathbf{x}) = \mathbb{I}_S$, where $\mathbb{I}$ is the indicator function taking value 1 for all $\mathbf{x} \in S$, and 0 otherwise. In rare event estimation, S is usually a set where the target $\widetilde{\pi}$ has few probability mass, and hence I is a small positive number. It is then not practical to simulate from the target, since most of the samples will not contribute to the estimator due to their evaluation in $\mathbb{I}_S(\mathbf{x})$ being zero. IS allows for sampling from a different distribution that will increase the efficiency of the method when $q(\mathbf{x})$ is close to $\mathbb{I}_S$. A recent MIS method called *ALOE* (at least one sample) is able to simulate from a mixture of proposals ensuring that all of them are in the integration region S in the case where $\widetilde{\pi}(\mathbf{x})$ is Gaussian and S is the union of half-spaces defined by a set of hyperplanes (linear constraints) [62]. As an example, the authors show successful results in a problem with 5772 constraints, in a 326-dimensional problem with a probability of $I \approx 10^{-22}$, with just $N = 10^4$ samples. ALOE has been recently applied for characterizing wireless communications systems through the estimation of the symbol error rate [63, 64].

3.3 Compressed and Distributed IS

In the past years, several works have focused on alleviating the computational complexity, communication, or storage in intensive IS methods. This computational burden appears often when the inferential problem is challenging and requires a large amount of simulated samples. This can happen because the adaptive schemes may require many iterations, because of the high-dimensional nature of the tackled problem, and/or because a high precision (low variance) is required in the estimate. In Ref. 65, several compressing schemes are proposed and theoretically analyzed for assigned importance weights to groups of samples for distributed or decentralized Bayesian inference. The framework is extended in Ref. 66, where a stronger theoretical support is given, and new deterministic and random rules for compression are given. The approach in Refs 67 and 68 considers the case of a single node that keeps simulating samples and assigning them an importance weight. The bottleneck here is the storage of the samples so one needs to decide at each time if the sample is stored or discarded. A compression algorithm is introduced for building a dictionary based on greedy subspace projections and a kernel density estimator of the targeted distribution with a limited number of samples. It is shown that asymptotic bias of this method is a tunable constant depending on the kernel bandwidth parameter and a compression parameter. Finally, some works have studied the combination of IS estimators in the distributed setting. For instance, in Ref. 69, Section 4, independent estimators are linearly combined with the combination weights being the inverse of the variance of each estimator. A similar approach is followed in Ref. 70, using the $\widehat{\text{ESS}}$ instead of the variance of the estimator (which is unknown in most practical problems). A Bayesian combination of Monte Carlo estimators is considered in Refs 71 and 72. Note that the MIS approach is, due to its own nature, an implicit linear combination of multiple estimators (each of them using samples from one or several proposals). This perspective is exploited for instance in Refs 73 and 74.

4 Adaptive Importance Sampling (AIS)

Since choosing a good proposal (or set of proposals) in advance is in general impossible, a common approach is the use of adaptive importance sampling (AIS) [75]. AIS algorithms are iterative methods for a gradual learning of one or multiple proposals that aim at approximating the target pdf. Algorithm alg: standard_AIS_alg describes a generic AIS algorithm through three basic steps: the simulation of samples from one or several proposals (sampling), the computation of the importance weight of each sample (weighting), and the update of the parameters that characterize the proposal(s) for repeating the previous steps in the next iteration (adaptation).

Most existing algorithms can be described in this framework that we describe with more detail. The generic AIS algorithm initializes N proposals $\{q_n(\mathbf{x}|\boldsymbol{\theta}_{n,1})\}_{n=1}^N$, parameterized each of them by a vector $\boldsymbol{\theta}_{n,1}$. Then, K samples are simulated from each proposal, $\mathbf{x}_{n,1}^{(k)}$, $n = 1, \ldots, N, k = 1, \ldots, K$, and weighted properly. Here again, many ways of sampling and weighting are possible, as described in Section 3.1. At the end of the weighting step, it is possible to approximate the integral of Equation (1) with either UIS or SNIS, and the target distribution with a discrete random measure, using the set of weighted samples $\{\mathbf{x}_{n,1}^{(k)}, w_{n,1}^{(k)}\}$, $n = 1, \ldots, N, k = 1, \ldots, K$. Finally, the parameters of the nth proposals are updated from $\boldsymbol{\theta}_{n,1}$ to $\boldsymbol{\theta}_{n,2}$. This three-step process is repeated until an iteration stoppage

Algorithm 1. Generic AIS algorithm

Input: Choose K, N, J, and $\{\boldsymbol{\theta}_{n,1}\}_{n=1}^N$
for $i = 1, \ldots, T$ **do**
  Sampling: Draw K samples from each of the N proposal pdfs, $\{q_{n,j}(\mathbf{x}|\boldsymbol{\theta}_{n,j})\}_{n=1}^N$, $\mathbf{x}_{n,j}^{(k)}, k = 1, \ldots, K,\ n = 1, \ldots, N$
  Weighting: Calculate the weights, $w_{n,j}^{(k)}$, for each of the generated KN samples.
  Adaptation: Update the proposal parameters $\{\boldsymbol{\theta}_{n,j}\}_{n=1}^N \longrightarrow \{\boldsymbol{\theta}_{n,j+1}\}_{n=1}^N$.
end for
Output: Return the KNJ pairs $\{\mathbf{x}_{n,j}^{(k)}, w_{n,j}^{(k)}\}$ for all $k = 1, \ldots, K, n = 1, \ldots, N, j = 1, \ldots, J$.

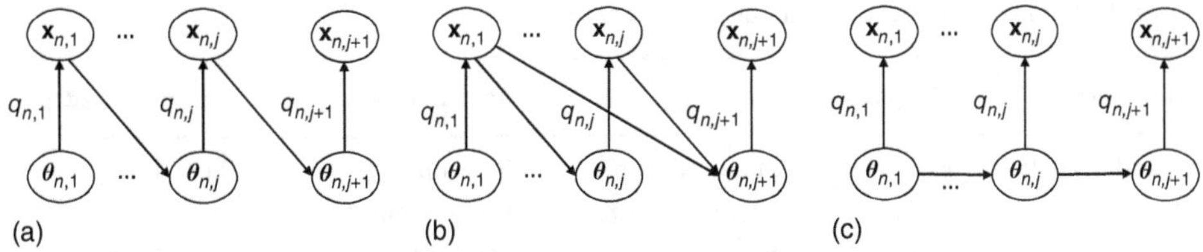

Figure 1 Graphical description of three possible dependencies between the adaptation of the proposal parameters $\theta_{n,t}$ and the samples. Note that $q_{n,t} \equiv q_{n,t}(\mathbf{x}|\theta_{n,t})$. (a) The proposal parameters are adapted using the last set of drawn samples (standard PMC, DM-PMC, N-PMC, M-PMC, and APIS). (b) The proposal parameters are adapted using all drawn samples up to the latest iteration (AMIS, CAIS, Daisee, EAMIS, and RS-CAIS). (c) The proposal parameters are adapted using an independent process from the samples (LAIS, GAPIS, GIS, IMIS, and SL-PMC).

criterion is met (e.g., a maximum number of iterations, J, is reached). Note that at the end, the estimators can use either all weighted samples from iterations 1 to J or only the samples from the last iteration.

The literature is vast in AIS methods, and a detailed description of all of them goes beyond the scope of this paper (see Ref. 75 for a thorough review). Most of the AIS algorithms can be classified within three categories, depending on how the proposals are adapted. Figure 1 shows graphically the three families of AIS algorithms, describing the dependencies for the adaptation of the proposal parameters and the simulation of the samples. Each subplot corresponds to each family, whose description and corresponding AIS algorithms of the literature are given below.

a) The proposal parameters are adapted using the last set of drawn samples (e.g., standard PMC [76], DM-PMC [77, 78], N-PMC [31], M-PMC [79], and APIS [80]).
b) The proposal parameters are adapted using all drawn samples up to the latest iteration (e.g., AMIS [81], CAIS [82], Daisee [83], EAMIS [84], and RS-CAIS [85]).
c) The proposal parameters are adapted using an independent process from the samples (LAIS [86, 87], GAPIS [88], GIS [89], IMIS [90], and SL-PMC [91]).

In Table 2, we describe some relevant AIS algorithms according to different features: the number of proposals; the weighting scheme (*nonlinear* corresponds to the clipping strategies of Section 2.5.1, *standard* is equivalent to Option 1 in Section 3.1, *spatial mixture* corresponds to Option 2 with $\psi(\mathbf{x}) = \sum_{i=1}^{N} q_{i,j}(\mathbf{x}|\theta_{i,j})$, and *temporal mixture* corresponds to Option 2 with $\psi(\mathbf{x}) = \sum_{\tau=1}^{j} q_{n,\tau}(\mathbf{x}|\theta_{n,\tau})$); and the parameters that are adapted (either location and scale or only location). In Table 3, we describe the computational complexity of the same algorithms according to the number of target evaluations, proposal evaluations, target evaluations per proposal, and proposal evaluations per proposal. In some AIS algorithms, the proposals converge with the number of iterations J, although proving this convergence (and the associated convergence rates) is in general a tough problem (see a recent result in Ref. 92). For many other AIS algorithms (e.g., DM-PMC, LAIS, and APIS), the proposals do not converge to any limiting distribution. Converge rates have been established only for simple classes of AIS algorithms which are based on optimized parametric proposals [92]. Note that AIS-based algorithms have also been used for optimization purposes [93, 94].

Table 2 Comparison of various AIS algorithms according to different features.

Algorithm	# Proposals	Weighting	Adaptation strategy	Parameters adapted
Standard PMC	$N > 1$	Standard	Resampling	Location
M-PMC	$N > 1$	Spatial mixture	Resampling	Location
N-PMC	Either	Nonlinear	Moment estimation	Location/scale
LAIS	$N > 1$	Generic mixture	MCMC	Location
DM-PMC	$N > 1$	Spatial mixture	Resampling	Location
AMIS	$N = 1$	Temporal mixture	Moment estimation	Location/scale
GAPIS	$N > 1$	Spatial mixture	Gradient process	Location/scale
APIS	$N > 1$	Spatial mixture	Moment estimation	Location

Table 3 Comparison of various AIS algorithms according to the computational complexity.

Algorithm	# Target evaluation	# Proposal evaluation	# Target evaluation/sample	# Proposal evaluation/sample
Standard PMC	NJ	NJ	1	1
N-PMC	NJ	NJ	1	1
M-PMC	KJ	KNJ	1	N
LAIS	$K(N+1)J$	KN^2J	$1 + 1/N$	N
DM-PMC	KNJ	KN^2J	1	N
AMIS	KJ	KJ^2	1	J
GAPIS	KNJ	KN^2J	1	N
APIS	KNJ	KN^2J	1	N

Acknowledgments

V.E. acknowledges support from the *Agence Nationale de la Recherche* of France under PISCES project (ANR-17-CE40-0031-01).

Notes

1 For the sake of easing the notation, from now on we use the same notation for denoting a random variable or one realization of a random variable.
2 From now on, we drop $\mathbf{y}$ to ease the notation, for example, $Z \equiv Z(\mathbf{y})$.

References

1 Bernardo, J.M. and Smith, A.F.M. (1994) *Bayesian Theory*, Wiley & sons, New York.
2 Robert, C.P. (2007) *The Bayesian Choice*, Springer, New York.

3 Box, G.E.P. and Tiao, G.C. (1973) *Bayesian Inference in Statistical Analysis*, Wiley & sons, New York.

4 Acton, F.S. (1990) *Numerical Methods That Work*, The Mathematical Association of America, Washington, DC.

5 Burden, R.L. and Faires, J.D. (2000) *Numerical Analysis*, Brooks Cole, Boston.

6 Kythe, P.K. and Schaferkotter, M.R. (2004) *Handbook of Computational Methods for Integration*, Chapman and Hall/CRC, Boca Raton, USA.

7 Plybon, B.F. (1992) *An Introduction to Applied Numerical Analysis*, PWS-Kent, Boston, MA.

8 Dunn, W.L. and Shultis, J.K. (2011) *Exploring Monte Carlo Methods*, Elsevier Science, Amsterdam.

9 Jaeckel, P. (2002) *Monte Carlo Methods in Finance*, Wiley, New York.

10 Gentle, J.E. (2004) *Random Number Generation and Monte Carlo Methods*, Springer, New York.

11 Kroese, D., Taimre, T., and Botev, Z. (2011) *Handbook of Monte Carlo Methods*, Wiley Series in Probability and Statistics, John Wiley and Sons, New York.

12 Liu, J.S. (2004) *Monte Carlo Strategies in Scientific Computing*, Springer, New York.

13 Robert, C.P. and Casella, G. (2004) *Monte Carlo Statistical Methods*, Springer, New York.

14 Kahn, H. (1950) Random sampling (Monte Carlo) techniques in neutron attenuation problems. *Nucleonics*, 6 (5), 27.

15 Hesterberg, T. (1995) Weighted average importance sampling and defensive mixture distributions. *Technometrics*, 37 (2), 185–194.

16 Tokdar, S.T. and Kass, R.E. (2010) Importance sampling: a review. *Wiley Interdiscip. Rev. Comput. Stat.*, 2 (1), 54–60.

17 Owen, A.B. (2013) *Monte Carlo theory, methods and examples*, Art Owen, Stanford, Palo Alto, USA.

18 Doucet, A. and Johansen, A.M. (2009) A tutorial on particle filtering and smoothing: fifteen years later. *Handbook of Non. Filt.*, 12 (656-704), 3.

19 Kong, A. (1992) A Note on Importance Sampling Using Standardized Weights. University of Chicago, Dept. of Statistics, *Tech. Rep, 348*.

20 Elvira, V., Martino, L., and Robert, C.P. (2018) Rethinking the effective sample size. *arXiv preprint arXiv:1809.04129*.

21 Martino, L., Elvira, V., and Louzada, F. (2017) Effective sample size for importance sampling based on discrepancy measures. *Signal Process.*, 131, 386–401.

22 Martino, L., Elvira, V., and Louzada, F. (2016) *Alternative Effective Sample Size Measures for Importance Sampling*. 2016 IEEE Statistical Signal Processing Workshop (SSP), pp. 1–5. IEEE.

23 Huggins, J.H. and Roy, D.M. (2019) Sequential Monte Carlo as approximate sampling: bounds, adaptive via ∞-ESS, and an application to particle Gibbs. *Bernoulli*, 25 (1), 584–622.

24 Elvira, V., Closas, P., and Martino, L. (2019) *Gauss-Hermite Quadrature for Non-gaussian Inference Via an Importance Sampling Interpretation*. 2019 27th European Signal Processing Conference (EUSIPCO), pp. 1–5. IEEE.

25 Elvira, V., Martino, L., and Closas, P. (2020) Importance Gaussian quadrature. *arXiv preprint arXiv:2001.03090*.

26 Agapiou, S., Papaspiliopoulos, O., Sanz-Alonso, D., *et al.* (2017) Importance sampling: intrinsic dimension and computational cost. *Stat. Sci.*, 32 (3), 405–431.

27 Chatterjee, S. and Diaconis, P. (2018) The sample size required in importance sampling. *Ann. Appl. Probab.*, 28 (2), 1099–1135.

28 Ryu, E.K. and Boyd, S.P. (2014) Adaptive importance sampling via stochastic convex programming. *arXiv preprint arXiv:1412.4845.*

29 Míguez, J. (2017) *On the Performance of Nonlinear Importance Samplers and Population Monte Carlo Schemes*. 2017 22nd International Conference on Digital Signal Processing (DSP), pp. 1–5. IEEE.

30 Ionides, E.L. (2008) Truncated importance sampling. *J. Comput. Graph. Stat.*, 17 (2), 295–311.

31 Koblents, E. and Míguez, J. (2015) A population Monte Carlo scheme with transformed weights and its application to stochastic kinetic models. *Stat. Comput.*, 25 (2), 407–425.

32 Miguez, J., Mariño, I.P., and Vázquez, M.A. (2018) Analysis of a nonlinear importance sampling scheme for Bayesian parameter estimation in state-space models. *Signal Process.*, 142, 281–291.

33 Vehtari, A., Gelman, A., and Gabry, J. (2015) Pareto smoothed importance sampling. *arXiv preprint arXiv:1507.02646.*

34 Martino, L., Elvira, V., Míguez, J., et al. (2018) *A Comparison of Clipping Strategies for Importance Sampling*. 2018 IEEE Statistical Signal Processing Workshop (SSP), pp. 558–562. IEEE.

35 Douc, R., Cappé, O., and Moulines, E. (2005) *Comparison of Resampling Schemes for Particle Filtering*. Proc. 4th Int. Symp. on Image and Signal Processing and Analysis, September 2005, pp. 64–69.

36 Li, T., Bolic, M., and Djuric, P.M. (2015) Resampling methods for particle filtering: classification, implementation, and strategies. *IEEE Signal Process. Mag*, 32 (3), 70–86.

37 Gordon, N., Salmond, D., and Smith, A.F.M. (1993) Novel approach to nonlinear and non-Gaussian Bayesian state estimation. *IEE Proc.-F Radar and Signal Process.*, 140, 107–113.

38 Doucet, A., De. Freitas, N., Murphy, K., and Russell, S. (2000) *Rao-Blackwellised Particle Filtering for Dynamic Bayesian Networks*. Proceedings of the Sixteenth conference on Uncertainty in artificial intelligence, pp. 176–183. Morgan Kaufmann Publishers Inc.

39 Pitt, M.K. and Shephard, N. (2001) Auxiliary variable based particle filters, in *Sequential Monte Carlo Methods in Practice*, Chap. 13 (eds A. Doucet., N. de Freitas., and N. Gordon), Springer, New York. pp. 273–293.

40 Kotecha, J. and Djurić, P.M. (2003) Gaussian particle filtering. *IEEE Trans. Signal Process.*, 51 (10), 2592–2601.

41 Djuric, P.M., Lu, T., and Bugallo, M.F. (2007) *Multiple Particle Filtering*. 2007 IEEE International Conference on Acoustics, Speech and Signal Processing-ICASSP'07, vol 3, pp III–1181. IEEE.

42 Elvira, V., Martino, L., Bugallo, M.F., and Djurić, P.M. (2018) *In Search for Improved Auxiliary Particle Filters*. 2018 26th European Signal Processing Conference (EUSIPCO), pp. 1637–1641. IEEE.

43 Elvira, V., Martino, L., Bugallo, M.F., and Djuric, P.M. (2019) Elucidating the auxiliary particle filter via multiple importance sampling [lecture notes]. *IEEE Signal Process Mag.*, 36 (6), 145–152.

44 Lee, A. and Whiteley, N. (2015) Variance estimation and allocation in the particle filter. *arXiv:1509.00394v1 [stat.CO].*

45 Bhadra, A. and Ionides, E.L. (2016) Adaptive particle allocation in iterated sequential Monte Carlo via approximating meta-models. *Stat. Comput.*, 26 (1-2), 393–407.

46 Elvira, V., Míguez, J., and Djurić, P. (2017) Adapting the number of particles in sequential monte carlo methods through an online scheme for convergence assessment. *IEEE Trans. Signal Process.*, 65 (7), 1781–1794.

47 Elvira, V., Míguez, J., and Djurić, P.M. (2019) New results on particle filters with adaptive number of particles. *arXiv preprint arXiv:1911.01383.*

48 Elvira, V., Martino, L., Luengo, D., and Bugallo, M.F. (2019) Generalized multiple importance sampling. *Stat. Sci.*, 34 (1), 129–155.

49 Martino, L., Elvira, V., and Louzada, F. (2016) *Weighting a Resampled Particle in Sequential Monte Carlo.* 2016 IEEE Statistical Signal Processing Workshop (SSP), pp. 1–5. IEEE.

50 Elvira, V., Martino, L., Luengo, D., and Bugallo, M.F. (2015) Efficient multiple importance sampling estimators. *IEEE Signal Proc. Let.*, 22 (10), 1757–1761, 2015.

51 Elvira, V., Martino, L., Luengo, D., and Bugallo, M.F. (2016) *Multiple Importance Sampling with Overlapping Sets of Proposals.* 2016 IEEE Statistical Signal Processing Workshop (SSP).

52 Elvira, V., Martino, L., Luengo, D., and Bugallo, M.F. (2016) Heretical multiple importance sampling. *IEEE Signal Process Lett.*, 23 (10), 1474–1478.

53 Medina-Aguayo, F.J. and Everitt, R.G. (2019) Revisiting the balance heuristic for estimating normalising constants. *arXiv preprint arXiv:1908.06514.*

54 Veach, E. and Guibas, L. (1995) *Optimally Combining Sampling Techniques for Monte Carlo Rendering.* SIGGRAPH 1995 Proceedings, pp. 419–428.

55 Owen, A. and Zhou, Y. (2000) Safe and effective importance sampling. *J. Am. Stat. Assoc.*, 95 (449), 135–143.

56 Sbert, M., Havran, V., and Szirmay-Kalos, L. (2018) Multiple importance sampling revisited: breaking the bounds. *EURASIP J. Adv. Signal Process.*, 2018 (1), 15.

57 Sbert, M. and Elvira, V. (2019) Generalizing the balance heuristic estimator in multiple importance sampling. *arXiv preprint arXiv:1903.11908.*

58 He, H.Y. and Owen, A.B. (2014) Optimal mixture weights in multiple importance sampling. *arXiv preprint arXiv:1411.3954.*

59 Sbert, M., Havran, V., and Szirmay-Kalos, L. (2016) Variance analysis of multi-sample and one-sample multiple importance sampling, in *Computer Graphics Forum*, vol. 35, Wiley Online Library, pp. 451–460.

60 Sbert, M. and Havran, V. (2017) Adaptive multiple importance sampling for general functions. *Visual Comput.*, 33 (6-8), 845–855.

61 Sbert, M., Havran, V., and Szirmay-Kalos, L. (2019) Optimal deterministic mixture sampling, in *Eurographics (Short Papers)*, pp. 73–76.

62 Owen, A.B., Maximov, Y., Chertkov, M., *et al.* (2019) Importance sampling the union of rare events with an application to power systems analysis. *Electron. J. Stat.*, 13 (1), 231–254.

63 Elvira, V. and Santamaría, I. (2019) *Efficient ser Estimation for Mimo Detectors via Importance Sampling Schemes*. 2019 Asilomar Conference on Signals, Systems and Computers, pp. 1–5. IEEE.

64 Elvira, V. and Santamaría, I. (2019) Multiple importance sampling for efficient symbol error rate estimation. *IEEE Signal Process Lett.*, 26 (3), 420–424.

65 Martino, L., Elvira, V., and Camps-Valls, G. (2018) Group importance sampling for particle filtering and mcmc. *Digital Signal Process.*, 82, 133–151.

66 Martino, L. and Elvira, V. (2018) Compressed Monte Carlo for distributed Bayesian inference. *viXra:1811.0505*.

67 Koppel, A., Bedi, A.S., Elvira, V., and Sadler, B.M. (2019) Approximate shannon sampling in importance sampling: nearly consistent finite particle estimates. *arXiv preprint arXiv:1909.10279*.

68 Bedi, A.S., Koppel, A., Elvira, V., and Sadler, B.M. (2019) *Compressed Streaming Importance Sampling for Efficient Representations of Localization Distributions*. 2019 Asilomar Conference on Signals, Systems and Computers, pp. 1–5. IEEE.

69 Douc, R., Guillin, A., Marin, J.M., and Robert, C.P. (2007) Minimum variance importance sampling via population Monte Carlo. *ESAIM Probab. Stat.*, 11, 427–447.

70 Nguyen, T.L.T., Septier, F., Peters, G.W., and Delignon, Y. (2014) *Improving smc Sampler Estimate by Recycling All Past Simulated Particles. Statistical Signal Processing (SSP)*, 2014 IEEE Workshop on, pp. 117–120. IEEE.

71 Luengo, D., Martino, L., Elvira, V., and Bugallo, M. (2015) *Bias Correction for Distributed Bayesian Estimators*. 2015 IEEE 6th International Workshop on Computational Advances in Multi-Sensor Adaptive Processing (CAMSAP), pp. 253–256. IEEE.

72 Luengo, D., Martino, L., Elvira, V., and Bugallo, M. (2018) Efficient linear fusion of partial estimators. *Digital Signal Process.*, 78, 265–283.

73 Havran, V. and Sbert, M. (2014) *Optimal Combination of Techniques in Multiple Importance Sampling*. Proceedings of the 13th ACM SIGGRAPH International Conference on Virtual-Reality Continuum and its Applications in Industry, pp. 141–150.

74 Sbert, M., Havran, V., Szirmay-Kalos, L., and Elvira, V. (2018) Multiple importance sampling characterization by weighted mean invariance. *Visual Comput.*, 34 (6-8), 843–852.

75 Bugallo, M.F., Elvira, V., Martino, L., *et al.* (2017) Adaptive importance sampling: the past, the present, and the future. *IEEE Signal Process. Mag.*, 34 (4), 60–79.

76 Cappé, O., Guillin, A., Marin, J.M., and Robert, C.P. (2004) Population Monte Carlo. *J. Comput. Graph. Stat.*, 13 (4), 907–929.

77 Elvira, V., Martino, L., Luengo, D., and Bugallo, M.F. (2017) Improving population Monte Carlo: alternative weighting and resampling schemes. *Sig. Process.*, 131 (12), 77–91.

78 Elvira, V., Martino, L., Luengo, D., and Bugallo, M.F. (2017) *Population Monte Carlo Schemes with Reduced Path Degeneracy*. Proc. IEEE Int. Work. Comput. Adv. Multi-Sensor Adap. Process. (CAMSAP 2017), pp. 1–5.

79 Cappé, O., Douc, R., Guillin, A., *et al.* (2008) Adaptive importance sampling in general mixture classes. *Stat. Comput.*, 18, 447–459.

80 Martino, L., Elvira, V., Luengo, D., and Corander, J. (2015) An adaptive population importance sampler: learning from the uncertanity. *IEEE Trans. Signal Process.*, 63 (16), 4422–4437.

81 Cornuet, J.M., Marin, J.M., Mira, A., and Robert, C.P. (2012) Adaptive multiple importance sampling. *Scand. J. Stat.*, 39 (4), 798–812.

82 El-Laham, Y., Elvira, V., and Bugallo, M.F. (2018) Robust covariance adaptation in adaptive importance sampling. *IEEE Signal Process Lett.*, 25 (7), 1049–1053.

83 Lu, X., Rainforth, T., Zhou, Y., *et al.* (2018) On exploration, exploitation and learning in adaptive importance sampling. *arXiv preprint arXiv:1810.13296*.

84 El-Laham, Y., Martino, L., Elvira, V., and Bugallo, M.F. (2019) *Efficient Adaptive Multiple Importance Sampling*. 2019 27th European Signal Processing Conference (EUSIPCO), pp. 1–5. IEEE.

85 El-Laham, Y., Elvira, V., and Bugallo, M.F. (2019) *Recursive Shrinkage Covariance Learning in Adaptive Importance Sampling*. Proc. IEEE Int. Work. Comput. Adv. Multi-Sensor Adap. Process. (CAMSAP 2019), pp. 1–5.

86 Martino, L., Elvira, V., Luengo, D., and Corander, J. (2017) Layered adaptive importance sampling. *Stat. Comput.*, 27 (3), 599–623.

87 Martino, L., Elvira, V., and Luengo, D. (2017) *Anti-tempered Layered Adaptive Importance Sampling*. 2017 22nd International Conference on Digital Signal Processing (DSP), pp. 1–5. IEEE.

88 Elvira, V., Martino, L., Luengo, L., and Corander, J. (2015) *A Gradient Adaptive Population Importance Sampler*. Proc. IEEE Int. Conf. Acoust., Speech Signal Process. (ICASSP 2015), April 19-24, 2015, pp. 4075–4079, Brisbane, Australia.

89 Schuster, I. (2015) Gradient importance sampling. Technical report. https://arxiv.org/abs/1507.05781.

90 Fasiolo, M., de Melo, F.E., and Maskell, S. (2018) Langevin incremental mixture importance sampling. *Stat. Comput.*, 28 (3), 549–561.

91 Elvira, V. and Chouzenoux, E. (2019) *Langevin-Based Strategy for Efficient Proposal Adaptation in Population Monte Carlo*. ICASSP 2019-2019 IEEE International Conference on Acoustics, Speech and Signal Processing (ICASSP), pp. 5077–5081. IEEE.

92 Akyildiz, Ö. D. and Míguez, J. (2019) Convergence rates for optimised adaptive importance samplers. arXiv preprint *arXiv:1903.12044*.

93 Moral, P.D., Doucet, A., and Jasra, A. (2006) Sequential Monte Carlo samplers. *J. R. Stat. Soc. Ser. B Stat. Methodol.*, 68 (3), 411–436.

94 Akyildiz, O.D., Marino, I.P., and Míguez, J. (2017) *Adaptive Noisy Importance Sampling for Stochastic Optimization*. IEEE 7th International Workshop on Computational Advances in Multi-Sensor Adaptive Processing (CAMSAP), pp. 1–5. IEEE.

Part III

Statistical Learning

10

Supervised Learning

Weibin Mo and Yufeng Liu

University of North Carolina at Chapel Hill, Chapel Hill, NC, USA

1 Introduction

Supervised learning is an important type of machine learning problems which focuses on the learning task using training data with both covariates and response variables. Supervised learning problems are commonly seen in practice. In finance, the future price of a stock can be forecast by the historical stock prices and many macroeconomic factors. The forecasting of the future stock price can help with the buy and sell decisions or the valuation of the underlying assets. In medicine, the patient's illness can be predicted by the patient's characteristics, symptoms, clinical test results, and the medical treatments received. The predictive model can help the physician to diagnose illness and decide whether to introduce a treatment therapy for a given patient. In a context-based recommender system, the contextual information such as time, location, and social connection can be used to predict the recipient's feedback, which can help to improve the effectiveness of the recommender. These applications involve the *covariate–response* data, also known as the *input–output* data. A common goal of these applications is to find a model that predicts the response from the covariates [1]. In contrast to the supervised learning problem, an *unsupervised learning* task does not involve the response variable, and the goals are typically related to dimension reduction or discovering useful patterns [2].

In this chapter, we mainly focus on supervised learning and specifically consider techniques that can be formulated as the optimization of "loss + penalty." In particular, the loss term keeps the fidelity of the resulting model to the data, while the penalty term penalizing the complexity can prevent the fitted model from overfitting [3]. Many existing supervised learning methods can be formulated in this framework [4–9]. In the supervised learning literature, there exist theoretical foundations for the penalized approaches with sharp generalization error bounds [10–12]. In modern high-dimensional applications, people may want to find a model that is not only predictive but also simple and interpretable [13]. In this case, using the penalties that produce sparse solutions in supervised learning problems can perform model estimation and model selection simultaneously [14, 15].

Our goal in this chapter is to provide a general overview of some commonly used methods under the "loss + penalty" framework. Some selected statistical models and computational algorithms for supervised learning are of our main focus. The organization of this chapter

Computational Statistics in Data Science.
Edited by Walter W. Piegorsch, Richard A. Levine, Hao Helen Zhang and Thomas C. M. Lee.
 ISBN 978-1-11956107-1

is as follows. In Section 2, we introduce the mathematical formulation of "loss + penalty" as the penalized empirical risk minimization (ERM) problem. In particular, in Sections 2.1 and 2.2, we discuss the basic questions of "why to penalize," the bias–variance trade-off, and "how to optimize," the first-order optimization methods. In Sections 3 and 4, we consider linear regression and classification as two main areas in supervised learning. We present some selected methods and specific computational algorithms. In Section 5, we discuss the extensions of these supervised learning methods for complex data, including nonlinear modeling and large-scale optimization. In Section 6, we summarize the chapter and provide some concluding remarks.

2 Penalized Empirical Risk Minimization

Suppose that we have the random vector $(\boldsymbol{X}, Y)$, where $\boldsymbol{X} \in \mathcal{X} \subseteq \mathbb{R}^p$ is a p-dimensional covariate vector, and $Y \in \mathcal{Y}$ is the response. Let $\ell : \mathcal{Y} \times \mathcal{Y} \to \mathbb{R}_+$ be a loss function. The goal of the supervised learning problem is to find $f : \mathcal{X} \to \mathcal{Y}$ that minimizes the *risk function*

$$\mathcal{R}(f) := \mathbb{E}\ell\,(Y, f(\boldsymbol{X}))$$

If the response takes continuous value, that is, $\mathcal{Y} = \mathbb{R}$, and the squared loss $\ell(y, \hat{y}) = (y - \hat{y})^2$ is considered, then the problem is a standard least-squared regression problem. If the response takes discrete value in $\mathcal{Y}$, and the 0–1 loss $\ell(y, \hat{y}) = \mathbb{1}(y \neq \hat{y})$ is considered, then the problem becomes a binary classification problem.

For learning practice, we usually specify a function class $\mathcal{F}$ on which f is optimized. Based on the training data $\mathcal{D}_n := \{(\boldsymbol{X}_i, Y_i)\}_{i=1}^n \overset{\text{iid}}{\sim} (\boldsymbol{X}, Y)$, we define the *empirical risk* of $f \in \mathcal{F}$ as

$$R_n(f) := \frac{1}{n} \sum_{i=1}^{n} \ell\left(Y_i, f(\boldsymbol{X}_i)\right)$$

Then, the general penalized *ERM* framework considers the following optimization problem:

$$\min_{f \in \mathcal{F}} \quad R_n(f) + J_\lambda(f) \tag{1}$$

Here, $J_\lambda(f)$ is a penalty term that controls the complexity of f with the tuning parameter λ. The penalized ERM formulation (1) can raise two basic problems: why to penalize and how to optimize. In Sections 2.1 and 2.2, we discuss the "why" and "how" questions from the statistical and computational points of view, respectively.

2.1 Bias–Variance Trade-Off

To motivate the introduction of the penalty term $J_\lambda(f)$ in the penalized ERM formulation (1), we first discuss two phenomena: the overoptimism of the empirical risk $R_n(\hat{f})$ and the bias–variance trade-off for the true risk $\mathcal{R}(\hat{f})$.

To assess the performance $\mathcal{R}(\hat{f})$ of a fitted model $\hat{f}$, we may use the *in-sample* and *out-of-sample* empirical risks based on the data, which are often referred as the *training* and *testing errors*, respectively, in the literature. In particular, the in-sample empirical risk $R_n(\hat{f}) := \frac{1}{n}\sum_{i=1}^{n}\ell(Y_i,\hat{f}(\boldsymbol{X}_i))$ utilizes the training sample $\mathcal{D}_n$ to estimate the true risk $\mathcal{R}(\hat{f})$. However, since the fitted model $\hat{f}$ depends on the training data $\mathcal{D}_n$, the in-sample empirical risk can be overoptimistic. For concrete discussion, assume that the data-generating process is $Y = \boldsymbol{X}^T\boldsymbol{\beta}^\star + \epsilon$, where $\boldsymbol{\beta}^\star \in \mathbb{R}^p$ is the true parameter vector, $\epsilon \sim \mathcal{N}(0,\sigma^2)$, and $\boldsymbol{X} \perp \epsilon$. Consider $\ell(y,\hat{y}) = (y-\hat{y})^2$ and $\mathcal{F} = \{\boldsymbol{x} \mapsto \boldsymbol{\beta}^T\boldsymbol{x} : \boldsymbol{\beta} \in \mathbb{R}^p\}$. For simplicity, we first assume $n > p$, and the training covariate matrix $\mathbb{X} \in \mathbb{R}^{n\times p}$ has a full column rank here. Suppose that $\hat{\boldsymbol{\beta}} \in \operatorname{argmin}_{\boldsymbol{\beta}\in\mathbb{R}^p}\frac{1}{n}\sum_{i=1}^{n}\ell(Y_i,\boldsymbol{X}_i^T\boldsymbol{\beta}) = \operatorname{argmin}_{\boldsymbol{\beta}\in\mathbb{R}^p}\|\boldsymbol{Y}-\mathbb{X}\boldsymbol{\beta}\|_2^2$, where $\boldsymbol{Y} \in \mathbb{R}^n$ is the training response vector. The resulting fitted model becomes $\hat{f}(\boldsymbol{x}) = \boldsymbol{x}^T\hat{\boldsymbol{\beta}}$. The in-sample empirical risk is $R_n(\hat{f}) = \frac{1}{n}\sum_{i=1}^{n}\ell\left(Y_i,\hat{f}(\boldsymbol{X}_i)\right) = \frac{1}{n}\|\boldsymbol{Y}-\mathbb{X}\hat{\boldsymbol{\beta}}\|_2^2$. Then, we have $\mathbb{E}R_n(\hat{f}) = \frac{n-p}{n}\sigma^2 < \sigma^2$. However, the true risk $\mathcal{R}(\hat{f}) = \mathbb{E}\ell\left(Y_{\text{out}},\hat{f}(\boldsymbol{X}_{\text{out}})\right) = \mathbb{E}(Y_{\text{out}} - \boldsymbol{X}_{\text{out}}^T\hat{\boldsymbol{\beta}})^2 = \sigma^2 + \mathbb{E}(\boldsymbol{X}_{\text{out}}^T(\hat{\boldsymbol{\beta}}-\boldsymbol{\beta}^\star))^2 > \sigma^2$, where $(\boldsymbol{X}_{\text{out}},Y_{\text{out}}) \perp \mathcal{D}_n$. It suggests that $\mathbb{E}R_n(\hat{f}) < \mathcal{R}(\hat{f})$, that is, the in-sample empirical risk is overoptimistic. Here, we call $\ell(Y_{\text{out}},\boldsymbol{X}_{\text{out}}^T\hat{\boldsymbol{\beta}})$ as the out-of-sample empirical risk since it evaluates at the out-of-sample point $(\boldsymbol{X}_{\text{out}},Y_{\text{out}})$.

The above discussion also informs a risk decomposition: $\mathcal{R}(\hat{f}) = \sigma^2 + \mathbb{E}(\boldsymbol{X}_{\text{out}}^T\hat{\boldsymbol{\beta}} - \mathbb{E}(\boldsymbol{X}_{\text{out}}^T\hat{\boldsymbol{\beta}}))^2 + (\mathbb{E}(\boldsymbol{X}_{\text{out}}^T\hat{\boldsymbol{\beta}}) - \mathbb{E}(\boldsymbol{X}_{\text{out}}^T\boldsymbol{\beta}^\star))^2 = \sigma^2 + \mathbf{Var}(\hat{f}(\boldsymbol{X}_{\text{out}})) + \mathbf{Bias}\,(\hat{f}(\boldsymbol{X}_{\text{out}}))^2$, for any estimated $\hat{\boldsymbol{\beta}}$. Assume that the underlying true coefficients are nonzero, that is, $\beta_j^\star \neq 0$ $(\forall 1 \le j \le p)$. For $0 \le q \le p$, we consider the restricted least-squared problem by setting the last $p-q$ coefficients being zero, $\min\{\|\boldsymbol{Y}-\mathbb{X}\boldsymbol{\beta}\|_2 : \boldsymbol{\beta} = \boldsymbol{\beta}_{(q)} \oplus \mathbf{0}_{p-q},\ \boldsymbol{\beta}_{(q)} \in \mathbb{R}^q\}$, which corresponds to the fitted model $\hat{f}^{(q)}$. We further assume that $\{\boldsymbol{X}_i\}_{i=1}^n, \boldsymbol{X}_{\text{out}} \overset{\text{iid}}{\sim} \mathcal{N}_p(\mathbf{0}_p,\mathbf{I}_p)$. Then, it can be shown that

$$\mathbf{Var}(\hat{f}^{(q)}(\boldsymbol{X}_{\text{out}})) = \frac{q\sigma^2}{n-q-1};\quad \mathbf{Bias}\,(\hat{f}^{(q)}(\boldsymbol{X}_{\text{out}}))^2 = \frac{n-1}{n-q-1}\|\boldsymbol{\beta}^\star_{((q+1):p)}\|_2^2$$

where $\boldsymbol{\beta}^\star_{((q+1):p)}$ is the subvector in $\boldsymbol{\beta}^\star$ corresponding to the indices from $q+1$ to p. That is, the variance of the fitted model increases with the number of nonzero variables q, while the bias generally decreases with q, especially when n is large. Therefore, the model complexity q trades-off between the variance and the bias of the fitted model $\hat{f}^{(q)}$. It is worthwhile to note that the full model with $q = p$ may not enjoy the lowest risk even though the linear model estimator is a *best linear unbiased estimator (BLUE)* of $\boldsymbol{\beta}^\star$. In contrast, the model complexity $q < p$ may lead to a biased estimator, but the corresponding fitted model $\hat{f}^{(q)}$ can have a smaller risk $\mathcal{R}(\hat{f}^{(q)})$ than the full model risk $\mathcal{R}(\hat{f}^{(p)})$.

In the previous discussion, we have shown that (i) the in-sample empirical risk, that is, the training error, can be overoptimistic and (ii) there is a general bias–variance trade-off when determining the model complexity. In the previous example, we want to select the model of complexity q that minimizes the true risk $\mathcal{R}(\hat{f}^{(q)})$, while the in-sample empirical risk $R_n(\hat{f}^{(q)})$ can be overoptimistic of $\mathcal{R}(\hat{f}^{(q)})$. In fact, using $R_n(\hat{f}^{(q)})$ to choose q will always prefer the full model with $q = p$, since $\mathbb{E}R_n(\hat{f}^{(q)}) = \frac{n-q}{n}\sigma^2 \ge \frac{n-p}{n}\sigma^2 = \mathbb{E}R_n(\hat{f}^{(p)})$. This motivates the introduction of the penalty term $J_\lambda(f)$ in the penalized ERM formulation (1) to perform estimation and model selection simultaneously. In particular, the term $J_\lambda(f)$ penalizing the complexity of f can prevent the fitted model from overfitting the training data. More in-depth theoretical foundation of penalization in supervised learning can be found in Barron *et al.* [10].

The tuning parameter λ plays a role that balances the emphasis on the training empirical risk $R_n(f)$ for model fidelity and the complexity penalty $J_\lambda(f)$ for not overfitting. The tuning parameter λ can be determined by the out-of-sample risks, the testing errors, of different models corresponding to various λs. The out-of-sample risks can be evaluated on a held-out dataset, known as the *validating set*, which is not used for the model training. When the available data are limited, the out-of-sample risks can also be determined by the *cross-validation (CV)*. To be specific, the training data $\mathcal{D}_n$ is divided into K-fold. For each fold of the data, we first use the remaining training data to fit the models. Then, we evaluate the out-of-sample risks on the targeting fold of the data. Finally, we aggregate the out-of-sample risks among all folds to tune the parameter λ.

2.2 First-Order Optimization Methods

In this section, we consider the first-order optimization techniques for solving the following penalized ERM problem, also known as the *composite convex minimization* [16]:

$$\min_{\theta\in\mathbb{R}^p} \quad \{Q(\theta) := R(\theta) + J(\theta)\} \tag{2}$$

where R and J are both continuous and convex functions. Moreover, R is assumed to be differentiable with ∇R being L_R-Lipschitz continuous, that is, $\| \nabla R(\theta_1) - \nabla R(\theta_2)\|_2 \le L_R \|\theta_1 - \theta_2\|_2$ for any $\theta_1, \theta_2 \in \mathbb{R}^p$. In this case, we say R is L_R*-Lipschitz gradient*.

First, we assume that J is also Lipschitz gradient, so that Q is L_Q-Lipschitz gradient. Consider an optimization upper bound of $Q(\theta)$ at $\theta_0 \in \mathbb{R}^p$:

$$\tilde{Q}_L(\theta;\theta_0) := Q(\theta_0) + \langle \nabla Q(\theta_0), \theta - \theta_0\rangle + (L/2)\,\|\theta - \theta_0\|_2^2$$

Then, for any $\theta \in \mathbb{R}^p$,

$$\begin{aligned}
&Q(\theta) - \tilde{Q}_{L_Q}(\theta;\theta_0)\\
&= Q(\theta) - Q(\theta_0) - \langle \nabla Q(\theta_0), \theta - \theta_0\rangle - (L_Q/2)\,\|\theta - \theta_0\|_2^2\\
&= \int_0^1 \langle \nabla Q\left((1-t)\theta_0 + t\theta\right) - \nabla Q(\theta_0), \theta - \theta_0\rangle \mathrm{d}t - (L_Q/2)\,\|\theta - \theta_0\|_2^2\\
&\quad \text{(by Mean Value Theorem)}\\
&\le L_Q \int_0^t \|(1-t)\theta_0 + t\theta - \theta_0\|_2 \mathrm{d}t \times \|\theta - \theta_0\|_2 - (L_Q/2)\,\|\theta - \theta_0\|_2^2\\
&\quad \text{(by } L_Q\text{-Lipschitz gradient)}\\
&= 0
\end{aligned}$$

That is, $Q(\theta) \le \tilde{Q}_{L_Q}(\theta;\theta_0)$ for any $\theta \in \mathbb{R}^p$. Moreover, $\nabla_\theta \tilde{Q}_L(\theta;\theta_0)|_{\theta=\theta_0} = \nabla Q(\theta_0)$. Therefore, $\tilde{Q}_{L_Q}(\cdot;\theta_0)$ is a convex upper bound of $Q(\cdot)$ such that $\theta^\star \in \operatorname{argmin}_{\theta\in\mathbb{R}^p} Q(\theta)$ if and only if $\theta^\star \in \operatorname{argmin}_{\theta\in\mathbb{R}^p} \tilde{Q}_{L_Q}(\theta;\theta^\star)$. Note that given θ_t, $\tilde{Q}_L(\cdot;\theta_t)$ is minimized at

$$\theta_{t+1} \leftarrow \theta_t - (1/L)\nabla Q(\theta_t) \tag{3}$$

Then, the *gradient descent algorithm (GDA)* iteratively updates (3) for a sufficiently large L until θ_{t+1} is close to θ_t or $\|\nabla Q(\theta_t)\|_2$ is small. It can be shown that for $L = L_Q$, we have optimization guarantee $Q(\theta_t) - Q(\theta^\star) \le \frac{2L_Q\|\theta_0 - \theta^\star\|^2}{t+1}$ (Nesterov [17]). If we further assume that Q

is μ_Q-*strongly convex*, that is, $Q(\theta) - (\mu_Q/2)\|\theta\|^2$ is still convex in θ for $0 < \mu_Q \leq L_Q$, then for $L = (L_Q + \mu_Q)/2$, we further have $Q(\theta_t) - Q(\theta^\star) \leq (L_Q/2)\, \|\theta_t - \theta^\star\|_2^2 \leq (L_Q/2)\left(\frac{L_Q-\mu_Q}{L_Q+\mu_Q}\right)^{2t} \|\theta_0 - \theta^\star\|_2^2$.

For now we consider the case that J may not be Lipschitz gradient, but it has a tractable *proximal operator*:

$$\text{prox}_J(\theta_0) := \underset{\theta\in\mathbb{R}^p}{\text{argmin}} \left\{(1/2)\, \|\theta - \theta_0\|_2^2 + J(\theta)\right\}$$

The following are some typical examples of the penalty term whose proximal operator can be derived analytically:

- (Breiman [18], Nonnegative Garrotte) Suppose that $J(\theta) = \lambda \sum_{j=1}^p \theta_j$ if $\theta_j \geq 0$ $(\forall 1 \leq j \leq p)$ and $+\infty$ otherwise. Then, $\text{prox}_J(\theta_0) = (\theta_0 - \lambda)_+$. Here, $(\cdot)_+$ is taken componentwise.
- (Tibshirani [19], LASSO) If $J(\theta) = \lambda\, \|\theta\|_1$, then $\text{prox}_J(\theta_0) = \mathbf{sign}(\theta_0) \odot (|\theta_0| - \lambda)_+$. Here, $\mathbf{sign}(\cdot)$, $|\cdot|$, and $(\cdot)_+$ are taken componentwise, and $\odot$ is the componentwise product.
- (Zou and Hastie [20], Elastic Net) If $J(\theta) = \lambda\left(\frac{1-\alpha}{2}\, \|\theta\|_2^2 + \alpha\, \|\theta\|_1\right)$, then $\text{prox}_J(\theta_0) = \frac{\mathbf{sign}(\theta_0)\odot(|\theta_0|-\lambda\alpha)_+}{1+\lambda(1-\alpha)}$.
- (Yuan and Lin [21], Grouped LASSO) If $J(\theta) = \lambda\, \|\theta\|_2$, then $\text{prox}_J(\theta_0) = \left(1 - \lambda/\, \|\theta_0\|_2\right)_+ \theta_0$.
- (Zhao *et al.* [22], iLASSO) If $J(\theta) = \lambda\|\theta\|_\infty$, then $\text{prox}_J(\theta_0) = \mathbf{0}_p$ if $\|\theta_0\|_1 \leq \lambda$ and $[\theta_0 \vee (-\eta)] \wedge \eta$ otherwise, for $\eta \geq 0$ satisfying $\sum_{j=1}^p (|\theta_{0j}| - \eta)_+ = 1$.
- (Yuan and Lin [21] and Zhao *et al.* [22], Grouped LASSO and CAP) Suppose that $J(\theta) = \sum_{g\in\mathcal{G}} J_g(\theta_g)$, where $g \subseteq [p] := \{1, 2, \dots, p\}$ is an index subset, θ_g is the corresponding subvector in θ, the collection $\mathcal{G}$ of the index subsets is a disjoint partition of (p), and $\theta = \oplus_{g\in\mathcal{G}}\theta_g$, then $\text{prox}_J(\theta_0) = \oplus_{g=1}^G \text{prox}_{J_g}(\theta_{0g})$. In particular, if $J_g(\theta_g) = \lambda\sqrt{|g|}\, \|\theta_g\|_2$, then the corresponding $J(\theta)$ is the grouped least absolute shrinkage and selection operator (LASSO) penalty. If $J_g(\theta_g) = \lambda\, \|\theta\|_{\gamma_g}^{\gamma_0}$ for $1 \leq \gamma_0, \gamma_g \leq \infty$, then the corresponding $J(\theta)$ is the *composite absolute penalty (CAP)*.
- (Jacob *et al.* [23], Overlap Norm) If $J(\theta) = \lambda \inf\left\{\sum_{g\in\mathcal{G}} \|\theta_g\|_2 : \theta = \sum_{g\in\mathcal{G}}\theta_g\right\}$ where $g \subseteq [p] := \{1, 2, \dots, p\}$ is an index subset, $\bigcup_{g\in\mathcal{G}} g = [p]$ while the index subsets among $\mathcal{G}$ need not be disjoint, then $\text{prox}_J(\theta_0) = \theta_0 - \Pi_\Theta(\theta_0)$, where Π_Θ is the Euclidean projection onto the set $\Theta := \bigcap_{g\in\mathcal{G}}\{\theta \in \mathbb{R}^p : \|\theta_g\|_2 \leq \lambda\}$.
- (Chernozhukov *et al.* [24], LAVA) If $J(\theta) = \lambda \inf\left\{\frac{1-\alpha}{2}\, \|\beta\|_2^2 + \alpha\, \|\delta\|_1 : \theta = \beta + \delta\right\}$, then $\text{prox}_J(\theta_0) = \frac{\theta_0+\lambda(1-\alpha)\mathbf{sign}(\theta_0)\odot\left[|\theta_0|-\lambda\alpha-\alpha/(1-\alpha)\right]_+}{1+\lambda(1-\alpha)}$.
- (Constrained Optimization Problems) Suppose that $\Theta \subseteq \mathbb{R}^p$ is a convex subset, $J(\theta) = 0$ if $\theta \in \Theta$ and $+\infty$ otherwise. Then, it becomes a constrained optimization problem $\min_{\theta\in\Theta} R(\theta)$, and $\text{prox}_J = \Pi_\Theta$ reduces to the Euclidean projection operator onto Θ. The following are the common constraints.
  - (Affine Subspace) If $\Theta = \{\theta \in \mathbb{R}^p : \langle \boldsymbol{a}, \theta\rangle = b\}$, then $\Pi_\Theta(\theta_0) = \theta_0 - \frac{\langle \boldsymbol{a},\theta_0\rangle - b}{\langle \boldsymbol{a},\boldsymbol{a}\rangle}\boldsymbol{a}$.
  - (Half Space) If $\Theta = \{\theta \in \mathbb{R}^p : \langle \boldsymbol{a}, \theta\rangle \leq b\}$, then $\Pi_\Theta(\theta_0) = \theta_0 - \frac{(\langle \boldsymbol{a},\theta_0\rangle - b)_+}{\langle \boldsymbol{a},\boldsymbol{a}\rangle}\boldsymbol{a}$. In particular, if $\Theta = [\boldsymbol{a}, \boldsymbol{b}] = \prod_{j=1}^p [a_j, b_j]$ is the box constraint, then $\Pi_\Theta(\theta_0) = (\theta_0 \vee \boldsymbol{a}) \wedge \boldsymbol{b}$. If $\Theta = \{\theta \in \mathbb{R}^p : \|\theta\|_\infty \leq \lambda\}$ is the ℓ^∞-ball, then $\Pi_\Theta(\theta_0) = [\theta_0 \vee (-\lambda)] \wedge \lambda$.
  - (ℓ^2-Ball) If $\Theta = \{\theta \in \mathbb{R}^p : \|\theta\|_2 \leq \lambda\}$, then $\Pi_\Theta(\theta_0) = \lambda\theta_0/(\|\theta_0\|_2 \vee \lambda)$.

- (ℓ^1-Ball) If $\Theta = \{\theta \in \mathbb{R}^p : \|\theta\|_1 \leq \lambda\}$ and $\|\theta_0\|_1 > \lambda$, then $\Pi_\Theta(\theta_0) = \mathbf{sign}(\theta_0) \odot (|\theta_0| - \eta)_+$ for $\eta \geq 0$ satisfying $\sum_{j=1}^{p} (|\theta_{0j}| - \eta)_+ = \lambda$. In particular, if $\Theta = \{\theta \in \mathbb{R}_+^p : \sum_{j=1}^{p} \theta_j = 1\}$ is the unit simplex, then $\Pi_\Theta(\theta_0) = (\theta_0 - \eta)_+$ for $\eta \in \mathbb{R}$ satisfying $\sum_{j=1}^{p} (\theta_{0j} - \eta)_+ = 1$. The fast algorithm for searching for the η can be found in Duchi *et al.* [25].

Based on the L_R-Lipschitz gradient of R and the proximal operator of J, we consider the following optimization upper bound of $Q(\theta) = R(\theta) + J(\theta)$ at $\theta_0 \in \mathbb{R}^p$:

$$\tilde{Q}_L(\theta; \theta_0) := R(\theta_0) + \langle \nabla R(\theta_0), \theta - \theta_0 \rangle + (L/2)\, \|\theta - \theta_0\|_2^2 + J(\theta) \tag{4}$$

Due to the previous discussion, we immediately have: (i) $Q(\theta) \leq \tilde{Q}_{L_R}(\theta; \theta_0)$ for any $\theta \in \mathbb{R}^p$ and (ii) $\partial Q(\theta_0) = \partial_\theta \tilde{Q}_L(\theta; \theta_0)\,|_{\theta=\theta_0}$. With the given θ_t, $\tilde{Q}_L(\cdot; \theta_t)$ is minimized at

$$\theta_{t+1} \in \text{prox}_{(1/L)J}\left(\theta_t - (1/L)\nabla R(\theta_t)\right) \tag{5}$$

Then, (5) provides the updating formula for the *iterative shrinkage-thresholding algorithm (ISTA)* [26, 27]. In particular, the step size $1/L$ can be determined by the *backtracking line search* at every iteration step. Specifically, at the tth step, we first initialize $L_0 < L_R$ and then search for the smallest $j \geq 0$ such that $L_j := 2^j L_0$ satisfying the well-known *Armijo condition*:

$$\tilde{\theta}_{t+1,j} \leftarrow \text{prox}_{(1/L_j)J}\left(\theta_t - (1/L_j)\nabla R(\theta_t)\right), \quad Q(\tilde{\theta}_{t+1,j}) \leq \tilde{Q}_{L_j}(\tilde{\theta}_{t+1,j}; \theta_t)$$

That is, the corresponding optimization upper bound $\tilde{Q}_{L_j}$ majorizes the true objective function at the next-step parameter $\tilde{\theta}_{t+1,j}$. Then, we set $\theta_{t+1} \leftarrow \tilde{\theta}_{t+1,j}$. The ISTA enjoys the same optimization guarantee as the GDA.

Both the GDA and the ISTA can be accelerated by the Nesterov's optimal-gradient methods [27, 28]. More materials on solving (2) by gradient-based approaches can be found in Nesterov [16, 17].

3 Linear Regression

In this section, we discuss the supervised learning problem with the response taking continuous value in $\mathcal{Y} = \mathbb{R}$. The most commonly used loss function for the continuous-valued response data is the squared loss $\ell(y, \hat{y}) = (y - \hat{y})^2$. In this case, the risk function $\mathcal{R}(f) = \mathbb{E}[Y - f(\mathbf{X})]^2$ is also known as the *prediction mean-squared error (PMSE)*. One simple statistical model to study is a linear model. We mainly focus on the linear regression problem, where the function class is $\mathcal{F} = \{\boldsymbol{x} \mapsto \boldsymbol{x}^T\boldsymbol{\beta} : \boldsymbol{\beta} \in \mathbb{R}^p\}$. The nonlinear extensions using the *kernel tricks* are left to Section 5.1.

3.1 Linear Regression and Ridge Regression

First, note that the ERM formulation (1) without the penalty term is equivalent to solving for the *least-squared estimator (LSE)* $\hat{\boldsymbol{\beta}}$ of the coefficient vector. When the data-generating process is assumed to be linear model with Normal error as in Section 2.1, the LSE is a BLUE and the *maximum-likelihood estimator (MLE)* of the problem. Furthermore, the confidence band for $\hat{f}(\boldsymbol{x}) = \boldsymbol{x}^T\hat{\boldsymbol{\beta}}$ can be determined using the F-test. In practice, the linear model can be easy to interpret. The t-tests on the significance of the coefficients can inform useful covariates for the fitted model. However, as discussed in Section 2.1, even though the underlying true coefficients are nonzero, the linear model based on all variables may not have the smallest PMSE. There are model selection techniques based on the information criteria for choosing the "best" model that targets on the PMSE directly, including the *best subset selection* and the *forward* and *backward selections* [29]. However, these methods can handle very limited numbers of variables. Recently, Bertsimas *et al.* [30] proposed to solve the best subset selection problem using the projection-gradient method, which shares the same structure as the ISTA in Section 2.2, but their ℓ^0-ball constraint is nonconvex. They showed the convergence of the algorithm and that the selection of thousands of variables can be handled.

In the linear model theory, when the training design matrix is ill-conditioned, the introduction of the penalty term can help to reduce the variance by sacrificing a small amount of bias. When the ℓ^2-penalty $J_\lambda(\boldsymbol{\beta}) = \frac{\lambda}{2} \|\boldsymbol{\beta}\|_2^2$ is used, the penalized ERM problem (1) becomes the *ridge regression*. Consider the *smoother matrix* for the ridge regression $\mathbf{S}_\lambda := \mathbb{X}(\mathbb{X}^T\mathbb{X} + n\lambda\mathbf{I}_n)^{-1}\mathbb{X}^T$. Then, the in-sample prediction of the response vector is $\mathbf{S}_\lambda\boldsymbol{Y}$. Let $\{d_j\}_{j=1}^p$ be the set of singular values of $\mathbb{X}$. Then, $\text{tr}(\mathbf{S}_\lambda) = \sum_{j=1}^p d_j^2/(d_j^2 + n\lambda)$ is the *effective dimension* of the ridge regression model. To better understand the bias–variance trade-off, assume the linear model with Normal error as in Section 2.1, and $\mathbb{X}^T\mathbb{X} = \text{diag}\{d_j^2\}_{j=1}^p$. Denote $\hat{f}_\lambda(\boldsymbol{x}) := \boldsymbol{x}^T\hat{\boldsymbol{\beta}}(\lambda)$ where $\hat{\boldsymbol{\beta}}(\lambda)$ is the ridge regression estimate. Then, for fixed $\boldsymbol{x}_0 \in \mathbb{R}^p$,

$$\textbf{Var}(\hat{f}_\lambda(\boldsymbol{x}_0)) = \sigma^2 \sum_{j=1}^p \frac{d_j^2}{(d_j^2 + n\lambda)^2} x_{0j}^2; \quad \textbf{Bias}\,(\hat{f}_\lambda(\boldsymbol{x}_0))^2 = \left(\sum_{j=1}^p \frac{n\lambda}{d_j^2 + n\lambda} x_{0j}\beta_j^\star \right)^2$$

When $\mathbb{X}^T\mathbb{X}$ is ill-conditioned, there exists some singular value $d_j \approx 0$. When $\lambda = 0$, $\hat{\boldsymbol{\beta}}(0)$ is the LSE with the effective dimension $\text{tr}(\mathbf{S}_0) = p$. The variance $\textbf{Var}(\hat{f}_0(\boldsymbol{x}_0)) = \sigma^2 \sum_{j=1}^p x_{0j}^2/d_j^2$ can be large due to the small denominator, while the bias is $\textbf{Bias}\,(\hat{f}_0(\boldsymbol{x}_0))^2 = 0$. As λ increases, the effective dimension and the variance decrease, while the bias increases. As $\lambda \to +\infty$, we have $\text{tr}(\mathbf{S}_\lambda) \to 0$, $\textbf{Var}(\hat{f}_\lambda(\boldsymbol{x}_0)) \to 0$, while $\textbf{Bias}(\hat{f}_\lambda(\boldsymbol{x}_0)) \to (\boldsymbol{x}_0^T\boldsymbol{\beta}^\star)^2$. In general, the best tuning parameter λ achieves the minimal $\textbf{Var}(\hat{f}_\lambda(\boldsymbol{x}_0)) + \textbf{Bias}\,(\hat{f}_\lambda(\boldsymbol{x}_0))^2$ at some $\lambda > 0$, giving the best PMSE.

3.2 LASSO

The ridge regression can trade-off the bias and variance by introducing the ℓ^2-penalty. However, the resulting coefficients are not sparse. In order to automatically perform variable selection when training the model, the ℓ^1-penalty can be used instead.

Consider the LASSO solution in the linear regression problem with $\hat{\boldsymbol{\beta}}(\lambda) \in \text{argmin}_{\boldsymbol{\beta}\in\mathbb{R}^p}\{(1/2)\|\boldsymbol{Y} - \mathbb{X}\boldsymbol{\beta}\|_2^2 + \lambda\|\boldsymbol{\beta}\|_1\}$, where $\boldsymbol{Y}$ is centered, and $\mathbb{X}$ is centered and standardized by column. Define $\boldsymbol{r} := \boldsymbol{Y} - \mathbb{X}\hat{\boldsymbol{\beta}}(\lambda)$ as the residual vector. Using the fact that $\hat{\boldsymbol{\beta}}(\lambda)$ is the fixed point of the ISTA iteration (5), we have

$$\begin{aligned}
\hat{\boldsymbol{\beta}}(\lambda) &= \text{prox}_{(\lambda/L)\|\cdot\|_1}\left(\hat{\boldsymbol{\beta}}(\lambda) + (1/L)\mathbb{X}^T\boldsymbol{r}\right) \\
&= \textbf{sign}\left(\hat{\boldsymbol{\beta}}(\lambda) + (1/L)\mathbb{X}^T\boldsymbol{r}\right) \odot \left(\left|\hat{\boldsymbol{\beta}}(\lambda) + (1/L)\mathbb{X}^T\boldsymbol{r}\right| - \lambda/L\right)_+
\end{aligned}$$

Denote $\boldsymbol{X}_j$ as the jth column vector in $\mathbb{X}$. If $\hat{\beta}_j(\lambda) \neq 0$, then $\hat{\beta}_j(\lambda) = \hat{\beta}_j(\lambda) + (1/L)\langle \boldsymbol{X}_j, \boldsymbol{r}\rangle - (\lambda/L)\mathbf{sign}[\hat{\beta}_j(\lambda)] \Leftrightarrow \langle \boldsymbol{X}_j, \boldsymbol{r}\rangle = \lambda\mathbf{sign}[\hat{\beta}_j(\lambda)]$. If $\hat{\beta}_j(\lambda) = 0$, then we have $|(1/L)\langle \boldsymbol{X}_j, \boldsymbol{r}\rangle| \leq \lambda/L \Leftrightarrow |\langle \boldsymbol{X}_j, \boldsymbol{r}\rangle| \leq \lambda$. If λ is sufficiently large, then there can be many js such that $|\langle \boldsymbol{X}_j, \boldsymbol{r}\rangle| < \lambda$, which correspond to the zero coefficients $\hat{\beta}_j(\lambda) = 0$. This explains the sparsity of the LASSO solution.

Define the active index set $\mathcal{A}(\lambda) := \{1 \leq j \leq p : |\langle \boldsymbol{X}_j, \boldsymbol{r}\rangle| = \lambda\}$. Fix $\lambda_0 \geq 0$. Denote $\mathcal{A}_0 := \mathcal{A}(\lambda_0)$, $\mathbb{X}_{\mathcal{A}_0} \in \mathbb{R}^{n\times|\mathcal{A}_0|}$ as the submatrix in $\mathbb{X}$ with column indices in $\mathcal{A}_0$ and $\hat{\boldsymbol{\beta}}_{\mathcal{A}_0}(\lambda)$ as the subvector in $\hat{\boldsymbol{\beta}}(\lambda)$ with indices in $\mathcal{A}_0$. Consider a small change from λ_0 to λ such that $\mathcal{A}(\lambda) = \mathcal{A}_0$. In that case, $\mathbf{sign}[\hat{\boldsymbol{\beta}}(\lambda)] = \mathbf{sign}[\hat{\boldsymbol{\beta}}(\lambda_0)]$. Denote $\boldsymbol{s}_0 := \mathbf{sign}[\hat{\boldsymbol{\beta}}_{\mathcal{A}_0}(\lambda_0)]$. Then, $-\mathbb{X}_{\mathcal{A}_0}^T\mathbb{X}_{\mathcal{A}_0}[\hat{\boldsymbol{\beta}}_{\mathcal{A}_0}(\lambda) - \hat{\boldsymbol{\beta}}_{\mathcal{A}_0}(\lambda_0)] = (\lambda - \lambda_0)\boldsymbol{s}_0$, which is equivalent to

$$\hat{\boldsymbol{\beta}}_{\mathcal{A}_0}(\lambda) = \hat{\boldsymbol{\beta}}_{\mathcal{A}_0}(\lambda_0) - (\lambda - \lambda_0)(\mathbb{X}_{\mathcal{A}_0}^T\mathbb{X}_{\mathcal{A}_0})^{-1}\boldsymbol{s}_0 \tag{6}$$

That is, the solution path $\hat{\boldsymbol{\beta}}(\lambda)$ is linear in λ when the active set $\mathcal{A}(\lambda)$ remains unchanged. In other words, the overall solution path is piecewise linear. This property characterizes the solutions to the LASSO problems for different λs. Based on this fact, Efron *et al.* [31] proposed the *least angle regression* (LAR) algorithm to solve for the solution path of $\hat{\boldsymbol{\beta}}(\lambda)$ from $\lambda = \lambda_{\max} := \max_{1\leq j\leq p}|\langle \boldsymbol{X}_j, \boldsymbol{Y}\rangle|$ down to $\lambda = 0$. In particular, starting from $\boldsymbol{\beta}_{\lambda_{\max}} = \boldsymbol{0}_p$, $\boldsymbol{r} = \boldsymbol{Y}$ and $\lambda_0 = \lambda_{\max}$, the algorithm proceeds with (6) until $\mathcal{A}(\lambda) \neq \mathcal{A}_0$, and then it updates $\lambda_0 \leftarrow \lambda$ and proceeds with (6) again. The piecewise linearity of the solution path can be helpful for developing algorithms analog to the LAR. Rosset and Zhu [32] systematically studied the general penalized ERM problem (1) and concluded that if R_n is quadratic or piecewise quadratic and J_λ is piecewise linear, then the solution path is piecewise linear.

When considering the LASSO problem in high dimensions, the path coordinate descent algorithm can enjoy computational efficiency [33, 34]. First consider a univariate LASSO solution $\operatorname{argmin}_{\beta\in\mathbb{R}}\{(1/2)\,\|\boldsymbol{Y} - \boldsymbol{X}\beta\|_2^2 + \lambda|\beta|\} = \operatorname{argmin}_{\beta\in\mathbb{R}}\{(1/2)(\beta - \langle \boldsymbol{X}, \boldsymbol{Y}\rangle)^2 + \lambda|\beta|\} = \operatorname{prox}_{\lambda|\cdot|}(\langle \boldsymbol{X}, \boldsymbol{Y}\rangle)$. Here, the proximal operator for $\lambda|\cdot|$ is obtained in Section 2.2. Then, suppose that for the p-variate LASSO problem, the jth coordinate is chosen to descent, given $\hat{\boldsymbol{\beta}}_{-j} := (\hat{\beta}_1, \hat{\beta}_2, \ldots, \hat{\beta}_{j-1}, \hat{\beta}_{j+1}, \ldots, \hat{\beta}_p)^T$. Denote $\mathbb{X}_{-j}$ as the submatrix in $\mathbb{X}$ ruling out the jth column, $\boldsymbol{r}_{-j} := \boldsymbol{Y} - \mathbb{X}_{-j}\hat{\boldsymbol{\beta}}_{-j}$. Then, the jth coordinate LASSO problem becomes $\hat{\beta}_j(\lambda) \in \operatorname{argmin}_{\beta_j\in\mathbb{R}}\{(1/2)\,\|\boldsymbol{r}_{-j} - \boldsymbol{X}_j\beta_j\|_2^2 + \lambda|\beta_j|\} = \operatorname{prox}_{\lambda|\cdot|}(\langle \boldsymbol{X}_j, \boldsymbol{r}_{-j}\rangle)$. The path coordinate descent algorithm is implemented as follows: first, it chooses the coordinate j cyclically through $1, 2, \ldots, p$; then, it solves the jth coordinate LASSO problem for the entire solution path $\hat{\beta}_j(\cdot)$. Notice that using the squared loss, the pathwise coordinate descent algorithm can be used for other penalties discussed in Section 2.2, since the coordinate problem turns out to be a proximal operator on the penalty term. When general Lipschitz-gradient losses are considered, we can perform coordinate descent on the quadratic optimization upper bound (4) in ISTA. Then it becomes a coordinate proximal GDA. For example, Friedman *et al.* [35] proposed the path coordinate descent algorithm for the MLE of the generalized linear model (GLM) with the elastic net penalty in the well-known `R` package `glmnet`.

The LASSO with the correctly chosen λ can enjoy the model selection consistency [36] and the $\sqrt{n}$-asymptotic normality [37]. However, Zou [38] pointed out that the choice of λ cannot be compatible to satisfy both properties simultaneously. Instead, he considered the *adaptive LASSO* penalty $J_\lambda(\boldsymbol{\beta}) = \sum_{j=1}^{p}\hat{w}_j|\beta_j|$, where the weight is determined adaptively $\hat{w}_j := 1/|\hat{\beta}_j|^\gamma$ by a consistent estimate of the coefficient vector $\hat{\boldsymbol{\beta}}$, for example, the LSE.

The adaptive LASSO can enjoy both the model selection consistency and the $\sqrt{n}$-asymptotic normality simultaneously for the appropriately chosen λ. Recent advances focus on the inference problem in high dimensions, including Refs 39–42.

In the literature, there exists some variants of the LASSO penalty that accounts for special structures, including the fused LASSO [43] that encourages similar patterns for successive coefficients, the grouped LASSO and CAP [21, 22] that encourage the group sparsity, the Dantzig selector [44] that recovers the compressed sparse signals, and the sparse regression incorporating graphical structure among predictors (SRIG) [45]. There are many other penalized regression methods using nonconvex penalties, including the ℓ^q-penalty for $q \in (0,1)$ [46], the *smoothly clipped absolute deviation (SCAD)* penalty [7], the hybrid of the ℓ^0 and ℓ^1 penalties [47], and the *minimax concave plus (MCP)* penalty [48]. We refer the readers to Refs 49 and 50 for more complete reviews.

4 Classification

In this section, we consider the classification problem that the response takes discrete values in $\mathcal{Y}$. We denote $d : \mathcal{X} \to \mathcal{Y}$ as the *decision rule*. Recall that the 0–1 loss $\ell(y, \hat{y}) = \mathbb{1}(y \neq \hat{y})$ is considered here. The corresponding risk function $\mathcal{R}(d) = \mathbb{P}[d(\boldsymbol{X}) \neq Y]$ is known as the *misclassification error* of d. We mainly discuss the binary classification problem $\mathcal{Y} = \{0, 1\}$ in Sections 4.1–4.4. In some scenario, we may use the sign coding of the response $\tilde{Y} = 2Y - 1 \in \{-1, 1\}$ for convenience. In Section 4.5, we consider the multicategory classification problem for $\mathcal{Y} = \{1, 2, \ldots, K\}$.

4.1 Model-Based Methods

In the binary classification problem where $\mathcal{Y} = \{0, 1\}$, we introduce the class conditional probability function as $\eta(\boldsymbol{x}) := \mathbb{P}(Y = 1|\boldsymbol{X} = \boldsymbol{x})$. Then, the *Bayes rule* of the classification problem is defined as $d_{\text{Bayes}}(\boldsymbol{x}) := \mathbb{1}[\eta(\boldsymbol{x}) \geq 1/2]$, with the corresponding *Bayes risk* defined as

$$
\begin{aligned}
\mathcal{R}_{\text{Bayes}} &:= \mathcal{R}(d_{\text{Bayes}}) \\
&= \mathbb{P}[d_{\text{Bayes}}(\boldsymbol{X}) \neq Y] \\
&= \mathbb{E}\left\{\eta(\boldsymbol{X})\mathbb{P}[d_{\text{Bayes}}(\boldsymbol{X}) \neq 1|\boldsymbol{X}] + [1 - \eta(\boldsymbol{X})]\mathbb{P}[d_{\text{Bayes}}(\boldsymbol{X}) \neq 0|\boldsymbol{X}]\right\} \\
&= \mathbb{E}\{\eta(\boldsymbol{X})\mathbb{1}[\eta(\boldsymbol{X}) < 1/2] + [1 - \eta(\boldsymbol{X})]\mathbb{1}[\eta(\boldsymbol{X}) \geq 1/2]\} \\
&= \mathbb{E}\{\eta(\boldsymbol{X}) \wedge [1 - \eta(\boldsymbol{X})]\}
\end{aligned}
$$

Then, for any decision rule $d : \mathcal{X} \to \{0, 1\}$, we have

$$
\begin{aligned}
\mathcal{R}(d) &= \mathbb{E}\{\eta(\boldsymbol{X})\mathbb{1}[d(\boldsymbol{X}) \neq 1] + [1 - \eta(\boldsymbol{X})]\mathbb{1}[d(\boldsymbol{X}) \neq 0|\boldsymbol{X}]\} \\
&\geq \mathbb{E}\{\eta(\boldsymbol{X}) \wedge [1 - \eta(\boldsymbol{X})]\} = \mathcal{R}_{\text{Bayes}}
\end{aligned}
$$

That is, the Bayes rule d_{Bayes} achieves the smallest possible risk $\mathcal{R}_{\text{Bayes}}$. Based on this fact, the model-based methods first impose model assumptions on $(\boldsymbol{X}, Y)$ and then estimate $\eta(\boldsymbol{x})$ using the likelihood-based approach based on the training data.

There are two popular model-based approaches for $(\boldsymbol{X}, Y)$. The first approach is to assume $\mathbb{P}(Y = y) = \pi_y \in (0, 1)$ and $\boldsymbol{X}|(Y = y) \sim \mathcal{N}_p(\boldsymbol{\mu}_y, \Sigma_y)$ for $y = 0, 1$. Then, using the Bayes formula $\frac{\eta(\boldsymbol{x})}{1-\eta(\boldsymbol{x})} = \frac{\pi_1 f_1(\boldsymbol{x})}{\pi_0 f_0(\boldsymbol{x})}$, where $f_y(\boldsymbol{x})$ is the density of $\boldsymbol{X}|(Y = y)$ for $y = 0, 1$, we have

$$\log\left(\frac{\eta(\boldsymbol{x})}{1-\eta(\boldsymbol{x})}\right) = \log\left(\frac{\pi_1}{\pi_0}\right) - \frac{1}{2}\log\det(\Sigma_0^{-1}\Sigma_1) - \frac{1}{2}\left[(\boldsymbol{x}-\boldsymbol{\mu}_1)^T\Sigma_1^{-1}(\boldsymbol{x}-\boldsymbol{\mu}_1) - (\boldsymbol{x}-\boldsymbol{\mu}_0)^T\Sigma_0^{-1}(\boldsymbol{x}-\boldsymbol{\mu}_0)\right] \tag{7}$$

Here, we remark that $\eta(\boldsymbol{x}) \geq 1/2$ if and only if $\log\left(\frac{\eta(\boldsymbol{x})}{1-\eta(\boldsymbol{x})}\right) \geq 0$. The formula (7) suggests that the Bayes rule $d_{\text{Bayes}}(\boldsymbol{x}) = \mathbb{1}\left[\log\left(\frac{\eta(\boldsymbol{x})}{1-\eta(\boldsymbol{x})}\right) \geq 0\right]$ has a quadratic decision boundary $\{\boldsymbol{x} \in \mathcal{X} : d_{\text{Bayes}}(\boldsymbol{x}) = 0\}$. Based on the training data $\mathcal{D}_n$, we estimate the parameters $(\pi_1, \pi_0, \boldsymbol{\mu}_1, \boldsymbol{\mu}_0, \Sigma_1, \Sigma_0)$ in (7) using the MLEs. This approach is known as the *quadratic discriminant analysis (QDA)*. In the Bayes formula (7), if we further assume that $\Sigma_1 = \Sigma_0 = \Sigma$, then it can be simplified to

$$\log\left(\frac{\eta(\boldsymbol{x})}{1-\eta(\boldsymbol{x})}\right) = \log\left(\frac{\pi_1}{\pi_0}\right) + (\boldsymbol{\mu}_1 - \boldsymbol{\mu}_0)^T\Sigma^{-1}\left(\boldsymbol{x} - \frac{\boldsymbol{\mu}_1 + \boldsymbol{\mu}_0}{2}\right) \tag{8}$$

The corresponding Bayes rule $d_{\text{Bayes}}(\boldsymbol{x})$ has a linear decision boundary in $\boldsymbol{x}$. We estimate the parameter $(\pi_1, \pi_0, \boldsymbol{\mu}_1, \boldsymbol{\mu}_0, \Sigma)$ in (8) by their MLEs based on the training data. This approach is called the *linear discriminant analysis (LDA)*.

The second approach is to assume $\log\left(\frac{\eta(\boldsymbol{x})}{1-\eta(\boldsymbol{x})}\right) = b + \boldsymbol{x}^T\boldsymbol{\beta}$, which is the logistic regression assumption for $(\boldsymbol{X}, Y)$. It results in the Bayes rule $d_{\text{Bayes}}(\boldsymbol{x}) = \mathbb{1}(b + \boldsymbol{x}^T\boldsymbol{\beta} \geq 0)$. Based on the training data $\mathcal{D}_n$, the parameters $(b, \boldsymbol{\beta})$ can be estimated by the MLEs for the logistic regression. We point out that the assumption of the logistic regression is weaker than the LDA, in the sense that the logistic regression only assumes the conditional model of $Y|\boldsymbol{X}$, while the LDA assumes the joint distribution of $(\boldsymbol{X}, Y)$. If we use the sign coding $\tilde{Y}_i = 2Y_i - 1 \in \{-1, 1\}$, then maximizing the likelihood of the logistic regression can be reformulated into the penalized ERM problem without penalty as follows:

$$\min_{b,\boldsymbol{\beta}} \quad \frac{1}{n}\sum_{i=1}^{n} \log\left\{1 + \exp[-\tilde{Y}_i(b + \boldsymbol{X}_i^T\boldsymbol{\beta})]\right\} \tag{9}$$

Here, we also define $\ell(y, \hat{y}) := \log(1 + e^{-y\hat{y}})$ as the *logistic loss*, also known as the deviance loss. Further extensions can also introduce the ℓ^2-penalty or ℓ^1-penalty of $\boldsymbol{\beta}$ as the penalty term in (9), which corresponds to the *penalized logistic regression (PLR)* approaches [6].

4.2 Support Vector Machine (SVM)

The model-based methods in Section 4.1 first estimate the class conditional probability function $\eta(\boldsymbol{x}) = \mathbb{P}(Y = 1|\boldsymbol{X} = \boldsymbol{x})$ and then induce the corresponding Bayes rule $d_{\text{Bayes}}(\boldsymbol{x}) = \mathbb{1}[\eta(\boldsymbol{x}) \geq 1/2]$. These approaches are often known as the *soft classifiers* [51]. However, if the class conditional probability function is hard to estimate in some complicated problems, then it is often more desirable to target on the decision rule that minimizes the risk directly [52]. Such classifiers are referred as the *hard classifiers*. Recall the sign coding $\tilde{Y} = 2Y - 1 \in \{-1, 1\}$. The goal of the hard classifiers is to find a *decision function* $f : \mathcal{X} \to \mathbb{R}$ which induces the decision rule $d(\boldsymbol{x}) := \mathbb{1}[f(\boldsymbol{x}) \geq 0]$, such that the misclassification error $\mathcal{R}(f) = \mathbb{P}[d(\boldsymbol{X}) \neq Y] = \mathbb{P}[\tilde{Y}f(\boldsymbol{X}) < 0]$ is minimized. Here, we abuse the notation $\mathcal{R}(f)$ to refer to $\mathcal{R}(d)$. In this section, we introduce the support vector machine (SVM) as one of the hard classifiers.

To begin with, we suppose that the training data $\mathcal{D}_n$ are linearly separable. That is, there exists $(b, \boldsymbol{\beta})$ such that $\tilde{Y}_i = \mathbf{sign}(b + \boldsymbol{X}_i^T\boldsymbol{\beta})$ for $1 \leq i \leq n$. In that case, the training misclassification error is 0. The SVM considers a separating hyperplane $\{\boldsymbol{x} \in \mathcal{X} : b + \boldsymbol{x}^T\boldsymbol{\beta} = 0\}$ that optimizes the following problem:

$$\begin{aligned} \max_{b,\boldsymbol{\beta},\gamma} \ & \gamma \\ \text{s.t.} \ & \tilde{Y}_i(b + \boldsymbol{X}_i^T\boldsymbol{\beta}) \geq \gamma; \ 1 \leq i \leq n \\ & \|\boldsymbol{\beta}\|_2 = 1 \end{aligned} \tag{10}$$

Here, the width γ of the *margin* $\{\boldsymbol{x} \in \mathcal{X} : |b + \boldsymbol{x}^T\boldsymbol{\beta}| \leq \gamma\}$ is maximized, such that $\boldsymbol{X}_i \in \{\boldsymbol{x} \in \mathcal{X} : |b + \boldsymbol{x}^T\boldsymbol{\beta}| \geq \gamma\}$ and $\tilde{Y}_i = \mathbf{sign}(b + \boldsymbol{X}_i^T\boldsymbol{\beta})$ for $1 \leq i \leq n$. Therefore, the SVM is also known as a *large-margin classifier*. The problem (10) can be shown equivalent to

$$\begin{aligned} \min_{b,\boldsymbol{\beta}} \ & \tfrac{1}{2} \|\boldsymbol{\beta}\|_2^2 \\ \text{s.t.} \ & \tilde{Y}_i(b + \boldsymbol{X}_i^T\boldsymbol{\beta}) \geq 1; \ 1 \leq i \leq n \end{aligned} \tag{11}$$

In particular, the solution to $\boldsymbol{\beta}$ in (11) corresponds to the margin width $\gamma = 1/\|\boldsymbol{\beta}\|_2$ in (10). We further introduce the Lagrange dual variables $\boldsymbol{\alpha} = (\alpha_1, \alpha_2, \dots, \alpha_n)^T$ for the inequality constraints. Then, the Lagrange dual problem of (11) becomes

$$\begin{aligned} \max_{\boldsymbol{\alpha}} \ & \sum_{i=1}^{n} \alpha_i - \tfrac{1}{2} \sum_{i=1}^{n} \sum_{i'=1}^{n} \alpha_i \alpha_{i'} \tilde{Y}_i \tilde{Y}_{i'} \langle \boldsymbol{X}_i, \boldsymbol{X}_{i'} \rangle \\ \text{s.t.} \ & \sum_{i=1}^{n} \alpha_i \tilde{Y}_i = 0 \\ & \alpha_i \geq 0; \qquad 1 \leq i \leq n \end{aligned} \tag{12}$$

The dual problem can be solved by the standard *quadratic programming (QP)* [53]. Moreover, the solution $\hat{\boldsymbol{\beta}}$ to the primal problem (12) relates to the solutions $(b, \hat{\boldsymbol{\alpha}})$ to the dual problem (11) through the *Karush–Kuhn–Tucker (KKT) conditions*: (i) $\hat{\boldsymbol{\beta}} = \sum_{i=1}^{n} \hat{\alpha}_i \tilde{Y}_i \boldsymbol{X}_i$ and (ii) $\begin{cases} \tilde{Y}_i(\hat{b} + \boldsymbol{X}_i^T\hat{\boldsymbol{\beta}}) = 1 \Rightarrow \hat{\alpha}_i \geq 0; \\ \tilde{Y}_i(\hat{b} + \boldsymbol{X}_i^T\hat{\boldsymbol{\beta}}) > 1 \Rightarrow \hat{\alpha}_i = 0. \end{cases}$ In other words, if $\hat{\alpha}_i > 0$, then $\tilde{Y}_i(\hat{b} + \boldsymbol{X}_i^T\hat{\boldsymbol{\beta}}) = 1$, so that $\boldsymbol{X}_i$ lies on the boundary of the margin and hence is called the *support vector* (SV). The solution to $\hat{b}$ in the primal problem (11) can be identified by the SVs with $\hat{b} = \tilde{Y}_i - \boldsymbol{X}_i^T\hat{\boldsymbol{\beta}}$.

When the training data $\mathcal{D}_n$ are not linearly separable, we introduce the slack variables $\xi = (\xi_1, \xi_2, \dots, \xi_n)^T$ and a cost parameter $C > 0$ for the misclassified sample points. The primal problem (11) can then be rewritten as

$$\begin{aligned} \min_{b,\boldsymbol{\beta},\xi} \ & \tfrac{1}{2} \|\boldsymbol{\beta}\|_2^2 + C \sum_{i=1}^{n} \xi_i \\ \text{s.t.} \ & \tilde{Y}_i(b + \boldsymbol{X}_i^T\boldsymbol{\beta}) \geq 1 - \xi_i; \ 1 \leq i \leq n \\ & \xi_i \geq 0; \qquad 1 \leq i \leq n \end{aligned} \tag{13}$$

Then, the corresponding dual problem becomes

$$\begin{aligned} &\max_{\alpha} \sum_{i=1}^{n} \alpha_i - \frac{1}{2}\sum_{i=1}^{n}\sum_{i'=1}^{n} \alpha_i \alpha_{i'} \tilde{Y}_i \tilde{Y}_{i'} \langle \boldsymbol{X}_i, \boldsymbol{X}_{i'} \rangle \\ &\text{s.t. } \sum_{i=1}^{n} \alpha_i \tilde{Y}_i = 0; \\ &\quad 0 \le \alpha_i \le C; \qquad 1 \le i \le n \end{aligned} \tag{14}$$

The second KKT condition becomes $\begin{cases} \tilde{Y}_i[\hat{b} + \boldsymbol{X}_i^T\hat{\beta}] > 1 \Rightarrow \hat{\alpha}_i = 0; \\ \tilde{Y}_i[\hat{b} + \boldsymbol{X}_i^T\hat{\beta}] < 1 \Rightarrow \hat{\alpha}_i = C; \\ \tilde{Y}_i[\hat{b} + \boldsymbol{X}_i^T\hat{\beta}] = 1 \Rightarrow 0 \le \hat{\alpha}_i \le C. \end{cases}$ In this case, we can use the SVs with $0 < \hat{\alpha}_i < C$ to identify $\hat{b}$.

Finally, we point out that the primal problem (13) can be reformulated as the penalized ERM problem [5]:

$$\min_{b,\beta} \quad \frac{1}{n}\sum_{i=1}^{n} [1 - \tilde{Y}_i(b + \boldsymbol{X}_i^T \boldsymbol{\beta})]_+ + \frac{\lambda}{2} \| \boldsymbol{\beta} \|_2^2 \tag{15}$$

Here, $(\cdot)_+ := \max\{\cdot, 0\}$. We define $\ell(y, \hat{y}) := (1 - y\hat{y})_+$ as the *hinge loss*. Further extensions can be obtained by replacing the ℓ^2-penalty of $\boldsymbol{\beta}$ by the ℓ^1-penalty to advocate sparsity, which corresponds to the ℓ^1-SVM [14].

4.3 Convex Surrogate Loss

In this section, we begin with a general discussion on the hard classification problem using the convex surrogate loss. Then, we introduce the *large-margin unified machines (LUMs)* that bridge the soft and hard classifiers.

4.3.1 Surrogate risk minimization

Recall that the goal of the hard classification problem is to directly minimize the misclassification error $\mathcal{R}(f) = \mathbb{E}\mathbb{1}[\tilde{Y}f(\boldsymbol{X}) < 0]$. However, the risk function $\mathcal{R}(f)$ is nonconvex and nonsmooth in f. The optimization of the penalized ERM problem based on $\mathcal{R}(f)$ can be difficult. Since the nonconvexity and nonsmoothness are due to the 0–1 loss, it can be more computationally tractable to replace the 0–1 loss by some surrogate losses that have better convexity and smoothness.

We first point out that the 0–1 loss can be viewed as the loss function $u \mapsto \mathbb{1}(u < 0)$ of the *functional margin* $\tilde{Y}f(\boldsymbol{X})$. Then, we can consider the general loss functions that measure the functional margin. For example, the SVM measures the functional margin by the hinge loss function $u \mapsto (1 - u)_+$. The squared loss $[\tilde{Y} - f(\boldsymbol{X})]^2 = [1 - \tilde{Y}f(\boldsymbol{X})]^2$ also measures the functional margin by $u \mapsto (1 - u)^2$. For further generality, we can use a nonnegative convex loss function $\phi : \mathbb{R} \to [0, +\infty)$ such that ϕ is differentiable at 0 and $\phi'(0) < 0$. It was shown in Bartlett *et al.* [54, Lemma 4] that there exists a $\gamma > 0$ such that $\gamma\phi(u) \ge \mathbb{1}(u < 0)$ for all $u \in \mathbb{R}$.

Based on the loss function ϕ, we define the *ϕ-risk* as $\mathcal{R}_\phi(f) := \mathbb{E}\phi[\tilde{Y}f(\boldsymbol{X})]$. Then, we have $\mathcal{R}(f) \le \gamma\mathcal{R}_\phi(f)$. That is, the ϕ-risk is an upper envelop of the misclassification error up to a scaling factor γ. Define the *empirical ϕ-risk* based on the training data $\mathcal{D}_n$ as $R_{\phi,n}(f) := \frac{1}{n}\sum_{i=1}^{n} \phi[\tilde{Y}_i f(\boldsymbol{X}_i)]$. Then, solving the ϕ-risk-based penalized ERM problem

$\min_{f \in \mathcal{F}} \{R_{\phi,n}(f) + J_\lambda(f)\}$ can be directly carried out based on the first-order optimization methods discussed in Section 2.2.

The validity of minimizing the empirical ϕ-risk can be justified. Suppose that $f_\phi^\star \in \operatorname{argmin}_{f:\mathcal{X}\to\mathbb{R}} \mathcal{R}_\phi(f)$ is the population minimizer of the ϕ-risk. Then, according to Bartlett *et al.* [54, Theorem 2], we have $\mathbf{sign}[f_\phi^\star(\boldsymbol{x})] = d_{\text{Bayes}}(\boldsymbol{x})$ on $\{\boldsymbol{x} \in \mathcal{X} : \eta(\boldsymbol{x}) \neq 1/2\}$. Such a property is known as the *Fisher consistency* [55]. The logistic regression using the logistic loss in Section 4.1 and the SVM using the hinge loss in Section 4.2 are both Fisher consistent. Suppose that $\hat{f}_n \in \operatorname{argmin}_{f\in\mathcal{F}} \{R_{\phi,n}(f) + J_\lambda(f)\}$ minimizes the ϕ-risk-based penalized ERM problem. Then, the *excessive risk* $\mathcal{R}(\hat{f}_n) - \mathcal{R}_{\text{Bayes}}$ can be bounded by model complexity + finite sample approximation error + $\mathcal{F}$-approximation error [54, 56].

4.3.2 Large-margin unified machines (LUMs)

In this section, we consider a specific family of the surrogate losses that unify the soft and hard classification methods. For $a > 0$ and $c \geq 0$, define the *LUM loss function* [3] as

$$V(u) := \begin{cases} 1 - u, & u < \frac{c}{1+c} \\ \frac{1}{1+c}\left(\frac{a}{(1+c)u - c + a}\right)^a, & u \geq \frac{c}{1+c} \end{cases} \tag{16}$$

In particular, when $c \to +\infty$ for some fixed $a > 0$, we have $V(u) \to (1-u)_+$, which corresponds to the hinge loss and the hard classifier SVM. When $c = 0$ and $a \to +\infty$, we have $V(u) = 1 - u$ for $u < 0$ and e^{-u} otherwise, which lies between the logistic loss $\log(1 + e^{-u})$ and $1 + \log(1 + e^{-u})$. Moreover, $\lim_{u\to+\infty}\{[\log(1+e^{-u})]/e^{-u}\} = 1$ and $\lim_{u\to-\infty}[\log(1+e^{-u}) - (-u)] = 0$. Therefore, the LUM behaves similar to the logistic regression as a soft classifier.

The LUM-based penalized ERM problem enjoys the following estimation properties. Consider $f_V^\star \in \operatorname{argmin}_{f:\mathcal{X}\to\mathbb{R}} \mathbb{E}V[\tilde{Y}f(\boldsymbol{X})]$. Then,

$$f_V^\star(\boldsymbol{x}) \begin{cases} = -\frac{1}{1+c}\left[\left(\frac{1-\eta(\boldsymbol{x})}{\eta(\boldsymbol{x})}\right)^{1/(a+1)} \times a - a + c\right], & 0 \leq \eta(\boldsymbol{x}) < 1/2 \\ \in \left[-\frac{c}{1+c}, \frac{c}{1+c}\right], & \eta(\boldsymbol{x}) = 1/2 \\ = +\frac{1}{1+c}\left[\left(\frac{\eta(\boldsymbol{x})}{1-\eta(\boldsymbol{x})}\right)^{1/(a+1)} \times a - a + c\right], & 1/2 < \eta(\boldsymbol{x}) \leq 1 \end{cases}$$

In particular, the induced decision rule $d_V^\star(\boldsymbol{x}) = \mathbf{sign}[f_V^\star(\boldsymbol{x})]$ is the same as $d_{\text{Bayes}}(\boldsymbol{x})$ on $\{\boldsymbol{x} \in \mathcal{X} : \eta(\boldsymbol{x}) \neq 1/2\}$. That is, the LUM loss is Fisher consistent. Moreover, since $f_V^\star(\boldsymbol{x})$ can be viewed as the monotone transform of the class conditional probability function $\eta(\boldsymbol{x})$, the estimation of $\eta(\boldsymbol{x})$ can be recovered by the inverse of this transformation. Under the *GLM* framework [57], the LUM with $c = 0$ can be cast as considering a family of link functions to the canonical parameter $\eta(\boldsymbol{x})$ for the Bernoulli distribution.

4.4 Nonconvex Surrogate Loss

Section 4.3 considers the convex loss function $\phi(u)$ as a surrogate of the 0–1 loss $\mathbb{1}(u < 0)$. However, $\lim_{u\to-\infty} \phi(u) = +\infty$, while $\lim_{u\to-\infty} \mathbb{1}(u < 0) = 1$. As a result, a sample point with large functional margin $\tilde{Y}f(\boldsymbol{X})$ can be highly influential under the ϕ-risk $\mathcal{R}_\phi$ [58] while less influential under the misclassification error $\mathcal{R}$. To prevent the unboundedness of the surrogate loss and align it more tightly with the 0–1 loss, we may truncate the surrogate loss function from above [59]. Specifically, define $\tilde{\phi}_s(u) := (s-u)_+$ and $\phi_s(u) := \tilde{\phi}_1(u) - \tilde{\phi}_s(u)$ for $s \leq 0$. Then, we have $\phi_s(u) = (1+|s|)\mathbb{1}(u < s) + (1-u)\mathbb{1}(s \leq u \leq 1)$. That is, ϕ_s truncates the hinge loss from above at $1+|s|$, so that the $\phi_s(u)$ remains constant for $u \leq s$. Wu and Liu [59] named ϕ_s as the *truncated hinge loss function*. The SVM based on the truncated hinge loss function is called the *robust SVM (RSVM)*.

Note that the truncated hinge loss function ϕ_s is nonconvex but rather *difference-of-convex (DC)*. Then, we consider the *DC algorithm (DCA)* to solve the corresponding penalized ERM problem $\min_{f \in \mathcal{F}} \left\{ R_{\phi_s,n}(f) + J_\lambda(f) \right\}$ [60]. Specifically, the objective function can be written as

$$\overbrace{R_{\phi_s,n}(f) + J_\lambda(f)}^{Q(f)} = \overbrace{R_{\tilde{\phi}_1,n}(f) + J_\lambda(f)}^{Q_1(f)} - \overbrace{R_{\tilde{\phi}_s,n}(f)}^{Q_2(f)}$$

where Q_1, Q_2 are convex in f. Then, we introduce the optimization upper bound of $Q(f)$ at f_0:

$$\tilde{Q}(f;f_0) := Q_1(f) - Q_2(f_0) - \langle \nabla_0, f - f_0 \rangle; \quad \text{for some } \nabla_0 \in \partial Q_2(f_0)$$

Due to the convexity of Q_1, Q_2, we have: (i) for fixed f_0, $\tilde{Q}(f;f_0)$ is convex in f; (ii) $Q(f) \leq \tilde{Q}(f;f_0)$ for all f; and (iii) $\partial_f \tilde{Q}(f;f_0)\mid_{f=f_0} \subseteq \partial Q(f_0)$. At the tth iteration, with the given f_t, the DCA minimizes the convex upper bound: $f_{t+1} \in \operatorname{argmin}_{f \in \mathcal{F}} \tilde{Q}(f;f_t)$. If f_{t+1} is close to f_t, then the DCA stops; otherwise, the DCA proceeds to the $(t+1)$th iteration. Liu *et al.* [60] showed that (i) the dual problem of $\min_{f \in \mathcal{F}} \tilde{Q}(f;f_0)$ is a QP problem; (ii) the DCA iterations terminate in finite steps; (iii) there exists an initial polytope such that the DCA converges to a global minimizer when the initial value is chosen from the polytope. In practice, we can train the standard SVM using the hinge loss to find the initial value; and (iv) the set of SVs in the RSVM is a subset of that in the standard SVM.

Instead of using the nonconvex surrogate loss to gain robustness, Wu and Liu [61] alternatively considered the adaptive weights for the hinge loss. Since the hinge loss $\phi(u) = (1-u)_+$ increases with $1+|u|$ as $u \to -\infty$, they introduced the weight function $w(u) := 1/(1+|u|)$ and target on the weighted hinge loss $\phi_w(u) := w(u)\phi(u) = \frac{(1-u)_+}{1+|u|} = \mathbb{1}(u<0) + \frac{1-u}{1+u}\mathbb{1}(0 \leq u \leq 1)$, which approximately equals the truncated hinge loss $\phi_0(u) = \mathbb{1}(u<0) + (1-u)\mathbb{1}(0 \leq u \leq 1)$. However, since ϕ_w is also nonconvex, they proposed to approximate the weight $w(u)$ adaptively. Suppose that $\hat{f}_{\text{SVM}}$ is the optimal solution of the standard SVM problem. The adaptive weight for the ith training sample is defined as $w_i := w[\tilde{Y}_i\hat{f}_{\text{SVM}}(\boldsymbol{X}_i)] = 1/[1+|\hat{f}_{\text{SVM}}(\boldsymbol{X}_i)|]$. The weighted SVM can be solved by the dual problem (14) by replacing the constraint $0 \leq \alpha_i \leq C$ by $0 \leq \alpha_i \leq Cw_i$ for the ith sample point.

4.5 Multicategory Classification Problem

In this section, we discuss the extensions to the multicategory classification problem. Suppose that $\mathcal{Y} = \{1, 2, \ldots, K\}$ for $K \geq 2$.

For the soft classification method, we first introduce the response vector $\vec{\boldsymbol{Y}} := (\mathbb{1}(Y = 1), \mathbb{1}(Y = 2), \ldots, \mathbb{1}(Y = K))^T$ and the class conditional probability vector $\mathbb{E}(\vec{\boldsymbol{Y}}|\boldsymbol{X} = \boldsymbol{x}) := \boldsymbol{p}(\boldsymbol{x})$. Then, the Bayes rule becomes $d_{\text{Bayes}}(\boldsymbol{x}) = \operatorname{argmax}_{1\leq k\leq K} p_k(\boldsymbol{x})$. Consider the polytomous response model using the *generalized log-linear model* specification: $\mathbf{C}\log[\mathbf{L}\boldsymbol{p}(\boldsymbol{x})] = \mathbf{x}\boldsymbol{\beta}$ [57, Section 6.5.4], where $\mathbf{C}, \mathbf{L}$ are linear transformations of compatible dimensions, the logarithm is taken componentwise, and $\mathbf{x}$ is a design matrix induced by the covariate vector $\boldsymbol{x}$. For example, if we let the Kth class be a reference class and specify $\log p_k(\boldsymbol{x}) - \log p_K(\boldsymbol{x}) = b_k + \boldsymbol{x}^T\boldsymbol{\beta}_k$ for $1 \leq k \leq K-1$, then it becomes the multinomial response model [57, Section 6.4.2]. If we let $q_k(\boldsymbol{x}) := \sum_{j=1}^{k} p_j(\boldsymbol{x})$ be the cumulative conditional probability and specify $\log q_k(\boldsymbol{x}) - \log[1 - q_k(\boldsymbol{x})] = \theta_k - \boldsymbol{x}^T\boldsymbol{\beta}$ for $1 \leq k \leq K-1$, then it becomes the *proportional-odds (PO)* model for the ordinal response [57, Section 5.2.2]. The soft classifier replaces the class conditional probability vector $\boldsymbol{p}(\boldsymbol{x})$ in the Bayes rule by its MLE based on the training data.

The hard classification directly targets on a K-dimensional-vector-valued decision function $\boldsymbol{f}(\boldsymbol{x}) := \left(f_1(\boldsymbol{x}), f_2(\boldsymbol{x}), \ldots, f_K(\boldsymbol{x})\right)^T$, which induces the decision rule $d(\boldsymbol{x}) = \operatorname{argmax}_{1\leq k\leq K} f_k(\boldsymbol{x})$. For identifiability, a sum-to-zero constraint $\sum_{k=1}^{K} f_k(\boldsymbol{x}) = 0$ is employed. Such a formulation reduces to the binary classification case as in Section 4.3 if $K = 2$. The multicategory classification problem can be cast to consider the multicategory surrogate loss $\ell(y, \boldsymbol{f})$ for $\boldsymbol{f} \in \mathbb{R}^K$ that characterizes the 0–1 loss $\mathbb{1}[d(\boldsymbol{X}) \neq Y]$. For example, Liu and Shen [62] introduced $\ell(y, \boldsymbol{f}) := \phi\left(f_y - \max_{k\neq y} f_k\right)$ for some univariate robust surrogate loss function ϕ. Liu and Yuan [63] proposed the combined loss $\ell(y, \boldsymbol{f}) := \gamma(K - 1 - f_y)_+ + (1-\gamma)\sum_{k\neq y}(1 + f_k)_+$ for some $\gamma \in [0, 1]$. Different choices of the multicategory surrogate losses and their Fisher consistencies were studied in Liu [64].

In the multicategory surrogate loss formulation, the vector-valued decision function $\boldsymbol{f}(\boldsymbol{x})$ has effective dimension $K-1$ due to the sum-to-zero constraint. In practice, dealing with the constraint can take more computational efforts. It can be more preferable to encode the decision function in the $(K-1)$-dimensional space directly and get rid of the sum-to-zero constraint. Zhang and Liu [65] considered a $(K-1)$-dimensional coding of the response $\boldsymbol{W}_Y \in \mathbb{R}^{K-1}$, with the unit-length arms $\{\boldsymbol{W}_k\}_{k=1}^K$ defined as

$$\boldsymbol{W}_k := \begin{cases} (K-1)^{-1/2}\mathbf{1}_{K-1}, & k = 1 \\ -\frac{1+\sqrt{K}}{(K-1)^{3/2}}\mathbf{1}_{K-1} + \sqrt{\frac{K}{K-1}}\boldsymbol{e}_{K-1,k-1}, & 2 \leq k \leq K \end{cases}$$

where $\mathbf{1}_{K-1} \in \mathbb{R}^{K-1}$ is the all-one vector, and $\boldsymbol{e}_{K-1,j} = (0, \ldots, 0, \overset{j\text{th}}{1}, 0, \ldots, 0)^T \in \mathbb{R}^{K-1}$ is the all-zero vector but with the jth component as 1. Note that the arms are chosen such that the angles between any two are equal, that is, $\langle \boldsymbol{W}_k, \boldsymbol{W}_{k'} \rangle = \mathbb{1}(k = k') - \frac{1}{K-1}\mathbb{1}(k \neq k')$. For a given $(K-1)$-dimensional-vector-valued decision function $\boldsymbol{f}(\boldsymbol{x})$, the induced decision rule is $d(\boldsymbol{x}) := \operatorname{argmin}_{1\leq k\leq K} \angle\left(\boldsymbol{W}_k, \boldsymbol{f}(\boldsymbol{x})\right) = \operatorname{argmax}_{1\leq k\leq K}\langle \boldsymbol{W}_k, \boldsymbol{f}(\boldsymbol{x})\rangle$. Moreover, the *multicategory functional margin* is defined as $\langle \boldsymbol{W}_Y, f(\boldsymbol{X})\rangle$, which is a multicategorical extension of the case in Section 4.3 when $K = 2$. Then, the large-margin classifiers based on the surrogate loss functions of the functional margin in Section 4.3 can be directly applied here. In particular, consider the LUM loss function V defined in (16), and $\boldsymbol{f}_V^\star \in \operatorname{argmin}_{f:\mathcal{X}\to\mathbb{R}^{K-1}} \mathbb{E}V[\langle \boldsymbol{W}_Y, \boldsymbol{f}(\boldsymbol{X})\rangle]$. Fix $\boldsymbol{x} \in \mathcal{X}$ such that $p_1(\boldsymbol{x}) > p_2(\boldsymbol{x}) > \cdots > p_K(\boldsymbol{x})$.

Then, the class conditional probability vector $\boldsymbol{p}(\boldsymbol{x})$ can be recovered as

$$\langle \boldsymbol{W}_k, \boldsymbol{f}_V^\star(\boldsymbol{x})\rangle = \begin{cases} +\frac{1}{c+1}\left[\left(\frac{p_k(\boldsymbol{x})}{p_K(\boldsymbol{x})}\right)^{1/(a+1)} \times a - a + c\right], & 1 \le k \le K-1 \\ -\frac{1}{c+1}\sum_{j=1}^{K-1}\left[\left(\frac{p_j(\boldsymbol{x})}{p_K(\boldsymbol{x})}\right)^{1/(a+1)} \times a - a + c\right], & k = K \end{cases}$$

5 Extensions for Complex Data

In this section, we consider some extensions of the supervised learning methods in the complex data scenarios. In Section 5.1, we consider nonlinear modeling using the well-known kernel tricks. In Section 5.2, we discuss the large-scale optimization problem when the sample size n and the dimension p can be huge.

5.1 Reproducing Kernel Hilbert Space (RKHS)

Suppose that $K : \mathcal{X} \times \mathcal{X} \to \mathbb{R}$ is a *positive semidefinite (PSD)* kernel function that satisfies: (i) K is symmetric and (ii) for any $n \in \mathbb{N}$ and $\{\boldsymbol{x}_i\}_{i=1}^n \subseteq \mathcal{X}$, the corresponding *kernel matrix* $\mathbf{K} := [K(\boldsymbol{x}_i, \boldsymbol{x}_{i'})]_{n\times n}$ is PSD. Then, the kernel function K induces a unique Hilbert space $\mathcal{H}_K$ of functions on $\mathcal{X}$, known as the *reproducing kernel Hilbert space (RKHS)* [66]. In particular,

$$\mathcal{H}_K = \left\{ \sum_{i=1}^n \alpha_i K(\boldsymbol{x}_i, \cdot) : n \in \mathbb{N},\ \{\alpha_i\}_{i=1}^n \subseteq \mathbb{R},\ \{\boldsymbol{x}_i\}_{i=1}^n \subseteq \mathcal{X} \right\}$$

Moreover, the RKHS $\mathcal{H}_K$ is equipped with an inner product $\langle \cdot, \cdot \rangle_K$ satisfying $\langle K(\boldsymbol{x}, \cdot), K(\boldsymbol{y}, \cdot)\rangle = K(\boldsymbol{x}, \boldsymbol{y})$ for any $\boldsymbol{x}, \boldsymbol{y} \in \mathcal{X}$. Such a property is known as the *reproducing property*. Suppose that K has the eigenexpansion $K(\boldsymbol{x}, \boldsymbol{y}) = \sum_{j=1}^\infty \gamma_j \phi_j(\boldsymbol{x})\phi_j(\boldsymbol{y})$ for $\boldsymbol{x}, \boldsymbol{y} \in \mathcal{X}$, $\gamma_j \ge 0$ and $\sum_{j=1}^\infty \gamma_j^2 < +\infty$. Then, for any $f \in \mathcal{H}_K$, there exists $\{\beta_j\}_{j=1}^\infty \subseteq \mathbb{R}$ such that $f(\cdot) = \sum_{j=1}^\infty \beta_j \phi_j(\cdot)$, and $\| f \|_K^2 := \langle f, f\rangle_K = \sum_{i=1}^\infty \beta_j^2/\gamma_j < +\infty$.

Let $\mathcal{G}_0$ be a finite-dimensional functional space on $\mathcal{X}$. For example, $\mathcal{G}_0 = \mathbf{span}\{1\}$ is the constant function on $\mathcal{X}$, typically accounting for an intercept. Consider the general RKHS-penalized ERM problem on $\mathcal{G}_0 \oplus \mathcal{H}_K$:

$$\min_{g_0 + h \in \mathcal{G}_0 \oplus \mathcal{H}_K} \frac{1}{n}\sum_{i=1}^n \ell\left(Y_i, g_0(\boldsymbol{X}_i) + h(\boldsymbol{X}_i)\right) + \lambda \| h \|_K^2 \tag{17}$$

The theoretical foundation of using the RKHS penalty to achieve a tight excessive risk bound can be found in Bartlett and Mendelson [11]. Moreover, even though $\mathcal{H}_K$ is an infinite-dimensional functional space, the optimization of (17) can be shown tractable through the following well-known Representer Theorem [67].

Theorem 1 (Representer Theorem). *The solution $\hat{f}_n$ to (17) has a representer of the form*

$$\hat{f}_n(\cdot) = g_0(\cdot) + \sum_{i=1}^n \alpha_i K(\boldsymbol{X}_i, \cdot)$$

for some $g_0 \in \mathbf{span}(\mathcal{G}_0)$ and $\{\alpha_i\}_{i=1}^n \subseteq \mathbb{R}$.

Theorem 1 has a direct connection to the dual problem (14) of the standard SVM. Consider the *linear kernel function* $K_{\text{lin}}(\boldsymbol{x},\boldsymbol{y}) = \langle \boldsymbol{x},\boldsymbol{y}\rangle = \sum_{j=1}^{p} x_j y_j$ for $\boldsymbol{x},\boldsymbol{y} \in \mathbb{R}^p$ and $\mathcal{G}_0 = \mathbf{span}\{1\}$. Then, $\mathcal{G}_0 \oplus \mathcal{H}_{K_{\text{lin}}} = \{\boldsymbol{x} \mapsto b + \sum_{j=1}^{p} \beta_j x_j : b, \beta_j \in \mathbb{R}\}$ is the space of linear functions on $\mathcal{X}$. For $h \in \mathcal{H}_{K_{\text{lin}}}$ such that $h(\boldsymbol{x}) = \sum_{j=1}^{p} \beta_j x_j$, $\|h\|_{K_{\text{lin}}}^2 = \sum_{j=1}^{p} \beta_j^2 = \|\boldsymbol{\beta}\|_2^2$. Therefore, (17) with the hinge loss becomes the standard SVM problem (15). In Section 4.2, we solve the SVM problem by its dual problem. The KKT condition informs that $\hat{\boldsymbol{\beta}} = \sum_{i=1}^{n} \hat{\alpha}_i \tilde{Y}_i \boldsymbol{X}_i$, where $\{\hat{\alpha}_i\}_{i=1}^{n}$ are the solutions to the dual variables. Then, the fitted decision function becomes $\boldsymbol{x} \mapsto b + \boldsymbol{x}^T \left(\sum_{i=1}^{n} \hat{\alpha}_i \tilde{Y}_i \boldsymbol{X}_i\right) = b + \sum_{i=1}^{n} \hat{\alpha}_i \tilde{Y}_i \langle \boldsymbol{X}_i, \boldsymbol{x}\rangle$, which coincides with the conclusion in Theorem 1.

In fact, using the duality to solve the SVM problem can be extended to the kernelized SVM problem. Let K be a general kernel function with the eigenexpansion $K(\boldsymbol{x},\boldsymbol{y}) = \sum_{j=1}^{\infty} \gamma_j \phi_j(\boldsymbol{x})\phi_j(\boldsymbol{y})$. Then, the kernelized SVM problem replaces the original covariates $\boldsymbol{X}_i$ by the induced features $\{\phi_j(\boldsymbol{X}_i)\}_{j=1}^{\infty}$, and the penalty term $\|\boldsymbol{\beta}\|_2^2$ by the RKHS-penalty $\sum_{j=1}^{\infty} \beta_j^2/\gamma_j$. As a result, the dual problem of the kernelized SVM (14) becomes

$$
\begin{aligned}
\max_{\boldsymbol{\alpha}} \ & \sum_{i=1}^{n} \alpha_i - \frac{1}{2} \sum_{i=1}^{n} \sum_{i'=1}^{n} \alpha_i \alpha_{i'} \tilde{Y}_i \tilde{Y}_{i'} \sum_{j=1}^{\infty} \gamma_j \phi_j(\boldsymbol{X}_i)\phi_j(\boldsymbol{X}_{i'}) \\
\text{s.t.} \ & \sum_{i=1}^{n} \alpha_i \tilde{Y}_i = 0 \\
& 0 \le \alpha_i \le C; \qquad 1 \le i \le n
\end{aligned}
\tag{18}
$$

Comparing the dual problems of the standard SVM (14) and the kernelized SVM (18), the only difference appears in the term $\langle \boldsymbol{X}_i, \boldsymbol{X}_{i'}\rangle$ and $\sum_{j=1}^{\infty} \gamma_j \phi_j(\boldsymbol{X}_i)\phi_j(\boldsymbol{X}_{i'}) = K(\boldsymbol{X}_i, \boldsymbol{X}_{i'})$. Then, the kernelized SVM can be obtained by replacing every $\langle\cdot,\cdot\rangle$ in the standard SVM by $K(\cdot,\cdot)$. In particular, we can use the QP to solve the dual problem of the kernelized SVM (18) and recover the primal solution as $b + \sum_{i=1}^{n} \hat{\alpha}_i \tilde{Y}_i K(\boldsymbol{X}_i, \cdot)$.

For the general RKHS-penalized ERM problem (17), Theorem 1 suggests that we can solve the following finite-dimensional optimization problem:

$$
\min_{g_0 \in \mathcal{G}_0, \boldsymbol{\alpha} \in \mathbb{R}^n} \quad \frac{1}{n} \sum_{i=1}^{n} \ell\left(Y_i, g_0(\boldsymbol{X}_i) + [\mathbf{K}]_{i\cdot}\boldsymbol{\alpha}\right) + \lambda \boldsymbol{\alpha}^T \mathbf{K} \boldsymbol{\alpha} \tag{19}
$$

where $\mathbf{K} := [K(\boldsymbol{X}_i, \boldsymbol{X}_{i'})]_{n \times n}$ is the kernel matrix at the training sample covariates, and $[\mathbf{K}]_{i\cdot}$ is the ith row vector in $\mathbf{K}$. For example, if $\mathcal{G}_0 = \mathbf{span}\{1\}$ and $\ell(y,\hat{y}) = (y-\hat{y})^2$ is the squared loss, then (19) in the matrix form becomes a *generalized ridge regression* problem $\min_{b,\boldsymbol{\alpha}}\{(1/n)\|\boldsymbol{Y} - b - \mathbf{K}\boldsymbol{\alpha}\|_2^2 + \lambda \boldsymbol{\alpha}^T \mathbf{K} \boldsymbol{\alpha}\}$.

Further extensions for nonlinear variable selections based on the RKHS can be found in Refs 68 and 69. We refer the readers to Hofmann *et al.* [70] for a comprehensive review on the RKHS with more modern applications.

5.2 Large-Scale Optimization

For large-scale problems, the dimension p and the sample size n can be potentially huge. Using the methods discussed in Section 2.2 may not be sufficient enough. Without loss of generality, consider the unpenalized ERM problem on the linear function class: $\min_{\boldsymbol{\beta} \in \mathbb{R}^p}\{Q(\boldsymbol{\beta}) := \frac{1}{n} \sum_{i=1}^{n} \ell\left(Y_i, \boldsymbol{X}_i^T \boldsymbol{\beta}\right)\}$. If the dimension p is large, then the full-dimensional parameter vector $\boldsymbol{\beta}$ and the gradient $\nabla Q(\boldsymbol{\beta})$ can require large amount of storage. Maintaining the full vector at a time can also be expensive. If the sample size n is large, then the gradient evaluation $\nabla Q(\boldsymbol{\beta}) = \frac{1}{n}\sum_{i=1}^{n} \nabla_{\boldsymbol{\beta}} \ell\left(Y_i, \boldsymbol{X}_i^T \boldsymbol{\beta}\right)$ can be time consuming, since the gradients at all sample points are required to compute.

When the dimension p is large, we can reduce the number of updating parameters per iteration by fixing most of the coordinates at their current iteration. Specifically, suppose that Q is coordinatewise Lipschitz gradient, that is, there exists $\{L_{Q,j}\}_{j=1}^{p} \subseteq \mathbb{R}_+$ such that

$$|\nabla_j Q(\boldsymbol{\beta} + h\boldsymbol{e}_{p,j}) - \nabla_j Q(\boldsymbol{\beta})| \leq L_{Q,j}|h|; \quad \boldsymbol{\beta} \in \mathbb{R}^p,\ h \in \mathbb{R},\ 1 \leq j \leq p$$

Define $p_\alpha(j) := L_{Q,j}^{\alpha} / \sum_{j'=1}^{p} L_{Q,j'}^{\alpha}$ for $\alpha \in \mathbb{R}$ and $1 \leq j \leq p$. Then, the *random coordinate descent method* (RCDM) [71] considers the following updates at the tth iteration:

1) Choose the coordinate index j_t from $[p]$ randomly with $\mathbb{P}(j_t = j | \mathcal{F}_{t-1}) = p_\alpha(j)$ for $1 \leq j \leq p$, where $\mathcal{F}_{t-1} := \sigma\{j_0, j_1, \cdots, j_{t-1}\}$ denotes the filtration generated by the historical random coordinate indices;
2) Update $\beta_{t+1,j_t} \leftarrow \beta_{t,j_t} - (1/L_{Q,j_t})\nabla_{j_t} Q(\boldsymbol{\beta}_t)$; $\beta_{t+1,j} \leftarrow \beta_{t,j}\ (\forall j \neq j_t)$.

It can be shown that the RCDM has optimization guarantee $\mathbb{E}[Q(\boldsymbol{\beta}_t)|\mathcal{F}_{t-1}] - Q(\boldsymbol{\beta}^\star) \leq \frac{2}{t+4}\left(\sum_{j=1}^{p} L_j^{\alpha}\right) r_{1-\alpha}^2(\boldsymbol{\beta}_0)$ for some radius $r_{1-\alpha}^2(\boldsymbol{\beta}_0)$. Here, the expectation $\mathbb{E}$ is taken over the random indices j_ts. The RCDM can be extended to the blockwise coordinate descent, where a block of coordinates are updated at each iteration.

When the sample size n is large, we can evaluate the gradient $\nabla Q(\boldsymbol{\beta}) = \frac{1}{n}\sum_{i=1}^{n} \nabla_{\boldsymbol{\beta}} \ell\left(Y_i, \boldsymbol{X}_i^T \boldsymbol{\beta}\right)$ by the stochastic version $\nabla_{\boldsymbol{\beta}} \ell\left(Y_\xi, \boldsymbol{X}_\xi^T \boldsymbol{\beta}\right)$, where the stochastic sample index ξ is chosen with $\mathbb{P}(\xi = i) = 1/n\ (\forall 1 \leq i \leq n)$. The *stochastic-gradient descent algorithm (SGDA)* [72, 73] updates at the tth iteration specifically as follows:

1) Choose the sample index i_t from $[n]$ randomly with $\mathbb{P}(i_t = i | \mathcal{F}_{t-1}) = 1/n$ for $1 \leq i \leq n$, where $\mathcal{F}_{t-1} := \sigma\{i_0, i_1, \ldots, i_{t-1}\}$ denotes the filtration generated by the historical random sample indices;
2) Update $\boldsymbol{\beta}_{t+1} \leftarrow \boldsymbol{\beta}_t - \alpha_t \nabla_{\boldsymbol{\beta}} \ell\left(Y_{i_t}, \boldsymbol{X}_{i_t}^T \boldsymbol{\beta}_t\right)$.

Assume that Q is L_Q-Lipschitz gradient, and μ_Q strongly convex, $M_1^2 := \sup_{\boldsymbol{\beta} \in \mathbb{R}^p} \frac{1}{n}\sum_{i=1}^{n} \|\nabla_{\boldsymbol{\beta}} \ell(Y_i, \boldsymbol{X}_i^T \boldsymbol{\beta})\|_2^2 < +\infty$. When the step size is chosen as $\alpha_t := \frac{M_0}{t+1}$ for $M_0 > 1/(2\mu_Q)$, we have optimization guarantee for the SGDA: $\mathbb{E}\|\boldsymbol{\beta}_t - \boldsymbol{\beta}^\star\|_2^2 \leq \frac{1}{t+1}\max\left\{\frac{M_0^2 M_1^2}{2\mu_Q M_0 - 1}, \|\boldsymbol{\beta}_0 - \boldsymbol{\beta}^\star\|_2^2\right\}$ and $\mathbb{E}Q(\boldsymbol{\beta}_t) - Q(\boldsymbol{\beta}^\star) \leq \frac{L_Q}{2(t+1)}\max\left\{\frac{M_0^2 M_1^2}{2\mu_Q M_0 - 1}, \|\boldsymbol{\beta}_0 - \boldsymbol{\beta}^\star\|_2^2\right\}$. Here, the expectation $\mathbb{E}$ is taken over the random indices i_ts.

There is a close connection between the random coordinate descent and the stochastic-gradient descent. That is, the stochastic-gradient descent on the primal space can be mimicked by the stochastic coordinate accent on the dual space. Such a correspondence motivates the methods of the *Stochastic Dual Coordinate Ascent (SDCA)* [74]. The coordinate descent/ascent algorithms can be generally simpler than the stochastic-gradient descent/ascent algorithms, since the coordinate subproblem is univariate and generally has a closed-form solution. A concrete example can be the path coordinate descent algorithm to solve for the LASSO solution path in Section 3.2.

6 Discussion

In this chapter, we review supervised learning under the penalized ERM (1) framework. We begin with the general discussion of the penalized ERM problem on "why to penalize" and "how to optimize." From the statistical point of view, we argue that the penalty term trades-off the bias and variance of the fitted model to perform model estimation and model selection simultaneously. From the computational point of view, we introduce the GDA and the ISTA to solve the penalized ERM as a composite convex minimization problem (2).

For linear regression, we highlight the bias–variance trade-off and emphasize the importance of model selection even when the true coefficients are all nonzero. Then, we discuss ridge regression whose tuning parameter λ trades-off the variance and bias explicitly to achieve the smallest risk, that is, the PMSE. In order to perform variable selection, we consider the LASSO problem with ℓ^1-penalty and discuss the LAR algorithm and the path coordinate descent algorithm when solving for the solution path.

For binary classification problems, we consider the soft classifiers that estimate the Bayes rule using model-based approaches and the hard classifiers that target on the decision rules and minimize the risk directly. We further discuss the convex surrogate risk minimization problem for hard classification and introduce the LUM that bridges the soft and hard classifiers. To robustify the hard classifiers with unbounded surrogate loss functions, we introduce the nonconvex surrogate loss and the corresponding DCA. For multicategory classification problems, we discuss the polytomous response modeling approaches for soft classification and the sum-to-zero-constrained multicategory surrogate losses and the angle-based representation for hard classification.

In the extensions for complex data, we first consider nonlinear modeling using the RKHS, where the Representer Theorem reduces the optimization problem to finite dimensions. Then, we discuss the large-scale optimization problem where the dimension p and the sample size n can be huge. We introduce the RCDM and the SGDA that can handle the large-p and large-n problems, respectively.

There are many other supervised learning methods that cannot be covered in this chapter. We refer the readers to the books and comprehensive reviews for smoothing techniques in Loader [75], generalized additive models in Hastie and Tibshirani [76], tree-based methods in Loh [77], ensemble methods such as boosting and random forest in Bühlmann [78], and deep learning in Refs 79 and 80.

References

1 Hastie, T., Tibshirani, R., and Friedman, J. (2009) *The Elements of Statistical Learning: Data Mining, Inference, and Prediction*, Springer Science & Business Media.

2 Bishop, C.M. (2006) *Pattern Recognition and Machine Learning*, springer.

3 Liu, Y., Zhang, H.H., and Wu, Y. (2011) Hard or soft classification? large-margin unified machines. *J. Am. Stat. Assoc.*, 106, 166–177.

4 Wahba, G. (1990) *Spline Models for Observational Data*, vol. 59, Siam.

5 Wahba, G. (1999) Support vector machines, reproducing kernel Hilbert spaces and the randomized GACV, *Adv. Kernel Methods-Support Vector Learn.*, 6, 69–87.

6 Lin, X., Wahba, G., Xiang, D. *et al.* (2000) Smoothing spline ANOVA models for large data sets with Bernoulli observations and the randomized GACV. *Ann. Stat.*, 28, 1570–1600.

7 Fan, J. and Li, R. (2001) Variable selection via nonconcave penalized likelihood and its oracle properties. *J. Am. Stat. Assoc.*, 96, 1348–1360.

8 Shen, X. and Ye, J. (2002) Adaptive model selection. *J. Am. Stat. Assoc.*, 97, 210–221.

9 Bühlmann, P. and Hothorn, T. (2007) Boosting algorithms: regularization, prediction and model fitting. *Stat. Sci.*, 22, 477–505.

10 Barron, A., Birgé, L., and Massart, P. (1999) Risk bounds for model selection via penalization. *Probab. Theory Relat. Fields*, 113, 301–413.

11 Bartlett, P.L. and Mendelson, S. (2002) Rademacher and Gaussian complexities: risk bounds and structural results. *J. Mach. Learn. Res.*, 3, 463–482.

12 Bartlett, P.L., Bousquet, O., and Mendelson, S. (2005) Local rademacher complexities. *Ann. Stat.*, 33, 1497–1537.

13 Fan, J. and Li, R. (2006) Statistical challenges with high dimensionality: feature selection in knowledge discovery. *arXiv preprint math/0602133.*

14 Zhu, J., Rosset, S., Tibshirani, R., and Hastie, T.J. (2004) *1-Norm Support Vector Machines*. Advances in Neural Information Processing Systems, pp. 49–56.

15 Witten, D.M. and Tibshirani, R. (2011) Penalized classification using Fisher's linear discriminant. *J. R. Stat. Soc.: Ser. B Stat. Methodol.*, 73, 753–772.

16 Nesterov, Y. (2013) Gradient methods for minimizing composite functions. *Math. Program.*, 140, 125–161.

17 Nesterov, Y. (2018) *Lectures on Convex Optimization*, vol.137, Springer.

18 Breiman, L. (1995) Better subset regression using the nonnegative garrote. *Technometrics*, 37, 373–384.

19 Tibshirani, R. (1996) Regression shrinkage and selection via the lasso. *J. R. Stat. Soc.: Ser. B Methodol.*, 58, 267–288.

20 Zou, H. and Hastie, T. (2005) Regularization and variable selection via the elastic net. *J. R. Stat. Soc.: Ser. B Stat. Methodol.*, 67, 301–320.

21 Yuan, M. and Lin, Y. (2006) Model selection and estimation in regression with grouped variables. *J. R. Stat. Soc.: Ser. B Stat. Methodol.*, 68, 49–67.

22 Zhao, P., Rocha, G., and Yu, B. (2009) The composite absolute penalties family for grouped and hierarchical variable selection. *Ann. Stat.*, 37, 3468–3497.

23 Jacob, L., Obozinski, G., and Vert, J.-P. (2009) *Group Lasso with Overlap and Graph Lasso*. Proceedings of the 26th Annual International Conference on Machine Learning, ACM, pp. 433–440.

24 Chernozhukov, V., Hansen, C., and Liao, Y. (2017) A lava attack on the recovery of sums of dense and sparse signals. *Ann. Stat.*, 45, 39–76.

25 Duchi, J., Shalev-Shwartz, S., Singer, Y., and Chandra, T. (2008) *Efficient Projections onto the l 1-Ball for Learning in High Dimensions*. Proceedings of the 25th international Conference on Machine Learning, ACM, pp. 272–279.

26 Combettes, P.L. and Wajs, V.R. (2005) Signal recovery by proximal forward-backward splitting. *Multiscale Model. Simul.*, 4, 1168–1200.

27 Beck, A. and Teboulle, M. (2009) A fast iterative shrinkage-thresholding algorithm for linear inverse problems. *SIAM J. Imag. Sci.*, 2, 183–202.

28 Nesterov, Y. (1983) *A Method for Unconstrained Convex Minimization Problem with the Rate of Convergence* $\mathcal{O}(1/k^2)$. Doklady AN USSR, vol. 269, pp. 543–547.

29 Miller, A. (2002) *Subset Selection in Regression*, Chapman and Hall/CRC.

30 Bertsimas, D., King, A., and Mazumder, R. (2016) Best subset selection via a modern optimization lens. *Ann. Stat.*, 44, 813–852.

31 Efron, B., Hastie, T., Johnstone, I., and Tibshirani, R. (2004) Least angle regression. *Ann. Stat.*, 32, 407–499.

32 Rosset, S. and Zhu, J. (2007) Piecewise linear regularized solution paths. *Ann. Stat.*, 35, 1012–1030.

33 Friedman, J., Hastie, T., Höfling, H., and Tibshirani, R. (2007) Pathwise coordinate optimization. *Ann. Appl. Stat.*, 1, 302–332.

34 Wu, T.T. and Lange, K. (2008) Coordinate descent algorithms for lasso penalized regression. *Ann. Appl. Stat.*, 2, 224–244.

35 Friedman, J., Hastie, T., and Tibshirani, R. (2010) Regularization paths for generalized linear models via coordinate descent. *J. Stat. Soft.*, 33, 1.

36 Zhao, P. and Yu, B. (2006) On model selection consistency of Lasso. *J. Mach. Learn. Res.*, 7, 2541–2563.

37 Knight, K. and Fu, W. (2000) Asymptotics for lasso-type estimators. *Ann. Stat.*, 28, 1356–1378.

38 Zou, H. (2006) The adaptive lasso and its oracle properties. *J. Am. Stat. Assoc.*, 101, 1418–1429.

39 Zhang, C.-H. and Zhang, S.S. (2014) Confidence intervals for low dimensional parameters in high dimensional linear models. *J. R. Stat. Soc.: Ser. B Stat. Methodol.*, 76, 217–242.

40 Van de Geer, S., Bühlmann, P., Ritov, Y., and Dezeure, R. (2014) On asymptotically optimal confidence regions and tests for high-dimensional models. *Ann. Stat.*, 42, 1166–1202.

41 Lu, S., Liu, Y., Yin, L., and Zhang, K. (2017) Confidence intervals and regions for the lasso by using stochastic variational inequality techniques in optimization. *J. R. Stat. Soc.: Ser. B Stat. Methodol.*, 79, 589–611.

42 Yu, G., Yin, L., Lu, S., and Liu, Y. (2019) Confidence intervals for sparse penalized regression with random designs. *J. Am. Stat. Assoc.*, 115, 1–38.

43 Tibshirani, R., Saunders, M., Rosset, S. *et al.* (2005) Sparsity and smoothness via the fused lasso. *J. R. Stat. Soc.: Ser. B Stat. Methodol.*, 67, 91–108.

44 Candes, E. and Tao, T. (2007) The Dantzig selector: statistical estimation when p is much larger than n. *Ann. Stat.*, 35, 2313–2351.

45 Yu, G. and Liu, Y. (2016) Sparse regression incorporating graphical structure among predictors. *J. Am. Stat. Assoc.*, 111, 707–720.

46 Frank, L.E. and Friedman, J.H. (1993) A statistical view of some chemometrics regression tools. *Technometrics*, 35, 109–135.

47 Liu, Y. and Wu, Y. (2007) Variable selection via a combination of the L0 and L1 penalties. *J. Comput. Graphical Stat.*, 16, 782–798.

48 Zhang, C.-H. (2010) Nearly unbiased variable selection under minimax concave penalty. *Ann. Stat.*, 38, 894–942.

49 Tibshirani, R. (2011) Regression shrinkage and selection via the lasso: a retrospective. *J. R. Stat. Soc.: Ser. B Stat. Methodol.*, 73, 273–282.

50 Bühlmann, P. and Van De Geer, S. (2011) *Statistics for High-Dimensional Data: Methods, Theory and Applications*, Springer Science & Business Media.

51 Wahba, G. (2002) Soft and hard classification by reproducing kernel Hilbert space methods. *Proc. Natl. Acad. Sci.*, 99, 16524–16530.

52 Wang, J., Shen, X., and Liu, Y. (2007) Probability estimation for large-margin classifiers. *Biometrika*, 95, 149–167.

53 Boyd, S. and Vandenberghe, L. (2004) *Convex Optimization*, University Press, Cambridge.

54 Bartlett, P.L., Jordan, M.I., and McAuliffe, J.D. (2006) Convexity, classification, and risk bounds. *J. Am. Stat. Assoc.*, 101, 138–156.

55 Lin, Y. (2004) A note on margin-based loss functions in classification. *Stat. Probab. Lett.*, 68, 73–82.

56 Boucheron, S., Bousquet, O., and Lugosi, G. (2005) Theory of classification: a survey of some recent advances. *ESAIM: Probab. Stat.*, 9, 323–375.

57 McCullagh, P. and Nelder, J. (1989) *Generalized Linear Models, Second Edition, Chapman & Hall/CRC Monographs on Statistics & Applied Probability*, Taylor & Francis.

58 Zhao, J., Yu, G., and Liu, Y. (2018) Assessing robustness of classification using an angular breakdown point. *Ann. Stat.*, 46, 3362–3389.

59 Wu, Y. and Liu, Y. (2007) Robust truncated hinge loss support vector machines. *J. Am. Stat. Assoc.*, 102, 974–983.

60 Liu, Y., Shen, X., and Doss, H. (2005) Multicategory ψ-learning and support vector machine: computational tools. *J. Comput. Graphical Stat.*, 14, 219–236.

61 Wu, Y. and Liu, Y. (2013) Adaptively weighted large margin classifiers. *J. Comput. Graphical Stat.*, 22, 416–432.

62 Liu, Y. and Shen, X. (2006) Multicategory ψ-learning. *J. Am. Stat. Assoc.*, 101, 500–509.

63 Liu, Y. and Yuan, M. (2011) Reinforced multicategory support vector machines. *J. Comput. Graphical Stat.*, 20, 901–919.

64 Liu, Y. (2007) Fisher consistency of multicategory support vector machines, in *Artificial Intelligence and Statistics* (eds M. Meila and X. Shen), PMLR, San Juan, Puerto Ricopp. pp. 291–298.

65 Zhang, C. and Liu, Y. (2014) Multicategory angle-based large-margin classification. *Biometrika*, 101, 625–640.

66 Aronszajn, N. (1950) Theory of reproducing kernels. *Trans. Am. Math. Soc.*, 68, 337–404.

67 Kimeldorf, G. and Wahba, G. (1971) Some results on Tchebycheffian spline functions. *J. Math. Anal. Appl.*, 33, 82–95.

68 Lin, Y. and Zhang, H.H. (2006) Component selection and smoothing in multivariate nonparametric regression. *Ann. Stat.*, 34, 2272–2297.

69 Zhang, H.H., Cheng, G., and Liu, Y. (2011) Linear or nonlinear? Automatic structure discovery for partially linear models. *J. Am. Stat. Assoc.*, 106, 1099–1112.

70 Hofmann, T., Schölkopf, B., and Smola, A.J. (2008) Kernel methods in machine learning. *Ann. Stat.*, 36, 1171–1220.

71 Nesterov, Y. (2012) Efficiency of coordinate descent methods on huge-scale optimization problems. *SIAM J. Optim.*, 22, 341–362.

72 Robbins, H. and Monro, S. (1951) A stochastic approximation method. *Ann. Math. Stat.*, 22, 400–407.

73 Bottou, L. (2010) *Large-Scale Machine Learning with Stochastic Gradient Descent*. Proceedings of COMPSTAT'2010, Springer, pp. 177–186.

74 Shalev-Shwartz, S. and Zhang, T. (2013) Stochastic dual coordinate ascent methods for regularized loss minimization. *J. Mach. Learn. Res.*, 14, 567–599.

75 Loader, C. (2012) Smoothing: local regression techniques, in *Handbook of Computational Statistics* (eds J. Gentle, W. Härdle, and Y. Mori), Springer, Berlin, Heidelberg, pp. 571–596.

76 Hastie, T.J. and Tibshirani, R.J. (1990) *Generalized Additive Models*, vol. 43, CRC press.

77 Loh, W.-Y. (2014) Fifty years of classification and regression trees. *Int. Stat. Rev.*, 82, 329–348.

78 Bühlmann, P. (2012) Bagging, boosting and ensemble methods, in *Handbook of Computational Statistics* (eds J.E. Gentle, W.K. Härdle, and Y. Mori), Springer, Berlin, Heidelberg, pp. 985–1022.

79 LeCun, Y., Bengio, Y., and Hinton, G. (2015) Deep learning. *Nature*, 521, 436–444.

80 Goodfellow, I., Bengio, Y., and Courville, A. (2016) *Deep Learning*, MIT Press, http://www.deeplearningbook.org.

11

Unsupervised and Semisupervised Learning

Jia Li and Vincent A. Pisztora

The Pennsylvania State University, University Park, PA, USA

1 Introduction

In scientific exploration, a natural question to ask before we have gained any insight is whether the cases under study fall into categories or classes. Each category has some distinct characteristics, and the variation within a category is low or ideally negligible comparing with that across different categories. Since the taxonomy of the classes is part of what we must find out, we face the challenge of unsupervised learning or clustering. For example, in single-cell data analysis [1–3], clusters identified computationally often motivate new hypothesis or substantiate the existing ones. They also enable researchers to decide which subgroups to drill into with more field experiments. Furthermore, clustering is intrinsically a data reduction mechanism, important especially in the era of big data. By representing every cluster, for instance, using the mean of the cluster members, the amount of data to be inspected can decrease tremendously. As a result, clustering is frequently carried out at the beginning of a data analysis pipeline. The applications of clustering span broadly across science, engineering, and commercial domains. In image processing and computer vision, a prominent paradigm for segmentation relies on clustering local features such as color components or results of convolution [4, 5]. In information retrieval, clustering is used to organize items in a database to improve efficiency [6, 7].

Classification (supervised learning) and clustering are two extreme ends of a whole spectrum. If all the training data are labeled, we have the problem of classification. If no label is given, the problem is clustering. We may have partially labeled datasets due to the high cost of acquiring class labels, for instance, expert diagnosis using medical images. In practice, we may have a much larger portion of the data being unlabeled. This leads us to the problem of semisupervised learning, which is reviewed in Section 3.

The rest of the chapter is organized as follows. Section 2 is on unsupervised learning. In Section 2.1, we introduce the conventional framework of mixture-model-based clustering. Next, relatively recent advances to tackle high dimensionality such as clustering by mode association (Section 2.1.2), hidden Markov model on variable blocks (HMM-VB)

Computational Statistics in Data Science.
Edited by Walter W. Piegorsch, Richard A. Levine, Hao Helen Zhang and Thomas C. M. Lee.
 ISBN 978-1-11956107-1

(Section 2.1.3), and variable selection (Section 2.1.4) are presented. In many modern applications, objects to be clustered are each represented by an unordered set of vectors assigned with weights, essentially a finite-support discrete distribution. In Section 2.2, we address a research topic that has attracted rapidly growing interest in the past decade, specifically, clustering under the Wasserstein metric between distributions. In Section 2.3, we cover the topic of assessing uncertainty of clustering results. In Section 3, we review semisupervised learning. Sections 3.2, 3.3, and 3.4 present overviews of foundational semisupervised approaches. Next, entropy minimization, consistency regularization, and mixup augmentation are presented in detail (Sections 3.5, 3.6, and 3.7). Section 3.8 then describes a state-of-the-art method, MixMatch, which utilizes a combination of these methods. Finally, conclusions are drawn in Section 4.

2 Unsupervised Learning

There are three major schools of approaches to clustering. The first type requires only pairwise distances between objects to be clustered. Such methods are appealing when the mathematical representation of the objects is complex or may not even be defined. The main drawback is the quadratic computational complexity for obtaining all the pairwise distances. Another accompanying issue is the lack of direct and compact description for each cluster. Only the memberships of objects are provided. Examples include linkage clustering [8] and spectral graph partitioning [9]. The second type of approach aims at optimizing a given merit function, which reflects the commonly accepted standards for good clustering. The basic principle is that objects in the same cluster should be similar to each other while those in different clusters should be as distinct as possible. The merit function is inevitably subjective depending on how the distances between objects are defined and the definition of the overall quality of clustering. K-means and k-center clustering [10] belong to this type. The third type of approach relies on statistical modeling [11], in particular, mixture models, which we focus on in this chapter.

2.1 Mixture-Model-Based Clustering

In the classic framework of mixture-model-based clustering [12], a mixture model is fitted first, usually by the EM algorithm, and then the posterior probability of each mixture component given a data point is computed. The component with the largest posterior is chosen for that point. Points associated with the same component form one cluster. In this section, we first introduce the basic approach with every cluster corresponding to one component in the mixture. Then, we discuss the drawbacks of associating every cluster with a single component and present methods to overcome them. As with many other statistical methodologies, in relatively recent literature, the challenge of high dimensionality has drawn primary attention. In particular, we present approaches for variable selection and approaches to estimating high-dimensional densities by exploiting latent graphical structures. At last, we briefly discuss clustering for sequential or spatial data which arise frequently in signal/image processing.

2.1.1 Gaussian mixture model

Suppose that the dataset $\mathbb{X} = \{\mathbf{x}_1, \ldots, \mathbf{x}_n\}, \mathbf{x}_i \in \mathbb{R}^d$, is an i.i.d. sample of a random vector $X = (X_1, \ldots, X_d)' \in \mathbb{R}^d$ with density function $f(\mathbf{x})$. Assume that sample points in each cluster follow a parametric distribution, usually referred to as the component distribution. Denote the kth component distribution by $\phi(\mathbf{x} \mid \theta_k)$, where θ_k is the parameter depending on the cluster identity and $k \in \{1, \ldots, M\}$. Let the prior probability of cluster k be α_k, $\sum_{k=1}^{M} \alpha_k = 1$. Denote the set of parameters θ_k, α_k, $k = 1, \ldots, M$, collectively as θ. Then, the density of X is

$$f(\mathbf{x} \mid \theta) = \sum_{k=1}^{M} \alpha_k \phi(\mathbf{x} \mid \theta_k) \tag{1}$$

To cluster $\mathbb{X}$, the mixture density $f(\mathbf{x} \mid \theta)$ is estimated first, usually by the maximum-likelihood criterion using the EM algorithm. Then, the posterior probabilities of the cluster labels of every point are computed, and the point is assigned to the label with the maximum posterior.

The most commonly used component distribution is Gaussian. We thus have the *Gaussian mixture model* (GMM) [12]. By imposing various structural constraints on the covariance matrices of the components, for example, diagonal, and requiring the components to share some aspects of the covariance matrices, for example, identical eigenvectors, many different versions of GMM are obtained for clustering [13, 14]. These different versions of GMM together with different numbers of components are evaluated by a model selection criterion, the best one being chosen for clustering. A popular R package, Mclust [15], provides functions to estimate such GMMs and perform clustering.

2.1.2 Clustering by mode association

In the basic setup of GMM-based clustering, each component is associated with one cluster. A thorough discussion on the limitations of treating components equivalently as clusters is available in Refs 16 and 17. One apparent drawback of the classic GMM paradigm is the implicit restriction of Gaussian-shaped clusters. In addition, the number of components in a GMM affects density estimation. The best number of components to achieve accurate density estimation is not necessarily a reasonable number of clusters. For instance, a single cluster may require a mixture model to capture its complexity or its deviation from the Gaussian shape. It is well known that mixture models with dramatically different numbers of components can yield quite similar density functions. This phenomenon is both a blessing and a curse. For the sake of density estimation, it means we can obtain good estimation without being too precise with the number of components. On the other hand, it is notoriously hard to decide the number of clusters by evoking a criterion that relies on the likelihood of the data.

To tackle the pitfalls of one-to-one correspondence between mixture components and clusters, Li *et al.* [17] proposed the framework of clustering by mode association, aptly called *modal clustering*. Theoretical results on the modes of GMMs are available in Ray *et al.* [18]. The feasibility of the framework relies on the modal EM (MEM) algorithm [17]. Given any density in the form of a mixture model, for example, kernel density estimator, MEM finds an ascending path from any point to a mode (local maxima). By mode association, data points that ascend to the same mode are grouped into one cluster. This criterion exploits the geometric characteristics of the density function. It is found that when the number of components in the mixture model changes significantly, the modes in the density are relatively stable, resulting in little change in the number of clusters. In fact, in their initial approach to clustering by mode association, Li *et al.* [17] simply use kernel density estimation, the number of mixture components being the same as the data size. By enlarging the kernel bandwidth gradually, hierarchical clustering is obtained. The clusters formed by mode association often take shapes very different from Gaussian. It is straightforward to apply modal clustering similarly to the general GMM as later studied by Lee and Li [19].

Consider the GMM in Equation (1) $f(\mathbf{x}) = \sum_{k=1}^{M} \alpha_k \phi(\mathbf{x} \mid \theta_k)$. Given any initial value $\mathbf{x}^{(0)}$, MEM solves a local maximum of the mixture by alternating the following two steps until a stopping criterion is met. Start with $r = 0$.

1. Let $p_k = \dfrac{\alpha_k \phi(\mathbf{x}^{(r)} \mid \theta_k)}{f(\mathbf{x}^{(r)})}, k = 1, \dots, M.$
2. Update $\mathbf{x}^{(r+1)} = \underset{\mathbf{x}}{\operatorname{argmax}} \displaystyle\sum_{k=1}^{M} p_k \log \phi(\mathbf{x} \mid \theta_k).$

The first step is the "Expectation" step where the posterior probability of each component k, $1 \leq k \leq M$, at the current point $\mathbf{x}^{(r)}$ is computed. The second step is the "Maximization" step. We assume that $\sum_{k=1}^{M} p_k \log \phi(\mathbf{x} \mid \theta_k)$ has a unique maximum, which is true for Gaussian components.

After estimating a GMM for the dataset, we can either apply MEM to each individual point and group those that ascend to the same mode into one cluster, as is done by Li *et al.* [17], or we can use MEM as a way to merge multiple components of the GMM into one cluster [19]. Specifically, this method of merging components into a cluster by MEM is called *Componentwise Mode Association Clustering (CMAC)* [19]:

1. Estimate a GMM for the dataset $\{\mathbf{x}_1, \dots, \mathbf{x}_n\}$, $f(\mathbf{x}) = \sum_{k=1}^{M} \alpha_k \phi(\mathbf{x} \mid \mu_k, \Sigma_k)$, where μ_k is the component mean, and Σ_k is the covariance matrix.
2. Apply MEM to each component mean μ_k. Let the number of distinct modes found by MEM be M', $M' \leq M$ in general. If the kth component mean μ_k is mapped to the mth mode, we denote the mapping by $\Lambda(k) = m$.
3. Partition $\mathbf{x}_i$ into M' clusters by first finding the component k with the maximum posterior probability given $\mathbf{x}_i$ and then mapping k to its mode:

$$\mathbf{x}_i \to \Lambda(\underset{k=1,\dots,K}{\operatorname{argmax}} \alpha_k \phi(\mathbf{x}_i \mid \mu_k, \Sigma_k))$$

2.1.3 Hidden Markov model on variable blocks

Mixture modeling, as a way of density estimation, encounters great obstacles when the dimension is high or merely moderate. A high-dimensional density tends to require a large number of mixture components to model. Generally speaking, more data are needed to estimate the Gaussian parameters (especially the covariance) of a single component in higher dimensions. Both tendencies necessitate a larger dataset for estimation, with size growing at a scale much faster than the linear order of the dimension. Furthermore, the computational intensity of estimating a GMM grows sharply with more components. Therefore, from the aspect of either estimation or computational efficiency, we are severely restricted in the number of components to assume for a GMM. To overcome the quandary, Lin and Li [20] proposed the HMM-VB. Variables are divided into groups called *variable blocks*, which are ordered as a chain. A latent state is assumed for each variable block, and these states are assumed to follow a Markov chain. Given the state of any variable block, the variables in this block are assumed to follow a Gaussian distribution with mean and covariance depending on the state. The graph structure of this latent state model is the same as the usual hidden Markov model (HMM). However, there is no real notion of "time," and any "time spot" corresponds to a particular variable block. As a result, there is a unique transition probability matrix at any time spot. Conceptually, HMM-VB is a special type of GMM with component means restricted on a lattice of the Cartesian product space of the variable blocks. In practice, if we cast HMM-VB as a GMM, the number of components is enormous, roughly exponential in the number of variable blocks and often much larger than the data size. Consequently, this link between HMM-VB and GMM cannot be exploited for estimation.

Suppose that the d-dimensional random vector X is partitioned into blocks $t = 1, 2, \ldots, T$, where T is the total number of blocks. Let the number of variables in block t be d_t, where $\sum_{t=1}^{T} d_t = d$. For brevity of discussion, assume that the d_1 variables in block 1 have indices before the d_2 variables in block 2, and so on. In general, obviously, such an ordering of variables may not hold. But this is only a matter of naming the variables and has no essential effect. Let $X^{(t)}$ denote the tth variable block. Without loss of generality, let $X^{(1)} = (X_1, X_2, \ldots, X_{d_1})'$ and $X^{(t)} = (X_{m_t+1}, X_{m_t+2}, \ldots, X_{m_t+d_t})'$, where $m_t = \sum_{\tau=1}^{t-1} d_\tau$, for $t = 2, \ldots, T$. Denote the underlying state of $X^{(t)}$ by s_t, $t = 1, \ldots, T$. Let the index set of s_t be $S_t = \{1, 2, \ldots, M_t\}$, where M_t is the number of mixture components for variable block $X^{(t)}$, $t = 1, \ldots, T$. Let the set of all possible sequences be $\hat{S} = S_1 \times S_2 \cdots \times S_T$. $|\hat{S}| = \prod_{t=1}^{T} M_t$. HMM-VB assumes:

1. $\{s_1, s_2, \ldots, s_T\}$ follow a Markov chain. Let $\pi_k = P(s_1 = k)$, $k \in S_1$. Let the transition probability matrix $A_t = (a_{k,l}^{(t)})$ between s_t and s_{t+1} be defined by $a_{k,l}^{(t)} = P(s_{t+1} = l | s_t = k)$, $k \in S_t$, $l \in S_{t+1}$.
2. Given s_t, $X^{(t)}$ is conditionally independent from the other $s_{t'}$ and $X^{(t')}$, $t' \neq t$. We also assume that given $s_t = k$, the conditional density of $X^{(t)}$ is the Gaussian distribution $\phi(X^{(t)} \mid \mu_k^{(t)}, \Sigma_k^{(t)})$.

Let $\mathbf{s} = \{s_1, \ldots, s_T\}$. A realization of X is denoted by $\mathbf{x}$, and a realization of $X^{(t)}$ is $\mathbf{x}^{(t)}$. To summarize, the density of HMM-VB is given by

$$f(\mathbf{x}) = \sum_{\mathbf{s} \in \hat{S}} \left(\pi_{s_1} \prod_{t=1}^{T-1} a_{s_t, s_{t+1}}^{(t)} \right) \cdot \prod_{t=1}^{T} \phi(\mathbf{x}^{(t)} | \mu_{s_t}^{(t)}, \Sigma_{s_t}^{(t)}) \tag{2}$$

Although HMM-VB can be viewed as a special GMM, the practice of identifying each mixture component as a cluster is apparently improper because of the huge number of components. Modal clustering is used instead. However, applying MEM directly to HMM-VB is computationally infeasible as the complexity of MEM is linear in the number of mixture components. It is discovered that the computational techniques of the Baum–Welch algorithm for estimating an HMM can be adapted to MEM, thus achieving complexity linear in the length of the chain. This new algorithm is called *Modal Baum–Welch* [20]. An R CRAN package called *HDclust* has been developed to estimate HMM-VB and to perform clustering based on it.

HMM-VB is further generalized to utilize a mixture model with a latent Bayesian network. This more general model allows for more complex dependence relationships among the latent states than a Markov chain would. The Baum–Welch and Modal Baum–Welch algorithms have been extended to this model [21]. We point out that some probabilistic graph models developed in signal/image processing two decades ago are examples of mixture models with latent Bayesian networks, for example, the spatial (so-called 2-D) HMM [22] and multiresolution 2-D HMM [23]. The 2-D HMMs have been used for image segmentation and classification. There is a subtle difference between using HMM-VB to cluster vectors and using HMMs to segment imagery or sequential data. For the former, each vector is an entire chain, and thus to cluster means dividing multiple chains into groups. While for the latter, each pixel (or local window around a pixel) is a state in the chain, and to cluster means partitioning the nodes on the chain (or mesh).

2.1.4 Variable selection

Variable selection methods for clustering can be categorized into roughly three types: methods that aim to remove redundancy among variables, those that seek to achieve high-quality clustering by a certain criterion, and those that improve fit of certain statistical models. Methods of the first type, often called *filter methods*, do not aim particularly at clustering. As a general data reduction strategy, they can be used for classification or visualization. The selection criteria include the maximum variance criterion [24, 25], principal components [26, 27], information gain [28, 29], entropy [30], and Laplacian score [31].

Methods of the second type originated from classification and are called *wrapper methods* in the literature. They usually search for a subset of variables under which the "best" clustering is achieved. In the case of classification, it is straightforward to measure the performance of the classifier, whereas the criterion to measure the quality of clustering is inevitably subjective since the true labels are unknown. Example criteria include the scatter separability [32], separability based on ridgelines [19], and Bayesian approaches [33]. Another line of ideas assumes that the clustering structure is most reliably revealed by considering all the dimensions. As a result, variable selection is conducted to best approximate the essential structure estimated from the full dimensions. Such a view point implies that variable selection is not meant to enhance clustering. Lee and Li [19] investigated such an approach along with a couple of alternative criteria and found that when the dimension is moderate, this approach can perform comparatively well. In Belkin and Niyogi [34], the full-dimensional data is used to identify a manifold which is assumed to retain the clustering structure, and then variable selection is conducted to best approximate the identified manifold.

The statistical modeling approaches rely largely on mixture models and fall into two subcategories. The first subcategory casts variable selection into a model selection problem by exploiting mixture models with specific formulations on how informative and noninformative variables relate to the membership of the mixture components, or equivalently, the cluster labels [11, 35–39]. The second subcategory of mixture-model-based clustering and variable selection methods exploits the mechanism of penalized modeling. A penalty term on the component means is added in the maximum-likelihood estimation of the mixture model, possibly shrinking the means across different components to a common value. If the component means of a variable are all equal, under certain setups of the mixture model, this variable becomes noninformative for the clustering structure. This line of research has been explored by Pan and Shen [40], Wang and Zhu [41], Xie *et al.* [42], Guo *et al.* [43], and Witten and Tibshirani [44, 45]. The clean formulation of penalized modeling that achieves simultaneous clustering and variable selection is elegant and appealing. However, these methods cannot be easily extended if the density model is not GMM. In addition, the penalty terms are designed to capture only a certain type of irrelevant variables for clustering and have an adverse effect on density estimation [40, 45].

2.2 Clustering of Distributional Data

In some applications, an instance/object is best characterized by a distribution. For different objects, the support sets of the distributions can differ. Ample examples arise in multimedia information retrieval and computer vision [46–48]. For instance, text documents are mathematically represented by the set of words (or more generally, terms) they contain, and each word is mapped to a point in a high-dimensional Euclidean space that captures the semantic distance between the words. The words are often assigned with weights indicating their significance to the document, for example, the frequency of occurring in the document adjusted by the rareness of the word. Such a mathematical representation is conveniently dubbed the "bag-of-word" model. Similarly, for images, "bag-of-word" representation is frequently used for various purposes. An image may be segmented first, and a feature vector extracted to characterize each segmented region, for which a weight proportional to the area size of the region would be assigned. Since the weights assigned to the "words" are usually normalized to yield unit sum, the "bag-of-word" representation is essentially a discrete distribution with finite support in a Euclidean space. In genomic sequence analysis, a sequence can be converted to a distribution on the combinations of symbols over a few positions. In this case, the support points of the distribution are symbolic.

In this section, we introduce a line of research for clustering distributional data based on the Wasserstein metric between distributions. We focus on the case of discrete distributions in the Euclidean space. For symbolic distributions, we can define a distance likewise by simply replacing the baseline distance in the Euclidean space with pairwise distances between the symbols. A thorough treatment of the Wasserstein metric and its applications in probability theory is given in Rachev [49]. The Wasserstein metric is well defined for distributions with different support points, an important difference from some popular distances such as K–L divergence.

In probability theory, Wasserstein distance is a metric defined for any two probability measures over a metric space, specifically $\mathbb{R}^d$ in our discussion.

Definition 1. *For $p \in [1, \infty)$ and Borel probability measures $\mathcal{P}$, $\mathcal{Q}$ on $\mathbb{R}^d$ with finite p-moments, the p-Wasserstein distance (Villani [50], Section 6) between $\mathcal{P}$ and $\mathcal{Q}$ is defined by*

$$W_p(\mathcal{P}, \mathcal{Q}) = \left[\inf_{\gamma \in \Pi(\mathcal{P},\mathcal{Q})} \int_{\mathbb{R}^d \times \mathbb{R}^d} \| \mathbf{x} - \mathbf{y} \|^p d\gamma(\mathbf{x}, \mathbf{y}) \right]^{1/p} \tag{3}$$

where $\Pi(f, g)$ is the collection of all joint probability measures on $\mathbb{R}^d \times \mathbb{R}^d$ whose marginals are $\mathcal{P}$ and $\mathcal{Q}$, respectively. More specifically, for all subsets $U \in \mathbb{R}^d$, $\gamma(U \times \mathbb{R}^d) = \mathcal{P}(U)$ and $\gamma(\mathbb{R}^d \times U) = \mathcal{Q}(U)$.

$\Pi(\mathcal{P},\mathcal{Q})$ is often called the *coupling set*, and its element γ the coupling distribution, which is also called a transport plan between $\mathcal{P}$ and $\mathcal{Q}$. We can regard γ as a matching matrix specifying the amount of mass at any support point in $\mathcal{P}$ that is transported to another support point in $\mathcal{Q}$. If the cost of transporting mass from a location x to y is $\| \mathbf{x}-\mathbf{y}\|^p$, then W_p^p is the minimum cost to move all the mass in $\mathcal{P}$ to that in $\mathcal{Q}$. See Villani [50] for theory on optimal transport (OT).

In particular, consider two discrete distributions $\mathcal{P}^{(a)} = \{(w_i^{(a)}, \mathbf{x}_i^{(a)}), i = 1,\dots,m_a\}$ and $\mathcal{P}^{(b)} = \{(w_j^{(b)}, \mathbf{x}_j^{(b)}), j = 1,\dots,m_b\}$, where $w_i^{(a)}$ (or $w_j^{(b)}$) is the probability assigned to support point $\mathbf{x}_i^{(a)}$ (or $\mathbf{x}_j^{(b)}$), $\mathbf{x}_i^{(a)}, \mathbf{x}_j^{(b)} \in \mathbb{R}^d$. Let $\mathcal{J}_a = \{1,\dots,m_a\}$, $\mathcal{J}_b = \{1,\dots,m_b\}$. The coupling γ is a joint probability mass function, $\gamma = (\pi_{i,j})_{i\in\mathcal{J}_a, j\in\mathcal{J}_b}$. Then

$$\begin{aligned}
&(W_p(\mathcal{P}^{(a)},\mathcal{P}^{(b)}))^p := \\
&\min_{\{\pi_{i,j}\geq 0\}} \sum_{i\in\mathcal{J}_a, j\in\mathcal{J}_b} \pi_{i,j} \| \mathbf{x}_i^{(a)} - \mathbf{x}_j^{(b)} \|_p^p \\
&\text{s.t.} \quad \sum_{i=1}^{m_a} \pi_{i,j} = w_j^{(b)}, \quad \forall j \in \mathcal{J}_b \\
&\qquad \sum_{j=1}^{m_b} \pi_{i,j} = w_i^{(a)}, \quad \forall i \in \mathcal{J}_a
\end{aligned} \tag{4}$$

In the following discussion, we use the L_2 norm and simply denote the Wasserstein metric by W (instead of W_2). The optimization problem (4) can be solved by linear programming (LP).

The *Wasserstein barycenter* of a set of distributions $\{\mathcal{P}^{(1)},\dots,\mathcal{P}^{(n)}\}$ is defined as a distribution on $\mathbb{R}^d$ that minimizes the sum of squared Wasserstein distances to these distributions. Denote the barycenter by $\overline{\mathcal{P}}$.

$$\overline{\mathcal{P}} := \operatorname*{argmin}_{\mathcal{P}} \sum_{i=1}^{n} W^2(\overline{\mathcal{P}},\mathcal{P}^{(i)})$$

It is proved that for discrete distributions with finite support, the barycenter is also a discrete distribution with finite support [51]. Computation of the Wasserstein barycenter has attracted much attention in the past decade. Most of the literature focuses on the case when the support points are fixed and shared among the $\mathcal{P}_i$'s. Hence, to solve $\overline{\mathcal{P}}$, we only need to solve the probabilities of all the support points. As a result, problem (4) is LP, which can be solved by various LP algorithms, for example, simplex, interior point method. However, the computational complexity grows rapidly when the support sizes of the distributions or the number of distributions increases. There are two major schools of approaches. The first type of approach achieves computational efficiency by adding an entropy regularization term on the transport plan (an approximation to the original problem) [52–54]. The second school of approaches aim at solving the exact problem [55, 56].

When the support points are not fixed, the optimization problem becomes substantially harder and is much less understood theoretically. A common practice is to preset the number of support points in the barycenter and then iteratively update the support points and their probabilities. Interestingly, this less-explored scenario was studied before the burst of interest on Wasserstein barycenter (and as far as the authors know, before the phrase Wasserstein barycenter was coined). Specifically, in their pursuit of creating a real-time automatic image annotation system, Li and Wang [57] proposed the D2-clustering algorithm for clustering discrete distributions with nonfixed support under the Wasserstein distance. In that algorithm, the technique of iteratively updating the support points and their probabilities (together with all the transport plans) has been proposed. As D2 clustering not only solves the barycenter of multiple distributions but also clusters them, it has an extra outer loop to iterate the update of the partition and the calculation of barycenter for each cluster. In the original D2 clustering, these probabilities are solved by a standard LP routine without satisfying scalability. Efforts have been devoted to improve the efficiency using both an ad hoc divide-and-conquer strategy [47] and the modern optimization technique of Bragman ADMM [56]. Recently, a state-of-the-art optimization method has been applied to solve the exact barycenter problem efficiently [58].

2.3 Uncertainty Analysis

When estimating the parameters of a statistical model, it is common practice to provide some form of uncertainty measures for the estimation, for example, standard deviation, confidence intervals. In scientific exploration, clusters may be treated as new discoveries or as evidence to substantiate hypotheses. It is natural to expect uncertainty assessment to be provided for the clustering result. However, this problem has not drawn due attention although there has been increasing interest in the literature recently.

A closely related topic to uncertainty assessment in cluster analysis is the measurement of result stability. We may want to distinguish the two at least in terms of their respective purposes. We can think of uncertainty as arising from the randomness of the data, while stability as arising from the nuances of the algorithms, for example, hyperparameters, initializations. However, as both involve evaluating various kinds of similarity between clustering results, these two concepts are blurred in the literature, often used interchangeably. Early work on defining stability measures focused on the similarity between overall partitions [59, 60]. These partitions are usually obtained from perturbed version of the original data, for example, by adding Gaussian noise or by bootstrap sampling. Various similarity measures between overall partitions have been proposed decades ago, the most commonly used being the Rand Index [61] or Adjusted Rand Index [62], along with others [63]. A straightforward idea to assess stability is to examine all the pairwise distances between partitions. A small average distance indicates high stability. More recently, efforts have been devoted to developing stability measures at the level of individual clusters or data points [64–66].

Li *et al.* [67] proposed a framework based on OT that unites ensemble clustering with uncertainty analysis. Ensemble clustering is a topic much explored in computer science [68–72]. In contrast to the previous methods for assessing stability, this method only requires the partitions of the points but not the points themselves and is thus purely combinatorial. Furthermore, the method reveals directly the set relationship between clusters such as one-to-one mapping, split, and merge instead of wrapping all the information in pairwise distances. When the pairwise distance is very small, it indicates that a one-to-one mapping is quite likely. But when the distance is moderate or high, it could be caused by different set relationships, for example, split, merge, or simply a lack of correspondence.

Consider two partitions $\mathcal{P}^{(m)} = \{C_1^{(m)}, \dots, C_{K_m}^{(m)}\}$, $m = 1, 2$, where $C_j^{(m)}$ is the jth cluster in the mth partition. The clusters in each partition follow a discrete distribution, $\mathcal{Q}^{(m)} = \{q_1^{(m)}, \dots, q_{K_m}^{(m)}\}$, for example, uniform or the empirical frequencies of the clusters.

A distance is defined between any pair of clusters, $d(C_i^{(1)}, C_j^{(2)})$, for example, the popular Jacaard index $d(C_i^{(1)}, C_j^{(2)}) = 1 - |C_i^{(1)} \cap C_j^{(2)}|/|C_i^{(1)} \cup C_j^{(2)}|$. We encode the matching between clusters in the two partitions by a matrix of matching weights, $\gamma = (\gamma_{i,j})_{i=1,\ldots,K_1, j=1,\ldots,K_2}$. The principle of cluster alignment is to minimize the sum of weighted distances between pairs of clusters in the two partitions. The weights are subject to certain constraints to guarantee that every cluster influences the matching to an extent proportional to its assigned probability. This is essentially the OT problem [50], occurring in the definition of the Wasserstein distance between distributions. Again, use $\Pi(Q^{(1)}, Q^{(2)})$ to denote the coupling set (see Section 2.2). We define the Wasserstein distance between two partitions by

$$D(\mathcal{P}^{(1)}, \mathcal{P}^{(2)}) := \min_{\gamma \in \Pi(Q^{(1)}, Q^{(2)})} \sum_{i=1}^{K_1} \sum_{j=1}^{K_2} \gamma_{i,j} d(C_i^{(1)}, C_j^{(2)}) \tag{5}$$

We characterize a partition $\mathcal{P}$ by the so-called cluster posterior matrix $P = (p_{l,i})_{l=1,\ldots,n, i=1,\ldots,K}$, where n is the data size, and K is the number of clusters. For the usual hard clustering, $p_{l,i} = 1$ if the lth data point belongs to the ith cluster and 0 otherwise. For mixture-mode-based clustering, we often compute $p_{l,i}$ as the posterior probability for cluster label i given $\mathbf{x}_l$ and convert to hard clustering by the maximum a posteriori criterion. Let the cluster posterior matrix of $\mathcal{P}^{(m)}$ be $P^{(m)}$. This matching matrix γ can be viewed as a "translator" between $\mathcal{P}^{(1)}$ and $\mathcal{P}^{(2)}$ such that they are subject to the same cluster labeling scheme. Let the rowwise normalized matrix of γ be Γ, called cluster *mapping matrix* from $\mathcal{P}^{(1)}$ to $\mathcal{P}^{(2)}$. Then, the aligned cluster-posterior matrix of $\mathcal{P}^{(1)}$ with respect to $\mathcal{P}^{(2)}$ is $P^{(1\to 2)} = P^{(1)}\Gamma$. If we treat one partition as the reference and compute the aligned cluster-posterior matrix of every other partition with respect to the reference, we can compute the average aligned cluster-posterior matrix $\overline{P}$ under the labeling scheme of the reference. The "mean partition" $\overline{\mathcal{P}}$ is then defined by $\overline{P}$. One apparent choice for the stability measure at the level of partition is the average Wasserstein distance between each partition and the mean partition.

After cluster alignment, we can label the clusters consistently across all the partitions. In Li *et al.* [67], several set relationships are defined including one-to-one match (or simply match), split, merge, and lack of correspondence. Roughly speaking, one-to-one match means that the two clusters can be considered as random realizations of the same cluster. It is possible that two partitions are so different that the match relation does not exist between any pair of clusters. Consider a collection of matching clusters S_i, $i = 1, \ldots, m$, each specified by the set of points it contains. Every S_i comes from one partition. The *covering point set* (CPS) S_α at a covering level α is defined as the smallest set of points such that at least $100(1-\alpha)\%$ of S_is are subsets of S_α. Use $|\cdot|$ to denote the cardinality of a set. $I(\cdot)$ is the indicator function that equals 1 if the argument is true and 0 otherwise. CPS is solved by the following optimization problem:

$$\begin{aligned} &\min_S |S| \\ &\text{s.t.} \sum_{i=1}^{m} I(S_i \subset S) \geq m(1-\alpha) \end{aligned} \tag{6}$$

The optimization problem is solved approximately by an algorithm called *Least Impact First Targeted removal* (LIFT). CPS plays a role for sets of points similar to the role of confidence intervals for numerical parameters.

Given the CPS S_α of a collection of clusters S_i, $i = 1, \dots, m$, the level of uncertainty for the clusters can be measured by the extent of agreement between S_is and S_α. Suppose that $S_i \subset S_\alpha$. Define *tightness ratio* by $R_t(S_i|S_\alpha) = \frac{|S_i|}{|S_\alpha|}$. For clusters not fully covered by S_α, that is, $S_i \not\subset S_\alpha$, define *coverage ratio* by $R_c(S_i|S_\alpha) = \frac{|S_i \cap S_\alpha|}{|S_i|}$. Let $\overline{R}_t(S_\alpha)$ be the average tightness ratio for the S_is that are covered, and $\overline{R}_c(S_\alpha)$ be the average coverage ratio for those not fully covered. To quantify the stability of a cluster, we must consider that in general only a proportion of bootstrap partitions contain a cluster matched with the base cluster. Let that proportion be ρ. Then, the stability of a base cluster can be defined as $\rho\overline{R}_t(S_\alpha)$. A pipeline for applying the uncertainty analysis for biomedical data is presented in Zhang *et al.* [73]. The method is implemented in an R package called *OTclust*, available on CRAN.

3 Semisupervised Learning

In contrast to unsupervised learning, in the semisupervised setting, both labeled and unlabeled observations are available for modeling. This setting has become increasingly important in response to changes in modern machine learning tasks and changes in data availability. Modern machine learning tasks (e.g., object detection [74] and text translation [75]) have become progressively more challenging, and thus their model training has become more data intensive. Simultaneously, large-scale unlabeled datasets have become more common and available as information is digitized and made available online [76, 77]. Often, however, it is unfeasible to fully label these large datasets due to constraints of time, cost, and expertise. These developments have naturally led to the question of whether a large unlabeled dataset could be paired with a smaller labeled one to improve a model's task performance without the need for additional labeled observations. Current state-of-the-art results on tasks such as image classification demonstrate significant benefits to such semisupervised approaches [78–80].

3.1 Setting

More formally, the semisupervised setting can be defined as follows. We consider our dataset to be $\mathcal{D} = (\mathcal{D}_L, \mathcal{D}_U)$, where $\mathcal{D}_L = (\mathcal{X}_L, \mathcal{Y}_L)$ comprises the labeled observations, and $\mathcal{D}_U = (\mathcal{X}_U)$ the unlabeled observations, often with $|\mathcal{X}_L| << |\mathcal{X}_U|$. Our data are generated with $\mathbf{x}_i \overset{\text{i.i.d.}}{\sim} p(\mathbf{x})$ and $(\mathbf{x}_i, \mathbf{y}_i) \sim p(\mathbf{x}, \mathbf{y})$. Although this general setting supports all manner of statistical learning tasks, we focus on estimation of the posterior class probability $p(\mathbf{y}|\mathbf{x})$ for classification.

The challenge of semisupervised learning is to now improve estimation of $p(\mathbf{y}|\mathbf{x})$ by incorporating the information contained in $\mathcal{D}_U$. A natural choice is to use these unlabeled observations to obtain a better estimate for the marginal distribution $p(\mathbf{x})$. This information, however, is beneficial only if there is an exploitable relationship between $p(\mathbf{x})$ and the desired $p(\mathbf{y}|\mathbf{x})$. To bridge this gap, a low-density separation assumption is often invoked[1] [81]. This assumption states that the decision boundary between classes should lie in a low-density region of the feature space. In other words, proximity of observations $\mathbf{x}_1$ and $\mathbf{x}_2$ in a high-density region of the feature space implies proximity of their respective conditional distributions, $p(\mathbf{y}|\mathbf{x}_1)$ and $p(\mathbf{y}|\mathbf{x}_2)$. Importantly, the low-density separation assumption is often paired with a second assumption, namely, the manifold assumption. This assumption states that the data lie approximately on a low-dimensional manifold embedded in the feature space [81]. Now, incorporating the manifold assumption, proximity between observations can be measured along the manifold instead of the feature space directly. Incorporating these assumptions into training provides extra structure for candidate models and utilizes all available observations.

Semisupervised modeling operationalizes these assumptions using several approaches. Here, we present a select overview with a focus on component approaches of the current state of the art. We begin with an introduction to three historically important classes of method, self-training [82], generative models, and graphical models [83]. We then review entropy minimization, consistency regularization, and mixup augmentation. Finally, we describe a state-of-the-art method, MixMatch, which combines these three approaches. For a more comprehensive review, see Refs 81, 84 and 85.

3.2 Self-Training

Among the first semisupervised learning approaches were self-training algorithms, developed first for pattern recognition [86, 87] and later for text classification [88]. In this paradigm, classifiers are trained iteratively. In each iteration, the model is first trained on labeled data and then used to predict labels for the unlabeled data. Those observations receiving a high-confidence prediction are then treated as labeled in succeeding iterations. Later developments in this area include multiview training algorithms such as cotraining [89] and tri-training [90], which apply self-training using various model ensembles and pseudolabeling rules.

3.3 Generative Models

Another early approach to semisupervised learning was the generalization of existing generative models, such as linear discriminant analysis [91], and later mixture models [92], to accommodate partially labeled data settings. In contrast to discriminative models that are satisfied to estimate the conditional distribution $p(\mathbf{y}|\mathbf{x})$, generative models obtain an estimate of the full joint distribution $p(\mathbf{x}, \mathbf{y})$, which is then used to build the classifier. While more descriptive of the data-generating process than discriminative models, generative models are also thus more complex and difficult to estimate [93]. Despite these drawbacks, generative modeling using neural networks has yielded competitive performance [94].

3.4 Graphical Models

Graphical models rely on representation of data as a graph, with each observation serving as a node and relationships between observations encoded as weighted edges. Application of such models to the semisupervised setting is done in two steps, graph construction and model training. Graph construction aims to map data from the feature space to a representative graph with weight matrix $\mathcal{W}$ using various mechanisms such as k-nearest neighbors, ϵ-neighborhoods, and Gaussian kernels [85]. More involved approaches perform kernel tuning [95], embedding [96, 97], or metric learning [98]. Classifiers are then trained to reduce a supervised loss $\mathcal{L}_S$ and use a second loss term to preserve similarity of labels predicted by the classifier h_θ parameterized by θ along the graph among related observations as measured by $\mathcal{W}$. In particular, this second loss term measures the dissimilarity $\mathcal{D}$ between pairs of predicted label distributions $h(x_i)$ and $h(x_j)$ for all observations $i, j \in \{1, 2, \dots, N\}$ in a sample of size N and then weighs each pair's dissimilarity by their edge weight $\mathcal{W}_{ij}$ (Equation 7). Methods vary in the form of this regularizing term [99–103].

$$\mathcal{L}_G = \mathcal{L}_S + \lambda \sum_{i,j} \mathcal{W}_{ij}\mathcal{D}(h(\mathbf{x}_i), h(\mathbf{x}_j)) \tag{7}$$

More recent work has combined neural network classifiers with graphical regularization [104] and has developed novel network architectures directly incorporating graph structure into the classifier's functional form [105].

3.5 Entropy Minimization

Although not inherently semisupervised, entropy minimization has been shown to be synergistic with many such models and has thus been utilized in various forms in recent work [78, 79, 106]. Motivating this technique is the intuition that a good classifier should produce high confidence predictions, that is, the conditional distribution $p(\mathbf{y}|\mathbf{x})$ should assign the correct class a probability of 1. Several approaches have been developed to encourage this type of "one-hot" prediction. The first entropy minimization approach introduces a loss regularizing the empirical conditional entropy of model predictions [93]. Mutual exclusivity loss instead penalizes a differentiable measure of a prediction's deviation from one-hot [107]. Temperature sharpening of the model prediction can also be used to achieve the entropy minimization effect in postprocessing [78].

3.6 Consistency Regularization

Regularization is a well-studied concept from supervised learning which proposes penalization of model complexity to discourage selection of overfitted models. In semisupervised learning, this approach has been extended to include model complexity measures incorporating density information provided by unlabeled data. These complexity measures are, in turn, used to bias model selection toward those meeting targeted semisupervised assumptions.

Consistency regularization is a method that aims to impose the low-density separation assumption on the classifier. First, existing observations $\mathbf{x}_i, i \in \{1, 2, \dots, N\}$ for a sample of size N are used to generate new observations $\tilde{\mathbf{x}}_i$ using various augmentation schemes [79, 108–110]. These augmented data $\tilde{\mathbf{x}}_i$ are generated to match their "parent" observations $\mathbf{x}_i$ in conditional distribution (i.e., $p(\mathbf{y}|\mathbf{x}_i) = p(\mathbf{y}|\tilde{\mathbf{x}}_i))^2$. These augmented $\tilde{\mathbf{x}}_i$ are then incentivized to share predicted conditional distributions ($h_\theta(\tilde{\mathbf{x}}_i)$) with a target T_i through a loss term measuring the dissimilarity $\mathcal{D}$ between them[3].

$$\mathcal{L}_{\text{CR}} = \sum_{i=1}^{N} \mathcal{D}(h_\theta(\tilde{\mathbf{x}}_i), T_i) \tag{8}$$

For some intuition, this target T_i can be thought of as the predicted conditional distribution of the parent observation (i.e., $h_\theta(\mathbf{x}_i)$). Thus, this loss can be seen as enforcing class consistency between parent and augmentation. As detailed below, the various implementations of consistency regularization differ in the choice of augmentation function, dissimilarity measure, and target.

When augmentations populate local regions around original observations, enforcement of class consistency pushes the decision boundary away from the original data and into lower density regions, producing higher margin classifiers. When augmentations are not local to the original observations in the feature space, this reasoning can be adapted to measure locality along the data manifold instead. In this case, the decision boundaries are repulsed from manifold-local regions. In either case, this loss term encourages selection of models with decision boundaries in low-density regions and incorporates the density information of both labeled and unlabeled data in doing so. The most promising results of such consistency regularization have come from its coupling with neural network classifiers.

Among the early pairings of consistency regularization with neural networks is the Π-model [111]. The Π-model generates augmented data ($\tilde{\mathbf{x}}_{i,1}$) using context-standard transformations of the input data[4] and adds in-network stochasticity using dropout. The predicted conditional distribution of an additional augmented observation $h_\theta(\tilde{\mathbf{x}}_{i,2})$ is used as the target. Finally, the dissimilarity is calculated using the squared L2 norm.

$$\mathcal{L}_{\Pi} = \sum_{i=1}^{N} ||h_\theta(\tilde{\mathbf{x}}_{i,1}) - h_\theta(\tilde{\mathbf{x}}_{i,2})||_2^2 \tag{9}$$

Temporal ensembling builds on the success of the Π-model by developing an improved consistency target while maintaining the same augmentation generation and dissimilarity measure [111]. Given that the Π-model target $h_\theta(\tilde{\mathbf{x}}_{i,2})$ varies significantly throughout training, the authors propose using an exponential moving average (EMA) of model outputs from past epochs instead (10). The resultant consistency loss is shown below[5].

$$\hat{h}_i = \alpha \hat{h}_i + (1-\alpha) h_\theta(\tilde{\mathbf{x}}_i) \tag{10}$$

$$\mathcal{L}_{\text{TE}} = \sum_{i=1}^{N} ||h_\theta(\tilde{\mathbf{x}}_i) - \hat{h}_i||_2^2 \tag{11}$$

This loss (11) has the effect of smoothing consistency learning as the EMA target has lower variability across training. It also has the more practical benefit of halving the number of model evaluations per epoch.

Further refinement of the consistency target is achieved by the Mean Teacher approach [80]. Here, the authors propose using the predicted class distribution of a model weighted by the EMA of past classifier weights as the target (12).

$$\hat{\theta}_i = \alpha\hat{\theta}_i + (1 - \alpha)\theta \tag{12}$$

$$\mathcal{L}_{\text{MT}} = \sum_{i=1}^{N} ||h_\theta(\tilde{\mathbf{x}}_i) - h_{\hat{\theta}_i}(\tilde{\mathbf{x}}_i)||_2^2 \tag{13}$$

This loss (13) provides several advantages over the temporal ensemble[6]. Firstly, the mean teacher provides a more accurate consistency target, benefitting from a higher level of information transfer through the "memory" weights $\hat{\theta}_i$, than the temporal ensemble does through its predicted class distribution "memory" $\hat{h}_i$. This target is also more frequently updated (once per training step as opposed to once per epoch) and thus provides a prediction incorporating more recent updates. While this method implements the standard data augmentations of past works, a more powerful in-network regularization, Shake–Shake [112], is also used to further perturb the data representations.

Unlike previously described approaches that focus on consistency target improvement, Virtual Adversarial Training (VAT) instead targets the augmentation procedure [79]. During training, augmentations are generated by shifting parent observations in the adversarial direction (14). The adversarial direction is, in this case, the direction in the feature space that causes the largest changes in the model's predictions. The consistency loss is then calculated using the Kullback-Leibler (KL) divergence between the current model's predicted class distribution on the parent observation and the adversarial augmentation ($\tilde{\mathbf{x}}_i$).

$$\tilde{\mathbf{x}}_i = \mathbf{x}_i + r_{\text{adv}} \tag{14}$$

$$\mathcal{L}_{\text{VAT}} = \sum_{i=1}^{N} \text{KL}(h_\theta(\mathbf{x}_i), h_\theta(\tilde{\mathbf{x}}_i)) \tag{15}$$

As with all consistency regularization approaches, VAT provides smoothing of the classifier around the data. However, an important advantage to this approach is that smoothing is applied more tactically than standard augmentation schemes – augmented data is only generated in the directions in which consistency learning is most needed. Adversarial augmentation is also more data-type agnostic, as it does not require explicit class-preserving transformations that are often data-type dependent.

3.7 Mixup

Mixup is a powerful data-type agnostic augmentation technique originally developed for supervised learning [113]. A new augmented observation $\tilde{\mathbf{x}}_{ij}$ is constructed as the convex combination of two existing data points, where the mixing weight λ is drawn from a Beta distribution with hyperparameter α. The label corresponding to the augmented observation $\tilde{\mathbf{y}}_{ij}$ is similarly defined as the corresponding convex combination of the labels of the two original observations.

$$\tilde{\mathbf{x}}_{ij} = \lambda\mathbf{x}_i + (1 - \lambda)\mathbf{x}_j \tag{16}$$

$$\tilde{\mathbf{y}}_{ij} = \lambda\mathbf{y}_i + (1 - \lambda)\mathbf{y}_j \tag{17}$$

When adapted to the semisupervised setting, mixup combinations involving unlabeled observations substitute pseudolabels for unknown true labels[7]. By injecting linearly interpolated examples into the training process, this augmentation procedure encourages model predictions to evolve linearly between observations, acting as a regularizer against sudden prediction changes between observations. This is in contrast to the previously discussed role of augmentation in consistency regularization, which seeks to maintain a constant decision function between observations. The interaction between these two mechanisms is exploited by the MixMatch [78] and Interpolation Consistency Training [114] methods.

3.8 MixMatch

Model training via MixMatch and similar methods combines multiple semisupervised methods to generate more-than-the-sum-of-their-parts classification performance [78, 114, 115]. MixMatch applies consistency regularization using two sources of augmentation. First, standard augmentation functions are applied to the data with unlabeled observations assigned a sharpened average prediction as their pseudolabel. The resulting dataset is then further jittered with mixup augmentation. The classifier is finally trained with a supervised loss and an L2 consistency regularizer term.

4 Conclusions

In this chapter, we have reviewed major schools of approaches to unsupervised and semisupervised learning. For semisupervised learning, the emphasis is on recent work in the machine learning community. For unsupervised learning, the methods described are heavily tilted toward our own work. We present several of the reviewed methods in much greater detail to serve as a handbook for practitioners.

Acknowledgment

Jia Li's research is partially supported by the National Science Foundation [DMS-2013905].

Notes

1 The low-density assumption is closely related to several other popular assumptions – the semisupervised smoothness assumption and the cluster assumption.
2 For simplicity, discussion is limited to the case of one augmentation per parent observation; however, inclusion of multiple augmentations is straightforward.
3 Common dissimilarity measures include the Kullback-Leibler (KL) divergence and the squared L2 norm.
4 Experiments are performed on image datasets using random translation and flipping – considered standard "weak" augmentations.
5 During training, the target is treated as constant with respect to the model parameters θ.

6 As with the temporal ensemble, the target is treated as constant with respect to the model parameters θ during training.
7 Pseudolabel generation varies between methods but often involves predictions from a teacher model.

References

1 Kvistborg, P., Gouttefangeas, C., Aghaeepour, N. *et al.* (2015) Thinking outside the gate: single-cell assessments in multiple dimensions. *Immunity*, 42 (4), 591–592.
2 Navin, N., Kendall, J., Troge, J. *et al.* (2011) Tumour evolution inferred by single-cell sequencing. *Nature*, 472 (7341), 90–94.
3 Kim, K.T., Lee, H.W., Lee, H.O. *et al.* (2016) Application of single-cell RNA sequencing in optimizing a combinatorial therapeutic strategy in metastatic renal cell carcinoma. *Genome Biol.*, 17, 80.
4 Malik, J., Belongie, S., Leung, T., and Shi, J. (2001) Contour and texture analysis for image segmentation. *Int. J. Comput. Vision*, 43 (1), 7–27.
5 Li, J. (2011) Agglomerative connectivity constrained clustering for image segmentation. *Stat. Anal. Data Min.: ASA Data Sci. J.*, 4 (1), 84–99.
6 Charikar, M., Chekuri, C., Feder, T., and Motwani, R. (2004) Incremental clustering and dynamic information retrieval. *SIAM J. Comput.*, 33 (6), 1417–1440.
7 Jardine, N. and van Rijsbergen, C.J. (1971) The use of hierarchic clustering in information retrieval. *Inf. Storage Retrieval*, 7 (5), 217–240.
8 Gower, J.C. and Ross, G.J.S. (1969) Minimum spanning trees and single linkage cluster analysis. *J. R. Stat. Soc.: Ser. C (Appl. Stat.)*, 18 (1), 54–64.
9 Pothen, A., Simon, H.D., and Liou, K-P. (1990) Partitioning sparse matrices with eigenvectors of graphs. *SIAM J. Matrix Anal. Appl.*, 11 (3), 430–452.
10 Gonzalez, T.F. (1985) Clustering to minimize the maximum intercluster distance. *Theor. Comput. Sci.*, 38, 293–306.
11 Fraley, C. and Raftery, A.E. (2002) Model-based clustering, discriminant analysis, and density estimation. *J. Am. Stat. Assoc.*, 97 (458), 611–631.
12 McLachlan, G.J. and Peel, D. (2004) *Finite Mixture Models*, John Wiley & Sons.
13 Banfield, J.D. and Raftery, A.E. (1993) Model-based gaussian and non-gaussian clustering. *Biometrics*, 49, 803–821.
14 Celeux, G. and Gérard, G. (1993) Gaussian Parsimonious Clustering Models. Technical Report RR-2028, INRIA. https://hal.inria.fr/inria-00074643.
15 Fraley, C., Raftery, A.E., Brendan Murphy, T., and Scrucca, L. (2012) mclust Version 4 for R: Normal Mixture Modeling for Model-Based Clustering, Classification, and Density Estimation. Technical Report.
16 Li, J. (2005) Clustering based on a multilayer mixture model. *J. Comput. Graph. Stat.*, 14 (3), 547–568.
17 Li, J., Ray, S., and Lindsay, B.G. (2007) A nonparametric statistical approach to clustering via mode identification. *J. Mach. Learn. Res.*, 8, 1687–1723.
18 Ray, S. and Lindsay, B.G. 2005 The topography of multivariate normal mixtures. *Ann. Stat.*, 33 (5), 2042–2065.

19 Lee, H. and Li, J. (2012) Variable selection for clustering by separability based on ridgelines. *J. Comput. Graph. Stat.*, 21 (2), 315–336.

20 Lin, L. and Li, J. (2017) Clustering with hidden Markov model on variable blocks. *J. Mach. Learn. Res.*, 18 (1), 3913–3961.

21 Li, J. and Lin, L. (2017) Baum–Welch algorithm on directed acyclic graph for mixtures with latent Bayesian networks. *Stat*, 6 (1), 303–314.

22 Li, J., Najmi, A., and Gray, R.M. (2000) Image classification by a two-dimensional hidden Markov model. *IEEE Trans. Signal Process.*, 48 (2), 517–533.

23 Li, J., Gray, R.M., and Olshen, R.A. (2000) Multiresolution image classification by hierarchical modeling with two-dimensional hidden Markov models. *IEEE Trans. Inf. Theory*, 46 (5), 1826–1841.

24 Ben-Hur, A. and Guyon, I. (2003) Detecting stable clusters using principal component analysis, in *Functional Genomics*, Springer, pp. 159–182.

25 Hackstadt, A.J. and Hess, A.M. (2009) Filtering for increased power for microarray data analysis. *BMC Bioinformatics*, 10 (1), 11.

26 Yeung, K. and Ruzzo, W. (2001) Principal component analysis for clustering gene expression data. *Bioinformatics*, 17, 763–774.

27 Chang, W.C. (1983) On using principal components before separating a mixture of two multivariate normal distributions. *Appl. Stat.*, 32, 267–275.

28 Liu, T., Liu, S., Chen, Z., and Ma, W-Y. (2003) *An Evaluation on Feature Selection for Text Clustering*. Proceedings of the 20th International Conference on Machine Learning (ICML-03), pp. 488–495.

29 Raileanu, L.E. and Stoffel, K. (2004) Theoretical comparison between the Gini index and information gain criteria. *Ann. Math. Artif. Intell.*, 41 (1), 77–93.

30 Yu, L. and Liu, H. (2003) *Feature Selection for High-Dimensional Data: A Fast Correlation-Based Filter Solution*. Proceedings of the 20th International Conference on Machine Learning (ICML-03), pp. 856–863.

31 He, X., Cai, D., and Niyogi, P. (2006) *Laplacian Score for Feature Selection*. Advances in Neural Information Processing Systems, pp. 507–514.

32 Dy, J.G. and Brodley, C.E. (2004) Feature selection for unsupervised learning. *J. Mach. Learn. Res.*, 5, 845–889.

33 Liu, J.S., Zhang, J.L., Palumbo, M.J., and Lawrence, C.E. (2003) Bayesian clustering with variable and transformation selections. *Bayesian Stat.*, 7, 249–275.

34 Belkin, M. and Niyogi, P. (2002) *Laplacian Eigenmaps and Spectral Techniques for Embedding and Clustering*. Advances in Neural Information Processing Systems, pp. 585–591.

35 Law, M.H.C., Figueiredo, M.A.T., and Jain, A.K. (2004) Simultaneous feature selection and clustering using mixture models. *IEEE Trans. Pattern Anal. Mach. Intell.*, 26 (9), 1154–1166.

36 Tadesse, M.G., Sha, N., and Vannucci, M. (2005) Bayesian variable selection in clustering high-dimensional data. *J. Am. Stat. Assoc.*, 100 (470), 602–617.

37 Raftery, A.E. and Dean, N. (2006) Variable selection for model-based clustering. *J. Am. Stat. Assoc.*, 101 (473), 168–178.

38 Maugis, C., Celeux, G., and Martin-Magniette, M-L. (2009) Variable selection for clustering with gaussian mixture models. *Biometrics*, 65 (3), 701–709.

39 Marbac, M. and Sedki, M. (2018) VarSelLCM: an R/C++ package for variable selection in model-based clustering of mixed-data with missing values. *Bioinformatics*, 35 (7), 1255–1257.

40 Pan, W. and Shen, X. (2007) Penalized model-based clustering with application to variable selection. *J. Mach. Learn. Res.*, 8, 1145–1164.

41 Wang, S. and Zhu, J. (2008) Variable selection for model-based high-dimensional clustering and its application to microarray data. *Biometrics*, 64 (2), 440–448.

42 Xie, B., Pan, W., and Shen, X. (2008) Penalized model-based clustering with cluster-specific diagonal covariance matrices and grouped variables. *Electron. J. Stat.*, 2, 168.

43 Guo, J., Levina, E., Michailidis, G., and Zhu, J. (2010) Pairwise variable selection for high-dimensional model-based clustering. *Biometrics*, 66 (3), 793–804.

44 Witten, D.M. and Tibshirani, R. (2010) A framework for feature selection in clustering. *J. Am. Stat. Assoc.*, 105 (490), 713–726.

45 Witten, D.M. and Tibshirani, R. (2013) sparcl: Perform sparse hierarchical clustering and sparse K-means clustering. *R Package Version*, 1 (3).

46 Wang, J.Z., Li, J., and Wiederhold, G. (2001) Simplicity: semantics-sensitive integrated matching for picture libraries. *IEEE Trans. Pattern Anal. Mach. Intell.*, 23 (9), 947–963.

47 Zhang, Y., Wang, J.Z., and Li, J. (2015) Parallel massive clustering of discrete distributions. *ACM Trans. Multimedia Comput., Commun. Appl. (TOMM)*, 11 (4), 1–24.

48 Ye, J., Li, Y., Wu, Z. *et al.* (2017) *Determining Gains Acquired from Word Embedding Quantitatively Using Discrete Distribution Clustering*. Proceedings of the 55th Annual Meeting of the Association for Computational Linguistics (Volume 1: Long Papers), pp. 1847–1856.

49 Rachev, S.T. (1985) The Monge–Kantorovich mass transference problem and its stochastic applications. *Theory Probab. Appl.*, 29 (4), 647–676.

50 Villani, C. (2008) *Optimal Transport: Old and New*, Vol. 338, Springer Science & Business Media.

51 Anderes, E., Borgwardt, S., and Miller, J. (2016) Discrete Wasserstein barycenters: optimal transport for discrete data. *Math. Meth. Oper. Res.*, 84 (2), 389–409.

52 Cuturi, M. (2013) *Sinkhorn Distances: Lightspeed Computation of Optimal Transport*. Advances in Neural Information Processing Systems, pp. 2292–2300.

53 Cuturi, M. and Doucet, A. (2014) *Fast Computation of Wasserstein Barycenters*. Proceeding of International Conference on Machine Learning, 685–693, Vancouver, Canada.

54 Benamou, J-D., Carlier, G., Cuturi, M. *et al.* (2015) Iterative Bregman projections for regularized transportation problems. *SIAM J. Sci. Comput.*, 37 (2), A1111–A1138.

55 Ye, J. and Li, J. (2014) *Scaling Up Discrete Distribution Clustering Using ADMM*. 2014 IEEE International Conference on Image Processing (ICIP), IEEE, pp. 5267–5271.

56 Ye, J., Wu, P., Wang, J.Z., and Li, J. (2017) Fast discrete distribution clustering using Wasserstein barycenter with sparse support. *IEEE Trans. Signal Process.*, 65 (9), 2317–2332.

57 Li, J. and Wang, J.Z. (2008) Real-time computerized annotation of pictures. *IEEE Trans. Pattern Anal. Mach. Intell.*, 30 (6), 985–1002.

58 Yang, L., Li, J., Sun, D., and Toh, K-C. (2018) A fast globally linearly convergent algorithm for the computation of Wasserstein barycenters. *arXiv preprint arXiv:1809.04249.*

59 Luxburg, U. (2010) Clustering stability: an overview. *Foundations and Trends® in Machine Learning*, 2 (3), 235–274.

60 Leisch, F. (2015) Resampling methods for exploring cluster stability, in *Handbook of Cluster Analysis*, Chapman and Hall/CRC, pp. 658–673.

61 Rand, W.M. (1971) Objective criteria for the evaluation of clustering methods. *J. Am. Stat. Assoc.*, 66, 846–850.

62 Hubert, L. and Arabie, P. (1985) Comparing partitions. *J. Classif.*, 2, 193–218.

63 Zhou, D., Li, J., and Zha, H. (2005) *A New Mallows Distance Based Metric for Comparing Clusterings.* Proceedings of the 22nd International Conference on Machine Learning, pp. 1028–1035.

64 Koepke, H. and Clarke, B. (2013) A Bayesian criterion for cluster stability. *Stat. Anal. Data Min.: ASA Data Sci. J.*, 6 (4), 346–374.

65 Hennig, C. (2007) Cluster-wise assessment of cluster stability. *Comput. Stat. Data Anal.*, 52 (1), 258–271.

66 Yu, H., Chapman, B., Di Florio, A. *et al.* (2019) Bootstrapping estimates of stability for clusters, observations and model selection. *Comput. Stat.*, 34 (1), 349–372.

67 Li, J., Seo, B., and Lin, L. (2019) Optimal transport, mean partition, and uncertainty assessment in cluster analysis. *Stat. Anal. Data Min.: ASA Data Sci. J.*, 12 (5), 359–377.

68 Strehl, A. and Ghosh, J. (2002) Cluster ensembles – a knowledge reuse framework for combining multiple partitions. *J. Mach. Learn. Res.*, 3, 583–617.

69 Fred, A. (2001) *Finding Consistent Clusters in Data Partitions.* Workshop on Multiple Classifier Systems, 309–318.

70 Vega-Pons, S. and Ruiz-Shulcloper, J. (2011) A survey of clustering ensemble algorithms. *Int. J. Pattern Recognit. Artif. Intell.*, 25, 337–372.

71 Amiri, S., Clarke, B.S., Clarke, J.L., and Koepke, H. (2019) A general hybrid clustering technique. *J. Comput. Graph. Stat.*, 28 (3), 540–551.

72 Amiri, S., Clarke, B.S., and Clarke, J.L. (2018) Clustering categorical data via ensembling dissimilarity matrices. *J. Comput. Graph. Stat.*, 27 (1), 195–208.

73 Zhang, L., Lin, L., and Li, J. (2020) CPS analysis: self-contained validation of biomedical data clustering. *Bioinformatics*, 03. ISSN 1367-4803. doi: 10.1093/bioinformatics/btaa165.

74 Zhao, Z., Zheng, P., Xu, S., and Wu, X. (2019) Object detection with deep learning: a review. *IEEE Trans. Neural Networks Learn. Syst.*, 30 (11), 3212–3232.

75 Young, T., Hazarika, D., Poria, S., and Cambria, E. (2018) Recent trends in deep learning based natural language processing [review article]. *IEEE Comput. Intell. Mag.*, 13, 55–75.

76 Lahiri, S. (2014) *Complexity of Word Collocation Networks: A Preliminary Structural Analysis.* Proceedings of the Student Research Workshop at the 14th Conference of the European Chapter of the Association for Computational Linguistics, Association for Computational Linguistics, Gothenburg, Sweden, pp. 96–105. http://www.aclweb.org/anthology/E14-3011.

77 Thomee, B., Shamma, D.A., Friedland, G. *et al.* (2016) YFCC100M: the new data in multimedia research. *Commun. ACM*, 59, 64–73.

78 Berthelot, D., Carlini, N., Goodfellow, I.G. *et al.* (2019) *Mixmatch: A Holistic Approach to Semi-Supervised Learning*. NeurIPS.

79 Miyato, T., Maeda, S., Koyama, M., and Ishii, S. (2019) Virtual adversarial training: a regularization method for supervised and semi-supervised learning. *IEEE Trans. Pattern Anal. Mach. Intell.*, 41, 1979–1993.

80 Tarvainen, A. and Valpola, H. (2017) *Mean Teachers are Better Role Models: Weight-Averaged Consistency Targets Improve Semi-Supervised Deep Learning Results*. Proceedings of the 31st International Conference on Neural Information Processing Systems. InNIPS'17, pp. 1195–1204. ISBN 9781510860964.

81 Chapelle, O., Bernhard, S., and Zien, A. (2006) *Semi-Supervised Learning*, MIT Press.

82 Triguero, I., García, S., and Herrera, F. (2013) Self-labeled techniques for semi-supervised learning: taxonomy, software and empirical study. *Knowl. Inf. Syst.*, 42, 245–284.

83 Subramanya, A. and Talukdar, P.P. (2014) Graph-Based Semi-Supervised Learning. *Synth. Lect. Artif. Intell. Mach. Learn.*, 8 (4), 1–25.

84 van Engelen, J.E. and Hoos, H.H. (2019) A survey on semi-supervised learning. *Mach. Learn.*, 109, 373–440.

85 Zhu, X.J. (2005) Semi-Supervised Learning Literature Survey. Technical Report. University of Wisconsin-Madison Department of Computer Sciences.

86 Scudder, H.J. (1965) Probability of error of some adaptive pattern-recognition machines. *IEEE Trans. Inf. Theory*, 11, 363–371.

87 Agrawala, A.K. (1970) Learning with a probabilistic teacher. *IEEE Trans. Inf. Theory*, 16, 373–379.

88 Yarowsky, D. (1995) *Unsupervised Word Sense Disambiguation Rivaling Supervised Methods*. ACL.

89 Blum, A. and Mitchell, T. (1998) *Combining Labeled and Unlabeled Data with Co-Training*. COLT' 98.

90 Zhou, Z-H. and Li, M. (2005) Tri-training: exploiting unlabeled data using three classifiers. *IEEE Trans. Knowl. Data Eng.*, 17, 1529–1541.

91 McLachlan, G.J. and Ganesalingam, S. (1982) Updating a discriminant function on the basis of unclassified data. *Commun. Stat.-Simul. Comput.*, 11 (6), 753–767.

92 Miller, D.J. and Uyar, H.S. (1996) *A Mixture of Experts Classifier with Learning Based on Both Labelled and Unlabelled Data*. NIPS.

93 Grandvalet, Y. and Bengio, Y. (2004) *Semi-Supervised Learning by Entropy Minimization*. CAP.

94 Kingma, D.P., Mohamed, S., Rezende, D.J., and Welling, M. (2014) *Semi-Supervised Learning with Deep Generative Models*. NIPS.

95 Zhang, X. and Lee, W.S. (2006) *Hyperparameter Learning for Graph Based Semi-Supervised Learning Algorithms*. NIPS.

96 Belkin, M. and Niyogi, P. (2002) Laplacian eigenmaps for dimensionality reduction and data representation. *Neural Comput.*, 15, 1373–1396.

97 Tenenbaum, J.B., de Silva, V., and Langford, J. (2011) A global geometric framework for nonlinear dimensionality reduction. *Science*, 290, 2319–2323.

98 Dhillon, P.S., Talukdar, P.P., and Crammer, K. (2010) *University of Pennsylvania Department of Computer and Information Science Technical Report No. MS-CIS-10-18.* Inference Driven Metric Learning (IDML) for Graph Construction.

99 Blum, A. and Chawla, S. (2001) *Learning from Labeled and Unlabeled Data Using Graph Mincuts.* ICML.

100 Zhu, X. and Ghahramani, Z. (2002) Learning from Labeled and Unlabeled Data with Label Propagation. Technical Report. Center for Automated Learning and Discovery School of Computer Science, Carnegie Mellon University.

101 Zhu, X., Ghahramani, Z., and Lafferty, J.D. (2003) *Semi-Supervised Learning Using Gaussian Fields and Harmonic Functions.* ICML.

102 Zhou, D., Bousquet, O., Lal, T.N. *et al.* (2003) *Learning with Local and Global Consistency.* NIPS.

103 Belkin, M., Niyogi, P., and Sindhwani, V. (2006) Manifold regularization: a geometric framework for learning from labeled and unlabeled examples. *J. Mach. Learn. Res.*, 7, 2399–2434.

104 Weston, J., Ratle, F., Mobahi, H., and Collobert, R. (2012) *Deep Learning via Semi-supervised Embedding*, Volume 7700, Springer, pp. 639–655.

105 Kipf, T. and Welling, M. (2017) Semi-supervised classification with graph convolutional networks. *ArXiv*, abs/1609.02907.

106 Sajjadi, M., Javanmardi, M., and Tasdizen, T. (2016) *Regularization with Stochastic Transformations and Perturbations for Deep Semi-Supervised Learning.* NIPS.

107 Sajjadi, M., Javanmardi, M., and Tasdizen, T. (2016) *Mutual Exclusivity Loss for Semi-Supervised Deep Learning.* 2016 IEEE International Conference on Image Processing (ICIP), pp. 1908–1912.

108 Xie, Q., Dai, Z., Hovy, E.H. *et al.* (2019) Unsupervised data augmentation for consistency training. *arXiv: Learning.*

109 Shorten, C. and Khoshgoftaar, T.M. (2019) A survey on image data augmentation for deep learning. *J. Big Data*, 6, 1–48.

110 Perez, L. and Wang, J. (2017) The effectiveness of data augmentation in image classification using deep learning. *ArXiv*, abs/1712.04621.

111 Laine, S. and Aila, T. (2017) Temporal ensembling for semi-supervised learning. *ArXiv*, abs/1610.02242.

112 Gastaldi, X. (2017) Shake-shake regularization. *ArXiv*, abs/1705.07485.

113 Zhang, H., Cissé, M., Dauphin, Y., and Lopez-Paz, D. (2018) Mixup: beyond empirical risk minimization. *ArXiv*, abs/1710.09412.

114 Verma, V., Lamb, A., Kannala, J. *et al.* (2019) *Interpolation Consistency Training for Semi-Supervised Learning.* IJCAI.

115 Sohn, K., Berthelot, D., Li, C-L. *et al.* (2020) Fixmatch: simplifying semi-supervised learning with consistency and confidence. *ArXiv*, abs/2001.07685.

12

Random Forests

Peter Calhoun[1], Xiaogang Su[2], Kelly M. Spoon[3], Richard A. Levine[4], and Juanjuan Fan[4]

[1]Jaeb Center for Health Research, Tampa, FL, USA
[2]Department of Mathematical Sciences, University of Texas, El Paso, TX, USA
[3]Computational Science Research Center, San Diego State University, San Diego, CA, USA
[4]Department of Mathematics and Statistics, San Diego State University, San Diego, CA, USA

1 Introduction

Random forest constructs a set of decision trees through binary splits of the data to create homogeneous groups relative to data inputs. Random forest has been identified as an excellent predictive tool relative to machine learning competitors; see, for example, Caruana and Niclulescu-Mizil [1], Caruana *et al.* [2], Fernandez-Delgado *et al.* [3], and He *et al.* [4]. Random forest, along with boosting, presents as a base ensemble learner whereby predictive accuracy is gained through averaging over or voting across predictions from a collection of trees. Random forest requires few statistical assumptions and handles a variety of data structures and outcomes. The approach can handle nonlinear relationships with interactions and missing records without the need for variable transformation nor variable selection. The algorithm is relatively simple and often requires less user input than other machine learning algorithms, such as neural networks and gradient boosting trees.

In Section 2, we briefly introduce random forest, to set the notation and terminology for this chapter to sit as a self-contained piece. In Section 3, we discuss extensions on the base random forest. We have a particular eye on improving predictive accuracy and variable selection bias. We highlight three extensions: extremely randomized trees (ERT), acceptance-rejection trees (ART), and conditional random forest (CRF). We also identify key articles in the literature detailing these methods, further extensions, and relevant statistical software in the R programming environment [5]. Random forest has enjoyed successful application in many scientific disciplines under a variety of outcome types and study designs. The survival forest, as the name suggests, extends random forest to (censored) survival data; this approach has an encyclopedic entry. In Sections 4 and 5, we highlight random forest extensions to precision medicine, personalized learning, and more generally applications in which individualized treatment effects (ITEs) and the corresponding subgroup analysis are desired. To this end, we discuss random forest of interaction trees (RFIT) for both randomized clinical trials and observational studies.

Computational Statistics in Data Science.
Edited by Walter W. Piegorsch, Richard A. Levine, Hao Helen Zhang and Thomas C. M. Lee.
 ISBN 978-1-11956107-1

As part of the discussion, we again identify the literature that provides further details and additional approaches to study differential treatment effects.

2 Random Forest (RF)

2.1 RF Algorithm

Random forest, proposed by Breiman [6], combines bagging [7] and random feature selection [8–10]. A random forest is an ensemble of classification or regression trees. Tree-based methods belong among supervised learning methods because they require a response variable. A tree is referred to as a classification or regression tree depending on whether the response variable is categorical or continuous. We refer the reader to Breiman *et al.* [11] for tree-related concepts and terminology. The basic concepts of a tree as well as some terminology are also explained at the end of this section using the tree in Figure 1 based on the headache data.

A tree is grown by reiteratively splitting the data into two child nodes while minimizing the heterogeneity. Let y_i and $\mathbf{x}_i$ denote, respectively, the response and a p-dimensional covariate vector for the ith observation. Let R_t, with $t = L$ or R, represent the left or right

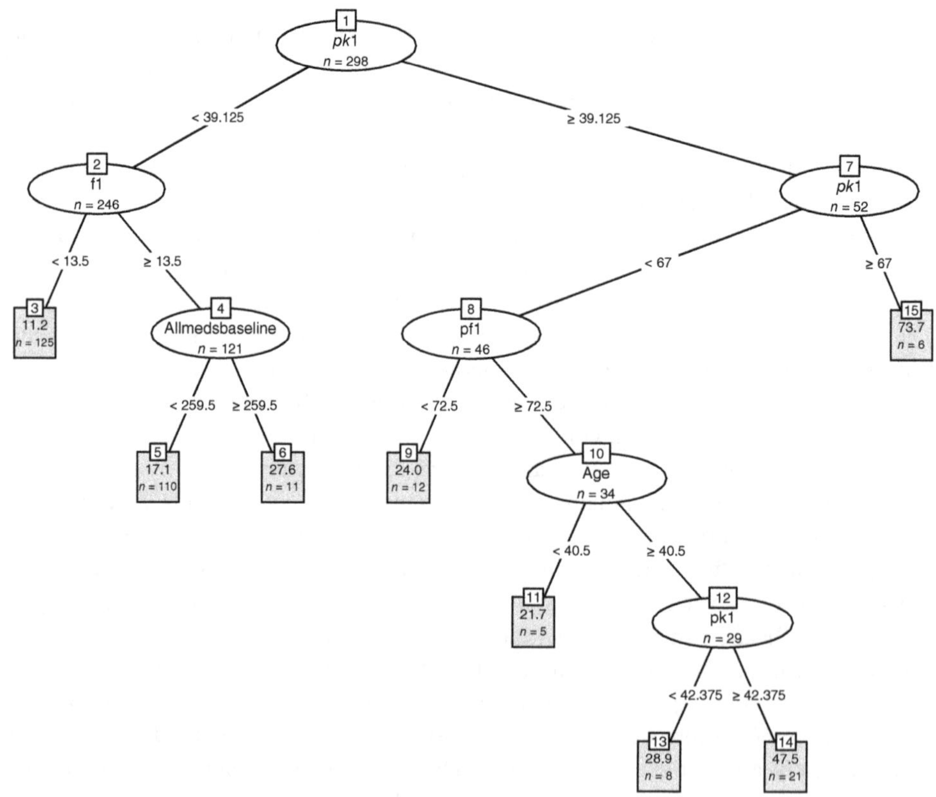

Figure 1 Decision tree for headache data.

region (child node). For the purpose of evaluating a possible split, let n be the sample size for the parent node, and n_t, with $t = L$ or R, the sample size for the left or right child node. Similarly, for regression problems, we let $\bar{y}_t$, with $t = L$ or R, be the mean response in the left or right child node, and for classification problems, we let $\hat{p}_{tk}$ be the proportion of class k observations in region R_t. The heterogeneity, often referred to as a splitting statistic, is a weighted average of the impurity measures:

$$p(R_L)i(R_L) + p(R_R)i(R_R) \tag{1}$$

where $p(R_t) = n_t/n$, with $t = L$ or R, is the proportion of observations in region R_t, and $i(R_t)$ is the impurity measure for region R_t. For regression problems, the impurity measure is often the mean squared error (MSE): $i(R_t) = 1/n_t \sum_{x_i \in R_t} (y_i - \bar{y}_t)^2$; for classification problems, this is often the Gini index: $i(R_m) = \sum_{k=1}^{K} \hat{p}_{tk}(1 - \hat{p}_{tk})$. There are many other impurity measures to split the data, such as entropy for classification problems and log-rank for survival problems. Further details regarding efficient splitting statistics and categorical predictors are described by Hastie *et al.* [12] and James *et al.* [13].

Under RF, trees are grown using bootstrap replicas or subsamples (i.e., sampling the original training set with or without replacement; observations not selected are called the out-of-bag sample). At each node, *mtry* variables are randomly selected, and the best split is determined by considering all distinct cut-points for each variable selected and taking the split that minimizes the heterogeneity, Equation (1). This process is repeated for each child node until a stopping criterion is reached, typically when few observations remain, or the node is pure. This algorithm can be implemented in the **randomForest** and **randomForestSRC** R packages [14, 15].

To illustrate tree-based methods, we consider the headache data collected from a randomized acupuncture trial [16, 17], available at https://trialsjournal.biomedcentral.com/articles/10.1186/1745-6215-7-15. We consider the 298 patients with chronic headache who were randomized to receive up to 12 acupuncture treatments over three months and completed the trial with no missing covariates. The primary outcome used here is the headache severity score at month 12. Also included in the analysis are 18 covariates, which are demographic, medical, or treatment variables measured at baseline. See Vickers [17] and Su *et al.* [18] for details. A regression tree is given in Figure 1. The root node contains all of the data, or 298 subjects in this example. The first split, which minimizes the heterogeneity measure (1) among all candidate splits, is based on the variable *pk*1 (headache severity score at baseline and higher scores indicate more severe headache), with subjects having *pk*1 scores less than 39.12 sent to the left child node (or left region), and subjects having *pk*1 scores greater than or equal to 39.12 sent to the right child node (or right region). The 246 subjects in the left child node are further split by *f*1 (headache frequency at baseline) with the cut-point at 13.5, while the 52 subjects in the right child node are further split by *pk*1 again, but this time with the cut-point at 67. Continuing with this process, every subject in the data ends up in one and only one terminal node, denoted by a rectangle in Figure 1. Inside each node are the mean severity score and the node size. This tree was built using the R package *rpart*, including a tree pruning step to avoid model overfitting. For random forest, each tree is built based on a bootstrap or subsample of the data, and the trees are typically not pruned; so, they are typically larger, with fewer subjects in terminal nodes. For prediction, new observations are sent down each tree, and depending on which terminal node they land, the node average is given as the prediction. The prediction from the random forest is the average of the predictions from all trees of the forest.

2.2 RF Advantages and Limitations

Random forest is an ensemble method that grows many trees and averages the predictions from each tree. The rationale is that using bootstrap replicas and randomly selecting variables for split rules produces diverse trees, and selecting the best split ensures that each tree is informative. However, other RF extensions have gone further by adding more randomization producing even more diverse trees – this is discussed in the following section. The goal is to reduce the variance by averaging many different trees and decrease the bias by growing only informative trees; this bias and variance trade-off is present in all supervised learning algorithms [13].

The RF algorithm is invariant to monotonic transformations of input variables. Specifically, the region formed by splitting a predictor variable X is equivalent to the region formed by splitting the predictor variable $f(X)$, where $f(\cdot)$ is a monotonic transformation. This is a useful property as users do not need to transform and standardize any input variables. RF also does not require selecting variables, although removing uninformative variables may improve accuracy. Finally, RF can utilize the out-of-bag (OOB) sample to estimate the error rate and determine variable importance. The error rate is estimated by sending each OOB observation down the tree to obtain a predicted outcome. For regression problems, the predicted response is averaged across all trees, and the overall error rate is determined. For classification problems, Breiman [6] proposed the majority vote method where the most common classification is taken as the predicted response, although a better method may be to average the classification probabilities across trees [19]. The MSE or misclassification rate is the usual error rate metrics, but other loss functions can also be used. The variable importance for a variable, X, can be determined by randomly permuting the values of X in the OOB sample and comparing the error rate before and after the permutation. This method is called the unscaled permuted variable importance, although other popular choices include scaled permuted variable importance and minimal depth [20].

A major strength of RF is the ability to handle interactions. Since the best split is found conditioning on the previous splits, the interaction of two predictor variables is explicitly assessed; this occurs for all decision trees. Growing a random forest with a depth of one will prevent any interactions, and comparing random forests with a depth of one versus two or more allows the user to assess if interactions are present. Interaction trees (ITs) are further discussed in Sections 4 and 5. Nonlinear relationships are also formed by partitioning a variable at various cut-points across its domain. The flexibility of RF to handle complex relationships and interactions demonstrates how RF often outperforms classical methods.

Despite random forest's many advantages, there remain four major barriers. First, RF favors variables with more distinct cut-points or categories. This bias is often due to the multiple comparison effect – a variable with many cut-points has more opportunities of finding a split with low heterogeneity than a variable with few cut-points. Strobl *et al.* [21] also find that RF can favor variables with more missing data and demonstrates that this is due to the bias and variance effects of the splitting statistic – Gini index favors variables with smaller sample sizes (i.e., more missing data). Second, growing many trees and using an exhaustive search to determine the best split is computationally intensive. Third, high prediction accuracy often requires optimizing parameters and may still yield worse results than other machine learning algorithms. Finally, the splitting statistic assumes that observations are independent. These barriers have led many researchers to propose modifications and extensions to the RF algorithm. The following section summarizes the more popular extensions that diminish the variable selection bias, reduce the computation time, and improve prediction accuracy. We refer to relevant articles that extend the algorithm to multivariate outcomes and repeated measurements.

3 Random Forest Extensions

Several extensions of random forest have been proposed to either alleviate the shortcomings of random forest or extend implementation to handle various outcomes. This section describes some of the popular extensions and newest ideas but does not include all of the existing algorithms. Random forest is considered a "black box" ensemble method where inputs are fitted, and accurate predictions are returned. To better understand the inner workings and interpretation, people have utilized partial dependence plots, proximity plots, variable interactions, and variable importance. However, understanding how RF achieves high predictive performance, and in particular, how changing the algorithm affects the performance, is still relatively unknown. Therefore, most of the extensions are assessed, for example, using real and synthetic data sets and demonstrating improvements in terms of prediction accuracy, computation time, memory size, or variable selection bias. Many changes appear trivial in nature but have profound effects on the algorithm's performance. The algorithms described below focus on improvements in regression and classification problems but would naturally extend to other outcomes. The first extension mentioned in the following subsection is a relatively simple modification to random forest but yields several important improvements.

3.1 Extremely Randomized Trees (ERT)

ERT is one of the more popular RF extensions [22]. The main revisions are ERT grows trees using the whole training set (instead of bootstrap replicas) and randomly selecting both the variables and cut-points at each node. Specifically, *mtry* variables are randomly selected and then *nsplit* cut-points are randomly chosen for each variable selected. The split that minimizes the heterogeneity, Equation (1), among the selected candidate splits is taken, and this process is repeated for each child node until a stopping criterion is reached (see Algorithm 1). The rationale behind ERT is that randomizing cut-points grows diverse trees that reduces the variance, while using the whole training set minimizes the bias. Several papers have found that this simple extension can yield greater predictive performance, particularly for classification problems [22, 23]. The improved accuracy of ERT is due to maximizing the individual tree strength and minimizing the correlation among trees in the forest [24]. Breiman [6] also shows that a low correlation between the individual trees yields a low upper bound of the generalization error of an ensemble. Additionally, choosing an equal number of candidate splits for each variable selected reduces the multiple comparison effect and improves computation time. Strobl *et al.* [25] illustrate a clear bias with RF where variables with many distinct values or categories are preferred, but this bias was reduced substantially when using ERT [23, 26]. Considering only a small number of candidate splits for each variable greatly improves computation time, although Geurts *et al.* [22] find that the tree depth is larger for ERT compared with random forest. This algorithm can be implemented in the **randomForestSRC** R package [15].

Algorithm 1. Algorithm for Extremely Randomized Trees (ERT)

```
1 for t = 1, 2, ..., ntree do
2     Pick mtry random variables in original training set
3     Pick nsplit random cut-points from each chosen variable
4     Calculate splitting statistic for each candidate split
5     Pick the split that minimizes the heterogeneity
6     Repeat steps 2–6 for data in each child node until a stopping criterion is reached
```

While ERT has many advantages, there are a few limitations too. First, the use of the whole training set does not provide an OOB sample that can be utilized to estimate accuracy and optimize parameters. One could employ cross-validation, but this is an extra step not needed with RF. Second, the ERT adds another parameter, *nsplit*, that may need to be optimized, although typically this is set to one. Finally, while Geurts *et al.* [22] found good performance in regression settings, Calhoun *et al.* [23] found that ERT had worse predictive performance compared with RF in all 10 regression data sets assessed with an overall MSE around 20% higher than the MSE with RF. A large-scale benchmark experiment is warranted comparing RF and ERT in regression problems.

3.2 Acceptance-Rejection Trees (ART)

Another RF extension is ART designed to improve prediction accuracy and reduce variable selection bias [23]. Under ART, trees are grown using bootstrap replicas or subsamples, and at each node, a random variable and then a random cut-point are chosen, and the candidate split is either accepted or rejected based on whether it satisfies a splitting criterion. Specifically, a p-value is calculated testing whether the two child node groups differ and the split is taken if the p-value is less than some minimum p-value threshold, *minpvalue* (see Algorithm 2). The p-value comes from a two-sample t-test for regression problems, and a chi-square test for classification problems. The rationale behind ART is similar to ERT where picking a single random variable and then cut-point grows diverse trees and reduces the multiple comparison effect, while the use of an acceptance-rejection criterion ensures that only informative trees are grown. The chosen minimum p-value threshold may need to be optimized, but the number of variables selected is no longer a tuning parameter. Calhoun *et al.* [23] find that a p-value threshold of 0.05–0.25 yields the best predictive performance for binary problems, and a threshold of 0.01 for regression problems. Importantly, the p-value threshold is used to reject bad splits but does not indicate significance.

Using a threshold of one is very similar to the totally randomized trees algorithm by Geurts *et al.* [22] and not recommended as many trees would not be informative.

Algorithm 2. Algorithm for acceptance-rejection trees (ART)

```
1 for t = 1, 2, ..., ntree do
2     Generate new training set, L_N, by sampling with replacement from the original data
3     Pick a random variable in L_N
4     Pick a random cut-point from the chosen variable
5     Calculate splitting statistic and its respective p-value
6     If p-value < minpvalue threshold, then take split; otherwise, repeat steps 3–6
7     Once a split is chosen, repeat steps 3–7 for data in each child node until a stopping
      criterion is reached
```

A comparison of 10 binary outcome data sets found that ART had the best overall accuracy compared against RF and ERT. The variable selection bias was also greatly reduced similar to ERT, and the computation time was much lower than RF and only marginally greater than ERT. This algorithm can currently only be implemented from GitHub source code [27].

The main disadvantages with ART are that it is a newer algorithm that has not been fully assessed or built using the faster C++ software, and the code does not currently include some of the features with random forest, such as the ability to handle missing data. The overall prediction accuracy for regression problems was also slightly worse for ART compared with RF. A comparison of four random forest algorithms found that the MSE for ART was lowest on 3 of 10 regression problems, with one data set reducing the error by almost 50%, but ART also had the highest error on 5 of 10 regression problems. Overall, ART is worth consideration for regression problems and a strong alternative for binary problems or variable importance.

3.3 Conditional Random Forest (CRF)

CRF takes a different approach by selecting one variable first and then choosing the best split [28]. Under CRF, trees are grown either using bootstrap replicas or subsamples, and *mtry* variables are randomly selected similar to random forest. However, at each node, the algorithm tests the global hypothesis of independence between any of the randomly selected covariates and the response. This is performed using a permutation test on a weighted mean or linear statistic by fixing the covariates and conditioning on all possible permutations of the response. The computed test statistic can be either the maximum of the absolute value of a standardized linear statistic or a quadratic statistic, and a *p*-value is calculated and compared against a minimum *p*-value threshold. If the global hypothesis is rejected, then the covariate with the smallest *p*-value is selected. Once a variable is selected, the algorithm splits the data using a two-sample linear statistic and evaluating all possible splits (see Algorithm 3). The logic for CRF is to select a single variable in an unbiased way and then take the best split to reduce the variance and produce informative trees. The global hypothesis test is an important step for a single decision tree to prevent overfitting but can also be useful in random forest to prevent growing large uninformative trees and to reduce computation time. This algorithm can be implemented in the **partykit** R package [29].

Algorithm 3. Algorithm for Conditional Random Forest (CRF)

1 **for** $t = 1, 2, \dots, ntree$ **do**
2 Generate new training set, L_N, by sampling with or without replacement from the original data
3 Pick *mtry* random variables in original training set
4 Test global hypothesis of independence using permutation test
5 If global hypothesis p-value is greater than pre-specified α, then stop partitioning node; otherwise, continue to steps 6–8
6 Select the variable with the smallest individual p-value
7 Select the best split using an exhaustive search on all distinct cut-points
8 Once a split is chosen, repeat steps 2–8 for data in each child node until a stopping criterion is reached

Strobl *et al.* [25] showed that the variable selection bias from random forest is greatly reduced when using CRF with subsampling. For this reason, the default recommendation is to use subsampling with CRF. Hothorn *et al.* [28] found that CRF had greater predictive performance compared with random forest for simulations where there was an uninformative variable with many categories.

The limitations of CRF are the computational intensity of performing a global hypothesis test, selecting a single variable at each node, and computing the best split using an exhaustive search. While the algorithm has shown good performance at variable selection, a large benchmark study is needed comparing the predictive performance and computation time of CRF and random forest. This algorithm is a strong alternative, particularly when variable selection or importance is desirable.

3.4 Miscellaneous

The three RF extensions described above share a common goal of giving each variable an equal chance of being selected (either by randomization or an unbiased selection criterion). ERT and ART also attempt to produce uncorrelated, informative trees to improve prediction accuracy. Other algorithms with the same goal exist. Random Uniform Forests (RUF [30]) is similar to ERT except that it uses bootstrap replicas or subsamples, randomly selects variables with replacement, and randomly chooses regions using a continuous uniform distribution. This often requires more computation time than ERT, and more research is needed to assess prediction accuracy; this algorithm is implemented in the **randomUniformForest** R package. Bharathidason and Jothi Venkataeswaran [24] take a more direct approach to reduce the correlation by building a random forest and removing trees with high error rates and large correlation to other trees.

Other researchers have suggested modifications to the splitting statistic. Smooth sigmoid surrogate (SSS) trees replace the indicator function with a smooth sigmoid function to determine the best split; this idea and how it can be applied for ITs is further discussed in Section 4. Ishwaran [31] evaluates weighted splitting rules and demonstrates that different weighting schemes can have major effects on performance. Strobl *et al.* [21] use the maximally selected Gini gain to provide unbiased split selection for classification problems. Oblique random forest (ORF) splits the data based on multivariate models, usually ridge regression, logistic regression, or support vector machines. This algorithm can be implemented using the **ObliqueRF** R package [32]. Finally, some researchers have made computational revisions to implement faster or memory efficient algorithms of random forest, which include R packages **ranger** and **Rborist** [33, 34].

All of the algorithms described above have focused on classification and regression problems, but random forests have been extended to handle survival [35] and ordinal outcomes [36]. These extensions require redefining the splitting statistic and modifying how the error rate, predictions, and variable importance are computed for these outcomes. Many algorithms described in Section 3 are applicable to these settings too. Additionally, random forest has been extended to handle multivariate and correlated outcomes, an important research area but outside the scope of this chapter. Importantly, using random forest ignoring the dependence can yield inaccurate predictions and flawed interpretations. The algorithms must also ensure that estimated error rates are unbiased as having repeated measurements may cause the OOB sample to be correlated with the in-bag sample. The interested reader should consider Multivariate Random Forest (MRF [37]), Random Effects/Expectation-Maximization (RE-EM) trees [38], random forest for correlated survival data [26, 39], Mixed-Effects Random Forest (MERF; [40]), and Repeated Measures Random Forest (RMRF [41]).

4 Random Forests of Interaction Trees (RFIT)

The RF methodology, together with its concomitant features or by-products such as variable importance ranking and partial dependence plots, has been extended for other uses with moderate modification. As we have seen in earlier sections, random forest is a perturb-and-combine ensemble learning method. While variants are available, data perturbation in the standard version of RF is facilitated by bootstrap resampling, where an additional step of randomly selecting the splitting variable is crucial for expediting the computation and for producing more heterogeneous and less correlated tree structures. In terms of combining results from individual learners, either averaging or majority voting is common, depending on the type of the target or response variable.

The entire engineering of the standard RF is ready for extension to other areas. One important area as such is precision medicine, where the aim is to account for the differential effects of a putative treatment and discover the optimal treatment regime.

4.1 Modified Splitting Statistic

Consider data $\{(y_i, T_i, \mathbf{x}_i) : i = 1, \ldots, n\}$ that consist of n iid copies of $(Y, T, \mathbf{X})$, where Y is an outcome variable that is assumed to be continuous, T is a 0/1-coded binary variable indicating the treatment assignment, and $\mathbf{X}$ is a p-dimensional covariate vector. Concerning the assessment of treatment effect, whether the treatment assignment is randomized or not is an important issue. We assume that a random mechanism is employed for assigning the treatment for now.

One quantity of keen interest to precision medicine is the ITE. ITE is best defined with the causal inference framework; see, for example, Rubin [42]. Let Y_1' and Y_0' denote the potential outcomes for a subject when assigned to the treated and untreated groups, respectively. The ITE of a subject with covariate vector $\mathbf{x}$ is defined as $\delta = \delta(\mathbf{x}) = E(Y_1' - Y_0' \mid \mathbf{X} = \mathbf{x})$. Random forests can be extended to make accurate estimation of ITE δ. To do so, one major modification is merely needed for the splitting rule. In the following paragraphs, we present the RFIT procedure by Su *et al.* [18], which is built on ITs [43] for subgroup analysis.

A binary split s induces the following 2×2 table, where n_{1L} denotes the number of treated subjects in the left child node, $\bar{y}_{1L}$ denotes the sample mean response for treated subjects in the left child node, and so on for notations in the other cells.

	Child node	
Treatment	**Left**	**Right**
0	$(\bar{y}_{0L}, n_{0L})$	$(\bar{y}_{0R}, n_{0R})$
1	$(\bar{y}_{1L}, n_{1L})$	$(\bar{y}_{1R}, n_{1R})$

A splitting statistic can be naturally derived from the test of $H_0: \beta_3 = 0$ in the interaction model:

$$y_i = \beta_0 + \beta_1 T_i + \beta_2 \Delta_i + \beta_3 T_i \cdot \Delta_i + \varepsilon_i \text{ with } \varepsilon_i \overset{IID}{\sim} N(0, \sigma^2) \tag{2}$$

where $\Delta_i = \Delta(x_{ij}; c) = I(x_{ij} \leq c)$ indicates the left/right child node membership. Note that $I(x_{ij} \leq c)$ is general enough to represent any split of the data regardless of the type of the covariate x_j, since a categorical covariate can be 'ordinal'ized by ranking its categories according to the estimated treatment effect within each category [43]. The least-squares estimate of β_3, given by $\hat{\beta}_3 = (\bar{y}_{1L} - \bar{y}_{0L}) - (\bar{y}_{1R} - \bar{y}_{0R})$, can be interpreted as the difference in differences (DID), corresponding to heterogeneity in treatment effects. If the Wald test is used, this amounts to

$$Q(c) = \frac{\{(\bar{y}_{1L} - \bar{y}_{0L}) - (\bar{y}_{1R} - \bar{y}_{0R})\}^2}{\hat{\sigma}^2 \, (1/n_{1L} + 1/n_{0L} + 1/n_{1R} + 1/n_{0R})} \tag{3}$$

where

$$\hat{\sigma}^2 = \frac{1}{n-4}\left(\sum_{i=1}^{n} y_i^2 - \sum_{k=0,1} \sum_{t \in \{L,R\}} n_{kt}\, \bar{y}_{kt}^2\right) \tag{4}$$

denotes the pooled estimator of σ^2.

For each covariate x_j, the best cut-off point $c^\star$ is sought via exhaustive search by maximizing $Q(c)$ over all permissible cut-off points. Nevertheless, the exhaustive search is computationally slow and yields erratic fluctuations owing to its discrete nature. Su *et al.* [18]

advocate an SSS method by replacing the indicator function Δ_i with a sigmoid function $s_i = s(x_{ij}; a, c) = [1 + \exp\{-a(x_{ij} - c)\}]^{-1}$ in model (2), where $a > 0$ is a shape parameter. For standardized x_j, a is recommended to be fixed at a value in [10, 50]. Other functions such as the rectified linear unit (ReLU) could be used as well. After the replacement, estimation of model (2) becomes a nonlinear least-squares problem, yet with multivariate decision variables. To further simplify the procedure, Su *et al.* [18] note that every component in $Q(c)$ could be written as a function of Δ_i, for example, $n_{1L} = \sum_{i=1}^{n} \Delta_i$. Therefore, an approximation of $Q(c)$ could be obtained by replacing Δ_i with s_i in each component. Let $\tilde{Q}(c)$ denote the approximated quantity, which is a function of c only. Maximizing $\tilde{Q}(c)$ with respect to c becomes a one-dimensional smooth optimization problem, which can be solved efficiently via, for example, the Brent method as implemented in the R [5] function `optimize`. Su *et al.* [18] have demonstrated the advantages of this SSS approach in both computational speed and accuracy in finding $c^{\star}$.

In terms of splitting rules, alternatives are available. Subgroup Identification based on Differential Effect Search (SIDES [44]) pursues subgroups with enhanced treatment effects, possibly taking into account both efficacy and toxicity. QUINT (QUalitative INteraction Trees [45]) tackles qualitative interactions. Loh *et al.* [46] develop splitting rules that are less prone to biased variable selection. One is referred to Lipkovich *et al.* [47] for a recent survey of these approaches.

4.2 Standard Errors

Given an individual with covariate vector $\mathbf{x}$, the standard paradigm of RF [6] is then used to obtain an ensemble estimate of the ITE $\delta(\mathbf{x})$. Take B bootstrap samples. For each bootstrap sample, a tree structure $\mathcal{T}_b$ is obtained by partitioning data with maximum $Q(c)$ and termed as an IT. Let t denote the terminal node of $\mathcal{T}_b$ that this individual with $\mathbf{x}$ falls into. An estimate of δ with $\mathcal{T}_b$ is given by $\hat{\delta}_b = \bar{y}_{1t} - \bar{y}_{0t}$. The ensemble estimate is then $\bar{\delta} = \sum_{b=1}^{B} \hat{\delta}_b / B$, the average of $\hat{\delta}_b$s.

One remarkable property of the ensemble estimator $\bar{\delta}$ is that its standard error (SE) can be obtained via an infinitesimal jackknife (IJ) approach proposed by Efron [48]. Let N_{bi} be the number of times that the ith observation appears in the bth bootstrap sample, for $i = 1, \ldots, n$ and $b = 1, \ldots, B$. Denote $\mathbf{N}_i = (N_{1i}, \ldots, N_{Bi})^T$ and $\hat{\boldsymbol{\delta}} = (\hat{\delta}_1, \ldots, \hat{\delta}_B)^T$, both being B-dimensional vectors. Let

$$\widehat{\text{cov}}(\mathbf{N}_i, \hat{\boldsymbol{\delta}}) = \frac{1}{B} \sum_{b=1}^{B} (N_{bi} - 1)(\hat{\delta}_b - \bar{\delta})$$

be the sample covariance between $\mathbf{N}_i$ and $\hat{\boldsymbol{\delta}}$. The IJ estimator of variance of $\bar{\delta}$ is given by

$$\widehat{var}(\bar{\delta}) = \sum_{i=1}^{n} \widehat{\text{cov}}(\mathbf{N}_i, \hat{\boldsymbol{\delta}})$$

The calculation of $\widehat{\text{cov}}(\mathbf{N}_i, \hat{\boldsymbol{\delta}})$ is illustrated in the table below, where the sample covariance is computed between each column and the last column. Further simplification in computing $\widehat{var}(\bar{\delta})$, as well as bias correction, is available; see Wager *et al.* [49] and Su *et al.* [18] for details.

	Observation in data				Estimated
Bootstrap	**1**	**2**	**...**	***n***	**ITE**
1	N_{11}	N_{12}	...	N_{1n}	$\hat{\delta}_1$
2	N_{21}	N_{22}	...	N_{2n}	$\hat{\delta}_2$
⋮	⋮	⋮	⋮	⋮	⋮
B	N_{B1}	N_{B2}	...	N_{Bn}	$\hat{\delta}_B$
				RFIT estimate:	$\bar{\delta}$

4.3 Concomitant Outputs

Since RF involves intensive computation and massive output from bootstrap samples, it is natural to make maximal use of the results and extract more information from the individual trees. Common by-products of RF include a proximity or distance matrix, the variable importance ranking, and partial dependence plots, which are important concomitant features that are useful for various purposes. It is worth noting that the splitting rule in RFIT is made in the spirit of goodness of split [50] by maximizing the between-node difference, which is a little different from minimizing the within-node impurity in conventional CART or RF as shown in Equation (1). Therefore, certain modifications may be necessary.

See Su *et al.* [43] for a modified procedure to obtain variable importance ranking with RFIT. There is little modification needed for obtaining the proximity matrix. An extension of partial dependence plot with RFIT is provided by Algorithm 4. The idea is to estimate ITE by varying the values of covariate X_j, while retaining the joint distribution of other covariates. To do so, a number of equal-distanced values $\{x_{mj}\}_{m=1}^{M}$ are taken from the range of X_j. Then, we replace the X_j column in the original data by each value x_{mj}, while keeping other covariates unchanged. This yields a new data set of the same size as the original data. Next, we send down this new data set to the RFIT forest and estimate ITE for each row. The average of these ITEs is computed and denoted as $\hat{\delta}_{mj}$. Then, the M pairs $\{(x_{mj}, \hat{\delta}_{mj}) : m = 1, \ldots, M\}$ are plotted against each other.

Algorithm 4. Partial Dependence Plot with RFIT

Data: Data $\mathcal{L} = \{(\mathbf{x}_i, y_i)\}_{i=1}^{n}$ and RFIT $\mathcal{F} = \{\mathcal{T}_b\}_{b=1}^{B}$
Result: Partial dependence plot for X_j

```
initialize Obtain equal-distanced points {x_mj : m = 1, ..., M} from the range of X_j ;
begin
    for m = 1 to M do
        Replace the X_j value in L with x_mj to form new data set L_m of size n;
        Send L_m down to each tree in the forest F and obtain RF-predicted ITEs {δ̂_imj}_{i=1}^n;
        Average δ̂_mj = Σ_{i=1}^n δ̂_imj / n
    Plot points (x_mj, δ̂_mj) for m = 1, ..., M.
end
```

4.4 Illustration of RFIT

To illustrate RFIT and its features, we again use the headache data [16, 17]. A basic description of the data is provided in Section 2.1. Here, the primary outcome (`pk1-pk5`) is taken as the difference in headache severity score between baseline and 12 months. Hence, a higher value indicates a larger improvement.

Figure 2a plots the estimated ITE with $B = 2000$, together with the error bars. The horizontal line in gray is the average of ITEs, yielding an estimate of the average treatment effect. Figure 2b plots the distance matrix obtained from RFIT. On the leftmost, subjects 43 and 151 are quite distant from others. These two patients are highlighted with brown diamonds in Figure 2a. Acupuncture seems to have a detrimental effect on them. On the right bottom part, subjects {248, 191, 15, 37} are clustered together. They are highlighted with triangles in Figure 2a. They represent a group of patients for whom acupuncture was highly beneficial.

Figure 3a plots the variable importance ranking. It can be seen that the baseline headache severity score `pk1` is the most important effect moderator, followed by `age` and baseline `f1`. In Figure 3b, the partial dependence plots are made for the three most important moderators. It can be seen that patients who have a high headache severity score at baseline (`pk1`) may benefit from the acupuncture treatment, similarly for the baseline `f1` score. In terms of `age`, it seems that acupuncture has a more positive effect on younger patients than older ones.

5 Random Forest of Interaction Trees for Observational Studies

In an experimental study, randomization ensures that the treatment groups would have similar distributions with respect to confounding and other background variables. However, for ethical or practical reasons, random assignment of treatment is not always feasible, in which case one may resort to observational or quasi-experimental studies. Observational studies are conducted in many fields. For example, educational interventions are often observational where students are free to choose different teaching methods or supplemental instruction classes. In public health, observational studies are often conducted when the "treatment" (e.g., smoking) is unethical to assign or when the effect of an exposure leads to a rare disease (case–control studies are observational in nature). The lack of randomization in observational studies may lead to the so-called self-selection bias where subjects with certain characteristics are more likely to be in a particular treatment group. Such imbalance in the data must be accounted for in the statistical analysis, otherwise inferences from observational studies may be biased.

5.1 Propensity Score

The propensity score is defined as the conditional probability of treatment given covariates,

$$P(T = 1|X) = E(T|X)$$

where T is the treatment indicator, while X denotes all covariates excluding the treatment indicator. The propensity score provides a scalar summary of all the covariates: under the

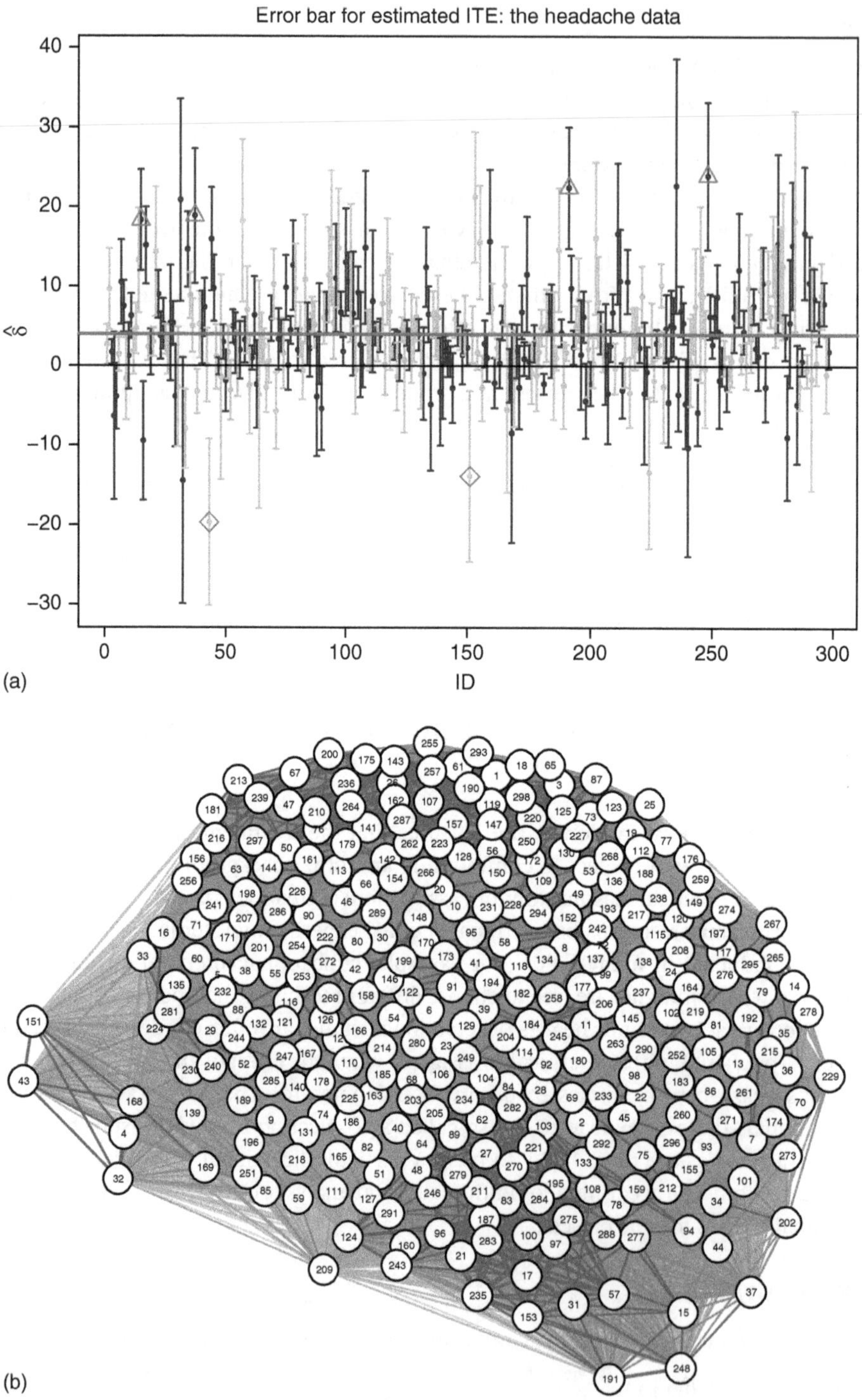

Figure 2 RFIT analysis of the headache data: (a) Estimated ITE with SE error bars; (b) Visualization of the distance matrix from RFIT via force-directed graph drawing.

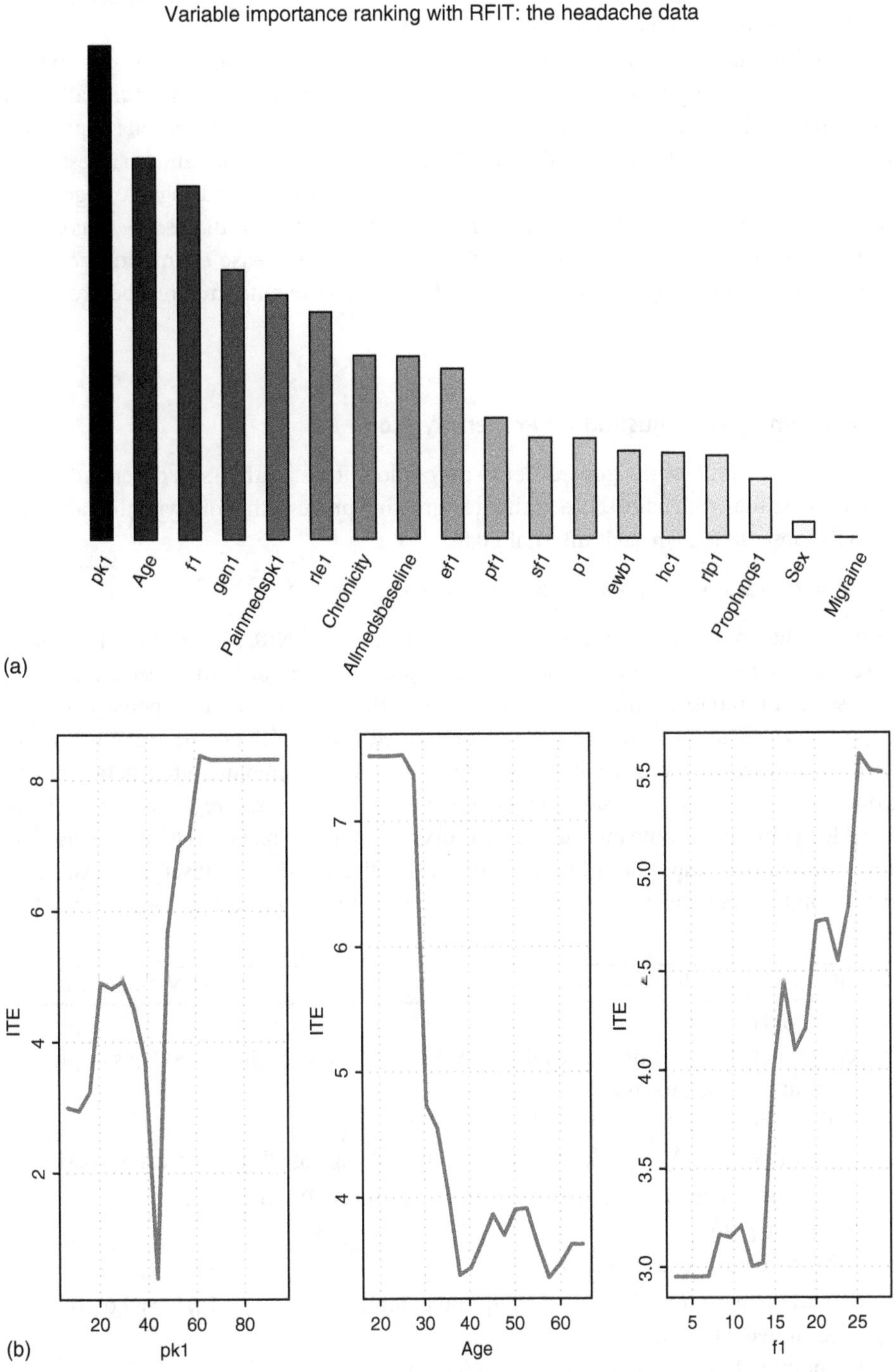

Figure 3 Exploring important effect moderators in the headache data: (a) Variable importance ranking; (b) Partial dependence plots for `pk1`, `age`, and `f1`.

assumption of strong ignorability, the distribution of X given the propensity score is balanced between the treated and control groups [51].

By including the propensity score as part of the tree-growing procedure, ITs may be applied to data from observational studies. While propensity scores have traditionally been calculated using parametric methods such as logistic regression, these methods require several assumptions about the data. Watkins *et al.* [52] used three different random forest-based methods to calculate propensity and found that they outperformed logistic regression, especially at controlling for confounding variables in observational studies. We thus choose to create our propensity scores using random forests. To increase estimation accuracy, propensity scores are computed based on all data in a separate random forest before ITs are constructed.

5.2 Random Forest Adjusting for Propensity Score

The splitting rule follows the general setup for random forest with ITs as described above. To incorporate the propensity score in the tree-growing process, the following linear model is fit to the observations in each internal node:

$$y_i = \beta_0 + \beta_1 T_i + \beta_2 \Delta_i + \beta_3 T_i \cdot \Delta_i + \beta_4 \cdot S_i + \varepsilon_i \tag{5}$$

where S_i is the propensity score for the treatment, ε_i are iid $N(0, \sigma^2)$, and $i = 1, \ldots, n$ for n observations. Since the purpose of RFIT is to predict the treatment effect for a particular observation rather than the outcome variable, the inclusion of a propensity score in the splitting rule should allow for more accurate predictions of the treatment effect for an observation. Though this model is written for a continuous outcome, other data types for the outcome variable can be accommodated. For example, a logistic regression model may be used for a binary outcome measure, and a survival model on hazard rate may be used for a time-to-event outcome. The splitting process is applied to all or a subset of the variables and cut-points under consideration, so that Breiman's random forest, as well as the other

Algorithm 5. Algorithm for variable importance within RFIT for observational data

for $b = 1$ *to* B **do**
- Let $\mathcal{T}_b$ index the tree under consideration and let L_b denote the bootstrap sample of data that was used to create tree $\mathcal{T}_b$;
- Send $\overline{L}_b$, the out-of-bag sample, down tree $\mathcal{T}_b$;
- Calculate $G(\mathcal{T}_b) = \sum_{h \in \mathcal{T}_b - \tilde{\mathcal{T}}_b} G(h)/|\mathcal{T}_b - \tilde{\mathcal{T}}_b|$, the average of all squared test statistics in the internal nodes, $\mathcal{T}_b - \tilde{\mathcal{T}}_b$, of tree $\mathcal{T}_b$, based on the OOB sample $\overline{L}_b$;
- **for** $j = 1$ *to* p **do**
  - Permute the values of $(X_j \cdot Z)$ in $\overline{L}_b$;
  - Send the permuted $\overline{L}_b$ down tree $\mathcal{T}_b$ and compute the new $G_j(\mathcal{T}_b)$ based on the permuted data set;
  - Compute $VI_j(\mathcal{T}_b) = \{G(\mathcal{T}_b) - G_j(\mathcal{T}_b)\}/G(\mathcal{T}_b)$
- Average $VI_j(\mathcal{T}_b)$ for a particular variable X_j across all B trees in the random forest

RF extensions discussed earlier, may be utilized. The variable and cut-point with the largest squared t test statistic for the coefficient β_3 on the interaction term,

$$G(h) = \max\ (t_{\beta_3})^2 \tag{6}$$

is chosen as the best split. All observations are then sent to a child node based on this best split, and the process is repeated on these child nodes until a stopping rule is satisfied.

The purpose of ITs is to divide the data into groups that are increasingly similar with respect to their treatment effect. To remove any residual confounding effects, the treatment effect needs to be estimated in each terminal node while adjusting for the propensity score. The overall treatment effect from the random forest is the average treatment effect from each tree.

5.3 Variable Importance

The advantage of a random forest over separate counterfactual models, in addition to its higher estimation accuracy, is that it allows for variable importance rankings on all the variables. To this end, we propose a variable importance algorithm, adapted from Breiman [6], but is designed to identify only those variables with strong interaction effects with treatment. Specifically, we propose permuting only the product of variable and treatment, allowing for the model to hold the main and treatment effects constant between the original data set and the permuted data set.

5.4 Simulation Study

To evaluate the performance of the propensity score addition in the IT splitting rule, we simulated a data set with 1500 observations and four covariates (X_1, X_2, X_3, and X_4). All the covariates were simulated from a discrete uniform distribution taking the following values: $\{0,\ 0.1,\ 0.2,\ \dots,\ 1.0\}$. A single covariate, X_2, was used to determine the likelihood of an observation being in the treatment group. Values in the lowest third $[0,\ 0.3]$ of the distribution of X_2 were associated with a probability of 0.1 of being in the treatment group, those in the middle third $[0.4,\ 0.6]$ with a probability of 0.5, and those in the upper third $[0.7,\ 1.0]$ with a probability of 0.9. Bernoulli draws using the previously described probability were then used to determine treatment group assignment:

$$T_i \sim \text{Bernoulli} \begin{cases} p = 0.1 & \text{if } X_2 \in [0,\ 0.3] \\ p = 0.5 & \text{if } X_2 \in [0.4,\ 0.6] \\ p = 0.9 & \text{if } X_2 \in [0.7,\ 1.0] \end{cases}$$

Two covariates, X_1 and X_2, were then used to create the outcome measure having both main and interaction effects:

$$Z_{ji} = \begin{cases} 1 & \text{if } X_{ji} \leq 0.5 \\ 0 & \text{if } X_{ji} > 0.5 \end{cases}$$

$$y_i = 10 \cdot T_i + 2 \cdot Z_{1i} + 2 \cdot Z_{2i} + 5 \cdot Z_{1i} \cdot T_i + 5 \cdot Z_{2i} \cdot T_i + \epsilon_i, \quad \epsilon_i \sim N(0,1)$$

The true treatment effect is given by: true treatment effect $= 10 + 5 \cdot Z_{1i} + 5 \cdot Z_{2i}$.

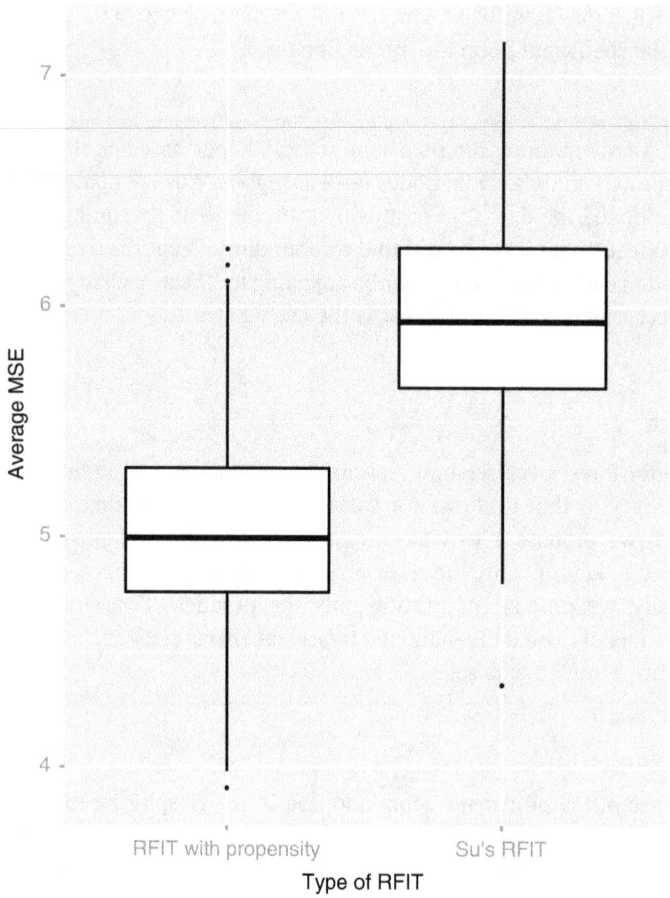

Figure 4 Comparison of MSE averaged over 1000 interaction trees using methods with and without correcting for the propensity score in the splitting rule.

The data was divided into a training set of 1000 observations and a test set of 500 observations. The training set was used to create an RFIT of 1000 ITs using both the method outlined by Su *et al.* [43] and our proposed method. The MSE for predicted treatment effects on the test set was calculated for both methods. This was repeated 250 times, and the results are displayed in Figure 4. As seen in the boxplot in Figure 4, the predictions when including the propensity score in the splitting rule model outperformed the original method proposed by Su *et al.* [43], which was designed for randomized clinical trials. For all 250 simulated samples, the RFIT with the propensity score outperformed the RFIT without the propensity score.

Several other approaches are available in extending tree models and random forests to explore differential treatment effects, especially when working with observational data. Su *et al.* [53] introduced the concept of facilitating score and proposed a causal inference tree procedure where the data are split in such a way that the simple difference of group means can be a valid estimate of the treatment effect within each terminal node. The virtual twins [54] approach estimates the two potential outcomes separately via ordinary RF. This approach is further explored by Lu *et al.* [55], where they combine results by varying different choices of tuning parameters such as *mtry* and minimum node size. Wager and Athey [56] proposed a causal forest with honest splitting rules. A comparison of causal effect random forest of interaction trees (CERFIT) incorporating the propensity score with other existing methods can be found in a submitted manuscript [57].

6 Discussion

In this chapter, we provide an introduction and comprehensive review of random forest algorithms, including the original RF algorithm by Breiman [6] and variations such as ERT by Geurts *et al.* [22], ART by Calhoun *et al.* [23], and conditional random forest by Hothorn *et al.* [28]. We discuss the advantages and limitations of these algorithms in terms of prediction accuracy for classification and regression problems, variable selection bias, computation time, and the availability of R packages and code. In addition, we provide brief discussions of other RF extensions including algorithms for survival and correlated responses. We provide references for all these extensions.

We also discuss RFIT and its applications to precision medicine. In particular, an analysis of headache data from a randomized acupuncture trial [16, 17] showcases the usefulness of the concomitant outputs from random forest. Random forest goes beyond a highly accurate black-box tool and is capable of informing the relationship between input variables and the response via variable importance rankings and partial dependence plots. The SE estimates for the ITE from RFIT allow statistical inferences for the ITE.

Random forest has also enjoyed much success in the analysis of observational data. With highly accurate estimates of the propensity score from RF, extending RFIT to learning analytics seems to be a natural idea. We provide a simulation study in this chapter to demonstrate that incorporating propensity score into the RFIT algorithm is effective in reducing the self-selection bias that is present in observational data. We are currently conducting research in this area geared toward evaluating student success programs and learning individualized treatment regimes.

References

1 Caruana, R. and Niclulescu-Mizil, A. (2006) *An Empirical Comparison of Supervised Learning Algorithms*. Proceedings of the 23rd International Conference on Machine Learning, pp. 161–168.

2 Caruana, R., Karampatziakis, N., and Yessenalina, A. (2008) An Empirical Evaluation of Supervised Learning in High Dimensions. Proceedings of the 25th International Conference on Machine Learning, pp. 96–103.

3 Fernandez-Delgado, M., Cernadas, E., Barro, S., and Amorim, D. (2014) Do we need hundreds of classifiers to solve real world classification problems? *J. Mach. Learn. Res.*, 15, 3133–3181.

4 He, L., Levine, R.A., Fan, J. *et al.* (2018) Random forest as a predictive analytics alternative to regression in institutional research. *Pract. Assess. Res. Evaluation*, 23 (1), 1–16. https://scholarworks.umass.edu/pare/vol23/iss1/1/

5 R Core Team (2020) *R: A Language and Environment for Statistical Computing*, R Foundation for Statistical Computing, Vienna, Austria. https://www.R-project.org/ (accessed 07 June 2021).

6 Breiman, L. (2001) Random forests. *Mach. Learn.*, 45, 5–32.

7 Breiman, L. (1996) Bagging predictors. *Mach. Learn.*, 24, 123–140.

8 Ho, T. (1995) *Random Decision Forest*. Proceedings of the 3rd International Conference on Document Analysis and Recognition, 1, pp. 278–282.

9 Ho, T. (1998) The random subspace method of constructing decision forests. *IEEE Trans. Patt. Anal. Mach. Intell.*, 20, 832–844.

10 Amit, Y. and Geman, D. (1997) Shape quantization and recognition with randomized trees. *Neural Comput.*, 9, 1545–1588.

11 Breiman, L., Friedman, J., Olshen, R., and Stone, C. (1984) *Classification and Regression Trees*, Wadsworth International Group, Belmont, CA.

12 Hastie, T., Tibshirani, R., and Friedman, J. (2009) *The Elements of Statistical Learning: Data Mining, Inference, and Prediction*, 2nd edn, Springer.

13 James, T., Witten, D., Hastie, T., and Friedman, J. (2013) *An Introduction to Statistical Learning with Applications in R*, Springer, New York.

14 Liaw, A. and Wiener, M. (2002) Classification and regression by randomForest. *R News*, 2 (3), 18–22.

15 Ishwaran, H. and Kogalur, U.B. (2007) Random survival forest for R. *R News*, 7 (2), 25–31.

16 Vickers, A.J., Rees, R.W., Zollman, C.E. *et al.* (2004) Acupuncture for chronic headache in primary care: large, pragmatic, randomised trial. *Br. Med. J., Primary Care*, 328, 744. doi: 10.1136/bmj.38029.421863.EB.

17 Vickers, A.J. (2006) Whose data set is it anyway? Sharing raw data from randomized trials. *Trials*, 7, 15. doi: 10.1186/1745-6215-7-15.

18 Su, X., Pena, A., Liu, L., and Levine, R. (2018) Random forests of interaction trees (RFIT) for estimating individualized treatment effects in randomized trials. *Stat. Med.*, 37 (17), 2547–2560.

19 Malley, J.D., Kruppa, J., Dasgupta, A. *et al.* (2012) Probability machines: consistent probability estimation using nonparametric learning machines. *Meth. Inf. Med.*, 51, 74–81.

20 Ishwaran, H., Kogalur, U.B., Gorodeski, E.Z. *et al.* (2015) High-dimensional variable selection for survival data. *J. Am. Stat. Assoc.*, 105, 205–217.

21 Strobl, C., Boulesteix, A., Zeileis, A., and Augustin, T. (2007) Unbiased split selection for classification trees based on Gini Index. *Comput. Stat. Data Anal.*, 52, 483–501.

22 Geurts, P., Ernst, D., and Wehenkel, L. (2006) Extremely randomized trees. *Mach. Learn.*, 63, 3–42.

23 Calhoun, P., Hallett, M.J., Su, X. *et al.* (2019) Random forest with acceptance-rejection trees. *Comput. Stat.*, 35, 983–999.

24 Bharathidason, S. and Jothi Venkataeswaran, C. (2014) Improving classification accuracy based on random forest model with uncorrelated high performing trees. *Int. J. Comp. Appl.*, 191 (13), 26–30.

25 Strobl, C., Boulesteix, A., Zeileis, A., and Hothorn, T. (2007) Bias in random forest variable importance measures: illustrations, sources and a solution. *BMC Bioinf.*, 8, 25–46.

26 Hallett, M.J., Fan, J., Levine, R.A., and Nunn, M.E. (2014) Random forest and variable importance rankings for correlated survival data, with applications to tooth loss. *Stat. Modell.*, 14 (6), 523–547.

27 Calhoun, P. (2020) ART and RMRF code. GitHub repository. https://github.com/pcalhoun1/AR-code (accessed 07 June 2021).

28 Hothorn, T., Hornik, K., and Zeileis, A. (2006) Unbiased recursive partitioning: a conditional inference framework. *J. Comput. Graph. Stat.*, 15 (3), 651–674.

29 Hothorn, T. and Zeileis, A. (2015) partykit: a modular toolkit for recursive partytioning in R. *J. Mach. Learn. Res.*, 16, 3905–3909.

30 Ciss, S. (2015) randomUniformForest: random uniform forest for classification, regression and unsupervised learning. R package. http://CRAN.R-project.org/package=randomUniformForest (accessed 07 June 2021).

31 Ishwaran, H. (2015) The effect of splitting on random forests. *Mach. Learn.*, 99, 75–118.

32 Menze, B. and Splitthoff, N. (2012) obliqueRF: oblique random forests from recursive linear model splits. R package. https://CRAN.R-project.org/package=obliqueRF (accessed 07 June 2021).

33 Wright, M.N. and Ziegler, A. (2017) ranger: a fast implementation of random forests for high dimensional data in C++ and R. *J. Stat. Soft.*, 77, 1–17.

34 Seligman, M. (2019) Rborist: extensible, parallelizable implementation of the random forest algorithm. R package. https://CRAN.R-project.org/package=Rborist (accessed 07 June 2021).

35 Ishwaran, H., Kogalur, U.B., Blackstone, E.H., and Lauer, M.S. (2008) Random survival forests. *Ann. Appl. Stat.*, 2 (3), 841–860.

36 Janitza, S., Tutz, G., and Boulesteix, A. (2016) Random forest for ordinal responses: prediction and variable selection. *Comput. Stat. Data Anal.*, 96, 57–73.

37 Segal, M. and Xiao, Y. (2011) Multivariate random forests. *WIREs Data Min. Knowl. Discovery*, 1, 80–87.

38 Sela, R.J. and Simonoff, J.S. (2012) RE-EM trees: a data mining approach for longitudinal and clustered data. *Mach. Learn.*, 86, 169–207.

39 Fan, J., Nunn, M.E., and Su, X. (2009) Multivariate exponential survival trees and their application to tooth prognosis. *Comput. Stat. Data Anal.*, 53, 1110–1121.

40 Hajjem, A., Bellevance, F., and Larocque, D. (2014) Mixed-effects random forest for clustered data. *J. Stat. Comput. Simul.*, 84, 1313–1328.

41 Calhoun, P., Levine, R.A., and Fan, J. (2021) Repeated measures random forest (RMRF): identifying factors associated with nocturnal hypoglycemia. *Biometrics*, 77 (1), 343–351.

42 Rubin, D.B. (2005) Causal inference using potential outcomes: design, modeling, decisions. *J. Am. Stat. Assoc.*, 100, 322–331.

43 Su, X., Tsai, C.-L., Wang, H. *et al.* (2009) Subgroup analysis via recursive partitioning. *J. Mach. Learn. Res.*, 10, 141–158.

44 Lipkovich, I., Dmitrienko, A., Denne, J., and Enas, G. (2011) Subgroup identification based on differential effect search (SIDES): a recursive partitioning method for establishing response to treatment in patient subpopulations. *Stat. Med.*, 30, 2601–2621.

45 Dusseldorp, E. and van Mechelen, I. (2014) Qualitative interaction trees: a tool to identify qualitative treatment-subgroup interactions. *Stat. Med.*, 33, 219–237.

46 Loh, W.-Y., He, X., and Man, M. (2015) A regression tree approach to identifying subgroups with differential treatment effects. *Stat. Med.*, 34, 1818–1833.

47 Lipkovich, I., Dmitrienko, A., and D'Agostino, R.B. (2017) Tutorial in biostatistics: data-driven subgroup identification and analysis in clinical trials. *Stat. Med.*, 36, 136–196.

48 Efron, B. (2014) Estimation and accuracy after model selection (with discussion). *J. Am. Stat. Assoc.*, 109, 991–1007.

49 Wager, S., Hastie, T., and Efron, B. (2014) Confidence intervals for random forests: the jackknife and the infinitesimal jackknife. *J. Mach. Learn. Res.*, 15, 1625—1651.

50 LeBlanc, M. and Crowley, J. (1993) Survival trees by goodness of split. *J. Am. Stat. Assoc.*, 88, 457–467.

51 Rosenbaum, P.R. and Rubin, D.B. (1983) The central role of the propensity score in observational studies for causal effects. *Biometrika*, 70, 41–55.

52 Watkins, S., Jonsson-Funk, M., Brookhart, M.A. *et al.* (2013) An empirical comparison of tree-based methods for propensity score estimation. *Health Serv. Res.*, 48 (5), 1798–1817.

53 Su, X., Kang, J., Fan, J. *et al.* (2012) Facilitating score and causal inference trees for large observational data. *J. Mach. Learn. Res. (JMLR)*, 13, 2955–2994.

54 Foster, J.C., Taylor, J.M.C., and Ruberg, S.J. (2011) Subgroup identification from randomized clinical trial data. *Stat. Med.*, 30, 2867–2880.

55 Lu, M., Sadiq, S., Feaster, D.J., and Ishwaran, H. (2018) Estimating individual treatment effect in observational data using random forest methods. *J. Comput. Graph. Stat.*, 27, 209–219.

56 Wager, S. and Athey, S. (2018) Estimation and inference of heterogeneous treatment effects using random forests. *J. Am. Stat. Assoc.*, 113, 1228–1242.

57 Li, L., Levine, R.A., Su, X., and Fan, J. (2020) Causal Effect Random Forest of Interaction Trees for Observational Data, Applied to Educational Interventions, (Submitted).

13

Network Analysis

Rong Ma and Hongzhe Li

University of Pennsylvania, Philadelphia, PA, USA

1 Introduction

Metabolic exchanges among the microbial communities play an important role in determining the microbial ecological dynamics. Microbial communities are intertwined by metabolic links as the latter determines the cellular activities with building blocks and energy. Thus, studying metabolomic interactions with microbial communities at the global level is important in the understanding of microbial ecology and evolution [1]. However, it is well known that detecting metabolite cross-links or microbe–metabolite interaction is difficult due to the intrinsically dynamic nature and the complexity of microbial communities [2, 3]. In microbiome–metabolomics studies, due to varying sequencing depths across different individual samples, the read counts are often normalized into proportions to provide quantification of microbial relative abundances. As a result, the intrinsic simplex constraint of the microbial counts data makes the analysis even more challenging. Moreover, due to the advanced sequencing technology and mass spectrometry, high-dimensional datasets are often generated that include abundance measurements of thousands of microbes and metabolites.

Compositional data, referring to quantitative representation of the proportions of fractions of some whole, appears in many applications including geochemical compositions in geology, portfolio compositions in stock marketing, and species compositions of biological communities in ecology. Instead of quantifying the actual values, compositional data only characterizes the relative amounts of the parts. Therefore, many existing statistical methods, if not examined and adjusted properly, can be inferentially unreliable when applied to compositional data [4]. In the context of microbiome–metabolomics studies, standard analysis of pairwise dependencies across microbes and metabolites often ignores the simplex constraint of the microbial relative abundances, which may lead to spurious discoveries [5]. Moreover, the compositional effects can be further magnified by the low diversity of the microbial taxa, that is, a few taxa make up the overwhelming majority of the microbiome [6].

Recently, Morton *et al.* [7] proposed a neural network approach combined with a Bayesian formulation to estimate the conditional probability that each molecule is present, given

Computational Statistics in Data Science.
Edited by Walter W. Piegorsch, Richard A. Levine, Hao Helen Zhang and Thomas C. M. Lee.
 ISBN 978-1-11956107-1

the presence of a specific microorganism. One advantage of this method is its biological interpretability, namely, the quantification of microbe–metabolite co-occurrence structure. However, it is also of interest to infer the conditional dependence of a given pair of metabolite and microbial species, given all the other metabolites and microbes. In this chapter, we adopt a graphical modeling framework to investigate the microbe–metabolite interaction network.

Graphical models [8–11] have proven to be useful for investigating the conditional dependence structure among a large number of random variables. Among them, the framework of Gaussian graphical model, where the conditional dependence structure is characterized by the inverse population covariance matrix or the precision matrix, has been widely considered for various applications. In particular, in light of the recent surge of large datasets in different fields such as genomics, finance, and social science, many interesting problems such as graphical model selection and statistical inference about the model parameters have been extensively studied under the high-dimensional setting where the number of variables p exceeds the number of observations n. For example, in Meinshausen and Bühlmann [12], a computationally efficient neighborhood selection method based on the Lasso was proposed and shown to be consistent for learning sparse high-dimensional graphs. In Refs 13 and 14, a Lasso-type ℓ_1-penalized Gaussian maximum-likelihood estimator for the precision matrix was proposed, whose theoretical guarantees under the high-dimensional setting were investigated by Ravikumar *et al.* [15]. Meanwhile, in Refs 16 and 17, based on a constrained ℓ_1 minimization approach, the so-called CLIME estimator for the high-dimensional precision matrix was proposed and carefully analyzed. More recently, other interesting problems such as large-scale hypothesis testing [18], construction of confidence intervals [19, 20], and joint estimation of multiple precision matrices [21] have been studied. See also Refs 17, 22–30 and the references therein for recent development. Nevertheless, the validity of most of the existing methods relies on some common distributional assumptions such as *i.i.d.* random variables with sub-Gaussian or polynomial tails, which are not likely to hold in the presence of compositional data.

In microbiome studies, although various methods have been developed for inferring the microbial interaction network based on compositional data [6, 31–34], their statistical properties are less well understood. Using the idea of centered log-ratio (CLR) transformation introduced by Aitchison [4], Cao *et al.* [35] proposed a consistent composition-adjusted thresholding estimator of the basis covariance matrix of the latent variables under high-dimensional settings. Despite the deep connections between the basis covariance matrix and the corresponding Gaussian graphical models, it remains unclear how to construct a consistent graphical model estimator in the presence of compositional data.

This chapter considers the problem of estimating the high-dimensional graphical models for mixed compositional and continuous data. The key idea of our approach is to apply CLR transformation to compositional data and then estimate the sparse precision matrix using the method of Cai *et al.* [16]. We present the rates of convergence of the resulting estimates under the matrix spectral norm. We observe that the estimation error is decomposed into sum of an approximation error and an estimation error. We also present an estimator that is consistent for graphical model selection. We emphasize that these theoretical properties are derived under the assumptions imposed on the distribution of true bacterial counts (after log-transformation) instead of on the compositional vector.

2 Gaussian Graphical Models for Mixed Partial Compositional Data

2.1 A Statistical Framework for Mixed Partial Compositional Data

Consider a random vector $(\mathbf{W}, \mathbf{Z}) \in \mathbb{R}^p$, where the subvectors $\mathbf{W} \in \mathbb{R}^{p_A}$ and $\mathbf{Z} \in \mathbb{R}^{p_B}$ with $p = p_A + p_B$ represent two sets of random variables of different types. Specifically, in light of our motivation about microbe–metabolite interaction network, we assume that $\mathbf{W}$ represents the true abundance of p_A bacterial taxa, which are not directly observable, but observable in terms of its relative abundances (proportions), and assume $\mathbf{Z}$ as the metabolomics measurements, whose realizations can be observed. In other words, if we have n realizations of $(\mathbf{W}, \mathbf{Z})$, for the realizations of the subvector $\mathbf{W} \in \mathbb{R}^{p_A}$, denoted as $\mathbf{W}_1, \mathbf{W}_2, \dots, \mathbf{W}_n$, we can only observe n scaled vectors $\mathbf{W'}_i = c_i\mathbf{W}_i$ for $i = 1, \dots, n$, with $c_i \in (0, 1)$ reflecting sequencing depth of the ith sample, and all c_is are unknown. In practice, $\mathbf{W}$ usually represents absolute count for some object of interest, and it is conventionally consensus to impose distributional assumption on its log-transformed version, say $\mathbf{Y} = \log(\mathbf{W})$. The other set of variables, $\mathbf{Z}$, usually in continuous measurement, is assumed to be potentially correlated with the set of variables $\mathbf{W}$ or $\mathbf{Y}$. In our applications, $\mathbf{Z}$, as the amount of different metabolites, may play important roles in regulating different aspects of the microbial community quantified by $\mathbf{W}$. It is then of great biological interest to study the pairwise dependency between different bacteria and metabolites.

Specifically, we assume that the random vector $(\mathbf{Y}, \mathbf{Z})$ follows a multivariate normal distribution $N_p(\mu, \mathbf{\Sigma}_0)$, where the covariance matrix $\mathbf{\Sigma}_0$ reflects the dependence structure among all the p components of $(\mathbf{Y}, \mathbf{Z})$. Thus, our original motivation of studying the conditional dependence structure among gut microbiome and metabolites can be translated into the problem of estimating the precision matrix $\mathbf{\Omega}_0 = \mathbf{\Sigma}_0^{-1}$ of the random vector $(\mathbf{Y}, \mathbf{Z})$, or its Gaussian graphical model, with n observations $(\mathbf{W'}_i, \mathbf{Z}_i)$, $i = 1, \dots, n$.

As argued by many works such as Refs 5, 35, and 36, researchers should work with a compositional version of $\mathbf{W'}$ instead of directly imposing distributional assumptions on $\mathbf{W'}$. In fact, it is straightforward that a compositional version of $\mathbf{W'}$ is invariant to the effects of c_i's and can be expressed as a direct transformation of $\mathbf{Y}$. Suppose that the random vector $\mathbf{X} = (X_1, \dots, X_n)^{\mathrm{T}}$ is a compositional version of $\mathbf{W'}$; in other words,

$$X_i = \frac{W'_i}{\sum_{i=1}^n W'_i} = \frac{W_i}{\sum_{i=1}^n W_i} = \frac{\exp(Y_i)}{\sum_{i=1}^n \exp(Y_i)}, \qquad \text{for all } i$$

where W'_i, W_i, and Y_i are the ith components of $\mathbf{W'}$, $\mathbf{W}$, and $\mathbf{Y}$, respectively. In the following section, we construct estimators of the precision matrix $\mathbf{\Omega}_0$ based on samples of $(\mathbf{X}, \mathbf{Z})$ under the high-dimensional setting where $p \gg n$.

Throughout our chapter, for a vector $\mathbf{a} = (a_1, \dots, a_p)^T \in \mathbb{R}^p$, we define $|\mathbf{a}| = \sum_{i=1}^p a_i$, and $|\mathbf{a}|_2 = \sqrt{\sum_{i=1}^p a_i^2}$. For a matrix $\mathbf{A} = (a_{ij}) \in \mathbb{R}^{p\times q}$, we define the spectral norm $||\mathbf{A}||_2 = \sup_{|\mathbf{x}|_2<1} |\mathbf{A}\mathbf{x}|_2$ or equivalently the largest singular value of $\mathbf{A}$. For a symmetric matrix $\mathbf{A}$, it is also the largest eigenvalue of $\mathbf{A}$. We also define the matrix l_1 norm $\|\mathbf{A}\|_{L_1} = \max_{1\le j\le q}\sum_{i=1}^{p}|a_{ij}|$, the elementwise l_1 norm $\|\mathbf{A}\|_1 = \sum_{i=1}^{p}\sum_{j=1}^{q}|a_{ij}|$, the elementwise l_∞ norm $\|\mathbf{A}\|_\infty = \max_{1\le i\le p,1\le j\le q}|a_{ij}|$, and the Frobenius norm $\|\mathbf{A}\|_F = \sqrt{\sum_{i=1}^{p}\sum_{j=1}^{q}a_{ij}^2}$. $\mathbf{I}$ is the $p\times p$ identity matrix. The notation $\mathbf{A}\succ 0$ denotes that $\mathbf{A}$ is positive definite. For any index sets $B\subseteq\{1,\dots,p\}, C\subseteq\{1,\dots,q\}$, we define the submatrix $\mathbf{A}_{BC}\in\mathbb{R}^{|B|\times|C|}$ by choosing from $\mathbf{A}$ the corresponding rows in the set B and the columns in the set C. For sequences $\{a_n\}$ and $\{b_n\}$, $a_n = o(b_n)$ means that $\lim_n a_n/b_n = 0$, and $a_n = O(b_n)$ or $a_n \lesssim b_n$ means there exists a constant C such that $a_n < Cb_n$ for all n.

2.2 Estimation of Gaussian Graphical Models of Mixed Partial Compositional Data

The covariance matrix $\mathbf{\Sigma}_0$ of the random vector $(\mathbf{Y},\mathbf{Z})\in\mathbb{R}^{p_A+p_B}$ can be partitioned into four submatrices, namely,

$$\mathbf{\Sigma}_0 = \begin{bmatrix}\mathbf{\Sigma}_{0,YY} & \mathbf{\Sigma}_{0,YZ}\\ \mathbf{\Sigma}_{0,ZY} & \mathbf{\Sigma}_{0,ZZ}\end{bmatrix}_{(p_A+p_B)\times(p_A+p_B)} \tag{1}$$

Cao et al. [35] showed that, when only the compositional version $\mathbf{X}$ of $\mathbf{Y}$ is observable, the population covariance matrix of $\mathbf{Y}$, namely, $\mathbf{\Sigma}_{0,YY}$, is not identifiable. However, they found that the covariance matrix of the CLR transformed $\mathbf{X}$, denoted in what follows as $\mathbf{\Gamma}_{0,XX}\in\mathbb{R}^{p_A\times p_A}$, can be a good approximation of $\mathbf{\Sigma}_{0,YY}$, especially under the high-dimensional sparse settings. To be specific, suppose that $\mathbf{\Gamma}_{0,XX} = (\gamma_{ij}^0)_{p_1\times p_A}$ is defined by $\gamma_{ij}^0 = \mathrm{Cov}\{\log(X_i/g(\mathbf{X}), \log(X_j/g(\mathbf{X})\}$, $i,j\in\{1,\dots,p_A\}$, where $g(\mathbf{X}) = (\Pi_{j=1}^{p_A}X_j)^{1/p_A}$, it was shown by Cao *et al.* [35] that

$$\|\mathbf{\Gamma}_{0,XX} - \mathbf{\Sigma}_{0,YY}\|_\infty \le 3p_A^{-1}\,\|\mathbf{\Sigma}_{0,YY}\|_{L_1} \tag{2}$$

The approximation is accurate when $\mathbf{\Sigma}_{0,YY}$ is sparse and the dimension p_A is relatively large.

Given the n observations $(\mathbf{X_i},\mathbf{Z_i}) = (X_{i1},\dots,X_{ip_A},Z_{i1},\dots,Z_{ip_B})^\top$, $i = 1\dots,n$, of $(\mathbf{X},\mathbf{Z})\in\mathbb{R}^{p_A+p_B}$, we define a modified sample covariance matrix by first taking the CLR transformation of the compositional data $\mathbf{X_i}$ and then compute the pairwise sample covariances. Specifically, we define

$$\hat{\mathbf{\Gamma}} = \begin{bmatrix}\hat{\mathbf{\Gamma}}_{XX} & \hat{\mathbf{\Gamma}}_{XZ}\\ \hat{\mathbf{\Gamma}}_{ZX} & \hat{\mathbf{\Gamma}}_{ZZ}\end{bmatrix}_{(p_A+p_B)\times(p_A+p_B)} \tag{3}$$

where the entries in the submatrix $\hat{\mathbf{\Gamma}}_{XX}$ are defined by $\hat{\gamma}_{ij} = \frac{1}{n}\sum_{k=1}^{n}(\gamma_{ki}-\overline{\gamma}_i)(\gamma_{kj}-\overline{\gamma}_j)$ for $i,j\in\{1,\dots,p_A\}$, with $\gamma_{ki} = \log(X_{ki}/g(\mathbf{X_k}))$ and $\overline{\gamma}_i = n^{-1}\sum_{k=1}^{n}\gamma_{ki}$; the entries in the submatrix $\hat{\mathbf{\Gamma}}_{XZ}$ are defined by $\hat{\gamma}_{ij} = \frac{1}{n}\sum_{k=1}^{n}(\gamma_{ki}-\overline{\gamma}_i)(Z_{kj-p_A}-\overline{Z}_{j-p_A})$ for $i\in\{1,\dots,p_A\}, j\in\{p_A+1,\dots,p_A+p_B\}$; the submatrix $\hat{\mathbf{\Gamma}}_{ZX} = \hat{\mathbf{\Gamma}}_{XZ}^\top$; and the entries in the submatrix $\hat{\mathbf{\Gamma}}_{ZZ}$ are defined by $\hat{\gamma}_{ij} = \frac{1}{n}\sum_{k=1}^{n}(Z_{k,i-p_A}-\overline{Z}_{i-p_A})(Z_{kj-p_1}-\overline{Z}_{j-p_A})$ for $i,j\in\{p_A+1,\dots,p_A+p_B\}$.

In Cao *et al.* [35], it was shown that a consistent sparse estimator of $\mathbf{\Sigma}_{0,YY}$ can be constructed based on $\hat{\mathbf{\Gamma}}_{XX}$. As a result, following the ideas of Refs 16 and 17 where a rate-optimal sparse precision matrix estimator (named as CLIME) was obtained by solving a constrained l_1 optimization problem based on the sample covariance matrix, we propose an estimator for the precision matrix $\mathbf{\Omega}_0$ of the mixed random variables $(\mathbf{Y}, \mathbf{Z})$ by solving the following optimization problem:

$$\min \|\mathbf{\Omega}\|_1 \qquad \text{subject to: } \|\hat{\mathbf{\Gamma}}\mathbf{\Omega} - \mathbf{I}\|_\infty \leq \lambda_n \tag{4}$$

The intuition behind the above formulation is that, if $\hat{\mathbf{\Gamma}}$ is a good estimate of $\mathbf{\Sigma}_0$, for some appropriately chosen small λ_n, the solution to (4) should be sufficiently close to the true precision matrix $\mathbf{\Omega}_0$. Suppose that the solution to (4) is $\hat{\mathbf{\Omega}}_1$, which is not necessarily symmetric, then our proposed estimator $\hat{\mathbf{\Omega}}$ for the precision matrix $\mathbf{\Omega}_0$ is defined by choosing the smaller one from every pair of symmetric entries in $\hat{\mathbf{\Omega}}_1$ as the value for both of the corresponding entries in $\hat{\mathbf{\Omega}}$. In other words, we define

$$\hat{\mathbf{\Omega}} = (\hat{\omega}_{ij})_{1\leq i,j\leq p}, \quad \text{where } \hat{\omega}_{ij} = \hat{\omega}_{ji} = \hat{\omega}^1_{ij} I\{\hat{\omega}^1_{ij} \leq \hat{\omega}^1_{ji}\} + \hat{\omega}^1_{ji} I\{\hat{\omega}^1_{ij} > \hat{\omega}^1_{ji}\} \tag{5}$$

Our final estimator $\hat{\mathbf{\Omega}}$ is symmetric, and the results in Section 3 indicate that it is positive definite with high probability.

Computationally, Cai *et al.* [16] showed that (4) is equivalent to the following sets of convex optimization problems:

$$\min|\beta|_1 \qquad \text{subject to } |\hat{\mathbf{\Gamma}}\beta - e_i|_\infty \leq \lambda_n, \quad i = 1, \ldots, p \tag{6}$$

where $\beta \in \mathbb{R}^p$, and $\{e_i\}_{i=1}^p$ is the canonical basis of the Euclidean space $\mathbb{R}^p$. Therefore, the solution to the above optimization problems $\{(\hat{\beta}_1, \hat{\beta}_2, \ldots, \hat{\beta}_p)\}$ is exactly $\hat{\mathbf{\Omega}}_1$. This fact would make both the implementation and the theoretical analysis of our procedure much easier. Moreover, for the selection of the tuning parameter λ_n, we recommend the existing approaches such as cross-validation and stability selection [37] as implemented by the R package `flare`. See also Section 5.

Alternatively, we can also consider solving the following Lasso-type problem:

$$\text{minimize} \left\{ \text{tr}(\hat{\mathbf{\Gamma}}_p \mathbf{\Omega}) - \log|\mathbf{\Omega}| + \lambda \sum_{j\neq k} |\omega_{jk}| \right\}, \qquad \text{where } \mathbf{\Omega} \geq 0 \tag{7}$$

with $\hat{\mathbf{\Gamma}}_p = \arg\min_{\mathbf{\Gamma}\geq 0} \|\hat{\mathbf{\Gamma}} - \mathbf{\Gamma}\|_\infty$. The estimator obtained from the above optimization problem is usually referred as the graphical Lasso estimator, which is based on the ℓ_1-penalized maximum-likelihood formulation [12–14]. Numerically, under the Gaussian graphical model, the graphical Lasso approach (7) would provide a sparse estimator having similar performance as the one from (4) [16]. Nevertheless, the following discussions will focus on the proposed estimator (5). The theoretical results for the graphical Lasso estimator can be derived similarly by following Ravikumar *et al.* [15], but under a more complicated "irrepresentable condition."

3 Theoretical Properties

In this section, we provide theoretical properties of our proposed precision matrix estimator $\hat{\mathbf{\Omega}}$, including the consistency and the rates of convergence under the spectral norm, l_1 norm, and l_∞ norm, in both probabilistic and risk-bound arguments. We also show that as long as the approximation error is negligible in comparison with the estimation error, our proposed estimator is minimax rate optimal.

3.1 Assumptions

We start with introducing the theoretical assumptions needed for our subsequent results. For simplicity of presentation, we denote A and B as index sets for the variables belonging to the compositional part and the general continuous part, respectively. In other words, $A = \{1, 2, \dots, p_A\}$, $B = \{p_A + 1, \dots, p_A + p_B\}$, and apparently $p = p_A + p_B = |A| + |B| = |A \cup B|$.

The assumptions are:

(C1) (Exponential-type tails) Let $\log p = o(n)$, and there exist $\alpha > 0$ and K such that

$$\max_{i \in A \cup B} \mathbb{E} \exp\{\alpha (Y_i - \mu_i)^2\} \leq K < \infty$$

without loss of generality, we take $\mu_i = 0$, $i = 1, \dots, p$ for our analysis.

(C2) The basis covariance matrix $\mathbf{\Sigma}_0$ belongs to the class

$$\mathcal{U}(q_1, s_1(p_A), M_1) = \left\{ \mathbf{\Sigma} : \mathbf{\Sigma} \succ 0, \max_{i \in A \cup B} \sigma_{ii} \leq M_1, \max_{i \in A \cup B} \sum_{j \in A} |\sigma_{ij}|^{q_1} \leq s_1(p_A) \right\}$$

where $s_1(p_A) = o(p_A)$, $0 \leq q_1 < 1$, and M_1 is some constant.

(C3) The basis precision matrix $\mathbf{\Omega}_0$ belongs to the class

$$\mathcal{V}(q_2, s_2(p), M_2) = \left\{ \mathbf{\Omega} : \mathbf{\Omega} \succ 0, \|\mathbf{\Omega}\|_{L_1} \leq M_2, \max_{i \in A \cup B} \sum_{j=1}^{p} |\omega_{ij}|^{q_2} \leq s_2(p) \right\}$$

where $s_2(p) = o(p)$, $0 \leq q_2 < 1$, and M_2 is some constant.

(C4) The proportion of compositional variables is a constant asymptotically, that is, $p_A/p \to c \in (0, 1)$.

In conditions (C2) and (C3), weak sparsity assumptions are imposed on both the basis covariance matrix and the precision matrix, which can be restrictive in some applications. However, we note that condition (C2) can actually be replaced by the following condition, imposed on the class of basis covariance matrices:

(C5) $$\mathcal{U}(s_1(p_A)) = \left\{ \mathbf{\Sigma} : \mathbf{\Sigma} \succ 0, \max_{i \in B} \left| \sum_{j \in A} \sigma_{ij} \right| \leq s_1(p_A) \right\}, \quad \text{where } s_1(p_A) = o(p_A)$$

When (C2) is replaced by (C5), all of the theoretical properties still hold only with some minor changes of the proof details. Comparing with (C2), one advantage of (C5) is that it does not necessarily enforce sparsity on the basis covariance matrix. In addition, our simulation study in Section 5 also shows good performance of our proposed estimator when the covariance matrix is dense.

3.2 Rates of Convergence

We first state the rate of convergence for our proposed estimator $\hat{\mathbf{\Omega}}$ under the spectral norm.

Theorem 1. *Under (C1), (C2), and (C3), let $\lambda_n = C_1\sqrt{\log p/n} + C_2 s_1(p_A)/p_A$. Then*

$$\|\hat{\mathbf{\Omega}} - \mathbf{\Omega}_0\|_2 \leq C_3 s_2(p) \left(C_1 \sqrt{\frac{\log p}{n}} + C_2 \frac{s_1(p_A)}{p_A} \right)^{1-q_2}$$

holds with probability greater than $1 - O(p^{-C})$, *where* C_1, C_2, C_3, *and* C *are positive constants depending only on* α, K, q_1, q_2, M_1, *and* M_2.

There are two general components in the above rate of convergence. The first one is due to the estimation error, whereas the second one is due to the approximation bias introduced by the CLR transformation. When the second component $s_1(p_A)/p_A$ is negligible compared to the estimation error $\sqrt{\log p/n}$, or

$$\log p \lesssim n \lesssim \frac{p_A^2 \log p}{s_1^2(p_A)}$$

the rate of convergence of our estimator would be the same as that of the standard CLIME estimator in Cai *et al.* [16]. It is promising in the sense that as long as the approximation procedure is guaranteed to be accurate, the estimation procedure would recover the rate of convergence as if we had the compositional data in its original scale, that is, the unobserved Y_is. Moreover, when M_2 does not depend on n, p, and $s_1(p_A)/p_A = o(\sqrt{\log p/n})$, the rate of convergence is of the order $s_2(p)(\log p/n)^{(1-q_2)/2}$, which matches the minimax lower bound under the spectral norm within the class of precision matrices $\mathcal{V}(q_2, s_2(p), M_2)$ as shown in Cai *et al.* [17], and in this case our proposed estimator is minimax rate optimal.

Next, we provide the expected rate of convergence $\sup_{\boldsymbol{\Omega}_0 \in \mathcal{U}} \mathbb{E}\, \|\hat{\boldsymbol{\Omega}} - \boldsymbol{\Omega}_0\|_2^2$ under the same sets of conditions. Since in general the expectation of $\|\hat{\boldsymbol{\Omega}} - \boldsymbol{\Omega}_0\|_2^2$ may not exist, we modify the problem in the spirit of Cai *et al.* [16]. Let $\{\hat{\boldsymbol{\Omega}}_{1\rho}\}$ be the solution to the optimization problem:

$$\min \|\boldsymbol{\Omega}\|_1 \qquad \text{subject to: } |\hat{\boldsymbol{\Gamma}}_\rho \boldsymbol{\Omega} - \mathbf{I}|_\infty \le \lambda_n \tag{8}$$

where $\hat{\boldsymbol{\Gamma}}_\rho = \hat{\boldsymbol{\Gamma}} + \rho \mathbf{I}$ with $\rho > 0$. Similarly, we symmetrize the initial solution $\{\hat{\boldsymbol{\Omega}}_{1\rho}\}$ to get our final estimator $\{\hat{\boldsymbol{\Omega}}_\rho\}$, whose expectation under spectral norm $\mathbb{E}\, \|\hat{\boldsymbol{\Omega}}_\rho - \boldsymbol{\Omega}_0\|_2^2$ is well defined. Therefore, we have the following result.

Theorem 2. *Under (C1), (C2), and (C3), let* $\lambda_n = C_1\sqrt{\log p/n} + C_2 s_1(p_A)/p_A$ *and* $\rho = \sqrt{\log p/n}$. *If* $p \ge n^\xi$ *for some* $\xi > 0$, *then*

$$\sup_{\boldsymbol{\Omega}_0 \in \mathcal{U}} \mathbb{E}\, \|\hat{\boldsymbol{\Omega}}_\rho - \boldsymbol{\Omega}_0\|_2^2 \le C_3 s_2^2(p) \left(C_1\sqrt{\frac{\log p}{n}} + C_2 \frac{s_1(p_A)}{p_A} \right)^{2-2q_2}$$

where C_1, C_2, *and* C_3 *are positive constants depending only on* α, K, q_1, q_2, M_1, *and* M_2.

In addition to the spectral norm considered above, similar results can be obtained for the matrix l_∞ norm and the Frobenius norm.

Theorem 3. *Under the conditions of Theorem 1, we have*

$$|\hat{\boldsymbol{\Omega}} - \boldsymbol{\Omega}_0|_\infty \le C_1\sqrt{\frac{\log p}{n}} + C_2 \frac{s_1(p_A)}{p_A}$$

$$\frac{1}{p}\, \|\hat{\boldsymbol{\Omega}} - \boldsymbol{\Omega}_0\|_F^2 \le C_3 s_2(p) \left(C_1\sqrt{\frac{\log p}{n}} + C_2 \frac{s_1(p_A)}{p_A} \right)^{2-q_2}$$

with probability at least $1 - O(p^{-C})$*, where* $C,\ C_1,\ C_2,$ *and* C_3 *are positive constants depending only on* $\alpha,\ K,\ q_1,\ q_2,\ M_1,$ *and* M_2.

Theorem 4. *Under the conditions of Theorem 2, we have*

$$\sup_{\Omega_0 \in \mathcal{U}} \mathbb{E}|\hat{\Omega}_\rho - \Omega_0|_\infty^2 \le C_3 \left(C_1 \sqrt{\frac{\log p}{n}} + C_2 \frac{s_1(p_A)}{p_A} \right)^2$$

$$\frac{1}{p} \sup_{\Omega_0 \in \mathcal{U}} \mathbb{E} \,\|\hat{\Omega}_\rho - \Omega_0\|_F^2 \le C_3 s_2(p) \left(C_1 \sqrt{\frac{\log p}{n}} + C_2 \frac{s_1(p_A)}{p_A} \right)^{2-q_2}$$

where $C_1,\ C_2,$ *and* C_3 *are positive constants depending only on* $\alpha,\ K,\ q_1,\ q_2,\ M_1,$ *and* M_2.

In Theorems 3 and 4, again, when the approximation error $s_1(p_A)/p_A$ is negligible in comparison with the estimation error $\sqrt{\log p/n}$, the rates of convergence of our proposed estimator are minimax optimal over the class of precision matrices $\mathcal{V}(q_2, s_2(p), M_2)$. In particular, the results concerning entrywise l_∞ norm will be used to construct the estimator with consistent graphical model selection property.

4 Graphical Model Selection

A good estimate of precision matrix can be useful in graphical model selection. However, extra care should be taken to obtain a consistent estimator of the support or the sign matrix of the true precision matrix. Denote the support of a matrix $A \in \mathbb{R}^{p\times p}$ as $S(A) = \{(i,j) : a_{ij} \ne 0, 1 \le i,j \le p\}$ and the sign matrix as $\mathcal{M}(A) = \{\mathrm{sgn}(a_{ij}), 1 \le i,j \le p\}$. To construct a precision matrix estimator that is consistent in terms of graphical model selection, we propose a hard-thresholding estimator $\tilde{\Omega} = \{\tilde{\omega}_{ij}\}$ based on $\hat{\Omega}$ in (5) as $\tilde{\omega}_{ij} = \hat{\omega}_{ij} I\{|\hat{\omega}_{ij}| \ge \tau_n\}$, where $\tau_n \ge 4M_2\lambda_n$ is a tuning parameter, and λ_n is given in Theorem 1.

Define $\mathcal{M}(\tilde{\Omega}) = \{\mathrm{sgn}(\tilde{\omega}_{ij}), 1 \le i,j \le p\}$, $\mathcal{M}(\Omega_0) = \{\mathrm{sgn}(\omega_{ij}^0), 1 \le i,j \le p\}$, $S(\Omega_0) = \{(i,j) : \omega_{ij}^0 \ne 0\}$, and $\theta_{\min} = \min_{(i,j)\in S(\Omega_0)} |\omega_{ij}^0|$. The following theorem shows that $\tilde{\Omega}$ is consistent for graphical model selection.

Theorem 5. *Under (C1), (C2), and (C3), if* $\theta_{\min} > 2\tau_n$*, then we have* $\mathcal{M}(\tilde{\Omega}) = \mathcal{M}(\Omega_0)$ *with probability at least* $1 - O(p^{-C})$ *for some constant* $C > 0$.

5 Analysis of a Microbiome–Metabolomics Data

We illustrate our proposed methods by analyzing a dataset from the Pediatric Longitudinal Study of Elemental Diet and Stool Microbiome Composition (PLEASE) study, a prospective cohort study to investigate the effects of inflammation, antibiotics, and diet as environmental stressors on the gut microbiome in pediatric Crohn's disease [38–40].

Our dataset contains both gut microbiome and fecal metabolite data from a set of 90 pediatric patients with Crohn's disease at baseline. To obtain the microbial relative abundance measurements, shotgun metagenomic sequencing was applied to the stool samples of each subject, leading to compositional data of 45 relative common bacterial genera for each sample. In addition, 335 different known metabolites were also measured on each subject at the baseline. We first filtered out the samples for whom only 10 or fewer bacteria were captured from the metagenomic sequencing and eliminated the bacterial genera that were observed in 30 or fewer individuals. For the metabolite measurements, the abundance for each metabolite was normalized so as to be approximately normally distributed across individuals. In our analysis, we only kept the biologically relevant metabolites that include amino acids, growth factors, and nucleosides. After the sample and metabolite filtering, the final dataset includes a total of 81 distinct patients, with 25 bacteria genera and 189 metabolites to be considered in the downstream network analysis. For the samples with zero count for certain bacterial genera, we imputed the proportions using nonparametric missing value imputation with random forests, as implemented in the R package `missForest`. After data imputation, the zero counts were in general substituted by some small positive numbers.

Figure 1 shows the microbe–metabolite interaction network based on the Gaussian graphical model estimated by our proposed estimator (5), where the selection of tuning parameter is based on stability selection method implemented in the R package `flare`. Since we are mainly interested in the microbe–metabolite interaction network, only such interactions are plotted in Figure 1, where the IDs and the names of the corresponding metabolites are given in Table 1. To better visualize the identified interactions, Figure 2 shows the marginal correlations between pairs of linked microbe and metabolite through scatter plots, where the levels of the metabolite are plotted against the central log-ratio-transformed microbial abundances, and from which both moderate and strong marginal correlations are discernible.

One interesting observation from the microbe–metabolite interaction network shown in Figure 1 is that most of the metabolites are associated with very few bacteria, and a given bacterium is associated with a small set of metabolites. This suggests relative simple associations between bacteria and metabolites in the guts of patients with Crohn's disease. However, despite the growing literature surrounding this topic [41, 42], existing results concerning the microbial metabolism are still relatively scarce, and more empirical evidences are needed in order to claim scientific validity of our numerical findings. In particular, it would be interesting to replicate such findings in other datasets.

6 Discussion

This chapter presents a method of estimating the conditional dependence structure across a set of compositional variables and continuous variables. The key idea of our method is based on the CLR transformation of compositional variables. By considering a convex optimization problem with the transformed data, we are able to estimate the latent sparse precision

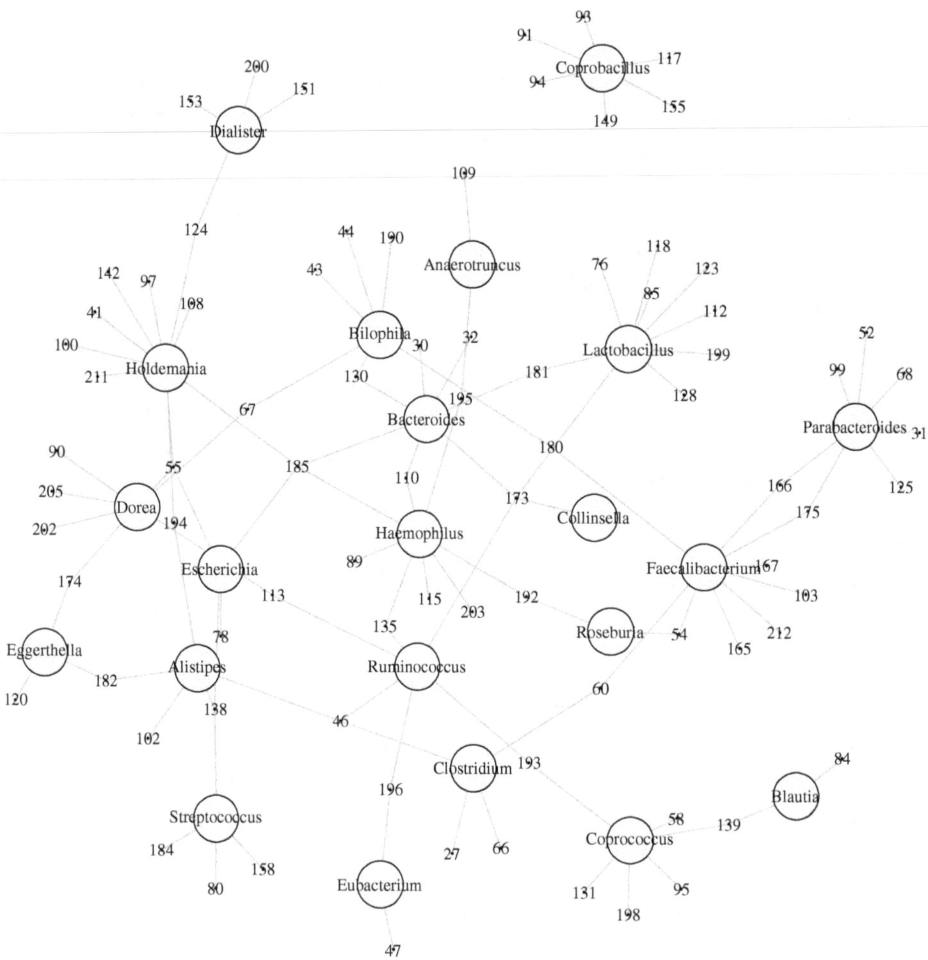

Figure 1 The metabolite–microbe interaction network. Only edges linking a metabolite and a microbe are presented, the IDs and the names of the corresponding metabolites are given in Table 1.

matrix of the basis variables under the assumptions of sparsity and high dimensionality. The method provides one solution to investigate microbe–metabolite interaction network in microbiome studies.

Excessive zeros in microbial compositions are often observed in microbiome studies, which may be due to undersampling or low sequencing depth or absence of bacteria in a given sample. Such zeros complicated the CLR transformation that we use in this chapter. Assuming that all zeros are due to undersampling, various imputation or model-based methods have been developed for estimating the compositions [5, 43]. It is an important future topic to develop flexible methods that can handle such excessive zeros in studying the microbe–metabolite interactions.

Table 1 List of metabolites shown in Figure 1.

Index	ID+Name	Index	ID+Name
27	HMDB11134 Palmitoleic.acid	30	HMDB01043 Linoleic.acid
31	HMDB00518 Oleic.acid	32	HMDB00619 Stearic.acid
41	HMDB03876 15-HETE	43	HMDB03073 11-HETE
44	HMDB05089 LTB4	46	HMDB01483 PGF2
47	HMDB00277 Sphingosine.1-Phosphate	52	HMDB00806 Glycolithocholic.acid
54	HMDB00220 Glycodeoxycholic.acid	55	HMDB03229 Glycoursodeoxycholic.acid
58	HMDB00036 Taurochenodesoxycholic.acid	60	HMDB00874 Taurohyodeoxycholic.acid
66	HMDB00101 2-deoxyadenosine	67	HMDB00014 2-deoxycytidine
68	HMDB01476 3-hydroxyanthranilic.acid	76	HMDB00462 allantoin
78	HMDB01906 aminoisobutyric.acid	80	HMDB01123 anthranilic.acid
84	HMDB00056 beta-alanine	85	HMDB00043 betaine
89	HMDB00904 citrulline	90	HMDB01046 cotinine
91	HMDB00064 creatine	93	HMDB00574 cysteine
94	HMDB00630 cytosine	95	HMDB00092 dimethylglycine
97	HMDB03345 glucose	99	HMDB00641 glutamine
100	HMDB00131 glycerol	102	HMDB00870 histamine
103	HMDB00177 histidine	108	HMDB00182 lysine
109	HMDB00696 methionine	110	HMDB02005 methionine.sulfoxide
112	HMDB00026 N-carbamoyl-beta-alanine	113	HMDB01406 niacinamide
115	HMDB00214 ornithine	117	HMDB00716 pipecolic.acid
118	HMDB00162 proline	120	HMDB00267 pyroglutamic.acid
123	HMDB00187 serine	124	HMDB00259 serotonin
125	HMDB01257 spermidine	128	HMDB00167 threonine
130	HMDB00248 thyroxine	131	HMDB00925 trimethylamine-N-oxide
135	HMDB00299 xanthosine	138	HMDB00824 C3.carnitine
139	HMDB02095 C3-DC.carnitine	142	HMDB13127 C4-OH.carnitine
149	HMDB00651 C10.carnitine	151	HMDB02250 C12.carnitine
153	HMDB05066 C14.carnitine	155	HMDB13331 C14:2.carnitine
158	HMDB00848 C18.carnitine	165	HMDB00510 2-aminoadipate
166	HMDB00017 4-pyridoxate	167	HMDB00072 aconitate
173	HMDB00124 hexose.monophosphate	174	HMDB00121 folate
175	HMDB00122 fructose/glucose/galactose	180	HMDB00714 hippurate
181	HMDB00676 homocystine	182	HMDB00130 homogentisate
184	HMDB00157 hypoxanthine	185	HMDB00195 inosine
190	HMDB00247 sorbitol	192	HMDB00262 thymine
193	HMDB00300 uracil	194	HMDB00289 urate
195	HMDB00296 uridine	196	HMDB00292 xanthine
198	HMDB00138 glycocholate	199	HMDB00036 taurocholate
200	HMDB00631 glycodeoxycholate	202	HMDB00626 deoxycholate
203	HMDB00893 suberate	205	HMDB00020 hydroxyphenylacetate
211	HMDB00694 2-hydroxyglutarate	212	HMDB00211 inositol

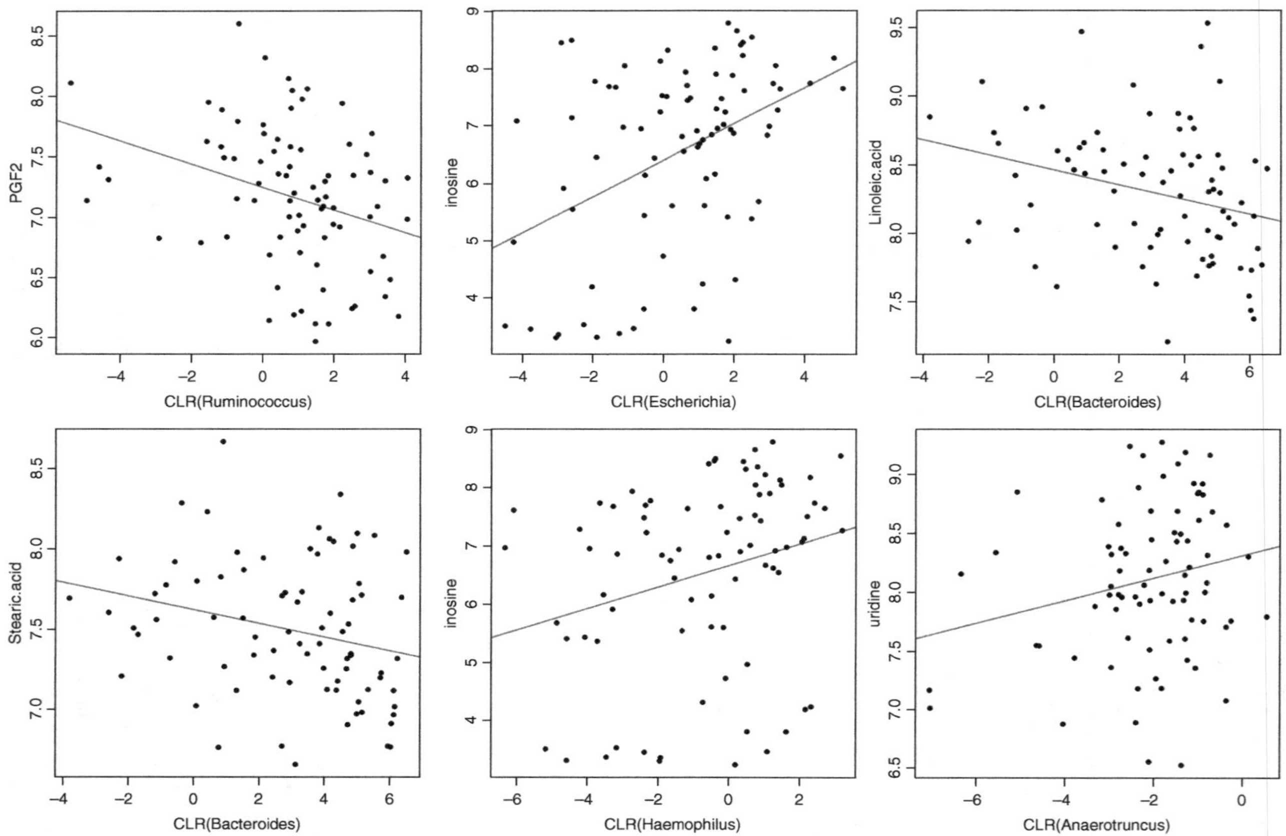

Figure 2 Scatter plots of microbe and metabolite pairs.

References

1 Chubukov, V., Gerosa, L., Kochanowski, K., and Sauer, U. (2014) Coordination of microbial metabolism. *Nat. Rev. Microbiol.*, 12 (5), 327–340.

2 Tang, J. (2011) Microbial metabolomics. *Curr. Genom.*, 12 (6), 391–403.

3 Ponomarova, O. and Patil, K.R. (2015) Metabolic interactions in microbial communities: untangling the gordian knot. *Curr. Opin. Microbiol.*, 27, 37–44.

4 Aitchison, J. (1982) The statistical analysis of compositional data. *J. R. Stat. Soc. Ser. B Methodol.*, 44, 139–177.

5 Cao, Y., Zhang, A., and Li, H. (2020) Multi-sample estimation of bacterial composition matrix in metagenomics data. *Biometrika*, 107 (1), 75–92.

6 Friedman, J. and Alm, E.J. (2012) Inferring correlation networks from genomic survey data. *PLoS Comput. Biol.*, 8 (9), e1002687.

7 Morton, J.T., Aksenov, A.A., Nothias, L.F. *et al.* (2019) Learning representations of microbe–metabolite interactions. *Nat. Meth.*, 16 (2), 1306–1314.

8 Lauritzen, S.L. (1996) *Graphical Models*, vol. 17, Clarendon Press.

9 Jordan, M.I. (2004) Graphical models. *Stat. Sci.*, 19 (1), 140–155.

10 Loh, P.-L. and Wainwright, M.J. (2013) Structure estimation for discrete graphical models: generalized covariance matrices and their inverses. *Ann. Stat.*, 41 (6), 3022–3049.

11 Drton, M. and Maathuis, M.H. (2017) Structure learning in graphical modeling. *Annu. Rev. Stat. Appl.*, 4, 365–393.

12 Meinshausen, N. and Bühlmann, P. (2006) High-dimensional graphs and variable selection with the lasso. *Ann. Stat.*, 34, 1436–1462.

13 Yuan, M. and Lin, Y. (2007) Model selection and estimation in the gaussian graphical model. *Biometrika*, 94, 19–35.

14 Friedman, J., Hastie, T., and Tibshirani, R. (2008) Sparse inverse covariance estimation with the graphical lasso. *Biostatistics*, 9 (3), 432–441.

15 Ravikumar, P., Wainwright, M.J., Raskutti, G., and Yu, B. (2011) High-dimensional covariance estimation by minimizing l_1-penalized log-determinant divergence. *Electron. J. Stat.*, 5, 935–980.

16 Cai, T., Liu, W., and Luo, X. (2011) A constrained l_1 minimization approach to sparse precision matrix estimation. *J. Am. Stat. Assoc.*, 106 (494), 594–607.

17 Cai, T.T., Ren, Z., and Zhou, H.H. (2016) Estimating structured high-dimensional covariance and precision matrices: optimal rates and adaptive estimation. *Electron. J. Stat.*, 10 (1), 1–59.

18 Liu, W. (2013) Gaussian graphical model estimation with false discovery rate control. *Ann. Stat.*, 41 (6), 2948–2978.

19 Ren, Z., Sun, T., Zhang, C.-H., and Zhou, H.H. (2015) Asymptotic normality and optimalities in estimation of large gaussian graphical models. *Ann. Stat.*, 43 (3), 991–1026.

20 Jankova, J. and Van de Geer, S. (2015) Confidence intervals for high-dimensional inverse covariance estimation. *Electron. J. Stat.*, 9 (1), 1205–1229.

21 Cai, T.T., Li, H., Liu, W., and Xie, J. (2016) Joint estimation of multiple high-dimensional precision matrices. *Stat. Sin.*, 26 (2), 445–464.

22 Liu, W. (2017) Structural similarity and difference testing on multiple sparse gaussian graphical models. *Ann. Stat.*, 45 (6), 2680–2707.

23 Cai, T.T. (2017) Global testing and large-scale multiple testing for high-dimensional covariance structures. *Annu. Rev. Stat.Appl.*, 4, 423–446.

24 Ni, Y., Müller, P., Zhu, Y., and Ji, Y. (2018) Heterogeneous reciprocal graphical models. *Biometrics*, 74 (2), 606–615.

25 Zhu, Y. and Li, L. (2018) Multiple matrix gaussian graphs estimation. *J. R. Stat. Soc. Ser. B, Stat. Methodol.*, 80 (5), 927.

26 Gan, L., Narisetty, N.N., and Liang, F. (2019) Bayesian regularization for graphical models with unequal shrinkage. *J. Am. Stat. Assoc.*, 114 (527), 1218–1231.

27 Neykov, M., Lu, J., and Liu, H. (2019) Combinatorial inference for graphical models. *Ann. Stat.*, 47 (2), 795–827.

28 Wang, Y., Segarra, S., and Uhler, C. (2020) High-dimensional joint estimation of multiple directed gaussian graphical models. *Electron. J. Stat.*, 14 (1), 2439–2483.

29 Kumar, S., Ying, J., de Miranda Cardoso, J.V., and Palomar, D.P. (2020) A unified framework for structured graph learning via spectral constraints. *J. Mach. Learn. Res.*, 21 (22), 1–60.

30 Solea, E. and Li, B. (2020) Copula gaussian graphical models for functional data. *J. Am. Stat. Assoc.*, 1–13.

31 Kurtz, Z.D., Müller, C.L., Miraldi, E.R. *et al.* (2015) Sparse and compositionally robust inference of microbial ecological networks. *PLoS Comput. Biol.*, 11 (5), e1004226.

32 Lovell, D., Pawlowsky-Glahn, V., Egozcue, J.J. *et al.* (2015) Proportionality: a valid alternative to correlation for relative data. *PLoS Comput. Biol.*, 11 (3), e1004075.

33 Yuan, H., He, S., and Deng, M. (2019) Compositional data network analysis via lasso penalized d-trace loss. *Bioinformatics*, 35 (18), 3404–3411.

34 Yoon, G., Gaynanova, I., and Müller, C.L. (2019) Microbial networks in spring-semi-parametric rank-based correlation and partial correlation estimation for quantitative microbiome data. *Front. Genet.*, 10, 516.

35 Cao, Y., Lin, W., and Li, H. (2019) Large covariance estimation for compositional data via composition-adjusted thresholding. *J. Am. Stat. Assoc.*, 114 (526), 759–772.

36 Mandal, S., Van Treuren, W., White, R.A. *et al.* (2015) Analysis of composition of microbiomes: a novel method for studying microbial composition. *Microb. Ecol. Health Dis.*, 26 (1), 27663.

37 Meinshausen, N. and Bühlmann, P. (2010) Stability selection. *J. R. Stat. Soc. Ser. B, Stat. Methodol.*, 72 (4), 417–473.

38 Lewis, J.D., Chen, E.Z., Baldassano, R.N. *et al.* (2015) Inflammation, antibiotics, and diet as environmental stressors of the gut microbiome in pediatric Crohn's disease. *Cell Host Microbe*, 18 (4), 489–500.

39 Lee, D., Baldassano, R.N., Otley, A.R. *et al.* (2015) Comparative effectiveness of nutritional and biological therapy in north American children with active Crohn's disease. *Inflamm. Bowel Dis.*, 21 (8), 1786–1793.

40 Ni, J., Shen, T.-C.D., Chen, E.Z. *et al.* (2017) A role for bacterial urease in gut dysbiosis and Crohn's disease. *Sci. Transl. Med.*, 9 (416), eaah6888.

41 Sung, J., Kim, S., Cabatbat, J.J.T. *et al.* (2017) Global metabolic interaction network of the human gut microbiota for context-specific community-scale analysis. *Nat. Commun.*, 8 (1), 1–12.

42 Kundu, P., Manna, B., Majumder, S., and Ghosh, A. (2019) Species-wide metabolic interaction network for understanding natural lignocellulose digestion in termite gut microbiota. *Sci. Rep.*, 9 (1), 1–13.

43 Martin-Fernandez, J.A., Palarea-Albaladejo, J., and Olea, R.A. (2011) Dealing with zeros, in *Compositional Data Analysis* (eds V. Pawlowsky-Glahn and A. Buccianti), John Wiley & Sons, 43–58.

14

Tensors in Modern Statistical Learning

Will Wei Sun[1], *Botao Hao*[2], *and Lexin Li*[3]

[1] *Purdue University, West Lafayette, IN, USA*
[2] *DeepMind, London, UK*
[3] *University of California, Berkeley, CA, USA*

1 Introduction

Tensors, also known as multidimensional arrays, are generalizations of vectors and matrices to higher dimensions. In recent years, tensor data are fast emerging in a wide variety of scientific and business applications, including, but not limited to, recommendation systems [1, 2], speech or facial recognitions [3, 4], networks analysis [5, 6], knowledge graphs, and relational learning [7, 8], among many others. Tensor data analysis is thus gaining increasing attention in statistics and machine-learning communities. In this survey, we provide an overview of tensor analysis in modern statistical learning.

We begin with a brief introduction of tensor notations, tensor algebra, and tensor decompositions. For more details on tensor basics, we refer the readers to Kolda and Bader [9]. We then divide our survey into four topics, depending on the nature of the learning problems: (i) tensor supervised learning, including tensor predictor regression and tensor response regression, (ii) tensor unsupervised learning, including tensor clustering and tensor graphical model, (iii) tensor reinforcement learning (RL), including low rank tensor bandit and low-rank Markov decision process (MDP), and (iv) tensor deep learning, including deep neural networks compression and deep learning theory via tensor formulation. For each topic, we start with the study goals and some motivating applications. We then review several key methods and some related solutions. We conclude each topic by a discussion of some open problems and potential future directions.

We also note that there have already been several excellent survey papers on tensor learning in statistics and machine learning, for instance, Rabanser *et al.* [10]; Sidiropoulos *et al.* [11]; Janzamin *et al.* [12]; Song *et al.* [13]; Bi *et al.* [14]. However, our review differs in terms of the focus and the organization of different tensor learning topics. Particularly, Rabanser *et al.* [10]; Sidiropoulos *et al.* [11]; Janzamin *et al.* [12] concentrated on tensor decomposition, which aims to dissolve tensors into separable representations, while Song *et al.* [13] reviewed tensor completion, which aims to impute the unobserved entries of a partially observed tensor. Tensor decomposition and tensor completion are both

Computational Statistics in Data Science.
Edited by Walter W. Piegorsch, Richard A. Levine, Hao Helen Zhang and Thomas C. M. Lee.
 ISBN 978-1-11956107-1

fundamental problems in tensor data analysis. However, given there are already fairly thorough reviews on these topics, we will not go over them in detail but instead refer to the aforementioned survey articles. Bi *et al.* [14] divided numerous tensor methods by three major application areas, that is, recommendation systems, biomedical imaging, and network analysis. We instead divide our review by different types of learning problems. Moreover, Bi *et al.* [14] only briefly mentioned some connections between tensor analysis and deep learning, while one of the focuses of our chapter is about more recent topics of tensor RL and tensor deep learning and their relations with tensor analysis.

Given fast development of tensor learning, it is inevitable that we will miss some important papers in this survey. Nevertheless, our goal is to provide a good entry point to the area of tensor data analysis, with emphasis on statistical models and properties as well as connections with other learning topics.

2 Background

We begin with a brief review of some basics of tensors. For more details, we refer the readers to Kolda and Bader [9] for an excellent review.

2.1 Definitions and Notation

The *order* of a tensor, also referred to as the *mode*, is the dimension of the array. A first-order tensor is a vector, a second-order tensor is a matrix, and tensors of order three and higher are referred to as high-order tensors, see Figure 1. The *fiber* of a tensor is defined by fixing all indices but one. For example, given a third-order tensor $\mathcal{X} \in \mathbb{R}^{p_1\times p_2\times p_3}$, its mode-1, -2, and -3 fibers are denoted as $\mathcal{X}_{:jk}$, $\mathcal{X}_{i:k}$, and $\mathcal{X}_{ij:}$, respectively.

2.2 Tensor Operations

Tensor *unfolding*, also known as tensor *matricization*, is a tensor operation that arranges tensor fibers into a matrix. Given a tensor $\mathcal{X} \in \mathbb{R}^{p_1\times p_2\times\ldots\times p_D}$, the mode-$d$ unfolding, denoted as $\mathcal{X}_{(d)}$, arranges the mode-d fibers to be the columns of the resulting matrix. For example, the mold-1 unfolding of a third-order tensor $\mathcal{X} \in \mathbb{R}^{p_1\times p_2\times d_3}$, denoted by $\mathcal{X}_{(1)}$, results in the matrix $[\mathcal{X}_{:11}, \ldots, \mathcal{X}_{:p_21}, \ldots, \mathcal{X}_{:p_2p_3}] \in \mathbb{R}^{p_1\times(p_2p_3)}$; see Figure 2 for a graphic illustration. Tensor *vectorization* is a tensor operation that arranges tensor fibers into a vector. The vectorization of tensor $\mathcal{X} \in \mathbb{R}^{p_1\times p_2\times\ldots\times p_D}$, denoted by vec($\mathcal{X}$), is the vector of length $\prod_{d=1}^{D} p_d$ that

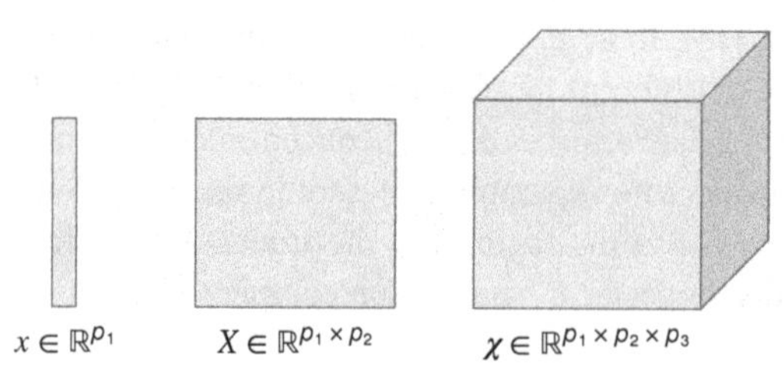

Figure 1 An example of first-, second-, and third-order tensors.

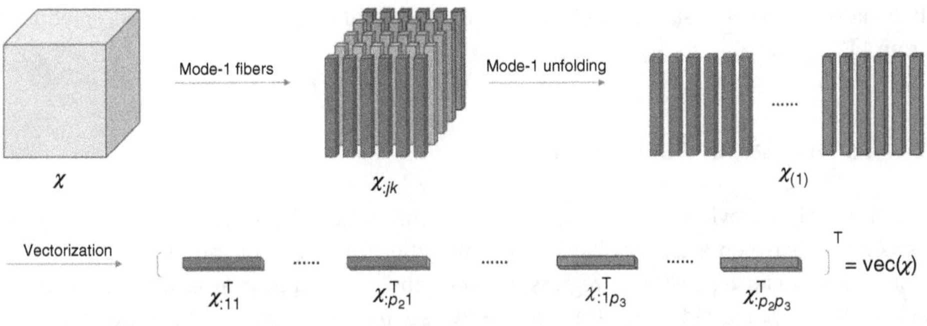

Figure 2 Tensor fibers, unfolding and vectorization.

is obtained by stacking the mode-1 fibers of $\mathcal{X}$. For example, given an order-three tensor $\mathcal{X} \in \mathbb{R}^{p_1 \times p_2 \times p_3}$, $\text{vec}(\mathcal{X}) = (\mathcal{X}^{\top}_{:11}, \ldots, \mathcal{X}^{\top}_{:p_2 1}, \ldots, \mathcal{X}^{\top}_{:p_2 p_3})^{\top}$; again see Figure 2 for an illustration.

For two tensors $\mathcal{X}, \mathcal{Y} \in \mathbb{R}^{p_1 \times p_2 \times \ldots \times p_D}$, their *inner product* is defined as $\langle \mathcal{X}, \mathcal{Y} \rangle = \sum_{i_1,\ldots,i_D} \mathcal{X}_{i_1,\ldots,i_D} \mathcal{Y}_{i_1,\ldots,i_D}$. For a tensor $\mathcal{X} \in \mathbb{R}^{p_1 \times p_2 \times \ldots \times p_M}$ and a matrix $\boldsymbol{A} \in \mathbb{R}^{J \times p_m}$, *the d-mode tensor matrix product*, denoted by $\times_d$, is defined as $\mathcal{X} \times_d \boldsymbol{A} \in \mathbb{R}^{p_1 \times \ldots \times p_{d-1} \times J \times p_{d+1} \times \ldots \times p_D}$. In this operation, each mode-d fiber of $\mathcal{X}$ is multiplied by the matrix $\boldsymbol{A}$, and elementwisely, we have $(\mathcal{X} \times_d \boldsymbol{A})_{i_1,\ldots,i_{d-1},j,i_{d+1},\ldots,i_D} = \sum_{i_d=1}^{p_d} \mathcal{X}_{i_1,\ldots,i_D} \boldsymbol{A}_{j i_d}$.

2.3 Tensor Decompositions

We next introduce two tensor decompositions that play fundamental roles in tensor data analysis.

The first is the *CANDECOMP/PARAFAC (CP)-decomposition*. For a Dth-order tensor $\mathcal{B}^*$, the rank-R CP decomposition of $\mathcal{B}^*$ is defined as

$$\mathcal{B}^* = \sum_{i=1}^{R} w_r^* \boldsymbol{\beta}_{r,1}^* \circ \cdots \circ \boldsymbol{\beta}_{r,D}^* \tag{1}$$

where $w_r^* \in \mathbb{R}$; $\boldsymbol{\beta}_{r,d}^* \in \mathbb{S}^{p_d}$, $r = 1, \ldots, R$, $d = 1, \ldots, D$, $\mathbb{S}^d = \{\mathbf{v} \in \mathbb{R}^d \mid \|\mathbf{v}\| = 1\}$; and $\circ$ denotes the outer product. The CP decomposition is sometimes abbreviated as $\mathcal{B}^* = [[\boldsymbol{W}^*; \boldsymbol{B}_1^*, \ldots, \boldsymbol{B}_D^*]]$, where $\boldsymbol{W}^* = \text{diag}(w_1^*, \ldots, w_R^*) \in \mathbb{R}^{R \times R}$ is a diagonal matrix, and $\boldsymbol{B}_d^* = [\boldsymbol{\beta}_{1,d}^*, \ldots, \boldsymbol{\beta}_{R,d}^*] \in \mathbb{R}^{p_d \times R}$ are the factor matrices. If $\mathcal{B}^*$ admits a CP structure (1), then the number of free parameters in $\mathcal{B}^*$ is reduced from $\prod_{d=1}^{D} p_d$ to $R \times \sum_{d=1}^{D} p_d$.

The second is the *Tucker decomposition*. For a Dth-order tensor $\mathcal{B}^*$, the rank-$(R_1, \ldots, R_D)$ Tucker decomposition of $\mathcal{B}^*$ is defined as

$$\mathcal{B}^* = \sum_{r_1=1}^{R_1} \cdots \sum_{r_D=1}^{R_D} w_{r_1,\ldots,r_D}^* \boldsymbol{\beta}_{r_1,1}^* \circ \cdots \circ \boldsymbol{\beta}_{r_D,D}^* \tag{2}$$

where $w_{r_1,\ldots,r_D}^* \in \mathbb{R}$ and $\boldsymbol{\beta}_{r_d,d}^* \in \mathbb{S}^{p_d}$, $r_d = 1, \ldots, R_d$, $d = 1, \ldots, D$. The Tucker decomposition is sometimes abbreviated as $\mathcal{B}^* = [[\boldsymbol{W}^*; \boldsymbol{B}_1^*, \ldots, \boldsymbol{B}_D^*]]$, where $\boldsymbol{W}^* = (w_{r_1,\ldots,r_D}^*) \in \mathbb{R}^{R_1 \times \ldots \times R_D}$ is the Dth-order core tensor, and $\boldsymbol{B}_d^* = [\boldsymbol{\beta}_{1,d}^*, \ldots, \boldsymbol{\beta}_{R_d,d}^*] \in \mathbb{R}^{p_d \times R_d}$ are the factor matrices.

If $\mathcal{B}^*$ admits a Tucker structure (2), then the number of free parameters in $\mathcal{B}^*$ is reduced from $\prod_{d=1}^{D} p_d$ to $\sum_{d=1}^{D} R_d \times p_d + \prod_{d=1}^{D} R_d$.

3 Tensor Supervised Learning

The first topic we review is tensor supervised learning, where the primary goal is to study the association between a tensor object and some other univariate or multivariate variables. The problem can be cast as a regression, and tensor can appear at either the predictor side or the response side. This leads to the two subtopics we review: the tensor predictor regression and the tensor response regression. The tensor supervised learning idea can also be generalized to involve multiple tensors on one side of the regression or having tensors showing up on both sides of the regression model.

3.1 Tensor Predictor Regression

3.1.1 Motivating examples

Neuroimaging data often take the form of tensors. For instance, electroencephalography (EEG) measures voltage value from numerous electrodes placed on scalp over time, and the resulting data is a two-dimensional matrix. Anatomical magnetic resonance imaging (MRI) measures brain structural features such as cortical thickness, and the data is a three-dimensional tensor. Figure 3 shows an example of 3D MRI at different slices and directions. It is often of great scientific interest to model the association between the tensor-valued images and the clinical outcomes such as diagnostic status or cognition and memory scores. This can be formulated as a tensor predictor regression problem, where the response is a binary or continuous scalar, the predictor is an imaging tensor, and the goal is to understand the change of the outcome as a function of the tensor.

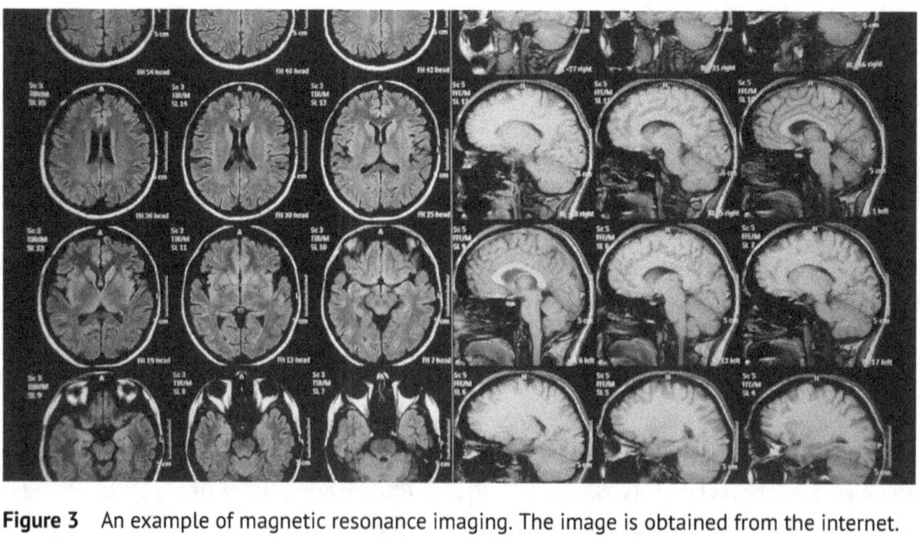

Figure 3 An example of magnetic resonance imaging. The image is obtained from the internet.

3.1.2 Low-rank linear and generalized linear model

Consider a Dth-order tensor predictor $\mathcal{X}_i \in \mathbb{R}^{p_1 \times \dots \times p_D}$ and a scalar response $y_i \in \mathbb{R}$, for i.i.d. data replications $i = 1, \dots, n$. Zhou *et al.* [15] considered the tensor predictor regression model of the form

$$y_i = \langle \mathcal{B}^*, \mathcal{X}_i \rangle + \epsilon_i \tag{3}$$

where $\mathcal{B}^* \in \mathbb{R}^{p_1 \times \dots \times p_D}$ denotes the coefficient tensor that captures the association between $\mathcal{X}_i$ and y_i and is of primary interest, and $\epsilon_i \in \mathbb{R}$ denotes the measurement error. Without loss of generality, the intercept term is set to zero to simplify the presentation. Model (3) is a direct generalization of the classical multivariate linear regression model. The issue, however, is that $\mathcal{B}^*$ involves $\prod_{d=1}^{D} p_d$ parameters, which is ultrahigh dimensional and far exceeds the typical sample size. To efficiently reduce the dimensionality, Zhou *et al.* [15] imposed the CP low-rank structure (1) on $\mathcal{B}^*$. Accordingly, the number of unknown parameters involved in $\mathcal{B}^*$ is reduced to $R \sum_{d=1}^{D} p_d$. They then proposed to estimate $\mathcal{B}^*$ via penalized maximal likelihood estimation, by solving

$$\min_{w_r, \boldsymbol{\beta}_{r,1}, \dots, \boldsymbol{\beta}_{r,D}} \sum_{i=1}^{n} \left(y_i - \left\langle \sum_{r=1}^{R} w_r \boldsymbol{\beta}_{r,1} \circ \dots \circ \boldsymbol{\beta}_{r,D}, \mathcal{X}_i \right\rangle \right)^2 + \sum_{d=1}^{D} \sum_{r=1}^{R} P_\lambda(|\boldsymbol{\beta}_{r,d}|) \tag{4}$$

under the additional constraints that $w_r > 0$ and $\|\boldsymbol{\beta}_{r,d}\|_2 = 1$ for all $r = 1, \dots, R$ and $d = 1, \dots, D$, and $P_\lambda(\cdot)$ is a sparsity-inducing penalty function indexed by the tuning parameter λ. This penalty helps to obtain a sparse estimate of $\boldsymbol{\beta}_{r,d}$, which translates to sparsity in the blocks of $\mathcal{B}^*$, and in turn facilitates the interpretation of $\mathcal{B}^*$. Denote the factor matrices $\mathbf{B}_d = [\boldsymbol{\beta}_{1,d}, \dots, \boldsymbol{\beta}_{R,d}] \in \mathbb{R}^{p_d \times R}$, for $d = 1, \dots, D$. Zhou *et al.* [15] proposed a block updating algorithm to solve (4) for each $\mathbf{B}_d$ while fixing all other $\mathbf{B}_{d'}$, $d' \neq d$. They further considered a generalized linear model formulation of (3) by introducing a link function so as to work with a binary or count type y_i.

Relatedly, Li *et al.* [16] extended (3) to multivariate response variables. Guhaniyogi *et al.* [17] formulated the tensor predictor regression (3) in a Bayesian setting and introduced a novel class of multiway shrinkage priors for tensor coefficients. Li *et al.* [18] considered the Tucker decomposition (2) for $\mathcal{B}^*$ and demonstrated its flexibility over the CP decomposition. Zhang *et al.* [19] extended (3) to the generalized estimating equation setting for longitudinally observed imaging tensors.

3.1.3 Large-scale tensor regression via sketching

A common challenge associated with the tensor predictor regression with a low-rank factorization is the high computational cost. This is especially true when the dimension of the tensor predictor is large. Sketching offers a natural solution to address this challenge and is particularly useful when the dimensionality is ultrahigh, the sample size is super large, or the data is extremely sparse.

Yu and Liu [20] introduced the subsampled tensor-projected gradient approach for a variety of tensor regression problems, including the situation when the response is a tensor too. Their algorithm was built upon the projected gradient method with fast tensor power iterations and leveraged randomized sketching for further acceleration. In particular, they used count sketch [21] as a subsampling step to generate a reduced data and then feed the data into tensor-projected gradient to estimate the final parameters.

Zhang *et al.* [22] utilized importance sketching for low-rank tensor regressions. They carefully designed sketches based on both the response and the low-dimensional structure of the parameter of interest. They proposed an efficient algorithm, which first used the high-order orthogonal iteration [23] to determine the importance sketching directions, then performed importance sketching and evaluated the dimension-reduced regression using the sketched tensors, and constructed the final tensor estimator using the sketched components. They showed that their algorithm achieves the optimal mean-squared error under the low-rank Tucker structure and randomized Gaussian design.

3.1.4 Nonparametric tensor regression

Although the linear tensor regression provides a simple and concise solution, the linearity assumption in (3) can be restrictive in numerous applications [24, 25]. For instance, Hao *et al.* [26] showed that, in a digital advertising study, the association between the click-through rate and the impression tensor of various ads on different devices is clearly nonlinear.

Hao *et al.* [26] proposed a nonparametric extension of model (3), by assuming

$$y_i = \sum_{j_1=1}^{p_1} \cdots \sum_{j_D=1}^{p_D} f^*_{j_1 \ldots j_D}([\mathcal{X}_i]_{j_1 \ldots j_D}) + \epsilon_i \tag{5}$$

where $[\mathcal{X}_i]_{j_1 \ldots j_D}$ denotes the $(j_1, \ldots, j_D)$th entry of the tensor $\mathcal{X}_i$, and $f^*_{j_1 \ldots j_D}(\cdot)$ is some smooth function that can be approximated by *B*-splines [27],

$$f^*_{jkl}([\mathcal{X}_i]_{jkl}) \approx \sum_{h=1}^{H} \beta^*_{j_1 \ldots j_D h} \psi_h([\mathcal{X}_i]_{j_1 \ldots j_D}),\ 1 \leq j_1 \leq p_1, \ldots, 1 \leq j_D \leq p_3$$

with the *B*-spline basis $\psi_{j_1 \ldots j_D h}$ and coefficients $\beta^*_{j_1 \ldots j_D h}$. Let $[\mathcal{F}_h(\mathcal{X}_i)]_{j_1 \ldots j_D} = \psi_{j_1 \ldots j_D h}([\mathcal{X}_i]_{j_1 \ldots j_D})$ and $[\mathcal{B}_h]_{j_1 \ldots j_D} = \beta^*_{j_1 \ldots j_D h}$. The compact tensor representation of their model is

$$y_i = \sum_{h=1}^{H} \langle \mathcal{B}_h, \mathcal{F}_h(\mathcal{X}_i) \rangle + \epsilon_i \tag{6}$$

In this model, $\mathcal{F}_h(\mathcal{X}_i) \in \mathbb{R}^{p_1 \times \ldots \times p_D}$ is the predictor tensor under the *B*-spline transformation, and $\mathcal{B}_h \in \mathbb{R}^{p_1 \times \ldots \times p_D}$ captures the association information. The linear tensor regression model (3) becomes a special case of (6), with $\psi_{j_1 \ldots j_D h}(x) = x$ and $H = 1$. By considering nonlinear basis functions, for example, trigonometric functions, model (6) is more flexible and has a better prediction power. Moreover, Hao *et al.* [26] imposed the CP structure (1) on $\mathcal{B}_h$ and a groupwise penalty to screen out the nuisance components. They proposed to solve the following penalized optimization problem:

$$\min_{\beta_{1hr}, \ldots, \beta_{Dhr}} \frac{1}{n} \sum_{i=1}^{n} \left(y_i - \sum_{h=1}^{H} \left\langle \sum_{r=1}^{R} \beta_{1hr} \circ \ldots \circ \beta_{Dhr}, \mathcal{F}_h(\mathcal{X}_i) \right\rangle \right)^2 + \lambda \sum_{d=1}^{D} \sum_{j=1}^{p_d} \sqrt{\sum_{h=1}^{H} \sum_{r=1}^{R} \beta^2_{dhrj}} \tag{7}$$

The optimization in (7) is done in a blockwise manner for β_{dhr}, $d = 1, \ldots, D$, and each block is solved by the backfitting algorithm for the standard sparse additive model [28]. The regularization parameter λ is tuned by cross-validation.

Relatedly, Zhou *et al.* [29] considered a broadcasted nonparametric tensor regression model where all entries of the tensor covariate are assumed to share the same function, which is a special case of (5).

3.1.5 Future directions

There are a number of open questions for tensor predictor regression. One is to integrate multiple tensor predictors, each of which represents a tensor measurement from a data modality, and there are multiple modalities of data collected for the same group of experimental subjects. Challenges include how to model the interactions between different tensors, and how to perform statistical inference. In addition, it is of interest to investigate how to speed up the computation in nonparametric tensor regression. One possible solution is to use the sketching idea, or the divide-and-conquer approach [30], when the data cannot fit into a single machine.

3.2 Tensor Response Regression

3.2.1 Motivating examples

While the tensor predictor regression focuses on understanding the change of a phenotypic outcome as the tensor varies, in numerous applications, it is important to study the change of the tensor as the covariates vary. One example is anatomical MRI, where the data takes the form of a 3D tensor, and voxels correspond to brain spatial locations. Another example is functional magnetic resonance imaging (fMRI), where the goal is to understand brain functional connectivity encoded by a symmetric matrix, with rows and columns corresponding to brain regions, and entries corresponding to interactions between those regions. In both examples, it is of keen scientific interest to compare the scans of brains, or the brain connectivity patterns, between the subjects with some neurological disorder and the healthy controls, after adjusting for additional covariates such as age and sex. Both can be formulated as a regression problem, with image tensor or connectivity matrix serving as the response, and the disease indicator and other covariates forming the predictors.

3.2.2 Sparse low-rank tensor response model

Consider a Dth-order tensor response $\mathcal{Y}_i \in \mathbb{R}^{p_1\times\cdots\times p_D}$ and a vector of predictors $\boldsymbol{x}_i \in \mathbb{R}^{p_0}$, for i.i.d. data replications $i = 1, \ldots, n$. Rabusseau and Kadri [31]; Sun and Li [32] considered the tensor response regression model of the form

$$\mathcal{Y}_i = \mathcal{B}^* \times_{m+1} \boldsymbol{x}_i + \mathcal{E}_i \tag{8}$$

where $\mathcal{B}^* \in \mathbb{R}^{p_1\times\cdots\times p_D\times p_0}$ is an $(D+1)$th-order tensor coefficient that captures the association between $\boldsymbol{x}_i$ and $\mathcal{Y}_i$, and $\mathcal{E}_i \in \mathbb{R}^{p_1\times\cdots\times p_D}$ is an error tensor that is independent of $\boldsymbol{x}_i$. Without loss of generality, the intercept term is set to zero to simplify the presentation.

Both Rabusseau and Kadri [31] and Sun and Li [32] imposed the rank-R CP structure (1) for the coefficient tensor $\mathcal{B}^*$, while Sun and Li [32] further incorporated the sparsity structure. Specifically, Sun and Li [32] proposed to solve

$$\min_{\substack{w_r,\boldsymbol{\beta}_{r,d}\\ r\in[R],d\in[D+1]}} \frac{1}{n}\sum_{i=1}^{n}\left\|\mathcal{Y}_i - \sum_{r=1}^{R} w_r(\boldsymbol{\beta}_{r,D+1}^{\top}\boldsymbol{x}_i)\boldsymbol{\beta}_{r,1}\circ\cdots\circ\boldsymbol{\beta}_{r,D}\right\|_F^2, \text{ subject to } \|\boldsymbol{\beta}_{r,d}\|_0 \le s_d \tag{9}$$

and $\|\boldsymbol{\beta}_{r,d}\|_2 = 1$, where s_j is the sparsity parameter. In (9), the sparsity of the decomposed components is encouraged via a hard-thresholding penalty. The optimization in (9) is utterly different from that of (4) for tensor predictor regression, which leads to a more complicated algorithm and a more subtle interplay between the computational efficiency and the statistical rate of convergence. To solve (9), Sun and Li [32] proposed an iterative updating algorithm consisting of two major steps. In the first step, the estimation of $w_r, \boldsymbol{\beta}_{r,1}, \dots, \boldsymbol{\beta}_{r,d}$ for $k \in [K]$, given $\boldsymbol{\beta}_{r,D+1}$, $r \in [R]$ and $w_{r'}, \boldsymbol{\beta}_{r',1}, \dots, \boldsymbol{\beta}_{r',d}$, $r' \neq r$, is reformulated as a sparse rank-1 tensor decomposition problem [33], while in the second step, the estimation of $\boldsymbol{\beta}_{r,D+1}$ for $r \in [R]$, given $w_r, \boldsymbol{\beta}_{r,1}, \dots, \boldsymbol{\beta}_{r,D}$, $r \in [R]$ and $\boldsymbol{\beta}_{r',D+1}$, $r' \neq r$, becomes a standard least-squares optimization problem and has a closed-form solution.

3.2.3 Additional tensor response regression models

Li and Zhang [34] proposed an envelope-based tensor response model, which utilized a generalized sparsity principle to exploit the redundant information in the tensor response, and seeked linear combinations of the response that are irrelevant to the regression. Raskutti *et al.* [35] developed a class of sparse regression models, under the assumption of Gaussian error, when either or both the response and predictor are tensors. Their approach required a crucial condition that the regularizer was convex and weakly decomposable, and the low rankness of the estimator was achieved via a tensor nuclear norm penalty. Later, Chen *et al.* [36] proposed a projected gradient descent algorithm to efficiently solve the nonconvex optimization in tensor response regression and provided the theoretical guarantees for learning high-dimensional tensor regression models under different low-rank structural assumptions. Motivated by longitudinal neuroimaging studies where image tensors are often missing, Zhou *et al.* [37] developed a regression model with partially observed dynamic tensor as the response and external covariates as the predictor vector. Their solution combined the tensor completion loss idea of a single partially observed tensor [38] with the tensor response regression model of Sun and Li [32] and developed an elementwise updating algorithm.

3.2.4 Future directions

There are a number of open questions for tensor response regression. One is how to obtain a consistent estimator of the rank R when the CP structure is employed. More importantly, it remains open to derive the corresponding convergence rate and combine the estimated rank with the subsequent estimator of $\mathcal{B}^*$ when studying the asymptotic properties. The existing solutions generally treat R as known in the asymptotic studies. Moreover, the current studies have primarily focused on parameter estimation, whereas parameter inference remains a challenging and open question for tensor response regression, especially when the sample size is limited.

4 Tensor Unsupervised Learning

The second topic we review is tensor unsupervised learning, which involves no external variables. We review two topics: tensor clustering and tensor graphical model. The former aims to identify clusters by studying the structure of tensor itself, whereas the latter aims to characterize the dependency structure of the individual mode of tensor-valued data.

4.1 Tensor Clustering

4.1.1 Motivating examples

Consider two motivating examples. One is a digital advertisement example consisting of the click-through rates for advertisements displayed on an internet company's webpages over weeks during the ad campaign. The data is a fourth-order tensor, recording the click-through rate of multiple users over a collection of advertisements by different publishers and published on different devices, and the data was aggregated across time. The goal is to simultaneously cluster users, advertisements, and publishers to improve user behavior targeting and advertisement planning. Another example is dynamic brain connectivity analysis based on fMRI data, where the data is in the form of brain region by region by time tensor, and the goal is to cluster over time, so as to better understand the interactions of distinct brain regions and their dynamic patterns over time. Both examples can be formulated as a tensor clustering problem. The prevalent clustering solutions, however, have mainly focused on clustering of vector- or matrix-valued data. Notably, biclustering extends the classical clustering along both the observations (rows) and the features (columns) of a data matrix [39, 40].

4.1.2 Convex tensor co-clustering

We first review a convex coclustering method that extends biclustering to tensor coclustering by solving a convex formulation of the problem. Specifically, without loss of generality, Chi *et al.* [41] considered a third-order tensor $\mathcal{X} \in \mathbb{R}^{p_1 \times p_2 \times p_3}$. They assumed that the observed data tensor is a noisy realization of an underlying tensor that exhibits a checkerbox structure modulo some unknown reordering along each of its modes. Suppose that there are K_1, K_2, and K_3 clusters along mode 1, 2, and 3, respectively. If the (i_1, i_2, i_3)th entry in $\mathcal{X}$ belongs to the cluster defined by the r_1th mode-1 group, r_2th mode-2 group, and r_3th mode-3 group, then the observed tensor element $x_{i_1 i_2 i_3}$ is

$$x_{i_1 i_2 i_3} = c^*_{r_1 r_2 r_3} + \epsilon_{i_1 i_2 i_3} \tag{10}$$

where $c^*_{r_1 r_2 r_3}$ is the mean of the cocluster defined by the r_1th mode-1 partition, r_2th mode-2 partition, and r_3th mode-3 partition, and $\epsilon_{i_1 i_2 i_3}$ is the noise. Consequently, the observed tensor $\mathcal{X}$ can be written as the sum of a mean tensor $\mathcal{U}^* \in \mathbb{R}^{p_1 \times p_2 \times p_3}$, whose elements are expanded from the cocluster means tensor $\mathcal{C}^* \in \mathbb{R}^{K_1 \times K_2 \times K_3}$ and a noise tensor $\mathcal{E} \in \mathbb{R}^{p_1 \times p_2 \times p_3}$. Figure 4 illustrates an underlying mean tensor $\mathcal{U}^*$ after permuting the slices along each of the modes to reveal a checkerbox structure. The coclustering model in (10) is the three-way analogue of the checkerboard mean model often employed in biclustering data matrices [39, 40].

Estimating model (10) consists of finding the partitions along each mode and finding the mean values of the $K_1 K_2 K_3$ coclusters. The challenge is the first step, that is, finding the partitions $\mathcal{G}_1, \mathcal{G}_2$, and $\mathcal{G}_3$, which denote the indices of the r_1th mode-1, r_2th mode-2, and r_3th mode-3 groups, respectively. Chi *et al.* [41] proposed to solve a convex relaxation to the original combinatorial optimization problem, by simultaneously identifying the partitions along the modes of $\mathcal{X}$ and estimating the cocluster means through the optimization of the following convex objective function:

$$F_\gamma(\mathcal{U}) = \frac{1}{2}\|\mathcal{X} - \mathcal{U}\|_F^2 + \underbrace{\gamma[R_1(\mathcal{U}) + R_2(\mathcal{U}) + R_3(\mathcal{U})]}_{R(\mathcal{U})} \tag{11}$$

Figure 4 A third-order tensor with a checkerbox structure.

where $R_1(\mathcal{U}) = \sum_{i<j} w_{1,ij} \|\mathcal{U}_{i::} - \mathcal{U}_{j::}\|_F$, $R_2(\mathcal{U}) = \sum_{i<j} w_{2,ij} \|\mathcal{U}_{:i:} - \mathcal{U}_{:j:}\|_F$, and $R_3(\mathcal{U}) = \sum_{i<j} w_{3,ij} \|\mathcal{U}_{::i} - \mathcal{U}_{::j}\|_F$. By seeking the minimizer $\hat{\mathcal{U}}_\gamma \in \mathbb{R}^{p_1 \times p_2 \times p_3}$ of (11), it casts coclustering as a signal approximation problem, modeled as a penalized regression, to estimate the true cocluster mean tensor $\mathcal{U}^*$. The quadratic term in (11) quantifies how well $\mathcal{U}$ approximates $\mathcal{X}$, while the regularization term $R(\mathcal{U})$ penalizes deviations away from a checkerbox pattern. The nonnegative parameter γ tunes the relative emphasis on these two terms and is selected via a BIC-type information criterion. The nonnegative weights $w_{d,ij}$ fine-tune the shrinkage of the slices along the dth mode. Chi *et al.* [41] showed that the solution $\hat{\mathcal{U}}$ for (11) produces an entire solution path of checkerbox coclustering estimates that varies continuously in γ, from the least smoothed model, where $\hat{\mathcal{U}} = \mathcal{X}$, and each tensor element occupies its own cocluster, to the most smoothed model, where all the elements of $\hat{\mathcal{U}}$ are identical, and all tensor elements belong to a single cocluster.

4.1.3 Tensor clustering via low-rank decomposition

We next review tensor clustering based on low-rank tensor decompositions [42, 43]. Unlike the convex tensor coclustering of Chi *et al.* [41] that targets a single tensor object, here we target the problem of clustering a collection of tensor samples.

Given N copies of the Dth-order tensors, $\mathcal{X}_1, \ldots, \mathcal{X}_N \in \mathbb{R}^{p_1 \times \cdots \times p_D}$, Papalexakis *et al.* [42]; Sun and Li [43] aimed to uncover the underlying cluster structures of the N samples, with K clusters, and an equal number of $l = N/K$ samples per cluster, for simplicity. Sun and Li [43] proposed to first stack all n tensor samples into a $(D+1)$th-order tensor $\mathcal{T} \in \mathbb{R}^{p_1 \times \cdots \times p_D \times N}$, then consider a structured decomposition of $\mathcal{T}$, and finally apply a usual clustering algorithm, for example, K-means, to the matrix from the tensor decomposition that corresponds to the last mode to obtain the cluster assignment. Figure 5 shows a schematic plot of this method. Specifically, assume that the tensor $\mathcal{T}$ is observed with noise, that is, $\mathcal{T} = \mathcal{T}^* + \mathcal{E}$, where $\mathcal{E}$ is an error tensor, and $\mathcal{T}^*$ is the true tensor with a rank-R CP decomposition structure, $\mathcal{T}^* = \sum_{r=1}^{R} w_r^* \boldsymbol{\beta}_{r,1}^* \circ \cdots \circ \boldsymbol{\beta}_{r,D+1}^*$, where $\boldsymbol{\beta}_{r,j}^* \in \mathbb{R}^{p_j}$, $\|\boldsymbol{\beta}_{j,r}^*\|_2 = 1$, $w_r^* > 0$, and $j = 1, \ldots, D + 1$, $r = 1, \ldots, R$. Then, the cluster structure of samples along the last mode of the tensor $\mathcal{T}$ is

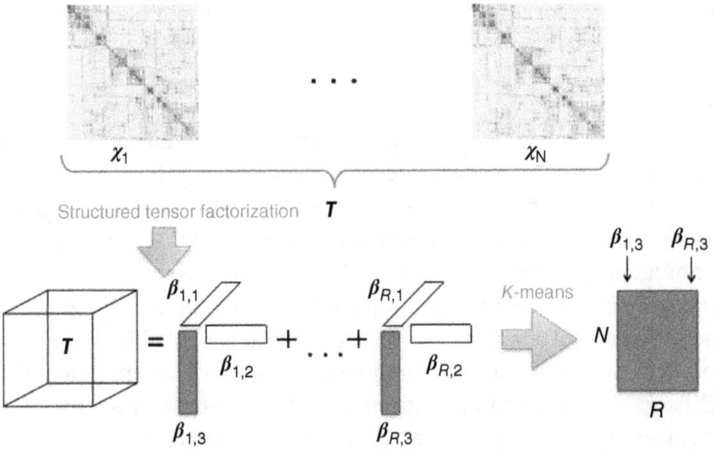

Figure 5 A schematic illustration of the low-rank tensor clustering method.

fully determined by the matrix that stacks the decomposition components, that is,

$$\boldsymbol{B}^*_{D+1} = (\boldsymbol{\beta}^*_{1,D+1}, \dots, \boldsymbol{\beta}^*_{R,D+1}) = (\underbrace{\boldsymbol{\mu}_1^{*\mathrm{T}}, \dots, \boldsymbol{\mu}_1^{*\mathrm{T}}}_{l\text{samples}}, \dots, \underbrace{\boldsymbol{\mu}_K^{*\mathrm{T}}, \dots, \boldsymbol{\mu}_K^{*\mathrm{T}}}_{l\text{samples}})^{\mathrm{T}} \in \mathbb{R}^{N\times R}$$

where $\boldsymbol{\mu}_k^* = (\mu^*_{1,k}, \dots, \mu^*_{R,k}) \in \mathbb{R}^R$, $k = 1, \dots, K$, indicates the cluster assignment. Accordingly, the true cluster means of the tensor samples $\mathcal{X}_1, \dots, \mathcal{X}_N$ can be written as

$$\underbrace{\mathcal{M}_1 := \sum_{r=1}^{R} w_r^* \boldsymbol{\beta}^*_{r,1} \circ \cdots \circ \boldsymbol{\beta}^*_{r,D} \mu^*_{r,1},}_{\text{cluster center 1}} \quad \dots, \quad \underbrace{\mathcal{M}_K := \sum_{r=1}^{R} w_r^* \boldsymbol{\beta}^*_{r,1} \circ \cdots \circ \boldsymbol{\beta}^*_{r,D} \mu^*_{r,K}}_{\text{cluster center K}}$$

This reveals the key structure, that is, each cluster mean is a linear combination of the outer product of R rank-1 basis tensors, and all the cluster means share the same R basis tensors.

Sun and Li [43] further introduced the sparsity and smoothness fusion structures in tensor decomposition to capture the sparsity and dynamic properties of the tensor samples. They proposed an optimization algorithm consisting of an unconstrained tensor decomposition step followed by a constrained optimization step. They established theoretical guarantee for their proposed dynamic tensor clustering approach, by deriving the corresponding nonasymptotic error bound, the rate of convergence, and the cluster recovery consistency.

4.1.4 Additional tensor clustering approaches

We briefly discuss some additional tensor clustering methods. Zhang *et al.* [44] unfolded tensor in each mode to construct an affinity matrix and then applied spectral clustering algorithm on this affinity matrix to obtain the cluster structure. Wu *et al.* [45] utilized super-spacey random walk to propose a tensor spectral coclustering algorithm for a nonnegative three-mode tensor. More recently, Luo and Zhang [46] studied high-order clustering with planted structures for testing whether a cluster exists and identifying the support of cluster.

4.1.5 Future directions

In the model $\mathcal{T} = \mathcal{T}^* + \mathcal{E}$ considered by Sun and Li [43], no distributional assumption is imposed on the error tensor $\mathcal{E}$. If one further assumes that $\mathcal{E}$ is a standard Gaussian tensor, then the method reduces to a tensor version of Gaussian mixture model with identity covariance matrix. One possible future direction is to consider a more general tensor Gaussian mixture model with nonidentity covariance matrix. The tensor cluster means and the covariance matrices can be estimated using a high-dimensional expectation–maximization algorithm [47] in which the maximization step solves a penalized weighted least squares. Moreover, in the theoretical analysis of all the aforementioned tensor clustering projects, the true number of clusters was assumed to be given. It is of great interest to study the property of the tensor clustering when the number of cluster is estimated [48].

4.2 Tensor Graphical Model

4.2.1 Motivating examples

Tensor graphical model aims to characterize the dependency structure of the individual mode of the tensor-valued data. As an example, consider the microarray study for aging [49], where multiple gene expression measurements are recorded on multiple tissue types of multiple mice with varying ages, which forms a set of third-order gene–tissue–age tensors. It is of scientific interest to study the dependency structure across different genes, tissues, and ages.

4.2.2 Gaussian graphical model

Similar to the vector-valued graphic model, He *et al.* [50]; Sun *et al.* [51] assumed that the Dth-order tensor $\mathcal{T} \in \mathbb{R}^{p_1 \times \cdots \times p_D}$ follows a tensor normal distribution with zero mean and covariance matrices $\mathbf{\Sigma}_1, \ldots, \mathbf{\Sigma}_D$. Denote it by $\mathcal{T} \sim \text{TN}(\mathbf{0}; \mathbf{\Sigma}_1, \ldots, \mathbf{\Sigma}_D)$, and its probability density function is given by

$$p(\mathcal{T}|\mathbf{\Sigma}_1, \ldots, \mathbf{\Sigma}_D) = (2\pi)^{-p/2} \left\{ \prod_{d=1}^{D} |\mathbf{\Sigma}_d|^{-p/(2p_d)} \right\} \exp(-\|\mathcal{T} \times \mathbf{\Sigma}^{-1/2}\|_F^2/2) \tag{12}$$

where $p = \prod_{d=1}^{D} p_d$ and $\mathbf{\Sigma}^{-1/2} = \{\mathbf{\Sigma}_1^{-1/2}, \ldots, \mathbf{\Sigma}_D^{-1/2}\}$. When $D = 1$, it reduces to the vector normal distribution with zero mean and covariance $\mathbf{\Sigma}_1$. Following Kolda and Bader [9], $\mathcal{T} \sim \text{TN}(\mathbf{0}; \mathbf{\Sigma}_1, \ldots, \mathbf{\Sigma}_D)$ if and only if $\text{vec}(\mathcal{T}) \sim \text{N}(\text{vec}(\mathbf{0}); \mathbf{\Sigma}_D \otimes \ldots \otimes \mathbf{\Sigma}_1)$, where $\otimes$ denotes the Kronecker product.

Given n copies of i.i.d. samples $\mathcal{T}_1, \ldots, \mathcal{T}_n$ from $\text{TN}(\mathbf{0}; \mathbf{\Sigma}_1^*, \ldots, \mathbf{\Sigma}_D^*)$, the goal of tensor graphical modeling is to estimate the true covariance matrices $\mathbf{\Sigma}_1^*, \ldots, \mathbf{\Sigma}_D^*$, and the corresponding true precision matrices $\mathbf{\Omega}_1^*, \ldots, \mathbf{\Omega}_D^*$, where $\mathbf{\Omega}_d^* = \mathbf{\Sigma}_d^{*-1}, d = 1, \ldots, D$. For identifiability, assume that $\|\mathbf{\Omega}_d^*\|_F = 1$ for $d = 1, \ldots, D$. This renormalization does not change the graph structure of the original precision matrix. A standard solution is the penalized maximum-likelihood estimation which minimizes

$$\frac{1}{p}\text{tr}[\mathbf{S}(\mathbf{\Omega}_D \otimes \ldots \otimes \mathbf{\Omega}_1)] - \sum_{d=1}^{D} \frac{1}{p_d} \log |\mathbf{\Omega}_d| + \sum_{d=1}^{D} P_{\lambda_d}(\mathbf{\Omega}_d) \tag{13}$$

where $\mathbf{S} = n^{-1}\sum_{i=1}^{n} \text{vec}(\mathcal{T}_i)\text{vec}(\mathcal{T}_i)^\top$, and $P_{\lambda_d}(\cdot)$ is a penalty function indexed by the tuning parameter λ_d. Adopting the usual lasso penalty used in vector graphical model, let $P_{\lambda_d}(\boldsymbol{\Omega}_d) = \lambda_d\|\boldsymbol{\Omega}_d\|_{1,\text{off}}$, where $\|\cdot\|_{1,\text{off}}$ means that the sparsity penalty is applied to the off-diagonal elements of the matrix. The problem reduces to the classical sparse vector graphical model [52, 53] when $D = 1$, and the sparse matrix graphical model [54–57] when $D = 2$. He *et al.* [50] showed that the global minimizer of (13) enjoys nice theoretical properties.

Note that the objective function in (13) is biconvex, in the sense that it is convex in terms of $\boldsymbol{\Omega}_d$ when the rest of $D-1$ precision matrices are fixed. Exploring this biconvex property, Sun *et al.* [51] proposed to solve (13) by alternatingly updating one precision matrix while fixing the rest, which is equivalent to minimizing

$$\frac{1}{p_d}\text{tr}(\mathbf{S}_d\boldsymbol{\Omega}_d) - \frac{1}{p_d}\log|\boldsymbol{\Omega}_d| + \lambda_d\|\boldsymbol{\Omega}_d\|_{1,\text{off}} \tag{14}$$

where $\mathbf{S}_d = p_d/(np)\sum_{i=1}^{n}\mathbf{V}_i^d\mathbf{V}_i^{d\top}$, $\mathbf{V}_i^d = [\mathcal{T}_i \times \{\boldsymbol{\Omega}_1^{1/2}, \dots, \boldsymbol{\Omega}_{d-1}^{1/2}, 1_{p_d}, \boldsymbol{\Omega}_{d+1}^{1/2}, \dots, \boldsymbol{\Omega}_D^{1/2}\}]_{(d)}$, $\times$ denotes the tensor product operation, and $[\cdot]_{(d)}$ denotes the mode-d matricization operation. Minimizing Equation (14) corresponds to estimating the vector-valued Gaussian graphical model, which can be efficiently solved [52, 53]. Sun *et al.* [51] further showed that the estimator of their tensor lasso algorithm is able to achieve the desirable optimal statistical rates. In particular, their estimator $\hat{\boldsymbol{\Omega}}_d$ satisfies

$$\|\hat{\boldsymbol{\Omega}}_d - \boldsymbol{\Omega}_d^*\|_F = O_P\left(\sqrt{\frac{p_d(p_d+s_d)\log p_d}{np}}\right);\ \|\hat{\boldsymbol{\Omega}}_d - \boldsymbol{\Omega}_d^*\|_\infty = O_P\left(\sqrt{\frac{p_d\log p_d}{np}}\right)$$

where $p = \prod_{d=1}^{D} p_d$, and the sparsity parameter s_d is the number of nonzero entries in the off-diagonal component of $\boldsymbol{\Omega}_d^*$. The above error bound implies that when the mode $D \geq 3$, the estimator from the tensor lasso algorithm can achieve estimation consistency even if we only have access to one observation, that is, $n = 1$. This is because the estimation of the dth precision matrix takes advantage of the information from all other modes of the tensor data. This phenomenon only exists in tensor graphical model when $D \geq 3$, which reveals an interesting blessing of dimensionality phenomenon. Moreover, this rate is minimax-optimal since it is the best rate one can obtain even when $\boldsymbol{\Omega}_j^*$ $(j \neq d)$ were known.

As a follow-up, Lyu *et al.* [58] further proposed a debiased statistical inference procedure for testing hypotheses on the true support of the sparse precision matrices and employed it for testing a growing number of hypothesis with false discovery rate (FDR) control. They also established the asymptotic normality of the test statistic and the consistency of the FDR-controlled multiple testing procedure.

4.2.3 Variation in the Kronecker structure

In addition to the Kronecker product structure considered in (12), Greenewald *et al.* [59] considered a Kronecker sum structure of $\Omega = \Psi_1 \oplus \Psi_2 = (\Psi_1 \otimes \mathbb{I}) + (\Psi_2 \otimes \mathbb{I})$. They showed that the new structure on the precision matrix leads to a nonseparable covariance matrix that provides a richer model than the Kronecker product structure. Alternatively, Wang *et al.* [60] proposed a Sylvester-structured graphical model to estimate precision matrices associated with tensor data and used a Kronecker sum model for the square root factor of the precision matrix.

4.2.4 Future directions

All the existing works have assumed that the tensor data follows a tensor normal distribution. A natural future direction is to relax this normal distribution requirement, extend to the higher order nonparanormal distribution [61], and utilize a robust rank-based likelihood estimation. When the order of the tensor is $D = 2$, it reduces to the semiparametric bigraphical model considered in Ning and Liu [62].

5 Tensor Reinforcement Learning

The third topic we review is tensor RL. RL is an area of machine learning that focuses on how an agent interacts with and takes actions in an environment in order to maximize the notion of cumulative rewards. It is a fast-growing field; see Sutton and Barto [63] for a review and the references therein. We highlight two topics that involve tensor learning in RL: stochastic low-rank tensor bandit and learning MDP via tensor decomposition. In both cases, tensor methods serve as a powerful dimension reduction tool, which efficiently reduces the complexity of the RL problems.

5.1 Stochastic Low-Rank Tensor Bandit

5.1.1 Motivating examples

The growing availability of tensor data provides an unique opportunity for decision-makers to efficiently develop multidimensional decisions for individuals [2, 13, 64, 65]. For instance, consider a marketer who wants to design an advertising campaign for products with promotion offers across different marketing channels and user segments. This marketer needs to estimate the probability of user i clicking offer j in channel k for any (i, j, k) combination so that the most relevant users will be targeted for a chosen product and channel. Figure 6 gives a graphic illustration.

Traditional static recommendation systems using tensor methods [2, 13, 65] do *not* interact with the environment to update the estimation. Besides, they usually suffer from cold start in the absence of information from new customers, new products, or new contexts. An interactive recommendation system for multidimensional decisions is urgently needed.

RL offers a dynamic and interactive policy of recommendations. One of the fundamental problems in RL is the exploration–exploitation trade-off, in the sense that the agent must balance between exploiting existing information to accrue immediate reward and investing in exploratory behavior that may increase future reward. Multiarmed bandit [66] can be viewed as a simplified version of RL that exemplifies this exploration–exploitation trade-off and itself has plenty of applications in online advertising and operations research [67]. We review the problem of stochastic low-rank tensor bandit, a class of bandits whose mean reward can be represented as a low-rank tensor.

5.1.2 Low-rank tensor bandit problem formulation

We begin with a brief introduction of basic notations and concepts of multiarmed bandit. For more details, we refer the readers to Lattimore and Szepesvári [66]. In the vanilla K-armed bandit, the agent interacts with the environment for n rounds. At round $t \in [n]$,

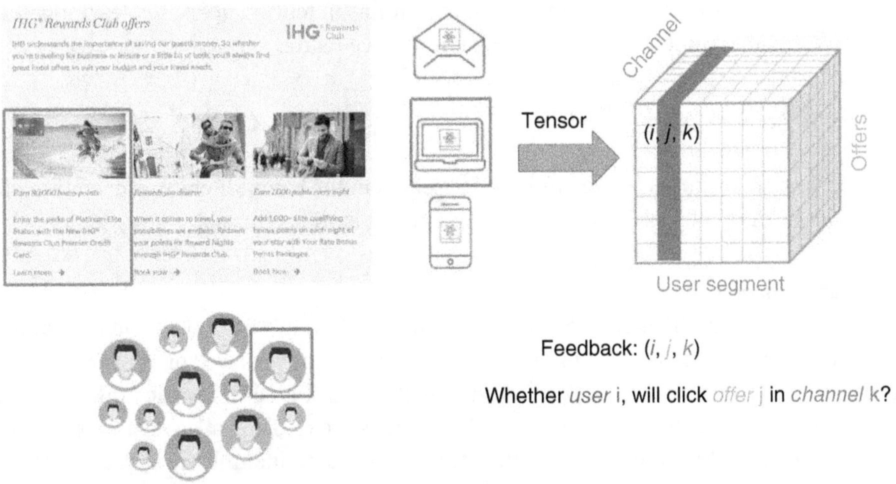

Figure 6 The tensor formulation of multidimensional advertising decisions.

the agent faces a multidimensional decision set $\mathcal{A} \subseteq \mathbb{R}^{p_1 \times \cdots \times p_d}$, and the cardinality of $\mathcal{A}$ can be either finite or infinite. The agent pulls an arm $I_t \in [K]$ and observes its reward y_{I_t}, which is drawn from a distribution associated with the arm I_t, denoted by P_{I_t} with an mean reward μ_{I_t}. It is important to point out that in multiarmed bandit problems, the objective is to minimize the expected cumulative regret, which is defined as

$$R_n = n \max_{k \in [K]} \mu_k - \mathbb{E}\left[\sum_{t=1}^{n} y_t\right] \tag{15}$$

where the expectation is with respect to the randomness in the environment and policy.

Next, we introduce the low-rank tensor bandit. The classical vanilla multiarmed bandit can be treated as a special case of tensor bandit where the order of the tensor is 1, and the action set $\mathcal{A}$ only consists of canonical basis vectors, for example, e_i that has 1 on its ith coordinate and 0 anywhere else. At round $t \in [n]$, based on historical information, the agent selects an action $\mathcal{A}_t$ from $\mathcal{A}$ and observes a noisy reward y_t, which can be written as

$$y_t = \langle \mathcal{X}, \mathcal{A}_t \rangle + \epsilon_t \tag{16}$$

where $\mathcal{X}$ is an unknown tensor parameter that admits a low-rank structure, and ϵ_t is a random noise. Model (16) can be viewed as a special case of the so-called stochastic linear bandit [68–71], where the mean reward can be parameterized into a linear form. However, naively implementing the existing linear bandit algorithms is to suffer high regret, since none of them utilizes the intrinsic low-rank structure of $\mathcal{X}$.

At a glance, the tensor bandit model (16) looks similar to the tensor predictor regression model (3) in tensor supervised learning. However, the two have some fundamental distinctions. First, (16) considers a sequential setting, in the sense that $\mathcal{A}_t$ has to be sequentially collected by the agent rather than given ahead. Consequently, $\mathcal{A}_t$ and $\mathcal{A}_{t-1}$ may be highly dependent, and the dependency structure is extremely difficult to characterize. By contrast, (3) can be viewed as corresponding to the offline setting where $\mathcal{A}_t$ is fully observed.

Second, instead of minimizing the mean square error as in tensor supervised learning, the objective in tensor bandit is to minimize the cumulative regret,

$$R_n = \sum_{t=1}^{n} \langle \mathcal{X}, \mathcal{A}^* \rangle - \sum_{t=1}^{n} \langle \mathcal{X}, \mathcal{A}_t \rangle \tag{17}$$

where $\mathcal{A}^* = \text{argmax}_{\mathcal{A}} \langle \mathcal{A}, \mathcal{X} \rangle$. As commonly observed in the bandit literature, even though $\mathcal{A}^*$ may not be optimally estimated, the optimal regret is still achievable.

5.1.3 Rank-1 bandit

Several existing RL methods can be categorized into the framework of (16), and they differ in terms of the structure of the action set $\mathcal{A}$ and the assumptions placed on $\mathcal{X}$. Particularly, Katariya *et al.* [72]; Katariya et al. [73]; Trinh et al. [74] considered stochastic rank-1 matrix bandit, where $\mathcal{X}$ is a rank-1 matrix, and vec$(\mathcal{A}_t)$ is a basis vector. The rank-1 structure greatly alleviates the difficulty of the problem, since one only needs to identify the largest values of the left- and right-singular vectors to find the largest entry of a nonnegative rank-1 matrix. Alternatively, Katariya *et al.* [72]; Katariya et al. [73] proposed special elimination-based algorithms, and Trinh *et al.* [74] viewed rank-1 bandit as a special instance of unimodal bandit [75]. However, neither of these solutions is applicable for general-rank matrices.

5.1.4 General-rank bandit

Kveton *et al.* [76]; Lu *et al.* [77] studied the extension of stochastic general low-rank matrix bandit, and Hao *et al.* [78] further generalized to stochastic low-rank tensor bandit. In particular, Kveton *et al.* [76] relied on a strong hot-topic assumption on the mean reward matrix, and their algorithm was computationally expensive. Lu *et al.* [77] utilized the ensemble sampling for low-rank matrix bandit but did not provide any regret guarantee due to the theoretical challenges in handling sampling-based exploration. Hao *et al.* [78] proposed a version of epoch-greedy algorithm [79] and a tensor elimination algorithm to handle both data-poor regime and -rich regime. The corresponding worse-case regret bounds were derived, though it is unclear if those bounds are optimal. In addition, Jun *et al.* [80]; Lu *et al.* [81] studied stochastic contextual low-rank matrix bandit, where vec$(\mathcal{A}_t)$ can be an arbitrary feature vector, and Hamidi *et al.* [82] considered linear contextual bandit with a low-rank structure.

5.1.5 Future directions

The key principle to design an algorithm for low-rank tensor bandit is to efficiently utilize the low-rank information while balancing the exploration–exploitation trade-off. Unfortunately, there is no consensus about what types of algorithm can explore the low-rank information in both a provable and practical manner. Actually, there is no direct upper confidence bound or Thompson sampling type algorithm for low-rank tensor bandit that is justified both empirically and theoretically for different structured bandit problems. The challenge is to construct a valid confidence bound or the posterior distribution of a nonconvex estimator in the sequential setting. In theory, although several regret upper bounds have been derived [78, 80, 81], the minimax lower bound of low-rank tensor bandit remains unestablished.

5.2 Learning Markov Decision Process via Tensor Decomposition

5.2.1 Motivating examples

We next turn to full RL of how an agent takes actions in an environment. A classical application is robotics, where a robot is to autonomously discover an optimal behavior through trial-and-error interactions with its environment; see Kober *et al.* [83] for a survey of RL in robotics. Particularly, Kober *et al.* [83] noted that a key challenge facing robotics RL is the high dimensionality of both the action space and the state space, due to many degrees of freedom of modern anthropomorphic robots. Tensor methods again offer useful dimension reduction tools.

5.2.2 Dimension reduction of Markov decision process

MDP is a fundamental model in RL that characterizes the interactions between an agent and an environment. We first briefly introduce some basic notations about MDP. For more details, we refer the readers to Puterman [84]. An instance of MDP $\mathcal{M}$ can be specified by a tuple $(\mathcal{S}, \mathcal{A}, \mathcal{P}, \boldsymbol{R})$, where $\mathcal{S}$ and $\mathcal{A}$ are the state and action spaces, $\mathcal{P} \in \mathbb{R}^{|\mathcal{S}|\times|\mathcal{A}|\times|\mathcal{S}|}$ is the transition probability tensor, and $\boldsymbol{R} \in \mathbb{R}^{|\mathcal{S}|\times|\mathcal{A}|}$ is a matrix whose entries represent the reward after taking a certain action under a certain state. A policy $\pi \in \mathbb{R}^{|\mathcal{S}|\times|\mathcal{A}|}$ is a set of probability distributions on actions conditioned on each state. In addition, $|\mathcal{S}| = p, |\mathcal{A}| = q$.

In most applications, for example, the robotics, the exact transition probability tensor of the MDP is unknown, and only a batch of empirical transition trajectories is available to the learner. Then, one of the key tasks is to efficiently estimate the MDP transition tensor from the batch data. A challenge, however, is the scale of the data, which makes both model estimation and policy optimization intractable [63].

Dimension reduction of MDP through matrix or tensor decompositions appears in a variety of RL solutions, including the MDP with rich observations [85], the state aggregation model [86–88], and the hidden Markov model [89], among others.

5.2.3 Maximum-likelihood estimation and Tucker decomposition

Ni and Wang [90] proposed a joint dimension reduction method for both the action and state spaces of the MDP transition tensor through the Tucker decomposition (2),

$$\mathcal{P} = [\![\tilde{\mathcal{P}}; \boldsymbol{U}_1, \boldsymbol{U}_2, \boldsymbol{U}_3]\!]$$

where $\tilde{\mathcal{P}} \in \mathbb{R}^{r_1\times r_2\times r_3}$ is the core tensor, and $\boldsymbol{U}_1 \in \mathbb{R}^{p\times r_1}$, $\boldsymbol{U}_2 \in \mathbb{R}^{q\times r_2}$, and $\boldsymbol{U}_3 \in \mathbb{R}^{p\times r_3}$ are the factor matrices. The Tucker rank (r_1, r_2, r_3) can be viewed as the intrinsic dimension of the MDP. When $q = 1$, the MDP reduces to a Markov chain, and the Tucker decomposition reduces to the spectral decomposition of the Markov chain [87, 91]. The factor matrices provide natural features for representing functions and operators on the action and state spaces, which can be applied together with feature-based RL methods [92].

A natural way to estimate the low-rank MDP transition tensor from the batch data is through maximum-likelihood estimation. Suppose that there are n independent state–action transition triplets $\{(s_k, a_k, s_k')\}_{k\in[n]}$. For $1 \le s, s' \le p, 1 \le a \le q$, define the empirical count as $n_{sas'} = \sum_{k=1}^{n} \mathbf{1}_{\{s_k=s,a_k=a,s_k'=s'\}}$. Given a fixed policy π, the negative log-likelihood based on the state–action transition triples $\{(s_k, a_k, s_k')\}_{k\in[n]}$ is

$$L(P) = -\sum_{s=1}^{p}\sum_{a=1}^{q}\sum_{s'=1}^{p} n_{sas'} \log(P_{(s,a,s')}) + C$$

where C is some constant unrelated with $\boldsymbol{\mathcal{P}}$. To estimate the MDP from sample transitions, Ni and Wang [90] proposed the following Tucker-constrained maximum-likelihood estimator:

$$\text{minimize } L(\boldsymbol{\mathcal{Q}}), \text{ such that } \boldsymbol{\mathcal{Q}}_{(\cdot,a,\cdot)}1_p = 1_p, \text{ Tuker-rank}(\boldsymbol{\mathcal{Q}}) \le (r_1, r_2, r_3), \text{and } a \in \mathcal{A}$$

Theoretically, Ni and Wang [90] showed that the maximum-likelihood estimator $\hat{\boldsymbol{\mathcal{P}}}$ satisfies the following bound with a high probability:

$$\|\hat{\boldsymbol{\mathcal{P}}} - \boldsymbol{\mathcal{P}}\|_F^2 \lesssim \left(\frac{\tilde{p}^2 \tilde{r}^2}{n} + q\sqrt{\frac{\log(\tilde{p})}{n}} \right) \tag{18}$$

where $\tilde{p} = \max(p, q), \tilde{r} = \max(r_1, r_2, r_3)$. The bound in (18) suggests that the estimation error is largely determined by the Tucker rank of the MDP instead of its actual dimension. This makes model compression possible with a limited number of data observations.

5.2.4 Future directions

Many questions in MDP remain open. For instance, it is unclear if the error-bound (18) is minimax optimal. After obtaining the low-rank representations of the MDP, it remains unclear how to embed them into the existing RL planning algorithms, and how the approximation error would affect the planning phase.

6 Tensor Deep Learning

The last topic we review is tensor deep learning. Deep learning represents a broad family of machine-learning methods based on artificial neural networks [93]. It has received enormous attention in recent years, thanks to its remarkable successes in a large variety of applications, including, but not limited to, image classification [94], speech recognition [95], and game playing [96]. We review two topics that connect tensors with deep learning: tensor-based compression of deep neural networks and deep learning theory through tensor representation.

6.1 Tensor-Based Deep Neural Network Compression

6.1.1 Motivating examples

Convolutional neural network (CNN) is perhaps the most common network structure in deep learning. It typically consists of a large number of convolutional layers, followed by a few fully connected layers. Therefore, it often requires a vast number of parameters and an enormous amount of training time even on the modern GPU clusters. For instance, the well-known VGG-19 network architecture [97] contains 10^8 parameters and requires over 15G floating-point operations to classify a single image. On the other hand, there is a growing interest to deploy CNNs on mobile devices, for example, smartphones and self-driving cars, to implement real-time image recognition and conversational system. Unfortunately, the expensive computational cost, in both time and memory, of the standard CNN architectures prohibits their deployments on such devices. For that reason, there have recently emerged some promising works to speed up CNNs through tensor-based dimension reduction.

Recurrent neural network (RNN) is another common network structure in deep learning [98]. It is particularly suitable for modeling temporal dynamics and has demonstrated excellent performance in sequential prediction tasks, for example, speech recognition [99] and traffic forecasting [100]. Despite their effectiveness for smooth and short-term dynamics, it is difficult to generalize RNN to capture nonlinear dynamics and long-term temporal dependency. Moreover, the standard version of RNN and its memory-based extension such as the long short-term memory (LSTM) network suffer from an excessive number of parameters, making it difficult to train and also susceptible to overfitting.

6.1.2 Compression of convolutional layers of CNN

Denton *et al.* [101]; Lebedev *et al.* [102]; Tai *et al.* [103] proposed low-rank approximations for the convolutional layers of CNN. Particularly, Lebedev *et al.* [102] applied the CP decomposition (1) for the convolutional layers, while Kim *et al.* [104] applied the Tucker decomposition (2) on the convolutional kernel tensors of a pretrained network and then fine-tuned the resulting network. Meanwhile, which decomposition is better depends on the application domains, tasks, network architectures, and hardware constraints. Recognizing this issue, Hayashi *et al.* [105] proposed to characterize a decomposition class specific to CNNs, by adopting a flexible hypergraphical notion in tensor networks. This class includes modern light-weight CNN layers, such as the bottleneck layers in ResNet [106], the depthwise separable layers in Mobilenet V1 [107], and the inverted bottleneck layers in Mobilenet V2 [108], among others. Moreover, this class can also deal with nonlinear activations by combining neural architecture search with the LeNet and ResNet architectures. Furthermore, Kossaifi *et al.* [109] introduced a tensor factorization framework for efficient multidimensional convolutions of higher order CNNs, with applications to spatiotemporal emotion estimation.

6.1.3 Compression of fully-connected layers of CNN

In a standard CNN architecture, the activation tensors of convolutional layers are first flattened and then connected to the outputs through fully connected layers. This step introduces a large number of parameters, and the flattening operation may also lose multimodal information. As an example, in the VGG-19 network architecture, about 80% of its parameters come from the fully connected layers [97]. Motivated by these observations, Novikov *et al.* [110] applied the tensor-train decomposition, Ye *et al.* [111] applied the block-term decomposition, and Kossaifi *et al.* [112] applied the Tucker decomposition, all focusing on reducing the number of parameters in the fully connected layers.

Figure 7 provides an outline of the tensor-based CNN compression strategy from Kossaifi *et al.* [112]. Built upon a standard CNN architecture, it consists of two new layers, a tensor contraction layer and a tensor regression layer, as the end-to-end trainable components of deep neural networks. After the standard convolutional layer and activation step, the tensor contraction layer reduces the dimensionality of the activation tensor $\mathcal{X}_i$ via a Tucker decomposition to obtain a dimension-reduced tensor $\mathcal{X}'_i$. The tensor regression layer then directly associates $\mathcal{X}'_i$ with the response y_i via a low-rank Tucker structure on the coefficient $\boldsymbol{\mathcal{B}}$, which helps avoid the flattening operation in the traditional fully connected layer. All the parameters can be efficiently learned via end-to-end backpropagation.

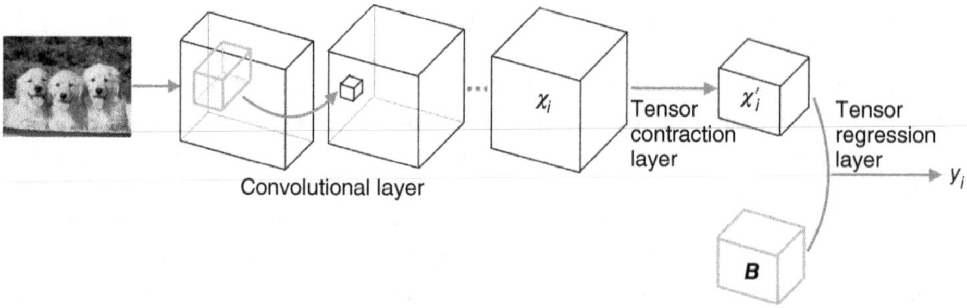

Figure 7 Illustration of the tensor-based CNN compression from Kossaifi *et al.* [112]. *Source*: Based on Kossaifi, J., Lipton, Z. C., Khanna, A., Furlanello, T. and Anandkumar, A. (2020). Tensor regression networks. Journal of Machine Learning Research 1–21.

6.1.4 Compression of all layers of CNN

In addition to compression of the convolutional layers and fully connected layers separately, there is the third category of compression methods targeting all layers. This enables to learn the correlations between different tensor dimensions. Moreover, the low-rank structure on the weight tensor acts as an implicit regularization and can substantially reduce the number of parameters. Specifically, Kasiviswanathan *et al.* [113] incorporated the randomized tensor sketching technique and developed a unified framework to approximate the operations of both the convolutional and fully connected layers in CNNs. Kossaifi *et al.* [114] proposed to fully parameterize all layers of CNNs with a single high-order low-rank tensor, where the modes of the tensor represent the architectural design parameters of the network, including the number of convolutional blocks, depth, number of stacks, and input features.

6.1.5 Compression of RNN

Yang *et al.* [115]; Yu *et al.* [116]; Su *et al.* [117] utilized the tensor-train decomposition to efficiently learn the nonlinear dynamics of RNNs, by directly using high-order moments and high-order state transition functions. In addition, Ye *et al.* [118] proposed a compact and flexible structure called the block-term tensor decomposition for dimension reduction in RNNs and showed that it is not only more concise but also able to attain a better approximation to the original RNNs with much fewer parameters.

6.1.6 Future directions

Although the tensor-based DNN compression methods have shown great empirical success, the theoretical properties are still not yet fully understood. Moreover, the existing solutions have been focusing on the low-rank structure for dimension reduction. It is potentially useful to consider the additional sparsity structure, for example, the sparse tensor factorization [33], to further reduce the number of parameters and to improve the interpretability of the tensor layers in CNNs or RNNs.

6.2 Deep Learning Theory through Tensor Methods

6.2.1 Motivating examples

Despite the wide empirical success of deep neural network models, their theoretical properties are much less understood. Next, we review a few works that use tensor representations to facilitate the understanding of the expressive power, compressibility, generalizability, and other properties of deep neural networks.

6.2.2 Expressive power, compressibility and generalizability

Cohen *et al.* [119] used tensor as an analytical tool to study the expressive power of deep neural networks, where the expressive power refers to the representation ability of a neural network architecture. They established an equivalence between the neural network and hierarchical tensor factorization and showed that a shallow network corresponds to a rank-1 CP decomposition, whereas a deep network corresponds to a hierarchical Tucker decomposition. Through this connection, they further proved that, other than a measure zero negligible set, all functions that can be implemented by a deep network of the polynomial order would require an exponential order shallow network to realize. Built on this general tensor tool, various recent works have extended the study of expressive power to the overlapping architecture of deep learning [120], RNNs with multiplicative recurrent cells [121], and RNNs with rectifier nonlinearities [122].

Li *et al.* [123] employed tensor analysis to derive a set of data-dependent and easily measurable properties that tightly characterize the compressibility and generalizability of neural networks. Specifically, the compressibility measures how much the original network can be compressed without compromising the performance on a training dataset more than a certain range. The generalizability measures the performance of a neural network on the unseen testing data. Compared to the generalization bounds via compression scheme [124], Li *et al.* [123] provided a much tighter bound for the layerwise error propagation, by exploiting the additional structures in the weight tensor of a neural network.

6.2.3 Additional connections

There are other connections between deep learning theory and tensors. Janzamin *et al.* [125] provided a polynomial-time algorithm based on tensor decomposition for learning one-hidden-layer neural networks with twice-differential activation function and known input distributions. Moreover, Ge *et al.* [126] considered learning a one-hidden-layer neural network and proved that the population risk of the standard squared loss implicitly attempts to decompose a sequence of low-rank tensors simultaneously. Mondelli and Montanari [127] also established connections between tensor decomposition and the problem of learning a one-hidden-layer neural network with activation functions given by low-degree polynomials. They provided evidence that in certain regimes, and for certain data distributions, the one-hidden-layer neural network cannot be learnt in polynomial time. So, similar to Ge *et al.* [126], they also considered the case when the data distribution is normal.

6.2.4 Future directions

Aforementioned works [125–127] provide theoretical foundations for the connection between tensor decomposition and learning one-hidden-layer neural network. It is of interest to study how such a connection can be extended to more general deep neural network architectures and more general data distributions. It is also of interest to investigate if the theoretical results of Li *et al.* [123] can be extended to study the compressibility and generalizability of more deep neural network architectures.

Acknowledgments

Sun's research was partially supported by Office of Naval Research grant N00014-18-1-2759. Li's research was partially supported by NSF grant DMS-1613137 and NIH grants R01AG034570 and R01AG061303.

References

1 Rendle, S. and Schmidt-Thieme, L. (2010) *Pairwise Interaction Tensor Factorization for Personalized Tag Recommendation*. International Conference on Web Search and Data Mining.

2 Bi, X., Qu, A., Shen, X. *et al.* (2018) Multilayer tensor factorization with applications to recommender systems. *Ann. Stat.*, 46, 3308–3333.

3 Vasilescu, M. and Terzopoulos, D. (2002) *Multilinear Analysis of Image Ensembles: Tensorfaces*. European Conference on Computer Vision.

4 Ma, X., Zhang, P., Zhang, S. *et al.* (2019) A tensorized transformer for language modeling. Advances in Neural Information Processing Systems.

5 Li, W., Liu, C.-C., Zhang, T. *et al.* (2011) Integrative analysis of many weighted co-expression networks using tensor computation. *PLoS Comput. Biol.*, 7, e1001106.

6 Ermis, B., Acar, E., and Cemgil, A.T. (2015) Link prediction in heterogeneous data via generalized coupled tensor factorization. Data Mining and Knowledge Discovery.

7 Trouillon, T., Dance, C.R., Gaussier, E. *et al.* (2017) Knowledge graph completion via complex tensor factorization. *J. Mach. Learn. Res.*, 18, 4735–4772.

8 Liu, Y., Yao, Q., and Li, Y. (2020) Generalizing Tensor Decomposition for N-ary Relational Knowledge Bases. Proceedings of The Web Conference 2020.

9 Kolda, T. and Bader, B. (2009) Tensor decompositions and applications. *SIAM Rev.*, 51, 455–500.

10 Rabanser, S., Shchur, O., and Gnnemann, S. (2017) Introduction to tensor decompositions and their applications in machine learning. *arXiv preprint arXiv:1711.10781*.

11 Sidiropoulos, N.D., De Lathauwer, L., Fu, X. *et al.* (2017) Tensor decomposition for signal processing and machine learning. *IEEE Trans. Signal Process.*, 65, 3551–3582.

12 Janzamin, M., Ge, R., Kossaifi, J., and Anandkumar, A. (2019) Spectral learning on matrices and tensors. *Found. Trends Mach. Learn.*, 12, 393–536.

13 Song, Q., Ge, H., Caverlee, J., and Hu, X. (2019) Tensor completion algorithms in big data analytics. *ACM Trans. Knowl. Discovery Data*, 13, 148.

14 Bi, X., Tang, X., Yuan, Y. *et al.* (2020) Tensor in statistics. *Ann. Rev. Stat. Appl.*, 8, 2.1–2.24.

15 Zhou, H., Li, L., and Zhu, H. (2013) Tensor regression with applications in neuroimaging data analysis. *J. Am. Stat. Assoc.*, 108, 540–552.

16 Li, Z., Suk, H.-I., Shen, D., and Li, L. (2016) Sparse multi-response tensor regression for Alzheimer's disease study with multivariate clinical assessments. *IEEE Trans. Med. Imaging*, 35, 1927–1936.

17 Guhaniyogi, R., Qamar, S., and Dunson, D.B. (2017) Bayesian tensor regression. *J. Mach. Learn. Res.*, 18, 2733–2763.

18 Li, X., Xu, D., Zhou, H., and Li, L. (2018) Tucker tensor regression and neuroimaging analysis. *Stat. Biosci.*, 10, 520–545.

19 Zhang, X., Li, L., Zhou, H., and Shen, D. (2019) Tensor generalized estimating equations for longitudinal imaging analysis. *Stat. Sin.*, 29, 1977–2005.

20 Yu, R. and Liu, Y. (2016) *Learning from Multiway Data: Simple and Efficient Tensor Regression*. International Conference on Machine Learning.

21 Clarkson, K.L. and Woodruff, D.P. (2017) Low-rank approximation and regression in input sparsity time. *J. ACM*, 63, 145.

22 Zhang, A., Luo, Y., Raskutti, G., and Yuan, M. (2020) Islet: fast and optimal low-rank tensor regression via importance sketching. *SIMODS*, 2, 444–479.

23 De Lathauwer, L., De Moor, B., and Vandewalle, J. (2000) On the best rank-1 and rank-(r_1, r_2, …, r_n) approximation of higher-order tensors. *SIAM J. Matrix Anal. Appl.*, 21, 1324–1342.

24 Kanagawa, H., Suzuki, T., Kobayashi, H. *et al.* (2016) *Gaussian Process Nonparametric Tensor Estimator and Its Minimax Optimality*. International Conference on Machine Learning.

25 Suzuki, T., Kanagawa, H., Kobayashi, H. *et al.* (2016) Minimax optimal alternating minimization for kernel nonparametric tensor learning. Advances in Neural Information Processing Systems.

26 Hao, B., Wang, B., Wang, P. *et al.* (2019) Sparse tensor additive regression. *arXiv preprint arXiv:1904.00479*.

27 Hastie, T.J. and Tibshirani, R.J. (1990) *Generalized Additive Models*, vol. 43, CRC Press.

28 Ravikumar, P., Lafferty, J., Liu, H., and Wasserman, L. (2009) Sparse additive models. *J. R. Stat. Soc., Ser. B*, 71, 1009–1030.

29 Zhou, Y., Wong, R.K.W., and He, K. (2020) Broadcasted nonparametric tensor regression. *arXiv preprint arXiv:2008.12927*.

30 Zhang, Y., Duchi, J., and Wainwright, M. (2015) Divide and conquer kernel ridge regression: a distributed algorithm with minimax optimal rates. *J. Mach. Learn. Res.*, 16, 3299–3340.

31 Rabusseau, G. and Kadri, H. (2016) Low-rank regression with tensor responses. Advances in Neural Information Processing Systems.

32 Sun, W. and Li, L. (2017) Sparse tensor response regression and neuroimaging analysis. *J. Mach. Learn. Res.*, 18, 4908–4944.

33 Sun, W., Lu, J., Liu, H., and Cheng, G. (2017) Provable sparse tensor decomposition. *J. R. Stat. Soc., Ser. B*, 79, 899–916.

34 Li, L. and Zhang, X. (2017) Parsimonious tensor response regression. *J. Am. Stat. Assoc.*, 112, 1131–1146.

35 Raskutti, G., Yuan, M., Chen, H. *et al.* (2019) Convex regularization for high-dimensional multiresponse tensor regression. *Ann. Stat.*, 47, 1554–1584.

36 Chen, H., Raskutti, G., and Yuan, M. (2019) Non-convex projected gradient descent for generalized low-rank tensor regression. *J. Mach. Learn. Res.*, 20, 172–208.

37 Zhou, J., Sun, W.W., Zhang, J., and Li, L. (2020) Partially observed dynamic tensor response regression. *arXiv preprint arXiv:2002.09735*.

38 Jain, P. and Oh, S. (2014) Provable tensor factorization with missing data. Advances in Neural Information Processing Systems.

39 Madeira, S.C. and Oliveira, A.L. (2004) Biclustering algorithms for biological data analysis: a survey. *IEEE/ACM Trans. Comput. Biol. Bioinf.*, 1, 24–45.

40 Chi, E.C., Allen, G.I., and Baraniuk, R.G. (2017) Convex biclustering. *Biometrics*, 73, 10–19.

41 Chi, E.C., Gaines, B.R., Sun, W.W. *et al.* (2018) Provable convex co-clustering of tensors. *arXiv preprint arXiv:1803.06518*.

42 Papalexakis, E.E., Sidiropoulos, N.D., and Bro, R. (2013) From K-means to higher-way co-clustering: multilinear decomposition with sparse latent factors. *IEEE Trans. Signal Process.*, 61, 493–506.

43 Sun, W.W. and Li, L. (2019) Dynamic tensor clustering. *J. Am. Stat. Assoc.*, 114, 1894–1907.

44 Zhang, C., Fu, H., Liu, S. *et al.* (2015) *Low-Rank Tensor Constrained Multiview Subspace Clustering*. Proceedings of the IEEE International Conference on Computer Vision.

45 Wu, T., Benson, A.R., and Gleich, D.F. (2016) General tensor spectral co-clustering for higher-order data. Advances in Neural Information Processing Systems.

46 Luo, Y. and Zhang, A.R. (2020) Tensor clustering with planted structures: statistical optimality and computational limits. *arXiv preprint arXiv:2005.10743*.

47 Hao, B., Sun, W.W., Liu, Y., and Cheng, G. (2018) Simultaneous clustering and estimation of heterogeneous graphical models. *J. Mach. Learn. Res.*, 18 (271), 1–58.

48 Wang, J. (2010) Consistent selection of the number of clusters via cross validation. *Biometrika*, 97, 893–904.

49 Zahn, J., Poosala, S., Owen, A. *et al.* (2007) AGEMAP: a gene expression database for aging in mice. *PLoS Genet.*, 3, 2326–2337.

50 He, S., Yin, J., Li, H., and Wang, X. (2014) Graphical model selection and estimation for high dimensional tensor data. *J. Multivar. Anal.*, 128, 165–185.

51 Sun, W., Wang, Z., Liu, H., and Cheng, G. (2015) Non-convex statistical optimization for sparse tensor graphical model. *Adv. Neural Inf. Process. Syst.*, 28, 1081–1089.

52 Yuan, M. and Lin, Y. (2007) Model selection and estimation in the gaussian graphical model. *Biometrika*, 94, 19–35.

53 Friedman, J., Hastie, H., and Tibshirani, R. (2008) Sparse inverse covariance estimation with the graphical Lasso. *Biostatistics*, 9, 432–441.

54 Leng, C. and Tang, C. (2012) Sparse matrix graphical models. *J. Am. Stat. Assoc.*, 107, 1187–1200.

55 Yin, J. and Li, H. (2012) Model selection and estimation in the matrix normal graphical model. *J. Multivar. Anal.*, 107, 119–140.

56 Tsiligkaridis, T., Hero, A.O., and Zhou, S. (2013) On convergence of Kronecker graphical Lasso algorithms. *IEEE Trans. Signal Process.*, 61, 1743–1755.

57 Zhou, S. (2014) Gemini: graph estimation with matrix variate normal instances. *Ann. Stat.*, 42, 532–562.

58 Lyu, X., Sun, W.W., Wang, Z. *et al.* (2019) Tensor graphical model: non-convex optimization and statistical inference. *IEEE Trans. Pattern Anal. Mach. Intell.*, 42, 2024–2037.

59 Greenewald, K., Zhou, S., and Hero III, A. (2019) Tensor graphical lasso (teralasso). *J. R. Stat. Soc., Ser. B*, 81, 901–931.

60 Wang, Y., Jang, B., and Hero, A. (2020) *The Sylvester Graphical Lasso (Syglasso).* International Conference on Artificial Intelligence and Statistics.

61 Liu, H., Lafferty, J., and Wasserman, L. (2009) The nonparanormal: semiparametric estimation of high dimensional undirected graphs. *J. Mach. Learn. Res.*, 10, 2295–2328.

62 Ning, Y. and Liu, H. (2013) High-dimensional semiparametric bigraphical models. *Biometrika*, 100, 655–670.

63 Sutton, R.S. and Barto, A.G. (2018) *Reinforcement Learning: An Introduction*, MIT press.

64 Ge, H., Caverlee, J., and Lu, H. (2016) *Taper: A Contextual Tensor-Based Approach for Personalized Expert Recommendation.* Proceedings of the 10th ACM Conference on Recommender Systems.

65 Frolov, E. and Oseledets, I. (2017) Tensor methods and recommender systems. *Wiley Interdisc. Rev. Data Min. Knowl. Discov.*, 7, e1201.

66 Lattimore, T. and Szepesvári, C. (2020) *Bandit Algorithms*, Cambridge University Press.

67 Li, L., Chu, W., Langford, J., and Schapire, R.E. (2010) *A Contextual-Bandit Approach to Personalized News Article Recommendation.* Proceedings of the 19th International Conference on World Wide Web.

68 Dani, V., Hayes, T.P., and Kakade, S.M. (2008) *Stochastic Linear Optimization Under Bandit Feedback.* 21st Annual Conference on Learning Theory, COLT 2008.

69 Rusmevichientong, P. and Tsitsiklis, J.N. (2010) Linearly parameterized bandits. *Math. Oper. Res.*, 35, 395–411.

70 Chu, W., Li, L., Reyzin, L., and Schapire, R. (2011) *Contextual Bandits with Linear Payoff Functions.* Proceedings of the Fourteenth International Conference on Artificial Intelligence and Statistics.

71 Abbasi-Yadkori, Y., Pál, D., and Szepesvári, C. (2011) Improved algorithms for linear stochastic bandits. Advances in Neural Information Processing Systems.

72 Katariya, S., Kveton, B., Szepesvari, C. *et al.* (2017) Stochastic rank-1 bandits. Artificial Intelligence and Statistics.

73 Katariya, S., Kveton, B., Szepesvári, C. *et al.* (2017) *Bernoulli Rank-1 Bandits for Click Feedback.* Proceedings of the 26th International Joint Conference on Artificial Intelligence.

74 Trinh, C., Kaufmann, E., Vernade, C., and Combes, R. (2020) Solving Bernoulli rank-one bandits with unimodal Thompson sampling. Algorithmic Learning Theory.

75 Combes, R. and Proutiere, A. (2014) *Unimodal Bandits: Regret Lower Bounds and Optimal Algorithms.* International Conference on Machine Learning.

76 Kveton, B., Szepesvári, C., Rao, A. *et al.* (2017) Stochastic low-rank bandits. *arXiv preprint arXiv:1712.04644.*

77 Lu, X., Wen, Z., and Kveton, B. (2018) *Efficient Online Recommendation Via Low-Rank Ensemble Sampling.* Proceedings of the 12th ACM Conference on Recommender Systems.

78 Hao, B., Zhou, J., Wen, Z., and Sun, W.W. (2020) Low-rank tensor bandits. *arXiv preprint arXiv:2007.15788.*

79 Langford, J. and Zhang, T. (2008) The epoch-greedy algorithm for multi-armed bandits with side information. Advances in Neural Information Processing Systems.

80 Jun, K.-S., Willett, R., Wright, S., and Nowak, R. (2019) Bilinear bandits with low-rank structure. *arXiv preprint arXiv:1901.02470.*

81 Lu, Y., Meisami, A., and Tewari, A. (2020) Low-rank generalized linear bandit problems. *arXiv preprint arXiv:2006.02948*.

82 Hamidi, N., Bayati, M., and Gupta, K. (2019) Personalizing many decisions with high-dimensional covariates. Advances in Neural Information Processing Systems.

83 Kober, J., Bagnell, J.A., and Peters, J. (2013) Reinforcement learning in robotics: a survey. *Int. J. Rob. Res.*, 32, 1238–1274.

84 Puterman, M.L. (2014) *Markov Decision Processes: Discrete Stochastic Dynamic Programming*, John Wiley & Sons.

85 Azizzadenesheli, K., Lazaric, A., and Anandkumar, A. (2016) *Reinforcement Learning of POMDPs Using Spectral Methods*. Proceedings of the 29th Annual Conference on Learning Theory (COLT2016).

86 Bertsekas, D.P., Bertsekas, D.P., Bertsekas, D.P., and Bertsekas, D.P. (2005) *Dynamic Programming and Optimal Control*, vol. 1, Athena Scientific, Belmont, MA.

87 Zhang, A. and Wang, M. (2019) Spectral state compression of Markov processes. *IEEE Trans. Inf. Theory*, 66, 3202–3231.

88 Duan, Y., Ke, T., and Wang, M. (2019) State aggregation learning from Markov transition data. Advances in Neural Information Processing Systems.

89 Hsu, D., Kakade, S.M., and Zhang, T. (2012) A spectral algorithm for learning hidden Markov models. *J. Comput. Syst. Sci.*, 78, 1460–1480.

90 Ni, C. and Wang, M. (2019) *Maximum Likelihood Tensor Decomposition of Markov Decision Process*. 2019 IEEE International Symposium on Information Theory (ISIT). IEEE.

91 Li, X., Wang, M., and Zhang, A. (2018) *Estimation of Markov Chain Via Rank-Constrained Likelihood*. 35th International Conference on Machine Learning, ICML 2018. International Machine Learning Society (IMLS).

92 Ernst, D., Geurts, P., and Wehenkel, L. (2005) Tree-based batch mode reinforcement learning. *J. Mach. Learn. Res.*, 6, 503–556.

93 LeCun, Y., Bengio, Y., and Hinton, G. (2015) Deep learning. *Nature*, 521, 436–444.

94 Krizhevsky, A., Sutskever, I., and Hinton, G.E. (2012) *ImageNet Classification with Deep Convolutional Neural Networks*. Proceedings of the 25th International Conference on Neural Information Processing Systems - Volume 1, NIPS'12.

95 Hinton, G., Deng, L., Yu, D. *et al.* (2012) Deep neural networks for acoustic modeling in speech recognition: the shared views of four research groups. *IEEE Signal Processing Mag.*, 29, 82–97.

96 Silver, D., Huang, A., Maddison, C.J. *et al.* (2016) Mastering the game of go with deep neural networks and tree search. *Nature*, 529, 484–489.

97 Simonyan, K. and Zisserman, A. (2015) *Very Deep Convolutional Networks for Large-Scale Image Recognition*. International Conference on Learning Representations.

98 Hochreiter, S. and Schmidhuber, J. (1997) Long short-term memory. *Neural Comput.*, 9, 1735–1780.

99 Graves, A., Mohamed, A.R., and Hinton, G. (2013) *Speech Recognition with Deep Recurrent Neural Networks*. IEEE International Conference on Acoustics, Speech and Signal Processing (ICASSP).

100 Li, Y., Yu, R., Shahabi, C., and Liu, Y. (2018) *Diffusion Convolutional Recurrent Neural Ntwork: Data-Driven Traffic Forecasting*. International Conference on Learning Representations.

101 Denton, E., Zaremba, W., Bruna, J. *et al.* (2014) *Exploiting Linear Structure Within Convolutional Networks for Efficient Evaluation.* Proceedings of the 27th International Conference on Neural Information Processing Systems, NIPS'14.

102 Lebedev, V., Ganin, Y., Rakhuba, M. *et al. Speeding-Up Convolutional Neural Networks Using Fine-Tuned CP-Decomposition.* International Conference on Learning Representations.

103 Tai, C., Xiao, T., Zhang, Y. *et al.* (2016) *Convolutional Neural Networks with Low-Rank Regularization.* International Conference on Learning Representations.

104 Kim, Y.-D., Park, E., Yoo, S. *et al.* (2016) *Compression of Deep Convolutional Neural Networks for Fast and Low Power Mobile Applications.* International Conference on Learning Representations.

105 Hayashi, K., Yamaguchi, T., Sugawara, Y., and Maeda, S.-i. (2019) Exploring unexplored tensor network decompositions for convolutional neural networks. Advances in Neural Information Processing Systems.

106 He, K., Zhang, X., Ren, S., and Sun, J. (2016) *Deep Residual Learning for Image Recognition.* Proceedings of the IEEE Conference on Computer Vision and Pattern Recognition.

107 Howard, A.G., Zhu, M., Chen, B. *et al.* (2017) Mobilenets: efficient convolutional neural networks for mobile vision applications. *arXiv preprint arXiv:1704.04861.*

108 Sandler, M., Howard, A., Zhu, M. *et al.* (2018) *Mobilenetv2: Inverted Residuals and Linear Bottlenecks.* Proceedings of the IEEE Conference on Computer Vision and Pattern Recognition.

109 Kossaifi, J., Toisoul, A., Bulat, A. *et al.* (2020) *Factorized Higher-Order CNNS with an Application to Spatio-Temporal Emotion Estimation.* Proceedings of the IEEE/CVF Conference on Computer Vision and Pattern Recognition.

110 Novikov, A., Podoprikhin, D., Osokin, A., and Vetrov, D. (2015) Tensorizing Neural Networks. Proceedings of the 28th International Conference on Neural Information Processing Systems - Volume 1, InNIPS'15.

111 Ye, J., Li, G., Chen, D. *et al.* (2020) Block-term tensor neural networks. *Neural Networks*, 130, 11–21.

112 Kossaifi, J., Lipton, Z.C., Khanna, A. *et al.* (2020) Tensor regression networks. *J. Mach. Learn. Res.*, 21, 1–21.

113 Kasiviswanathan, S.P., Narodytska, N., and Jin, H. (2018) *Network Approximation Using Tensor Sketching.* Proceedings of the 27th International Joint Conference on Artificial Intelligence.

114 Kossaifi, J., Bulat, A., Tzimiropoulos, G., and Pantic, M. (2019) *T-Net: Parametrizing Fully Convolutional Nets with a Single High-Order Tensor.* Proceedings of the IEEE Conference on Computer Vision and Pattern Recognition.

115 Yang, Y., Krompass, D., and Tresp, V. (2017) *Tensor-Train Recurrent Neural Networks for Video Classification.* International Conference on Machine Learning.

116 Yu, R., Zheng, S., Anandkumar, A., and Yue, Y. (2019) Long-term forecasting using higher-order tensor RNNS. *arXiv preprint arXiv:1711.00073v2.*

117 Su, J., Byeon, W., Huang, F. *et al.* (2020) Convolutional tensor-train LSTM for Spatio-temporal learning. *arXiv preprint arXiv:2002.09131.*

118 Ye, J., Wang, L., Li, G. *et al.* (2018) *Learning Compact Recurrent Neural Networks with Block-Term Tensor Decomposition*. Proceedings of the IEEE Conference on Computer Vision and Pattern Recognition.

119 Cohen, N., Sharir, O., and Shashua, A. (2016) On the Expressive Power of Deep Learning: A Tensor Analysis. Conference on Learning Theory.

120 Sharir, O. and Shashua, A. (2018) *On the Expressive Power of Overlapping Architectures of Deep Learning*. International Conference on Learning Representations.

121 Khrulkov, V., Novikov, A., and Oseledets, I. (2018) *Expressive Power of Recurrent Neural Networks*. International Conference on Learning Representations.

122 Khrulkov, V., Hrinchuk, O., and Oseledets, I. (2019) *Generalized Tensor Models for Recurrent Neural Networks*. International Conference on Learning Representations.

123 Li, J., Sun, Y., Su, J. *et al.* (2020) *Understanding Generalization in Deep Learning Via Tensor Methods*. International Conference on Artificial Intelligence and Statistics.

124 Arora, S., Ge, R., Neyshabur, B., and Zhang, Y. (2018) *Stronger Generalization Bounds for Deep Nets Via a Compression Approach*. 35th International Conference on Machine Learning, ICML 2018.

125 Janzamin, M., Sedghi, H., and Anandkumar, A. (2015) Beating the perils of non-convexity: guaranteed training of neural networks using tensor methods. *arXiv preprint arXiv:1506.08473*.

126 Ge, R., Lee, J.D., and Ma, T. (2018) *Learning One-Hidden-Layer Neural Networks with Landscape Design*. 6th International Conference on Learning Representations, ICLR 2018.

127 Mondelli, M. and Montanari, A. (2019) *On the Connection between Learning Two-Layer Neural Networks and Tensor Decomposition*. The 22nd International Conference on Artificial Intelligence and Statistics.

15

Computational Approaches to Bayesian Additive Regression Trees

Hugh Chipman[1], Edward George[2], Richard Hahn[3], Robert McCulloch[3], Matthew Pratola[4], and Rodney Sparapani[5]

[1] *Acadia University Wolfville, Nova Scotia, Canada*
[2] *The Wharton School, University of Pennsylvania, Philadelphia, PA, USA*
[3] *The School of Mathematical and Statistical Sciences, Arizona State University, Tempe, AZ, USA*
[4] *The Ohio State University, Columbus, OH, USA*
[5] *Institute for Health and Equity, Medical College of Wisconsin, Milwaukee, WI, USA*

1 Introduction

Our problem is to predict a target variable y, given the information in a vector of predictor variables $\mathbf{x}$. Approaches based on trees play a large role in the development of predictive methodology. The classic CART work [1], which uses a single tree, is still a very important part of our toolkit. Ensemble methods, which use many trees, such as random forests [2] and boosting [3], have proven remarkably effective. The XGBoost approach to boosting [4] is heavily used in applications. In this chapter, we explore some of the modeling and computational issues involved in an approach to a Bayesian analysis of tree-based models.

The Bayesian approach offers some attractive features. Perhaps, most fundamentally, priors can be used to express interesting beliefs about complex models. Computation of the posterior motivates interesting explorations of the model space and helps us assess our inferential uncertainty. Multiple Bayesian tree-based models may be embedded in larger models using the standard hierarchical modeling framework.

A Bayesian approach requires us to formulate a tree model as a parameter, place a prior on the parameter, and define a computable likelihood. Section 2 reviews the approach developed in Ref. 5 (hereafter CGM98). Section 3 reviews the Markov chain Monte Carlo (MCMC) algorithm of CGM98 for computing the posterior. The presentation in this chapter spells out some important details not readily discernible from CGM98. These algorithmic details underlie the implementations in the `R` packages `BART` and `BayesTree` that are both available on the Comprehensive R Archive Network (CRAN): https://cran.r-project.org. Section 3 reviews the more recent advances in tree model MCMC due to Pratola [6]. These algorithms are used in the `R` package `rbart` that is also on CRAN. All three of these `R` packages are based on code written in C++, which is then called from `R`.

Computational Statistics in Data Science.
Edited by Walter W. Piegorsch, Richard A. Levine, Hao Helen Zhang and Thomas C. M. Lee.
 ISBN 978-1-11956107-1

Building upon CGM98 and boosting (in particular Ref. 7), Section 4 reviews the Bayesian approach to ensemble tree modeling developed in Ref. 8 (hereafter CGM10). The model developed in CGM10 is known as Bayesian Additive Regression Trees (BART). BART is applied to the classic Boston housing values and air pollution data set in Section 5. Section 6 reviews the MCMC algorithm for posterior computation.

In Section 7, we review two modeling approaches to illustrate the power of Bayesian tree modeling beyond the basic development of CGM98 and CGM10. We make no attempt at a comprehensive review of Bayesian tree models but highlight two examples that we find compelling. Section 7.1 explores the fundamental issues of sparsity and variable selection. Section 7.2 develops computational and modeling approaches that dramatically improve the computational speed of Bayesian approaches, making inference with large numbers of observations and predictors feasible. Note that the two articles that use multiple BART models to address issues in causal inference are Richard Hahn *et al.* [9] and McCulloch *et al.* [10]. Section 8 concludes.

2 Bayesian CART

We begin by laying out the structure of a tree model. This follows the general structure of the usual thing seen in a CART type model, but our notation and discussion is targeted toward our ultimate goal of a Bayesian analysis. Our Bayesian approach requires a specification of a prior on a tree based model and an MCMC algorithm for posterior computation.

Note that while we first cover the basics of modeling and computation for a model based on a single tree (Sections 2 and 3), this methodology underlies the more powerful BART approach, as a complete understanding of the single-tree material is needed to understand BART (Sections 4 and 6).

2.1 A Single-Tree Model

A binary tree consists of a set of internal nodes and a set of terminal nodes. We also call the terminal nodes bottom nodes. Each internal node has a binary decision rule associated with it. Internal nodes spawn left and right children, each of which in turn may be internal nodes with decision rules or terminal nodes. Each terminal node has a parameter associated with it. We let $\mathcal{T}$ denote the tree including the decision rules at the interior nodes. Let $\Theta = (\theta_1, \theta_2, \dots, \theta_b)$ denote the set of parameters at the b bottom nodes.

Given a predictor vector $\mathbf{x}$, you "drop it down the tree" using each decision rule to send $\mathbf{x}$ left or right to the left child node or the right child node. When $\mathbf{x}$ finally lands in a terminal node, there is a parameter value awaiting it.

Figure 1 depicts a simple example; the rest of this section will discuss this illustration. The tree has four internal nodes with labels $\{1, 2, 3, 5\}$ and five terminal nodes with labels $\{4, 10, 11, 6, 7\}$. The left child of node i is labeled $2i$, and the right child is labeled $2i + 1$.

Each decision rule is based on a single predictor x_i. The decision rule in node 1 uses x_2. The form of the decision rule depends on whether x_i is numeric or categorical. With numeric variables, we choose a cut-point c and then go left if $x_i \le c$, and right otherwise. The decision rule in node 2 uses x_1 and $c = 3$. With categorical variables, a decision rule

Figure 1 A Bayesian tree.

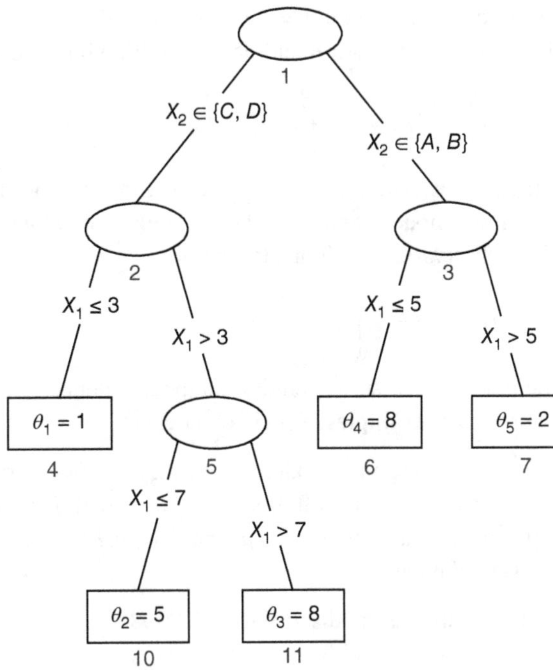

specifies which categories go left, and the rest go right. For example, x_2 is categorical with possible values $\{A, B, C, D\}$. The decision rule in node 1 of Figure 1 sends categories $\{C, D\}$ left and categories $\{A, B\}$ right.

It is convenient to have a linear integer index for the bottom node parameters. Our convention is that we number the bottom nodes "left to right." For example, $\Theta = (1, 5, 8, 8, 2)$. We have $\theta_2 = 5$ even though this corresponds to the bottom node with integer label 10. Each predictor vector $\mathbf{x}$ has a corresponding bottom node, and we let $\zeta(\mathbf{x})$ be the linear index of the bottom node corresponding to $\mathbf{x}$. So, if $\mathbf{x} = (x_1, x_2) = (4, B)$, then $\zeta(\mathbf{x}) = 4$.

2.2 Tree Model Likelihood

The parameter of our model is $(\mathcal{T}, \Theta)$. To obtain our likelihood, we start from a parametric model $Y \sim f(y \mid \theta)$. The idea is that given $(\mathcal{T}, \Theta)$ and $\mathbf{x}$, we drop $\mathbf{x}$ down the tree and then use the θ value in the terminal node $\mathbf{x}$ lands in. If we let $\zeta(\mathbf{x})$ be the index in Θ of the terminal node corresponding to $\mathbf{x}$, then

$$Y \mid \mathbf{x}, (\mathcal{T}, \Theta) \sim f(y \mid \theta_{\zeta(\mathbf{x})})$$

Given data $(y^k, \mathbf{x}_k)$, $k = 1, 2, \dots, n$, we can let $\theta^k = \theta_{\zeta(\mathbf{x}_k)}$, so that for $y = (y^1, y^2, \dots, y^n)$ and $\mathbf{x} = (\mathbf{x}_1, \mathbf{x}_2, \dots, \mathbf{x}_n)$,

$$f(y \mid \mathbf{x}, (\mathcal{T}, \Theta)) = \prod_{k=1}^{n} f(y^k \mid \theta^k)$$

where we assume that the Y^k are independent, given the $\{\mathbf{x}_k\}$.

It is convenient to let $y_i = \{y^k : \theta^k = \theta_i\}$, the set of y assigned to the ith terminal node. Then, we can write our likelihood by multiplying across terminal nodes,

$$f(y \mid \mathbf{x}, (\mathcal{T}, \Theta)) = \prod_{i=1}^{b} f(y_i \mid \theta_i)$$

Using $y_i = (y_{i1}, y_{i2}, \dots, y_{iv}, \dots, y_{in_i})$, where n_i is the number of observations assigned to the ith terminal node, we can also write our likelihood for a given terminal node by multiplying across observations assigned to that node,

$$f(y_i \mid \theta_i) = \prod_{v=1}^{n_i} f(y_{iv} \mid \theta_i)$$

where we again assume conditional independence.

Three basic examples of such a model are

1. the binary response model $\theta = p$ with $f(y \mid p) \sim \text{Bernoulli}(p)$,
2. the mean-variance shift model $\theta = (\mu, \sigma)$ with $f(y \mid \mu, \sigma) \sim N(\mu, \sigma^2)$,
3. the mean-shift model $\theta = \mu$ and $f(y \mid \mu, \sigma) \sim N(\mu, \sigma^2)$, with a common σ across all terminal nodes.

These examples are discussed in CGM98.

2.3 Tree Model Prior

To complete our Bayesian approach, we need to specify a prior on the model parameter $(\mathcal{T}, \Theta)$. Fundamental to our prior is the decomposition

$$p(\mathcal{T}, \Theta) = p(\mathcal{T})\, p(\Theta \mid \mathcal{T})$$

This decomposition greatly facilitates the prior choice. In particular, note that the dimension of Θ depends on $\mathcal{T}$. Since $\mathcal{T}$ captures the partitioning, and Θ captures parameters within partitions, it seems reasonable to think about $\mathcal{T}$ first and then Θ conditional on $\mathcal{T}$.

2.3.1 $p(\mathcal{T})$

We specify $p(T)$ by describing the process by which a tree $\mathcal{T}$ may be drawn from $p(\mathcal{T})$. We start with a tree that consists of a single node. We then recursively grow the tree by specifying:

- $p_{\text{SPLIT}}(\eta, \mathcal{T})$: the probability we *split* the terminal node η of tree $\mathcal{T}$ so that it gains left and right node children.
- $p_{\text{RULE}}(\eta, \mathcal{T})$: a distribution over the decision rules assignable to the current terminal node η of tree $\mathcal{T}$, should we decide to split it into left and right children.

Given $p_{\text{SPLIT}}(\eta, \mathcal{T})$ and $p_{\text{RULE}}(\eta, \mathcal{T})$, we randomly grow the tree, recursively, until we have decided not to split each bottom node. Each time we split, we assign the rule by drawing from $p_{\text{RULE}}(\eta, \mathcal{T})$. We choose $p_{\text{SPLIT}}(\eta, \mathcal{T})$ and $p_{\text{RULE}}(\eta, \mathcal{T})$ so that they only depend on the part of $\mathcal{T}$ above the current terminal node η. This ensures that our bottom node splitting process does not depend on the order in which we consider the bottom nodes for splitting.

We let $p_{\text{SPLIT}}(\eta, \mathcal{T})$ have the form

$$p_{\text{SPLIT}}(\eta, \mathcal{T}) = \frac{\alpha}{(1 + d_\eta)^\beta} \tag{1}$$

where d_η is the depth of node η in tree $\mathcal{T}$, and $0 < \alpha < 1$ and $\beta > 0$ are the hyperparameters. A single-node tree has depth zero. This allows us to express the idea that it gets harder to split as the tree grows. This plays a crucial role in the BART model where we need to express a prior preference for smaller trees. In a single-tree model, a value of $\beta = 0.5$ would be reasonable, while in BART, $\beta = 2$ is a common default. Interesting alternative enhancements of these choice for $p(\mathcal{T})$ have been proposed by Linero [11], Rockova and Saha [12], and Rockova and van der Pas [13].

We now turn to the choice of $p_{\text{RULE}}(\eta, \mathcal{T})$. Essentially, the basic default choice is uniform but taking into account which variables and rules are *available*, given η and $\mathcal{T}$. Recall that a predictor is considered to be either numeric or categorical.

For a given categorical x and current bottom node η, the set of available categories are all the categories that have been sent to that bottom node. For example in Figure 1 the categories $\{C, D\}$ are available in bottom nodes 4, 10, and 11. A categorical variable is said to be available in a bottom node if there are at least two categories available in the node.

For a numeric x, a rule is determined by the choice of cut-point. For each x_i, we initially choose a discrete set of possible cut-points. Typically we base our choice on the observed values in the training data. Basic choices would be a set of unique x values or quantiles or a uniform grid of value between the min and max. At a bottom node η, a subset of the possible splits are available for forming a new rule. For example in Figure 1, you would not consider a split value less than or equal to 5 for x_1 in terminal node 7, since observations in that bottom node are already restricted to have $x_1 > 5$. Given our initial set of discrete cut-points, a choice of numeric predictor, and a bottom node, we can determine the set of available cut-points. The numeric variable is said to be available if the set of available cut-points is nonempty.

We can now define $p_{\text{RULE}}(\eta, \mathcal{T})$ by drawing uniformly from the set of available predictors and then uniformly from the set of available rules, given the choice of predictor. The `R` package `BayesTree` uses this prior specification for numeric and categorical predictors, and much of detail in the underlying C++ code is devoted to determining the availability of variables and rules. The `R` packages `BART` and `rbart` only allow numeric predictors. With only numeric predictors, a categorical variable must be encoded with dummy variables with consequences for the implied prior. Note that unlike in the linear model, K dummies are included for a variable with K categories.

There are many interesting alternative specifications. See, for example, Section 7.1.

With a discrete set of cut-points for each numeric variable, $\mathcal{T}$ belongs to a large but discrete set. MCMC steps involving draws of $\mathcal{T}$ will be sampling from a discrete parameter space and will rely on Metropolis–Hastings (MH) proposals (Section 3).

2.3.2 $p(\Theta \mid \mathcal{T})$

Recall that $\Theta = (\theta_1, \theta_2, \dots, \theta_b)$, where b is the number of bottom nodes in the tree $\mathcal{T}$. A simplifying assumption is prior independence across bottom nodes,

$$p(\Theta \mid \mathcal{T}) = \prod_{i=1}^{b} p(\theta_i)$$

The θ values for the bottom nodes are IID $\theta_i \sim p(\theta)$. With this assumption, we only have to choose the distribution $p(\theta)$.

Our model is (suppressing $\mathbf{x}$)

$$p(y, \mathcal{T}, \Theta) = p(T) \left[\prod_{i=1}^{b} p(\theta_i)\right] \left[\prod_{i=1}^{b} p(y_i \mid \theta_i)\right] = p(T) \prod_{i=1}^{b} p(\theta_i)\, p(y_i \mid \theta_i)$$

The basic computations are then simplified by choosing $p(y \mid \theta)$ from a standard family and $p(\theta)$ from the corresponding conjugate prior. For example, in BART, θ is just a normal mean so that the (conditionally) conjugate prior is just a univariate normal.

3 Tree MCMC

In this section, we outline MCMC approaches for drawing from

$$p(\mathcal{T}, \Theta \mid y) \propto p(\mathcal{T})\, p(\Theta \mid \mathcal{T})\, p(y \mid \mathcal{T}, \Theta)$$

where we have again suppressed $\mathbf{x}$.

Our basic strategy is to integrate out Θ and then use a variety of MH transitions to propose changes to $\mathcal{T}$.

To integrate out Θ, first note

$$p(\mathcal{T} \mid y) \propto p(\mathcal{T})\, p(y \mid \mathcal{T})$$

Then, $p(y \mid \mathcal{T})$ can be computed as

$$\begin{aligned} p(y \mid \mathcal{T}) &\propto \int p(\Theta \mid \mathcal{T})\, p(y \mid \mathcal{T}, \Theta)\, \mathrm{d}\Theta \\ &= \int \left[\prod_{i=1}^{b} p(\theta_i)\, p(y_i \mid \theta_i)\right] \mathrm{d}\theta_1 \mathrm{d}\theta_2 \dots \mathrm{d}\theta_b \\ &= \prod_{i=1}^{b} \int p(\theta_i)\, p(y_i \mid \theta_i)\, \mathrm{d}\theta_i \\ &= \prod_{i=1}^{b} p(y_i) \end{aligned}$$

With the choice of a conjugate prior, each $p(y_i)$ is computable in closed form. It is just the joint predictive density (or probability mass function) for the subset of observations assigned to bottom node i of the tree $\mathcal{T}$.

This decomposition has important computational benefits. We will draw from $p(\mathcal{T} \mid y)$ using various MH schemes, each of which propose changes to a current tree $\mathcal{T}$. When just a part of $\mathcal{T}$ changes, some individual observations will move from one terminal node to another. That is, only a subset of the y_i will change, and only the corresponding subset of the integrals $\int p(\theta_i)\, p(y_i \mid \theta_i)\, \mathrm{d}\theta_i$ have to be recomputed.

Below, we detail the MH proposals used in CGM98. We will have

- A pair of complementary BIRTH/DEATH moves. In a BIRTH move, we propose adding a rule and pair of children to a terminal node. In DEATH move, we propose killing a pair of children, so that their parent becomes a terminal node.

- CHANGE Rule move. We propose changing the rule at an interior node.
- SWAP Rule move. We propose swapping the rules for a parent/child pair of interior nodes.

These moves are used in the R package `BayesTree`. At each MCMC iteration, the BIRTH/DEATH move is chosen with probability 0.5, the CHANGE Rule move is chosen with probability 0.4, and the SWAP Rule is chosen with probability 0.1. Within a BIRTH/DEATH move, BIRTH or DEATH is chosen at random with equal probability, unless one of these moves is not possible (e.g., DEATH for a tree with a single bottom node). Probabilities of BIRTH, DEATH, CHANGE, and SWAP are hard coded into the procedure.

Notably, the R package `BART` only uses the BIRTH/DEATH move in the marginal $\mathcal{T}$ space and redraws each θ_i at each MCMC step and still works remarkably well. This is because the BART MCMC works much better than the single-tree MCMC.

All of our moves construct a proposed Markov transition. Let $\mathcal{T}_0$ be the current tree, and $\mathcal{T}^*$ be the proposed tree which is some modification of $\mathcal{T}_0$. We accept the proposal with MH probability

$$\alpha = \min\left\{1, \frac{P(\mathcal{T}^* \mid y)\, P(\mathcal{T}^* \to \mathcal{T}_0)}{P(\mathcal{T}_0 \mid t)\, P(\mathcal{T}_0 \to \mathcal{T}^*)}\right\} \tag{2}$$

where $P(\mathcal{T}_0 \mid y)$ and $P(\mathcal{T}^* \mid y)$ are the posterior probabilities of trees $\mathcal{T}_0$ and $\mathcal{T}^*$, respectively. Thus, $P(T \mid y) \propto p(T)\, p(y \mid T)$. $P(P \to \mathcal{T}_0)$ is the probability of proposing $\mathcal{T}_0$ while at $\mathcal{T}^*$, and $P(\mathcal{T}_0 \to \mathcal{T}^*)$ is the probability of proposing $\mathcal{T}^*$ while at $\mathcal{T}_0$. $P(\mathcal{T}_0 \mid y)$ and $P(\mathcal{T}^* \mid y)$ will depend on both the likelihood and our prior, while the transition probabilities depend on the mechanics of our proposal.

Given $\mathcal{T}$, we can easily draw Θ using

$$p(\Theta \mid \mathcal{T}, y) \propto \prod_{i=1}^{b} p(y_i \mid \theta_i) p(\theta_i)$$

Hence, each θ_i may be drawn independently. With the choice of a standard likelihood and conjugate prior, methods for making these draws are typically readily available.

Clearly, the fundamental moves are the BIRTH/DEATH moves. These moves allow trees to grow and shrink in size.

3.1 The BIRTH/DEATH Move

In a BIRTH proposal, a bottom node of the current tree is chosen, and we propose to give it a pair of children. A *nog* node of a tree is a node that has children but no grandchildren. Thus, both children of a nog node are bottom nodes. In a DEATH proposal, we choose a nog node from the current tree, and we propose "killing its children." In Figure 1, we might propose a BIRTH at any of the bottom nodes 4, 10, 11, 6, and 7. We could propose a DEATH move at the two nog nodes 5 and 3.

We first describe the BIRTH move in detail. Let $\mathcal{T}_0$ denote the *current* tree, and $\mathcal{T}^*$ denote the *proposed* tree. Thus, $\mathcal{T}^*$ differs from $\mathcal{T}_0$ only in that one of the bottom nodes of $\mathcal{T}_0$ has given birth to a pair of children in $\mathcal{T}^*$.

First, we discuss the likelihood contribution. As noted above,

$$p(y \mid \mathcal{T}) = \prod_{i=1}^{b} p(y_i \mid \mathcal{T}) \tag{3}$$

Thus, the contribution of the likelihood to the ratio $P(\mathcal{T}^* \mid y)/P(\mathcal{T}_0 \mid y)$ in (2) is just

$$\frac{p(y_l, y_r \mid \mathcal{T}^*)}{p(y_l, y_r \mid \mathcal{T}_0)} = \frac{p(y_l \mid \mathcal{T}^*)\, p(y_r \mid \mathcal{T}^*)}{p(y_l, y_r \mid \mathcal{T}_0)} \tag{4}$$

where y_l denotes the observations in the new left child in $\mathcal{T}^*$, and y_r denotes the observations in the new right child in $\mathcal{T}^*$. All other contributions to the likelihoods cancel out because of the product form of (3).

As with the likelihood, much of the prior contributions to the posterior ratio cancel out since the trees differ only at the two new bottom nodes, and our stochastic tree growing prior draws tree components independently at different "places" of the tree. Hence, the prior contribution to the $P(\mathcal{T}^* \mid y)/P(\mathcal{T}_0 \mid y)$ ratio is

$$\frac{(PG)\,(1 - PGl)\,(1 - PGr)\,P(\text{rule})}{(1 - PG)} \tag{5}$$

where:

- *PG*: prior probability of growing at chosen bottom node of $\mathcal{T}_0$,
- *PGl*: prior probability of growing at new left child in $\mathcal{T}^*$,
- *PGr*: prior probability of growing at new right child in $\mathcal{T}^*$, and
- *P*(rule): prior probability of choosing the rule defining the new children in $\mathcal{T}^*$, given by p_{RULE}.

We draw the candidate rule for $\mathcal{T}^*$ by drawing from the prior so that *P*(rule) is given by $p_{\text{RULE}}(\eta, \mathcal{T}_0)$, where η is the bottom node we have randomly chosen for a potential birth.

Finally, the ratio $P(\mathcal{T}^* \to \mathcal{T}_0)/P(\mathcal{T}_0 \to \mathcal{T}^*)$ is given by

$$\frac{(PD)\,(P\text{nog})}{(PB)\,(P\text{bot})\,P(\text{rule})} \tag{6}$$

where

- *PD*: probability of choosing the death proposal at tree $\mathcal{T}^*$.
- *P*nog: probability of choosing the nog node that gets you back $\mathcal{T}_0$.
- *PB*: probability of choosing a birth proposal at $\mathcal{T}_0$.
- *P*bot: probability of choosing the $\mathcal{T}_0$ bottom node such that a birth gets you to $\mathcal{T}^*$.
- *P*(rule): probability of drawing the new splitting rule to generate $\mathcal{T}^*$'s children.

Our proposal draw of the new rule generating the two new bottom nodes is a draw from the prior. It is in this draw that variable selection (or, perhaps, variable proposal) occurs! Note that since our proposal for the *rule* is a draw from the prior, it cancels out in the ratio (2).

The final MH ratio used for BIRTH is

$$\min\left\{1, \frac{(PG)(1 - PGl)(1 - PGr)}{(1 - PG)}\, \frac{(PD)(P\text{nog})}{(PB)(P\text{bot})}\, \frac{p(y_l \mid \mathcal{T}^*)\, p(y_r \mid \mathcal{T}^*)}{p(y_l, y_r \mid \mathcal{T}_0)}\right\}$$

The formulas given above correspond exactly to the C++ source code in the `R` packages `BayesTree` and `BART`.

For a DEATH move, we choose a nog node of $\mathcal{T}_0$ and propose killing the two children to create $\mathcal{T}^*$. The MH acceptance probability is

$$\min\left\{1, \frac{(1-PG)(PB)(Pbot)}{(PG)(1-PGl)(1-PGr)(PD)(Pnog)} \frac{p(y_l, y_r \mid \mathcal{T}^*)}{p(y_l \mid \mathcal{T}_0)\, p(y_r \mid \mathcal{T}_0)}\right\}$$

where

- *PG*: prior probability of spawning children at the proposed new bottom node of $\mathcal{T}^*$ (nog node of $\mathcal{T}_0$).
- *PB*: probability of a BIRTH move at $\mathcal{T}^*$.
- *P*bot: probability of choosing the bottom node of $\mathcal{T}^*$ such that a birth gets you back to $\mathcal{T}_0$.
- *PGl*: prior probability of adding children at the proposed left child of $\mathcal{T}_0$.
- *PGr*: prior probability of adding children at the proposed right child of $\mathcal{T}_0$.
- *PD*: probability of a DEATH move at $\mathcal{T}_0$.
- *P*nog: probability of choosing the nog node at $\mathcal{T}_0$.

3.2 CHANGE Rule

The CHANGE Rule move picks an interior node and then modifies the current tree by changing the decision rule at the chosen node. Our transition $P(\mathcal{T}_0 \to \mathcal{T}^*)$ is made up of the steps:

1. Draw node η from $\mathcal{T}_0$ by drawing uniformly from the set of interior nodes.
2. Draw a rule from $p_{\text{RULE}}(\eta, \mathcal{T}_0)$.
3. Replace the decision rule at node η of $\mathcal{T}_0$ with the rule drawn in the second step to obtain $\mathcal{T}^*$.

After we draw $\mathcal{T}^*$, we check that the resulting tree has nonzero prior probability. For example, our prior does not allow logically empty bottom nodes since rules are always checked to be drawn using available variables. If $\mathcal{T}^*$ is such that $p(\mathcal{T}^*)$ is 0, then we can immediately reject the move without further computation.

The number of interior nodes in $\mathcal{T}_0$ and $\mathcal{T}^*$ are the same, and each interior node of each tree clearly has available variables (otherwise it could not have a splitting rule). Also recall that $p_{\text{RULE}}(\eta, \mathcal{T})$ only depends on the part of $\mathcal{T}$ above η in $\mathcal{T}$. Hence, we have the property that $P(\mathcal{T}_0 \to \mathcal{T}^*) = P(\mathcal{T}^* \to \mathcal{T}_0)$, so that the ratio cancels in the MH acceptance ratio.

$$\alpha = \min\left\{1, \frac{p(\mathcal{T}^*)\, p(y \mid \mathcal{T}^*)}{p(\mathcal{T}_0)\, p(y \mid \mathcal{T}_0)}\right\}$$

To compute $p(y \mid \mathcal{T})$ for either of $\mathcal{T}_0$ or $\mathcal{T}^*$, we only have to consider observations in bottom nodes below η since the contributions for other bottom nodes will cancel.

3.3 SWAP Rule

In the SWAP Rule step, we randomly pick a parent–child pair that are both internal nodes. We then swap their splitting rules. If both children have the identical rule, we swap the splitting rule of the parent with both children.

Similar to the CHANGE Rule proposal, a key observation is that the proposal step for SWAP is symmetric. The general expression of the MH acceptance probability is as in (2). For SWAP, the proposal distributions $P(T_0 \rightarrow T^*)$ and $P(T^* \rightarrow T_0)$ will cancel in (2). Only the likelihood and prior terms need to be calculated.

The proposal for SWAP is a draw (with equal probability) from the list of interior nodes having at least one child that is nonterminal. This list constitutes the parents of the swap. For each parent, there will be at least one child with a rule that could be swapped with the parent. Once a parent is chosen, the two children are inspected. If only one child is nonterminal, that child will be the one chosen for the SWAP. If both children are nonterminal and have different rules, then one of the two children will be chosen (with equal probability) for the swap. If both children have identical rules, then the parent rule and the child rules are swapped, and both children get the parent rule.

Once the proposal has chosen a parent–child pair to swap, the rules are swapped and the resulting tree checked to determine if the swap produces any necessarily empty terminal nodes. If there are necessarily empty terminal nodes, this corresponds to a proposed tree T^* with prior probability 0, and thus, the MH step will not accept. This check can be carried out without referring to the data, since only the rules of T_0 and T^* need to be checked.

Assuming that the proposal does not have 0 prior probability, the prior probabilities for T^* and T_0 are calculated for the entire trees. Although there is cancelation in the ratio of prior terms for parts of the tree that do not change, the prior computation is relatively quick and so is simply carried out for the full trees.

The calculation of likelihood for T^* requires reassignment of data among all bottom nodes that are below the parent. The likelihoods can be calculated for subsets of T^* and T_0, for all bottom nodes below the parent of the proposal. The two likelihood values and two prior values are sufficient to evaluate α in (2).

If the SWAP proposal is not accepted, then the tree is restored to T_0. If the proposal is accepted, the change to the tree has already been made (to allow computation of prior and likelihood at T^*).

3.4 Improved Tree Space Moves

As is well known, the proposal distribution is a key user-specified parameterization of the MH MCMC algorithm that has a large effect on how well, and how efficient, MH sampling can be performed. In the best-case scenario, draws from the true posterior are directly available, giving an acceptance ratio of 1. In practice, a distribution that is simple to draw from is used as the proposal. This leads to an algorithm that is practically implementable but uses a proposal having only moderate accuracy (often only locally) to the true posterior, leading to many rejected (i.e., wasted) samples and slower convergence. Nonetheless, the practical usefulness of MH has led to its widespread adoption.

The situation becomes more challenging in the modern setting where one is interested in performing Bayesian inference for complex, high-dimensional models. In CGM98 and CGM10, a pragmatic approach for the case of Bayesian regression trees (a complex, high-dimensional model) was taken by designing the proposals described above that explore tree space by incrementally making the model just slightly more or less complex (via BIRTH or DEATH at a single terminal or nog node, respectively) or just slightly adjusting an existing tree's ruleset (via CHANGE or SWAP at a single node or pair of nodes, respectively). However, in some settings, it has been recognized that this proposal distribution may lead to slow convergence and/or inaccurate sampling – an issue of eminent practical relevance even if the required properties for the asymptotic convergence of MH sampling are satisfied.

Good alternatives to the CGM98 algorithm are not necessarily obvious since one would like to retain the simplicity, locality, and efficiency of the algorithm. Recent work has provided some alternatives and refinements at moderate increases in computational cost when a problem demands more effective sampling of the posterior. Pratola [6] introduces a new ROTATE proposal, defines a PERTURB proposal as a refined version of CHANGE, and also revises the basic MCMC loop, as described in Algorithm 1. Algorithm 1 is for the mean-shift model, which will be a building block for BART in Section 4.

Algorithm 1. Updated Bayesian CART MCMC Algorithm

```
procedure BAYESIAN CART-ITERATION(y, X, num_trees)
output An approximate draw from the tree posterior
    Draw T|σ², y via BIRTH, DEATH, or ROTATE at one random eligible internal node
    Set num_internal = number of internal nodes of tree T
    Set num_terminal = number of terminal nodes of tree T
    for j in 1 to num_internal do
        Draw rule (v_j, c_j)|T, σ², y via PERTURB
    for j in 1 to num_terminal do
        Draw μ_j|T, y via Gibbs
    Draw σ²|T, Θ, y via Gibbs
return
```

3.4.1 Rotate

Similar to SWAP, ROTATE maps the existing internal structure of a tree into a plausible alternative (i.e., one that could have been generated by a longer sequence of BIRTH/DEATH proposals). But while SWAP only considers one or two possible alternatives, ROTATE generates a larger (stochastic) set of possible transitions. Unlike SWAP, ROTATE also considers the descendants of the ROTATE node in forming the possible transitions, and the further up (down) the tree, the more (less) possible ROTATE transitions there are. If one thinks of BIRTH/DEATH as the simplest possible rearrangement of a tree, ROTATE can then be thought of as generalizing the ideas of SWAP, BIRTH, and DEATH in an elegant way to arbitrary internal locations of a tree. For instance, while BIRTH/DEATH involves the likelihood contributions for y_l, y_r, ROTATE involves the likelihood contributions for the data involved in the left/right subtrees of the ROTATE proposal, say y_{T_l}, y_{T_r}. Heuristically then, ROTATE is a less local proposal than BIRTH/DEATH and more diverse than SWAP but not so global nor so diverse as to be too inefficient. Finally, ROTATE is its own inverse, making application of this algorithmically generated proposal distribution practically tangible.

3.4.2 Perturb

Similar to CHANGE, PERTURB aims to update the rules in an existing tree. This is done in two ways: updating the cut-points or updating the (variable,cut-point) pairs. Note that

PETURB is applied to all nodes in a tree, leading to more efficient exploration of this aspect of the posterior distribution. This is made possible by more efficient generation of cut-point proposals, which are conditioned on both the ancestral and descendant parts of the tree for the node being updated. Similarly, variable proposals are made more efficient using a preconditioned proposal distribution; Pratola [6] suggests using a correlation metric such as Spearman rank correlation to form the preconditioned transition matrix, although other choices are possible. Note also that both variants of PETURB can simultaneously update all internal nodes that are at the same tree depth, thereby exploiting parallelism to make such computations more efficient.

3.4.3 The complex mixtures that are tree proposals

Modifying the individual proposals as described above only goes part of the way to ameliorating Bayesian Tree MCMC algorithms. Part of the tale is in how smartly these proposals are used. Recall that for BIRTH/DEATH, the particular proposal selected from either of these choices is determined by the flip of an equally weighted coin. And the corresponding terminal or nog node selected for the chosen move is also randomly drawn with equal weight. But why not prefer a BIRTH in shallower parts of tree space or a DEATH in deeper parts of tree space? Similarly, in the BIRTH/DEATH/ROTATE mixture, should these be equally weighted, or should one proposal be preferred depending on the state of the tree? Such issues are very much nontrivial and would lead away from the simple, pragmatic, proposal distributions that have seen so much success. One alternative is to leverage parallel computation to explore a large set of possible transitions to avoid devising a clever strategy to determine what the mixture ought to be at any given iteration of the algorithm. Such is the strategy of Mohammadi *et al.* [14], who use the BD–MCMC algorithm to select among all possible BIRTH/DEATH moves (or BIRTH/DEATH/ROTATE moves) at a rate that is proportional to their posterior probability rather than the default (weighted) mixture. While this increases the number of required computations needed at each step of the MCMC, such computations can be largely hidden via effective parallelization, resulting in more efficient sampling of the posterior per unit time.

4 The BART Model

BART (CGM10) builds on the Bayesian analysis of a single tree to consider an ensemble of trees. BART is inspired by Friedman's work [7] on boosting but uses the power of the Bayesian machinery.

To transition from the single-tree development of Section 2, we start with a single tree but let $\theta = \mu$ be a single mean parameter. Rather than using Θ to denote the collection of bottom node parameters, we switch notation to $\mathcal{M} = (\mu_1, \mu_2, \ldots, \mu_b)$, a collection of mean parameters for the bottom nodes.

We then define the function $g(\mathbf{x}; \mathcal{T}, \mathcal{M})$ to be $\mu_{\zeta(\mathbf{x})}$, where ζ is as in Section 2. That is, we drop $\mathbf{x}$ down the tree $\mathcal{T}$ until it lands in a bottom node and finds a μ_i awaiting it, which is then the value of g. Clearly g looks like a step function corresponding to the classic regression tree of classic CART.

We can turn a single-tree model indexed by parameter $(\mathcal{T}, \mathcal{M})$ into a probability model with a likelihood by adding an error term,

$$Y = g(\mathbf{x}; \mathcal{T}, \mathcal{M}) + \epsilon, \quad \epsilon \sim N(0, \sigma^2)$$

BART follows Friedman (and more generally the boosting literature) by replacing the single-tree mean model $g(\mathbf{x}; \mathcal{T}, \mathcal{M})$ with a sum of m trees $f(\mathbf{x}) = \sum_{j=1}^{m} g(\mathbf{x}; \mathcal{T}_j, \mathcal{M}_j)$.

$$Y = f(\mathbf{x}) + \epsilon, \quad \epsilon \sim \mathcal{N}(0, \sigma^2) \quad \text{where } f \overset{\text{prior}}{\sim} \text{BART} \tag{7}$$

As in Section 2, each $\mathcal{T}_j$ is a recursive binary regression tree. $\mathcal{M}_j$ contains the terminal node constants μ_{ij}, for which $g(\mathbf{x}; \mathcal{T}_j, \mathcal{M}_j)$ is the step function that assigns $\mu_{ij} \in \mathcal{M}_j$ to $\mathbf{x}$ according to the sequence of splitting rules in $\mathcal{T}_j$.

For each value of $\mathbf{x}$, under (7), $E(Y \mid \mathbf{x})$ is equal to the sum of all the terminal node μ_{ij}s assigned to $\mathbf{x}$ by the $g(\mathbf{x}; \mathcal{T}_j, \mathcal{M}_j)$s. Thus, the sum-of-trees function is flexibly capable of approximating a wide class of functions from R^n to R, especially when the number of trees m is large. Note also that the sum-of-trees representation is simply the sum of many simple multidimensional step functions from R^n to R, namely the $g(\mathbf{x}; \mathcal{T}_j, \mathcal{M}_j)$, rendering it much more manageable than basis expansions with more complicated elements such as multidimensional wavelets or multidimensional splines.

The BART model specification is completed by introducing a prior distribution over all the parameters of the sum-of-trees model, namely $(\mathcal{T}_1, \mathcal{M}_1), \ldots, (\mathcal{T}_m, \mathcal{M}_m)$ and σ. Note that $(\mathcal{T}_1, \mathcal{M}_1), \ldots, (\mathcal{T}_m, \mathcal{M}_m)$ entail all the bottom node parameters as well as the tree structures and splitting rules, a very large number of parameters, especially when m is large. To cope with this parameter explosion, we use a "regularization" prior that effectively constrains the fit by keeping each of the individual tree effects from being unduly influential. Without such a regularizing influence, large tree components would overwhelm the rich structure of (7), thereby limiting its scope of fine structure approximation.

4.1 Specification of the BART Regularization Prior

To simplify the specification of this regularization prior, we restrict attention to symmetric independence priors of the form

$$p((\mathcal{T}_1, \mathcal{M}_1), \ldots, (\mathcal{T}_m, \mathcal{M}_m), \sigma) = \left[\prod_j \left(\prod_i p(\mu_{ij} \mid \mathcal{T}_j)\right) p(\mathcal{T}_j)\right] p(\sigma) \tag{8}$$

where $\mu_{ij} \in \mathcal{M}_j$, thereby reducing prior specification to the choice of prior forms for $p(\mathcal{T}_j)$, $p(\mu_{ij} \mid \mathcal{T}_j)$, and $p(\sigma)$. To simplify matters further, we use identical prior forms for every $p(\mathcal{T}_j)$ and for every $p(\mu_{ij} \mid \mathcal{T}_j)$. As detailed in the following paragraphs, each of these prior forms are controlled by just a few interpretable hyperparameters that can be calibrated to yield surprisingly effective default specifications for regularization of the sum-of-trees model.

For $p(\mathcal{T}_j)$, we use the prior developed in Section 2. Note, however, that the values for α and β are typically very different in BART. In BART, we often use $\alpha = 0.95$ and $\beta = 2$, whereas with a single tree, we use a much smaller β. This expresses the idea that we do not expect the individual trees to be large.

For $p(\mu_{ij} \mid \mathcal{T}_j)$, we use the conjugate normal distribution $\mathcal{N}(\mu_\mu, \sigma_\mu^2)$, which allows μ_{ij} to be marginalized out as in Section 3, vastly simplifying MCMC posterior calculations.

To guide the specification of the hyperparameters μ_μ and σ_μ, we note that under (7), it is highly probable that $E(Y \mid \mathbf{x})$ lies between $y_{\min}$ and $y_{\max}$, the minimum and maximum of the observed values of Y in the data, and that the prior distribution of $E(Y \mid \mathbf{x})$ is $\mathcal{N}(m\ \mu_\mu, m\ \sigma_\mu^2)$ (because $E(Y \mid \mathbf{x})$ is the sum of m independent μ_{ij}s under the sum-of-trees model). Based on these facts, we use the informal empirical Bayes strategy of choosing μ_μ and σ_μ so that $\mathcal{N}(m\ \mu_\mu, m\ \sigma_\mu^2)$ assigns substantial probability to the interval $(y_{\min}, y_{\max})$. This is conveniently done by choosing μ_μ and σ_μ so that $m\ \mu_\mu - k\ \sqrt{m}\ \sigma_\mu = y_{\min}$ and $m\ \mu_\mu + k\ \sqrt{m}\ \sigma_\mu = y_{\max}$ for some preselected value of k such as 1, 2, or 3. For example, $k = 2$ would yield a 95% prior probability that $E(Y \mid \mathbf{x})$ is in the interval $(y_{\min}, y_{\max})$. The goal of this specification strategy for μ_μ and σ_μ is to ensure that the implicit prior for $E(Y \mid \mathbf{x})$ is in the right "ballpark" in the sense of assigning substantial probability to the entire region of plausible values of $E(Y \mid \mathbf{x})$ while avoiding overconcentration and -dispersion of the prior with respect to the likelihood. As long as this goal is met, BART seems to be very robust to variations of these specifications.

For $p(\sigma)$, we also use a conjugate prior, here the inverse chi-square distribution $\sigma^2 \sim \nu\ \lambda/\chi_\nu^2$. Here again, we use an informal empirical Bayes approach to guide the specification of the hyperparameters ν and λ, in this case to assign substantial probability to the entire region of plausible values of σ while avoiding overconcentration and overdispersion of the prior. Essentially, we calibrate the prior df ν and scale λ with a "rough data-based overestimate" $\hat{\sigma}$ of σ. Two natural choices for $\hat{\sigma}$ are (i) a "naive" specification, the sample standard deviation of Y, and (ii) a "linear model" specification, the residual standard deviation from a least-squares linear regression of Y on all the predictors. We then pick a value of ν between 3 and 10 to get an appropriate shape, and a value of λ so that the qth quantile of the prior on σ is located at $\hat{\sigma}$, that is $P(\sigma < \hat{\sigma}) = q$. We consider large values of q such as 0.75, 0.90, and 0.99 to center the distribution below $\hat{\sigma}$.

5 BART Example: Boston Housing Values and Air Pollution

Here, we demonstrate BART with the classic Boston housing example [15]. This data is based on the 1970 US Census where each observation represents a Census tract in the Boston Standard Metropolitan Statistical Area. For each tract, there was a localized air pollution estimate, the concentration of nitrogen oxides, `nox`, based on a meteorological model that was calibrated to monitoring data. Restricted to tracts with owner-occupied homes, there are $N = 506$ observations. We will predict the median value of owner-occupied homes (in thousands of dollars), `mdev`, by 13 covariates including `nox` which is our primary interest.

However, BART does not directly provide a summary of the effect of a single covariate, or a subset of covariates, on the outcome. Friedman's partial dependence function [7] can be employed with BART to summarize the marginal effect due to a subset of the covariates, $\boldsymbol{x}_S$, by aggregating over the complement covariates, $\boldsymbol{x}_C$, that is, $\boldsymbol{x} = [\boldsymbol{x}_S, \boldsymbol{x}_C]$. The marginal dependence function is defined by fixing $\boldsymbol{x}_S$ while aggregating over the observed settings of the complement covariates in the data set: $f(\boldsymbol{x}_S) = N^{-1} \sum_{i=1}^{N} f(\boldsymbol{x}_S, \boldsymbol{x}_{iC})$. For example, suppose that we want to summarize `mdev` by `nox` while aggregating over the other 12 covariates in the Boston housing data. In Figure 2, we demonstrate the marginal estimate and its 95%

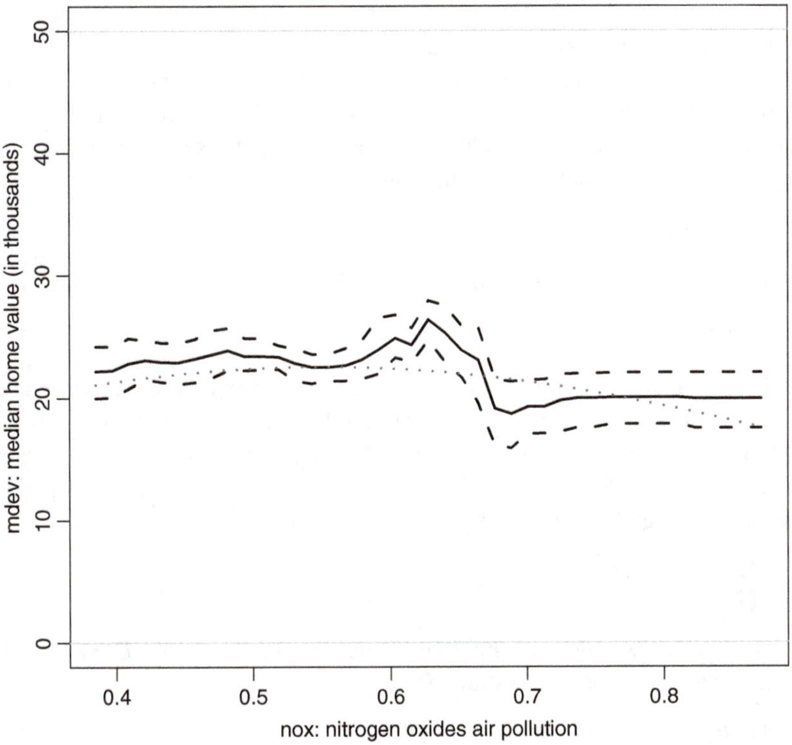

Figure 2 The Boston housing data was compiled from the 1970 US Census, where each observation represents a Census tract in Boston with owner-occupied homes. For each tract, we have the median value of owner-occupied homes (in thousands of dollars), `mdev`, and 13 other covariates including a localized air pollution estimate, the concentration of nitrogen oxides `nox`, which is our primary interest. We summarize the marginal effect of `nox` on `mdev` while aggregating over the other covariates with Friedman's partial dependence function. The marginal estimate and its 95% credible interval are shown. The line with short dashes comes from the linear regression model of Harrison and Rubinfeld [15] where a quadratic effect of `nox` with respect to the logarithm of `mdev` is assumed. *Source:* Based on Harrison and Rubinfeld [15].

credible interval: notice that BART has discerned a complex nonlinear relationship between `mdev` and `nox` from the data. Note that this example including data and source code can be found in the `BART R` package [16] as the `nox.R` demonstration program.

6 BART MCMC

Combining the regularization prior with the likelihood, $L((\mathcal{T}_1, \mathcal{M}_1), \dots, (\mathcal{T}_m, \mathcal{M}_m), \sigma \mid y)$ induces a posterior distribution

$$p((\mathcal{T}_1, \mathcal{M}_1), \dots, (\mathcal{T}_m, \mathcal{M}_m), \sigma \mid y) \tag{9}$$

over the full sum-of-trees model parameter space. Here, y is the observed $n \times 1$ vector of Y values in the data which are assumed to be independently realized. Note also that here and below we suppress explicit dependence on $\mathbf{x}$ as we assume $\mathbf{x}$ to be fixed throughout.

Although analytically intractable, the following backfitting MCMC algorithm can be used to very effectively simulate samples from this posterior.

This algorithm is a Gibbs sampler at the outer level. Let $\mathcal{T}_{(j)}$ be the set of all trees in the sum *except* $\mathcal{T}_j$, and similarly define $\mathcal{M}_{(j)}$, so that $\mathcal{T}_{(j)}$ will be a set of $m-1$ trees, and $\mathcal{M}_{(j)}$ the associated terminal node parameters. A Gibbs sampling strategy for sampling from (9) is obtained by m successive draws of $(\mathcal{T}_j, \mathcal{M}_j)$ conditionally on $(\mathcal{T}_{(j)}, \mathcal{M}_{(j)}, \sigma)$:

$$(\mathcal{T}_j, \mathcal{M}_j) \mid T_{(j)}, \mathcal{M}_{(j)}, \sigma, y \tag{10}$$

$j = 1, \dots, m$, followed by a draw of σ from the full conditional:

$$\sigma \mid \mathcal{T}_1, \dots \mathcal{T}_m, \mathcal{M}_1, \dots, \mathcal{M}_m, y \tag{11}$$

The draw of σ in (11) is simply a draw from an inverse gamma distribution, which can be straightforwardly obtained by routine methods. More subtle is the implementation of the m draws of $(\mathcal{T}_j, \mathcal{M}_j)$ in (10). This can be done by taking advantage of the following simplifying reduction. First, observe that the conditional distribution $p(\mathcal{T}_j, \mathcal{M}_j \mid \mathcal{T}_{(j)}, \mathcal{M}_{(j)}, \sigma, y)$ depends on $(\mathcal{T}_{(j)}, \mathcal{M}_{(j)}, y)$ only through $\mathbf{R}_j = (r_{j1}, \dots, r_{jn})'$, the $n \times 1$ vector of partial residuals

$$r_{ji} \equiv y_i - \sum_{k \neq j} g(\mathbf{x}_i; T_k, \mathcal{M}_k) \tag{12}$$

obtained from a fit that excludes the jth tree. Thus, the m draws of $(\mathcal{T}_j, \mathcal{M}_j)$ given $(\mathcal{T}_{(j)}, M_{(j)}, \sigma, y)$ in (10) are equivalent to m draws from

$$(\mathcal{T}_j, \mathcal{M}_j) \mid \sigma, \mathbf{R}_j \tag{13}$$

$j = 1, \dots, m$. Each of these draws is then done using the methods along the lines of those discussed in Section 3. We margin out $\mathcal{M}_j$ and then use MH proposals to modify $\mathcal{T}_j$. Given $\mathcal{T}_j$, we can draw $\mathcal{M}_j$.

The R package `BayesTree` uses all MH tree proposals in CGM98 and Section 3 for BART estimation. The R package `BART` just uses the BIRTH/DEATH step and redraws all the μ_{ij} at each MCMC iteration. This very simple approach works remarkably well in practice. The R package `rbart` implements BART (and a heteroskedastic version) using the more sophisticated tree moves of Section 3.4.

We initialize the chain with m simple single-node trees and then repeat iterations until satisfactory convergence is obtained. Fortunately, this backfitting MCMC algorithm appears to mix very well as we have found that different restarts give remarkably similar results even in difficult problems. At each iteration, each tree may increase or decrease the number of terminal nodes by 1, or change one or two splitting rules. The sum-of-trees model, with its abundance of unidentified parameters, allows the "fit" to glide freely from one tree to another. Because each move makes only small incremental changes to the fit, we can imagine the algorithm as analogous to sculpting a complex figure by adding and subtracting small dabs of clay.

For inference based on our MCMC sample, we rely on the fact that our backfitting algorithm is *ergodic*. Thus, the induced sequence of sum-of-trees functions

$$f^*(\cdot) = \sum_{j=1}^{m} g(\cdot\,; \mathcal{T}_j^*, \mathcal{M}_j^*) \tag{14}$$

from the sequence of draws $(\mathcal{T}_1^*, \mathcal{M}_1^*), \dots, (\mathcal{T}_m^*, \mathcal{M}_m^*)$, is converging to $p(f \mid y)$, the posterior distribution of the "true" $f(\cdot)$. Thus, by running the algorithm long enough after a suitable burn-in period, the sequence of f^* draws, say $f_1^*, \dots, f_K^*$, may be regarded as an approximate, dependent sample of size K from $p(f \mid y)$. Bayesian inferential quantities of interest can then be approximated with this sample as follows.

To estimate $f(\mathbf{x})$ or predict Y at a particular $\mathbf{x}$, in sample or out of sample, a natural choice is the average of the after burn-in sample $f_1^*, \dots, f_K^*$,

$$\frac{1}{K}\sum_{k=1}^{K} f_k^*(\mathbf{x}) \tag{15}$$

which approximates the posterior mean $E(f(\mathbf{x}) \mid y)$. Posterior uncertainty about $f(\mathbf{x})$ may be gauged by the variation of $f_1^*(\mathbf{x}), \dots, f_K^*(\mathbf{x})$. For example, a natural and convenient $(1 - \alpha)\%$ posterior interval for $f(\mathbf{x})$ is obtained as the interval between the upper and lower $\alpha/2$ quantiles of $f_1^*(\mathbf{x}), \dots, f_K^*(\mathbf{x})$.

7 BART Extentions

In this section, we mention some BART extensions. The Bayesian formulation and the corresponding MCMC approaches provide a rich environment for model and algorithm enhancement. We do not attempt to survey developments in Bayesian trees but point to two very powerful examples of extending or modifying the BART approach. In Section 7.1, the BART prior is modified to enhance search for models that use a small number of predictors. In Section 7.2, the computational and modeling approach is extensively modified to enable a "BART-like" inference, which is much faster and can handle much larger data sets.

7.1 The DART Sparsity Prior

Various Bayesian variable selection techniques applicable to BART have been studied [8, 11, 17–21]. Here, we focus on the sparse variable selection prior of Linero [11] for which we use the acronym DART (where "D" stands for the Dirichlet distribution). Let us represent the variable selection probabilities by s_j, where $j = 1, \dots, P$. Now, replace the uniform variable selection prior in BART with a Dirichlet prior as $[s_1, \dots, s_P] \mid \theta \overset{\text{prior}}{\sim} \text{D}(\theta/P, \dots, \theta/P)$. The prior for θ is induced via $\theta/(\theta + \rho) \overset{\text{prior}}{\sim} \text{Beta}(a,\ b)$. The typical settings are $b = 1$ and $\rho = P$. The distribution of θ controls the sparsity of the model: $a = 0.5$ induces a sparse posture, while $a = 1$ is not sparse and similar to the uniform prior with probability $s_j = P^{-1}$. If additional sparsity is desired, then you can set ρ to a value smaller than P.

The key to understanding the inducement of sparsity is the distribution of the arguments to the Dirichlet prior: θ/P. It can be shown that $\theta/P \sim F(a, b, \rho/P)$, where $F(.)$ is the beta prime distribution scaled by ρ/P [22]. The nonsparse setting is $(a, b, \rho/P) = (1, 1, 1)$. As you can see in Figure 3 [16], sparsity is promoted by reducing ρ, reducing a, or even further by reducing both.

Now, let us turn our attention to the posterior computation of the Dirichlet sparse prior. For a Dirichlet prior placed on the variable-splitting probabilities, $\mathbf{s}$, its posterior samples

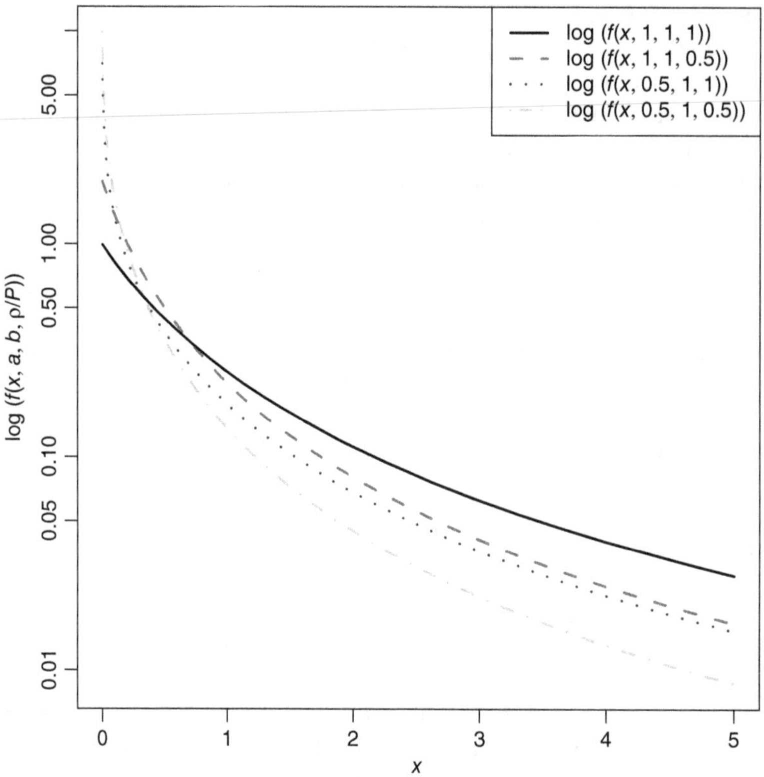

Figure 3 The distribution of θ/P and the sparse Dirichlet prior [16]. The key to understanding the inducement of sparsity is the distribution of the arguments to the Dirichlet prior: $\theta/P \sim F(a, b, \rho/P)$, where $F(.)$ is the beta prime distribution scaled by ρ/P. Here, we plot the natural logarithm of the scaled beta prime density, $f(.)$, at a nonsparse setting and three sparse settings. The nonsparse setting is $(a, b, \rho/P) = (1, 1, 1)$ (solid black line). As you can see in the figure, sparsity is promoted by reducing ρ (long dashed line), reducing a (short dashed line), or even further by reducing both (mixed dashed gray line). *Source*: Sparapani *et al.* [16].

are drawn via Gibbs sampling with conjugate Dirichlet draws. The Dirichlet parameter is updated by adding the total variable branch count over the ensemble, m_j, to the prior setting, $\frac{\theta}{P}$, that is, $[\frac{\theta}{P} + m_1, \dots, \frac{\theta}{P} + m_P]$. In this way, the Dirichlet prior induces a "rich get richer" variable selection strategy. The sparsity parameter, θ, is drawn on a discrete grid of values [11]: this draw only depends on $[s_1, \dots, s_P]$.

7.1.1 Grouped variables and the DART prior

Here, we take the opportunity to address a common pitfall of a Dirichlet prior for variable selection with a so-called grouped variable. Suppose that we have P variables, but Q variables of the covariates correspond to a grouped variable such as a series of dummy indicators encoded for a single categorical variable (suppose that these are the first Q variables without loss of generality): $x_1, \dots, x_Q$. N.B. Obviously, these developments apply to multiple grouped variables; however, for brevity, a single grouped variable will suffice to elucidate the problem and a solution. We denote the variable selection probabilities for all covariates as $\boldsymbol{s} = [s_1, \dots, s_P]$. There are two other probabilities of interest: the collapsed probabilities, $\boldsymbol{p} = [s_1 + \cdots + s_Q, s_{Q+1}, \dots, s_P]$, and the rescaled probabilities, $\boldsymbol{q} = [\tilde{s}_1, \dots, \tilde{s}_Q]$, where $\tilde{s}_j \propto s_j$ such that $\sum_{j=1}^{Q} \tilde{s}_j = 1$. If we blindly use Dirichlet variable selection probabilities on data such as this, then we arrive at the following:

$$\boldsymbol{s}|\theta \overset{\text{prior}}{\sim} \mathrm{D}_P(\theta/P, \dots, \theta/P)$$

where the subscript P is the order of the Dirichlet

$$\boldsymbol{p}|\theta \overset{\text{prior}}{\sim} \mathrm{D}_{\tilde{P}}(Q\theta/P, \theta/P, \dots, \theta/P)$$

where $\tilde{P} = P - Q + 1$

$$\boldsymbol{q}|\theta \overset{\text{prior}}{\sim} \mathrm{D}_Q(\theta/P, \dots, \theta/P)$$

The distribution of p_1, the first element of p, puts more prior weight on the grouped variable than the others. And now, the solution to the problem is trivial: rescale $\boldsymbol{q}$ by Q^{-1} while naturally redefining $\boldsymbol{p}$ and $\boldsymbol{s}$ as follows:

$$\boldsymbol{p}|\theta \overset{\text{prior}}{\sim} \mathrm{D}_{\tilde{P}}(\theta/\tilde{P}, \dots, \theta/\tilde{P})$$

$$\boldsymbol{q}|\theta \overset{\text{prior}}{\sim} \mathrm{D}_Q(Q^{-1}\theta/\tilde{P}, \dots, Q^{-1}\theta/\tilde{P})$$

$$\boldsymbol{s}|\theta \overset{\text{prior}}{\sim} \mathrm{D}_P(Q^{-1}\theta/\tilde{P}, \dots, Q^{-1}\theta/\tilde{P}, \theta/\tilde{P}, \dots, \theta/\tilde{P})$$

$$\overset{\text{prior}}{\sim} \mathrm{D}_P((\boldsymbol{q}|\theta), (\boldsymbol{p}|\theta))$$

7.2 XBART

Markov chain algorithms based on independent local modifications to individual trees, or even just nodes of trees, are potentially slow to explore the immense space of binary trees. In some respects, it is remarkable that randomly selecting a variable to split on and a cut-point to split at work as well as it does! Greedy procedures based on recursive partitioning and exhaustive search, such as those used in CART, may be able to more rapidly converge to local modes, especially when sample sizes are large and deep trees are required to approximate the response surface. However, optimization-based procedures produce a single output even when quite different trees fit the data essentially equally well. The XBART algorithm (for "Xcellerated," or "accelerated," BART) is a hybrid approach, which borrows elements of recursive partitioning by exhaustive search with elements of stochastic likelihood-weighted posterior sampling. The result is a stationary Markov chain that can be used to define its own estimator of the response surface, or draws from which can be used to initialize BART MCMC algorithms, reducing burn-in time. This section describes the XBART algorithm, with a special focus on the computational innovations this hybrid approach facilitates. For theoretical discussion and extensive simulation evidence, see He *et al.* [23] and He and Hahn [24].

7.2.1 The XBART algorithm and `GrowFromRoot`

At a high level, the XBART algorithm proceeds according to a series of iterative parameter updates, much like the original BART Gibbs sampler. Indeed, the sampling steps for σ and the leaf parameters μ are exactly the same as the full conditional updates from BART backfitting. Likewise, XBART's tree updates are based on the residualized response, given the other trees in the *collection* and their parameters. Where XBART differs is that individual trees are regrown anew at each update, rather than being modified incrementally. That is, rather than making a single transition to each tree, the current tree is deleted and regrown in full according to a recursive, but stochastic, growing process (individual branches stop growing stochastically). The main algorithm is presented in 2; the key subroutine, `GrowFromRoot`, is shown in 3. Although samples from this algorithm do not constitute draws from a bona fide Bayesian posterior, Monte Carlo averages may still be computed to define various estimators, specifically predictions for new observations.

Algorithm 2. Accelerated Bayesian Additive Regression Trees (XBART)

procedure XBART($\mathbf{y}, \mathbf{X}, C, L$, num_samples)
output Samples of forest
  $p \leftarrow$ number of columns of $\mathbf{X}$
  $N \leftarrow$ number of rows of $\mathbf{X}$
  Initialize $r_l^{(0)} \leftarrow \mathbf{y}/L$.
  for k in 1 to num_samples **do**
    for l in 1 to L **do**
      Calculate partial residual $r_l^{(k)}$ as shown in CGM10.
      if $k < I$ **then**
        GrowFromRoot($r_l^{(k)}$,$\mathbf{X}$)
      else
        GrowFromRoot($r_l^{(k)}$,$\mathbf{X}$)
      $\sigma^2 \sim$ Inverse-Gamma($N + \alpha, r_l^{(k)t} r_l^{(k)} + \eta$)
  return

Algorithm 3. GrowFromRoot

procedure GrowFromRoot(r, $\mathbf{X}$) ▷ Fit a tree to response vector r with predictors $\mathbf{X}$.
output A tree T_l.
  $N \leftarrow$ number of rows of r, $\mathbf{X}$
  $p \leftarrow$ number of columms of r, $\mathbf{X}$
  Evaluate expression [11] for C evenly spaced cut-points for each of p predictors.
  Sample a cut-point with probabilities given in expression [21].
  if sample no-split option **then**
    Sample leaf parameter according to $\mu \sim N\left(\sum \mathrm{r}/\left[\sigma^2\left(\frac{1}{\tau}+\frac{N}{\sigma^2}\right)\right], 1/\left[\frac{1}{\tau}+\frac{N}{\sigma^2}\right]\right)$. **return**
  else
    Partition data according to the selected cut-point.
    GrowFromRoot(y_{left},$\mathbf{X}_{\text{left}}$)
    GrowFromRoot(y_{right},$\mathbf{X}_{\text{right}}$)

The `GrowFromRoot` subroutine can be conceptualized as a sequence of draws from the posterior of "local Bayesian agents." At each node of the tree, the local Bayesian agent who "lives" at that node is given the data from the node above and updates her prior over a finite set of parameters, corresponding to partitions of the data. The likelihood used by these agents is the same as that from the BART model, but the local parameter set consists only of the available local partitions, irrespective of the previous or subsequent structure of the tree. Accordingly, the "local posterior" at each node is computed as a simple application of Bayes rule to a discrete parameter set. All available divisions are considered at each step, making the XBART algorithm comparatively fast at isolating partitions that are strongly indicated by the data. Formally, each local agent is tasked with partitioning the data into two parts (or leave it unpartitioned). Observations in the same partition are assumed to have the same, unknown, location parameter; therefore, the prior predictive distribution – obtained by integrating out the partition-specific mean – is a mean-zero multivariate normal distribution with covariance

$$\mathbf{V} = \tau \mathrm{J}\mathrm{J}^t + \sigma^2 \mathbf{I}$$

where τ is the prior variance of the leaf-specific mean parameter, σ^2 is the variance of the additive error, and J is a column vector of all ones. The prior predictive density of $\mathrm{y} \sim \mathcal{N}(0, \mathbf{V})$ is

$$p(\mathrm{y} \mid \tau, \sigma^2) = (2\pi)^{-n/2} \det(\mathbf{V})^{-1/2} \exp\left(-\frac{1}{2}\mathrm{y}^t\mathbf{V}^{-1}\mathrm{y}\right)$$

which can be simplified, using the matrix inversion Algorithm, to

$$\mathbf{V}^{-1} = \sigma^{-2}\mathbf{I} - \frac{\tau}{\sigma^2(\sigma^2 + \tau n)}\mathrm{J}\mathrm{J}^t$$

Sylvester's determinant theorem applied to $\det \mathbf{V}^{-1}$ yields a log-predictive likelihood of

$$-\frac{n}{2}\log(2\pi) - n\log(\sigma) + \frac{1}{2}\log\left(\frac{\sigma^2}{\sigma^2 + \tau n}\right)$$
$$-\frac{1}{2}\frac{\mathrm{y}^t\mathrm{y}}{\sigma^2} + \frac{1}{2}\frac{\tau}{\sigma^2(\sigma^2 + \tau n)}s^2$$

where $s \equiv \mathrm{y}^t\mathrm{J} = \sum_i y_i$, so that $\mathrm{y}^t\mathrm{J}\mathrm{J}^t\mathrm{y} = (\sum_i y_i)^2 = s^2$. Considering both partitions, $b \in \{\text{left}, \text{right}\}$, gives a combined log-predictive likelihood of

$$\sum_b \left\{ -\frac{n_b}{2}\log(2\pi) - n_b\log(\sigma) + \frac{1}{2}\log\left(\frac{\sigma^2}{\sigma^2 + \tau n_b}\right) \right.$$
$$\left. -\frac{1}{2}\frac{\mathrm{y}_b^t\mathrm{y}_b}{\sigma^2} + \frac{1}{2}\frac{\tau}{\sigma^2(\sigma^2 + \tau n_b)}s_b^2 \right\}$$
$$= -n\log(2\pi) - n\log(\sigma) - \frac{1}{2}\frac{\mathrm{y}^t\mathrm{y}}{\sigma^2}$$
$$+\frac{1}{2}\sum_b \left\{ \log\left(\frac{\sigma^2}{\sigma^2 + \tau n_b}\right) + \frac{\tau}{\sigma^2(\sigma^2 + \tau n_b)}s_b^2 \right\}$$

The first three terms are not functions of the partition yielding a "local likelihood" proportional to

$$\sum_b \left\{ \log\left(\frac{\sigma^2}{\sigma^2 + \tau n_b}\right) + \frac{\tau}{\sigma^2(\sigma^2 + \tau n_b)}s_b^2 \right\} \tag{16}$$

where n_b and s_b are functions of the partition (which is defined by the cut-point). These formulae have been written in terms of data y to emphasize the "local" interpretation/justification of the model. In the implementation, however, the data are the partial residuals.

Selection of a variable to split on, and a cut-point to split at, is then a sample according to Bayes rule:

$$\pi(v,c) = \frac{\exp(\ell(c,v))\kappa(c)}{\sum_{v'=1}^{p}\sum_{c'=0}^{C}\exp(\ell(c',v'))\kappa(c')} \tag{17}$$

where

$$\begin{aligned}\ell(v,c) &= \frac{1}{2}\left\{\log\left(\frac{\sigma^2}{\sigma^2+\tau n(\leq,v,c)}\right)+\frac{\tau}{\sigma^2(\sigma^2+\tau n(\leq,v,c))}s(\leq,v,c)^2\right\}\\ &+\frac{1}{2}\left\{\log\left(\frac{\sigma^2}{\sigma^2+\tau n(>,v,c)}\right)+\frac{\tau}{\sigma^2(\sigma^2+\tau n(>,v,c))}s(>,v,c)^2\right\}\end{aligned}$$

for $c \neq 0$. The partition size is denoted $n(\leq, v, c)$, which is the number of observations such that $x_v \leq c$; similarly, $s(\leq, v, c)$ is the sum of the residual $r_l^{(k)}$ of those same observations. The complement quantities, $n(>, v, c)$ and $s(>, v, c)$, are defined analogously. A uniform prior is applied to the cut-points, so that $\kappa(c \neq 0) = 1$.

Stochastic termination of the growing process is achieved by including a "no split" option in the local agents' parameter sets, effectively corresponding to a cut location that lies outside of the range of the data. The prior on this parameter can be chosen such that the XBART prior predictive (the algorithm applied to no data) corresponds to the usual BART prior predictive. Formally, for $c = 0$, which corresponds to no split,

$$\ell(v,c) = \frac{1}{2}\left\{\log\left(\frac{\sigma^2}{\sigma^2+\tau n}\right)+\frac{\tau}{\sigma^2(\sigma^2+\tau n)}s^2\right\}$$

and $\kappa(0) = \frac{1-\alpha(1+d)^{-\beta}}{\alpha(1+d)^{-\beta}}$. With this weight, the probability of *splitting* is the complement set of not splitting:

$$p_{\text{SPLIT}} = 1 - \frac{|C|(\alpha^{-1}(1+d)^{\beta}-1)}{|C|(\alpha^{-1}(1+d)^{\beta}-1)+|C|} = \alpha(1+d)^{-\beta}$$

just as in the original BART prior.

Relative to BART MCMC samplers, XBART has higher per-iteration cost because it must evaluate the likelihood at $|C|$ points at each node during `GrowFromRoot`. The benefits of this higher cost are (usually) improved posterior exploration leading to dramatically fewer required iterations. Still, any improvement to the per-iteration computational burden is beneficial, and the recursive structure of XBART permits a number of helpful improvements.

Two particular innovations deserve to be highlighted: presorting the predictor variables and using cut-point values based on local quantiles (as opposed to using all valid cut-points at each node).

Presorting predictor variables

Because the BART marginal likelihood depends only on partition sums, the sufficient statistics for *all* cut-points at a given node can be calculated with a single pass through the data at each variable by computing a cumulative sum, provided that the response values (in the form of the partial residual) are accessible in sorted order (for each predictor). More formally, define the cumulative sums in terms of a matrix of indices, $\mathbf{O}$, with elements o_{vh}

denoting the index of the hth largest observation of the x_vth variable in the original data matrix. In terms of **O**, the partition sums can be expressed as

$$s(\leq, v, c) = \sum_{h \leq c} r_{O_{vh}} \tag{18}$$

and

$$s(>, v, c) = \sum_{h=1}^{n} r_{lh} - s(\leq, v, c) \tag{19}$$

where r denotes the vector of partial residuals from the other trees. These sums are the inputs to the `GrowFromRoot` split criterion. To perform a similar operation at the subsequent node, the variable sorting must be maintained; fortunately, this can be achieved efficiently by "sifting" the variables. After a variable v and cut-point c are drawn, the algorithm partitions **O** into two matrices $\mathbf{O}^{\leq}$ and $\mathbf{O}^{>}$, which are populated sequentially by evaluating each element of **O** in turn and sending it to the next element of either $\mathbf{O}^{\leq}$ or $\mathbf{O}^{>}$, according to whether the corresponding element has $x_j \leq c$ or not. By populating each row of $\mathbf{O}^{\leq}$ and $\mathbf{O}^{>}$ by sequentially scanning the rows of **O**, the ordering is preserved for the next step of the recursion.

Adaptive nested cut-points

The discrete Bayes rule calculation at the heart of the stochastic `GrowFromRoot` procedure is computationally intensive when sample sizes are large (especially at early stages of the growing process, such as the split at the root), because each data point defines a valid cutting location. In some measure, this is why the BART MCMC implementations favor a predefined grid of candidate cut locations (perhaps based on marginal quantiles or a uniform grid). The recursive nature of the `GrowFromRoot` algorithm permits an "adaptive" alternative, where a nonexhaustive set of quantiles can be considered at each node, where the quantiles are computed relative to the available data at the present node. Conveniently, these quantiles need never be computed explicitly; instead, one simply evaluates the likelihood at "strides" by skipping a fixed number of observations (in sorted order) when calculating the marginal likelihood split criterion. All of the cumulative sums must still be computed, but the sampling is performed among a much smaller subset of cut-points, saving significant computational effort on both the likelihood evaluations and the random variable generation. This approach does not reduce the expressivity of the model, as any cut-point can eventually be selected, just perhaps further down the tree.

Thus, there is a trade-off between coarser/sparser cut-point candidates and eventual tree depth. In practice, using tens or hundreds of cut-points (rather than thousands or more) seems to gives good performance. Intuitively, the adaptive cut-point strategy will work well when there are large regions of covariate space where the function is relatively flat and others where it is comparatively variable. Coarser cut-point sets permit rapid identification of the flat regions, while simultaneously growing deeper trees in regions of high response surface variation. A function that oscillates rapidly uniformly over the predictor space may be more efficiently fit with a denser set of candidate cut-points.

7.2.2 Warm-start XBART

An especially appealing aspect of the XBART algorithm is its use in conjunction with traditional MCMC BART, by initializing independent Markov chains at draws from XBART.

This approach combines XBART's ability to rapidly find potentially large trees that fit the data well with the valid posterior uncertainty assessment that MCMC provides. Provided that each draw from XBART is from a starting location in a high probability region of the BART posterior, burn-in times are negligible for each chain, leading to substantially lower run times. Meanwhile, the diversity of the various starting locations results in wider credible intervals for quantities of interest, such as point-wise predictions. Nearness of the tree draws (according to various metrics) from the separate chains may also be used as a gauge of mixing, although in practice simply appending the separate draws appears to yield conservatively wide intervals, which has its own appeal. Simulation results indicate that warm-start XBART is faster and has better point-wise coverage of the mean function compared to either XBART or MCMC BART [24].

8 Conclusion

Bayesian tree modeling is a rich area of ongoing research, with challenges ranging from fundamental modeling to the construction of computational algorithms. The Bayesian approach offers many advantages, for example, BART infers the depth of each tree rather than having to tune it using cross-validation as in most non-Bayesian boosting approaches. But there is cost to the Bayesian advantages. Not everyone wants to choose a prior, and not everyone wants MCMC draws.

As empirical analysis continues to take center stage today, we see a growing variety of applications in data science with many different kinds of objectives. We believe that the fundamentals of Bayesian thinking will continue to play a role in the development of methodology that is relevant to real-world decision-making, and Bayesian tree models will continue to be a useful part of that bigger picture.

References

1 Brieman, L., Friedman, J., Olshen, R., and Stone, C. (1993) *Classification and Regression Trees*, Chapman & Hall.

2 Breiman, L. (2001) Random forests. *Mach. Learn.*, 45 (1), 5–32.

3 Freund, Y. and Schapire, R.E. (1997) A decision-theoretic generalization of on-line learning and an application to boosting. *J. Comput. Syst. Sci.*, 55 (1), 119–139.

4 Chen, T. and Guestrin, C. (2016) *XGBoost: A Scalable Tree Boosting System*. Proceedings of the 22nd ACM SIGKDD International Conference on Knowledge Discovery and Data Mining, KDD '16, pp. 785–794, New York, NY. Association for Computing Machinery.

5 Chipman, H.A., George, E.I., and McCulloch, R.E. (1998) Bayesian CART model search. *J. Am. Stat. Assoc. U.S.A.*, 93 (443), 935–948.

6 Pratola, M.T. (2016) Efficient metropolis–hastings proposal mechanisms for Bayesian regression tree models. *Bayesian Anal.*, 11, 885–911.

7 Friedman, J.H. (2001) Greedy function approximation: a gradient boosting machine. *Ann. Stat.*, 29 (5), 1189–1232.

8 Chipman, H.A., George, E.I., and McCulloch, R.E. (2010) BART: Bayesian additive regression trees. *Ann. Appl. Stat.*, 4 (1), 266–298.

9 Richard Hahn, P., Murray, J.S., and Carvalho, C.M. (2020) Bayesian regression tree models for causal inference: regularization, confounding, and heterogeneous effects (with discussion). *Bayesian Anal.*, 15 (3), 965–1056.

10 McCulloch, R.E., Sparapani, R.A., Logan, B.R., and Laud, P.W. (2021) Causal inference with the instrumental variable approach and Bayesian nonparametric machine learning. *arXiv preprint*, 2102.01199.

11 Linero, A. (2018) Bayesian regression trees for high dimensional prediction and variable selection. *J. Am. Stat. Assoc. U.S.A.*, 113 (522), 626–36.

12 Ročková, V. and Saha, E. (2019) On theory for BART, in *Proceedings of Machine Learning Research*, vol. 89 (eds K. Chaudhuri and M. Sugiyama), PMLR, pp. 2839–2848.

13 Ročková, V. and van der Pas, S. (2020) Posterior concentration for Bayesian regression trees and forests. *Ann. Stat.*, 48 (4), 2108–2131.

14 Mohammadi, R., Pratola, M., and Kaptein, M. (2020) Continuous-time birth-death MCMC for Bayesian regression tree models. *J. Mach. Learn. Res.*, 21 (201), 1–26.

15 Harrison Jr, D. and Rubinfeld, D.L. (1978) Hedonic housing prices and the demand for clean air. *J. Environ. Econ. Manage.*, 5 (1), 81–102.

16 Sparapani, R., Spanbauer, C., and McCulloch, R. (2021) Nonparametric machine learning and efficient computation with Bayesian additive regression trees: the BART R package. *J. Stat. Soft.*, 97 (1), 1–66.

17 Chipman, H.A., George, E.I., and McCulloch, R.E. (2013) Bayesian regression structure discovery, in *Bayesian Theory and Applications* (eds P. Damien, P. Dellaportas, N. Polson, and D. Stephens), Oxford University Press, Oxford, UK.

18 Bleich, J., Kapelner, A., George, E.I., and Jensen, S.T. (2014) Variable selection for BART: an application to gene regulation. *Ann. Appl. Stat.*, 8 (3), 1750–1781.

19 Hahn, P.R. and Carvalho, C.M. (2015) Decoupling shrinkage and selection in Bayesian linear models: a posterior summary perspective. *J. Am. Stat. Assoc. U.S.A.*, 110 (509), 435–448.

20 McCulloch, R.E., Carvalho, C., and Hahn, R. (2015) *A General Approach to Variable Selection Using Bayesian Nonparametric Models*. Joint Statistical Meetings, Seattle, 09 August 2015 to 13 August 2015

21 Liu, Y. and Ročková, V. (2021) Variable selection via Thompson sampling. *J. Am. Stat. Assoc. U.S.A.*, 1–18.

22 Lloyd Johnson, N., Kotz, S., and Balakrishnan, N. (1995) *Continuous Univariate Distributions*, vol. 2, 2nd edn, John Wiley & Sons, New York.

23 He, J., Yalov, S., and Richard Hahn, P. (2019) *XBART: Accelerated Bayesian Additive Regression Trees*. The 22nd International Conference on Artificial Intelligence and Statistics, pp. 1130–1138.

24 He, J. and Richard Hahn, P. (2021) Stochastic tree ensembles for regularized nonlinear regression. *J. Am. Stat. Assoc. U.S.A.*, 1–61.

Part IV

High-Dimensional Data Analysis

16

Penalized Regression

Seung Jun Shin[1] *and Yichao Wu*[2]

[1] *Korea University, Seoul, South Korea*
[2] *University of Illinois at Chicago, Chicago, IL, USA*

1 Introduction

Regression is a classical and important problem in statistics and data science to study how a response variable Y depends on covariates or predictor variables $\mathbf{X} = (X_1, X_2, \ldots, X_p)^T$. Given a random sample $\{(y_i, \mathbf{x}_i), i = 1, 2, \ldots, n\}$ of *i.i.d.* copies of $(Y, \mathbf{X})$, the regression seeks to estimate the unknown regression function $f(\cdot)$ defined as $f(\mathbf{x}) = E(Y|\mathbf{X} = \mathbf{x})$. To estimate f, one can minimize the empirical squared error loss functional

$$\min_{f \in \mathcal{F}} \sum_{i=1}^{n} \{y_i - f(\mathbf{x}_i)\}^2 \tag{1}$$

with respect to f in some function space $\mathcal{F}$, the choice of which depends on the model context. For example, the linear regression assumes the unknown regression function to have a linear form $f(\mathbf{x}) = \beta_0 + \boldsymbol{\beta}^T\mathbf{x}$. In this case, the minimization is over the involved regression parameters, namely, $\min_{\beta_0, \boldsymbol{\beta}} \sum_{i=1}^{n} (y_i - \beta_0 - \boldsymbol{\beta}^T\mathbf{x}_i)^2$. In this chapter, we focus on the mean regression with the squared error loss for the sake of brevity, but the extension to general loss functions such as the check loss for the quantile regression is straightforward.

Interpretation and prediction are two fundamental goals of regression and follow naturally after the estimation of regression function f. The linear regression is particularly useful to interpret the association between the response and predictors but often too poor for the prediction due to its restrictive linearity assumption which may not be valid in practical applications. It is possible to allow f to be far more flexible, but it may yield a trivial solution, namely, interpolation, which is extremely poor for both interpretation and prediction.

Penalized regression provides a natural way to compromise these two extremes and has gained a great popularity in contemporary applications in statistics and data science. It improves both interpretability and prediction accuracy by introducing penalty to control the complexity of the regression function in a data-adaptive manner. One of the earliest form of the penalization in statistical communities is the ridge regression [1] that is originally proposed to mitigate multicolinearity in the linear regression. The ridge

Computational Statistics in Data Science.
Edited by Walter W. Piegorsch, Richard A. Levine, Hao Helen Zhang and Thomas C. M. Lee.
 ISBN 978-1-11956107-1

regression is a version of Tikhonov regularization [2], a general tool for handling ill-posed problems. The ridge regression is also closely related to the Stein estimator [3] from which shrinkage estimation is originated. Regularization and shrinkage estimation are well-known synonyms of the penalization.

The penalized regression is defined in a general form as follows:

$$\hat{f}_\lambda = \underset{f \in \mathcal{F}}{\operatorname{argmin}} \sum_{i=1}^{n} (y_i - f(\mathbf{x}_i))^2 + \lambda J(f) \tag{2}$$

where J is the penalty that measures roughness of f, and a nonnegative constant $\lambda \geq 0$ denotes the regularization or tuning parameter that controls the balance between the fitness of data measured by the squared error loss and the model complexity measured by the penalty functional J. The choice of λ is crucial in practice and to be discussed later in great detail. Through the method of Lagrange multipliers, one can prove under certain conditions that (2) can be equivalently rewritten as the following constraint optimization problem:

$$\min_{f \in \mathcal{F}} \sum_{i=1}^{n} (y_i - f(\mathbf{x}_i))^2 \qquad \text{subject to} \qquad J(f) \leq \rho$$

for some $\rho \geq 0$ with a one-to-one correspondence between λ and ρ. This constraint form is useful to understand why the penalized regression can outperform its unpenalized counterpart. It is well known that the squared error risk $E(\hat{f}_\lambda - f)^2$ can be decomposed as the sum of the variance and squared bias of $\hat{f}_\lambda$. The constraint $J(f) \leq \rho$ increases model bias but reduces estimation variance. For a properly chosen ρ (or λ), the penalized regression estimator can beat the best unbiased estimator which is known to be optimal in classical statistical theories. Moreover, the penalized regression controls the complexity of $\hat{f}_\lambda$ at a certain level, resulting in better interpretable estimates.

In statistics, there are two domains where the penalization has been canonical. One is nonparametric regression where the complexity must be restricted to obtain a sensible estimator, and the other is the linear regression with high-dimensional predictors where the variable selection is crucial to have an interpretable model with improved accuracy. In this chapter, we provide a selective overview of penalized regression as follows. Section 2 is devoted for the penalized nonparametric regression, while Section 3 is for the penalized linear regression with high-dimensional predictors. We also describe how to select the tuning parameter in the penalized regression in Section 4, which is particularly important in practice.

2 Penalization for Smoothness

Penalization is a canonical tool in nonparametric regression that assumes f living on an infinite dimensional space of functions, in which case without appropriate penalization one can easily run into overfitting. Let us start with a univariate regression function $f(x)$ with $x \in \mathbb{R}$. The smoothing spline [4–7] is one of the most popular penalized methods in nonparametric regression. Smoothing spline estimator solves

$$\min_{f \in S_2} \sum_{i=1}^{n} (y_i - f(x_i))^2 + \lambda \int [f''(t)]^2 dt \tag{3}$$

where $f'(t) = \frac{d}{dt}f(t)$ and $f''(t) = \frac{d^2}{dt^2}f(t)$ denote the first- and second-order derivatives, respectively, and S_2 is the second-order Sobolev space defined as $S_2 = \{f : f$ and f' are absolutely continuous, and $\int \{f''(g)\}^2 dt < \infty\}$. Namely, the smoothing spline (3) seeks a smooth function f which explains its name. The solution of (3) is known to be a natural cubic spline with knots at all distinct x_is. Consequently, the smoothing spline estimator can be represented in a closed-form expression as $f(x) = \theta_0 + \theta^T \mathbf{B}(x)$, where $\theta = (\theta_1, \dots, \theta_n)^T$, and $\mathbf{B}(x) = \{B_1(x), \dots, B_n(x)\}^T$ denotes the set of basis functions such as B-spline (assuming all x_is are distinct). The smoothing spline (3) is thus equivalent to solve

$$\min_{\theta} \sum_{i=1}^{n} \{y_i - \theta^T \mathbf{B}(x_i)\}^2 + \lambda \theta^T \mathbf{K} \theta \tag{4}$$

where $\mathbf{K} = \{\int B_i''(t) B_j''(t) dt\}_{i,j=1}^{n} \in \mathbb{R}^{n \times n}$. This reveals a direct connection of (4) to a kernel ridge regression (KRR) defined as follows:

$$\min_{f \in \mathcal{H}_K} \sum_{i=1}^{n} (y_i - f(x_i))^2 + \lambda \|f\|_{\mathcal{H}_K}^2 \tag{5}$$

where $\mathcal{H}_K$ denotes the reproducing kernel Hilbert spaces (RKHSs) generated by a nonnegative kernel K, and $\|\cdot\|_{\mathcal{H}_K}^2$ is the squared RKHS norm. In fact, the smoothing spline can be cast into a penalized regression on RKHS, which provides an elegant way to analyze the smoothing spline based on RKHS theory [8, 9].

There are several ways to extend (3) to the multivariate regression function. First, the penalty term can be generalized to multidimensional functions. For example, the thin-plate spline [10] employs the following functional to penalize a bivariate regression function:

$$J(f) = \int_{\mathbb{R}^2} \left[\left(\frac{\partial f(\mathbf{x})}{\partial x_1^2} \right)^2 + 2 \left(\frac{\partial f(\mathbf{x})}{\partial x_1 \partial x_2} \right)^2 + \left(\frac{\partial f(\mathbf{x})}{\partial x_2^2} \right)^2 \right] d\mathbf{x}$$

and the extension to a higher dimension is also possible. Thin-plate spline shares many properties with the aforementioned univariate case and has a closed-form solution as (3) does.

For the multivariate case, another popular way is to apply the ANOVA decomposition to the regression function, known as the smoothing spline ANOVA (SS-ANOVA) model [9]. For example, the SS-ANOVA model with main effects only assumes $f(\mathbf{x}) = b + \sum_{j=1}^{p} f_j(x_j)$ and $\mathcal{F} = \{1\} \oplus \mathcal{F}^1 \oplus \cdots \oplus \mathcal{F}^p$, where $\mathcal{F}^j$ denotes the second-order Sobolev space for X_j, and $\oplus$ denotes the direct sum operator. Employing the RKHS, the SS-ANOVA model solves

$$\min_{f \in \mathcal{H}_K} \sum_{i=1}^{n} \{y_i - f(\mathbf{x}_i)\}^2 + \lambda \sum_{j=1}^{p} \theta_j^{-1} \|P^j f\|_{\mathcal{H}_K}^2 \tag{6}$$

where $P^j f$ denotes the orthogonal projection of f onto $\mathcal{H}_K^j$ that is analogously defined as $\mathcal{F}^j$. The additional tuning parameter $\theta_j > 0$ is compounded with λ but introduced for computational purposes.

Representer Theorem [8] states that the KRR with a given kernel K has a finite form solution as $f(\mathbf{x}) = b + \sum_{i=1}^{n} \theta_i K(\mathbf{x}, \mathbf{x}_i)$. This yields a straightforward extension of (5) for the multivariate case. The choice of the kernel is crucial in practice. Popular kernel functions include, for example, the linear kernel $K(\mathbf{x}, \mathbf{x}') = \mathbf{x}^T\mathbf{x}'$ and the radial (or Gaussian) kernel $K(\mathbf{x}, \mathbf{x}') = \exp\{-\gamma\|\mathbf{x} - \mathbf{x}'\|^2\}$. If the linear kernel is adopted, it essentially leads to a penalized linear regression.

3 Penalization for Sparsity

Penalized regression is also very popular in linear regression, especially for the case with high-dimensional predictors. Although the ridge regression can still be employed to improve prediction accuracy, the interpretability suffers severely when there are too many predictors in the model. In such high-dimensional regression, it is essential to select informative variables, and the sparsity of the regression coefficient estimator is thus highly desirable. Toward variable selection, L_0-penalty ($\|\boldsymbol{\beta}\|_0 = \sum_{j=1}^{p} I(\beta_j = 0)$, where $I(\cdot)$ denotes the indicator function) is a natural choice due to its equivalency to the best subset selection. However, L_0-penalized regression is an NP-hard combinatorial problem. Bridge regression [11] penalized by L_q-norm ($\|\boldsymbol{\beta}\|_q = (\sum_{j=1}^{p} |\beta_j|^q)^{1/q}$) with $0 < q < 2$ was proposed as an intermediate solution.

Tibshirani [12] proposed the least absolute shrinkage and selection operator (LASSO) that employs L_1-norm as a penalty, $\|\boldsymbol{\beta}\|_1 = \sum_{j=1}^{p} |\beta_j|$. The LASSO solves

$$\min_{\beta_0, \boldsymbol{\beta}} \sum_{i=1}^{n} (y_i - \beta_0 - \boldsymbol{\beta}^T\mathbf{x})^2 + \lambda\|\boldsymbol{\beta}\|_1 \tag{7}$$

for some $\lambda > 0$. Because of the geometry of the L_1-norm, the LASSO estimator is sparse and is capable of performing variable selection. In terms of computation, the L_1-norm can be viewed as a convex relaxation of the L_0-norm, which makes the LASSO easier to optimize. For example, the coordinate decent algorithm [13] can be easily implemented and is very efficient as the coordinatewise optimizer of the LASSO turns out to be a soft thresholding of the corresponding ordinary least square (OLS) estimator. The LARS algorithm [14] provides an elegant way to compute the entire regularization solution path of the LASSO estimate by exploiting the piecewise linearity of the LASSO solution as a function of λ.

There are numerous extensions of the LASSO. The LASSO suffers when predictors are highly correlated, and Zou and Hastie [15] tackles this by proposing the elastic net penalty, a hybrid version of the LASSO and the ridge regression. To be more precise, it solves

$$\min_{\beta_0, \boldsymbol{\beta}} \sum_{i=1}^{n} (y_i - \beta_0 - \boldsymbol{\beta}^T\mathbf{x})^2 + \lambda\left\{\alpha\|\boldsymbol{\beta}\|_1 + \frac{1}{2}(1-\alpha)\|\boldsymbol{\beta}\|_2^2\right\}$$

where $\|\boldsymbol{\beta}\|_2 = \sqrt{\boldsymbol{\beta}^T\boldsymbol{\beta}}$, and $\alpha \in [0, 1]$ controls the balance between the LASSO and ridge penalties. The elastic net penalty reduces to the LASSO when $\alpha = 1$ and to the ridge regression when $\alpha = 0$.

The group LASSO [16] extends the LASSO to the case with grouped variables that share an identical sparsity structure. Suppose that there are G groups of predictors $\mathbf{x}_{i,1}, \mathbf{x}_{i,2}, \ldots, \mathbf{x}_{i,G}$ such that $\mathbf{x}_i = (\mathbf{x}_{i,1}^T, \mathbf{x}_{i,2}^T, \ldots, \mathbf{x}_{i,G}^T)^T$ without loss of generality, where $\mathbf{x}_{i,g} \in \mathbb{R}^{p_g}$ denotes the covariates of the gth group with group size $p_g \geq 1, g = 1, \ldots, G$ and $\sum_{g=1}^{G} p_g = p$. Now, the group LASSO solves

$$\min_{\beta_0,\boldsymbol{\beta}} \sum_{i=1}^{n} \left(y_i - \beta_0 - \sum_{g=1}^{G} \boldsymbol{\beta}_g^T \mathbf{x}_{i,g} \right)^2 + \lambda \sum_{g=1}^{G} \|\boldsymbol{\beta}_g\|_2$$

where $\boldsymbol{\beta}_g$ denotes the coefficient vector corresponding to the gth group variables. The group LASSO shrinks the regression coefficients vector of the gth group, $\boldsymbol{\beta}_g$ to zero simultaneously. Each group of coefficients is either all in or all out. It reduces to the LASSO when $G = p$.

The LASSO estimator is biased even when $|\beta_j|$ is large. This results in the variable selection inconsistency of the LASSO estimator unless certain conditions are satisfied [17]. Zou [18] and Zhang and Lu [19] proposed the adaptive LASSO as a simple remedy to remove the bias of the LASSO. The adaptive LASSO solves

$$\min_{\beta_0,\boldsymbol{\beta}} \sum_{i=1}^{n} (y_i - \beta_0 - \boldsymbol{\beta}^T \mathbf{x})^2 + \lambda \sum_{j=1}^{p} w_j |\beta_j|$$

where the weight w_j is chosen to be inversely proportional to $|\beta_j|^\gamma$ for some $\gamma > 0$ or its estimate. Thus, the adaptive LASSO reduces the bias of the LASSO by less penalizing more informative variables with larger $|\beta_j|$.

Fan and Li [20] rigorously analyzed the penalized linear regression and identified desired properties that a good penalty function should possess: unbiasedness, sparsity, and continuity. They then proposed smoothly clipped absolute deviance (SCAD) penalty that possesses the aforementioned properties. It was shown that the SCAD-penalized linear regression estimator enjoys the oracle property that the estimator behaves as if the true model were known, when n is large. Minimax concave penalty (MCP [21]) is another popular penalty that shares a similar sprit with the SCAD penalty. By construction, both penalties are nonconvex (Figure 1), which makes the corresponding optimization nontrivial. However, Breheny and Huang [22] showed that the coordinate decent algorithm can solve these nonconvex penalization problems very efficiently.

As a generalization of the sparsity, Ke *et al.* [23] introduced *homogeneity* to refer clustering structures under which the coefficients belonging to the same cluster share an identical value. The fused LASSO [24] is one of the earliest attempts to pursue the homogeneity in regression, and it solves

$$\min_{\beta_0,\boldsymbol{\beta}} \sum_{i=1}^{n} (y_i - \beta_0 - \boldsymbol{\beta}^T \mathbf{x})^2 + \lambda_1 \sum_{j=1}^{p} |\beta_j| + \lambda_2 \sum_{j=2}^{p} |\beta_j - \beta_{j-1}|$$

Note that the fused LASSO penalizes the first-order absolute difference, which encourages adjacent coefficients to be identical. Tibshirani [25] proposed the trend filtering that extends the fused LASSO by replacing the first-order difference with higher order ones. Ke *et al.* [23] developed a hybrid pairwise penalty as a compromise between the fused LASSO and total variation penalty [26] to explore more complex homogeneity structure in regressions.

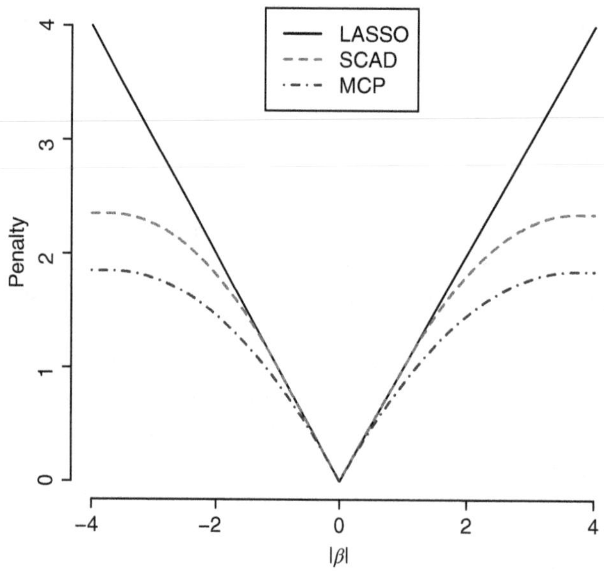

Figure 1 LASSO and nonconvex penalties: both SCAD and MCP do not penalize the regression coefficient β when $|\beta|$ is large and yield (nearly) unbiased estimators.

4 Tuning Parameter Selection

Tuning parameter λ plays an important role in the penalized regression, and its selection is crucial in practice. It is desirable to choose λ that minimizes the prediction error (PE) for independent observations, while keeping f as simple as possible, that is, $\lambda^* = \text{argmin}_\lambda \text{ PE}(\hat{f}_\lambda)$, where

$$\text{PE} = \text{PE}(\hat{f}_\lambda) = E\{Y - \hat{f}_\lambda(\mathbf{X})\}^2 \tag{8}$$

Here, $(Y, \mathbf{X})$ denotes a random observation independent of $(y_i, \mathbf{x}_i), i = 1, \dots, n$ used for training $\hat{f}_\lambda$. Estimation of PE (8) is not trivial since the expectation is with respect to $(Y, \mathbf{X})$ unknown in the training step. A naive empirical version of PE is training error (TE) defined as

$$\text{TE} = \text{TE}(\hat{f}_\lambda) = \frac{1}{n} \sum_{i=1}^{n} \{y_i - \hat{f}_\lambda(\mathbf{x}_i)\}^2$$

but always underestimates (8).

There are various ways to estimate PE. Efron [27] categorized them into two broad classes: (i) cross-validation (CV) and (ii) penalty methods. The K-fold -CV randomly splits the data $(y_i, \mathbf{x}_i), i = 1, \dots, n$ into K folds and then computes an empirical PE rate of $\hat{f}_\lambda^{(-k)}$ for the kth fold, $k = 1, \dots, K$, where $\hat{f}_\lambda^{(-k)}$ denotes the model fitted from data except the kth fold. The CV estimates (8) by averaging these error estimates over all $k = 1, \dots, K$. The leave-one-out CV (LOO-CV) refers the case when $K = n$, and is shown to be an unbiased estimator of PE, which justifies the use of K-fold CV. However, CV is often computationally intensive.

The generalized CV (GCV [8]) that approximates LOO-CV estimator without repetitions is a popular alternative since it substantially reduces the computational cost.

The penalty methods estimate (8) from TE by adjusting its bias via penalization. Various information criterions such as Akaike's information criterion (AIC) and Bayesian information criterion (BIC) belong to this class. Generalized information criterion (GIC [28]) is defined as

$$\mathrm{GIC}(\lambda) = \mathrm{TE}(\hat{f}_\lambda) + \kappa_n \, \mathrm{df}(\hat{f}_\lambda)$$

where $\mathrm{df}(\hat{f}_\lambda)$ denotes the effective degrees of freedom (EDF) of $\hat{f}_\lambda$. The GIC adjusts the bias of TE by adding a penalty term proportional to the EDF of the model, and a constant κ_n controls the amount of penalty. Note that the GIC reduces to AIC when $\kappa_n = 2$ and BIC when $\kappa_n = \log n$.

It is thus essential to compute the EDF of $\hat{f}_\lambda$ for the penalty methods. Stein [29] established a rigorous definition of the EDF of f with Gaussian error $\epsilon_i \sim N(0, \sigma^2)$, and Ye [30] further generalized it as follows:

$$\mathrm{EDF}(\hat{f}_\lambda) = \frac{1}{\sigma^2} \sum_{i=1}^{n} \mathrm{Cov}\{\hat{f}_\lambda(\mathbf{x}_i), y_i\} \tag{9}$$

This provides a general way to compute the EDF of $\hat{f}_\lambda$. For example, many L_2-penalized regression estimators such as ridge regression and smoothing spline are linear smoothers that can be represented as $\hat{\mathbf{f}}_\lambda = \mathbf{H}_\lambda \mathbf{y}$ for some smoothing matrix $\mathbf{H}_\lambda \in \mathbb{R}^{n\times n}$ depending on λ, where $\mathbf{f}_\lambda = \{\hat{f}_\lambda(\mathbf{x}_1), \dots, \hat{f}_\lambda(\mathbf{x}_n)\}^T$ and $\mathbf{y} = (y_1, y_2, \dots, y_n)^T$. By (9), we have $\mathrm{EDF}(\hat{f}_\lambda) = \mathrm{tr}(\mathbf{H}_\lambda)$, which coincides with the conventional definition of degrees of freedom of the linear regression. Zou *et al.* [31] showed that the EDF of LASSO estimate is the number of nonzero coefficients, and this, combined with the LARS algorithm, substantially simplifies the tuning procedure of LASSO.

References

1 Hoerl, A.E. and Kennard, R.W. (1970) Ridge regression: biased estimation for nonorthogonal problems. *Technometrics*, 12, 55–67.

2 Tikhonov, A.N., Goncharsky, A., Stepanov, V., and Yagola, A.G. (2013) *Numerical Methods for the Solution of Ill-Posed Problems*, vol. 328, Springer Science & Business Media.

3 James, W. and Stein, C. (1961) Estimation with quadratic loss. *Proc. Fourth Berkeley Symp. Math. Statist. Probab.*, 1, 361–379.

4 Reinsch, C.H. (1967) Smoothing by spline functions. *Numerische Mathematik*, 10, 177–183.

5 Kimeldorf, G.S. and Wahba, G. (1970) A correspondence between Bayesian estimation on stochastic processes and smoothing by splines. *Ann. Math. Stat.*, 41 (2), 495–502.

6 Kimeldorf, G.S. and Wahba, G. (1971) Some results on Tchebycheffian spline functions. *J. Math. Anal. Appl.*, 33 (1), 82–95.

7 de Boor, C. (1978) *A Practical Guide to Splines*, Springer-Verlag.

8 Wahba, G. (1990) *Spline Models for Observational Data*, SIAM.

9 Gu, C. (2013) *Smoothing Spline ANOVA Models*, vol. 297, Springer Science & Business Media.

10 Wood, S.N. (2003) Thin plate regression splines. *J. R. Stat. Soc., Ser. B*, 65 (1), 95–114.

11 Frank, L.E. and Friedman, J.H. (1993) A statistical view of some chemometrics regression tools. *Technometrics*, 35 (2), 109–135.

12 Tibshirani, R. (1996) Regression shrinkage and selection via the lasso. *J. R. Stat. Soc., Ser. B*, 58 (1), 267–288.

13 Wu, T.T. and Lange, K. (2008) Coordinate descent algorithms for lasso penalized regression. *Ann. Appl. Stat.*, 2 (1), 224–244.

14 Efron, B., Hastie, T., Johnstone, I., and Tibshirani, R. (2004) Least angle regression. *Ann. Stat.*, 32 (2), 407–499.

15 Zou, H. and Hastie, T. (2005) Regularization and variable selection via the elastic net. *J. R. Stat. Soc., Ser. B*, 67 (2), 301–320.

16 Yuan, M. and Lin, Y. (2006) Model selection and estimation in regression with grouped variables. *J. R. Stat. Soc., Ser. B*, 68 (1), 49–67.

17 Zhao, P. and Yu, B. (2006) On model selection consistency of lasso. *J. Mach. Learn. Res.*, 7, 2541–2563.

18 Zou, H. (2006) The adaptive lasso and its oracle properties. *J. Am. Stat. Assoc.*, 101 (476), 1418–1429.

19 Zhang, H.H. and Lu, W. (2007) Adaptive lasso for cox's proportional hazards model. *Biometrika*, 94 (3), 691–703.

20 Fan, J. and Li, R. (2001) Variable selection via nonconcave penalized likelihood and its oracle properties. *J. Am. Stat. Assoc.*, 96 (456), 1348–1360.

21 Zhang, C.-H. (2010) Nearly unbiased variable selection under minimax concave penalty. *Ann. Stat.*, 38 (2), 894–942.

22 Breheny, P. and Huang, J. (2011) Coordinate descent algorithms for nonconvex penalized regression, with applications to biological feature selection. *Ann. Appl. Stat.*, 5 (1), 232.

23 Ke, Z.T., Fan, J., and Wu, Y. (2015) Homogeneity pursuit. *J. Am. Stat. Assoc.*, 110 (509), 175–194.

24 Tibshirani, R., Saunders, M., Rosset, S. *et al.* (2005) Sparsity and smoothness via the fused lasso. *J. R. Stat. Soc., Ser. B*, 67 (1), 91–108.

25 Tibshirani, R.J. (2014) Adaptive piecewise polynomial estimation via trend filtering. *Ann. Stat.*, 42 (1), 285–323.

26 Harchaoui, Z. and Lévy-Leduc, C. (2010) Multiple change-point estimation with a total variation penalty. *J. Am. Stat. Assoc.*, 105 (492), 1480–1493.

27 Efron, B. (2004) The estimation of prediction error: covariance penalties and cross-validation. *J. Am. Stat. Assoc.*, 99 (467), 619–632.

28 Zhang, Y., Li, R., and Tsai, C.-L. (2010) Regularization parameter selections via generalized information criterion. *J. Am. Stat. Assoc.*, 105 (489), 312–323.

29 Stein, C.M. (1981) Estimation of the mean of a multivariate normal distribution. *Ann. Stat.*, 9, 1135–1151.

30 Ye, J. (1998) On measuring and correcting the effects of data mining and model selection. *J. Am. Stat. Assoc.*, 93 (441), 120–131.

31 Zou, H., Hastie, T., and Tibshirani, R. (2007) On the "degrees of freedom" of the lasso. *Ann. Stat.*, 35 (5), 2173–2192.

17

Model Selection in High-Dimensional Regression

Hao H. Zhang

University of Arizona, Tucson, AZ, USA

1 Model Selection Problem

In statistical data analysis, we sample observations and use them to infer the generation process underlying the data and make future predictions. Toward this, a theoretical assumption is usually made on the unknown true model, call f, which governs or characterizes the process. Typically, we assume that f belongs to some model class $\mathcal{M}$, which can be linear models, piecewise linear models without jumps, or smooth nonlinear models. Then, we use the data to evaluate and compare all the models in $\mathcal{M}$, in the hope of discovering the true f or finding the best approximation within $\mathcal{M}$; this is the practice of model selection.

Example: In multivariate regression problems, we observe the p-dimensional covariates $\mathbf{X} \in \mathbb{R}^p$ and the response variable $Y \in \mathbb{R}$, and the goal is to learn their relationship $E(Y|\mathbf{X}) = f(\mathbf{X})$ with $f \in \mathcal{M}$ from the data. The following are three examples of $\mathcal{M}$:

- $\mathcal{M}$ = {linear functions} $= \{f | f = \beta_0 + \mathbf{X}^T\boldsymbol{\beta}\}$, assuming Y is linearly related to X_js.
- $\mathcal{M} = \{f | f = \beta_0 + \sum_{j=1}^{p} \beta_j X_j + \sum_{j,k=1}^{p} \beta_{jk} X_j X_k\}$, allowing interaction effects between X_js.
- $\mathcal{M}$ = {piecewise linear functions with multiple jumps}, allowing f to be nonlinear and discontinuous.

The choice of $\mathcal{M}$ is usually subjective and made by analysts, by taking into account historical data and information, prior knowledge and experience, or just computational convenience. Since the true f is unknown, it is possible that the model class $\mathcal{M}$ is not chosen properly or not large enough to even contain f. In this case, the goal of model selection is to find one model in $\mathcal{M}$ that can approximate the truth in some optimal sense. The British statistician George E. Box had a famous quote "All models are wrong, but some are useful" [1].

To implement model selection, there are two essential elements:

1. *selection criterion*: evaluate the quality of all models in $\mathcal{M}$ to compare or rank them;
2. *search algorithm*: search over $\mathcal{M}$ to find the best one.

Computational Statistics in Data Science.
Edited by Walter W. Piegorsch, Richard A. Levine, Hao Helen Zhang and Thomas C. M. Lee.
 ISBN 978-1-11956107-1

A good model selection procedure should balance between a model's complexity and its generalization performance on future data prediction. If two models have similar prediction accuracy, then the simpler model is always preferred. A variety of model selection criteria have been developed for model selection in the literature. The commonly used are information criteria including Akaike's Information Criterion (AIC) [2, 3], Bayesian Information Criterion (BIC) [4], the deviance information criterion (DIC) [5, 6], and Mallow's C_p [7, 8] and their modified versions [9–12]. Bayesian model selection approaches take into account model uncertainty by putting a prior distribution on each candidate model and its parameters [13–16].

There is a rich literature on variable selection in linear regression models; see Linhart and Zucchini [17], Rao and Wu [18], and Miller [19] for a comprehensive review. Popular methods include the best subset selection and stepwise selection. The best subset search, also known as the exhaustive search, compares all models in $\mathcal{M}$ and selects the best one based on a certain criteria. It is theoretically appealing and often regarded as the "holy grail" of methods for model selection. However, the best subset selection is not scalable for high-dimensional data. For example, when $p = 20$, the model space $\mathcal{M}$ contains more than 1 million models, which makes the exhaustive search computationally expensive even infeasible. Using the software R, the best subset selection can be performed by the *leaps* package using a branch-and-bound algorithm [20]. Bertsimas *et al.* [21] propose to formulate the best subset selection as a mixed-integer optimization (MIO) problem, which can then be implemented by highly optimized MIO solvers. This idea greatly speeds up the best subset selection and makes it feasible for handling high-dimensional data with p in the hundreds and even thousands. Alternatively, greedy procedures such as forward selection and backward elimination visit only a subset of candidate models within $\mathcal{M}$ in a certain order to select the best model from the subset. These procedures are faster and popular in practice, but their results are suboptimal from the theoretical perspective as they cannot guarantee to identify the true model even with infinitely many observations. Furthermore, their discrete selection process often leads to unstable results with high variance, and it is hard to study their asymptotic properties [9, 10, 22].

Recent progress in computers and technologies enables the collection of massive and high-dimensional data. There are a variety of scenarios for the relative scale of p versus the sample size n: (i) p is fixed but comparable n, (ii) p diverges with n at a certain rate, say, $p \propto n^{\alpha}$ with $0 < \alpha < 1$, and (iii) $p \gg n$. Model selection is even more important in high-dimensional data analysis, as it is a critical step to reduce the data dimension, extract signals, and get rid of noises in the data. When p is larger than n, standard methods such as ordinary least squares (OLS) and the maximal-likelihood estimator (MLE) cannot be directly applied to the raw data, but they can be used after the low-dimensional true model is identified. The number of candidate models in the model class $\mathcal{M}$ is enormous when p is large. For example, the simple linear regression model class $\mathcal{M}$ contains totally 2^p models, which grows exponentially fast with p. Searching for the best model or the true model in such a large model class is challenging, just like "finding a needle in the haysack"; this is the so-called curse of dimensionality. Over the past two decades, a variety of modern methods and theory and scalable algorithms have been developed to tackle computational and theoretical challenges in model selection. This chapter provides a review on some recent results for model selection in high-dimensional linear regression, regression models with interaction effects, and nonparametric regression models.

Notations: We focus on the regression models in this chapter. Assume that we take a sample of size n observations $(\mathbf{x}_i, y_i), i = 1, \ldots, n$, from the random pair $(\mathbf{X}, Y) \sim Pr(\mathbf{X}, Y)$, where $\mathbf{X} \in \mathbb{R}^p$ is the p-vector regressors, and $Y \in \mathbb{R}$ is the real-valued response variable. In ordinary regression models, the goal is to estimate the relationship between $\mathbf{X}$ and Y from the data points, expressed as $Y_i = f(\mathbf{X}_i) + \epsilon_i$, where f is the function form of the regression relationship, and the error terms ϵ_i are i.i.d. with zero mean and a constant variance σ^2. For any vector $\mathbf{a} \in \mathbb{R}^p$, define its l_2 norm as $||\mathbf{a}|| = \sqrt{\mathbf{a}^T\mathbf{a}}$ and its l_1 norm as $||\mathbf{a}||_1 = \sum_{j=1}^{p} |a_j|$. For nonparametric regression, we assume the function $f(\mathbf{x}) \in \mathcal{F}$, which is some function space defined on the domain $\mathcal{X} = [0,1]^p$.

2 Model Selection in High-Dimensional Linear Regression

Linear regression models assume

$$Y_i = \beta_0 + \sum_{j=1}^{p} \beta_j X_{ij} + \epsilon_i \tag{1}$$

where ϵ_is are i.i.d. errors with mean zero and finite variance. In model (1), we call the predictors with nonzero coefficients as "important" variables. Define $A = \{j : \beta_j \neq 0, j = 1, \ldots, p\}$. The goal of model selection is to identify A, which is also known as the problem of variable selection. For high-dimensional data with a large p and the sparse true model $|A| = p_0 \ll p$, variable selection can effectively remove noise variable and therefore greatly enhance the model's prediction performance and interpretability. In the following sections, we review a variety of recently developed variable selection methods for high-dimensional linear regression.

2.1 Shrinkage Methods

Using matrix notations, we denote the response vector $\mathbf{y} = (y_1, \ldots, y_n)^T$, the vector of regression coefficients $\boldsymbol{\beta} = (\beta_1, \ldots, \beta_p)^T$, and the $n \times p$ design matrix $\mathbf{X}$, with the (i,j)th entry being x_{ij} for $i = 1, \ldots, n$ and $j = 1, \ldots, p$. For shrinkage methods, assume that both $\mathbf{y}$ and the columns in $\mathbf{X}$ are centered, so that the intercept β_0 can be omitted in the model fitting. We also standardize the columns in $\mathbf{X}$ to make the regression coefficients comparable.

The main idea of the shrinkage methods is to solve a penalized OLS problem by imposing penalties on regression coefficients and shrinking them toward zero. In general, shrinkage methods solve the following optimization problem:

$$\min_{\boldsymbol{\beta}} ||\mathbf{y} - \mathbf{X}\boldsymbol{\beta}||^2 + J_\lambda(\boldsymbol{\beta}) \tag{2}$$

where the penalty function $J_\lambda(\boldsymbol{\beta})$ is designed to shrink small coefficients to zero exactly and achieve the solution sparsity. The tuning parameter $\lambda > 0$ controls the amount of shrinkage: the larger the λ, the greater the amount of shrinkage on regression coefficients. A variety of penalty functions have been proposed in the literature, and popular examples include

- $J_\lambda(\beta) = \lambda|\beta|_1 = \lambda \sum_{j=1}^{p} |\beta_j|$ (LASSO) [23]
- $J'_\lambda(\beta) = \sum_{j=1}^{p}[I(|\beta_j| \leq \lambda) + \frac{(a\lambda-|\beta|)_+}{(a-1)\lambda} I(|\beta_j| > \lambda)]$ (SCAD) [22]
- $J_\lambda(\beta) = \lambda \sum_{j=1}^{p}[(1-a)|\beta_j|^2 + a|\beta_j|]$ (elastic net) [24]
- $J_\lambda(\beta) = \lambda \sum_{j=1}^{p} w_j|\beta_j|$ (adaptive LASSO) [25, 26]
- $J_\lambda(\beta) = \lambda \sum_{j=1}^{p} w_j[(1-a)|\beta_j|^2 + a|\beta_j|]$ (adaptive elastic net) [27],

where a is an additional tuning parameter in smoothly clipped absolute deviation (SCAD) and the elastic net, and w_j are the weights in the adaptive least absolute shrinkage and selection operator (LASSO) and adaptive elastic net penalties chosen adaptively from the data. See the references for details regarding the selection of λ and a as well as the construction of w_js. Different penalty functions have their own strengths and limitations.

The LASSO a soft-thresholding penalty and can produce exactly zeros in the solution. It is widely used in practice, due to its simple implementation and effective performance in real problems. Computationally, it enjoys convex programming, global convergence guarantee, and scalable path-finding algorithms. When $p \gg n$, the LASSO can select at most n predictors before it saturates. In practice, the LASSO may produce noticeable bias for large regression coefficients. This bias issue can be mitigated by the adaptive LASSO, which adjusts the penalty by weights w_js such that smaller penalties are imposed on important variables than unimportant ones. In practice, given a root-n consistent estimator $\tilde{\beta}$, the weight can be constructed as

$$w_j = 1/|\tilde{\beta}_j|^\gamma, \quad j = 1, \ldots, p$$

for some $\gamma > 0$. Zou [25] shows that, if the weights w_js are data dependent and chosen properly, then the adaptive lasso estimates are asymptotically unbiased and have optimal variance.

The SCAD penalty can be derived from its derivative function given the above, resulting in a nonconcave piecewise quadratic function form [22]. It is continuously differentiable except at the origin and can produce spare and approximately unbiased coefficients. There are two tuning parameters, λ and a with $(a > 1)$, and Fan and Li [22] showed that $a = 3.7$ worked well in various contexts of linear models.

The elastic net penalty encourages group selection by combining the ridge penalty and the LASSO penalty. For high-dimensional regression problems, the predictors are often highly correlated; the LASSO tends to select one arbitrary variable from the group and ignore the remaining ones, while the elastic net is shown to improve LASSO by selecting strongly correlated predictors together as groups [24]. The adaptive elastic net shares the same spirit as the adaptive LASSO by assigning different weights on the coefficients in the elastic net.

Other well-known penalty functions include the nonnegative garrote [10], the least angle regression (LARS) [28], the Dantzig selector [29], the minimax concave penalty (MCP) [30], and Lv and Fan [31]. All of these penalty functions apply continuous shrinkage to coefficients, making small coefficients become zero in the solution, and hence produce a parsimonious model.

2.2 Sure Screening Methods

For ultrahigh-dimensional data, the number of predictors p may grow at a much faster rate than n, for example, $\log(p) = O(n^\kappa)$ with $\kappa > 0$. In these settings, it is useful to first reduce p

to a moderate scale by a prescreening procedure, before a refined variable selection method is applied. Toward this, Fan and Lv [32] propose a sure screening method, called Sure Independence Screening (SIS), to reduce dimensionality from high to a relatively large scale below n. Define

$$\mathbf{w} = \mathbf{X}^T\mathbf{y}$$

where $\mathbf{X}$ is first standardized columnwisely. The SIS method first sorts the predictors based on $\mathbf{w}$ in a decreasing order and then produces the submodel

$$\hat{\mathcal{A}} = \{j : |w_j| \text{ is among the first } [n\gamma] \text{ largest of all}\}$$

for some $\gamma \in (0,1)$. Fan and Lv [32] show that SIS has the sure screening property for even exponentially growing dimension under some regularity conditions. Furthermore, an iterate SIS (ISIS) is proposed to enhance the finite sample performance.

In the literature, other sure screening methods were also developed, including forward regression screening [33], sure screening in the context of classification problems [Features annealed independence rules (FAIR)] [34], nonparametric independence screening (NIS) [35], and sure screening for interaction effects [36].

2.3 Model Selection Theory

We discuss a variety of theoretical properties of linear model selection procedures in terms of their variable selection and model prediction performance in the asymptotic sense. Let $\hat{\mathcal{A}}$ be the set of variables selected by a procedure, and let $\hat{\boldsymbol{\beta}}$ be the estimated linear model regression coefficient.

In general, a model selection procedure is said to be *model selection consistent* if it can identify the true model when n goes to infinity. For linear regression models, a model selection procedure is *selection consistent* if it can identify the true set of important variables $\mathcal{A}$ correctly with probability going to 1 as n increases to infinity, that is,

$$\Pr(\hat{\mathcal{A}} = \mathcal{A}) \to 1 \quad \text{as} \quad n \to \infty$$

A variable selection procedure has oracle properties [22] if it satisfies the following:

i) It can asymptotically identify the collect model $\mathcal{A}$;
ii) the estimator $\hat{\boldsymbol{\beta}}$ has the optimal rate and estimates nonzero parameters in $\mathcal{A}$ as efficiently as if the true $\mathcal{A}$ were known. In other words,

$$\sqrt{n}(\hat{\boldsymbol{\beta}}_{\mathcal{A}} - \boldsymbol{\beta}_{\mathcal{A}}) \to_d N(\mathbf{0}, \boldsymbol{\Sigma})$$

where $\boldsymbol{\Sigma}$ is the convariance matrix knowing $\mathcal{A}$.

A variable screening method is said to be *screening consistent* if

$$\Pr(\hat{\mathcal{A}} \subset \mathcal{A}) \to 1 \quad \text{as} \quad n \to \infty$$

Theoretical properties of the LASSO were studied by various researchers, including Donoho and Huo [37], Meinshausen and Bühlmann [38], Yuan and Lin [39], Zhao and Yu [40], Zhang and Huang [41], and Zou [25]. In particular, Donoho *et al.* [42] proved the near-minimax optimality for the LASSO with orthogonal predictors. Zhao and Yu [40] show that the LASSO is model selection consistent if the underlying model satisfies some nontrivial conditions such as the Irrepresentable Condition (IC). The LASSO does not possess the oracle properties due to its conflict requirement on the rate of λ to satisfy optimal prediction and model selection consistency [22, 43, 38].

Fan and Li [22] point out that the oracle properties are closely related to the superefficiency phenomenon [44] and show that the SCAD has oracle properties. Zou [25] shows that the adaptive solution is continuous in beta, establishes the oracle properties of the adaptive LASSO, and derives an oracle inequality to show that the adaptive LASSO is near-minimax optimal. The data-dependent nature of w_j is critical to assure that, as $n \to \infty$, the weights for unimportant variable increase to infinity, whereas those for important variables converge to a finite constant. Similarly, the elastic net does not hold oracle properties, and the adaptive elastic net has oracle properties when the weights are chosen properly.

2.4 Tuning Parameter Selection

Tuning parameter selection is important for shrinkage methods to assure their optimal performance in practice. Various selection criteria were proposed to choose λ adaptively from the data, including AIC, BIC, extended BIC (EBIC) [45], and generalized information criterion (GIC) [46]. Alternatively, K-fold cross-validation (CV) is also used to select the tuning parameter.

Assuming that p is fixed, Wang *et al.* [47] consider selecting λ in the SCAD and show that the tuning parameter λ obtained by BIC can identify the true model consistently, whereas AIC and CV may fail to guarantee model selection consistency. When p diverges with n, Wang *et al.* [48] show that a modified BIC still works for tuning parameter selection.

Fan and Tang [49] study the tuning parameter selection for ultrahigh-dimensional data, by allowing p to grow exponentially with n, that is, $\log(p) = O(p^{\kappa})$ for some $\kappa > 0$. They propose to select the tuning parameter by the GIC with an appropriate model complexity penalty. To ensure the model selection consistency, they consider a range for the model complexity penalty in the GIC and find that this model complexity penalty should diverge at the rate of some power of $\log(p)$, depending on the tail probability behavior of Y. Furthermore, they propose a uniform choice of the model complexity penalty to consistently identify the true model with asymptotic probability 1.

2.5 Numerical Computation

The LASSO, adaptive LASSO, and elastic net all solve convex programming, which makes the computation easy with a guarantee on global convergence. Their entire solution paths can be obtained using the LARS algorithm [28]. The computational cost is of order $O(np^2)$, which is the same as that of computing a single OLS fit. The solution paths greatly facilitate the tuning based on K-fold CV, which can be conveniently implemented in R, for example, using the functions cv.lars() for the LASSO tuning in the "LARS" package and cv.enet() for tuning the elastic net in the "elasticnet" package.

The SCAD penalty leads to the minimization problems with a nondifferentiable and nonconcave objective function, which makes computation challenging. Fan and Li [22] proposed a local quadratic approximation (LQA) algorithm to solve the problem iteratively. Zou and Li [50] propose a local linear approximation (LLA) algorithm for maximizing the penalized likelihood for a broad class of concave penalty functions and establish its convergence and theoretical properties.

3 Interaction-Effect Selection for High-Dimensional Data

3.1 Problem Setup

In regression problems, the predictors often work together and including their interactions can further improve prediction. Applications include gene–gene interaction (epistatic) effects in the genome-wide association studies. Interaction-effect selection is an important problem in high-dimensional data analysis, and we review some recent works.

Consider the linear model with two-way interaction effects

$$Y = \beta_0 + \beta_1 X_1 + \cdots + \beta_p X_p + \gamma_{11} X_1^2 + \gamma_{12} X_1 X_2 + \cdots + \gamma_{pp} X_p^2 + \epsilon \tag{3}$$

where β_0 is the intercept, $\boldsymbol{\beta} = (\beta_1, \ldots, \beta_p)^T$ is the p-vector of main-effect coefficients, and $\boldsymbol{\gamma} = (\gamma_{11}, \gamma_{12}, \ldots, \gamma_{pp})^T$ is the p^2-vector of interaction-effect coefficients. In addition to the design matrix for main effects $\mathbf{X} = (\mathbf{X}_1, \ldots, \mathbf{X}_p)$, there is a much larger design matrix for two-way interactions, denoted by

$$\mathbf{X}^{\circ 2} = \mathbf{X} \circ \mathbf{X} = (\mathbf{X}_1 \star \mathbf{X}_1, \mathbf{X}_1 \star \mathbf{X}_2, \ldots, \mathbf{X}_p \star \mathbf{X}_p)$$

where $\star$ denotes the entrywise product of two column vectors.

For a linear model with main effects only, X_j is regarded *important* if and only if $\beta_j \neq 0$. However, Hao and Zhang show that this definition does not hold any more for two-way interaction effect models as it is not invariant of a simple location-scale transformation for the predictors. They further provide a valid definition: in the quadratic model (3), X_j is *important* if and only if $\beta_j^2 + \sum_{k=1}^{p} \gamma_{jk}^2 > 0$, and $X_j X_k$ is *important* if $\gamma_{jk} \neq 0$. When p is large, interaction-effect selection has a couple of major challenges described as follows:

The number of terms in model (3) is 2^d in total, with $d = (p^2 + 3p)/2$. Even for a moderate p, the number of candidate models is enormously large. For example, if $p = 10$, there are approximately 3.6×10^{19} models to choose from, making the search for the true model much more difficult than the main-effect selection.

In model (3), there is a built-in hierarchical structure among the covariates: the interaction term $X_j X_k$ is the child of X_j and X_k. In the literature, it has been advocated that interaction selection should be performed subject to hierarchical constraints, that is, interaction effects are selected only if their corresponding main effects are selected [51, 52]. When p is large, hierarchy preserving induces additional challenges for model selection.

In the literature, the hierarchical structure is often mathematically formulated as heredity conditions [53–55]. In particular, the strong-heredity condition is

$$\gamma_{jk} \neq 0 \quad \text{only if} \quad \beta_j \beta_k \neq 0 \; \forall 1 \leq j, k \leq p \tag{4}$$

The weak-heredity condition is

$$\gamma_{jk} \neq 0 \quad \text{only if} \quad \beta_j^2 + \beta_k^2 \neq 0, \; \forall \quad 1 \leq j, k \leq p \tag{5}$$

These heredity conditions are usually used as constraints on the parameters when a model selection procedure or a computational algorithm is developed.

3.2 Joint Selection of Main Effects and Interactions

The joint selection method selects main effects and interaction effects simultaneously, subject to the hierarchical model structure. It is also called one-stage analysis. A variety of penalized regression methods have been proposed, including Zhao *et al.* [52], Yuan *et al.* [56], Choi *et al.* [57], and Bien *et al.* [58]. In order to satisfy the heredity conditions (4) or (5), some asymmetric penalty functions or inequality constraints are introduced in the optimization problem. For example, the Strong Heredity Interaction Model (SHIM) method proposed by Choi *et al.* [57] solves the following problem:

$$\min_{\beta,\gamma} \sum_{i=1}^{n} (y_i - g(\mathbf{x}_i))^2 + \lambda_\beta \sum_{j=1}^{p} |\beta_j| + \lambda_\gamma \sum_{j<k} |\gamma_{jk}|$$

where $g(\mathbf{x}) = \beta_0 + \sum_{j=1}^{p} \beta_j x_{ij} - \sum_{j<k} \gamma_{jk}\beta_j\beta_k(x_{ij}x_{ik})$, and $\beta_{jk} = \gamma_{jk}\beta_j\beta_k$ for $j < k$. When $p < n$ and fixed, under certain regularity conditions, these methods enjoy nice theoretical properties such as model selection consistency and oracle properties. However, they are typically computationally expensive for large p and not scalable for $p \gg n$ problems, as their optimization requires the operation of the full design matrix $\mathbf{X}$ and $\mathbf{X}^{\circ 2}$.

3.3 Two-Stage Approach

Two-stage selection methods, first considered by Efron *et al.* [28], separate the selection of the main effects and the interaction terms into two stages. For strong hierarchy, they conduct the following:

i) fit a linear model with main effects only (ignoring interaction effects) and select a model $\hat{\mathcal{A}}$;
ii) add interactions with both parents in $\hat{\mathcal{A}}$ to the model $\hat{\mathcal{A}}$ and select interactions while forcing main effects in $\hat{\mathcal{A}}$ in the final model.

For weak hierarchy, one can modify (ii) by requiring only one parent in $\hat{\mathcal{A}}$. Two-stage methods are popular for high-dimensional data analysis due to its computational efficiency: the number of interactions is greatly reduced after stage (i); they do not need to deal with the entire matrix $\mathbf{X}^{\circ 2}$; the hierarchical structure can be preserved without involving complex constraints as in the joint analysis. However, two-stage analysis was generally seen as a heuristic approach, as its theoretical foundation was not clearly understood. Since two-stage methods conduct variable selection under a misspecified model (by intentionally leaving out interactions) at stage (i), its selection consistencies were questioned in the literature.

Recently, Hao and Zhang [36] studied the problem "Under what conditions on the data distribution would two-stage methods work and have theoretical guarantees to discover the true two-way interaction model?" This question was answered under the hierarchical model assumption. The key to answer this question is to understand whether and when two-stage methods can identify all the important main effects at stage (i), so that all important interactions are included for selection at stage (ii). By analyzing the covariance structure between the main effects and the interaction effects, Hao and Zhang [36] established the necessary and sufficient conditions for two-stage methods to work properly. They further proposed new methods for interaction selection: two-stage forward selection [Interaction Selection by Two-stage Forward (iFORT)] and two-stage LASSO. Under regularity conditions and the strong-heredity condition, the iFORT is shown to have the sure screening property for interactions even when $p \gg n$. The computational complexity of the iFORT is linear in p, and its implementation is scalable since it does not require storage or operation of the augmented matrix. In one example of Hao and Zhang [36] with $p = 10\ 000$ and $n = 400$, it takes iFOR fewer than 30 s to complete the entire process of interaction selection.

Two-stage LASSO approach implements the LASSO at both stages:

i) fit the standard LASSO to a linear model containing the main effects only,

$$\min_{\beta} \sum_{i=1}^{n} \left(Y_i - \beta_0 - \sum_{j=1}^{p} \beta_j x_{ij} \right)^2 + \lambda \sum_{j=1}^{p} |\beta_j|$$

and denote the solution by $\hat{\boldsymbol{\beta}}^{\text{lasso}}$. Define $\hat{\mathcal{A}} = \{j : \hat{\beta}_j^{\text{lasso}} \neq 0, j = 1, \ldots, p\}$.

ii) solve

$$\sum_{i=1}^{n} \left(Y_i - \beta_0 - \sum_{j \in \hat{\mathcal{A}}} \beta_j X_j - \sum_{j,k \in \hat{\mathcal{A}}} \gamma_{jk} X_j X_k \right)^2 + \lambda \sum_{j,k \in \hat{\mathcal{A}}} |\gamma_{jk}|$$

Under certain regularity conditions, two-stage LASSO is shown to be sign consistent [36].

3.4 Regularization Path Algorithm under Marginality Principle (RAMP)

For penalized linear regression, solution-path algorithms provide the state-of-the-art computational tools for variable selection in high-dimensional data analysis. Popular algorithms include LARS [28], its extensions [59–61], and the coordinate descent algorithm (CDA) [62–65].

Recently, Hao *et al.* [66] developed a Regularization Algorithm under Marginality Principle (RAMP) algorithm to compute the whole solution path, while preserving model hierarchy, for interaction-effect selection. For the entire range of a tuning parameter λ, the RAMP utilizes CDA to compute the l_1 regression coefficients for main effects and interactions subject to model hierarchy. Figure 1 illustrates the RAMP solution paths under the strong and weak heredity conditions, respectively. The data $X_{ij} \overset{\text{i.i.d.}}{\sim} N(0,1)$, with $n = 500, p = 100$, and the error term $\epsilon \sim N(0,1)$. The true model is $Y = X_1 + 3X_6 + 4X_1X_3 + 5X_1X_6 + \epsilon$, satisfying the weak hierarchy but not the strong hierarchy. In the left plot, the strong-heredity RAMP selects X_1 and X_6 before picking up X_1X_6 on the solution path, and the interaction X_1X_3 is not selected until at a very late stage. In the right plot, the weak-heredity RAMP selects X_6, X_1X_6, X_1 and X_1X_3 sequentially.

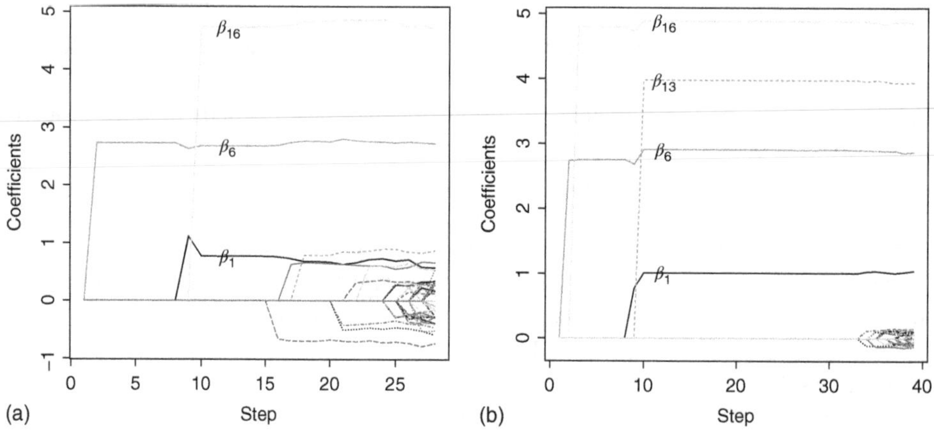

Figure 1 Hierarchy-preserving solution paths by RAMP. (a) Strong hierarchy; (b) Weak hierarchy.

4 Model Selection in High-Dimensional Nonparametric Models

Nonparametric methods play a fundamental role in statistics and machine learning due to their flexibility and ability to discover nonlinear and complex relationships between variables. A nonparametric regression models assumes

$$Y_i = f(\mathbf{X}) + \epsilon_i, \quad i = 1, \dots, n$$

where f has an unspecified function form, $\mathbf{X} \in [0,1]^p$, and the error terms ϵ_is have mean zero and finite variance. We assume that each $X_j \in [0,1]$ for notational convenience; otherwise, one can always standardize each predictor. To facilitate the estimation, we decompose the p-variate function $f(X_1, \dots, X_p)$ into the sum of functional components as

$$f(\mathbf{X}) = b + \sum_{j=1}^{p} f_j(X_j) + \sum_{j<k}^{p} f_{jk}(X_j, X_k) + \cdots \tag{6}$$

where b is a constant, f_js are the main-effect terms, and f_{jk}s are the two-way interactions. The identifiability of the terms in (6) is assured by imposing side conditions.

In practice, the sequence (6) is often truncated by retaining only low-order terms for easy estimation and interpretation. For example, if only the main effects are preserved, we obtain the additive model

$$Y_i = b + \sum_{j=1}^{p} f_j(X_{ij}) + \epsilon_i, \quad i = 1, \dots, n \tag{7}$$

where the identifiability is assured if we center f_js over the samples by $\sum_{i=1}^{n} f_j(x_{ij}) = 0$ for $j = 1, \dots, p$. The f_js can be estimated by smoothing techniques such as kernel estimators [67–69], local weighted polynomial regression [70, 71], generalized additive models (GAMs) [72–75], smoothing splines [76–82], and multivariate adaptive regression splines (MARS) [83, 84].

4.1 Model Selection Problem

We consider the problem of model selection for the additive model (7). Model selection is needed only if Y depends on $\mathbf{X}$ only through its subset, that is, not all X_js have effects on Y. In other words, the true model is

$$f(\mathbf{X}) = b + \sum_{f_j \neq 0} f_j(X_j) + \sum_{f_j \equiv 0} 0(X_j)$$

where only those predictors with nonzero f_js contribute to the prediction of Y. A function g is a zero function on [0,1], that is, $g \equiv 0$, if and only if $g(t) = 0 \forall t \in [0,1]$. Define $\mathcal{A} = \{j : f_j \neq 0, j = 1, \dots, p\}$, the index set of *important* variables. The goal of model selection or variable selection is to estimate $\mathcal{A}$ from the data.

Traditional variable selection methods for nonparametric regression were either based on hypothesis tests or stepwise approaches. For example, model selection for additive models can be formulated as hypothesis testing problems

$$f_j(x_j) = 0 \quad \text{vs} \quad f_j(x_j) \neq 0, \quad \text{for } j = 1, \dots, p$$

and variables with large p-values are regarded as "unimportant." One well-known work is the generalized likelihood ratio [75], where the test statistics are shown to follow rescaled chi-squared distributions asymptotically. Alternative methods build f in a stepwise fashion by adding or deleting basis functions, such as Classification and Regression Tree (CART) [86], MARS [83], and Stone *et al.* [87], which is similar to forward selection and backward elimination in linear models. For additive regression splines, knot-selection methods such as TURBO [84] and BRUTO [85] are developed to select knots for constructing the basis functions. In the context of smoothing spline Analysis of Variance (ANOVA), Gu [82] suggests a model checking tool based on cosine diagnostics after model fitting.

For high-dimensional data, the problem of model selection needs to perform dimension reduction while estimating smoothing functions for important effects. The early works are based on adaptive estimation such as projection pursuit, CART [86], and MARS [83]. In recent years, a variety of penalization approaches are proposed to achieve smooth and sparse estimation simultaneously in additive models, including the basis pursuit [88, 89], the group-LASSO methods [90], COSSO [91–93], and sparse additive models (SpAMs) [94]. In the context of local polynomial regression, the rodeo [95] is proposed and studied. For partially linear models, the linear and nonlinear discover method (LAND) [96] is designed for selecting linear and nonlinear terms simultaneously. Based on the nature of their penalty functions, these penalized methods fall into two categories: penalty on basis coefficients and function soft-thresholding methods.

Penalty on basis coefficients

The idea is to first represent each f_j in terms of a basis expansion and then apply a shrinkage penalty on the basis coefficients. This type of method is a direct extension of linear regression shrinkage to nonparametric model settings. Examples include the basis pursuit [88, 89] and the group-LASSO methods [90].

Function soft-thresholding methods

The idea is to apply a soft-thresholding operator on the function space $\mathcal{F}$ and shrink the function components f_j toward the zero function. Examples include the COSSO-type methods [91–93] and the SpAM [94].

In the following sections, we provide a selective review on these works in the context of additive models (7).

4.2 Penalty on Basis Coefficients

In model (7), assume that each component $f_j \in \mathcal{F}_j[0,1]$, which is expanded by a finite number M of preselected basis functions. Then, we have

$$f_j(x_j) = \sum_{m=1}^{M} \beta_{jm}\phi_m(x_j), \quad j = 1, \dots, p \tag{8}$$

where ϕ_ms are the basis functions such as polynomial basis or B-splines, and β_{jm}s are the coefficients of the basis functions. Define $\boldsymbol{\beta}_j = (\beta_{j1}, \dots, \beta_{jM})^T, j = 1, \dots, p$ and $\boldsymbol{\beta} = (\boldsymbol{\beta}_1^T, \dots, \boldsymbol{\beta}_p^T)^T$.

Polynomial basis example

Consider a partition of [0,1] by the knots $0 = \xi_0 < \xi_1 < \dots < \xi_K < \xi_{K+1} = 1$ into K subintervals, $I_{Kt} = [\xi_t, \xi_{t+1}), t = 0, \dots, K-1$ and $I_{KK} = [\xi_K, \xi_{K+1}]$, where the positive integer $K \equiv K_n = n^\nu$ with $0 < \nu < 0.5$ such that $\max_{1\le k\le K+1}|\xi_k - \xi_{k-1}| = O(n^{-\nu})$. Let $\mathcal{F}_j[0,1]$ be the space of polynomial splines of degree $l \ge 1$ consisting of functions s satisfying: (i) s is a polynomial degree of l on I_{Kt}, for $1 \le t \le K$; (ii) for $l \ge 2$ and $0 \le l' \le l-2$, s is l' times continuously differentiable on [0,1]. Define $m_n = K_n + l$. Then, there exists a normalized B-spline basis $\{\phi_k, 1 \le k \le m_n\}$ for $W_n^l[0,1]$ [97].

Since each component f_j is fully determined by its basis coefficient group $\boldsymbol{\beta}_j$, the problem of selecting nonzero components f_js amounts to identifying nonzero groups $\boldsymbol{\beta}_j$s. To shrink function components toward zero, one can impose the soft-thresholding penalty on their coefficients

$$\min_{b,\boldsymbol{\beta}} \sum_{i=1}^{n} \left\{ y_i - b - \sum_{j=1}^{p}\sum_{m=1}^{M} \beta_{jk}\phi_m(x_{ij}) \right\}^2 + \lambda \sum_{j=1}^{p} J(\boldsymbol{\beta}_j) \tag{9}$$

where $\lambda > 0$ and M are the smoothing parameters controlling the model fit and sparsity, and $J(\boldsymbol{\beta}_j)$ is the penalty function.

Basis pursuit

The basis pursuit [88] and the likelihood basis pursuit [89] employ l_1-type penalty on the coefficients,

$$J(\boldsymbol{\beta}_j) = \sum_{m=1}^{M} |\beta_{jm}|, \; j = 1, \dots, p$$

respectively, under the context of wavelets and smoothing splines. In practice, the model fit is not very sensitive to the exact value of M as long as it is sufficiently large to provide adequate flexibility.

Adaptive group LASSO

Huang *et al.* [90] propose to apply the adaptive group lasso penalty on the coefficients to select functional components, by solving

$$\min_{\beta} \sum_{i=1}^{n} \left\{ y_i - b - \sum_{j=1}^{p} \sum_{m=1}^{M} \beta_{jm}\phi_k(x_{ij}) \right\}^2 + \lambda \sum_{j=1}^{p} w_j \|\beta_j\|_2$$

$$\text{subject to} \sum_{i=1}^{n} \sum_{m=1}^{M} \beta_{jm}\phi_m(x_{ij}) = 0, \quad j = 1, \ldots, p \tag{10}$$

where $\lambda > 0$ is a tuning parameter, $\|\beta_j\|_2$ is the l_2 norm of $\beta_j \in \mathbb{R}^M$, and the weights $\mathbf{w} = (w_1, \ldots, w_p)^T \geq 0$ are given constants. The weights suggested by Huang *et al.* [90] are

$$w_j = \|\tilde{\beta}_j\|_2^{-1}, \quad j = 1, \ldots, p$$

where $\tilde{\beta}$ is obtained by solving a standard group lasso problem with $w_1 = \cdots = w_p = 1$. The parameter λ can be selected by BIC [4] or EBIC [45]. Huang *et al.* [90] proved that, under certain regularity conditions, if the group lasso estimator is used as the initial estimator, then the adaptive group lasso estimator can select nonzero components correctly with probability approaching one asymptotically, and it also achieves the optimal rate of convergence for nonparametric estimation of additive models; the results also hold for $p > n$.

4.3 Component Selection and Smoothing Operator (COSSO)

One convenient and unified framework of fitting the additive model (7) is through the reproducing kernel Hilbert space framework (RKHS) [78]. Assume that $f_j \in 1 \oplus \mathcal{H}_j$, which is an RKHS on [0,1], for all $j = 1, \ldots, p$. For example, $\mathcal{H}_j$ is the second-order Sobolev space on [0,1], $\mathcal{S}^2[0,1] = \{g : g, g'$ are absolutely continuous, $\int_0^1 [g''(t)]^2 < \infty\}$. When endowed with the inner product

$$< g, h > = \int_0^1 g(t)dt \int_0^1 h(t)dt + \int_0^1 g'(t)dt \int_0^1 h'(t)dt + \int_0^1 g''(t)h''(t)dt,$$
$$\forall g, h \in \mathcal{S}^2[0,1]$$

the space $\mathcal{S}^2[0,1] = \{1\} \oplus \overline{\mathcal{S}}^2[0,1]$ is an RKHS associated with the reproducing kernel $\mathcal{K}(s,t) = 1 + k_1(s)k_1(t) + k_2(s)k_2(t) - k_4(|s-t|)$, where $k_1(s) = s - 0.5$, $k_2(s) = [k_1(s)^2 - 1/12]/2$, and $k_4(s) = [k_1^4(s) - k_1^2(s)/2 + 7/240]/24$. Define $\mathcal{H} = \{1\} \oplus_{j=1}^{p} \overline{\mathcal{H}}_j$, which is an RKHS over $[0,1]^p$.

The COSSO [91, 92] imposes a soft-thresholding functional to shrink function components to exactly the zero function and therefore achieve sparsity. In particular, the COSSO solves

$$\min_{f \in \mathcal{H}} \frac{1}{n} \sum_{i=1}^{n} \{y_i - f(\mathbf{x}_i)\}^2 + \lambda \sum_{j=1}^{p} \|P^j f\|_{\mathcal{H}} \tag{11}$$

where P^j is the projection of f onto $\mathcal{H}_j$, and λ is the smoothing parameter. The penalty term $\sum_{j=1}^{p} \|P^j f\|_{\mathcal{H}}$ is a sum of RKHS norms. Lin and Zhang [91] show that the penalty (11) has the soft-thresholding property, which encourages both smoothness and sparsity of the solution in function estimation. When f is truly a linear function of X_js, then the COSSO penalty reduces to the LASSO penalty.

Computationally, the COSSO function a convex functional, and the solution to (11) is guaranteed to exist and has a finite-dimensional representation. Lin and Zhang [91] suggest an iterative algorithm to the COSSO optimization problem, which alternatively fits a traditional smoothing spline and solves a nonnegative garrote problem until convergence. They further propose an efficient one-step algorithm and develop an *R* package *COSSO* to implement the one-step algorithm. For very large datasets, the package implements a parsimonious basis method to reduce the number of parameters.

The COSSO estimator has nice asymptotic properties for function estimation and model selection. Under certain regularity conditions, the COSSO estimator is shown to achieve the optimal rate of convergence $n^{-2/5}$ if λ converges to zero at a proper rate. In the special case of a tensor product design with periodic functions, the COSSO can select the correct model structure with probability tending to 1 as n goes to infinity [91].

4.4 Adaptive COSSO

The COSSO treats all the functional components equally, which can induce large bias in estimation. To resolve this, Storlie *et al.* [93] suggest to penalize function components differently according to their relative importance: important function components are penalized less than unimportant components. Toward this, the adaptive COSSO (ACOSSO) is proposed to solve a weighted COSSO-penalty problem

$$\min_{f \in \mathcal{H}} \frac{1}{n} \sum_{i=1}^{n} \{y_i - f(\mathbf{x}_i)\}^2 + \lambda \sum_{j=1}^{p} w_j \|P^j f\|_{\mathcal{H}} \tag{12}$$

where $w_j > 0$ are chosen adaptively for each functional component, for example,

$$w_j = \|P^j \tilde{f}\|_{L_2}^{-\gamma}, \quad j = 1, \ldots p$$

where $\tilde{f}$ is some initial estimator of f, $\|P^j \tilde{f}\|_{L_2}$ is the L_2 norm of $P^j \tilde{f}$, and $\gamma > 0$ is a prespecified constant. In practice, the initial estimator $\tilde{f}$ can be either the traditional smoothing spline solution or the COSSO solution. The ACOSSO can be solved by modifying the COSSO algorithm.

Nonparametric oracle property

A nonparametric estimator $\hat{f}$ has the nonparametric-oracle property if $\|\hat{f} - f\|_n \to 0$ at the optimal rate, and $\hat{f}_j \equiv 0$ for all $j \notin \mathcal{A}$ with probability tending to 1 as n goes to infinity. Here, $\|f\|_n^2 = \frac{1}{n} \sum_{i=1}^{n} \{f(\mathbf{x}_i)\}^2$ is the squared norm of f evaluated at the design points.

The ACOSSO estimator has the nonparametric oracle (np-oracle) property when the weights are chosen properly. Assume that the input $\mathbf{X}$ follows a tensor product design. Let $f \in \mathcal{H}$ with $\mathcal{H} = \{1\} \oplus \bar{S}^2_{\text{per},1} \oplus \cdots \oplus \bar{S}^2_{\text{per},p}$, where $S^2_{\text{per},j} = \{1\} \oplus \bar{S}^2_{\text{per},j}$ is the second-order Sobolev space of periodic functions of X_j defined on [0,1]. Assume that the error terms ϵ_is are independent, mean zero, and uniformly sub-Gaussian. Define the weights, $w_{j,n} = \|P^j \tilde{f}\|_{L_2}^{-\gamma}$, where $\tilde{f}$ is given by the traditional smoothing spline with $\tau_0 \sim n^{-4/5}$, and $\gamma > 3/4$. Storlie *et al.* [93] show that, if $\lambda_n \sim n^{-4/5}$, then the ACOSSO estimator has the np-oracle property.

4.5 Sparse Additive Models (SpAM)

Assume that $f_j \in \mathcal{H}_j$, a Hilbert space of measurable functions $f_j(x_j)$ such that $\mathbb{E}(f_j(X_j)) = 0$, $\mathbb{E}(f_j^2(X_j)) < \infty$, with the inner product $< f_j, f_j' > = \mathbb{E}(f_j(X_j)f_j'(X_j))$. Ravikumar *et al.* [94] propose the SpAM method which conducts model selection by imposing the L_2 norm penalty on nonparametric components

$$\begin{aligned} &\min_{f_1,\ldots,f_p} \mathbb{E}\left(Y - \sum_{j=1}^{p} f_j(X_j)\right)^2 \\ &\text{subject to } \sum_{j=1}^{p} \sqrt{\mathbb{E}(f_j^2(X_j))} \leq L \\ &\qquad \mathbb{E}(f_j) = 0, \quad j = 1, \ldots, p \end{aligned} \tag{13}$$

where L is the tuning parameter. A backfitting algorithm is applied to solve the convex optimization problem (13).

Persistent property

Define the risk function of f by $R(f) = \mathbb{E}(Y - f(\mathbf{X}))^2$. An estimator $\hat{f}_n$ is said to be *persistent* relative to a class of functions $\mathcal{M}_n$ if

$$R(\hat{f}_n) - R(f_n^*) \to_p 0$$

where $f_n^* = \arg\min_{f \in \mathcal{M}_n} R(f)$ is the predictive oracle.

Ravikumar *et al.* [94] establish theoretical properties of the SpAM estimator in terms of its risk consistency and model selection consistency. Under some regularity conditions, the SpAM is persistent relative to the class of additive models when the tuning parameter L is chosen properly. And it is also model selection consistent (called *sparsistent* in the paper), that is, $P(\hat{A}_n = A) \to 1$ as $n \to \infty$ when L satisfies a proper rate.

4.6 Sparsity-Smoothness Penalty

For each $j = 1, \ldots, p$, define $\|f_j\|_n^2 = \frac{1}{n}\sum_{i=1}^{n} f_j^2(x_{ij})$ and the smoothness measure $I^2(f_j) = \int_0^1 [f_j''(t)]^2 dt$. Meier *et al.* [98] propose the following sparsity-smooth penalty function:

$$J_{\lambda_1,\lambda_2}(f_j) = \lambda_1 \sqrt{\|f_j\|_n^2 + \lambda_2 I^2(f_j)}$$

where $\lambda_1, \lambda_2 \geq 0$ are tuning parameters to control the degree of penalty on functional components, and fit the additive model (7) by solving

$$\min_{f \in \mathcal{H}} \frac{1}{n} \sum_{i=1}^{n} \{y_i - f(\mathbf{x}_i)\}^2 + \sum_{j=1}^{p} J_{\lambda_1,\lambda_2}(f_j) \tag{14}$$

Computationally, one can show that the solution of (14) are natural cubic splines with knots at $x_{ij}, i = 1, \ldots, n$. Therefore, each function component f_j can be expanded with a set of cubic B-spline basis, $f_j(x_j) = \sum_{m=1}^{M} \beta_{jm}\phi_m(x_j)$, as in (8), where ϕ_k are B-spline basis. One typical choice is that $m_n - 4 \asymp \sqrt{n}$. For each $j = 1, \ldots p$, let $B_{ij} = (\phi_1(x_{ij}), \ldots, \phi_{m_n}(x_{ij}))^T$, which consists of the values of basis functions evaluated at x_{ij}. Let $\mathbf{B}_j = (B_{1j}, \dots, B_{nj})^T$ and $\mathbf{B} = (\mathbf{B}_1, \dots, \mathbf{B}_p)$. This will lead to the following equivalent: optimization problem (14)

$$\arg\min_{\beta} \|\mathbf{y} - \mathbf{B}\beta\|^2 + \lambda_2 \sum_{j=1}^{p} \sqrt{\frac{1}{n}\beta_j^T \mathbf{B}_j^T \mathbf{B}_j \beta_j + \lambda_2 \beta_j^T \Omega_j \beta_j} \tag{15}$$

where Ω_j is an $M \times M$ matrix with its kl entry equal to $\int_0^1 \phi_k''(x)\phi_l''(x)\mathrm{d}x$, with $1 \le k, l \le M$. The optimization problem (15) can be seen as a group lasso problem and solved by the CDAs. Meier *et al.* [98] derive an oracle inequality for the estimator under the compatibility condition.

4.7 Nonparametric Independence Screening (NIS)

Fan *et al.* [35] extend the idea of SIS from linear models to nonparametric models and propose the NIS. The basic idea of NIS is to fit the marginal nonparametric model for each predictor X_j and then rank their relative importance by the goodness of fit of marginal models. In particular, consider the marginal nonparametric regressions problem,

$$\min_{f_j \in L_2(P)} E\{Y - f_j(X_j)\}^2, \quad j = 1, \dots, p \tag{16}$$

where P is the joint distribution of $(\mathbf{X}, Y)$, and $L_2(P)$ is the class of square integrable functions under the measure P. The minimizer of (16) is given by $f_j = E(Y|X_j)$.

For convenience, assume $f_j \in W_n^l[0,1]$, the space of polynomial splines of degree $l \ge 1$ defined in Section 3.4, expressed by a linear combination of a set of B-spline basis as (8). Then, we can estimate f_j by

$$\hat{f}_j = \arg \min_{f_j \in W_n^l[0,1]} \frac{1}{n} \sum_{i=1}^{n} \{y_i - f_j(x_{ij})\}^2, \quad j = 1, \dots, p$$

The NIS ranks the magnitude of marginal estimators $\hat{f}_j$s and selects the subset of top-ranked variables by

$$\widehat{\mathcal{M}}_{\nu_n} = \{j : \|\hat{f}_j\|_n^2 \ge \nu_n, j = 1, \dots, p\}$$

where $\|\hat{f}_j\|_n^2 = \frac{1}{n}\sum_{i=1}^n \hat{f}_j^2(x_{ij})$, and ν_n is a prespecified thresholding value. In this way, the NIS can greatly reduce the data dimensionality from p to $|\widehat{\mathcal{M}}_n|$, which is typically much smaller than p.

Under some technical conditions, Fan *et al.* [35] show that the SIS has the sure screening property,

$$P(A \subset \widehat{\mathcal{M}}_{\nu_n}) \to 1$$

when ν_n is selected properly. This sure screening result holds even if p grows at an exponential rate of the sample size n. Furthermore, the false selection rate is shown to converge to zero at an exponential rate. An iterative NIS (INIS) can be implemented to further reduce the false positive rate and increase the stability of the standard NIS.

5 Concluding Remarks

Model selection plays a critical role in all kinds of statistical analysis problems, including regression, classification, density estimation, clustering, and network analysis. For complex problems, where the data are heterogeneous, or from multiple sources, or partially observed, the underlying model is dynamical, the estimator is overparameterized (such as deep learning), and the problem of model selection is more challenging and less studied. They demand new methods and theoretical development in the future.

One important topic relevant to model selection is postselection inference, that is, how to take into account the uncertainty introduced by the selection process and make valid inference on the final model. The validity of the classical statistical inference, for example, statistical tests and confidence intervals, is based on the assumption that model selection and data analysis are two separate processes using independent data. However, the common practice is to perform data-driven model selection first and then derive statistical inference from the resulting model. A series of early work on postselection inference were done by Leeb [99] and Leeb and Potscher [100, 101], and recent works include Berk *et al.* [102] and Belloni *et al.* [103].

References

1 Box, G.E.P. (1976) Science and statistics. *J. Am. Stat. Assoc.*, 71, 791–799.

2 Akaike, H. (1973) Maximum likelihood identification of Gaussian autoregressive moving average models. *Biometrika*, 60, 255–265.

3 Akaike, H. (1977) On entropy maximization principle, in *Application of Statistics* (ed. P.R. Krishnaiah), North Holland, Amsterdam, pp. 27–41.

4 Schwarz, G. (1978) Estimating the dimension of a model. *Ann. Stat.*, 6, 461–464.

5 Spiegelhalter, D., Best N., Carlin B., and Linde A. (2002) Bayesian measures of model complexity and fit. *J. R. Stat. Soc. B*, 64, 583–639.

6 Berg, A., Meyer, R., and Yu, J. (2004) Deviance information criterion for comparing stochastic volatility models. *J. Bus. Econ. Stat.*, 22, 107–120.

7 Mallows, C.L. (1973) Some comments on C_p. *Technometrics*, 15, 661–675.

8 Mallows, C.L. (1995) More comments on C_p. *Technometrics*, 37, 362–372.

9 Breiman, L. and Spector, P. (1992) Subset selection and evaluation in regression: the X-random case. *Int. Stat. Rev.*, 60, 291–319.

10 Breiman, L. (1995) Better subset selection using the non-negative garrote. *Technometrics*, 37, 373–384.

11 Shao, J. (1993) Linear model selection by cross-validation. *J. Am. Stat. Assoc.*, 88, 486–494.

12 Shao, J. (1996) Bootstrap model selection. *J. Am. Stat. Assoc.*, 91, 655–665.

13 George, E.I. and McCulloch, R.E. (1993) Variable selection via Gibbs sampling. *J. Am. Stat. Assoc.*, 88, 881–889.

14 George, E.I. and McCulloch, R.E. (1997) Approaches to Bayesian variable selection. *Stat. Sin.*, 7, 339–373.

15 Chipman, H., George, E.I., and McCulloch, R.E. (2001) The practical implementation of Bayesian model selection (with discussion), in *Institute of Mathematical Statistical Lecture Notes - Mono- graph Series*, vol. 38 (ed. P. Lahiri), Institute of Mathematical Statistics (IMS), pp. 65–134.

16 Berger, J.O. and Pericchi, L.R. (2001) Objective Bayesian methods for model selection: introduction and comparison (with discussion), in *Institute of Mathematical Statistical Lecture Notes - Monograph Series*, vol. 38 (ed. P. Lahiri), Institute of Mathematical Statistics (IMS), pp. 135–207.

17 Linhart, H. and Zucchini, W. (1986) *Model Selection*, Wiley, New York.

18 Rao, C.R. and Wu, Y. (2001) On model selection (with discussion), in *Institute of Mathematical Statistical Lecture Notes - Monograph Series*, vol. 38 (ed. P. Lahiri), Institute of Mathematical Statistics (IMS), pp. 1–64.

19 Miller, A.J. (2002) *Subset Selection in Regression*, Chapman and Hall, London.

20 Furnival, G. and Wilson, R. (1974) Regressions by leaps and bounds. *Technometrics*, 16, 499–511.

21 Bertsimas, D., King, A., and Mazumder, R. (2016) Best subset selection via a modern optimization lens. *Ann. Stat.*, 44, 813–852.

22 Fan, J. and Li, R. (2001) Variable selection via nonconcave penalized likelihood and its oracle property. *J. Am. Stat. Assoc.*, 96, 1348–1360.

23 Tibshirani, R. (1996) Regression shrinkage and selection via the Lasso. *J. R. Stat. Soc. B*, 58, 147–169.

24 Zou, H. and Hastie, T. (2005) Regularization and variable selection via the elastic net. *J. R. Stat. Soc. B*, 67, 301–320.

25 Zou, H. (2006) The adaptive lasso and its oracle properties. *J. Am. Stat. Assoc.*, 101, 1418–1429.

26 Zhang, H.H. and Lu, W. (2007) Adaptive-LASSO for Cox's proportional hazard model. *Biometrika*, 94, 691–703.

27 Zou, H. and Zhang, H.H. (2009) On the adaptive elastic-net with a diverging number of parameters. *Ann. Stat.*, 37, 1733–1751.

28 Efron, B., Hastie, T., Johnstone, I., and Tibshirani, R. (2004) Least angle regression. *Ann. Stat.*, 32, 407–451.

29 Candes, E. and Tao, T. (2007) The Dantzig selector: statistical estimation when p is much larger than n. *Ann. Stat.*, 35, 2313–2351.

30 Zhang, C.- H. (2010) Nearly unbiased variable selection under minimax concave penalty. *Ann. Stat.*, 38, 894–942.

31 Lv, J. and Fan, Y. (2009) A unified approach to model selection and sparse recovery using regularized least squares. *Ann. Stat.*, 37, 3498–3528.

32 Fan, J. and Lv, J. (2008) Sure independence screening for ultrahigh dimensional feature space. *J. R. Stat. Soc. B*, 70, 849–911.

33 Wang, H. (2009) Forward regression for ultra-high dimensional variable screening. *J. Am. Stat. Assoc.*, 104, 1512–1524.

34 Fan, J. and Fan, Y. (2008) High-dimensional classification using features annealed independence rules. *Ann. Stat.*, 36, 2605–2637.

35 Fan, J., Feng, Y., and Song, R. (2011) Nonparametric independence screening in sparse ultra-high-dimensional additive models. *J. Am. Stat. Assoc.*, 106, 544–557.

36 Hao, N. and Zhang, H.H. (2014) Interaction screening for ultra-high dimensional data. *J. Am. Stat. Assoc.*, 109, 1285–1301.

37 Donoho, D. and Huo, X. (2002) Uncertainty principles and ideal atomic decompositions. *IEEE Trans. Inf. Theory*, 47, 2845–2863.

38 Meinshausen, N. and Bühlmann, P. (2006) Variable selection and high dimensional graphs with the Lasso. *Ann. Stat.*, 34, 1434–1462.

39 Yuan, M. and Lin, Y. (2007) On the nonnegative garrotte estimator. *J. R. Stat. Soc. B*, 69, 143–161.

40 Zhao, P. and Yu, B. (2006) On model selection of lasso. *J. Mach. Learn. Res.*, 7, 2541–2563.

41 Zhang, C.H. and Huang, J. (2006) The sparsity and bias of the Lasso selection in high-dimensional linear regression. *Ann. Stat.*, 36, 1567–1594.

42 Donoho, D., Johnstone, I., Kerkyacharian, G., and Picard, D. (1995) Wavelet shrinkage: asymptopia? (with discussion). *J. R. Stat. Soc. B*, 57, 301–337.

43 Leng, C., Lin, Y., and Wahba, G. (2004) A note on the Lasso and related procedures in model selection. *Stat. Sin.*, 16, 1273–1284.

44 Lehmann, E.L. and Casella, G. (1998) *Theory of Point Estimation*, Springer.

45 Chen, J. and Chen, Z. (2008) Extended Bayesian information criteria for model selection with large model space. *Biometrika*, 95, 759–771.

46 Zhang, Y., Li, R., and Tsai, C.L. (2010) Regularization parameter selections via generalized information criterion. *J. Am. Stat. Assoc.*, 105, 312–323.

47 Wang, H., Li, R., and Tsai, C.L. (2007) Tuning parameter selectors for the smoothly clipped absolute deviation method. *Biometrika*, 94, 553–568.

48 Wang, H., Li, B., and Leng, C. (2009) Shrinkage tuning parameter selection with a diverging number of parameters. *J. R. Stat. Soc. B*, 71, 671–683.

49 Fan and Tang (2013) Tuning parameter selection in high dimensional penalized likelihood. *J. R. Statist. Soc. B*, 75, 531–552.

50 Zou, H. and Li, R. (2008) One-step sparse estimates in nonconcave penalized likelihood models. *Ann. Stat.*, 36, 1509–1533.

51 Nelder, J.A. (1977) A reformulation of linear models. *J. R. Stat. Soc. A*, 140, 48–77.

52 Zhao, P., Rocha, G., and Yu, B. (2009) The composite absolute penalties family for grouped and hierarchical variable selection. *Ann. Stat.*, 37, 3468–3497.

53 Hamada, M. and Wu, C.F.J. (1992) Analysis of designed experiments with complex aliasing. *J. Qual. Technol.*, 24, 130–137.

54 Chipman, H. (1996) Bayesian variable selection with related predictors. *Can. J. Stat.*, 24, 17–36.

55 Chipman, H., Hamada, M., and Wu, C.F.J. (1997) A Bayesian variable-selection approach for analyzing designed experiments with complex aliasing. *Technometrics*, 39, 372–381.

56 Yuan, M., Joseph, V.R., and Zou, H. (2009) Structured variable selection and estimation. *Ann. Appl. Stat.*, 3, 1738–1757.

57 Choi, N.H., Li, W., and Zhu, J. (2010) Variable selection with the strong heredity constraint and its rracle property. *J. Am. Stat. Assoc.*, 105, 354–364.

58 Bien, J., J. Taylor, R. Tibshirani, *et al.* (2013) A lasso for hierarchical interactions. *Ann. Stat.*, 41, 1111–1141.

59 Park, M.Y. and Hastie, T. (2007) L1-regularization path algorithm for generalized linear models. *J. R. Stat. Soc. B*, 69, 659–677.

60 Wu, Y. (2011) An ordinary differential equation-based solution path algorithm. *J. Nonparametr. Stat.*, 23, 185–199.

61 Zhou, H. and Wu, Y. (2014) A generic path algorithm for regularized statistical estimation. *J. Am. Stat. Assoc.*, 109, 686–699.

62 Wu, T.T. and Lange, K. (2008) Coordinate descent algorithms for Lasso penalized regression. *Ann. Appl. Stat.*, 2, 224–244.

63 Friedman, J.H., Hastie, T., Hofling, H., and Tibshirani, R. (2007) Pathwise coordinate optimization. *Ann. Appl. Stat.*, 1, 302–332.

64 Friedman, J., Hastie, T., and Tibshirani, R. (2010) Regularization paths for generalized linear models via coordinate descent. *J. Stat. Soft.*, 33, 1–22.

65 Yu, Y. and Feng, Y. (2014) Apple: approximate path for penalized likelihood estimators. *Stat. Comput.*, 24, 803–819.

66 Hao, N., Feng, Y., and Zhang, H.H. (2014) Model selection for high-dimensional quadratic regression via regularization. *J. Am. Stat. Assoc.*, 113, 615–625.

67 Nadaraya, E. (1964) On estimating regression. *Theory Probab. Appl.*, 9, 141–142.

68 Altman, N.S. (1990) Kernel smoothing of data with correlated errors. *J. Am. Stat. Assoc.*, 85, 749–759.

69 Tsybakov, A.B. (2009) *Introduction to Nonparametric Estimation*, Springer, New York.

70 Cleveland, W. (1979) Robust locally weighted fitting and smoothing scatterplots. *J. Am. Stat. Assoc.*, 74, 829–836.

71 Fan, J. and Gijbels, I. (1996) *Local Polynomial Modeling and Its Applications*. Chapman and Hall.

72 Friedman, J.H. and Stuetzle, W. (1981) Projection pursuit regression. *J. Am. Stat. Assoc.*, 76, 817–823.

73 Buja, A., Hastie, T.J., and Tibshirani, R.J. (1989) Linear smoothers and additive models. *Ann. Stat.*, 17, 453–555.

74 Hastie, T.J. and Tibshirani, R.J. (1990) *Generalized Additive Models*, Chapman and Hall.

75 Fan, J. and Jiang, J. (2005) Nonparametric inference for additive models. *J. Am. Stat. Assoc.*, 100, 890–907.

76 Kimeldorf, G. and Wahba, G. (1971) Some results on Tchebycheffian spline functions. *J. Math. Anal. Appl.*, 33, 82–95.

77 de Boor, C. (1978) *A Practical Guide to Splines*, Springer, New York.

78 Wahba, G. (1990) *Spline Models for Observational Data*. SIAM CBMS-NSF Regional Conference Series in Applied Mathematics, vol. 59.

79 Green, P. and Silverman, B. (1994) *Nonparametric Regression and Generalized Linear Models: A Roughness Penalty Approach*, Chapman and Hall, Boca Raton.

80 Stone, C., Buja, A., and Hastie, T. (1994) The use of polynomial splines and their tensor-products in multivariate function estimation. *Ann. Stat.*, 22, 118–184.

81 Mammen, E. and van de Geer, S. (1997) Locally adaptive regression splines. *Ann. Stat.*, 25, 387–413.

82 Gu, C. (2002) *Smoothing Spline ANOVA Models*, Springer-Verlag.

83 Friedman, J.H. (1991) Multivariate adaptive regression splines (invited paper). *Ann. Stat.*, 19, 1–141.

84 Friedman, J.H. and Silverman, B.W. (1989) Flexible parsimonious smoothing and additive modeling. *Technometrics*, 31, 3–39.

85 Hastie, Tibshirani, and Friedman (2009) *The Elements of Statistical Learning: Data Mining, Inference, and Prediction*, Springer.

86 Breiman, L., Friedman, J.H., Stone, C., and Olshen, R.A. (1984) *Classification and Regression Trees.* , Taylor & Francis

87 Stone, C., Hansen, M., Kooperberg, C., and Truong, Y. (1997) Polynomial splines and their tensor products in extended linear modeling. *Ann. Stat.*, 25, 1371–1425.

88 Chen, S., Donoho, D.L., and Saunders, M.A. (1999) Atomic decomposition by basis pursuit. *SIAM J. Sci. Comput.*, 20 (1), 33–61.

89 Zhang, H.H., Wahba, G., Lin, Y. *et al.* (2004) Variable selection and model building via likelihood basis pursuit. *J. Am. Stat. Assoc.*, 99, 659–672.

90 Huang, J., Horovitz, J., and Wei, F. (2010) Variable selection in nonparametric additive models. *Ann. Stat.*, 38, 2282–2313.

91 Lin, Y. and Zhang, H.H. (2006) Component selection and smoothing in multivariate nonparametric regression. *Ann. Stat.*, 34, 2272–2297.

92 Zhang, H.H. and Lin, Y. (2006) Component selection and smoothing for nonparametric regression in exponential families. *Stat. Sin.*, 16, 1021–1042.

93 Storlie, C., Bondell, H., Reich, B., and Zhang, H.H. (2011) The adaptive COSSO for nonparametric surface estimation and model selection. *Stat. Sin.*, 21, 679–705.

94 Ravikumar, P., Liu, H., Lafferty, J., and Wasserman, L. (2009) Sparse additive models. *J. R. Stat. Soc. B.*, 71, 1009–1030.

95 Lafferty, J. and Wasserman, L. (2008) RODEO: sparse, greedy nonparametric regression. *Ann. Stat.*, 36, 28–63.

96 Zhang, H.H., Cheng, G., and Liu, Y. (2011) Linear or nonlinear? Automatic structure discovery for partially linear models. *J. Am. Stat. Assoc.*, 106, 1099–1112.

97 Schumaker, L. (1981) *Spline Functions: Basic Theory*, Cambridge Mathematical Library.

98 Meier, L., Van De Geer, S., and Buhlmann, P. (2009) High-dimensional additive modeling. *Ann. Stat.*, 37, 3779–3821.

99 Leeb, H. (2006) The Distribution of a Linear Predictor After Model Selection: Unconditional Finite-Sample Distributions and Asymptotic Approximations. Optimality. Institute of Mathematical Statistics Lecture Notes - Monograph Series, vol. 49, pp. 291–311.

100 Leeb, H. and Potscher, B. (2003) The finite-sample distribution of post-model-selection estimators and uniform versus nonuniform approximations. *Econ. Theory*, 19, 100–142.

101 Leeb, H. and Potscher, B.M. (2005) Model selection and inference: facts and fiction. *Econ. Theory*, 21, 21–59.

102 Berk, R., Brown, L., Buja, A. *et al.* (2013) Valid post-selection inference. *Ann. Stat.*, 41, 802–837.

103 Belloni, A., Chernozhukov, V., and Wei, Y. (2016) Post-selection inference for generalized linear models with many controls. *J. Bus. Econ. Stat.*, 34, 606–619.

18

Sampling Local Scale Parameters in High-Dimensional Regression Models

Anirban Bhattacharya[1] *and James E. Johndrow*[*2]

[1] *Texas A&M University, College Station, TX, USA*
[2] *University of Pennsylvania, Philadelphia, PA, USA*

1 Introduction

Consider a Gaussian linear model with likelihood

$$L(z \mid W\beta, \sigma^2) = (2\pi\sigma^2)^{-N/2} e^{-\frac{1}{2\sigma^2}(z-W\beta)'(z-W\beta)} \tag{1}$$

where W is a $N \times p$ matrix of covariates, $\beta \in \mathbb{R}^p$ is assumed to be a sparse vector, and $z \in \mathbb{R}^N$ is an N-vector of response observations. A popular hierarchical Bayesian approach to this problem chooses a "global–local" prior [1] for β, which is a Gaussian scale-mixture prior of the form

$$\begin{gathered} \beta_j \mid \sigma^2, \eta, \xi \overset{\text{iid}}{\sim} \mathrm{N}(0, \sigma^2 \xi^{-1} \eta_j^{-1}), \quad \eta_j^{-1/2} \overset{\text{iid}}{\sim} \upsilon_L, \quad j = 1, \ldots, p, \\ \xi^{-1/2} \sim \upsilon_G, \quad \sigma^2 \sim \mathrm{InvGamma}(\omega/2, \omega/2) \end{gathered} \tag{2}$$

where υ_{L} and υ_{G} are densities on $\mathbb{R}_+$. The choices of υ_{L} and υ_{G} commonly employed in the literature result in induced marginal densities on β_j that have singularities at zero and tails at least as heavy as an exponential [2–6] which results in a continuous approximation to the spike and slab prior [7]. See Figure 1 for an illustration of the behavior of some of these densities near the origin and in the tails. The choice of a prior that results in exponential or heavier tails for the slab component dates at least to Johnstone and Silverman [8], where it is shown to be necessary for minimaxity of empirical Bayes estimates.

In this chapter, we chronicle some of our recent efforts [9, 10] to scale up posterior sampling in the high-dimensional linear model with the horseshoe prior of Carvalho *et al.* [3]. We strive to minimize repeating what is already the prime focus there and instead aim to shed light on some important aspects of our algorithm development with a broader appeal beyond the immediate application. Specifically, our main focus is in sampling of the local scales $\eta_j^{-1/2}$ in blocked Gibbs sampling. While in this chapter we focus on the linear model, we expect that most of the discussion is relevant to blocked sampling for generalized linear models since these parameters are conditionally independent of data given β. Through

[*]Authors are listed alphabetically.

Computational Statistics in Data Science.
Edited by Walter W. Piegorsch, Richard A. Levine, Hao Helen Zhang and Thomas C. M. Lee.
 ISBN 978-1-11956107-1

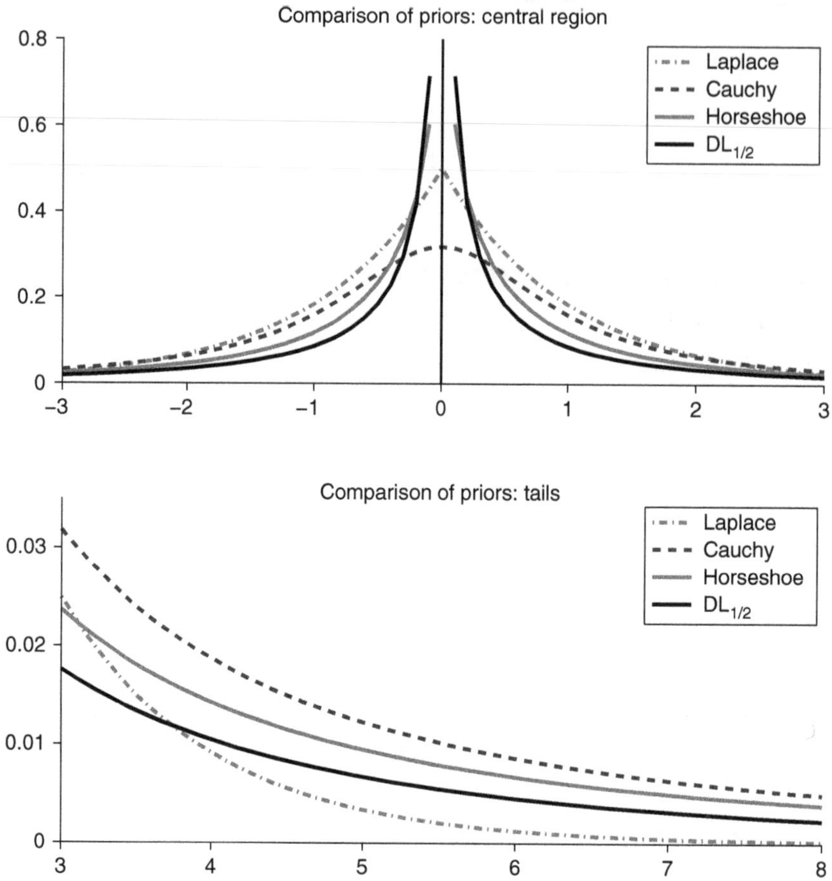

Figure 1 Marginal prior of $\beta_1 \mid \sigma = 1, \xi = 1$ for different choices of v_L. *Source*: Adapted from Bhattacharya, A., Pati, D., Pillai, N.S., and Dunson, D.B. (2015) Dirichlet–Laplace priors for optimal shrinkage. *J. Am. Stat. Assoc.*, **110** (512), 1479–1490.

this discussion, we also highlight the importance of delicately handling numeric issues that routinely arise in these applications.

2 A Blocked Gibbs Sampler for the Horseshoe

One of our motivations to develop more efficient algorithms for the horseshoe originated from the observation of Polson *et al.* [11, Supplement] that the global precision parameter ξ tends to mix very slowly in MCMC algorithms, especially when p is large. The approach in Polson *et al.* [11] to sample from the joint posterior of $(\beta, \sigma^2, \eta, \xi) \mid z$ is to use a Gibbs update rule of the form

$$\begin{aligned} &\sigma^2, \beta \mid \xi, \eta, z \\ &\quad \xi \mid \eta, \beta, \sigma^2 \\ &\quad \eta \mid \xi, \sigma^2, \beta \end{aligned} \tag{3}$$

The conditional posteriors of ξ and η in the second and third steps do not admit a standard form and Polson *et al.* [11] recommended slice sampling for these updates. The joint sampling of $\sigma^2, \beta \mid \xi, \eta, z$ in the first step can be carried out by first sampling $\sigma^2 \mid \xi, \eta, z$ from an inverse-gamma distribution and then sampling $\beta \mid \sigma^2, \xi, \eta, z$ from a p-variate Gaussian distribution. This algorithm was found to exhibit very high autocorrelation and poor mixing for ξ in the normal means setting. The authors also tried a parameter-expanded Gibbs sampler, with limited success.

Our basic approach in Johndrow *et al.* [10] to improve the mixing is to make more extensive use of block updating. Specifically, our proposed update rule takes the form

$$\begin{aligned} &\xi, \sigma^2, \beta \mid \eta, z \\ &\qquad \eta \mid \xi, \sigma^2, \beta \end{aligned} \tag{4}$$

It is a common folklore among practitioners of Gibbs sampling that blocking more parameters together usually improves mixing, although this is not universally true. Moreover, the resulting updates from such additional blocking often get more complicated and may contribute to additional per-iteration cost. However, in the present scenario, the blocking strategy above leads to improved mixing without sacrificing any computational efficiency. The joint update of $\xi, \sigma^2, \beta \mid \eta, z$ in the first step of (4) is carried out by sequentially sampling

$$\begin{aligned} &\xi \mid \eta, z \\ &\sigma^2 \mid \eta, \xi, z \\ &\beta \mid \eta, \xi, \sigma^2, z \end{aligned} \tag{5}$$

The update of β remains the same as in Equation (3), while for σ^2 and ξ, we take advantage of the conditionally conjugate nature of the model prior to perform analytical marginalization.

2.1 Some Highlights for the Blocked Algorithm

Before describing the details of the various steps within the algorithm (4), we provide a quick highlight of some of the salient features. Figure 2 shows autocorrelations at lags 1–100 for $\log(\xi)$ from the old algorithm (3) compared to the algorithm in (4) for a simulation with problem size $N = 2000$, $p = 20\ 000$. It is immediately evident that autocorrelations at all lags are much lower in the new algorithm. Moreover, a feature not exposed by looking at autocorrelations alone is that the old algorithm actually does not converge in most cases; ξ just drifts toward infinity as the chain extends; see bottom panel of Figure 2.

Even though ξ mixes poorly within the old algorithm (3), an argument can be made that it is the β_js that we care about, and all is okay as long as those elements exhibit reasonable mixing. However, we show in Johndrow *et al.* [10] that the statistical performance of the new algorithm is also superior in terms of a number of different metrics. A particularly interesting phenomenon takes place for "intermediate-sized" signals as explained below. Figure 3 shows trace plots from a path of length 20 000 for the 10th entry of β, whose true value of $2^{-1/4} \approx 0.84$ is slightly less than half the true residual standard deviation of $\sigma = 2$. The Horseshoe posterior for such intermediate-sized signals is apparently bimodal, with a mode at zero and a second one away from zero, a fact that has heretofore received little attention (see, e.g., the brief comment at Datta and Ghosh [12, p. 114]). This is of inferential interest, as it aptly reflects the posterior uncertainty associated with such

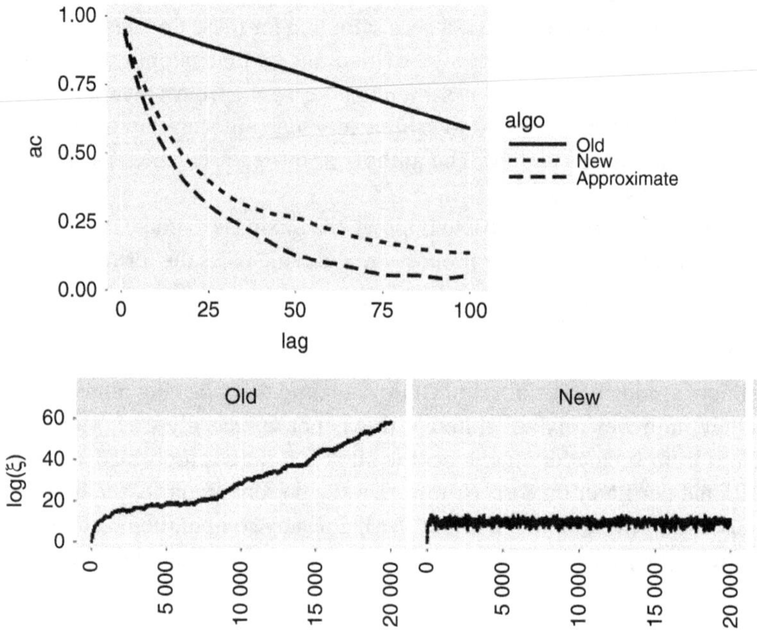

Figure 2 Estimated autocorrelations for log(ξ) for the three algorithms. Approximate refers to a more computationally efficient version of the exact blocked sampler developed in Johndrow *et al.* [10]. *Source*: Modified from Johndrow, J., Orenstein, P., and Bhattacharya, A. (2020) Scalable Approximate MCMC Algorithms for the Horseshoe Prior. *J. Mach. Learn. Res.*, **21**, 1–61. Available at: https://jmlr.org/papers/v21/19-536.html.

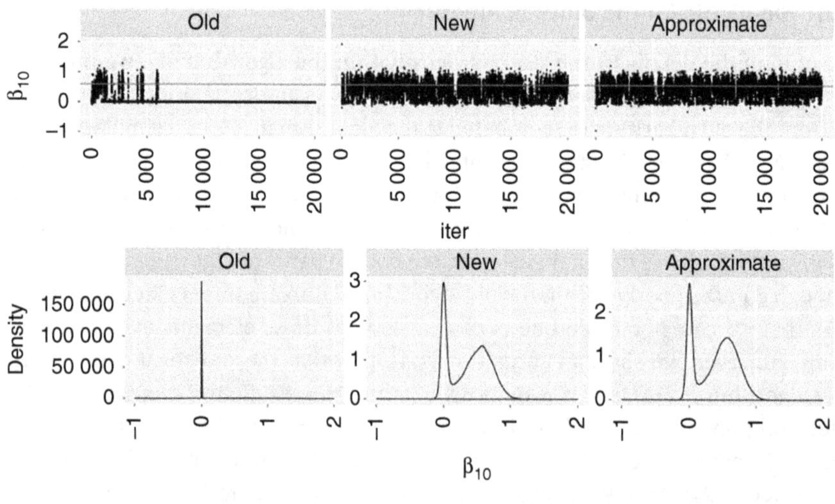

Figure 3 Trace plots (with true value indicated) and density estimates for one entry of β. *Source*: Modified from Johndrow, J., Orenstein, P., and Bhattacharya, A. (2020) Scalable Approximate MCMC Algorithms for the Horseshoe Prior. *J. Mach. Learn. Res.*, **21**, 1–61. Available at: https://jmlr.org/papers/v21/19-536.html.

intermediate signals and is also a nice illustration of how well the Horseshoe posterior approximates the posterior under sharp priors that place nonzero mass on the event $\beta_j = 0$ *a priori*. It is not surprising that this feature has not been a focus, since the old algorithm massively underestimates the relative sizes of the two modes and places most of its mass near the origin. The bimodality creates two deep potential wells in the target, and the old algorithm gets stuck in the larger mode during the second half of the run, as is apparent from the top left panel of Figure 3. This has inferential consequences, as any thresholding procedure based on the old algorithm will unequivocally declare β_{10} as a noise coefficient. The new algorithm is apparently more successful at crossing the potential "hill" between these wells.

A more complete analysis of the operating characteristics of the new algorithm can be found in Johndrow *et al.* [10] which we do not wish to repeat here. Rather, we hope to share some insights acquired from this exercise that possess a more wide-ranging appeal in high-dimensional Bayesian problems involving continuous shrinkage priors. We also point the readers to Refs 13–15 for some alternative algorithms for the horseshoe that have appeared in recent years.

3 Sampling (ξ, σ^2, β)

We make some remarks about the sampling steps within Equation (5). Define $D = \text{diag}(\eta_j^{-1})$, $\Gamma = \xi^{-1} D$, and $M_\xi = I_N + W\Gamma W'$; these quantities appear repeatedly going forward.

3.1 Sampling ξ

In Johndrow *et al.* [10], we show that the marginal posterior $\xi \mid \eta, z$, integrating over β and σ^2, is

$$p(\xi \mid \eta, z) = |M_\xi|^{-1/2} \left(\frac{\omega}{2} + \frac{1}{2} z' M_\xi^{-1} z \right)^{-(N+\omega)/2} \frac{1}{\sqrt{\xi}(1+\xi)} \tag{6}$$

Since this is not a standard density, we used a Metropolis algorithm to sample ξ. Since the conditional posterior of ξ has polynomial tails, it is not advisable to directly use a Metropolis random walk on ξ. We instead perform a normal random walk on $\log(\xi)$; the proposal standard deviation s can be easily tuned to achieve an acceptance rate of around 25%. For $p \gg n$ settings, we found $s = 0.8$ to be a good default choice.

3.2 Sampling σ^2

The conditional posterior of σ^2, integrating over β, is shown in Johndrow *et al.* [10] to be

$$\text{InvGamma}\left(\frac{n+\omega}{2}, \frac{z^T M_\xi^{-1} z + \omega}{2} \right)$$

In Johndrow *et al.* [10], we assumed a proper inverse-gamma prior on σ^2 as in Equation (2) instead of the more commonly used improper right Haar prior $\pi(\sigma^2) \propto (\sigma^2)^{-1}$. We have found the use of proper priors on the residual variance, an important constituent to obtaining convergence and for numerical stability of the algorithm. For example, we have observed in simulations that the residual variance sometimes converges to an apparent stationary distribution that puts almost no mass near its true value when the improper prior is used.

3.3 Sampling β

The full-conditional distribution of β is $\mathrm{N}((W'W+\Gamma^{-1})^{-1}W'z, \sigma^2(W'W+\Gamma^{-1})^{-1})$. Naively sampling from this p-variate Gaussian distribution has $O(p^3)$ complexity. We developed an alternative algorithm in Bhattacharya *et al.* [9] whose steps are as follows:

$$
\begin{aligned}
&\text{sample } u \sim \mathrm{N}(0,\Gamma) \text{ and } f \sim \mathrm{N}(0, I_N) \text{ independently}\\
&\qquad \text{set } v = Wu + f, \quad v^* = M_\xi^{-1}(z/\sigma - v)\\
&\qquad \text{set } \beta = \sigma(u + \Gamma W' v^*)
\end{aligned}
\tag{7}
$$

When $p > N$, the most expensive step above is the computation of the matrix product WDW' to form M_ξ, which has computational cost $O(N^2p)$. This is substantially smaller than $O(p^3)$ when $p \gg N$. A Gaussian full conditional as above routinely appears in many other high-dimensional applications such as reduced rank regression, factor models, and matrix factorizations, and the sampling trick (7) applies to all these settings.

Interestingly, observe that the matrix M_ξ appears in all the update steps above. The main focus of our work [10] is to further alleviate the computational cost associated with computing M_ξ for large N using a natural thresholding step exploiting the structure present in the problem. We refer the interested reader to Johndrow *et al.* [10] for further details on this approximate MCMC algorithm.

4 Sampling η

4.1 The Slice Sampling Strategy

The full conditional of η is proportional to

$$
\prod_{j=1}^{p} \frac{e^{-m_j\eta_j}}{1+\eta_j} \mathbb{1}_{(0,\infty)}(\eta_j) \tag{8}
$$

where $m_j = \beta_j^2\xi/(2\sigma^2)$. Clearly, the components η_j are conditionally independent of each other, and sampling from η amounts to independently sampling the η_js from their respective full conditionals. While this seems straightforward in principle, this basic sampling step from a univariate distribution needs to be repeated a very large number of times – for example, if $p \sim 100\,000$, and we run the MCMC for 10 000 iterates, then we are already sampling from this density 10^9 times – and even minor differences in accuracy will have a major overall repercussion. Many Bayesian hierarchical models share this feature of having a large number of scale parameters with conditionally independent posteriors, and much of the message below thus applies more generally.

Polson *et al.* [11] recommended a slice sampling algorithm for the above which proceeds as follows:

- sample $u_j \sim \text{Unif}\left(0, \frac{1}{1+\eta_j}\right)$,
- sample η_j from an exponential distribution with rate m_j truncated to the interval $(0, r_j)$, where $r_j = \frac{1-u_j}{u_j}$.

The above scheme equivalently performs Gibbs sampling on $\{(u_j, \eta_j)\}_{j=1}^{p}$ with joint density proportional to

$$\prod_{j=1}^{p} \mathbb{1}\left(0 < u_j < \frac{1}{1+\eta_j}\right) e^{-m_j\eta_j} \, \mathbb{1}_{(0,\infty)}(\eta_j)$$

since the density in Equation (8) is observed as the marginal density of η in the above display upon integrating over u.

While the two-step slice sampler looks entirely innocuous, it routinely runs into numerical issues unless care is exercised. To understand the cause of such numerical issues, focus on the rate parameter m_j which scales quadratically in β_j. If the truth is sparse, then a large fraction of the m_js assume very small values on any iteration of the sampler once the algorithm begins to converge. Now, the sampling of η_j from the truncated exponential distribution using the inverse cdf method is done by sampling $v_j \sim \text{Unif}(0, 1)$ and setting

$$\eta_j = -\frac{\log[1 - \{1 - \exp(-m_j r_j)\}v_j]}{m_j} \tag{9}$$

The numerical evaluation of the log and exp using built-in functions runs into numerical issues when $m_j r_j$ is close to machine precision, which results in `NaN` being returned for η_j. To avoid this, earlier implementations [9, 11] rely on truncation steps which replace a sampled value with a fixed numeric value (between 10^{-8} and 10^{-10}) if the sampled value is smaller than the specified threshold.

However, this is a band-aid on the real problem, which is that the expression in (9) needs to be evaluated with some additional care when m_j is small. We can rewrite it suggestively as

$$\eta_j = -\frac{1}{m_j} \log[1 + (\exp(-m_j r_j) - 1)v_j]$$

making it clear that we need to evaluate both $\log(1+x)$ and $e^x - 1$ for small x whenever m_j is small. A "robust" version of the inverse CDF sampling function for the truncated exponential that replaces calls to the generic functions `log(1+x)` and `exp(x)-1` with `log1p(x)` and `expm1(x)` when x is smaller than machine precision completely fixes all of the numerical problems and obviates the need for arbitrary replacement of small values of η_j. These latter two functions are well-known variants of `log(1+x)` and `exp(x)-1` that are robust to small values of `x` (i.e. values of x near machine precision). What is remarkable is that changing only this one function is enough to result in an algorithm that produces samples for ξ like those shown in the right panel of Figure 2. That is, this simple numerical issue seems to have been responsible for the Markov chain failing to converge for ξ.

It is worth noting that in most applications where truncated exponential sampling is desired, the inverse CDF method works just fine without the need to use robust versions of the logarithm and exponential functions. What is really at issue here is the combination of a prior that aggressively shrinks some of the parameters toward zero with moderate to large sample sizes and even larger dimension, which tends to make the posterior highly concentrated near zero for most of the coordinates of β. This in turn causes most of the m_j to be very small in (9). So while in this particular instance the numerical problems were, in retrospect, reasonably obvious and simple to fix, we view this as a representative example of how numerical stability can become a more important and delicate issue in writing MCMC algorithms targeting high-dimensional distributions whose "width" in many dimensions is extremely small.

4.2 Direct Sampling

We now describe some alternative strategies to directly sample η without introducing any auxiliary variable. There are a number of reasons that this strategy can be attractive compared to the slice sampler. For example, it makes theoretical analysis of the algorithm easier, both because it reduces the number of sampling steps in the blocked Gibbs sampler – since we no longer need to sample the u_j's – and because it allows direct analysis of the η dynamics without the need to marginalize over u. In Johndrow *et al.* [10], we exploited this to establish the geometric ergodicity of our blocked Gibbs sampler for $p \leq N$. Here, we describe two strategies for direct sampling from the full conditional for η: rejection and the inverse CDF method.

Consider the density

$$h_m(t) = C_m \frac{e^{-mt}}{1+t} \mathbb{1}_{(0,\infty)}(t) \tag{10}$$

where the normalizing constant $C_m = e^{-m}/E_1(m)$ with

$$E_1(x) = \int_x^\infty \frac{e^{-t}}{t} \, dt$$

denoting the exponential integral function. With this notation, the full conditional of η is $\bigotimes_{j=1}^p h_{m_j}$, with $m_j = \beta_j^2 \xi/(2\sigma^2)$ defined as before.

For large m, the density h_m is closely approximated by an Exponential(m) density. However, for small m, the picture is rather different – the exponential tails kick in only beyond $1/m$, and the polynomial part $(1+t)^{-1}$ dominates for smaller values. Thus, although h_m technically has exponential tails, it behaves like a heavy-tailed density for all practical purposes. These heuristics can be quantified as follows. In Johndrow *et al.* [10], we showed that the following steps produce a sample from h_m:

i) Sample $s \sim$ Exponential(m) and $\omega \sim$ Unif(0, 1) independently,
ii) If $s < \omega/(1-\omega)$, accept s as a sample from h_m, otherwise return to step (i).

The probability that a sample from Expo(m) is accepted as a sample from h_m is then given by

$$P\left(s < \frac{\omega}{1-\omega}\right) = \int_0^\infty m e^{-mt} \frac{1}{1+t} \, dt = m \, \exp(m) \, E_1(m)$$

For $m = 10$, this probability is already as large as 0.92; however, it is only 0.04 for $m = 10^{-2}$. Accounting for the fact that we need to sample p many of these η_js at any step of the Markov chain with p in the tens of thousands or more and most of the m_js on the smaller side, this simple-minded rejection scheme is rendered extremely ineffective. This led us to explore more efficient samplers, two of which we describe below.

4.2.1 Inverse-cdf sampler

We first explore sampling from h_m using the inverse cdf method. To that end, let us first calculate the cdf H_m of h_m. We have, for $x > 0$,

$$H_m(x) = C_m \int_0^x \frac{e^{-mt}}{1+t}\,\mathrm{d}t = C_m\, e^m \int_1^{x+1} \frac{e^{-mt}}{t}\mathrm{d}t = 1 - \frac{E_1(m(x+1))}{E_1(m)}$$

Hence, the inverse cdf method takes the form: sample $u \sim \mathrm{Unif}(0, 1)$ and solve

$$E_1(m(1+s)) = E_1(m) \times (1-u) \text{ for } s \tag{11}$$

The random variable s then has distribution h_m. The main obstacle at this point is solving the Equation (11), which requires numerical inversion of the exponential integral function. Somewhat surprisingly, high-level languages such as `Matlab` or `R` do not provide a built-in function for this purpose. For example, the `Matlab` function `gammaincinv`, inverse of the incomplete gamma function, does not return anything meaningful when its second argument is zero, which corresponds to the exponential integral function. Also, default equation solvers run into numerical instabilities or encounter issues with parallelizing p such equations. Due to the lack of default functionalities, we design a careful approach to solve the Equation (11). Depending on the value of m, we branch into two different solvers.

Small and moderate m ($m < 300$). For this case, we solve the equation $f(z) = 0$ using the Newton–Raphson algorithm, where $f(z) = E_1(z) - E_1(m) \times (1-u)$, and set $s = z/m - 1$. The details regarding the iterates, initialization, and stopping criterion are provided in the Appendix. This algorithm is very stable for $m \ll 1$ and works fine up to $m \sim 500$ but encounters numerical issues for larger values of m. We handle such values of m separately as follows.

Large m ($m > 300$). Here, our guiding principle is that a draw from $\mathrm{Expo}(m)$ has a very high probability (greater than 0.999) of getting accepted as a draw from h_m, and hence given u, the solution s to Equation (11) should not be too far away from $-\log(1-u)/m$.

It is well known that the function E_1 satisfies the inequalities

$$\frac{1}{2}e^{-x}\log\left(1+\frac{2}{x}\right) < E_1(x) < e^{-x}\log\left(1+\frac{1}{x}\right), \quad x > 0 \tag{12}$$

Motivated by the above, define, for $q \in [1/2, 1]$, the function

$$f_q(x) = \frac{q\, e^{-x}\log\left(1+\frac{(1/q)}{x}\right)}{E_1(x)}$$

with the aim to identify a q^* such that $f_{q^*}(x) \approx 1$ for large x. Repeatedly refining using the bisection method, we find $q^* := 0.5015$ provides an approximation up to five decimal

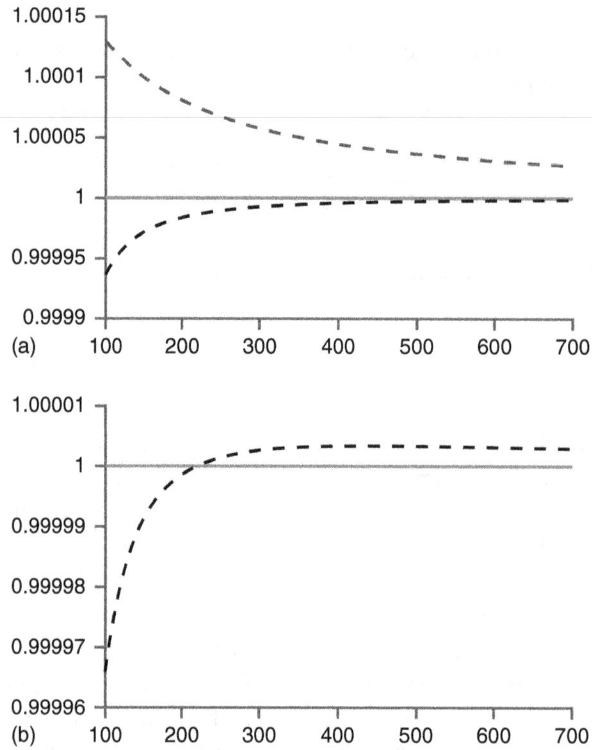

Figure 4 (a) Plots f_q for $q = 0.51$ and 0.50 (in dashed gray and dashed black, respectively). (b) The plot for $q^* = 0.5015$.

places; see Figure 4b. We now exploit the fact that f_{q^*} is flat for all practical purposes for $x > 300$. Recall that we are aiming to solve the equation $E_1(m + ms) = E_1(m) \times (1 - u)$ for large m. Set $z = ms$. Making the assumption that $f_{q^*}(m + z) \approx f_{q^*}(m)$, we obtain

$$q^* \, e^{-(m+z)} \log\left(1 + \frac{(1/q^*)}{m+z}\right) \approx q^* \, e^{-m} \log\left(1 + \frac{(1/q^*)}{m}\right) \times (1 - u)$$

After cancellations, we thus focus on solving the equation $g(z) = c$, where

$$g(z) = e^{-z} \log\left(1 + \frac{(1/q^*)}{m+z}\right), \quad c = \log\left(1 + \frac{(1/q^*)}{m}\right) \times (1 - u)$$

As an illustration, with $m = 10^{-6}$ and $u = 0.99$, the solution to Equation (11) is obtained as s =1.314826090972595e+06, while with $m = 600$ and $u = 0.80$, the solution s = 0.002677946618164, up to a user-defined tolerance level $\delta = 10^{-12}$ in either case (see the Appendix for details of how the tolerance is used in defining a stopping rule for the algorithm). The Newton–Raphson algorithm takes 15 and 4 iterations to converge in these two cases, respectively.

Rejection sampler. In Johndrow *et al.* [10], we developed a careful rejection sampler to sample from h_m for values of $m < 1$ by constructing an accurate upper envelope for h_m.

For values of m larger than one, we continued to use the simpler rejection sampler outlined in Section 4.2. Let

$$f_m(t) = mt + \log(1 + t), \quad t > 0$$

be the negative log-density (up to constants) corresponding to h_m. The observation that f_m is increasing and concave on $(0, \infty)$ is exploited to build a lower envelope to f_m, which translates to an upper envelope on the (unnormalized) density. To be specific, fix $0 < a < 1 < b$, and let

$$f_{\ell,m}(t) := \begin{cases} \log(1+t) & t \in [0, a/m) \\ A + \lambda_2(t - a/m) & t \in [a/m, 1/m) \\ I + \lambda_3(t - b/m) & t \in [1/m, b/m) \\ mt + \log(1 + b/m) & t \geq b/m \end{cases}$$

In the above display,

$$A = f_m(a/m),\ I = f_m(1/m),\ B = f_m(b/m),\ \lambda_2 = \frac{I - A}{(1-a)/m},\ \lambda_3 = \frac{B - I}{(b-1)/m}$$

In our implementation, we use default values $a = 1/5$ and $b = 10$. Observe that $f_{\ell,m}$ is an increasing function which is identical to $\log(1 + t)$ on $[0, a/m)$, linearly interpolates between (i) $f_m(a/m)$ and $f_m(1/m)$ on $[a/m, 1/m)$ and (ii) $f_m(1/m)$ and $f_m(b/m)$ on $[1/m, b/m)$, and equals $mt + \log(1 + b/m)$ on $[b/m, \infty)$. By construction, $f_{\ell,m} \leq f_m$ on $[0, a/m)$ and $(b/m, \infty)$, and the concavity of f_m implies $f_{\ell,m} \leq f_m$ on $[a/m, b/m]$. Together, we have

$$f_{\ell,m}(t) \leq f_m(t)\ \forall\ t \in (0, \infty)$$

Now, define a density $h_{\ell,m}$ on $(0, \infty)$ with $h_{\ell,m}(t) \propto e^{-f_{\ell,m}(t)}$ for $t > 0$. Since $e^{-f_{\ell,m}(t)} \leq e^{-f_m(t)}$ for all t, we used a version of rejection sampling for unnormalized densities (see, for example, Theorem 4.5 in Owen [16]) to propose the following sampler for h_m in Johndrow *et al.* [10]:

i) draw $s \sim h_{\ell,m}$ and $u \sim \text{Unif}(0, 1)$ independently.
ii) accept s as a sample from h_m if $u < e^{-(f_m - f_{\ell,m})(s)}$. Otherwise, back to step (i).

The density $h_{\ell,m}$ can be easily sampled from as it can be written as a mixture of four simple densities, three of which are truncated exponentials, and another one a compactly supported density with an easily invertible cdf; details can be found in Johndrow *et al.* [10].

We now analyze the acceptance rate of the above algorithm. From the proof of Theorem 4.5 of Owen [16], the acceptance probability $\alpha \equiv \alpha_m$ of the rejection sampler above is

$$\alpha_m = \frac{\int_0^\infty e^{-f_m(t)} dt}{\int_0^\infty e^{-f_{\ell,m}(t)} dt} \tag{13}$$

Figure 5 shows a plot of α_m as a function of $\log_{10}(m)$. Specifically, we choose a uniform grid for $\log_{10}(m)$ between -12 and 0, and for each m, calculate α_m in analytic closed form; the default values of a and b mentioned above were used to compute α_m. For such a wide range of values of m, our rejection sampler uniformly possesses excellent acceptance rates. This empirical observation is made rigorous in the following theorem.

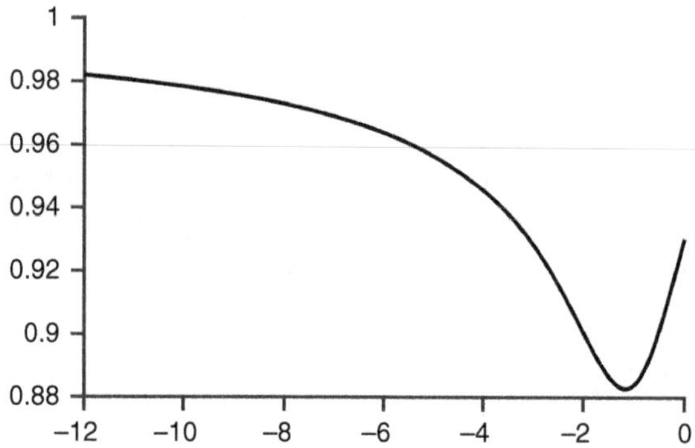

Figure 5 Plot of α_m as a function of $\log_{10}(m)$, where m varies between 10^{-12} and 1.

Theorem 1. *Let α_m be as defined in (13). Then, $\inf_{m\in(0,1)} \alpha_m > c$, where $c \in (0,1)$ is some fixed constant.*

Proof. We remain contented with the uniform result and do not wish to optimize c here in this proof. Write

$$\alpha_m = \frac{\int_0^{b/m} e^{-f_m(t)}\mathrm{d}t + \int_{b/m}^{\infty} e^{-f_m(t)}\mathrm{d}t}{\int_0^{b/m} e^{-f_{\ell,m}(t)}\mathrm{d}t + \int_{b/m}^{\infty} e^{-f_{\ell,m}(t)}\mathrm{d}t} := \frac{U+V}{W+Z}$$

A direct calculation yields $Z = e^{-b}/(b+m)$, which is bounded between $e^{-b}/(b+1)$ and e^{-b}/b as m varies over $(0,1)$. Next, by construction, it can be shown that there exists a constant $\kappa > 0$ independent of m such that

$$f_m(t) < f_{\ell,m}(t) + \kappa \;\forall\; t \;\in (0, b/m) \;\&\; \forall\; m \in (0,1)$$

This implies $U = \int_0^{b/m} e^{-f_m(t)}\mathrm{d}t \geq e^{-\kappa}W$. Now, bound $(U+V)/(W+Z) \geq U/(W+Z) \geq U/(e^{\kappa}U + Z)$. The proof is completed by observing that $U = E_1(m) - E_1(b+m)$ can be bounded from below by a constant uniformly over $m \in (0,1)$.

We now make some comments about parallel implementation of the rejection sampler. Since the η_j are all conditionally independent given the other parameters, it is clear that one can develop an embarassingly parallel implementation of this rejection sampler. What is perhaps less clear is that the gains from parallelization can be very large when p is large. To see this, first note that the waiting time τ_j until the first sample is accepted is a geometric random variable with mean α_j^{-1}, where α_j is given by (13) with arguments depending on m_j. This implies that a serial implementation has expected waiting time $\sum_j \alpha_j^{-1}$ to sample all p of the η_j's. In contrast, the waiting time for a näive parallel implementation will be $\mathbb{E}[\max_j \tau_j] \leq \sum_j \mathbb{E}[\tau_j]$. In the case where the α_j are similar across j, the expectation of the maximum will be much smaller than the expectation of the sum, which can result in large speedups. However, even this understates the advantage of parallelization. Ignoring communication costs, every time a sample of η_j is accepted for some j, the corresponding worker becomes available to begin rejection sampling for one of the other components of η that has not yet had an acceptance event. Thus, more resources become available to work on the

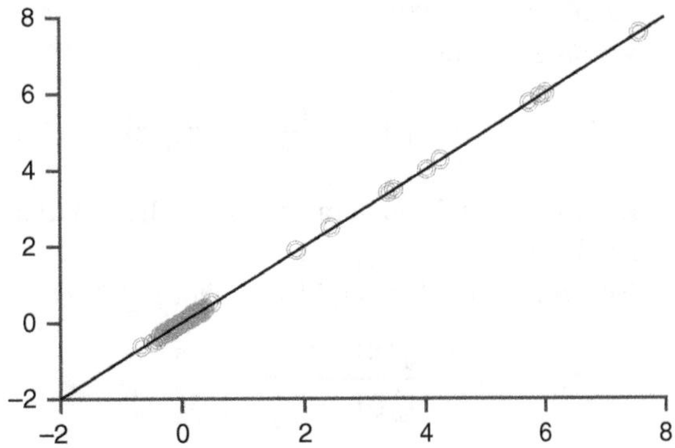

Figure 6 The posterior mean of β_j in a normal means problem: the x-axis and y-axis, respectively, correspond to the rejection sampler and the inverse cdf sampler being used to sample the η_js.

remaining components of η_j as samples are accepted. The gains are most pronounced in practice when communication costs are low but sampling costs high, such as when sampling many η_j in a multiprocessor or multicore environment.

Finally, we perform a small simulation study to compare the overall performance of the rejection sampler and the inverse cdf sampler. Figure 6 plots the posterior means $\hat{\beta}_j$s obtained from the inverse-cdf method (y-axis) versus the corresponding estimates from the rejection sampler (x-axis) in a normal means problem of dimension 200, where the 10 nonzero entries of β are sampled randomly between 3 and 8. We see excellent agreement between the two approaches, with the rejection sampler being the more computationally efficient. We note here that the Newton–Raphson iterates for the inverse-cdf sampler were also run in parallel, and hence the time comparison is fair.

5 Appendix: A. Newton–Raphson Steps for the Inverse-cdf Sampler for η

Small and moderate m **(**$m < 300$**).** For this case, we record here that $f'(z) = E_1'(z) = e^{-z}/z$. The Newton–Raphson iterates take the form:

Initialize. $z_0 = m$.

Iterate. For $t > 0$, set

$$z_{t+1} = z_t - \frac{f(z_t)}{f'(z_t)} = z_t + z_t\, e^{z_t}\, (E_1(z_t) - E_1(m) \times (1 - u))$$

Stopping criterion. We stop at $t = T$ when (for the first time)

$$\frac{|z_T - z_{T-1}|}{(z_{T-1} \wedge 1)} < \delta$$

for some prespecified tolerance δ. This criterion uses an absolute difference measure for values larger than one and a relative error for values smaller than one. The initial value $z_0 = m$ is chosen so that the corresponding starting value for s is 0. We also attempted Halley's method, which is a second-order extension of the Newton–Raphson method, with less success as it led to practically slower convergence in the above metric.

Large m **(**$m > 300$**).**
We note that for

$$g_{a,b}(z) := e^{-z} \log\left(1 + \frac{a}{b+z}\right), \quad g'_{a,b}(z) = -e^{-z}\left[\log\left(1 + \frac{a}{b+z}\right) + \frac{a}{(b+z)(a+b+z)}\right]$$

The Newton–Raphson iterates to solve the equation $g(z) = c$ take the form:
Initialize. $z_0 = -\log(1-u)$.
Iterate. Set $a = 1/q^*$ and $b = m$. For $t > 0$, set

$$z_{t+1} = z_t - \frac{g_{a,b}(z_t) - c}{g'_{a,b}(z_t)}$$

Stopping criterion. We stop at $t = T$ when (for the first time)

$$\frac{|z_T - z_{T-1}|}{(z_{T-1} \wedge 1)} < \delta$$

for some prespecified tolerance δ. The initialization $z_0 = -\log(1-u)$ reflects that $-\log(1-u)/m$ should be a very good initialization in the s-scale. We have noticed rapid convergence with this initialization.

Acknowledgment

Dr. Bhattacharya acknowledges NSF CAREER 1653404 for supporting this project.

References

1 Polson, N.G. and Scott, J.G. (2010) Shrink globally, act locally: Sparse Bayesian regularization and prediction, in *Bayesian Statistics*, vol. 9 (eds J. Bernardo, M. Bayarri, and J. Berger), Scholarship Online, Oxford, pp. 501–538. doi: 10.1093/acprof:oso/9780199694587.001.0001.

2 Griffin, J.E. and Brown, P.J. (2010) Inference with normal-gamma prior distributions in regression problems. *Bayesian Anal.*, 5 (1), 171–188.

3 Carvalho, C.M., Polson, N.G., and Scott, J.G. (2010) The horseshoe estimator for sparse signals. *Biometrika*, 97 (2), 465–480.

4 Armagan, A., Dunson, D.B., and Lee, J. (2013) Generalized double Pareto shrinkage. *Stat. Sin.*, 23 (1), 119.

5 Bhattacharya, A., Pati, D., Pillai, N.S., and Dunson, D.B. (2015) Dirichlet–Laplace priors for optimal shrinkage. *J. Am. Stat. Assoc.*, 110 (512), 1479–1490.

6 Zhang, Y., Reich, B.J., and Bondell, H.D. (2016) High dimensional linear regression via the R2-D2 shrinkage prior. *arXiv preprint arXiv:1609.00046*.

7 Scott, J.G. and Berger, J.O. (2010) Bayes and empirical-Bayes multiplicity adjustment in the variable-selection problem. *Ann. Stat.*, 38 (5), 2587–2619.

8 Johnstone, I.M. and Silverman, B.W. (2004) Needles and straw in haystacks: Empirical bayes estimates of possibly sparse sequences. *Ann. Stat.*, 32 (4), 1594–1649.

9 Bhattacharya, A., Chakraborty, A., and Mallick, B.K. (2016) Fast sampling with Gaussian scale mixture priors in high-dimensional regression. *Biometrika*, 103 (4), 985–991.

10 Johndrow, J., Orenstein, P., and Bhattacharya, A. (2020) Scalable Approximate MCMC Algorithms for the Horseshoe Prior. *J. Mach. Learn. Res.*, 21, 1–61. Available at: https://jmlr.org/papers/v21/19-536.html

11 Polson, N.G., Scott, J.G., and Windle, J. (2014) The Bayesian bridge. *J. R. Stat. Soc. Series B*, 76 (4), 713–733.

12 Datta, J. and Ghosh, J.K. (2013) Asymptotic properties of Bayes risk for the horseshoe prior. *Bayesian Anal.*, 8 (1), 111–132.

13 Makalic, E. and Schmidt, D.F. (2016) A simple sampler for the horseshoe estimator. *IEEE Signal Process Lett.*, 23 (1), 179–182.

14 Hahn, P.R., He, J., and Lopes, H.F. (2018) Efficient sampling for Gaussian linear regression with arbitrary priors. *J. Comput. Graph. Stat.*, 28 (1), 142–154. doi: 10.1080/10618600.2018.1482762.

15 Nishimura, A. and Suchard, M.A. (2018) Prior-preconditioned conjugate gradient for accelerated gibbs sampling in "large n & large p" sparse bayesian logistic regression models. *arXiv preprint arXiv:1810.12437*.

16 Owen, A.B. (2013) *Monte Carlo theory, methods and examples*. Available at: https://statweb.stanford.edu/~owen/mc/

19

Factor Modeling for High-Dimensional Time Series

Chun Yip Yau

Chinese University of Hong Kong, Shatin, Hong Kong

1 Introduction

Consider a high-dimensional time series $\{Y_t\}_{t=1,\dots,n}$, where $Y_t = (Y_{1,t}, Y_{2,t} \dots, Y_{p,t})' \in \mathbb{R}^p$ is a p-dimensional vector observed at time t. We call $\mathbf{y}_i = (Y_{i,1}, Y_{i,2}, \dots, Y_{i,n})'$ the ith component of the high-dimensional time series. For example, if $Y_{i,t}$ represents the price of stock i at time t, then Y_t collects the prices of all p stocks at time t, and $\mathbf{y}_i$ collects the historical prices of the ith stock from time 1 to n. The whole dataset is denoted by the $n \times p$-dimensional matrix $Y = (\mathbf{y}_1, \mathbf{y}_2, \dots, \mathbf{y}_p)$.

When p is large, it is difficult to simultaneously analyze the cross-correlations among the p components and the serial dependence across time. For example, the simplest vector autoregressive (VAR) model of order 1,

$$Y_t = \Phi Y_{t-1} + \epsilon_t \;,\; \epsilon_t \sim N(0, \Sigma_\epsilon)$$

already involves p^2 unknown parameters in the coefficient matrix $\Phi \in \mathbb{R}^{p\times p}$. Therefore, some structure has to be imposed for feasible estimation. For a small integer r, the *factor model* of the form

$$Y_t = \Lambda F_t + \epsilon_t \tag{1}$$

where $\Lambda \in \mathbb{R}^{p\times r}$ and $F_t \in \mathbb{R}^r$ is a latent low-dimensional stationary process, which provides a parsimonious option for modeling high-dimensional data. Intuitively, (1) states that the serial and cross-sectional dependence of the p-dimensional time series Y_t is captured by the r-dimensional latent common factor F_t. The error term $\epsilon_t \in \mathbb{R}^p$, also known as the *idiosyncratic component*, is an orthogonal white noise, that is, $\text{Cov}(\epsilon_t, \epsilon_s) = \Delta 1_{\{s=t\}}$, where Δ is a diagonal matrix. Also, $\text{Cov}(F_t, \epsilon_s) = 0$ for all t, s. Equivalently, (1) can be expressed as

$$Y' = \Lambda F + \epsilon \tag{2}$$

where $F = (F_1 \; F_2 \; \cdots \; F_n) \in \mathbb{R}^{r\times n}$ and $\epsilon = (\epsilon_1 \; \epsilon_2 \; \cdots \; \epsilon_n) \in \mathbb{R}^{p\times n}$.

Factor models can be classified as *static* or *dynamic*, according to whether time dependence is described in the model. For example, (1) is a static model since all variables appear

Computational Statistics in Data Science.
Edited by Walter W. Piegorsch, Richard A. Levine, Hao Helen Zhang and Thomas C. M. Lee.
 ISBN 978-1-11956107-1

at the same time t. On the other hand, dynamic factor models capture the dependence of Y_t on previous lags of the factor process and take the form

$$Y_t = \Lambda_0 F_t + \Lambda_1 F_{t-1} + \cdots + \Lambda_s F_{t-s} + \epsilon_t \tag{3}$$

where $\Lambda_j \in \mathbb{R}^{p\times r}$, $j = 1, \ldots, s$. Indeed, by defining $F_t = (F_t', F_{t-1}', \ldots, F_{t-s}')'$ and $\Lambda = (\Lambda_0, \Lambda_1, \ldots, \Lambda_s)$, the dynamic factor model (3) with an r-dimensional factor and lag order s can be expressed as a static factor model (1) with an $r(1+s)$-dimensional factor. Although static models cover dynamic models, dynamic model has a merit in explicitly describing the time dependence of the factor process on the data.

Factor model can also be classified as *exact* or *approximate*. If the noise ϵ_t has no cross and serial correlation, that is, Δ is diagonal and $\text{Cov}(\epsilon_t, \epsilon_s) = 0$ if $s \neq t$, then the model is exact. Relaxing this assumption to allow for cross or serial-correlation results in the approximate factor model. Since the factor loadings and factor process are usually of primary interest, for simplicity, we focus on the exact factor model unless otherwise specified. Indeed, many methods reviewed in this chapter also apply to the approximate factor model.

This chapter reviews the methodological and computational aspects of various estimation methods of factor models for high-dimensional time series. We introduce the identifiability issue of factor modeling in Section 2. In Section 3, we discuss estimation methods for the factor loadings and factor process under a known number of factors. Finally, determination of the number of factors is briefly discussed in Section 4.

2 Identifiability

In (1), as both Λ and F are unobserved, there is an issue of identifiability. Specifically, for any invertible matrix $H \in \mathbb{R}^{r\times r}$, (2) can be expressed as $Y' = (\Lambda H)(H^{-1}F) + \epsilon$. Thus, the sets (Λ, F) and $(\Lambda H, H^{-1}F)$ correspond to the same model. Since it requires r^2 parameters to completely specify an $r \times r$ invertible matrix, to ensure identifiability, we need r^2 restrictions on Λ and F.

Some common identifiability conditions include $\Lambda'\Lambda = I_r$, $\Lambda'\Lambda/p = I_r$, $FF' = I_r$, or $FF'/n = I_r$, which assert that either the columns of Λ or rows of F are orthonormal; see Bai and Ng [1]. However, ensuring the orthonormality of r vectors only requires $r(r+1)/2$ constraints. Thus, an additional set of $r(r-1)/2$ constraints has to be imposed to completely ensure the identifiability. For simplicity, we consider the following identifiability constraints:

$$\text{(i) } \Lambda'\Lambda = I_r, \qquad \text{(ii) } \lambda_{ij} = 0 \text{ for } 1 \le i < j \le r, \qquad \text{(iii) } \lambda_{ii} > 0 \text{ for } i = 1, \ldots, r \tag{4}$$

Although (ii) in (4) provides the required $r(r-1)/2$ constraints for identifiability, the sign of each column of Λ is still unidentified since (i) only ensures that the column norm is 1. Therefore, (iii) is needed to fix the signs of the columns. Given an arbitrary Λ satisfying (i), one can multiply Λ by a suitable rotation matrix to ensure that (ii) holds.

For other choices of identifying restrictions, see Bai and Li [2] for details.

3 Estimation of High-Dimensional Factor Model

3.1 Least-Squares or Principal Component Estimation

One way to estimate Λ is to minimize the least-squares criterion

$$S(\Lambda, F) = \sum_{t=1}^{n} |Y_t - \Lambda F_t|^2 = \mathrm{tr}((Y' - \Lambda F)'(Y' - \Lambda F)) \tag{5}$$

with, respectively, to Λ and F, subject to the identifiability constraints. Let $(\hat{\Lambda}, \hat{F})$ be the minimizer of S in (5). Differentiating S with respect to F, we have $-2\hat{\Lambda}'Y' + 2\hat{\Lambda}'\hat{\Lambda}\hat{F} = 0$. Combining with the identifiability constraint $\hat{\Lambda}'\hat{\Lambda} = I_r$ gives

$$\hat{F} = \hat{\Lambda}'Y' \tag{6}$$

Substituting (6) back to (5), we have $S(\hat{\Lambda}, \hat{F}) = \mathrm{tr}(YY') - \mathrm{tr}(\hat{\Lambda}'Y'Y\hat{\Lambda})$. In other words, $\hat{\Lambda}$ maximizes the quantity $\mathrm{tr}(\hat{\Lambda}'Y'Y\hat{\Lambda})$. From standard matrix algebra, we have

$$\hat{\Lambda} = (V_1\ V_2\ \cdots\ V_r) \tag{7}$$

where $V_i \in \mathbb{R}^p$ is the eigenvector corresponding to the ith largest eigenvalue of $Y'Y$. Note that as $Y'Y$ is a nonnegative definitive symmetric matrix, $V_i'V_j = 1_{\{i=j\}}$. Thus, the constraint $\hat{\Lambda}'\hat{\Lambda} = I_r$ is automatically satisfied.

If the noises $\{\epsilon_t\}$ are heteroscedastic or autocorrelated, Choi [3] suggests a weighted least-squares approach, which minimizes $\mathrm{tr}(\hat{\Delta}^{-1}(Y' - \Lambda F)'(Y' - \Lambda F))$, where the estimator $\hat{\Delta}$ of Δ may be obtained from the residuals of the ordinary least-squares estimator.

3.2 Factor Loading Space Estimation

This approach, developed by Lam *et al.* [4], considers estimating the column space of the factor loading matrix Λ. Specifically, since $\{\epsilon_t\}$ is a sequence of white noises uncorrelated with $\{F_t\}$, we have, for $k \geq 1$,

$$\Sigma_y(k) := \mathrm{Cov}(Y_{t+k}, Y_t) = \Lambda \mathrm{Cov}(F_{t+k}, F_t)\Lambda' =: \Lambda\Sigma_f(k)\Lambda' \tag{8}$$

Clearly, (8) implies that the columns of $\Sigma_y(k)$ lie in the column space of Λ. To pool together the information from different time lags without cancellations, one can consider

$$M := \sum_{k=1}^{k_0} \Sigma_y(k)\Sigma_y'(k) = \Lambda\left(\sum_{k=1}^{k_0} \Sigma_f(k)\Sigma_f'(k)\right)\Lambda' \tag{9}$$

where k_0 is a prespecified constant. By construction, the column spaces of M and Λ are the same. Since the column spaces of a matrix can be captured by the eigenvectors of the matrix, one can estimate Λ by

$$\hat{\Lambda} = (V_1\ V_2\ \cdots\ V_r) \tag{10}$$

where $V_i \in \mathbb{R}^p$ is the eigenvector corresponding to the ith largest eigenvalue of $\hat{M} = \sum_{k=1}^{k_0} \hat{\Sigma}_y(k)\hat{\Sigma}_y'(k)$, the sample version of M. Note that as $\hat{M}$ is a nonnegative definitive symmetric matrix, $V_i'V_j = 1_{\{i=j\}}$. Thus, the constraint $\hat{\Lambda}'\hat{\Lambda} = I_r$ is automatically satisfied.

Given $\hat{\Lambda}$, one can use (6) to estimate the factor process by $\hat{F} = \hat{\Lambda}'Y'$. Simulation evidences in Lam *et al.* [4] suggest that the choice of k_0 is not sensitive, and usually $k_0 \leq 5$ provides satisfactory results.

We remark that capturing the information of factor loading space is not limited to $\hat{M}$. Gao and Tsay [5] consider an alternative of $\hat{M}$ based on canonical correlation analysis; Pan and Yao [6] propose to estimate the factor loading space using an innovation expansion algorithm which requires solving a sequence of nonlinear optimization problems.

3.2.1 Improved Estimation of Factor Process

From (2), the estimated factor can be expressed as $\hat{F} = \hat{\Lambda}'Y' = \hat{\Lambda}'\Lambda F + \hat{\Lambda}'\epsilon$. Even if $\hat{\Lambda} \approx \Lambda$, the term $\hat{\Lambda}'\epsilon$ could contribute to a large bias on the estimation of F_t, especially when p is large. To tackle this problem, Gao and Tsay [7] propose a procedure to estimate the factor process by first eliminating the effect of ϵ. Specifically, assume that

$$Y' = \Lambda F + \epsilon = \Lambda F + \Gamma \mathbf{e} \tag{11}$$

where $\Gamma \in \mathbb{R}^{p\times(p-r)}$ satisfies $\Gamma'\Gamma = I_{p-r}$, $\mathbf{e} = (e_1\ e_2\ \cdots\ e_n) \in \mathbb{R}^{(p-r)\times n}$, $e_t \in \mathbb{R}^{p-r}$ is a white noise with covariance matrix Σ_e, and the largest K eigenvalues of Σ_e are diverging.

To remove the effect of $\epsilon = \Gamma\mathbf{e}$ in estimating F, the idea is to construct a matrix $\hat{B}$ such that

(i) $\hat{B}'\Gamma\mathbf{e}$ is negligible; (ii) $\hat{B}'\hat{\Lambda}$ is invertible.

If (i) and (ii) hold, then multiplying $\hat{B}'$ on both sides of (11) gives $\hat{B}'Y' \approx \hat{B}'\Lambda F$, and thus F can be estimated by

$$(\hat{B}'\hat{\Lambda})^{-1}\hat{B}'Y' \tag{12}$$

To achieve (i), one can estimate the orthogonal complement of $\Gamma\mathbf{e}$, which is equivalent to the orthogonal complement of $\Gamma\mathbf{e}\mathbf{e}'\Gamma' \approx \Gamma\Sigma_e\Gamma'$. Following the idea in (8) and (9), define

$$\Sigma_y := \text{Cov}(Y_t, Y_t) = \Lambda\text{Cov}(F_t, F_t)\Lambda' + \Gamma\text{Cov}(e_t, e_t)\Gamma' =: \Lambda\Sigma_f\Lambda' + \Gamma\Sigma_e\Gamma' \tag{13}$$

To extract the information about Γ, eliminate Λ in (13) by defining

$$S := \Sigma_y\Lambda_c\Lambda_c'\Sigma_y = \Gamma\Sigma_e\Gamma'\Lambda_c\Lambda_c'\Gamma'\Sigma_e\Gamma \tag{14}$$

where $\Lambda_c \in \mathbb{R}^{p\times(p-r)}$ is the orthogonal complement of Λ. By construction, S is symmetric, nonnegative definite with columns lying in the column space of $\Gamma\Sigma_e\Gamma'$. As Σ_e contains K diverging eigenvalues, the orthogonal complement of $\Gamma\Sigma_e\Gamma'$ can be estimated by the eigenvectors corresponding to the $p - K$ smallest eigenvalues of S.

Next, we discuss the estimation of S. In view of (10), the orthogonal complement of Λ, that is, Λ_c, can be estimated by the eigenvectors corresponding to the $p - r$ smallest eigenvalues of $\hat{M}$, that is,

$$\hat{\Lambda}_c = (V_{r+1}\ V_{r+2}\ \cdots\ V_p) \tag{15}$$

Thus, the sample version of S is given by $\hat{S} = \hat{\Sigma}_y\hat{\Lambda}_c'\hat{\Lambda}_c\hat{\Sigma}_y$. Given $\hat{S}$, one can define $B_* \in \mathbb{R}^{p\times(p-K)}$ as the matrix containing the eigenvectors corresponding to the $p - K$ smallest eigenvalues of $\hat{S}$, so that $B_*'\Gamma\mathbf{e}$ is negligible.

However, setting $\hat{B} = B_*$ does not fulfill ii) since $B_*'\hat{\Lambda} \in \mathbb{R}^{(p-K)\times r}$ is not a square matrix. This can be fixed by constructing a matrix $\hat{R} \in \mathbb{R}^{(p-K)\times r}$ so that $\hat{R}'B_*'\hat{\Lambda} \in \mathbb{R}^{r\times r}$ is invertible. This $\hat{R}$ can be defined as the matrix whose columns are the r eigenvectors corresponding to the r largest eigenvalues of $B_*'\hat{\Lambda}\hat{\Lambda}'B_*$. In conclusion, with $\hat{B} := B_*\hat{R}$, i) and ii) are fulfilled, and the improved estimator of F can be computed by (12).

3.3 Frequency-Domain Approach

The frequency-domain approach, developed by Forni *et al.* [8] and Forni *et al.* [9], addresses the dynamic factor model (3), with the assumption $\mathrm{Var}(\epsilon_t) = \Delta$ relaxed to a symmetric matrix instead of a diagonal matrix. The idea is based on the following result on frequency-domain time series:

Theorem 1. *(Theorem 9.3.1 of Brillinger [10]. Consider a zero-mean p-dimensional stationary time series $\{Y_t\}$ with absolutely summable autocovariance function $\Sigma_y(k)$ and spectral density matrix*

$$f_Y(\omega) := \frac{1}{2\pi}\sum_{k=-\infty}^{\infty} e^{-i\omega k}\Sigma_y(k)$$

Let $\{\mathbf{b}(u)\}$ and $\{\mathbf{c}(u)\}$, respectively, be $r \times p$- and $p \times r$-dimensional filters such that

$$E\left[\left(Y_t - \sum_{u=-\infty}^{\infty}\mathbf{c}(t-u)\xi_u\right)'\left(Y_t - \sum_{u=-\infty}^{\infty}\mathbf{c}(t-u)\xi_u\right)\right] \tag{16}$$

achieves its minimum value among all possible $r \times p$- and $p \times r$-dimensional filters, where

$$\xi_t = \sum_{u=-\infty}^{\infty}\mathbf{b}(t-u)Y_t$$

Then, $\{\mathbf{b}(u)\}$ and $\{\mathbf{c}(u)\}$ are, respectively, given by

$$\mathbf{b}(u) = \frac{1}{2\pi}\int_0^{2\pi}\mathbf{B}(\alpha)e^{iu\alpha}\,d\alpha, \qquad \mathbf{c}(u) = \frac{1}{2\pi}\int_0^{2\pi}\mathbf{C}(\alpha)e^{iu\alpha}\,d\alpha$$

where $\mathbf{B}(\omega) = (V_1(\omega)\ V_2(\omega)\cdots\ V_r(\omega))$, $\mathbf{C}(\omega) = \mathbf{B}(\omega)'$, and $V_j(\omega)$ is the eigenvector corresponding to the jth largest eigenvalue of $f_Y(\omega)$.

From Theorem 1, the r-dimensional filtered process ξ_t can be viewed as the factors of Y_t in the sense that $Y_t - \sum_{u=-\infty}^{\infty}\mathbf{c}(t-u)\xi_u$ is small. However, $\{\mathbf{c}(u)\}$ is a "two-sided" filter, and it requires future ξ_u's to approximate Y_t. Thus, ξ_t cannot serve as the factor F_t in model (3) since only past F_u's are required to approximate Y_t. To tackle this problem, Forni *et al.* [9] return to the time domain using the spectral densities of the process $X_t = \sum_{u=-\infty}^{\infty}\mathbf{c}(t-u)\xi_u$ and the idiosyncratic component $\epsilon_t = Y_t - X_t$, denoted by $f_X(\omega)$ and $f_\epsilon(\omega)$, respectively. Specifically, the covariance matrices $\Sigma_X(k)$ and $\Sigma_\epsilon(k)$ are first computed from $f_X(\omega)$ and $f_\epsilon(\omega)$. Then, for $j = 1, \ldots, p$, consider the generalized eigenvalue λ_j, which satisfies

$$\lambda_j = \arg\max_{\lambda\in\mathbb{R}^p} \lambda'\Sigma_X(0)\lambda \text{ s.t. } \lambda'\Sigma_\epsilon(0)\lambda = 1 \text{ and } \lambda'\Sigma_\epsilon(0)\lambda_i = 0 \tag{17}$$

for $i = 1, \ldots, j-1$. Intuitively, $\lambda_j Y_t = \lambda_j X_t + \lambda_j \varepsilon_t$ has $\lambda_j X_t$ maximized with $\lambda_j \varepsilon_t$ bounded and orthogonal to $\lambda_i \varepsilon_t$ for $i = 1, \ldots, j-1$. Thus, the linear combination $\lambda_j Y_t$ is close to the factor space, and $(\lambda_1 Y_t, \lambda_2 Y_t, \ldots, \lambda_r Y_t)'$ can be used as an estimate for the common factor F_t.

The computation of the frequency-domain estimation is summarized as follows:

Step 1: Spectral densities estimation

a) Estimate the spectral density matrix of Y_t by

$$\hat{f}_Y(\omega_h) = \frac{1}{2\pi} \sum_{k=-M}^{M} \left(1 - \frac{|k|}{M+1}\right) \hat{\Sigma}_y(k) e^{-ik\omega_h}$$

where $\omega_h = \frac{2\pi h}{2M+1}$, $h = 0, 1, \ldots, 2M$ are frequencies. The tuning parameter M has to satisfy $M \to \infty$ and $M/n \to \infty$ as $n \to \infty$, and a rule-of-thumb choice is $M = \frac{2}{3} n^{1/3}$.

b) For $h = 0, 1, \ldots, 2M$, compute the eigenvectors $\hat{V}_j(\omega_h)$ corresponding to the j-largest eigenvalues $\hat{a}_j$ of $\hat{f}_Y(\omega_h), j = 1, \ldots, r$.

c) Estimate the spectral densities of X_t and ε_t by

$$\hat{f}_X(\omega_h) = \sum_{j=1}^{r} \hat{a}_j \hat{V}_j(\omega_h) \hat{V}_j^*(\omega_h), \text{ and } \hat{f}_\varepsilon(\omega_h) = \hat{f}(Y)(\omega_h) - \hat{f}(X)(\omega_h)$$

where v^* stands for the transpose of the complex conjugate of a vector v.

d) Compute the sample covariance matrices of X_t and ε_t by the inverse Fourier transform

$$\hat{\Sigma}_X(k) = \frac{1}{2M+1} \sum_{k=-M}^{M} \hat{f}_X(\omega_h) e^{ik\omega_h},$$

$$\hat{\Sigma}_\varepsilon(k) = \frac{1}{2M+1} \sum_{k=-M}^{M} \hat{f}_\varepsilon(\omega_h) e^{ik\omega_h},$$

evaluated at $k = 0$

Step 2: Generalized eigenvalue estimation

a) Compute the generalized eigenvalue $\hat{\lambda}_j$ (see Theorem A.2.4 of Anderson [11]), which satisfies

$$\hat{\lambda}_j = \arg\max_{\lambda \in \mathbb{R}^p} \lambda' \hat{\Sigma}_X(0) \lambda \text{ s.t. } \lambda' \hat{\Sigma}_\varepsilon(0) \lambda = 1 \text{ and } \lambda' \hat{\Sigma}_\varepsilon(0) \lambda_i = 0$$

for $j = 1, \ldots, p$ and $i = 1, \ldots, j-1$.

b) Set $\hat{F}_t = (\hat{\lambda}_1 Y_t, \hat{\lambda}_2 Y_t, \ldots, \hat{\lambda}_r Y_t)'$. The estimated factor loading $\hat{\Lambda}_j, j = 0, \ldots, s$ can be obtained by regressing Y_t against $\hat{F}_t$ using (3).

3.4 Likelihood-Based Estimation

Maximum-likelihood estimation (MLE) is a popular approach in statistics as it is efficient in many classical statistics problems [12]. To conduct maximum-likelihood estimation, a parametric model has to be imposed. Therefore, in view of (1) and (3), one has to impose a model for the factor process $\{F_t\}$, and a distribution for ϵ_t. The multivariate normal distribution

$N(\mathbf{0}, \Delta)$ is a natural candidate for ϵ_t. Also, since $\{F_t\}$ is a low-dimensional process, it is natural to further assume that $\{F_t\}$ follows a vector ARMA (VARMA) model, that is,

$$\Phi(B)F_t = \Theta(B)z_t \tag{18}$$

where $z_t \sim N(\mathbf{0}, \Sigma_z)$, $\Phi(B) = I - \Phi_1 B - \cdots - \Phi_{\tilde{p}} B^{\tilde{p}}$, $\Theta(B) = I + \Theta_1 B + \cdots + \Theta_{\tilde{q}} B^{\tilde{q}}$, Σ_z, Φ_j, and Θ are $r \times r$-dimensional matrices. On the other hand, the likelihood function can be formulated conditional on $\{F_t\}$, that is, treating F_t as fixed numbers [2]. In this section, we review several likelihood-based methods for the estimation of factor models.

3.4.1 Exact likelihood via Kalman filtering

Consider the factor model (1) with VARMA($\tilde{p}, \tilde{q}$) factor process (18). In Jungbacker and Koopman [13], the noise is also allowed to be a VAR process

$$\epsilon_t = \Psi_1 \epsilon_{t-1} + \cdots + \Psi_{\tilde{s}} \epsilon_{t-\tilde{s}} + e_t \tag{19}$$

where $e_t \sim N(\mathbf{0}, \Sigma_e)$ is the white noise. Without loss of generality, assume $\tilde{s} \geq \tilde{p}$. Otherwise, one can regard the factor process as a VARMA($\tilde{s}, \tilde{q}$) process with the last $\tilde{s} - \tilde{p}$ autoregressive coefficient matrices equaling zero. Denote $\Psi(B) = (I_p - \Psi_1 B - \cdots - \Psi_{\tilde{s}} B^{\tilde{s}})$, where B is the backshift operator, so that $\Psi(B)\epsilon_t = e_t$. Multiplying $\Psi(B)$ to both sides of (1), together with the VAR(1) representation of the VARMA process

$$\alpha_t := \begin{pmatrix} F_t \\ F_{t-1} \\ \vdots \\ F_{t-\tilde{p}+1} \end{pmatrix} = \begin{pmatrix} \Phi_1 & \Phi_2 & \cdots & \Phi_{\tilde{p}} \\ I_r & \mathbf{0} & \cdots & \mathbf{0} \\ \mathbf{0} & I_r & \cdots & \mathbf{0} \\ \vdots & \ddots & & \vdots \\ \mathbf{0} & \cdots & \mathbf{0} & I_r \end{pmatrix} \begin{pmatrix} F_{t-1} \\ F_{t-2} \\ \vdots \\ F_{t-\tilde{p}} \end{pmatrix} + \begin{pmatrix} I_r & \Theta_1 & \cdots & \Theta_{\tilde{q}} \\ \mathbf{0} & \mathbf{0} & \cdots & \mathbf{0} \\ \vdots & \ddots & & \vdots \\ \mathbf{0} & \mathbf{0} & \cdots & \mathbf{0} \end{pmatrix} \begin{pmatrix} z_t \\ z_{t-1} \\ \vdots \\ z_{t-\tilde{q}} \end{pmatrix} =: H\alpha_{t-1} + R\eta_t \tag{20}$$

one obtains the *state space representation*

$$\begin{aligned} \Psi(B)Y_t &= Z\alpha_t + e_t \\ \alpha_t &= H\alpha_{t-1} + R\eta_t \,, \quad \eta_t \sim N(\mathbf{0}, Q) \end{aligned} \tag{21}$$

where $Z = (\Lambda, -\Psi_1\Lambda, \ldots, -\Psi_{\tilde{s}}\Lambda, \mathbf{0}) =: (\overline{\Lambda}, \mathbf{0}) \in \mathbb{R}^{p\times\tilde{p}r}$, with $\overline{\Lambda} \in \mathbb{R}^{p\times(\tilde{s}+1)r}$, $\mathbf{0} \in \mathbb{R}^{p\times(\tilde{p}-\tilde{s}-1)r}$ is a zero matrix, and Q is defined by the Kronecker product $Q = I_{\tilde{q}+1} \bigotimes \Sigma_z$. To enhance computations, Jungbacker and Koopman [13] construct a full rank matrix $A = \begin{pmatrix} A_L \\ A_H \end{pmatrix} \in \mathbb{R}^{p\times p}$, where $A_L = (\overline{\Lambda}'\Sigma_e^{-1}\overline{\Lambda})^{-1}\overline{\Lambda}'\Sigma_e^{-1} \in \mathbb{R}^{(\tilde{s}+1)r\times p}$ and $A_H \in \mathbb{R}^{(\tilde{p}-\tilde{s}-1)r\times p}$, such that

$$\text{(i) } A_L\Sigma_e A_H' = 0, \text{ (ii) } A_H Z = 0, \text{ (iii) } |A_H\Sigma_e A_H'| = I_{p-(\tilde{s}+1)r} \tag{22}$$

With (22), multiplying A on both sides of the first equation of (21) yields

$$Y_t^L := A_L\Psi(B)Y_t = A_L Z\alpha_t + A_L e_t =: \tilde{Z}\alpha_t + e_t^L \tag{23}$$

$$Y_t^H := A_H\Psi(B)Y_t = A_H e_t =: e_t^H \tag{24}$$

$$\alpha_t = H\alpha_{t-1} + R\eta_t \tag{25}$$

where $e_t^L = A_L e_t$, and $e_t^H = A_H e_t$ satisfies $\begin{pmatrix} e_t^L \\ e_t^H \end{pmatrix} \sim N\left(\mathbf{0}, \begin{pmatrix} (\overline{\Lambda}' \Sigma_e^{-1} \overline{\Lambda})^{-1} & \mathbf{0} \\ \mathbf{0} & A_H \Sigma_e A_H' \end{pmatrix}\right) =: \begin{pmatrix} \Sigma_H & \mathbf{0} \\ \mathbf{0} & \Sigma_L \end{pmatrix}$. Since A is of full rank, the likelihood functions of $\{Y_1, \ldots, Y_n\}$ and $\{AY_1, \ldots, AY_n\}$ only differ by a Jacobian term $\log |A|^n$. Together with the independence of e_t^H and e_t^L, the log-likelihood function is

$$\ln L(Y_1, \ldots, Y_n) = \ln L(Y_1^L, \ldots, Y_n^L) + \ln L(Y_1^H, \ldots, Y_n^H) + n \ln |A| \tag{26}$$

Now, we evaluate each of the three terms on the right side of (26). First, as $\{Y_t^{(L)}\}_{t=1,\ldots,n}$ follows a low-dimensional state space model (23) and (25), its likelihood can readily be computed by

$$\ln L(Y_1^L, \ldots, Y_n^L) = -\frac{(\tilde{s}+1)rn}{2} \ln 2\pi - \frac{1}{2} \sum_{t=1}^{n} \ln |D_t| - \frac{1}{2} \sum_{t=1}^{n} v_t' D_t v_t \tag{27}$$

where the quantities v_t and D_t are computed via Kalman filtering,

$$\begin{aligned} v_t &= y_t^L - \tilde{Z} a_{t|t-1} \\ D_t &= \tilde{Z} P_{t|t-1} \tilde{Z}' + \Sigma_L \\ K_t &= H P_{t|t-1} \tilde{Z}' D_t^{-1} \\ a_{t+1|t} &= H a_{t|t-1} + K_t v_t \\ P_{t+1|t} &= H P_{t|t-1} H' - K_t D_t K_t' + RQR' \end{aligned}$$

with initial values $a_{1|0} = E(\alpha_1) = \mathbf{0}$ and $P_{1|0} = \mathrm{Var}(\alpha_1)$; see, for example, Brockwell and Davis [14] for details. Second, from (24),

$$\begin{aligned} \ln L(Y_1^H, \ldots, Y_n^H) &= -\frac{(\tilde{p}-\tilde{s}-1)rn}{2} \ln 2\pi - \frac{1}{2} \sum_{t=1}^{n} (Y_t^H)' \Sigma_H^{-1} Y_t^H \\ &= -\frac{(\tilde{p}-\tilde{s}-1)rn}{2} \ln 2\pi - \frac{1}{2} \sum_{t=1}^{n} \tilde{e}_t' \Sigma_e^{-1} \tilde{e}_t \end{aligned} \tag{28}$$

where $\tilde{e}_t = (I_p - \Sigma_e A_L' (A_L \Sigma_e A_L')^{-1} A_L) \Psi(B) Y_t$, and the last equality follows from (22) and the fact that

$$A_H' (A_H \Sigma_e A_H')^{-1} A_H \Sigma_e + A_L' (A_L \Sigma_e A_L')^{-1} A_L \Sigma_e = I_p$$

since the two terms on the left side are orthogonal projection matrices spanning $\mathbb{R}^p$. Note that explicit formula of A_H is not required to compute (28). Finally, note from (22) (iii) that

$$|A|^2 = |\Sigma_e|^{-1} |A \Sigma_e A'| = |\Sigma_e|^{-1} |A_L \Sigma_e A_L'| |A_H \Sigma_e A_H'| = |\Sigma_e|^{-1} ||\Sigma_L|$$

which implies

$$n \ln |A| = -\frac{1}{2} \ln \frac{|\Sigma_e|}{|\Sigma_L|} \tag{29}$$

Combining (27), (28), and (29), the likelihood function can be computed efficiently. Jungbacker and Koopman [13] developed an EM algorithm to optimize the likelihood function.

Remark 1. Alternative to the exact likelihood approach, two-step procedures, which estimate the factor F_t first and then the model of the factor process (18), have been developed. Specifically, Doz *et al.* [15] obtain principle component estimates, denoted as $\{\hat{F}_t^{(1)}\}$ and $\hat{\Lambda}^{(1)}$, and fit a VAR model on $\{\hat{F}_t^{(1)}\}$ in the first step. In the second step, using $\hat{\Lambda}^{(1)}$ and the estimated parameters of the VAR model, Kalman filter is employed to update the estimate for the factor $\{F_t\}$. Bai and Li [16] use a similar second step, but with the first step replaced by the estimation method in Section 3.4.3. On the other hand, Doz *et al.* [17] employ the same first step as in Doz *et al.* [15] and propose a different second step, which estimates the model of factor process using quasi-likelihood.

3.4.2 Exact likelihood via matrix decomposition

Instead of using Kalman filter and EM algorithm, Ng *et al.* [18] employ matrix decomposition techniques to efficiently compute the log-likelihood, score function, and Fisher information, so that Newton–Raphson method can be directly employed to compute the maximum-likelihood estimator. They consider the model

$$\begin{aligned} Y_t &= \Lambda F_t + \epsilon_t \\ F_t &= \Phi F_{t-1} + z_t \end{aligned} \tag{30}$$

where $\eta_t \sim N(0, \Delta)$, $\epsilon_t \sim N(0, \Sigma_z)$, and Δ is a diagonal matrix with diagonal elements $\overline{\Delta} = (\Delta_1, \ldots, \Delta_p)'$. The parameter set is defined at $\theta = (\Lambda, \Delta, \Phi)$. Define

$$F = \begin{pmatrix} F_1 \\ F_2 \\ \vdots \\ F_n \end{pmatrix}, \quad Y = \begin{pmatrix} Y_1 \\ Y_2 \\ \vdots \\ Y_n \end{pmatrix}, \quad L = \begin{pmatrix} \Lambda & 0 & \ldots & 0 \\ 0 & \Lambda & \ldots & 0 \\ \vdots & \vdots & \ddots & \vdots \\ 0 & 0 & \ldots & \Lambda \end{pmatrix}, \quad D = \begin{pmatrix} \Delta & 0 & \ldots & 0 \\ 0 & \Delta & \ldots & 0 \\ \vdots & \vdots & \ddots & \vdots \\ 0 & 0 & \ldots & \Delta \end{pmatrix}$$

$S = YY^T$, $\Psi = \text{Var}(F)$, and $\Omega = \text{Var}(Y) = L\Psi L^T + D$. The negative log-likelihood function (ignoring the constant term) is

$$Q(\theta) = \frac{1}{2}[\text{tr}(\Omega^{-1}S) + \log|\Omega|] \tag{31}$$

The major challenge in (31) is the computation of the inverse and determinant of the $np \times np$-dimensional matrix Ω. To tackle this problem, Ng *et al.* [18] employ the matrix identities

$$\Omega^{-1} = D^{-1} - D^{-1}L(\Psi^{-1} + J)^{-1}L^T D, \text{ and } |\Omega| = |\Delta|^n |\Psi^{-1}|^{-1} \cdot |\Psi^{-1} + L^T DL|$$

to express $Q(\theta)$ as

$$Q(\theta) = \frac{1}{2}(\log|\Delta|^n \cdot |\Psi| \cdot |\Upsilon|^{-1} + Y^T D^{-1} Y - (LD^{-1}Y)^T \Upsilon (L^T D^{-1} Y)) \tag{32}$$

where $\Upsilon := (\Psi^{-1} + L^T D^{-1} L)^{-1}$. Since Ψ is the covariance matrix of the VAR process, it is a block Toeplitz matrix, and it can be shown that Ψ^{-1} is a tridiagonal block matrix. This observation substantially simplifies the computations involving Υ and thus $Q(\theta)$. Moreover, the score function and the Fisher information matrix can also be expressed in terms of Υ. This allows the whole estimation procedure to be completed in $O(np)$ steps in each iteration of Newton–Ralphson algorithm.

3.4.3 Bai and Li's Quasi-likelihood estimation

If the factor process $\{F_t\}$ and the noise $\{\epsilon_t\}$ are normally distributed, then we have from (1) that $Y_t \sim N(\mathbf{0}, \Sigma_y)$, where [2]

$$\Sigma_y := \text{Var}(Y_t) = \Lambda \text{Var}(F_t)\Lambda' + \Delta =: \Lambda\Sigma_F\Lambda' + \Delta \tag{33}$$

Thus, the marginal distribution of Y_t is

$$f(Y_t) = \frac{1}{\sqrt{(2\pi)^n|\Sigma_y|}} e^{-\frac{1}{2}Y_t'\Sigma_y^{-1}Y_t} \tag{34}$$

Based on (34), Bai and Li [2] formulate the log-likelihood function as

$$\begin{aligned} \ln L &= -\frac{1}{2n}\ln|M_y| - \frac{1}{2n}\sum_{t=1}^{n} Y_t'M_y^{-1}Y_t \\ &= -\frac{1}{2n}\ln|M_y| - \frac{1}{2n}\text{tr}(\hat{\Sigma}_y M_y^{-1}) \end{aligned} \tag{35}$$

where $\hat{\Sigma}_y = \sum_{t=1}^{T} Y_tY_t'/n$, $M_y := \Lambda M_F \Lambda' + \Delta$, and $M_F = FF'/n$. Note that (35) is a quasi-likelihood since it ignores the serial dependence of Y_t and approximates the true variance Σ_y of Y_t by M_y. The use of M_y, which involves the sample moment $M_F = FF'/n$ instead of the population moment Σ_F, can gain computational efficiency under certain identifiability conditions. For example, Bai and Li [2] consider the identifiability conditions

$$FF'/n = I_r, \quad \text{and} \quad \frac{1}{N}\Lambda'\Delta^{-1}\Lambda \text{ is a diagonal matrix,}$$

so that $M_F = I_r$, and the parameters to be estimated reduce to Λ and Δ.

Differentiating (35) with respect to Λ and Δ, the maximum-likelihood estimates $\hat{\Lambda}$ and $\hat{\Delta}$ satisfy

$$\left.\frac{\partial \ln L}{\partial \Lambda}\right|_{\Lambda=\hat{\Lambda}} = \hat{\Lambda}'\hat{M}_y^{-1}(\hat{S}_y - \hat{M}_y) = 0 \tag{36}$$

$$\left.\frac{\partial \ln L}{\partial \text{diag}(\Delta)}\right|_{\Delta=\hat{\Delta}} = \text{diag}(\hat{M}_y^{-1}) - \text{diag}(\hat{M}_y^{-1}\hat{\Sigma}_y\hat{M}_y^{-1}) = 0 \tag{37}$$

where $\hat{M}_y = \hat{\Lambda}'\hat{\Lambda} + \hat{\Delta}$, and diag$(A)$ is a vector containing the diagonal elements of a square matrix A. The high-dimensional system of equations (36) and (37) can be readily solved by EM algorithm, see Rubin and Thayer [19] and Bai and Li [2] for details. Recently, Bai and Liao [20] consider an extension to cover sparse covariance matrix for ϵ by introducing a penalty term in the quasi-likelihood function.

3.4.4 Breitung and Tenhofen's Quasi-likelihood estimation

Breitung and Tenhofen [21] consider model (1) with the noise process satisfying the autoregressive model

$$\epsilon_{it} = \sum_{j=1}^{q_i} \rho_{j,i}\epsilon_{i,t-j} + e_{it} \tag{38}$$

where $e_{it} \sim WN(0, \sigma_i^2)$ for $i = 1, \ldots, p$ and $t \in \mathbb{Z}$. Considering the distribution of ϵ_{it}, the quasi-log-likelihood function is given by

$$\ln L(\theta) = -\sum_{i=1}^{p} \frac{n - q_i}{2} \log \sigma_i^2 - \sum_{i=1}^{p} \sum_{t=p_i+1}^{n} \frac{(e_{it} - \rho_{1,i} e_{i,t-1} - \cdots - \rho_{q_i,i} e_{i,t-q_i})^2}{2\sigma_i^2} \tag{39}$$

where $e_{it} = Y_{it} - \sum_{j=1}^{p} \Lambda_{ij} F_{jt}$ and $\theta = (F, \Lambda, \rho_{1,1}, \ldots, \rho_{1,q_1}, \ldots, \rho_{p,q_p}, \sigma_1^2, \ldots, \sigma_p^2)$ are unknown parameters. The $\ln L(\theta)$ is a quasi-likelihood in the sense that (38) is only a "working" model, and misspecification is allowed. In contrast to Bai and Li [2], the quasi-likelihood here involves F as unknown parameters and thus induces higher computational burden. Therefore, it is infeasible to obtain the maximum-likelihood estimator by simultaneously solving the score functions

$$\frac{\partial \ln L}{\partial \Lambda} = \frac{1}{\sigma_i^2} \sum_{t=q_i+1}^{n} e_{it} \left[F_t - \sum_{j=1}^{q_i} \rho_{i,j} F_{t-j} \right] = \mathbf{0} \tag{40}$$

$$\frac{\partial \ln L}{\partial F_t} = \sum_{i=1}^{p} \frac{1}{\sigma_i^2} \left(e_{it} - \sum_{j=1}^{q_i} \rho_{i,j} e_{i,t+j} \right) \Lambda_{i.}' = \mathbf{0} \tag{41}$$

$$\frac{\partial \ln L}{\partial \rho_{k,i}} = \frac{1}{\sigma_i^2} \sum_{t=q_i+1}^{n} e_{it} (Y_{i,t-k} - \Lambda_{i.} F_{t-k}) = 0 \tag{42}$$

$$\frac{\partial \ln L}{\partial \sigma_i^2} = \frac{1}{\sigma_i^4} \sum_{t=q_i+1}^{n} e_{it}^2 - \frac{n - q_i}{2\sigma_i^2} = 0$$

where $\Lambda_{i.}$ is the ith column of Λ, and $e_{is} = 0$ for $s > n$.

To tackle this problem, Breitung and Tenhofen [21] suggest a two-step estimation as follows. In the first step, the principle component estimators $\hat{\Lambda}^{(1)}$ and $\hat{F}^{(1)}$ are obtained based on (6) and (7). Then, in the second step, some of the score functions are employed to compute the estimated parameters. Specifically, with the estimates from the first step, define $\hat{e}_{i,t} = Y_{i,t} - \hat{\Lambda}_{i.}^{(1)} \hat{F}_t^{(1)}$. Then, using (42), for each $i = 1, \ldots, p$, one can solve for $(\hat{\rho}_{1,i}, \ldots, \hat{\rho}_{q_i,i})$ from

$$\sum_{t=q_i+1}^{n} (\hat{e}_{i,t} - \hat{\rho}_{1,i} \hat{e}_{i,t-1} - \cdots - \hat{\rho}_{q_i,i} \hat{e}_{i,t-q_i}) \hat{e}_{i,t-k} = 0, \text{ for } k = 1, \ldots, q_i \tag{43}$$

Note that solving (43) is equivalent to computing the least-squares estimator from the regression model $\hat{e}_{i,t} = \rho_{1,i} \hat{e}_{i,t-1} + \cdots + \rho_{q_i,i} \hat{e}_{i,t-q_i} + e_t$. Next, using (40), we solve for $\hat{\Lambda}_{i.}$ from

$$\sum_{t=q_i+1}^{n} \left[(Y_{it} - \hat{\Lambda}_{i.} \hat{F}_t^{(1)}) - \sum_{j=1}^{q_i} \hat{\rho}_{i,j} (Y_{it} - \hat{\Lambda}_{i.} \hat{F}_t^{(1)}) \right] \left[\hat{F}_t^{(1)} - \sum_{j=1}^{q_i} \hat{\rho}_{i,j} \hat{F}_{t-j}^{(1)} \right] = 0 \tag{44}$$

Finally, to gain computational efficiency, (41) is modified as

$$\sum_{i=1}^{p} \frac{1}{\hat{\omega}_i^2} (Y_{it} - \hat{\Lambda}_{i.} F_t) \hat{\Lambda}_{i.}' = 0 \tag{45}$$

to solve for $\hat{F}_t$, where $\hat{\omega}_i^2 := \frac{1}{T} \sum_{t=1}^{T} \hat{e}_{i,t}^2$. Intuitively, solving for $\hat{F}_t$ in (45) is equivalent to minimizing the weighted sum of squares $\sum_{i=1}^{p} (Y_{it} - \hat{\Lambda}_{i.} F_t)^2 / \hat{\omega}_i^2$ in which $\hat{\omega}_i^2$ is estimating the variance of $Y_{it} - \hat{\Lambda}_{i.}F_t$. Although Breitung and Tenhofen [21] do not consider the estimation of σ_i^2 since (38) is only a working model, the estimator can be defined by

$$\hat{\sigma}_i^2 = \frac{1}{n - q_i} \sum_{t=q_i+1}^{n} (\hat{\epsilon}_{i,t} - \hat{\rho}_{1,i}\hat{\epsilon}_{i,t} - \cdots - \hat{\rho}_{q_i,i}\hat{\epsilon}_{i,t-q_i})^2$$

Note that each of (43), (44), and (45) involves low-dimensional root solving and thus can be computed efficiently.

3.4.5 Frequency-domain (Whittle) likelihood

Fiorentini *et al.* [22] propose a frequency-domain likelihood for the estimation of dynamic factor model (3) with the factor process following a VARMA model (18). Moreover, each component of the noise process $\{\epsilon_{it}\}_{t=1,\dots}$ follows a univariate ARMA model

$$\alpha_i(B)\epsilon_{it} = \beta_i(B)e_{it}, \text{ where } e_{it} \sim N(0, \phi_i) \tag{46}$$

Denote the parameter vector as $\theta = (\boldsymbol{\phi}, \theta_f, \theta_\epsilon, \Lambda)$, where $\boldsymbol{\phi} = (\phi_1, \dots, \phi_p)$, θ_f is the parameters associated with the VARMA model (18), and θ_ϵ is the parameters associated with the ARMA models (46).

Denote $f_{\epsilon_i}(\omega)$ and $f_F(\omega)$ as the spectral density matrices of $\{\epsilon_{it}\}_{t=1,\dots}$ and $\{F_t\}_{t=1,\dots}$, respectively, evaluated at frequency ω. Assuming that the latent factors process $\{F_t\}$ are observed, then the independence of $\{\epsilon_{it}\}_{t=1,\dots}$ across $i = 1, \dots, p$ implies that the components $\mathbf{y}_1, \dots, \mathbf{y}_p$ are independent given the factor process. Thus, the "complete data" frequency-domain log-likelihood has a simple decomposition

$$\begin{aligned}\ln L_\theta(Y, F) &= \ln L_\theta(Y|F) + \ln L_\theta(F) = \sum_{i=1}^{p} \ln L_\theta(\mathbf{y}_i|F) + \ln L(F)\\ &= \sum_{i=1}^{p} \mathrm{WL}_\theta(\{\mathbf{y}_{it} - \Lambda_0 F_t - \cdots \Lambda_s F_{t-s}\}_{t=1,\dots,n}; f_{\epsilon_i}) + \mathrm{WL}_\theta(\{F_t\}_{t=1,\dots,n}; f_F)\end{aligned} \tag{47}$$

where

$$WL_\theta(\{z_t\}_{t=1,\dots,n}; f_z) = -\frac{n}{2}\ln(2\pi) - \frac{1}{2}\sum_{j=0}^{n-1} \ln|f_z(\omega_j)| - \frac{2\pi}{2}\sum_{j=0}^{n-1} \mathrm{tr}(f_z^{-1}(\omega_j)I_z(\omega_j))$$

is the Whittle likelihood (see, e.g., Brockwell and Davis [14]) of a time series $z_1, \dots, z_n$ with spectral density f_z and periodogram I_z, and $\omega_j = \frac{2\pi j}{n}, j = 1, \dots, n-1$ are the Fourier frequencies.

In practice, $\{F_t\}$ are not observed. Nevertheless, parameter estimates can be obtained using the generalized EM principle, which asserts that for a given $\theta^{(n)}$, any increase in $E(\ln L_\theta(Y, F)|Y, \theta^{(n)})$ must represent an increase in $\ln L_\theta(Y, F)$. In other words, the sequence $\{\theta^{(n)}\}_{n=1,\dots,}$ where

$$\theta^{(n+1)} = \arg\max_\theta E(\ln L_\theta(Y, F)|Y, \theta^{(n)}) \tag{48}$$

guarantees that $\ln L_{\theta^{(n)}}(Y, F)$ increases with n. To compute $\{\theta^{(n)}\}_{n=1,\dots,}$, Fiorentini *et al.* [22] derive an expression of $E_\theta(\theta^{(n)}) := E(\ln L_\theta(Y, F)|Y, \theta^{(n)})$ (E-step) and conduct the maximization (M-step) in (48) by a zig-zag procedure, which solves $\frac{\partial E_\theta(\theta^{(n)})}{\partial \phi} = 0$, $\frac{\partial E_\theta(\theta^{(n)})}{\partial \theta_f} = 0$,

$\frac{\partial E_\theta(\theta^{(n)})}{\partial \theta_\epsilon} = 0$, and $\frac{\partial E_\theta(\theta^{(n)})}{\partial \Lambda} = 0$ successively until convergence and sets the resulting parameter vector as $\theta^{(n+1)}$.

4 Determining the Number of Factors

The estimation methods discussed in Section 3 require a prespecified number of factors, r. In this section we briefly summarize the existing methods for estimating r.

4.1 Information Criterion

As in many model selection problems, using information criterion is a popular approach to select r. Under this approach, the estimated number of factor $\hat{r}$ is the minimizer of an information criterion over a range of values of r, say $r = 0, 1, \dots, r_{\max}$ for some prespecified $r_{\max}$. Typically, this choice of r strikes a good balance between a certain lack of fit measure and a model complexity penalty in the criterion. For example, Bai and Ng [1], Alessi *et al.* [23], and Li *et al.* [24] consider information criterion of the form

$$IC(r) = \ln\left[\frac{1}{np}\sum_{t=1}^{n}(Y_t - \hat{\Lambda}\hat{F}_t)^2\right] + r \times P(n,p) \tag{49}$$

where $P(n,p)$ is a function depending on n and p. Some examples include $P(n,p) = c\frac{np}{n+p}, c\frac{np}{n+p}\ln\left(c\frac{np}{n+p}\right), \frac{\ln\min(\sqrt{n},\sqrt{p})}{\min(\sqrt{n},\sqrt{p})}$, where c is a positive constant. Choi and Jeong [25] systematically compare the empirical performance of the above ICs with some classical information criteria such as AIC, BIC, and Hannan and Quinn's criterion. Other classical information criterion such as final prediction error is also studied in Chan *et al.* [26].

4.2 Eigenvalues Difference/Ratio Estimators

As we have seen in Sections 3.1 and 3.2, estimation of factor models is highly connected to the largest eigenvalues of the sample covariance matrix. In particular, many estimators are developed based on the fact that if the number of factor is r, then the r-largest eigenvalues of the sample covariance matrix would be substantially greater than the rest in magnitude. Therefore, r corresponds to the index where a large value is observed in the ratio or difference of adjacent eigenvalues. For example, Lam and Yao [27] suggest that

$$\hat{r} = \arg\min_{1\le i\le R}\frac{\hat{\lambda}_{i+1}}{\hat{\lambda}_i} \tag{50}$$

where the upper bound R may be taken as $p/2$ or $p/3$. Independently, Ahn and Horenstein [28] considered

$$\hat{r} = \arg\max_{1\le i\le R}\frac{\hat{\lambda}_i}{\hat{\lambda}_{i+1}} \text{ and } \hat{r} = \arg\max_{1\le i\le R}\frac{\ln(1+\hat{\lambda}_i/\sum_{k=i+1}^{M}\hat{\lambda}_k)}{\ln(1+\hat{\lambda}_{i+1}/\sum_{k=i+2}^{M}\hat{\lambda}_k)} \tag{51}$$

Xia *et al.* [29] modify (50) as the *contribution ratio* (CR) estimator

$$\hat{r} = \arg\min_{1 \leq i \leq R} \frac{\hat{\lambda}_{i+1} / \sum_{k=i+1}^{M} \hat{\lambda}_k}{\hat{\lambda}_i / \sum_{k=i}^{M} \hat{\lambda}_k} \tag{52}$$

where $M = \min\{n, p\}$. Alternatively, Li *et al.* [30] propose

$$\hat{r} = \left\{ \text{The first } i \geq 1 \text{such that } \frac{\hat{\lambda}_{i+1}}{\hat{\lambda}_i} > 1 - d_n \right\} - 1 \tag{53}$$

where d_n is a threshold parameter that can be calibrated by simulating Gaussian vectors.

Besides the ratios, differences of eigenvalues can be employed to determine r. Onatski [31] proposes the estimator $\hat{r} = \max\{i \leq r_{\max}^n : \hat{\lambda}_i - \hat{\lambda}_{i+1} \geq \delta\}$, where $r_{\max}^n \to \infty$, $r_{\max}^n/n \to 0$, and δ is a constant that needs to be calibrated based on the eigenvalues.

4.3 Testing Approaches

Gao and Tsay [7] adopted methods in testing high-dimensional white noise from Chang *et al.* [32] and Tsay [33] for estimating the number of factors. The idea is as follows. Recall in Section 3.2 that the estimated factor is $\hat{F} = \hat{\Lambda}'Y'$, where $\hat{\Lambda} = (V_1\ V_2\ \cdots\ V_r)$ defined in (10) are the eigenvectors corresponding to the rth largest eigenvalues of the matrix $\hat{M}$. In other words, denoting $\hat{G} = (V_1\ V_2\ \cdots\ V_p)$ and $\hat{\mathbf{u}}_t = (\hat{u}_{1t}, \dots, \hat{u}_{pt})' := \hat{G}'Y_t$, the first r component of $\hat{\mathbf{u}}_t$ is $\hat{F}_t$, and the remaining components $\hat{\mathbf{w}}_{r,t} = (\hat{u}_{r+1,t}, \dots, \hat{u}_{pt})$ should behave like a high-dimensional white noise if r is greater than the true number of factor. Therefore, one can test the null hypothesis that $\{\hat{\mathbf{w}}_{i,t}\}_{t=1,\dots,n}$ is a white noise sequentially for $i = 1, 2, \dots$, and set $\hat{r} = i$ if the ith test is the first one that does not reject the null hypothesis. Note that when $p > n$, the eigenvectors $V_{n+1}, \dots, V_p$ are degenerate. In this case, Gao and Tsay [7] suggest using $\hat{G} = (V_1\ V_2\ \cdots\ V_{p_*})$ with $p_* = \epsilon n$ for a small number $\epsilon \in (0, 1)$.

A testing procedure based on eigenvalues is developed by Kapetanios [34]. The test statistic is given by $\hat{T}(r) = \hat{\tau}_{n,p}^r(\hat{\lambda}_{r+1} - \hat{\lambda}_{r_{\max}+1})$, where $\hat{\tau}_{n,p}^r$ is a normalizing constant determined by subsampling, and $r_{\max}$ is a prespecified positive integer. The critical value is also estimated by sumsampling. The test is applied for $i = 1, 2, \dots$ sequentially, and the estimator $\hat{r}$ is defined as the first r such that $\hat{T}(r)$ does not exceed the critical value.

4.4 Estimation of Dynamic Factors

Since dynamic factor models (3) contain additional dynamic structure compared to (1), more delicate methods are required to estimate the dimension of the factor process, r. In Bai and Ng [35], Amengual and Watson [36], and Breitung and Pigorsch [37], a static factor model is first fitted to the data to obtain a factor process. Then, a VAR model is fitted to the factor process, and the number of dynamic factors is estimated based on some information criteria involving the estimated factor and the fitted VAR model. On the other hand, Hallin and Liska [38] and Onatski [39] directly estimate the number of dynamic factors using the eigenvalues of the periodogram, in the form of information criteria and a testing procedure, respectively.

Acknowledgment

Supported in part by HKSAR-RGC Grants CUHK 14308218, 14305517, 14302719.

References

1 Bai, J. and Ng, S. (2002) Determining the number of factors in approximate factor models. *Econometrica*, 70 (1), 191–221.

2 Bai, J. and Li, K. (2012) Statistical analysis of factor models of high dimension. *Ann. Stat.*, 40, 436–465.

3 Choi, I. (2012) Efficient estimation of factor models. *Econometric Theory*, 28, 274–308.

4 Lam, C., Yao, Q. and Bathia, N. (2011) Estimation of latent factors for high-dimensional time series. *Biometrika*, 98, 901–918.

5 Gao, Z. and Tsay, R.S. (2019a) A structural-factor approach for modeling high-dimensional time series and space-time data. *J. Time Ser. Anal.*, 40, 343–362.

6 Pan, J. and Yao, Q. (2008) Modelling multiple time series via common factors. *Biometrika*, 95, 365–379.

7 Gao, Z., and Tsay, R.S. (2019b) Structural-factor modeling of high-dimensional time series: Another look at factor models with diverging eigenvalues. *arXiv:1808.07932*, 1–38.

8 Forni, M., Giannone, D., Lippi, M., and Reichlin, L. (2000) The generalized dynamic factor model: identification and estimation. *Rev. Econ. Stat.*, 82 (4), 540–554.

9 Forni, M., Giannone, D., Lippi, M., and Reichlin, L. (2005) The generalized dynamic factor model: one-sided estimation and forecasting. *J. Am. Stat. Assoc.*, 100, 830–840.

10 Brillinger, D.R. (1981) *Time Series: Data Analysis and Theory*, Rinehart and Winston, Inc., Holt.

11 Anderson, T.W. (1984) *An Introduction to Multivariate Statistical Analysis*, Wiley, New York.

12 Shao, J. (2003). *Mathematical Statistics*, 2nd edn, Springer Science & Business Media, Springer-Verlag, New York.

13 Jungbacker, B. and Koopman, S.J. (2014) Likelihood-based dynamic factor analysis for measurement and forecasting. *Econ. J.*, 18 (2), 1–21.

14 Brockwell, P.J. and Davis, R.A. (1991) *Time Series: Theory and Method*, Springer, New York.

15 Doz, C., Giannone, D., and Reichlin, L. (2011) A two-step estimator for large approximate dynamic factor models based on Kalman filtering. *Journal of Econometrics*, 164, 188–205.

16 Bai, J. and Li, K. (2016) Maximum likelihood estimation and inference for approximate factor models of high dimension. *Rev. Econ. Stat.*, 98 (2), 298–309.

17 Doz, C., Giannone, D., and Reichlin, L. (2012) A quasi maximum likelihood approach for large approximate dynamic models. *Rev. Econ. Stat.*, 94, 1014–1024.

18 Ng, C.T., Yau, C.Y., and Chan, N.H. (2015) Likelihood inferences for high dimensional dynamic factor analysis with applications in finance. *J. Comput. Graph. Stat.*, 24 (3), 866–884.

19 Rubin, D.B. and Thayer, D.T. (1982) EM algorithms for ML factor analysis. *Psychometrika*, 47, 69–76.

20 Bai, J. and Liao, Y. (2016) Efficient estimation of approximate factor models via penalized maximum likelihood. *J. Econom.*, 191, 1–18.

21 Breitung, J. and Tenhofen, J. (2011) GLS estimation of dynamic factor model. *J. Am. Stat. Assoc.*, 106, 1150–1166.

22 Fiorentini, G., Galesi, A., and Sentana, E. (2018) A spectral EM algorithm for dynamic factor models. *J. Econom.*, 205, 249–279.

23 Alessi, L., Barigozzi, M., and Capasso, M. (2010) Improved penalization for determining the number of factors in approximate factor models. *Stat. Probab. Lett.*, 80, 1806–1813.

24 Li, H., Li, Q., and Shi, Y. (2017) Determining the number of factors when the number of factors can increase with sample size. *J. Econ.*, 197 (1), 76–86.

25 Choi, I. and Jeong, H. (2019) Model selection for factor analysis: some new criteria and performance comparisons. *Econom. Rev.*, 38 (6), 577–596.

26 Chan, N.H., Lu, Y., and Yau, C.Y. (2017) Factor modelling for high-dimensional time series: inference and model selection. *J. Time Ser. Anal.*, 38 (2), 285–307.

27 Lam, C. and Yao, Q. (2012) Factor modeling for high-dimensional time series: inference for the number of factors. *Ann. Stat.*, 40 (2), 694–726.

28 Ahn, L. and Horenstein, A.R. (2013) Eigenvalue ratio test for the number of factors. *Econometrica*, 81, 1203–1227.

29 Xia, Q., Liang, R., Wu, J., and Wong, H. (2018) Determining the number of factors for high-dimensional time series. *Stat. Interface.*, 11, 307–316.

30 Li, Z., Wang, Q., and Yao, Q. (2017) Identifying the number of factors from singular values of a large sample auto-covariance matrix. *Ann. Stat.*, 45 (1), 257–288.

31 Onatski, A. (2010) Determining the number of factors from empirical distribution of eigenvalues. *Rev. Econ. Stat.*, 92 (4), 1004–1016.

32 Chang, J., Yao, Q., and Zhou, W. (2017) Testing for high-dimensional white noise using maximum cross-correlations. *Biometrika*, 104 (1), 111–127.

33 Tsay, R. (2020) Testing for serial correlations in high-dimensional time series via extreme value theory. *J. Econom.*, 216(1), 106–117.

34 Kapetanios, G. (2010) A testing procedure for determining the number of factors in approximate factor models with large dataset. *J. Bus. Econ. Stat.*, 28 (3), 397–409.

35 Bai, J. and Ng, S. (2007) Determining the number of primitive shocks in factor models. *J. Bus. Econ. Stat.*, 25, 52–60.

36 Amengual, D. and Watson, M.W. (2007) Consistent estimation of the number of dynamic factors in a large *N* and *T* panel. *J. Bus. Econ. Stat.*, 25 (1), 91–96.

37 Breitung, J. and Pigorsch, U. (2011) A canonical correlation approach for selecting the number of dynamic factors. *Oxford B. Econ. Stat.*, 75, 23–36.

38 Hallin, M. and Liska, R. (2007) Determining the number of factors in the general dynamic factor model. *J. Am. Stat. Assoc.*, 102, 603–617.

39 Onatski, A. (2009) Testing hypotheses about the number of factors in large factor models. *Econometrica*, 77 (5), 1447–1479.

Part V

Quantitative Visualization

20

Visual Communication of Data: It Is Not a Programming Problem, It Is Viewer Perception

Edward Mulrow and Nola du Toit

NORC at the University of Chicago, Chicago, IL, USA

1 Introduction

1.1 Observation

It is common to hear people talk about telling a story with data visualization. Effectively communicating the story within a dataset can be complex. Many visualization packages exist to help make the process easier, but there are still many choices needed to produce visuals that are understandable. One should not be misled into thinking that visualization packages provide effective graphics without the need for critical review. Programs that produce visuals do not think. The graphic developer[1] working with the program does the thinking and should evaluate the output of the program in order to create more effective graphics.

A lot of ineffective graphics are found on the web and in publications. Sometimes they are created by novices with no visualization training. In some cases, well-intentioned graphic developers do not understand the data and instead concentrate on visuals that are aesthetically pleasing. Other times, the developers jump into the problem because they see an interesting programming problem. In many of these situations the developer gets caught up in the software and in making graphics that are neat, cool, and pleasing.

Having neat, cool, and pleasing visuals is a good objective, but thought should be given to the data and how well it is visualized with different graphical styles. Most importantly, developers should think about how well the graphic is perceived by viewers.

1.2 Available Guidance

There are many research papers and books that provide tools for helping one think through the process of evaluating visual displays. In their seminal papers on graphical perception, Cleveland and McGill [1, 2] theorized that, when creating data graphics, viewers were expected to perform a set of perceptual tasks, such as compare length, angle, direction, area, volume, and color saturation. These elementary perceptual tasks allowed viewers to decode the quantitative data encoded into graphics. Cleveland and McGill went on to rank perceptual tasks by testing the accuracy of users' ability to correctly understand

Computational Statistics in Data Science.
Edited by Walter W. Piegorsch, Richard A. Levine, Hao Helen Zhang and Thomas C. M. Lee.
 ISBN 978-1-11956107-1

the underlying information. Consequently, they produced a set of graphical formats that developers can use to determine the most appropriate graphic. Cleveland followed up on this work with his books *Visualizing Data* (1993) [3] and *The Elements of Graphing Data* (1994) [4]. These books further explain the graphical perception model and provide a host of well-designed graphics that emphasize making data the primary focus of a graphic.

Another luminary in the field of data graphics is Edward Tufte. His work provides wonderful examples of data visualizations produced through the ages as well as critiques of common errors in modern graphics. Most notably, his book titled *The Visual Display of Quantitative Information* (2001) [5] includes useful direction, such as maximizing the data-to-ink ratio and how to avoid distorting the data with "chartjunk." Tufte also reminds graphic developers to "Induce the viewer to think about the substance rather than about methodology, graphic design, the tech of graphic production, or something else."

Similarly, the work of Naomi Robbins provides practical guidance that can be implemented daily. In her book, "Creating more effective graphs" (2005) [6], Robbins includes bad graphics alongside her improved alternatives and offers readers many tips and techniques that are readily implemented. Specifically, Robbins presents an exhaustive checklist of items that developers of graphics should always include, such as checking the accuracy of scales and including a legend.

Another valuable resource is Alberto Cairo whose three books, *The Functional Art* (2013) [7], *The Truthful Art* (2016) [8], and *How Charts Lie: Getting Smarter About Visual Information* (2019) [9], have been an inspiration for modern-day data analysts. Drawing upon his experiences as a data journalist, Cairo brings a unique perspective to the field of data graphics and, specifically, infographics. Cairo gives readers tips on how to critically evaluate data visualizations as well as step-by-step guidance on how to blend computer science, statistics, and design into successful visualizations. Moreover, he explains how to produce charts and graphs that are accessible across different audiences and produce a story rich in data.

While Cleveland and McGill, Cleveland, Tufte, Robbins, and Cairo are indispensable references for graphic developers, there are many other authors worth noting for their approaches to creating effective graphics. For example, Stephanie Evergreen and Jessica Hullman both contribute greatly to the field of data graphics. Evergreen [10] provides useful, plain language guidance. Hullman *et al.* [11, 12] and Kay *et al.* [13] have begun to researching the often overlooked area of visualizing uncertainty.

1.3 Our Message

We examine two case studies exemplifying practices that lead to ineffective graphics. After reviewing each case and the resulting poor graphic, we suggest that graphic developers employ a common feedback tool, known as *StAR*, which the developer can use to judge the effectiveness of the graphic. We then return to each case study to imagine how the graphic would have turned out if StAR had been used to guide the development process. We also suggest that developers review their work products with colleagues and have colleagues use the StAR model as a way to consider effective alternatives. Again, we return to each case study to imagine how this review would turn out. Finally, we provide additional advice on iterating through the development and review process to arrive at an effective, if not pleasing, graphic.

2 Case Studies Part 1

We present two case studies as examples of the poor choices developers make when they do not give much thought to viewer perception. The styles of poor graphics we illustrate are common place; however, the stories are fictional and we have made up the characters. Data from the same source, state-level household-based estimates from the 2017 American Community Survey (ACS), are used in both examples.

2.1 Imogene: A Senior Data Analyst Who Becomes Too Interested in the Program

Imogene has worked for five years at a renowned think tank. She has been tasked with examining a household characteristic. As she prepares for a conference presentation, she decides that it would be good to have state population estimates for the eight states of interest: Alaska, California, Indiana, Missouri, New Jersey, Texas, and Wyoming.

For one slide in the presentation, she wants to compare the number of households across states. She believes that a simple bar chart of state estimates of the number of household from the 2017 ACS would be sufficient. She is able to quickly produce Figure 1. Upon revision, she realizes that the estimates for three states, Vermont, Alaska, and Wyoming, are difficult to see. While the graphic accurately depicts the relative sizes of each state's population, Imogene worries that a viewer cannot make a reasonable guess as to the population estimate for each state.

Imogene recalls seeing a bar chart on a website that used a broken axis that allowed viewers to better determine the estimate of each value. To her dismay, the software she is using does not have a procedure in place to automatically produce such a plot. But this does not

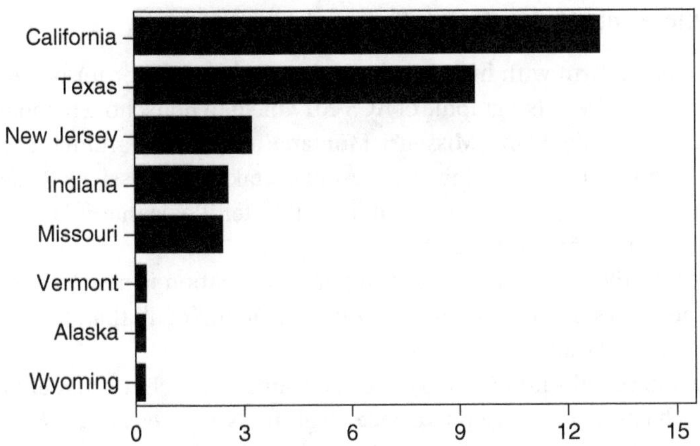

Figure 1 ACS 2017 state estimates of the number of households (millions).

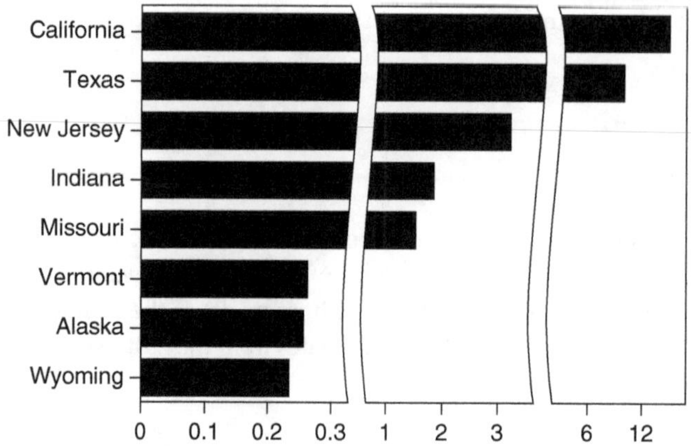

Figure 2 ACS 2017 state estimates of the number of households (millions). A broken axis is used so that it is easier to discern estimates for each state.

deter her, and in fact it excites her because there is a new programming problem to solve. She spends the next two days working on the program that produces Figure 2. Imogene is very proud of this plot. It was difficult to make, but she pulled it off. And now viewers can more easily determine the value of each state's estimate.

She shows it off to her colleagues and emphasizes that this is a programming achievement. However, one colleague is not impressed. He explains that the purpose of a bar chart is to visually judge the length of each bar from a common starting point. Using a broken axis for a bar chart creates a "visual lie." Imogene proceeds to argue that the graphic is appropriate and emphasizes the amount of time that was needed to produce such a graphic. Her colleague then points out the reason why the software does not readily have a procedure to create such a graphic: this graphic is a bad idea. Imogene is crestfallen. She is told to redo the graphic so that it is not a visual lie. She is not sure what to do. Her critics are not much help to her either. While they point out the poor graphical design, they do not offer her any alternative suggestions.

2.2 Regis: An Intern Who Wants to Get the Job Done Quickly

Regis is a new intern at research firm with high standards. The team manager for one of the projects to which he is assigned wants a graphic of ACS 2017 median household income estimates for some key states – Florida, Idaho, Missouri, Montana, and North Carolina. The estimates are being compared for an analysis the team has undertaken. She assigns Regis the task of visualizing the state estimates. She makes it a point to tell Regis that it is very important to show the uncertainty of the estimates.

Regis is a little perplexed. While he is a good programmer, visualization is new to him. He is also not sure how one shows the uncertainty of estimates. On top of all that, he has plans for the evening and does not want to work too late.

A friend, who is a fellow intern, tells him it is easy; it can be done in Excel in a matter of minutes. After a quick search based on the friend's advice, Regis finds what he needs. With a small amount of effort, he is able to produce Figure 3. His task is done. He is able to finish up some other tasks and can now enjoy the evening.

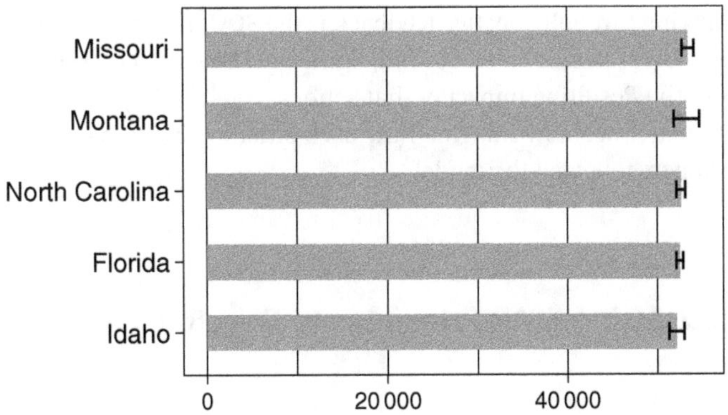

Figure 3 ACS 2017 median household income (USD) with 95% confidence intervals for five states.

When Regis returns to work the next day, he receives some bad news from the team manager. Evidently, she cannot get a good idea from the graphic that there are any differences in median household income across the states. When she did some comparison tests, she noticed that there were meaningful differences in the median household income for some pairs of states. This is not noticeable in Figure 3, so the graphic is not helpful. Regis needs to do something else.

Regis still does not understand the purpose for the graphic, and he is not being given any instructions on how to proceed. He is starting to question his desire to become a data analyst at this research firm.

3 Let StAR Be Your Guide

Reviewing the events that lead to the poorly designed graphics in our two case studies, certain things are apparent. The graphic developers used designs that, while commonly seen in the literature or on websites, resulted in displays that can be hard to interpret. Little thought was given to how a viewer would perceive the visual. In the first example, the developer chose a poor design and then got caught up in the programming puzzle. While a lot had to be learned, and a considerable amount of work went into the graphic, none of this mattered for the final results; it was a visual lie. In the second example, the developer chose something that was easy to produce. No thought was given as to the best way to display uncertainty, and the resulting bar chart contained useless error bars that are profligate and amount to chartjunk [14]. In both examples, as well as many other graphic development situations, most of the graphical perception problems could have been avoided if the developer took time to: (i) think through the situation and its tasks, (ii) carefully consider the actions needed, and (iii) assess the results of the actions. It is also the case that colleagues and supervisors could have been more effective if their feedback was more thoughtful.

The graphic developers lacked good feedback on the design and implementation. It is always good to solicit feedback from colleagues, but one can also run through some simple feedback guidelines initially to help with the process. An easy way to remember how to be effectively specific when thinking through effective feedback is the StAR feedback model.[2] First, think through the *S*ituation or *t*ask at hand. Second, figure out the *A*ction that should be taken. Finally, think about the *R*esulting impact of that action.

Using the StAR feedback model, we now imagine how each situation could have been different if the graphic developers followed the model.

4 Case Studies Part 2: Using StAR Principles to Develop Better Graphics

4.1 StAR Method: Imogene Thinks through and Investigates Changing Scales

Imogene needs to develop a graphic of ACS 2017 state estimates of the number of households for a presentation. She creates a simple bar chart (Figure 1) and realizes that the estimates for three of the states of interest are hard to discern. Thinking about the situation, she decides that for a presentation to an audience, this graphic is not the best way to present the data. She realizes that the problem is that the estimate values extend over more than one order of magnitude. She begins looking for ways to display the data that would allow her audience to discern the estimates for all the states in the display. She searches the web and finds a number of ideas. One is to use a broken axis to produce a graphic that allows viewers to better determine the estimate of each value in the bar chart. She does a little more research on this and reads that bar charts with truncated bars, as would be the case with Figure 2, are lie because the data are supposed to be encoded into the *length* of the bar [9].

The same reference [9] notes that a log scale could be used. Alternatively, one bar chart with all the state estimates on the same scale could be used with a second companion bar chart that zooms in on the low population states. Imogene finds both of these ideas to be fascinating and also easy to produce. Her software makes it very easy to change to a logarithmic scale (Figure 4), and producing two bar charts and placing them in a presentation is not time consuming either (Figure 5).

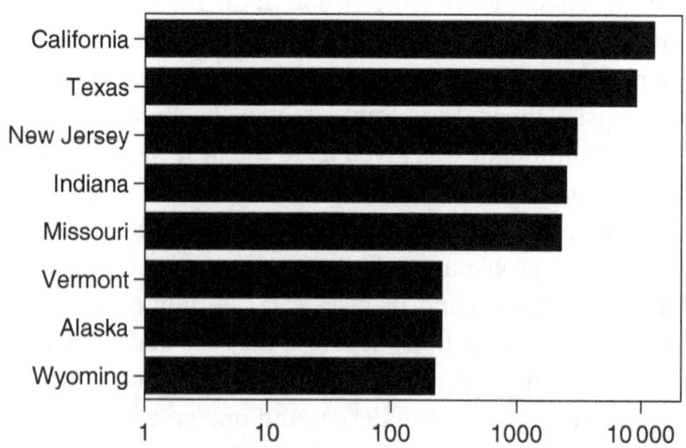

Figure 4 Log 10 US ACS 2017 state estimates of the number of households (per 1000 households).

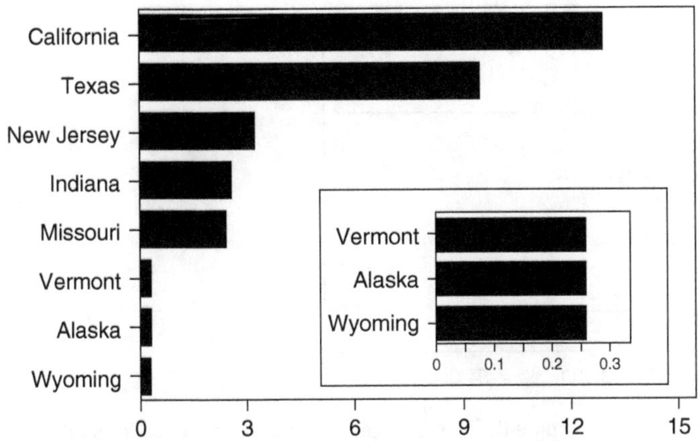

Figure 5 ACS 2017 state estimates of the number of households (millions), with insert.

As Imogene reviews her graphics, she considers which of the graphics would be best to show in her presentation. The bar chart with the log scale is simple, but she wonders if people will understand it. She knows that scientific minded viewers would be comfortable with a log scale, but she is going to present to a general audience. After thinking it through, she concludes that the two-graphic scheme, with one bar chart zoomed in on the small states, was the better choice because she will achieve her goals of preserving scale consistency, while allowing viewers to discern the values of each state estimate.

Imogen practices her presentation with work colleagues. Overall, the presentation is well received, and everyone understands her graphics. But one colleague asks some questions: "Did you *have* to use bar charts? Did you consider any other graphical form?" Imogene is confused by the questions; was a bar chart not the best way to view these data?

4.2 StAR Method: Regis Thinks through and Discovers an Interesting Way to Depict Uncertainty

Regis is an intern. His project team manager asks him to produce a graphic of ACS 2017 median household income estimates for some key states that are being compared. The manager tells Regis that it is very important to show the uncertainty of the estimates. Regis is not sure what to do and speaks with a friend, another intern, about it. The friend suggests he use a bar chart that includes confidence intervals at the end of the bars. Regis is able to produce Figure 3 quickly. Before he sends it to the team manager, he takes a good look at the graphic and notices that a viewer cannot really tell much about the uncertainty of the estimates with this graphic. He does a little research and finds that there are different reasons to consider uncertainty; one might want to make sure that viewers understand that the estimates are not exact. It could also be the case that a viewer might want the display to provide visual evidence to help make decisions. This notion intrigues Regis. He starts to think visualization is more interesting; it is more than just creating a pretty picture.

As Regis continues his research he learns that a dot plot is an effective alternative to a bar chart. He finds some R ggplot2 code to create such a plot and learns that it is easy to add confidence intervals to the plot. So, he gives it a try and produces Figure 6.

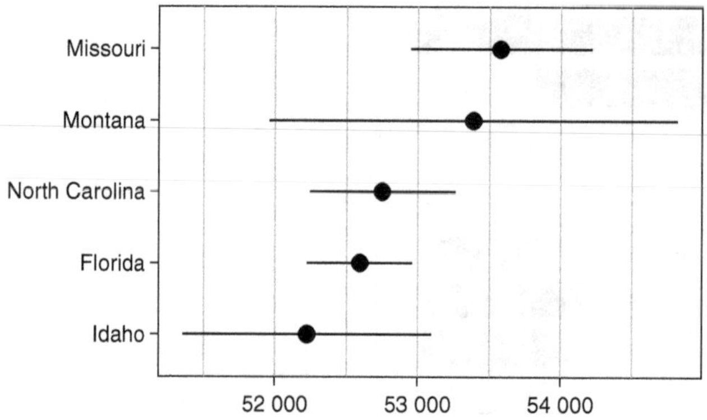

Figure 6 2017 ACS household median income (USD) estimates with 95% confidence intervals.

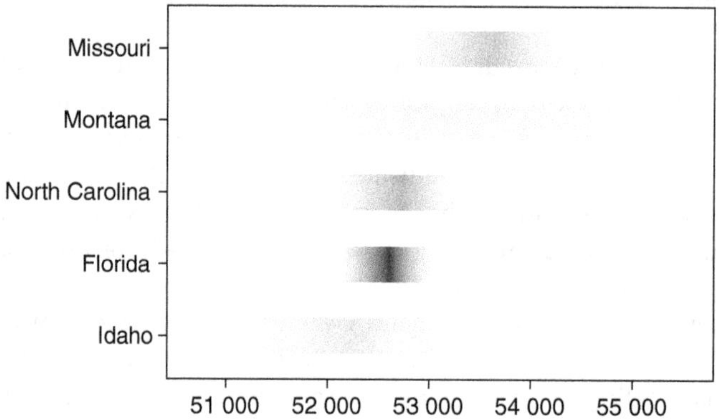

Figure 7 Sloppy plot of 2017 ACS household median income (USD) estimates.

Regis considers this new plot. He can certainly see the confidence intervals now. It helps that a dot plot does not need axes to start at 0 since a viewer compares the position of each dot. But he starts to wonder whether this style of plot really suggests uncertainty. The plots and intervals looked pretty certain to him. He is really into this now and decides to dig a little deeper.

He discovers many different visualization techniques related to showing the variation of the data, for example, box plots, violin plots, and raindrop plots. The one idea that intrigues him the most is the density strip [15], which is designed to be a display of uncertainty. Regis creates Figure 7, his own version of the density display using the ACS 2017 household median data. Regis thinks the display looks a little sloppy or dirty. He does a web search for "sloppy dirty graphics" and finds out about a wonderful book, *Picturing an Uncertain World* by Howard Wainer [16]. The author states that this style of plot is "exactly analogous to the too often ignored practice of writing down numerical values only to the number of digits that are significant."

As Regis thinks about both of the graphics, he cannot decide which one to send to the team manager. He sees that she is in her office and decides to see if she has time to help him. He shows her the two graphics and explains them to her. She likes both graphics and

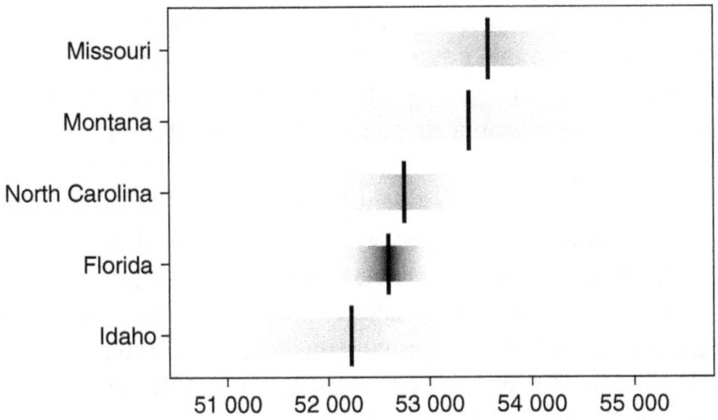

Figure 8 Sloppy plot of 2017 ACS household median income (USD) estimates with point estimate marker.

notes that she was hoping that the graphic would provide visual evidence of the results of hypothesis test calculations she has done. Figure 6 does not go along with her results, which tell her that some of the state estimates are statistically different. Figure 7 on the other hand might do the trick. To her it looks like the median household income for Missouri is different from all the other states except Montana, one of her key findings. She likes the plot, but tells Regis she would like it if the point estimates were highlighted in the plot.

Regis goes back to work to determine how he can highlight the point estimates. After thinking it through a little, he comes up with Figure 9, which is the Figure 8 sloppy plot with vertical bar point markers located at the point estimate value. He sends this version to the team manager. It has taken him most of the day to figure this out, but he is pleased with the plot. He has also learned quite a bit about graphics.

When Regis returns to work the next day, he sees the team manager and decides to check with her to make sure about the graphic. The team manager thanks Regis for all his efforts. She knows that it took a lot of effort on his part to create the graphic. But, as she has thought through the problem, she has concerns about the statistical tests she has done. She did several pairwise tests, but did not take into consideration that several tests were being done simultaneously with data. She asks Regis if he has heard about multiple testing problems; he has not. The manager would like a graphic that she is sure takes into account the multiple testing problem.

5 Ask Colleagues Their Opinion

Reviewing the Part 2 case studies, we see that both Imogene and Regis have each taken time to consider the task at hand, each thinks through the actions needed to create an effective graphic, and each evaluates the results of actions. Overall, both created nice graphics. But neither tried to think through a variety of alternatives.

It is difficult for one person to be able to think about a wide variety of alternatives. Therefore, one should seek out advice from others. Here you are looking for good advice. "Looks good to me," is not the type of feedback you want. It is best to get well-reasoned feedback.

As you approach a colleague for feedback, make sure you describe the situation so that they fully understand the task at hand. Check to make sure they have time to work with you. If so, let them know you are looking for alternatives. Go over the actions you have taken and ask if there are other approaches. Once a colleague gives you advice, ask if they are willing to look at the results with you.

When choosing a colleague for advice, see if you can identify someone with special knowledge that may be helpful for your situation. You probably would not ask a hair dresser about a plumbing problem, so do not ask for feedback from a novice (at least at this stage). Seek out colleagues who have developed effective graphics themselves. You may need to consult with a few colleagues together if there is more than one specialty involved, for example, visualization and multiple testing. The bottom line here is that you should use all the resources at your disposal. This includes trusted colleagues. In the following section, we consider how each of the Part 2 case studies can be taken further when the graphic developer seeks out a trusted colleague and obtains alternative ideas via the StAR feedback process.

6 Case Studies: Part 3

6.1 Imogene Gets Advice on Using Dot Plots

As in Section 4.1, Imogene has considered her task to create a presentation visualization of the population of nine states. Taking advice she found in Cairo, 2019, she creates a bar chart using a log scale, and a display that has one bar chart using the original scale, along with a companion "zoomed-in" bar chart of the three smallest states. Both displays provide good presentation visualizations, but she thinks the two-chart display with one bar chart zoomed in on the small states is the better choice because she is giving her presentation to a general audience. She decides to get a second opinion from her colleague Art, a research scientist with a good eye for visualization.

Imogene explains her problem to Art. He asks her to fully describe her presentation so that he understands the situation. Imogene explains why she needs a display of the population counts for the nine states and goes through the visualization issues. Art is impressed by her process and likes the graphics Imogene has created. But he feels that he should suggest an alternative graphical form just in case it provides a different look that an audience might like. He asks "Have you considered other ways to plot the data?"

Imogene never thought to consider anything other than a bar chart, and she was intrigued by the thought. She asks Art to explain his thoughts about alternatives. Art suggests that Imogene see what her display looks like using a dot plot. This allows a viewer to compare category summary statistics based on the position of the plotted dot along the scale. There is no need to have the scale start as zero for this graphical form as there is with a bar chart. Art notes that this would probably not solve all her problems, but it might make it easier to discern the population count value for the small states since the scale could start at a value closer to that of the smallest state population count. Imogene gives this a try and produces Figure 9. She does not think it improves much upon the bar charts. One could now guess that the three smallest state population estimates were at least 500 000, but it was still hard to discern the population estimates for these states. It was a little better than the bar chart,

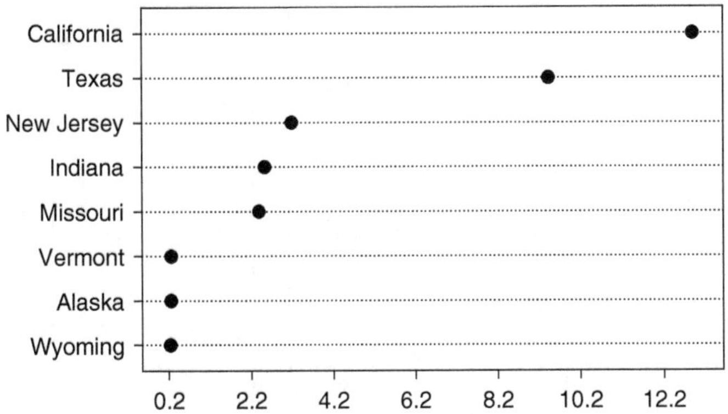

Figure 9 ACS 2017 state estimates of the number of households (millions).

but it still needed a companion zoom-in for the small states. She goes back to Art to show him the result.

Art agrees with her but has another suggestion. He has been looking through some of the visualization books he has on his shelf. He has a book on the R package *Lattice* [17]. In it the author has an example with a similar visualization problem to Imogene's. The solution is to have two dot plots, side by side, with different scales. Categories are ordered from bottom to top – smallest category value at the bottom. The set of categories with smaller values is in the left-side plot, and larger values are on the right side. The scales of the two plots are different so that it was easier to discern the values of the smaller categories. Imogene protests that the scale break is bad practice. Art notes that it was not really all that different from what she is doing with the zoomed-in bar chart. If a bar chart is used with a break in scale, it creates a visual lie because the viewer is trying to compare the length of the bars, and the full length of each bar is not visible with scale breaks. However, with dot plots you are *visualizing position along a scale*. As long as it is clear that there is a change in scale between the two plots, there should not be a problem. Imogene decides to give it a try and creates Figure 10.

She takes it to Art, and the two of them compare the graphics Imogene has created. While Art is partial to the dot plot, Imogene preferred the bar chart with a zoom-in for the small states. Art agrees that it is better for Imogene's presentation. It does not matter that Imogene does not agree with Art's preference; what matters is that Imogene thought through the situation and took appropriate action that led to a good result. She sought feedback from others and tried the suggested alternative. Her final choice of graphic was based on an examination of a few alternatives. She also learned more about visualization.

6.2 Regis Gets Advice on Visualizing in the Presence of Multiple Tests

As in Section 4.2, Regis has considered his task to create a visualization of household median incomes for five states. He created a dot plot with 95% confidence intervals around the plotted points as well as a density strip/sloppy plot. He likes the plots he has developed, and he talks them over with the team manager. She likes the plots as well, but wonders if they properly depict the situation's multiple testing setting. Regis realizes he needs

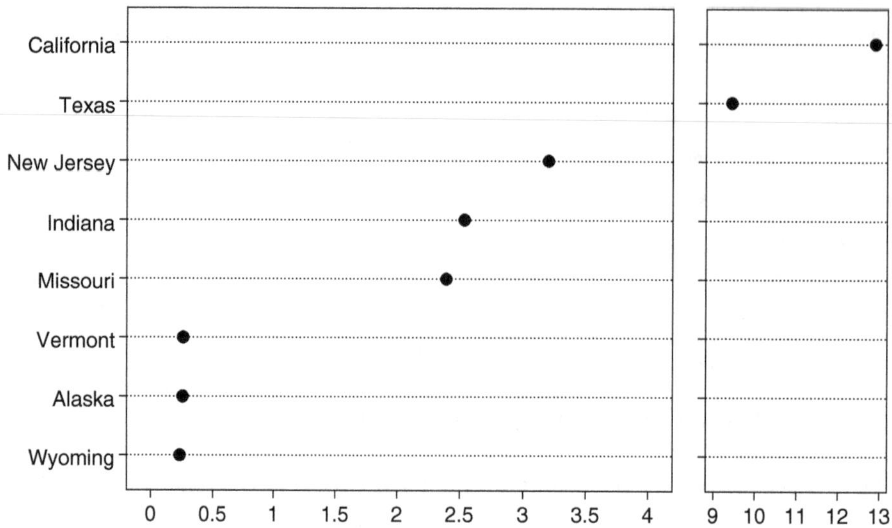

Figure 10 ACS 2017 state estimates of the number of households (millions). Two plots with different scaling are combined so that smaller values are easier to discern.

more advice from someone with experience that he does not have. He recalls meeting a senior fellow, a statistician with many years of experience, during his first week with the organization. Simone had offered to talk with him if he ever had a problem he needed help with. Regis suggests to his team manager that he talk with Simone about the problem. The team manager likes the idea and asks Regis to fill her in as he gets advice.

Simone is pleased that Regis has come to her with the problem. Regis carefully walks her through his thought process. He shows her each graphic he created. She is pleased with what he has done and notes that she has read many of Wainer's papers. In fact in one of the papers, "Depicting Error" (1996) [14], Wainer mentions that Figure 3 is a style that was outdated by 1996. It seems odd that over 20 years later that this bad form of depicting error is still used. Additionally, on the topic of multiple testing, Wainer supplies his thoughts on better ways to visualize this in the same paper! Simone goes on to explain the problem of multiple tests in general terms.

Regis asks Simone for the Wainer paper so that he can apply the technique to his situation. "I do better than that," Simone tells him, "*The American Statistician* recently published a paper on the topic by a group from the Census Bureau [18]."

Regis cannot believe his good luck. Asking Simone about the problem was the right move. He is about to head out to implement the techniques, but Simone asks him to talk a little longer. Multiple testing is complicated, and she wants to make sure he understands the paper. She walks him through the various techniques described in the paper. It turns out that there is more than one choice, and you need to pick the one that is best for your problem. As they talk about it, Simone advises him to use the Goldstein–Healy method to correct the significance level of the test for multiple testing. Using this method, the average confidence level across the 10 pairwise tests (5 states choose 2 in 10 tests she explains) will be 95%. Simone also points out that the group at the Census Bureau created an R package, RankingProject [19], to help with this type of visualization.

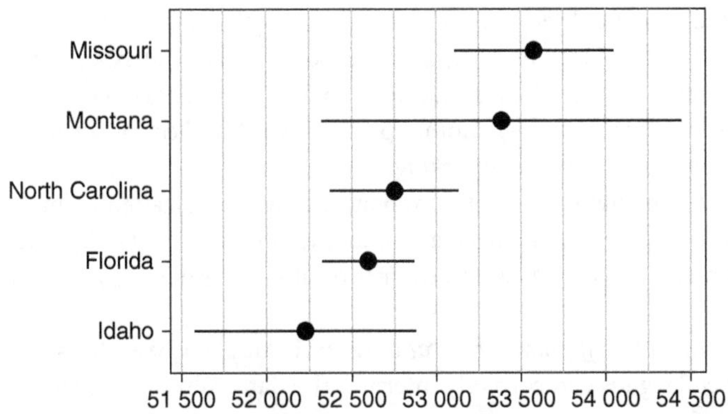

Figure 11 2017 ACS household median income (USD) estimates with confidence intervals. Confidence intervals are constructed using Goldstein and Healy method to correct for multiple testing. The average confidence level is approximately 95%; however, the confidence level of each interval is approximately 93%.

Regis reports to his team manager that the talk with Simone went well. He tells her about Simone's suggestions, and says he will have a new graphic done soon. He goes back to work on it and decides that he would like to try to implement Goldstein–Healy on his own. It is nice that there is an R package, but he feels that he will understand the problem better if he can write his own program. He produces Figure 11 and is pleased with the results. He double checks his work by also producing a version using RankingProject. He is convinced he has it right and emails it to the team manager.

The next day he stops by his team manager's office to check on things. As luck would have it, Simone sees him and joins Regis and the team manager. Regis explains that the graphic is designed so that the viewer can conclude that there is a meaningful difference in state household income between any two states that do not have overlapping intervals. The team manager is thrilled with the plot. It shows exactly what she wants. Regis is also happy. Not just about the graphic, though. He has learned a lot during this process and now views Simone as a mentor.

7 Iterate

The process we have outlined can be time consuming. It may not be possible to spend so much time discovering different alternatives. But when you have time, it can be very useful to get feedback from several colleagues.

Often, we get submerged in the process of creating a good graphic for too long and get stuck in the weeds; getting an outside perspective presents an opportunity to see the graphic with new eyes. Present draft versions to others, including coworkers, stakeholders, and even family members, and ask for feedback on the clarity of the graphic and its purpose. You do not need to use the StAR feedback process at every step. Save that for those with expertise in graphics for the particular field of application pertinent to the graphic.

For others, ask them what they understand from viewing the graphic. Ask what comparisons they are making. See if what they tell you aligns with Cleveland and McGill perceptual tasks you implemented. *A useful tip for determining the utility of a graphic is to point to a data point and ask someone to interpret it.* If they do not give the response you would expect, then you should consider an alternative format. However, bear in mind that you are not required to do anything with their feedback; it is simply another pair of eyes. *Also, become an expert on the differences between useful feedback and nitpicking.*

Furthermore, iterate, iterate, and iterate. Do not stop creating after the first attempt; make different versions of the same graphics and then solicit feedback. Find out which one serves the best purpose, and then revise and edit. Most likely, the first attempt is never going to be the best version. Keep going.

Another useful tip is to create a cutoff point for iterations. For example, if your boss says it is good enough, or your colleagues can correctly interpret the data. You have to stop at some point; figure out when it is good enough.

8 Final Thoughts

Do you always have to go through a long, drawn-out process to create an effective graphic? No, but if you are confused about a graphic's effectiveness, or the graphic's purpose, the viewer will be as well. It takes practice to develop good graphics. As one gains more experience, the process will go more quickly. But, avoid taking too many shortcuts. It will show in the visuals you produce.

Throughout our discussion, we have concentrated on basic graphics using relatively small-sized data sources. We have not addressed visualizing big data, nor any issues related to interactive graphics. As our colleague Naomi Robbins often suggests, if you learn good practices for visualizing small data, you will be able to apply the concepts to any size dataset and produce clear, effective graphics. Regardless of the data set you have, and the task at hand, the StAR process is scalable. Think through your situation, take appropriate actions, and evaluate the results of you actions.

Visualizing Data [3] ends stressing the notion that "Tools Matter." We wholeheartedly agree! It is not just about the graphical tools, though. It is also about the tools your use to develop and evaluate your graphic.

Notes

1 We use the phrase "graphic developer" to encompass analyst, programmer, designer, or whomever is creating the graphic.

2 While there are many references for StAR, they tend to be on the websites of consulting firms. We do not wish to endorse one over another, so we suggest performing a web search of the term "star feedback."

References

1 Cleveland, W. and McGill, R. (1984) Graphical perception: theory, experimentation, and application to the development of graphical methods. *J. Am. Stat. Assoc.*, 79, 807–22.

2 Cleveland, W. and McGill, R. (1987) Graphical perception: the visual decoding of quantitative information on graphical displays of data. *J. R. Stat. Soc. Ser. A*, 150, 192–229.

3 Cleveland, W. (1993) *Visualizing Data*, Hobart Press, Summit, NJ.

4 Cleveland, W. (1994) *The Elements of Graphing Data*, Revised edn, Hobart Press, Summit, NJ.

5 Tufte, E. (2001) *The Visual Display of Information*, 2nd edn, Graphics Press LLC, Cheshire, CT.

6 Robbins, N. (2013) *Creating More Effective Graphs*, Chart House, Wayne, NJ.

7 Cairo, A. (2013) *The Functional Art: An Introduction to Information Graphics and Visualization*, New Riders. www.newriders.com.

8 Cairo, A. (2016) *The Truthful Art: Data, Charts and Maps for Communication*, New Riders. www.newriders.com.

9 Cairo, A. (2019) *How Charts Lie*, W. W. Norton & Company. https://wwnorton.com/.

10 Evergreen, S. (2013) *Presenting Data Effectively: Communicating Your Findings for Maximum Impact*, SAGE, Thousand Oaks, CA.

11 Hullman, J., Adar, E., and Shah, P. (2011) *The Impact of Social Information on Visual Judgments*. Proceedings of the SIGCHI Conference on Human Factors in Computing Systems, pages 1461–1470. ACM.

12 Hullman, J., Resnick, P., and Adar, E. (2015) Hypothetical outcome plots outperform error bars and violin plots for inferences about reliability of variable ordering. *PLoS One*, 10, 11. http://idl.cs.washington.edu/papers/hops.

13 Kay, M., Kola, T., Hullman, J. R., and Munson, S. A. (2016) *When (ish) Is My Bus?: User-centered Visualizations of Uncertainty in Everyday, Mobile Predictive Systems*. Proceedings of the 2016 CHI Conference on Human Factors in Computing Systems (pp. 5092–5103). ACM.

14 Wainer, H. (1996) Depicting error. *Am. Stat.*, 50, 101–11.

15 Jackson, C.H. (2008) Displaying uncertainty with shading. *Am. Stat.*, 62 (4), 340–7.

16 Wainer, H. (2009) *Picturing the Uncertain World: How to Understand, Communicate, and Control Uncertainty through Graphical Display*, Princeton University Press. ISBN: 978-0-691-15267-7.

17 Sarkar, D. (2008) *Lattice: Multivariate Data Visualization with R*, Springer, New York. ISBN: 978-0-387-75968-5.

18 Wright, T., Klein, M., and Wieczorek, J. (2019) A Primer on visualizations for comparing populations, including the issue of overlapping confidence intervals. *Am. Stat.*, 73 (2), 165–78. doi: 10.1080/00031305.2017.1392359.

19 Wieczorek J. (2017) RankingProject: The Ranking Project: Visualizations for Comparing Populations. R package version 0.1.1. https://CRAN.R-project.org/package=RankingProject.

21

Uncertainty Visualization

Lace Padilla[1], Matthew Kay[2], and Jessica Hullman[2]

[1] *University of California, Merced, CA, USA*
[2] *Northwestern University, Evanston, IL, USA*

1 Introduction

Uncertainty is inherent to most data and can enter the analysis pipeline during the measurement, modeling, and forecasting phases [1]. Effectively communicating uncertainty is necessary for establishing scientific transparency. Further, people commonly assume that there is uncertainty in data analysis, and they need to know the nature of the uncertainty to make informed decisions [2]. However, understanding even the most conventional communications of uncertainty is highly challenging for novices and experts alike [3], which is due in part to the abstract nature of probability and ineffective communication techniques. Reasoning with uncertainty is unilaterally difficult, but researchers are revealing how some types of visualizations can improve decision-making in a variety of diverse contexts, from hazard forecasting [4, 5] to healthcare communication [6], to everyday decisions about transit [7].

Scholars have distinguished different types of uncertainty, including *aleatoric* (irreducible randomness inherent in a process), *epistemic* (uncertainty from a lack of knowledge that could theoretically be reduced given more information), and *ontological* uncertainty (uncertainty about how accurately the modeling describes reality, which can only be described subjectively) [8]. The term *risk* is also used in some decision-making fields to refer to quantified forms of aleatoric and epistemic uncertainty, whereas *uncertainty* is reserved for potential error or bias that remains unquantified. In this chapter, we use the term uncertainty to refer to quantified uncertainty that can be visualized, most commonly a probability distribution.

This chapter begins with a brief overview of the common uncertainty visualization techniques and then elaborates on the cognitive theories that describe how the approaches influence judgments. The goal is to provide readers with the necessary theoretical infrastructure to critically evaluate the various visualization techniques in the context of their own audience and design constraints. Importantly, there is no one-size-fits-all uncertainty visualization approach guaranteed to improve decisions in all domains, nor even guarantees that presenting uncertainty to readers will necessarily improve judgments or trust.

Computational Statistics in Data Science.
Edited by Walter W. Piegorsch, Richard A. Levine, Hao Helen Zhang and Thomas C. M. Lee.
 ISBN 978-1-11956107-1

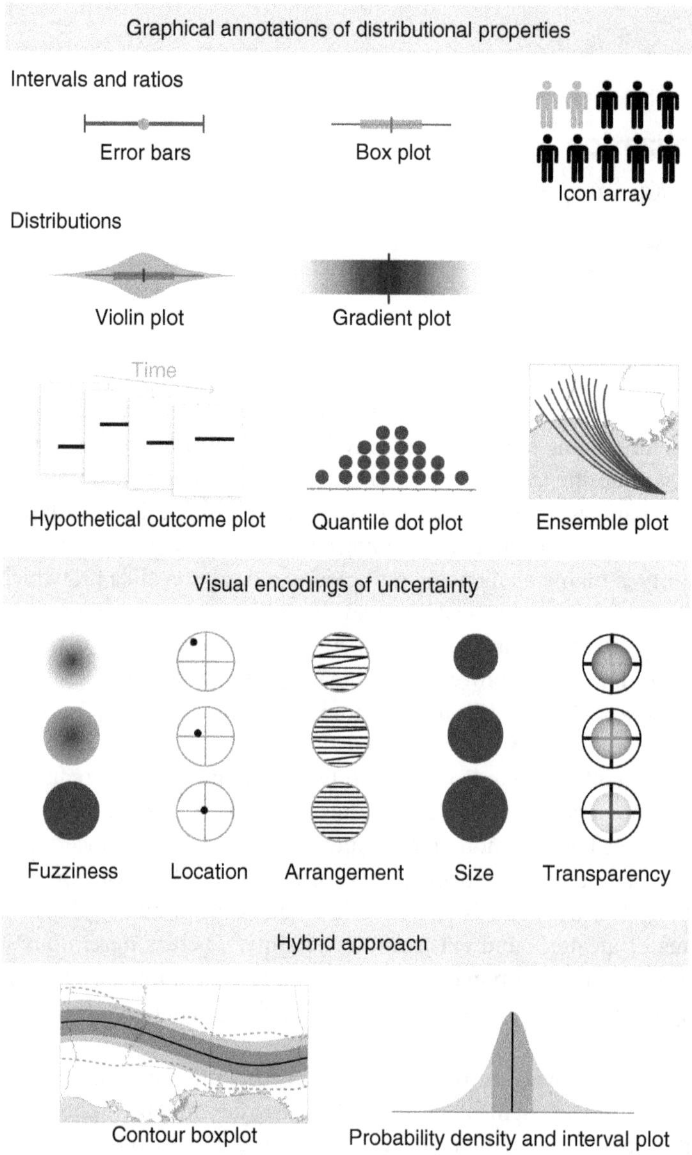
Graphical annotations of distributional properties
Intervals and ratios
Error bars
Box plot
Icon array
Distributions
Violin plot
Gradient plot
Time
Hypothetical outcome plot
Quantile dot plot
Ensemble plot
Visual encodings of uncertainty
Fuzziness
Location
Arrangement
Size
Transparency
Hybrid approach
Contour boxplot
Probability density and interval plot

Therefore, visualization designers must think carefully about each of their design choices or risk adding more confusion to an already difficult decision process.

1.1 Uncertainty Visualization Design Space

There are two broad categories of uncertainty visualization techniques, as shown in Figure 1. The first are graphical annotations that can be used to show properties of a distribution, such as the mean, confidence/credible intervals, and distributional moments. Numerous visualization techniques use the composition of marks (i.e., geometric primitives, such as dots, lines, and icons [16]) to display uncertainty directly, as in error bars depicting confidence or credible intervals. Other approaches use marks to display uncertainty implicitly as an inherent property of the visualization [17]. For example, hypothetical outcome plots (HOPs) [10] are random draws from a distribution that are presented in an animated sequence, allowing viewers to form an intuitive impression of the uncertainty as they watch. The second category of techniques focuses on mapping probability or confidence to a visual encoding channel (for reviews, see Refs 14, 18, 19). Visual encoding channels define the appearance of marks using controls such as color, position, and transparency [16]. Techniques that use encoding channels have the added benefit of adjusting a mark that is already in use, such as making a mark more transparent if the uncertainty is high. Marks and encodings that both communicate uncertainty can be combined to create hybrid approaches, such as in contour box plots [20] and probability

Figure 1 A subset of the graphical annotations used to show properties of a distribution and mappings of probability/confidence to visual variables. The visual variables that require color printing were excluded (e.g., color hue, color value, and color saturation). The examples are adapted from prior work: violin and gradient plots [9], hypothetical outcome plots [10], quantile dotplot [11], ensemble plot [12], icon array [13], fuzziness – transparency [14], contour boxplot [15], and probability density and interval plot [7]. *Source:* M. Correll and M. Gleicher, "Error bars considered harmful: Exploring alternate encodings for mean and error," IEEE transactions on visualization and computer graphics, vol. 20, no. 12, pp. 2142–2151, 2014; J. Hullman, P. Resnick, and E. Adar, "Hypothetical outcome plots outperform error bars and violin plots for inferences about reliability of variable ordering," PloS one, vol. 10, no. 11, p. e0142444, 2015; M. Kay, T. Kola, J. R. Hullman, and S. A. Munson, "When (ish) is my bus?: User-centered visualizations of uncertainty in everyday, mobile predictive systems," in Proceedings of the 2016 CHI Conference on Human Factors in Computing Systems, 2016: ACM, pp. 5092–5103; L. Liu et al., "Uncertainty Visualization by Representative Sampling from Prediction Ensembles," IEEE Transactions on Visualization and Computer Graphics, 2016 2016; B. J. Zikmund-Fisher et al., "Blocks, ovals, or people? Icon type affects risk perceptions and recall of pictographs," Medical decision making, vol. 34, no. 4, pp. 443–453, 2014; A. M. MacEachren, R. E. Roth, J. O'Brien, B. Li, D. Swingley, and M. Gahegan, "Visual Semiotics amp;amp; Uncertainty Visualization: An Empirical Study," IEEE Transactions on Visualization and Computer Graphics, vol. 18, no. 12, pp. 2496–2505, 2012/12// 2012; M. Mirzargar, R. T. Whitaker, and R. M. Kirby, "Curve boxplot: Generalization of boxplot for ensembles of curves," IEEE Transactions on Visualization and Computer Graphics, vol. 20, 2014 2014; M. Fernandes, L. Walls, F39 S. Munson, J. Hullman, and M. Kay, "Uncertainty displays using quantile dotplots or cdfs improve transit decision-making," in Proceedings of the 2018 CHI Conference on Human Factors in Computing Systems, 2018: ACM, p. 144.

density and interval plots [7] (Figure 1). In Figure 1, the contour box plot shows 95% intervals with greater transparency than 50% intervals and includes a mean line (black line) and outliers (dotted lines). Also, the probability density and interval plot in Figure 1 shows the shape of a density function with 50% intervals in darker gray and a mean line in black.

Some of the most common techniques in scientific communication are those that display intervals (see Figure 1, error bars and box plots). Despite their everyday use, visualizations of intervals have widely documented issues [3, 9, 21], such as provoking viewers to incorrectly think of distributional data as categorical [22]. For example, when summary information about the location of a natural disaster is plotted on a map with a contour line, people incorrectly interpret the area within the contour as the danger zone and locations just outside as safe [21] (see Section 2.3). Visualizations of intervals are generally hard for both experts and novices to use [3], and errors persist even with extensive instructions [23]. Rather than visualizing intervals, some research finds that using more expressive visualization techniques (e.g., violin and gradient plots in Figure 1) [9] can help people understand the uncertainty in the data more effectively. More expressive visualizations provide a fuller picture of the data by depicting more properties, such as the nature of the distribution and outliers, which can be lost with intervals.

Other work proposes that showing distributional information in a frequency format (e.g., 1 out of 10 rather than 10%) more naturally matches how people think about uncertainty and can improve performance (e.g., quantile dotplot and icon arrays in Figure 1 [11, 13]) (see Section 2.1). Visualizations that represent frequencies tend to be highly effective communication tools, particularly for individuals with low numeracy (e.g., inability to work with numbers) [24], and can help people overcome various decision-making biases [6]. Some approaches even *require* viewers to account for the uncertainty in making judgments of summary statistics (e.g., HOPs) [10], which can be useful because uncertainty information is commonly ignored or mentally substituted for simpler information (see Section 2.2).

Researchers have dedicated a significant amount of work to examining which visual encodings are most appropriate for communicating uncertainty, notably in geographic information systems and cartography [14, 18, 19] (see Visual Encodings in Figure 1 and Section 2.4). One goal of these approaches is to evoke a sensation of uncertainty, for example, using fuzziness, fogginess, or blur. Other work that examines uncertainty encodings also seeks to make looking-up values more difficult when the uncertainty is high, such as value-suppressing color pallets [25]. Given that there is no one-size-fits-all technique, in the following sections, we detail the emerging cognitive theories that describe *how* and *why* each visualization technique functions.

2 Uncertainty Visualization Theories

The empirical evaluation of uncertainty visualizations is challenging [26]. Many user experience goals (e.g., memorability [27], engagement [5], and enjoyment [28]) and performance metrics (e.g., speed, accuracy, and cognitive load [29]) can be considered when evaluating uncertainty visualizations [26]. Beyond identifying the metrics of evaluation, even the most simple tasks have countless configurations. As a result, it is hard for any single study to sufficiently test the effects of a visualization to ensure that it is appropriate to use in

Table 1 Summary of uncertainty visualization theory detailed in this chapter.

Theory	Summary	Visualization techniques
Frequency Framing [30] (Section 2.1)	Uncertainty is more intuitively understood in a frequency framing (1 out of 10) than in a probabilistic framing (10%)	Icon array [13] Quantile dotplot [11] Hypothetical outcome plots [16]
Attribute Substitution [31] Deterministic Construal Error [32] (Section 2.2)	If given the opportunity, viewers will mentally substitute uncertainty information for data that are easier to understand	Hypothetical outcome plots [16]
Visual Boundaries = Cognitive Categories [21] (Section 2.3)	Ranges that are represented by boundaries lead people to believe that data inside and outside the boundary are categorically different	Ensemble display [12] Error bar alternatives [7, 9]
Visual Semiotics [14] (Section 2.4)	Some encoding techniques naturally map onto uncertainty	Fuzziness, transparency, location, etc. [14] Value-suppressing color pallet [25]

all cases. Visualization guidelines based on a single or small set of studies are potentially incomplete. Theories can help bridge the gap between visualizations studies by identifying and synthesizing converging evidence, with the goal of helping scientists make predictions about how a visualization will be used. Understanding foundational theoretical frameworks will empower designers to think critically about the design constraints in their work and generate optimal solutions for their unique applications. The theories detailed in the following sections are only those that have mounting support from numerous evidence-based studies in various contexts. As an overview, Table 1 provides a summary of the dominant theories in uncertainty visualization, along with proposed visualization techniques.

2.1 Frequency Framing

The frequency-framing hypothesis was initially proposed by Gerd Gigerenzer [30] in response to popular theories, which argued that human reasoning systematically deviates from rational choice according to mathematical rules [33]. Gigerenzer hypothesized that our decisions seem flawed when we are provided with confusing information, such as probabilities communicated as percentiles (e.g., 10% chance). However, individuals can make rational choices if provided with information in a format they can understand easily, such as in frequencies or ratios (e.g., 1 out of 10). Gigerenzer argued that percentiles do not match the way people encounter probability in the world, and therefore lead to errors. Instead, it is more intuitive to depict probability as a frequency, as we have more exposure to these types of ratios (e.g., I hit traffic on this road 7 out of 10 times. I will take a different route tomorrow.) The frequency-framing hypothesis has substantial support from studies that find we can relatively automatically and accurately understand frequency formats, whereas probabilities are time consuming and highly error prone (for review and caveats, see Ref. 34).

One of the most effective ways to implement frequency framing of uncertainty information is with visualizations, and in this section we detail two promising frequency-framing techniques. Researchers, predominantly in healthcare communication, have extensively studied the use of icon arrays (Figure 1) to display ratios and have found strong evidence that they are useful for communicating forecasted probabilities of event outcomes. The second notable use of frequency formats in visualization is within the emerging study of quantile dotplots (Figure 1). While quantile dotplots are relatively new and have not received as much examination as icon arrays, they capitalize on the theoretical benefits of frequency framing and have demonstrated positive results in laboratory studies.

2.1.1 Icon arrays

A substantial body of research demonstrates that icon arrays are one of the most effective ways to communicate a single probabilistic value and can outperform textual descriptions of probabilities and frequencies [27, 35–42]. One of the key benefits of icon arrays is that they offload cognition by allowing a viewer's visual system to compare the denominator and the numerator in a frequency probability format. Visual comparisons of this nature are easier and faster than numerical calculations.

The difficulty in comparing ratios can produce common errors, such as individuals focusing on the numerator of each ratio and neglecting the denominator, called *denominator neglect* (for review, see Ref. 43). For example, when comparing a cancer with a mortality rate of *1286 of 10 000 people* to a cancer with a mortality rate of *24 of 100 people*, participants in a laboratory study incorrectly reported that the former cancer was riskier [44]. Researchers propose that individuals pay more attention to the relative differences in numerators (in this case, 1286 vs 24 deaths), even though they should consider the relative ratios (12.86% vs 24% mortality) [43, 44]. Several studies have found that icon arrays can reduce denominator neglect by allowing people to compare relative ratios visually [13, 42, 45, 46]. Additionally, other studies have found that people trust icon arrays more than other common visualization techniques [35], and they can reduce decision-making biases, including anecdotal evidence bias [27], side effect aversion [38, 47], and risk aversion [48].

The positive impacts of icon arrays, particularly on medical decision-making, are relatively consistent across studies that use various types of icons. However, if designers are interested in optimizing their icon selections, they should consider showing part-to-whole comparisons (i.e., both the denominator and the numerator). Designers should avoid showing only the numerator with icons and adding the denominator in text because viewers will make their judgments by considering the numerator and ignoring the denominator [46]. Icon arrays function by directing the viewer's attention to the information in the icons, so all the relevant information must be shown. Further, it is important to arrange the icons systematically in a grid that is easy to count. Various studies have found that icon arrays that are not arranged systematically are challenging to use [37], particularly for those with low numeracy [49, 50]. If two or more arrays will be compared, they should use the same denominator for each array, which will make the comparison easier.

2.1.2 Quantile dotplots

Icon arrays may be useful for communicating discrete data where only a small number of outcomes are possible (e.g., a positive or negative test result) [51]. When it comes to visualizing continuous variables, common approaches include probability density plots, which map probability to height (and by extension, area). However, users may have difficulty determining the exact density for any value because they need to visually calculate the integral under the curve [11]. Kay *et al.* [11] created the quantile dotplot as a frequency-framed alternative for displaying uncertainty for a continuous variable. As seen in Figure 2, a quantile dotplot represents a distribution where dots are sampled proportional to the quantiles of the distribution. In this case, each dot depicts a 5% probability. Using this figure as an illustration, imagine that the viewer's task is to determine if a bus will arrive in 8 min or later. With the quantile dotplot, the viewer can count the dots to determine that there is a 90% chance that the bus will in arrive 8 min or later.

Quantile dotplots have been tested in several empirical studies, which have found that they reduce the variance of probabilistic estimates compared to density plots [11] and improve recall of distributional data [52]. Other studies have found that quantile dotplots are more useful for decisions with risk compared to interval and density plots and are significantly better than textural descriptions of uncertainty [7]. Figure 2 illustrates the process of generating a quantile dotplot from a log-normal distribution.

Note that another way people can interpret both quantile dotplots and icon arrays is to make a visual area judgment. If viewers were to make an area judgment, they would not be utilizing the frequency information. Hence, icon arrays and quantile dotplots support both frequency- and non-frequency-based inferences. HOPs (Figure 3) are another example of visualizations that can use frequency framing in a way that viewers cannot fall back on non-frequency-based inferences. HOPS are described in the following section (Section 2.2) because they have the added benefit of requiring the viewer to consider uncertainty, which is an archetypal example of the theory detailed in that section.

2.2 Attribute Substitution

Reasoning with uncertainty is classically challenging, and one strategy that people unconsciously use to deal with difficult information is substitution [31]. Individuals will substitute a hard mental computation for an easier one. Researchers have studied this process extensively and termed it the *attribute substitution heuristic* [31]. A heuristic is a rule of thumb that people use to make decisions quickly, which can be beneficial if the heuristic produces a correct judgment or detrimental [53], as is the case with the *deterministic construal error* in visualizations [32].

The deterministic construal error is when individuals attempt to substitute visual uncertainty information for deterministic information. For example, Joslyn and LeClerc [32] found that when participants viewed mean temperature forecasts that included 95% confidence intervals depicted as bars with end caps, they incorrectly believed that the error bars represented high and low temperatures. The participants maintained this belief even when Joslyn and LeClerc tested a condition where the correct way to interpret the forecast was shown prominently in a key to the side of the display [32]. The authors proposed that viewers were substituting the complex uncertainty information for high- and low-temperature

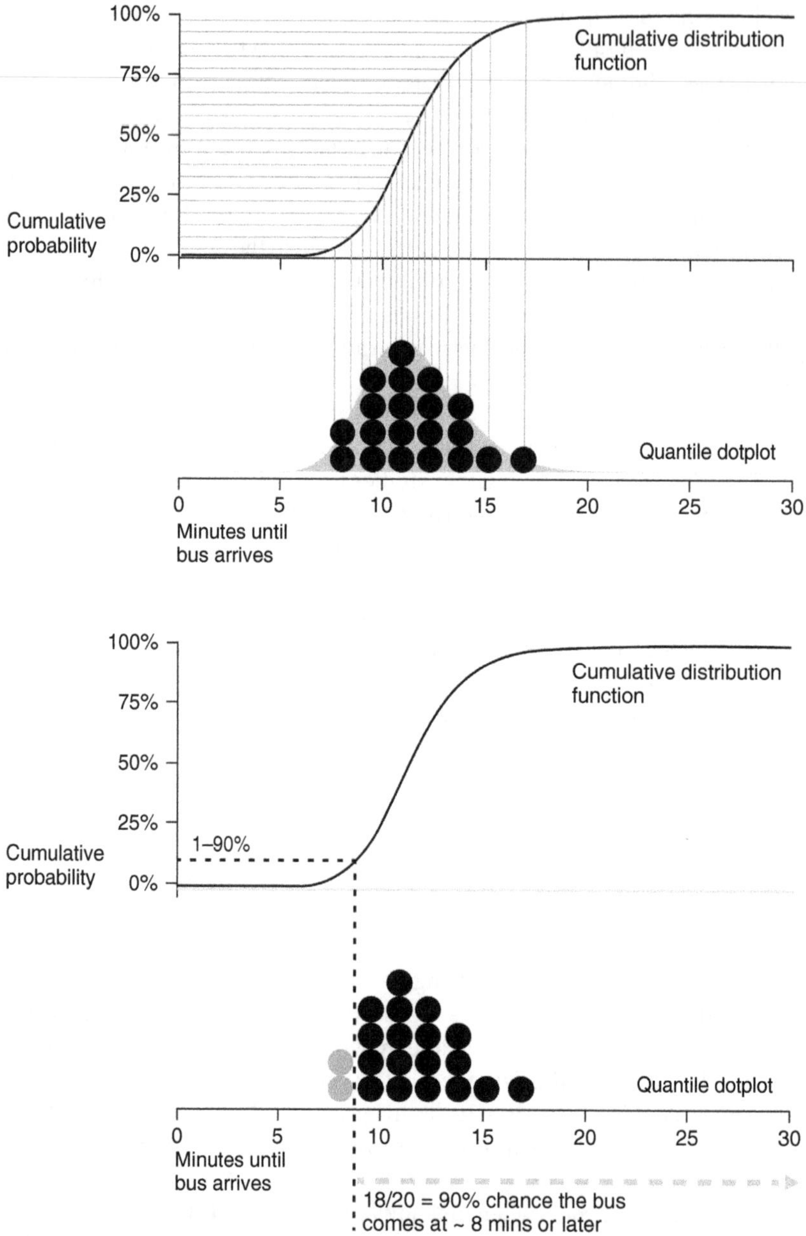

Figure 2 The process of generating a quantile dotplot from a log-normal distribution [11]. Tutorial in R can be found at https://github.com/mjskay/when-ish-is-my-bus/blob/master/quantile-dotplots.md. *Source*: Based on M. Kay, T. Kola, J. R. Hullman, and S. A. Munson, "When (ish) is my bus?: User-centered visualizations of uncertainty in everyday, mobile predictive systems," in Proceedings of the 2016 CHI Conference on Human Factors in Computing Systems, 2016: ACM, pp. 5092–5103.

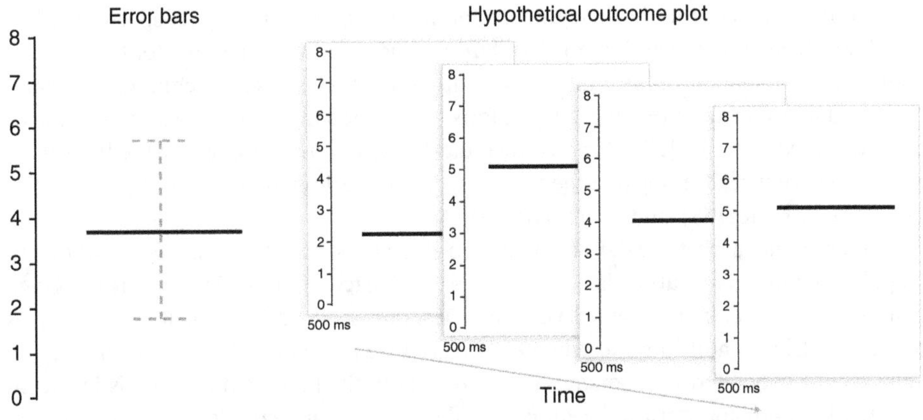

Figure 3 Illustration of HOPs compared to error bars from the same distribution [10]. *Source:* Based on J. Hullman, P. Resnick, and E. Adar, "Hypothetical outcome plots outperform error bars and violin plots for inferences about reliability of variable ordering," PloS one, vol. 10, no. 11, p. e0142444, 2015.

forecasts that were easier to understand [22, 32]. Other research with static visualizations has reported similar findings in which, if given the opportunity, viewers interpret uncertainty information incorrectly as deterministic information [3, 54–56].

Note that such a deterministic construal error has been found with visualizations and not in textual descriptions of the same information [32], meaning that the deterministic construal error might be a *visual-spatial bias*, which is a bias that is produced by the visual system [21]. In a paper that illustrates a cognitive model of decision-making with visualizations, Padilla *et al.* [21] proposed that visual-spatial biases are a unique form of bias that occurs early in the decision-making process (during visual encoding), making these type of biases particularly challenging to overcome as they influence all the downstream processes. Emerging research supports this hypothesis by demonstrating that even with extensive instructions, viewers' judgments are still influenced by deterministic construal errors, even though they are able to report the correct strategy at the end of the study [23, 55].

2.2.1 Hypothetical outcome plots

When viewers may be likely to ignore uncertainty in favor of simpler heuristics, any static visualization that encodes summary statistics runs the risk of allowing them to discount uncertainty in their judgments. One promising approach that can help to reduce the discounting of uncertainty is HOPs [10]. HOPs use random draws from a distribution and animate the draws over time. Figure 3 presents an example set of frames in which each frame is one random pull from the distribution on the left. The frames are shown in a random sequence for a short time (i.e., <500 ms), which creates an animation that can give viewers an intuitive sense of the uncertainty in the true mean.

A crucial theoretical contribution of HOPs is that they require viewers to build up a representation of distributional information in their mind. This approach has no one outcome that viewers can fixate on. Instead, they are forced by the visualization technique to both

(i) account for uncertainty in their understanding of the data and (ii) recognize that less probable outcomes do fall within the distribution. The second point is vital for hazard forecasting, where members of the public may be upset when a less likely event occurs because they failed to understand the full range of forecasted outcomes (e.g., Hurricane Katrina or the L'Aquila earthquake [57]). Empirical studies provide some support for the benefits of HOPs for lay viewers, finding that they can outperform static error bars [10, 58], icon arrays [59], line ensembles [58], and violin plots [10].

Another crucial aspect of HOPs compared to static uncertainty displays is that they can be applied to most distribution data and visual encoding techniques relatively easily, as long as animation is not already used. To create HOPs, one must first be able to draw samples from the distribution of interest, whether univariate (as in Figure 3) or multivariate. The draws can be generated via bootstrapping, referring to the large family of statistical techniques, which are appropriate for numerous data types. For example, bootstrapping can be used for generating hypothetical samples from an observed dataset, including parametric approaches (i.e., a model is fit to observed data and then samples are drawn from the model), and nonparametric approaches (i.e., resampling with replacement from observed data, which relaxes distributional assumptions). A large number of these samples can then be animated in sequence, with each sample appearing for only a short period of time. Research suggests that frame rates of 400–500 ms tend to perform best. HOPs are particularly useful in the case of complex visualizations where the distribution to be conveyed is a joint distribution with potential dependencies between variables. When visualizations already use the most accurate visual properties to show the data, for example, a choropleth map that uses position to show geographic location and color to show the value of a variable, conveying uncertainty may be difficult because it requires adding another visual property to an already complex visualization. As long as a visualization is not already animated, HOPs can be used without requiring the designer to choose another encoding for uncertainty and naturally display joint probabilities. This ability has inspired visualization researchers to use probabilistic animation to show uncertainty in geospatial data [60] as well as complex visualizations such as parallel coordinates plots [61].

2.3 Visual Boundaries = Cognitive Categories

Padilla *et al.* [21] proposed the data visualization theory that when visual boundaries, such as isocontours and error bars, are used for continuous data, the boundaries lead people to conceptualize the data as categorical. This theory is based on work by Barbra Tversky [62] in which she proposed that visual-spatial communications are intrinsically related to thought. To illustrate how concepts such as *containment* influence how we understand visual information, she writes, "Framing a picture is a way of saying that what is inside the picture has a different status from what is outside the picture" (p. 522 [62]). Padilla *et al.* [63] demonstrated how people perform the same task differently when presented with continuous or categorical depictions of the same data. The authors found that, in some cases, binned 2D scalar fields produce faster and more accurate judgments than continuous encodings of the same data, which may lead some individuals to conclude that they should always bin continuous data. However, the theory proposed by Padilla *et al.* [21] (visual boundaries = cognitive categories) emphasizes that visual boundaries force people

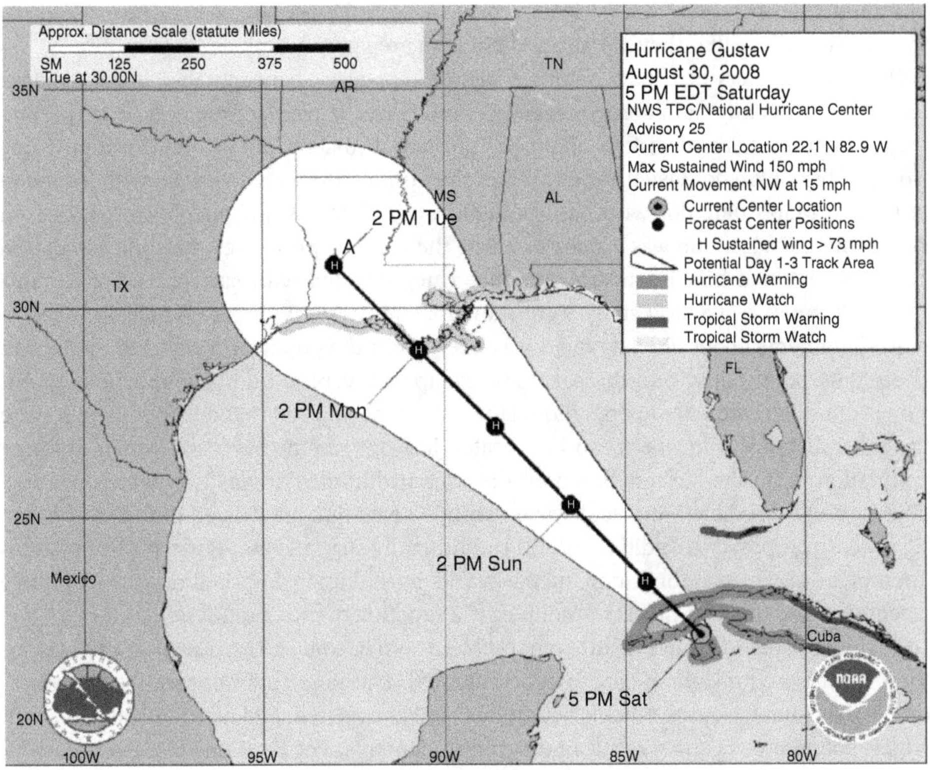

Figure 4 Example Cone of Uncertainty produced by the National Hurricane Center [5]. *Source*: L. Padilla, I. T. Ruginski, and S. H. Creem-Regehr, "Effects of ensemble and summary displays on interpretations of geospatial uncertainty data," Cognitive research: principles and implications, vol. 2, no. 1, p. 40, 2017. Licensed under CCBY 4.0.

to think *differently* about the data than a continuous encoding, which can be good or bad, depending on the nature of their decision.

The issue for uncertainty visualization is that most uncertainty data types are continuous. When a designer processes uncertainty data into a categorical format (e.g., mean values, ranges, or intervals), it fundamentally changes the way that a user forms an understanding of the data. For example, in hurricane forecasting, the most common way to represent the uncertainty in the storm's path is with the Cone of Uncertainty (see Figure 4). The Cone of Uncertainty is the current method used by the National Hurricane Center, and the border of the cone represents a 66% confidence interval around the mean predicted path. In numerous studies, researchers have found that viewers believe that areas inside the cone are categorically different from areas outside the cone [5, 23, 56]. When visualized with boundaries, viewers cannot ascertain that there is a distribution of uncertainty in the storm's path. The cognitive category created by the border of the cone makes viewers believe that areas inside the cone are in the danger zone, and areas outside are relatively safe [5, 56]. Participants' decisions are influenced by a subconscious categorical interpretation of the cone even when they are given instructions about how to interpret the visualization correctly and they can report the correct judgments at the end of the study [23]. The result of this inaccurate and persistent categorization may cause some people who reside just outside the cone to believe that they are safe and not take preparatory actions.

Other researchers also have found evidence that users conceptualize areas inside a boundary differently than areas outside [64, 65]. For example, McKenzie *et al.* [64] examined how users make decisions about the positional uncertainty in their location using Google Map's blue dot visualization. When the authors presented viewers with a version of Google Map's blue dot with hard boundaries, individuals' judgments were based on determining if a location was inside or outside the boundary [64]. Newman and Scholl [65] also demonstrated how boundaries produce categorization with bar charts. Participants in the Newman and Scholl study were shown mean values with bar charts and asked to make judgments about if a data point was likely from the population depicted by the bar. If the data point fell within the bar, the participants were more likely to believe that it came from the represented population. However, they believed that data points that were the same distance from the mean but located just outside the bar were from a different population [65]. The authors proposed that this *within-the-bar-bias* is due to perceptual object grouping, where our visual system groups items that are located near one another. The theory proposed in Padilla *et al.* [21] additionally suggests that our cognitive systems attempt to organize information in the world by grouping information into cognitive categories, and that this process is not purely a function of the visual system.

Cognitive categories have no inherent problem, and in some cases, designers might want their viewers to think about data categorically [63]. The concern for uncertainty visualization is that sometimes the boundaries are not well considered, and different choices about which boundaries to show result in different judgments. For the Cone of Uncertainty, in particular, there is no longer a justification for why the boundary is located at 66% (i.e., Why not 95% or 75%?). By plotting a hard boundary, viewers assume that the scientists are suggesting that the specific value of the boundary is important. Viewers understandably assume the value of a boundary is meaningful, particularly when the information about how the visualization was generated is insufficient, which is the case with hurricane news forecasts. In an analysis of the 20 most viewed television forecasts for Hurricane Irma in 2017, Padilla *et al.* [55] found that zero newscasters detailed how the Cone of Uncertainty was created or how to interpret it correctly, and the average length of the forecast was merely 1 : 52 min. Viewers have no choice but to assume that the scientists who made the forecast are indicating an important distinction with the boundary of the cone.

2.3.1 Ensemble displays

There are several alternatives to interval displays, such as the previously detailed HOPs. However, animations are not feasible in some cases. For example, in hurricane forecasting, static visualizations may be needed for printed reports or for regions that might not have access to high-speed Internet. Further, for hurricanes and other hazards, the time course of the hazard is uncertain. It is possible that when viewing HOPs of data where time information is critical, such as a natural disaster, viewers may incorrectly assume that the animation is depicting an event unfolding over time. Ensemble displays (see Figure 5) are another alternative to summary visualizations that researchers have tested extensively in the context of hurricane forecasting [5, 12, 55, 56, 66]. Ensemble displays are traditionally generated by making perturbations to a model's parameters and plotting the resulting

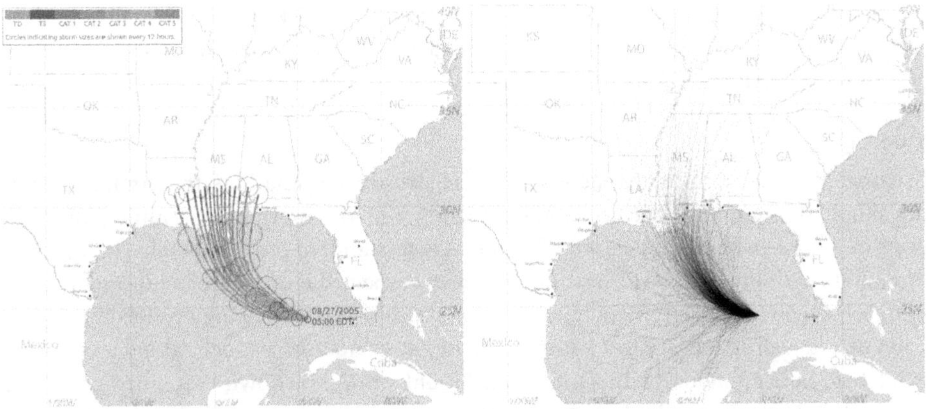

(a) (b)

Figure 5 (a) An example of an ensemble hurricane path display that utilizes a path-reconstruction procedure detailed in Liu *et al.* [66] and that also shows the intensity of the storm in the path color and the size of the storm with circle glyphs. (b) An earlier version of the ensemble display examined in Padilla *et al.* [5], Liu *et al.* [12], and Ruginski *et al.* [56] that does not use the path-reconstruction procedure. *Source*: L. Liu, L. Padilla, S. H. Creem-Regehr, and D. House, "Visualizing uncertain tropical cyclone predictions using representative samples from ensembles of forecast tracks," IEEE Transactions on Visualization Computer Graphics Forum, vol. 25, no. 1, pp. 882–891, © 2019 IEEE.

runs on a static display [12]. The result is a visualization that intrinsically shows the uncertainty in the storm's path. Early versions of the ensemble display outperformed the Cone of Uncertainty and other visualization techniques of the storm's path in laboratory studies [5, 12, 56].

Ensemble hurricane paths have received some criticism because older versions can look confusing, like a plate of spaghetti (a.k.a., spaghetti plots, see Figure 5b). Researchers have addressed concerns by developing a method for reconstructing the predicted paths from runs of the model [66] (see Figure 5a). The paths of the reconstructed ensembles have good separation and still communicate the uncertainty in the trajectory of the storm. The other benefit of using a path-reconstruction procedure is that fewer lines are needed to show the full range of outcomes, which declutters the display. In the case of hurricane forecasting, decluttering the display allows researchers to make the paths thicker and represent the intensity of the storm in color and the size of the storm with a glyph. Liu *et al.* [66] found that their study participants could effectively make decisions that incorporated the path, size, and intensity of the storm when it was visualized as an ensemble display.

Ensembles outperform all other versions of hurricane path visualizations, but they have also received significant alternative hypothesis testing by their creators, which has produced some caveats [5, 55]. In visualization research, alternative hypothesis testing is when researchers attempt to discover cases when their technique is not effective rather than focusing on optimal use cases. Researchers have found that people overreact when they see one ensemble member impacting their point of interest, such as their town [5, 55]. The same people do not overreact when an ensemble member barely misses their point of interest. Further, this effect is influenced by the number of ensemble members shown. For example, people overreact more when 1 out of 9 ensembles appears to be hitting their point of interest compared to 1 out of 33. This effect can be reduced only partially with training on how to interpret the displays correctly [55].

2.3.2 Error bars

In cases where researchers are interested in categorical interpretations, the summary statistics should be considered because of the difficulty we have interpreting sampling distributions of the mean shown as confidence intervals or standard error intervals [52, 67]. The sampling distribution is the distribution of means expected if one were to repeatedly draw samples of a given size *n* for a population. For example, when viewing results of an evaluation of a new drug relative to a control, one might wonder how much taking a new drug is likely to help a randomly drawn patient. Recent work has shown that when error bars are used to denote a standard error range of a control and treatment effect, laypeople are willing to (over)pay more for the treatment and overestimate the size of the effect compared to when the error bars show a standard deviation range [68]. Further, the relationship between statistical significance and whether or not two error bars overlap is often misunderstood: when two frequentist 95% confidence interval error bars do not overlap, it is correct to assume that the difference between the two quantities is significant at an alpha level of 0.05. However, when the two intervals do not overlap, it is incorrect to assume, as even researchers have been shown to do [3], that the difference between the two quantities is not significant.

In contrast to denoting uncertainty through boundary graphical marks meant to separately encode variance from central tendency and other distributional moments, approaches that map probability to a visual variable make uncertainty intrinsic to the presentation of other properties of a distribution like the mean. Correll and Gleicher [9] found that violin plots (mapping probability to area or width at a given *y* position) and gradient plots (mapping probability to opacity) lead to more intuitive assessments of value likelihood and "ability to surprise," which are more closely aligned with statistical definitions of uncertainty. However, the findings of other studies are less clear regarding how violin and density plots compare to error bars. Hullman *et al.* [10] observed little difference between judgments about probabilities from single distributions and multiple distributions from violin plots and error bars showing a standard deviation range. Fernandes *et al.* [7] found that a density plot leads to better quality decisions in a transportation context than in an interval, but users who used a hybrid density plot with overlaid Bayesian 50% and 95% credible intervals made better decisions after practice with the display than users of either encoding in isolation.

2.4 Visual Semiotics of Uncertainty

The final theory we detail in this chapter is the proposal that uncertainty encoding techniques that utilize visual metaphors for uncertainty, such as *graying out* using color saturation [69], *out of focus* using blur [70], *fogginess* using transparency [71, 72], adding noise using texture [73], or sketchiness [74], are more intuitive ways to communicate uncertainty (see Figure 1 for examples and Refs 14, 18, 19 for excellent reviews). The theory of visual semiotics of uncertainty, proposed by MacEachren *et al.* [14], suggests that visual encodings that prompt appropriate metaphors are easier to map onto the corresponding aspects of the information. This theory proposes that features that viewers spontaneously interpret as conveying uncertainty will be more effective than features that do not evoke uncertainty associations. For example, MacEachren *et al.* [14] asked viewers to judge the intuitiveness of numerous visual encodings of uncertainty (see Figure 1). They found that fuzziness, location, value, arrangement, size, and transparency were rated as highly intuitive. The theory of visual semiotics of uncertainty [14] has inspired numerous applications of metaphoric uncertainty visualization from cultural collections [75] to educational reporting systems [76].

In addition to the metaphoric association of uncertainty, some of the visualizations in this class of approaches map probability to visual properties and are designed to inhibit the viewer from resolving the value of a datum when the uncertainty is too high. For example, the location of a point on a map can be blurred proportional to the uncertainty in the position, such that the viewer cannot resolve an exact location [64]. Value-suppressing uncertainty palettes [25], which similarly attempt to make perception difficult in proportion to uncertainty, attempt to improve upon bivariate color maps that plot uncertainty to a separate color dimension from value, by making value judgments more difficult for uncertainty values. A viewer may be able to judge the value of a datum separately from its uncertainty using the bivariate color map by focusing only on hue, but the value-suppressing palette blends hues denoting value with gray proportional to how uncertain they are. As a result, the most uncertain values all appear as the same shade of gray. Correll, Moritz, and Heer [25] found that when applied to choropleth maps, users weigh uncertainty more heavily using the value-suppressing palettes in a decision task compared to the bivariate color map. The two key contributions of these approaches are that they elicit metaphoric associations with uncertainty and they restrict viewers from making overly precise judgments when uncertainty is high. In the following paragraphs, we discuss the implications of these contributions in turn.

The theory for why it is beneficial for uncertainty visualization to metaphorically depict uncertainty has to do with the concept of *natural mappings* [21, 28]. Natural mappings suggest that there are ways to display information that closely aligns with how people naturally think about the data. The importance of the alignment between an individual's mental representation of the data and the visual depiction of the data was initially described by Pinker [28] and expanded into a decision-making framework by Padilla *et al.* [21]. The theory suggests that when a visual representation matches how people think about the data, they will use their cognitive effort reserves to complete the task effectively. In contrast, if the discrepancy between how the information is presented and how people conceptualize it is large, they will first transform the visual variables in their minds to match their mental representation [28]. The transformation step uses some of the viewer's limited amount of mental effort, and less effort is left for the task. Uncertainty visualizations that naturally map onto how we conceptualize uncertainty may improve performance because viewers may not need to do superfluous mental transformations.

The theory of naturalness describes why metaphoric encodings of uncertainty may be helpful, but a variety of open questions concerning the exact nature of naturalness remain. Chief among them is the problem of determining how people conceptualize data. Without understanding a viewer's mental representation of data, attempts to naturally match how we think about data are guesses. Although educated guesses about how we mentally represent information are a good start, additional research is needed that more scientifically identifies our mental schemas for each type of data and context. Additionally, we have no clear way to determine the degree of relatedness between our conceptualization and the visual encoding. A range of relatedness likely impacts the extent of mental transformations required.

A more concrete contribution of metaphoric uncertainty encodings is that some techniques do not allow viewers to precisely look up values when uncertainty is high. Using a visualization technique that nudges viewers toward incorporating uncertainty in their decision-making process is a clever way of indirectly requiring them to use the uncertainty information. On the other hand, for tasks requiring viewers to look up specific values, metaphoric uncertainty can produce worse performance, simply because looking up values can be difficult. We recommend that designers think carefully about the nature of the tasks they are working with and weigh the pros and cons of using metaphoric encodings. Further, as detailed in Hullman *et al.* [26], researchers need to test uncertainty visualizations with a variety of tasks so that they do not come to incorrect conclusions about the efficacy of a visualization. For example, testing the use of *blur* with only a point-based look-up task might suggest that blur is a poor visualization choice. However, if a trend or area task were used, blur might prove to be a highly successful technique.

3 General Discussion

There are no one-size-fits-all uncertainty visualization approaches, which is why visualization designers must think carefully about each of their design choices or risk adding more confusion to an already difficult decision process. This chapter overviews many of the common uncertainty visualization techniques and the cognitive theory that describes *how* and *why* they function, to help designers think critically about their design choices. We focused on the uncertainty visualization methods and cognitive theories that have received the most support from converging measures (e.g., the practice of testing hypotheses in multiple ways), but there are many approaches not covered in this chapter that will likely prove to be exceptional visualization techniques in the future.

There is no single visualization technique we endorse, but there are some that should be critically considered before employing them. Intervals, such as error bars and the Cone of Uncertainty, can be particularly challenging for viewers. If a designer needs to show an interval, we also recommend displaying information that is more representative, such as a scatterplot, violin plot, gradient plot, ensemble plot, quantile dotplot, or HOP. Just showing an interval alone could lead people to conceptualize the data as categorical.

As alluded to in the prior paragraph, combining various uncertainty visualization approaches may be a way to overcome issues with one technique or get the best of both worlds. For example, each animated draw in a hypothetical outcome plot could leave a trace that slowly builds into a static display such as a gradient plot, or animated draws could be used to help explain the creation of a static technique such as a density plot, error bar, or quantile dotplot. Media outlets such as the New York Times have presented animated dots in a simulation to show inequalities in wealth distribution due to race [77]. More research is needed to understand if and how various uncertainty visualization techniques function together. It is possible that combining techniques is useful in some cases, but new and undocumented issues may arise when approaches are combined.

In closing, we stress the importance of empirically testing each uncertainty visualization approach. As noted in numerous papers [21–23, 55], the way that people reason with uncertainty is nonintuitive, which can be exacerbated when uncertainty information is communicated visually. Evaluating uncertainty visualizations can also be challenging, but it is necessary to ensure that people correctly interpret a display [26]. A recent survey of uncertainty visualization evaluations offers practical guidance on how to test uncertainty visualization techniques [26].

References

1 Pang, A.T., Wittenbrink, C.M., and Lodha, S.K. Approaches to uncertainty visualization, (in en). *Vis. Comput.*, 13 (8), 370–390. doi: 10.1007/s003710050111.

2 Joslyn, S. and Savelli, S. (2010) Communicating forecast uncertainty: public perception of weather forecast uncertainty. *Meteorol. Appl.*, 17 (2), 180–195.

3 Belia, S., Fidler, F., Williams, J., and Cumming, G. (2005) Researchers misunderstand confidence intervals and standard error bars. *Psychol. Methods*, 10 (4), 389.

4 Padilla, L., Creem-Regehr, S.H., and Thompson, W. (2020) The powerful influence of marks: visual and knowledge-driven processing in hurricane track displays. *JEP: Applied*, 26 (1), 1–15. doi: 10.1037/xap0000245.

5 Padilla, L., Ruginski, I.T., and Creem-Regehr, S.H. (2017) Effects of ensemble and summary displays on interpretations of geospatial uncertainty data. *Cognit. Res. Principles Implications*, 2 (1), 40.

6 Fagerlin, A., Zikmund-Fisher, B.J., and Ubel, P.A. (2011) Helping patients decide: ten steps to better risk communication. *J. Natl. Cancer Inst.*, 103 (19), 1436–1443.

7 Fernandes, M., Walls, L., Munson, S., et al. (2018) *Uncertainty Displays Using Quantile Dotplots or cdfs Improve Transit Decision-Making*. Proceedings of the 2018 CHI Conference on Human Factors in Computing Systems, ACM, p. 144.

8 Spiegelhalter, D. (2017) Risk and uncertainty communication. *Annu. Rev. Stat. Appl.*, 4, 31–60.

9 Correll, M. and Gleicher, M. (2014) Error bars considered harmful: exploring alternate encodings for mean and error. *IEEE Trans. Vis. Comput. Graph.*, 20, 2142–12, 2151.

10 Hullman, J., Resnick, P., and Adar, E. (2015) Hypothetical outcome plots outperform error bars and violin plots for inferences about reliability of variable ordering. *PLoS One*, 10 (11), e0142444.

11 Kay, M., Kola, T., Hullman, J.R., and Munson, S.A. (2016) *When (ish) Is My Bus?: User-Centered Visualizations of Uncertainty in Everyday, Mobile Predictive Systems*. Proceedings of the 2016 CHI Conference on Human Factors in Computing Systems, ACM, pp. 5092–5103.

12 Liu, L., Boone, A.P., Ruginski, I.T. et al. (2016) Uncertainty visualization by representative sampling from prediction ensembles. *IEEE Trans. Vis. Comput. Graph.*, 23 (9), 2165–2178.

13 Zikmund-Fisher, B.J., Witteman, H.O., Dickson, M. et al. (2014) Blocks, ovals, or people? Icon type affects risk perceptions and recall of pictographs. *Med. Decis. Mak.*, 34 (4), 443–453.

14 MacEachren, A.M., Roth, R.E., O'Brien, J. et al. Visual semiotics & uncertainty visualization: an empirical study. *IEEE Trans. Vis. Comput. Graph.*, 18 (12), 2496–2505. doi: 10.1109/TVCG.2012.279.

15 Mirzargar, M., Whitaker, R.T., and Kirby, R.M. (2014) Curve boxplot: generalization of boxplot for ensembles of curves. *IEEE Trans. Vis. Comput. Graph.*, 20, 2014. doi: 10.1109/TVCG.2014.2346455.

16 Munzner, T. (2014) *Visualization Analysis and Design*, CRC Press.

17 Deitrick, S. and Wentz, E.A. (2015) Developing implicit uncertainty visualization methods motivated by theories in decision science. *Ann. Assoc. Am. Geogr.*, 105 (3), 531–551.

18 Kinkeldey, C., MacEachren, A.M., and Schiewe, J. (2014) How to assess visual communication of uncertainty? A systematic review of geospatial uncertainty visualisation user studies. *Cartogr. J.*, 51 (4), 372–386.

19 Kinkeldey, C., MacEachren, A.M., Riveiro, M., and Schiewe, J. (2017) Evaluating the effect of visually represented geodata uncertainty on decision-making: systematic review, lessons learned, and recommendations. *Cartogr. Geogr. Inf. Sci.*, 44 (1), 1–21. doi: 10.1080/15230406.2015.1089792.

20 Whitaker, R.T., Mirzargar, M., and Kirby, R.M. (2013) Contour boxplots: a method for characterizing uncertainty in feature sets from simulation ensembles. *IEEE Trans. Visual. Comput. Graphics*, 19 (12), 2713–2722.

21 Padilla, L., Creem-Regehr, S., Hegarty, M., and Stefanucci, J. (2018) Decision making with visualizations: a cognitive framework across disciplines. *Cognit. Res. Principles Implications*, 3, 29.

22 Joslyn, S. and LeClerc, J. (2013) Decisions with uncertainty: the glass half full. *Curr. Dir. Psychol. Sci.*, 22 (4), 308–315.

23 Boone, A., Gunalp, P., and Hegarty, M. (2018) The influence of explaining graphical conventions on interpretation of hurricane forecast visualizations. *J. Exp. Psychol. Appl.*, 24 (3), 275.

24 Galesic, M., Garcia-Retamero, R., and Gigerenzer, G. (2009) Using icon arrays to communicate medical risks: overcoming low numeracy. *Health Psychol.*, 28 (2), 210.

25 Correll, M., Moritz, D., and Heer, J. (2018) *Value-Suppressing Uncertainty Palettes*. Proceedings of the 2018 CHI Conference on Human Factors in Computing Systems, pp. 1–11.

26 Hullman, J., Qiao, X., Correll, M. et al. (2018) In pursuit of error: a survey of uncertainty visualization evaluation. *IEEE Trans. Vis. Comput. Graph.*, 25 (1), 903–913.

27 Fagerlin, A., Wang, C., and Ubel, P.A. (2005) Reducing the influence of anecdotal reasoning on people's health care decisions: is a picture worth a thousand statistics? *Med. Decis. Mak.*, 25 (4), 398–405.

28 Pinker, S. (1990) A theory of graph comprehension, in *Artificial Intelligence and the Future of Testing* (ed. R. Frele), Erlbaum, Hillsdale, NJ, pp. 73–126.

29 Padilla, L.M., Castro, S.C., Quinan, P.S. et al. (2019) Toward objective evaluation of working memory in visualizations: a case study using pupillometry and a dual-task paradigm. *IEEE Trans. Vis. Comput. Graph.*, 26 (1), 332–342.

30 Gigerenzer, G. (1996) The psychology of good judgment: frequency formats and simple algorithms. *Med. Decis. Mak.*, 16 (3), 273–280.

31 Kahneman, D. and Frederick, S. (2002) Representativeness revisited: attribute substitution in intuitive judgment, in *Heuristics and Biases: The Psychology of Intuitive Judgment* (eds T. Gilovich, D. Griffin, and D. Kahneman), Cambridge University Press, Cambridge, MS.

32 Joslyn, S.L. and LeClerc, J.E. (2012) Uncertainty forecasts improve weather-related decisions and attenuate the effects of forecast error. *J. Exp. Psychol. Appl.*, 18 (1), 126.

33 Tversky, A. and Kahneman, D. (1974) Judgment under uncertainty: heuristics and biases. *Science*, 185 (4157), 1124–1131.

34 Visschers, V.H.M., Meertens, R.M., Passchier, W.W.F., and De Vries, N.N.K. (2009) Probability information in risk communication: a review of the research literature. *Risk Anal.*, 29 (2), 267–287.

35 Hawley, S.T., Zikmund-Fisher, B., Ubel, P. et al. (2008) The impact of the format of graphical presentation on health-related knowledge and treatment choices. *Patient Educ. Couns.*, 73 (3), 448–455.

36 Tait, A.R., Voepel-Lewis, T., Zikmund-Fisher, B.J., and Fagerlin, A. (2010) The effect of format on parents' understanding of the risks and benefits of clinical research: a comparison between text, tables, and graphics. *J. Health Commun.*, 15 (5), 487–501.

37 Feldman-Stewart, D., Brundage, M.D., and Zotov, V. (2007) Further insight into the perception of quantitative information: judgments of gist in treatment decisions. *Med. Decis. Mak.*, 27 (1), 34–43.

38 Waters, E.A., Weinstein, N.D., Colditz, G.A., and Emmons, K. (2006) Formats for improving risk communication in medical tradeoff decisions. *J. Health Commun.*, 11 (2), 167–182.

39 Waters, E.A., Fagerlin, A., and Zikmund-Fisher, B.J. (2016) Overcoming the many pitfalls of communicating risk, in *Handbook of Health Decision Science* (eds M.A. Diefenbach, S. Miller-Halegoua, and D.J. Bowen), Springer, New York, pp. 265–277.

40 Garcia-Retamero, R. and Galesic, M. (2009) Communicating treatment risk reduction to people with low numeracy skills: a cross-cultural comparison. *Am. J. Public Health*, 99 (12), 2196–2202.

41 Garcia-Retamero, R. and Galesic, M. (2009) Trust in healthcare, in *Encyclopedia of Medical Decision Making*, Kattan Ed., Sage, pp. 1153–1155.

42 Garcia-Retamero, R., Galesic, M., and Gigerenzer, G. (2010) Do icon arrays help reduce denominator neglect? *Med. Decis. Mak.*, 30 (6), 672–684.

43 Garcia-Retamero, R., Okan, Y., and Cokely, E.T. (2012) Using visual aids to improve communication of risks about health: a review. *Sci. World J.*, 2012.

44 Yamagishi, K. (1997) When a 12.86% mortality is more dangerous than 24.14%: implications for risk communication. *Appl. Cogn. Psychol.*, 11 (6), 495–506.

45 Okan, Y., Garcia-Retamero, R., Cokely, E.T., and Maldonado, A. (2012) Individual differences in graph literacy: overcoming denominator neglect in risk comprehension. *J. Behav. Decis. Mak.*, 25 (4), 390–401.

46 Stone, E.R., Sieck, W.R., Bull, B.E. et al. (2003) Foreground: background salience: explaining the effects of graphical displays on risk avoidance. *Organ. Behav. Hum. Decis. Process.*, 90 (1), 19–36.

47 Waters, E.A., Weinstein, N.D., Colditz, G.A., and Emmons, K.M. (2007) Reducing aversion to side effects in preventive medical treatment decisions. *J. Exp. Psychol. Appl.*, 13 (1), 11.

48 Schirillo, J.A. and Stone, E.R. (2005) The greater ability of graphical versus numerical displays to increase risk avoidance involves a common mechanism. *Risk Anal.*, 25 (3), 555–566.

49 Ancker, J.S., Weber, E.U., and Kukafka, R. (2011) Effect of arrangement of stick figures on estimates of proportion in risk graphics. *Med. Decis. Mak.*, 31 (1), 143–150.

50 Zikmund-Fisher, B.J., Witteman, H.O., Fuhrel-Forbis, A. et al. (2012) Animated graphics for comparing two risks: a cautionary tale. *J. Med. Internet Res.*, 14 (4), e106.

51 Kay, M., Morris, D., Schraefel, M., and Kientz, J.A. (2013) *There's No Such Thing as Gaining a Pound: Reconsidering the Bathroom Scale User Interface.* Proceedings of the 2013 ACM International Joint Conference on Pervasive and Ubiquitous Computing, pp. 401–410.

52 Hullman, J., Kay, M., Kim, Y.-S., and Shrestha, S. (2017) Imagining replications: graphical prediction & discrete visualizations improve recall & estimation of effect uncertainty. *IEEE Trans. Vis. Comput. Graph.*, 24 (1), 446–456.

53 Gigerenzer, G. and Gaissmaier, W. (2011) Heuristic decision making. *Annu. Rev. Psychol.*, 62, 451–482.

54 Grounds, M.A., Joslyn, S., and Otsuka, K. (2017) Probabilistic interval forecasts: an individual differences approach to understanding forecast communication. *Adv. Meteorol.*, 2017.

55 Padilla, L.M., Creem-Regehr, S.H., and Thompson, W. (2019) The powerful influence of marks: visual and knowledge-driven processing in hurricane track displays. *J. Exp. Psychol. Appl.*

56 Ruginski, I.T., Boone, A.P., Padilla, L.M. et al. (2016) Non-expert interpretations of hurricane forecast uncertainty visualizations. *Spat. Cogn. Comput.*, 16 (2), 154–172.

57 Cartlidge, E. (2012) Prison terms for L'Aquila experts shock scientists. *Science*, 338 (6106), 451–452.

58 Kale, A., Nguyen, F., Kay, M., and Hullman, J. (2018) Hypothetical outcome plots help untrained observers judge trends in ambiguous data. *IEEE Trans. Vis. Comput. Graph.*, 25 (1), 892–902.

59 Kim, Y.-S., Walls, L.A., Krafft, P., and Hullman, J. (2019) *A Bayesian Cognition Approach to Improve Data Visualization.* Proceedings of the 2019 CHI Conference on Human Factors in Computing Systems, ACM, p. 682.

60 Ehlschlaeger C. (1998) *Exploring Temporal Effects in Animations Depicting Spatial Data Uncertainty.* Association of American Geographers Annual Conference, Boston, MS, USA.

61 Feng, D., Kwock, L., Lee, Y., and Taylor, R. (2010) Matching visual saliency to confidence in plots of uncertain data. *IEEE Trans. Vis. Comput. Graph.*, 16 (6), 980–989.

62 Tversky, B. (2005) Visuospatial reasoning, in *The Cambridge Handbook of Thinking and Reasoning* (eds K. Holyoak and R. Morrison), Cambridge University Press, Cambridge, pp. 209–240.

63 Padilla, L., Quinan, P.S., Meyer, M., and Creem-Regehr, S.H. (2017) Evaluating the impact of binning 2d scalar fields. *IEEE Trans. Vis. Comput. Graph.*, 23 (1), 431–440.

64 McKenzie, G., Hegarty, M., Barrett, T., and Goodchild, M. (2016) Assessing the effectiveness of different visualizations for judgments of positional uncertainty. *Int. J. Geogr. Inf. Sci.*, 30 (2), 221–239.

65 Newman, G.E. and Scholl, B.J. (2012) Bar graphs depicting averages are perceptually misinterpreted: the within-the-bar bias. (in eng). *Psychon. Bull. Rev.*, 19 (4), 601–607. doi: 10.3758/s13423-012-0247-5.

66 Liu, L., Padilla, L., Creem-Regehr, S.H., and House, D. (2019) Visualizing uncertain tropical cyclone predictions using representative samples from ensembles of forecast tracks. *IEEE Trans. Visual. Comput. Graphics Forum*, 25 (1), 882–891.

67 Chance, B., del Mas, R., and Garfield, J. (2004) Reasoning about sampling distribitions, in *The Challenge of Developing Statistical Literacy, Reasoning and Thinking* (eds D. Ben-Zvi and J. Garfield), Kluwer Academic Publishers, Dordrecht, The Netherlands, pp. 295–323.

68 Hofman, J.M., Goldstein, D.G. and Hullman, J. (2020) *How Visualizing Inferential Uncertainty can Mislead Readers about Treatment Effects in Scientific Results*. Proceedings of the 2020 CHI Conference on Human Factors in Computing Systems (pp. 1–12).

69 Hengl T. (2003) *Visualisation of Uncertainty Using the HSI Colour Model: Computations with Colours*. 7th International Conference on GeoComputation.

70 Jiang, B., Ormeling, F. and Kainz, W. (1995) *Visualization Support for Fuzzy Spatial Analysis*. In Proc., ACSM/ASPRS Conference, Citeseer.

71 Rhodes, P.J., Laramee, R.S., Bergeron, R.D., and Sparr, T.M. (2003) Uncertainty visualization methods in isosurface rendering, in *Eurographics 2003, Short Papers*, (eds P. Armitage and T. Colton), The Eurographics Association, Sweden, pp. 1–5.

72 Maceachren, A.M., Robinson, A., Gardner, S. et al. (2005) Visualizing geospatial information uncertainty: what we know and what we need to know. *Cartogr. Geogr. Inf. Sci.*, 32, 160.

73 Howard, D. and MacEachren, A.M. (1996) Interface design for geographic visualization: tools for representing reliability. *Cartogr. Geogr. Inf. Syst.*, 23 (2), 59–77.

74 Boukhelifa, N., Bezerianos, A., Isenberg, T., and Fekete, J.-D. (2012) Evaluating sketchiness as a visual variable for the depiction of qualitative uncertainty. *IEEE Trans. Vis. Comput. Graph.*, 18 (12), 2769–2778.

75 Windhager, F., Salisu, S., and Mayr, E. (2019) Exhibiting uncertainty: visualizing data quality indicators for cultural collections. *Informatics*, 6 (3), 29.

76 Epp, C.D. and Bull, S. (2015) Uncertainty representation in visualizations of learning analytics for learners: current approaches and opportunities. *IEEE Trans. Learn. Technol.*, 8 (3), 242–260.

77 Badger, E., Miller, C.C., Pearce, A., and Quealy, K. (2018) Extensive data shows punishing reach of racism for black boys. *New York Times*, 19.

22

Big Data Visualization

Leland Wilkinson[1,2]

[1] *H_2O.ai, Mountain View, California, USA*
[2] *University of Illinois at Chicago, Chicago, IL, USA*

1 Introduction

Big data is a meaningless term. We hear endlessly, at least in the commercial realm, about massive files that must be conquered in order to take up the machine learning challenges *du jour*. And in the inflational rhetoric of the machine learning world, we periodically encounter escalating terms for file sizes (giga-, tera-, peta-, exa-, zetta-, etc.) as if they were a metric for the formidable computational problems confronting analytic software.

Big data is a meaningless term because it depends on context, and even with that restriction the term requires numerous qualifications to be meaningful. With rectangular data, for example, bigness depends on the number of rows and columns, the precision of numerical columns, the presence or absence of string columns, the length of stored strings, the number of distinct strings in each string column (cardinality), the storage format (CSV, binary, etc.), the sparsity of the data (percent of zeros or missing values), and other characteristics. Knowing the size of a file in bytes tells us nothing about these critical factors. For text data, bigness depends on how many distinct words appear in a file, the average number of characters in each word, how many languages appear in the file, how many distinct characters are there in each language, and so on. For image data, bigness depends on the resolution of the images, the number of images, the color model, pixel depth, and so on. For graph data, bigness depends on the number of nodes and edges, node and edge metadata, and so on. For streaming data, bigness depends on the rate of streaming and the complexity (volume) of the streaming packets.

More generally, data storage descriptors are seldom confined to single files. In document analysis, for example, single documents are often stored in single files. These files might be relatively small, but the number of such files can be enormous. In general, files of all types tend to be stored in massive distributed databases tailored for parallel computation. The challenges involved in analyzing these distributed databases hinge as much on the complexity of the database frameworks and the computing environments (on-premise, cloud, etc.), as on the size of the data resource.

Computational Statistics in Data Science.
Edited by Walter W. Piegorsch, Richard A. Levine, Hao Helen Zhang and Thomas C. M. Lee.
 ISBN 978-1-11956107-1

Not with standing these reservations, there indeed exist data sources for which the usual methods for analyzing or visualizing the data are not feasible. With rectangular files, for example, it is not unusual to encounter data sources with billions of rows or millions of columns. Some computer scientists imagine that these applications can be solved with enough memory, enough processors, or enough other machine resources to accommodate traditional algorithms. These advocates are wrong for a number of reasons.

At least for visualization, massive data resources present particular and daunting problems. There are several classes of these problems. First, human factors (perception and cognition) limit the number of stimuli that can be processed in a single glance or even in mediated perceptual or higher cognitive processes. Second, display resolution or bandwidth limits the size of data blobs that can form the basis of a visualization (the interface and internet handshaking instantiate a chokepoint). Third, real-time performance requirements of a visual analytics system often preclude the kind of responses available to massive data systems. Fourth, the curse of dimensionality drives distances between points in a space toward a constant as the number of dimensions approaches infinity. Finally, we run out of real estate when trying to plot big data in a single display area, even if we incorporate megapixel displays.

Potential solutions to these problems involve several different approaches. First, architecture considerations can dictate the type of designs that foster interactivity and exploration into large data sources. Second, data wrangling can produce data sources that are more amenable to exploration. And third, statistical graphics typologies can facilitate multivariate displays that encourage exploration and hypothesis testing.

The remainder of this chapter will address these three problems. We review basic architectures, data transformations, and graphics types that can help to ameliorate the difficulties encountered in analyzing rectangular big data resources. We cover issues and algorithms that lead in the right direction and that can be implemented in various software environments. Because of space requirements, we cover only data resources that are transformable to rectangular configurations. This strategy should lead to ideas for transforming other data structures into amenable forms for analysis.

This chapter will not cover specific application software for analyzing big data, in part because off-the-shelf visualization applications do not implement these methods. Furthermore, while some systems such as Plotly, Matplotlib, D3, and ggplot2 are popular among statisticians, engineers, and designers, they cannot handle big data on the scale we are considering. Nevertheless, it is possible for end users of systems such as *R* or Python to program solutions based on these methods. For examples of combining databases, statistical apps, and visualization apps to attack big data problems, see https://toddwschneider.com/. Also, see Ref. 1 for more extensive coverage.

2 Architecture for Big Data Analytics

Figure 1 illustrates a typical visualization dataflow as realized on a single platform (laptop, desktop, etc.). All the code and data reside on one device. While a dataflow like this can be implemented in a distributed environment (multiprocessor, cloud, etc.), it is not a design

Figure 1 Classic dataflow visualization architecture.

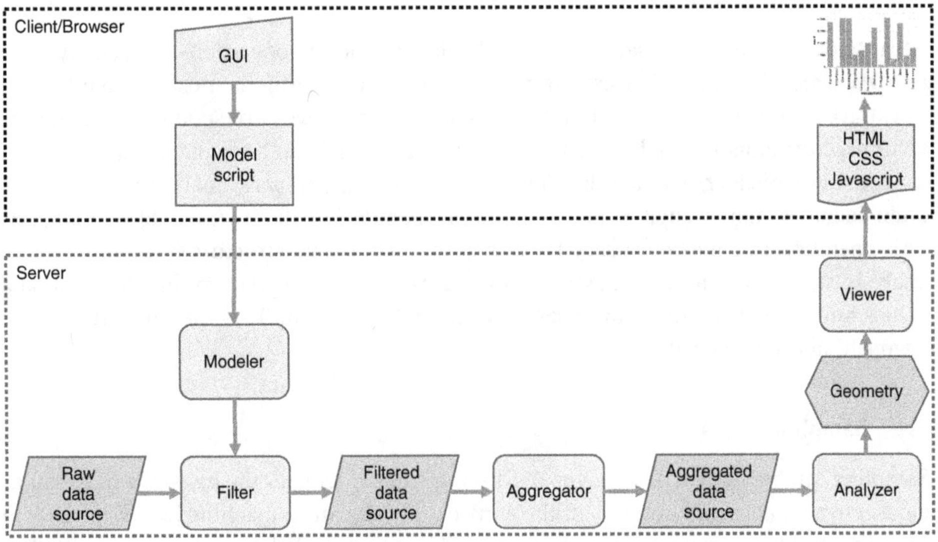

Figure 2 Client–server visualization architecture.

tailored to the capabilities of these more advanced systems. If all the data to be analyzed can fit in a local store, however, a single linear dataflow has its advantages, namely speed and simplicity.

Figure 2, by contrast, shows a design tailored for a distributed environment. Early versions of this design were called client–server systems, but the one illustrated here is designed for a Web environment. There are several considerations implicit in this design. First, the raw data can be arbitrarily large and the pieces may not be found in a single location; some massive visualization data resources are streaming, so they cannot be stored statically. Second, the end user interacts with this system through a scripting language. While some applications present a GUI view to the user, the back ends of these applications translate user interactions into scripts (usually JSON or XML) before communicating with the server. Third, the system includes a filter and an aggregator, which are components that distill large file inputs into more manageable datasets (e.g., a billion raw rows → a thousand aggregated rows). An aggregator is necessary whenever data are too massive to allow conventional statistical and graphical processing. An aggregator is necessary for another reason, however. Even if big data can be processed in a large distributed database such as Hadoop or Spark, the resulting graphic requires additional processing before transporting to the client browser. In the simplest case, a scatterplot on a billion points is too large to transmit all the points to a browser, too large to be processed in a typical browser, and too dense to view in a single display. An aggregator uses various weighted statistical methods to ameliorate this difficulty. Finally, an aggregator is designed to handle interactivity. While some systems construct bitmaps on a server and then send them to a browser for display, bitmaps are not suitable for brushing and linking.

3 Filtering

Because online data are usually stored in databases, filtering operations are usually done in SQL, which lends itself to scripting from a browser. Filtering is most frequently used to extract subsets of rows or columns of a data matrix (rows corresponding to males or columns corresponding to dates, etc.). An example using *R* can be found at https://github.com/toddwschneider/nyc-taxi-data/blob/master/analysis/prepare_analysis.sql. Filtering often includes more general data preparation, popularly called *wrangling* or *munging*. This operation becomes necessary for transforming one data structure to another before analysis (transposing a matrix, extracting n-grams from text, etc.) or for handling missing values and anomalies. In many cases, filtering reduces the bulk of the raw data so that aggregation is not needed.

3.1 Sampling

Sampling can be thought of as a type of filtering when it reduces the rows of a rectangular dataset to a manageable number. In the early days of statistical machine learning, random sampling to filter a dataset into manageable size was regarded as a cheap shortcut. The claim was that estimates of a model based on a random sample were not as "accurate" as estimates based on the whole "population" (batch). Nowadays, however, random sampling is at the heart of many machine learning algorithms, particularly in feature generation and cross-validation.

Random sampling can be especially valuable in visualization when estimates of error can help us to evaluate the suitability of a model. With massive datasets, conventional asymptotic confidence intervals are often impractical because they are deflated by huge n. Furthermore, joint distributions of points in real datasets are rarely bivariate normal; sometimes they do not even plausibly fit the exponential family or other familiar statistical distributions. In these cases, the bootstrap [2] is especially useful. Furthermore, we can often benefit by plotting sample estimates directly instead of pooling them to create confidence intervals. Figure 3 shows an example that plots bootstrapped piecewise regressions on a dataset from Gonnelli *et al.* [3]. While this example involves a relatively small batch, large datasets can be handled by plotting points after aggregation and plotting regression lines separately for each bootstrapped sample.

4 Aggregating

The function of an aggregator is to reduce a rectangular matrix $X_{n \times p}$ to a smaller rectangular matrix $X^*_{k \times d}$, where $k \ll n$ and $d \ll p$. We assume that all the elements of X are real (no strings), although we discuss ways to handle categorical values. The algorithms for performing this operation differ by dimensionality, so each will be outlined in a subsection.

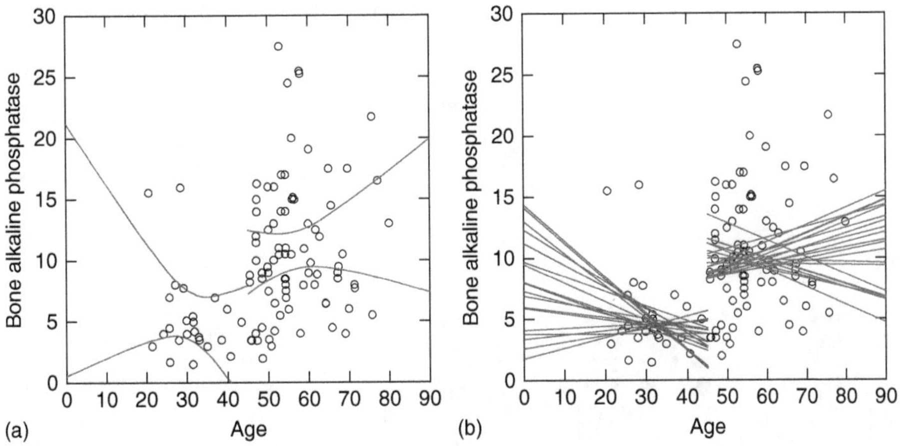

Figure 3 (a) Piecewise linear confidence intervals and (b) bootstrapped regressions [4].
Source: Wilkinson, L. The Grammar of Graphics, New York: Springer-Verlag, 2nd. © 2005 Springer Nature.

4.1 1D Continuous Aggregation

The simplest, and probably oldest, form of aggregation involves a single variable. While this problem involves a rather broad field of algorithms called *vector quantization*, there are at least two simple methods available. The first involves histogramming:

1. Choose a small bin width ($k = 500$ bins works well for most display resolutions).
2. Bin rows in one pass through the data.
3. When finished, average the values in each bin to get a single centroid value.
4. Delete empty bins and return centroids and counts in each bin.

The choice of k is based on the display resolution rather than n, as in ordinary histograms. An alternative method is based on dot plots [5]:

> Figure 4 shows how data (vertical stripes), dots, and histogram bars differ on a single set of data. Dot stacks in dot plots correspond more closely to the location of data values, but they are more expensive to compute, especially on large batches of data.

4.2 1D Categorical Aggregation

This algorithm is simple because it rests on a hash table object that is common in most computer language libraries.

1. Create a hash table to store the values of the categorical column.
2. Create a list whose entries will hold frequencies associated with each String value in the hash table.
3. For $i = 1$ to n: add the String value of x_i to the hash table and increment the associated frequency list entry.

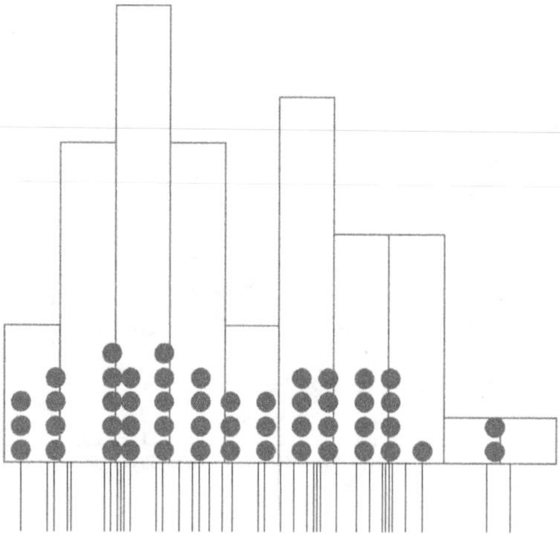

Figure 4 Dot plot and histogram.

We can visualize the result in a bar chart or dot plot [6]. There is a practical issue associated with categorical variables in big data, however. Many large datasets contain categorical values with high cardinality. This happens, for example, with product IDs or user names or Internet addresses. While a hash table can handle millions of category values, plotting the result can be difficult. One approach is to sort the categories by frequency and plot only the top 50 or 100 categories.

4.3 2D Aggregation

Two-dimensional aggregation is a simple extension of the one-dimensional histogram algorithm. We take a pair of columns to get (x, y) tuples and then bin them into a $k \times k$ rectangular grid. After binning, we delete empty bins and return centroids based on the averages of the coordinates of members in each grid cell. Figure 5 shows an example in which 100 000 points are binned into 1000. The two plots are almost indistinguishable. Figure 6 shows an example of 2D binning of clustered data.

Even though aggregated datasets are much smaller than the originals, it helps in plotting to use symbol size or opacity based on bin counts to render each point. That way points are less likely to occlude other nearby points. If we use opacity, then the joint density of the point configuration will be more apparent.

Although a little more expensive to compute, *hexagonal bins* [7, 8] are preferable to rectangular binning. With square bins, the distance from the bin center to the farthest point on the bin edge is larger than that to the nearest point in the neighboring bin. The square bin shape leads to local anisotropy and creates visible Moiré patterns. Hexagonal bins reduce this effect. Simple examples of hex binning can be found at https://everydayanalytics.ca/2014/09/5-ways-to-do-2d-histograms-in-r.html. Figure 7 shows an example involving hex binning of an enormous dataset.

4.3.1 2D binning on the surface of a sphere

The surface of a sphere is a two-dimensional object. Consequently, we can bin (x, y) tuples on a globe. Engineers at Facebook have plotted their entire network on the globe using

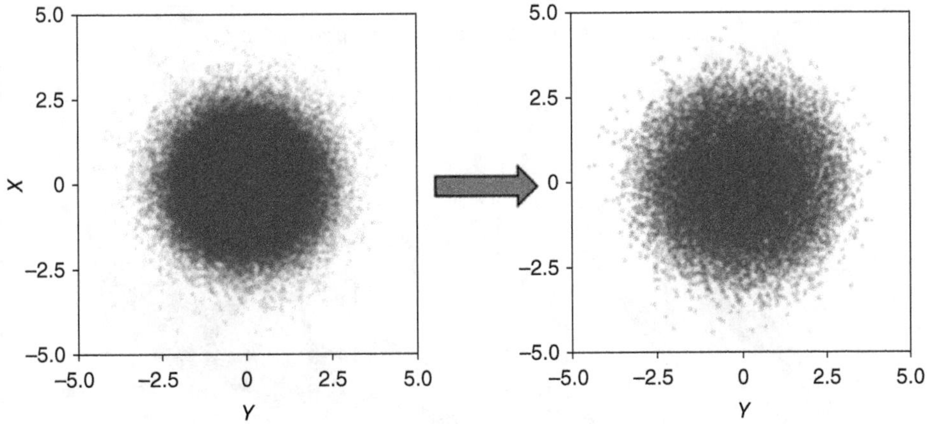

Figure 5 2D binning of 100 000 points.

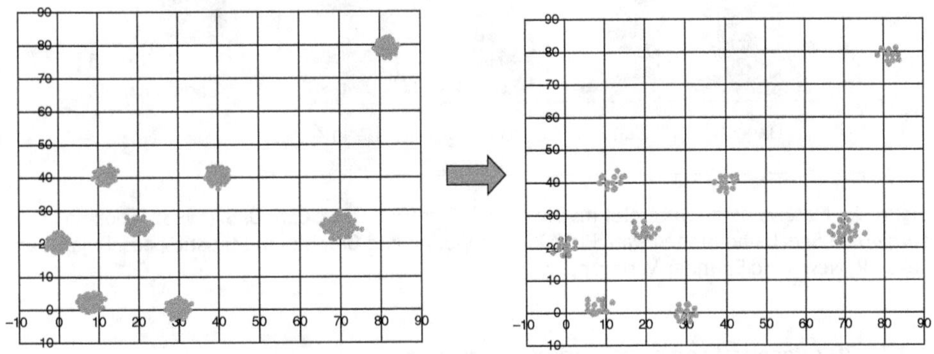

Figure 6 2D binning of thousands of clustered points.

binning. Ideally, we should use a vector quantizer such as hexagons to tile the globe, but a complete tiling of the sphere with hexagons is impossible. Compromises are available, however. Carr *et al.* [10] and Kimerling *et al.* [11] discuss this in more detail.

4.3.2 2D categorical versus continuous aggregation

Binning (x, y) tuples where x or y is categorical requires a different algorithm.

1. Create a hash table to store the values of the categorical column (assumed to be x here).
2. For $i = 1$ to n: add the String value of x_i to the hash table and add the value of y_i to a list of values associated with x_i.
3. When finished, average the values in each list element to get a single centroid value for each hash table entry.

The cardinality problems mentioned above for 1D categorical aggregation apply in this case as well.

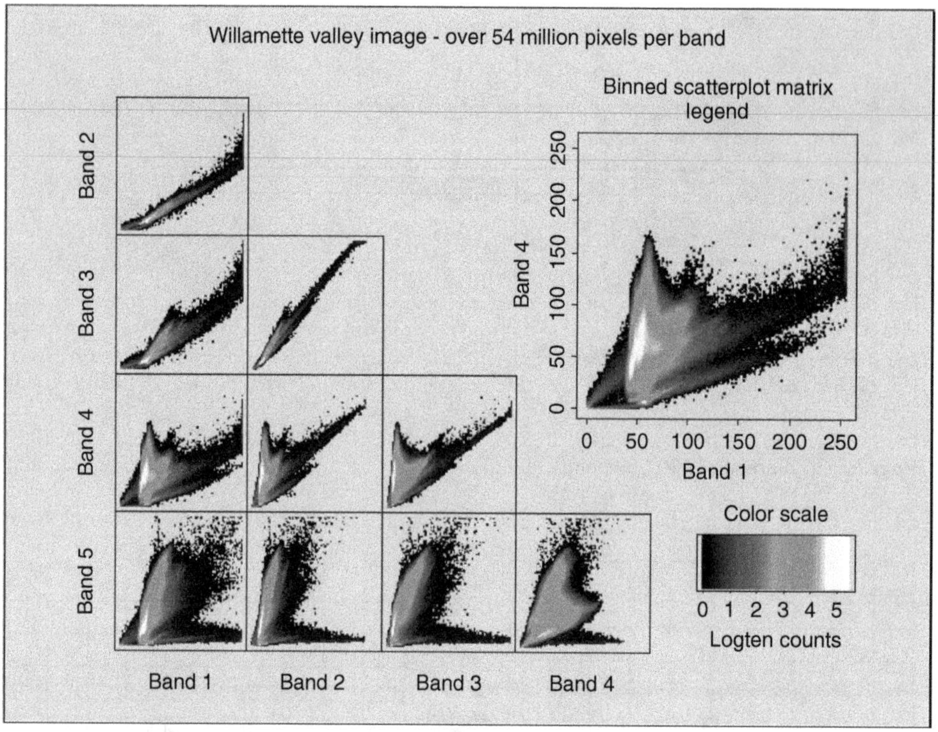

Figure 7 Massive data scatterplot matrix by Dan Carr [9]. *Source*: Carr, D. B. (1993), "Looking at Large Data Sets Using Binned Data Plots," in Computing and Graphics in Statistics, eds. Buja, A. and Tukey, P., New York: Springer-Verlag, pp. 7, 39.

4.3.3 2D categorical versus categorical aggregation

Binning into a two-way table involves a linked hash table. While the details of the algorithm are relatively simple, we face the prospect of having an enormous cross-tab as a result. We can sort the rows and columns by the marginal frequencies and select only the top 50 or 100 categories to display. If we are interested in outliers, we might select the *last* 50 categories in the list.

4.4 nD Aggregation

Higher dimensional aggregation is necessary when creating multivariate graphics such as parallel coordinate plots [12]. We must remember that points close together in a 2D projection (e.g., first two principal components or two raw variables) are not necessarily close together in higher dimensions. The nD aggregation algorithm merges points according to their distances in higher dimensional space.

The nD aggregation algorithm was first published in Wilkinson [13]. It is based on Hartigan's Leader fast-clustering algorithm [14].

1. If there are any categorical variables in the dataset, convert each categorical variable to a continuous variable using Correspondence Analysis [15, 16].

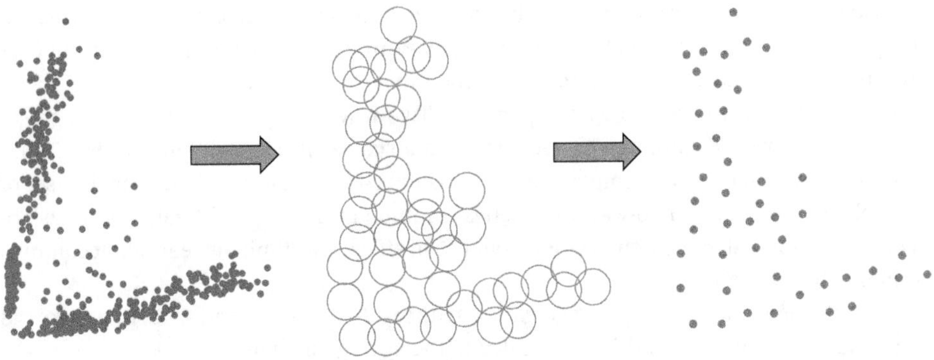

Figure 8 nD aggregator illustrated with 2D example.

2. Normalize the columns of the resulting n by p matrix X so that each column is on the unit interval.
3. Let *row*(i) be the ith row of X.
4. Let $\delta = 0.1/(\log n)^{1/p}$.
5. Initialize *exemplars*, a list of exemplars with initial entry [*row*(1)].
6. Initialize *members*, a list of lists; each exemplar has its own list of affiliated member indices.
7. For $i = 1$ to n: find $d =$ distance to closest *exemplar*(j); if $d < \delta$ add i to *members*(j); else add *row*(i) to *exemplars*(i) and i to *members*(i).

Figure 8 illustrates how the nD aggregator works. The circles all are the same size and cover all the instances in the raw dataset. Obviously, the nD aggregator algorithm will work in two dimensions, but we use specific binning aggregators for 1D and 2D aggregation because they are faster and simpler.

Some have suggested using k-means clustering to do the same job in n dimensions. There are several problems with that approach. First of all, k-means clusters are convex but are not radially symmetric; distances between clusters can be substantially affected by the shape of the clusters, so they are not representative for all the members in each cluster. Second, k-means does not scale well over a large number of clusters.

4.5 Two-Way Aggregation

A frequent problem when visualizing datasets with many columns is displaying many variables in a single plot. The most frequently recommended solution to this problem is to project the variables into a low-dimensional (usually 2D) space and base a graphic on that space. Recommended methods include principal components [17], multidimensional scaling [18], and manifold learning [19].

There are several drawbacks to this approach. First, distances between points in a high-dimensional space are not likely to correspond proportionally to distances in a 2D space. Projections like these tend to violate the triangle inequality. Second, we are left having to interpret the dimensions in the projection. With principal components, for example, we have a potentially large map involving linear coefficients when what we usually want is to see joint distributions on individual variables. Sparse principal components [20] can ameliorate this problem by setting some coefficients to zero, but we are still left with linear combinations of variables that can be difficult to interpret.

An alternative is to use two-way aggregation. Two-way aggregation applies nD aggregation to both rows and columns of a rectangular matrix. As nD aggregation clusters similar rows into a smaller set of exemplar rows, it clusters similar columns into a smaller set of exemplar columns. This two-way approach is common in the cluster literature but has not been widely applied in the visualization domain. In effect, we eliminate redundant columns from the final visualization.

Figure 9 shows how two-way aggregation reduces 30 000 rows and 25 columns to 148 rows and 15 columns. An outlier is revealed in the aggregated plot.

5 Analyzing

Using aggregation means that all statistics on an aggregated file must be computed by frequency-weighted algorithms. Frequency weights are ubiquitous in statistics packages such as SAS, SPSS, Stata, and SYSTAT but are not available in every function in *R* or Python. Ignoring frequency weights can produce erroneous results in basic statistics, regression, and other functions. Figure 10 shows how weights enter the computations in common statistical functions. The weight variable is highlighted in red. On the left are simple moments computed with frequency weights. On the right is a code snippet from Cleveland's *loess* smoother.

6 Big Data Graphics

This section features graphics that are suited or ill-suited for big data visualization. Aggregated data can be plotted with a wide variety of graphics, but a few present special problems.

6.1 Box Plots

Tukey designed the box plot (he called it a *schematic plot*) to be drawn by hand on a small batch of numbers [21]. The whiskers were designed not to enable outlier detection but to locate the display on the interval that supports the bulk of the values. Consequently, he chose the *Hspread* to correspond roughly to three standard deviations on normally distributed data. This choice led to two consequences: (i) it does not apply to skewed distributions, which constitute the instance many advocates think is the best reason for using a box plot in the first place and (ii) it does not include sample size in its derivation, which means that the box plot will falsely flag outliers on larger samples. As Dawson [22] shows, "regardless of size, at least 30% of samples drawn from a normally distributed population will have one or more data flagged as outliers." Figure 11 illustrates this problem for a sample of 100 000 normally distributed numbers. Thousands of points are denoted as outliers in the display.

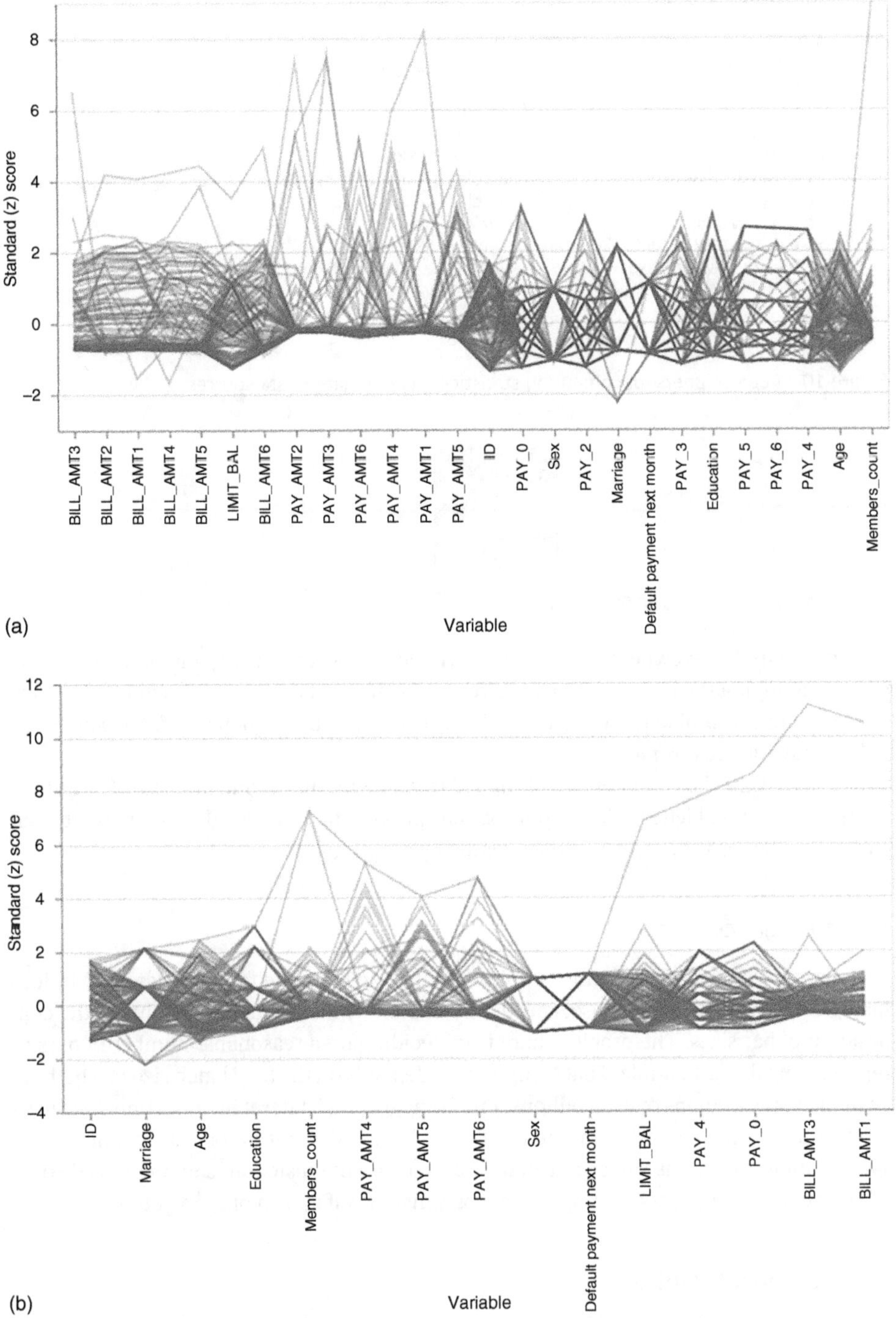

Figure 9 (a) Parallel coordinate plots of all columns and (b) aggregated columns. *Source*: UCI Credit Card Dataset, https://archive.ics.uci.edu/ml/datasets/default+of+credit+card+clients.

Moments (Python)

```
for x in data:
  if weights != None:
    wt = weights[i]
  if wt > 0:
    if x != None:
      xCount += 1
      xWeightedCount += wt
      xSum += x * wt
      xd = (x - xMean) * wt
      xMean += xd / xWeightedCount
      xSS += (x - xMean) * xd
```

Loess (Java)

```
for (int k = left; k <= right; k++) {
  double xk = x[k];
  double yk = y[k];
  double dist = xk - xi;
  if (k < i)
    dist = xi - xk;
  double wt = tricube(dist * denom) * weights[k] * frequencies[k];
  double xkw = xk * wt;
  sumWeights += wt;
  sumX += xkw;
  sumXSquared += xk * xkw;
  sumY += yk * wt;
  sumXY += yk * xkw;
}
```

Figure 10 Code snippets for computing statistics on aggregated data sources.

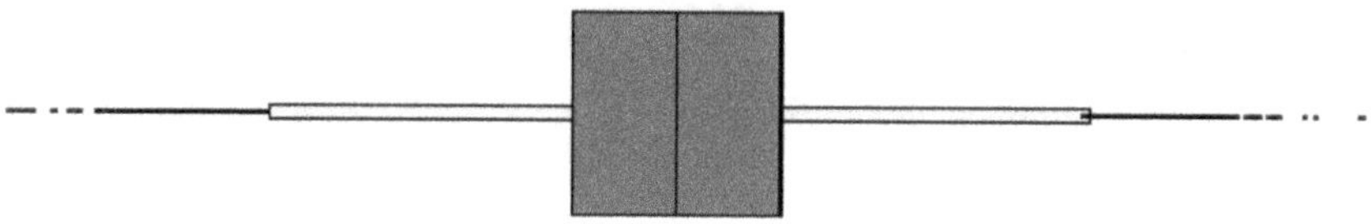

Figure 11 Box plots of 100 000 Gaussians.

To deal with the skewness problem, Hubert and Vandervieren [23] and others have suggested modifying the fences rule using a robust estimate of skewness. By contrast, Tukey's approach for this problem involved transforming the data through his *ladder of powers* [21] before drawing the box plot.

The letter-value box plot [24] was designed to deal with the second problem. The authors compute additional letter values (splitting the splits) until a statistical measure of fit is satisfied. Each letter-value region is represented by a rectangle.

6.2 Histograms

Histograms would seem to be the simplest graphics to compute for large datasets. Indeed, histograms on aggregated datasets need only to include the frequency weights in the computation of bar sizes. The problem here is in deciding on a reasonable number of bars for representing the full sample. That computation depends on the total batch size n, which can be in the many millions or even billions. But the aggregated dataset includes only $k \ll n$ distinct values. Using k to estimate the number of bars will result in too coarse a histogram. The solution? Do not use an aggregated dataset. Compute histograms in one pass through the raw data. After all, computing raw histograms is itself a form of aggregation.

6.3 Scatterplot Matrices

Scatterplot matrices (SPLOMs) were originally developed to display results of clustering algorithms [25]. They have since proved to be one of the most powerful visualizations for multivariate data. Like other multivariate displays, they suffer from two drawbacks. First,

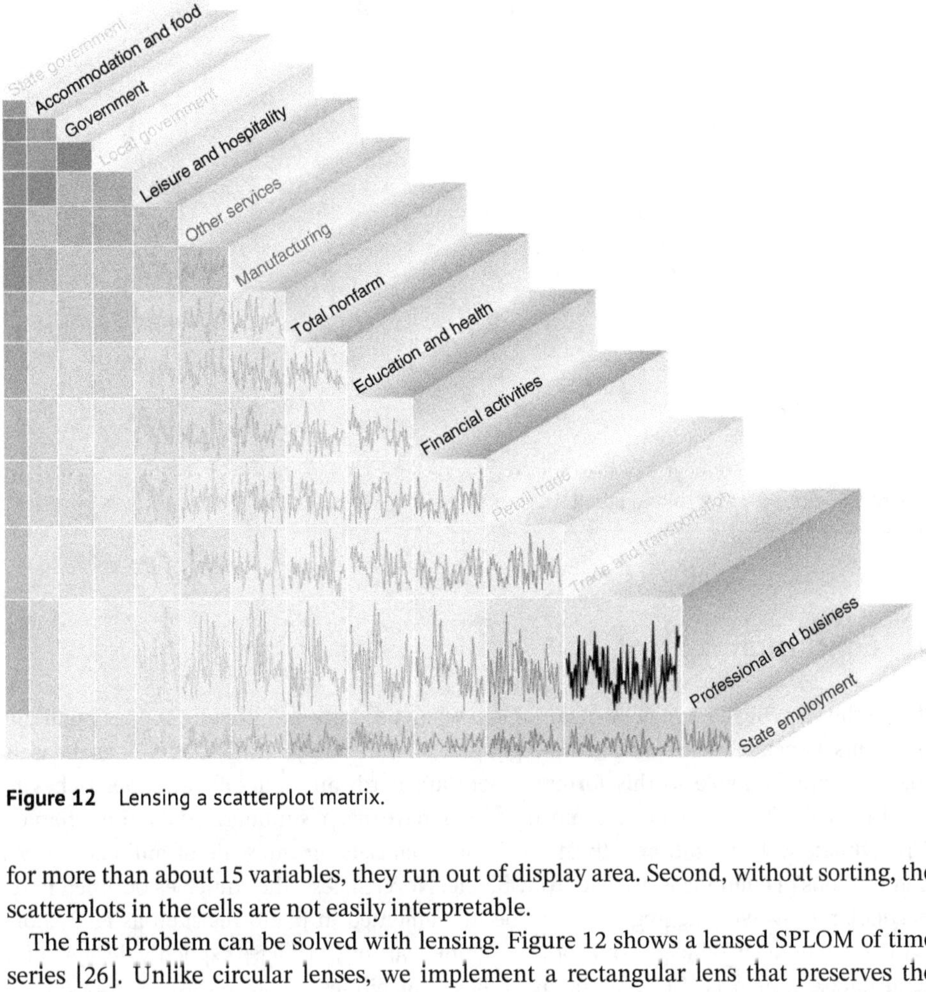

Figure 12 Lensing a scatterplot matrix.

for more than about 15 variables, they run out of display area. Second, without sorting, the scatterplots in the cells are not easily interpretable.

The first problem can be solved with lensing. Figure 12 shows a lensed SPLOM of time series [26]. Unlike circular lenses, we implement a rectangular lens that preserves the rectangular layout of the cells. As cells become too small to render the points, we simply color them with a single color. The second problem can be addressed with sorting the rows/columns. For this purpose, we can sort the variables on the basis of the first principal component loadings.

6.4 Parallel Coordinates

Parallel coordinates are a favorite multivariate display among visualization researchers. They can quickly become cluttered as the number of columns (variables) increase. We can apply the same methods we use for SPLOMs to remedy these issues. Figure 9 shows an example that exploits two-way aggregation to reduce clutter. Alternatively, we can lens the profiles to allow drill-down into details or we can add a horizontal scroll bar, as in Figure 13. This figure also includes the result of background clustering and outlier detection to facilitate interpretation.

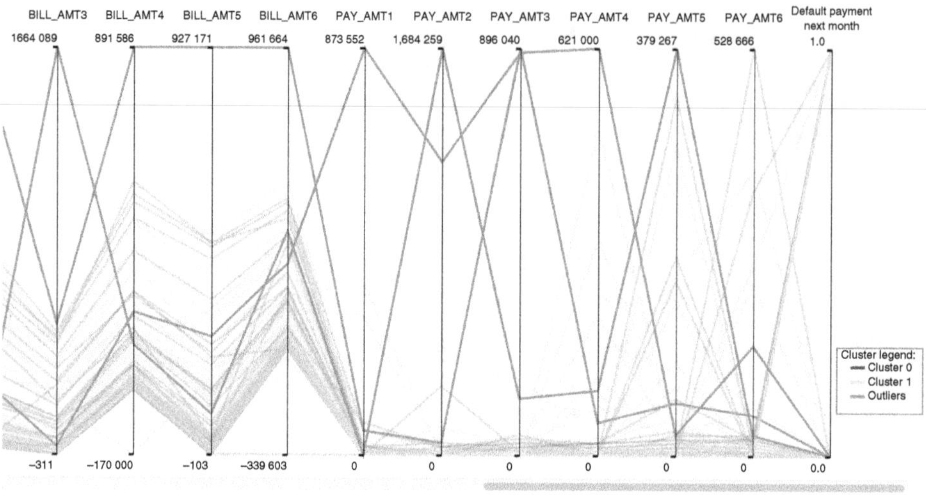

Figure 13 Sorted and scrolling parallel coordinates [27]. *Source*: H2O.ai (2019), AutoViz Driverless, https://www.h2o.ai/.

7 Conclusion

While the term big data may be meaningless when used unconditionally, it does highlight problems associated with visualizing big n and big p rectangular files, as well as datasets that are transformable to this format. There are hardware and software approaches to dealing with these problems, some of which have been summarized in this chapter. Unfortunately, visualization software to handle datasets on the scale of billions of rows and millions of columns is not widely available. Nevertheless, the strategies outlined here, particularly involving aggregation, can be implemented in platforms such as *R*, Python, SAS, Stata, or even Tableau. And some machine learning packages explicitly designed for large datasets, such as H_2O, offer many of these capabilities.

References

1 Unwin, A., Theus, M., and Hofmann, H. (2007) *Graphics of Large Datasets: Visualizing a Million*, Springer-Verlag, New York.
2 Efron, B. and Tibshirani, R. (1993) *An Introduction to the Bootstrap*, Chapman & Hall, New York.
3 Gonnelli, S., Cepollaro, C., Montagnani, A., *et al.* (1996) Bone alkaline phosphatase measured with a new immunoradiometric assay in patients with metabolic bone diseases. *Eur. J. Clin. Invest.*, 26, 391–396.
4 Wilkinson, L. (2005) *The Grammar of Graphics*, 2nd edn, Springer-Verlag, New York.
5 Wilkinson, L. (1999) Dot plots. *Am. Stat.*, 53, 276–281.
6 Cleveland, W.S. (1985) *The Elements of Graphing Data*, Hobart Press, Summit, NJ.

7 Kosugi, Y., Ikebe, J., Shitara, N., and Takakura, K. (1986) Graphical presentation of multidimensional flow histogram using hexagonal segmentation. *Cytometry*, 7, 291–294.

8 Carr, D.B., Littlefield, R.J., Nicholson, W.L., and Littlefield, J.S. (1987) Scatterplot matrix techniques for large N. *J. Am. Stat. Assoc.*, 82, 424–436.

9 Carr, D.B. (1993) Looking at large data sets using binned data plots, in *Computing and Graphics in Statistics*, (eds A. Buja and P. Tukey), Springer-Verlag, New York, pp. 7–39.

10 Carr, D., Kahn, R., Sahr, K., and Olsen, A.R. (1997) ISEA discrete global grids. *Stat. Comput. Graph. Newslett.*, 8, 31–39.

11 Kimerling, J.A., Sahr, K., White, D., and Song, L. (1999) Comparing geometrical properties of global grids. *Cartogr. Geogr. Inf. Sci.*, 26, 271–288.

12 Inselberg, A. (2009) *Parallel Coordinates: Visual Multidimensional Geometry and Its Applications*, Springer-Verlag, New York.

13 Wilkinson, L. (2018) Visualizing big data outliers through distributed aggregation. *IEEE Trans. Vis. Comput. Graph.*, 24, 56–66.

14 Hartigan, J. (1975) *Clustering Algorithms*, John Wiley & Sons, New York.

15 Greenacre, M. (1984) *Theory and Applications of Correspondence Analysis*, Academic Press, London.

16 Greenacre, M. and Blasius, J. (2006) *Multiple Correspondence Analysis and Related Methods*, Chapman & Hall/CRC, Boca Raton, FL.

17 Jolliffe, I. (2002) *Principal Component Analysis*, 2nd edn, Springer-Verlag, New York.

18 Borg, I. and Groenen, P.J.F. (2005) *Modern Multidimensional Scaling: Theory and Applications*, Springer, New York.

19 van der Maaten, L. and Hinton, G. (2008) Visualizing high-dimensional data using t-SNE. *J. Mach. Learn. Res.*, 9, 2579–2605.

20 Zou, H., Hastie, T., and Tibshirani, R. (2006) Sparse principal components. *J. Comput. Graph. Stat.*, 15, 265–286.

21 Tukey, J.W. (1977) *Exploratory Data Analysis*, Addison-Wesley Publishing Company, Reading, MA.

22 Dawson, R. (2011) How significant is a boxplot outlier? *J. Stat. Educ.*, 19, 1–12.

23 Hubert, M. and Vandervieren, E. (2008) An adjusted boxplot for skewed distributions. *Comput. Stat. Data Anal.*, 52, 5186–5201.

24 Hofmann, H., Kafadar, K., and Wickham, H. (2017) Letter-value plots: boxplots for large data. *J. Comput. Graph. Stat.*, 26, 469–477.

25 Hartigan, J.A. (1975) Printer graphics for clustering. *J. Stat. Comput. Simul.*, 4, 187–213.

26 Dang, T., Anand, A., and Wilkinson, L. (2013) TimeSeer: scagnostics for high-dimensional time series. *IEEE Trans. Vis. Comput. Graph.*, 19, 470–483.

27 H_2O.ai (2019) AutoViz – Driverless AI, https://h2o.ai/.

23

Visualization-Assisted Statistical Learning

Catherine B. Hurley and Katarina Domijan

Maynooth University, Maynooth, Ireland

1 Introduction

Statistical learning is a set of approaches for estimating a function of predictor variables. The goals of learning models are prediction, understanding how predictors impact on the estimated function, and inference. For linear models, predictor understanding, inference, and model comparison are easily obtained using standard statistical techniques. Regression diagnostics visualizations provide us with techniques for assessing model fit and may suggest ways of model improvement. Statistical learning algorithms are far more complex in structure, and understanding predictor impact, assessing model fit, and comparison of models are far more challenging. Most of the techniques for understanding and assessing statistical learning fits rely on visualization.

In recent years, we have made a number of contributions to the topic of visualization-assisted statistical learning through our work on the condvis project, which looks at interactive slice visualization for exploring machine learning models [1, 2] and accompanying recent R package [3]. Though not directly related to modeling per se, we have also investigated ordering or seriation algorithms which improve data visualizations by highlighting features and structure [4, 5]. Such visualizations assist in data exploration as a precursor to model fitting and can be considered part of the statistical learning pipeline.

Our goal in the current chapter is to give an overview of some techniques we have developed for visualization in the context of statistical learning. In Section 2, we explore the role of seriation in building data visualizations prior to modeling. In Section 3, we discuss PD plots [6], perhaps the most commonly used visualization technique for understanding predictor effects in machine learning fits. PD plots are a good starting point for moving on to interactive slice visualizations, which is the subject of Section 4. Throughout we focus on examples, rather than presenting technical details that can be found elsewhere. We conclude with some discussion.

Computational Statistics in Data Science.
Edited by Walter W. Piegorsch, Richard A. Levine, Hao Helen Zhang and Thomas C. M. Lee.
 ISBN 978-1-11956107-1

2 Better Visualizations with Seriation

The first dataset we look at is the Pima Indians dataset (provided as PimaIndiansDiabetes2 in package mlbench [7], originally from the UCI repository). The goal here is to predict occurrence of diabetes from eight numeric predictors. An initial display of the data will be useful in assessing how diabetes relates to the predictors. We return to this dataset in Section 4.2, where we interactively explore statistical learning fits for diabetes prediction.

Figure 1 is a parallel coordinate plot of the predictors, colored by the diabetes indicator. In this plot, we have chosen a predictor ordering to highlight group differences. Glucose is the single best predictor, scored by the ratio of the between- and within-group standard deviations on the linear discriminant variable, and (glucose, age) is the best predictor pair. The diabetic-positive groups have higher predictor values across all variables, but the group difference decreases moving from left to right across the plot.

To arrive at a predictor ordering, we used a seriation algorithm based on hierarchical clustering. See Earle and Hurley [4] for details and our package DendSer [5] for an implementation. First, form a matrix whose (i,j) entry is a score obtained as the ratio of the between- and within-group standard deviations on the linear discriminant variable calculated from the linear discriminant analysis (LDA) relating diabetes to predictors i and j. We wish to place pairs of variables with high scores adjacently so that the parallel coordinate plot highlights group differences. A hierarchical clustering using the scores as similarities will achieve this. Recall that the dendrogram from a hierarchical clustering of k objects gives an ordering of these objects, but this ordering is not unique: there are 2^{k-1} different orderings of the k objects consistent with the dendrogram. One of the seriation options in DendSer finds which of these orderings places the objects in an arrangement where the scores are closest to decreasing. Here, the scores used are LDA scores for individual predictors. Figure 2 shows the algorithm in action: The heatmap on the left uses the ordering of variables as they appear in the dataset, while the variable ordering in the second heatmap is

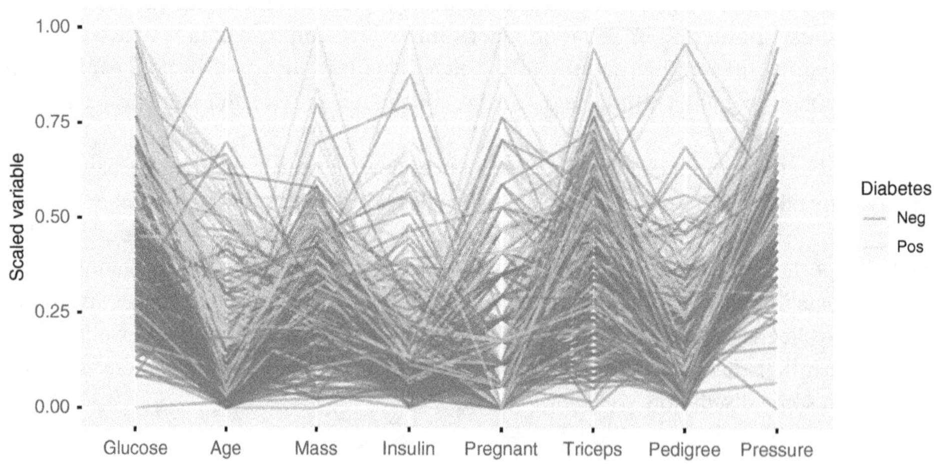

Figure 1 Parallel coordinate plot of the Pima data, colored by the diabetes indicator. Predictors that discriminate between the groups are shown on the left-hand side.

dataset are aged four to eight, and none of them are smokers. The ICE curves for smoke for these kids require predicting their FEV, were they to smoke, an extrapolation of the fitted models. For evaluating the effect of smoking, PD/ICE comparisons are better confined to children in higher age groups.

3.3 Interactive Conditional Visualization

In a previous paper [1] we outlined a paradigm for interactive visualization of statistical models, which is further developed in a more recent work [2]. Similar to ICE plots, our visualizations show how $\hat{f}$ varies across levels of a designated *section predictor* x_1 while the remaining predictors are held fixed at a value $\boldsymbol{x}_{(-1)} = \boldsymbol{u}$. The fitted curve $\hat{f}$ is superimposed on a subset of the observations $(x_{i,1}, y_i)$ which are "near" the section point $\boldsymbol{x}_{(-1)} = \boldsymbol{u}$. We call this a *section plot*. Points on the section plot are faded as their distance from the section point increases. The section plot is accompanied by a second visualization showing the predictors $\boldsymbol{x}_{(-1)}$, where the section point $\boldsymbol{u}$ is clearly indicated. We call these predictor plots *condition selector plots*. This methodology is implemented in the shiny-based R package condvis2 [3], which is applicable to a broad range of supervised and unsupervised learning fits.

Our condvis2 software is highly interactive, which adds tremendous powers of model investigation to the visualizations. Sifting through lots of visualizations exploring the model is quick and easy, and so the information about predictor impact, model assessment, and model comparison becomes immediately accessible. The main interactive choices are: First, pick one or two section predictors via a menu. Secondly, choose the section point $\boldsymbol{u}$. This may be set by clicking on the condition selector plots or by selecting a touring algorithm which will aim to present the analyst with an interesting series of section points. Such automatic touring algorithms are particularly relevant as the dimension of predictor space increases. Choice is not limited to observed predictor values, so "out-of-sample" section plots are available. Finally, choose a distance measure and a similarity threshold. Only observations whose distance are within this threshold are deemed near the section point.

Similar to ICE plots, our visualizations may show how $\hat{f}$ varies over predictor combinations $(x_1 = x, \boldsymbol{x}_{i\,-1} = \boldsymbol{u})$ that do not occur in the data. However, this will be obvious because there will be no superimposed observations at this $x_1 = x$ setting. Wild predictions resulting from extrapolations of $\hat{f}$ will also be obvious, reducing the likelihood of misleading the analyst.

Our condition selector plots of the predictors $\boldsymbol{x}_{(-1)}$ and the section point $\boldsymbol{u}$ show the slice of the predictor space that appears in the section plot. In the ICE plots of Figure 3, coloring the ICE curves by $\hat{y}$ gave clues as to its settings of predictors $\boldsymbol{x}_{(-1)}$ but with more complex datasets and in particular more predictors figuring out the relationship between the ICE curves and $\boldsymbol{x}_{(-1)}$ requires some detective work.

4 Condvis2 Case Studies

4.1 Interactive Exploration of FEV Regression Models

We revisit the FEV data of Section 3.2, with the linear model and random forest fits relating FEV to the predictors. As there are just a few predictors, model exploration is relatively straightforward. However, the example demonstrates how condvis is useful for comparing

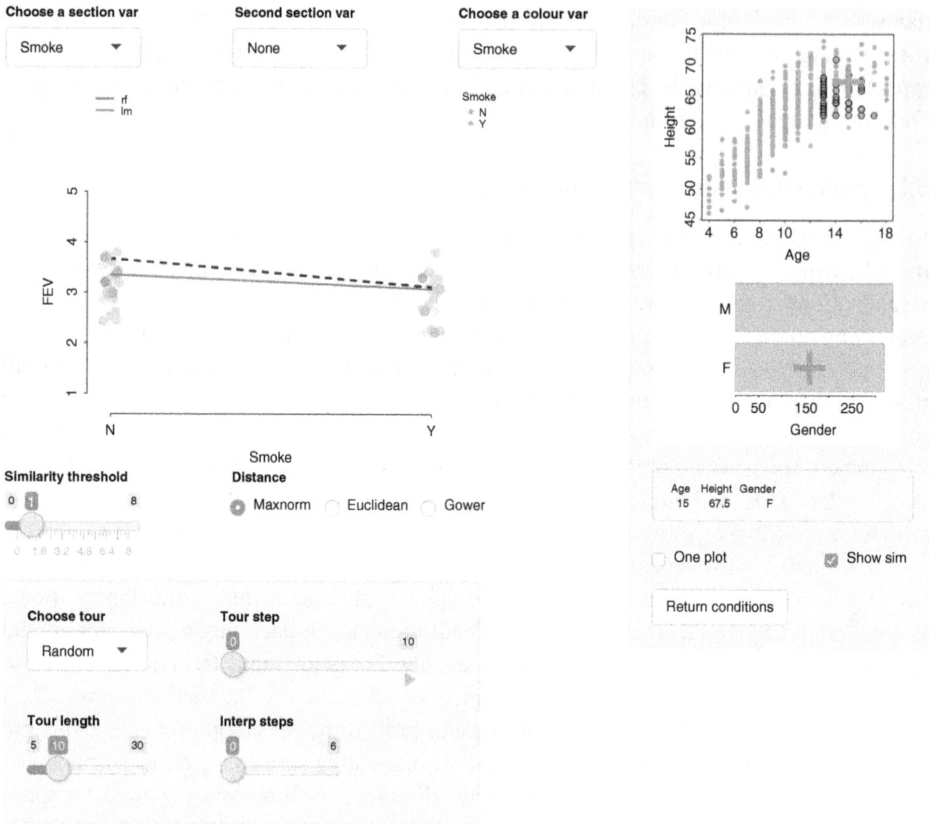

Figure 4 Condvis2 screenshot for a linear model and random forest fit to the FEV data, with smoke as the section variable. The random forest fit is in solid line, and the linear model in dashed line.

fits, assessing lack of fit, presence of extrapolation, and understanding predictor effects in a streamlined way.

A condvis2 screenshot of the two fits is given in Figure 4. The section point indicated by the gray cross is set to gender = F, age = 15, and height = 67.5. The dashed and solid lines on the section plot show the smoking effect for this condition, for the linear and random forest fits, respectively. Both fits show a negative impact of smoking on FEV, an effect which is more pronounced for the linear fit. However, the predicted response for the linear fit for nonsmokers does not agree with nearby observations, the darker gray filled circles. Clicking on the bar for males produces a section plot with the smoking effects for males of the same age and height (not shown), and again there is a negative smoking impact. Clicking again on the (age, height) condition plot to set age to 6 and height to 61 shows that there are no smokers for these conditions, so neither fit is to be trusted.

Figure 5 has another view of the fits, this time with two section predictors, age and smoke. We have clipped the bottom section of the display which has controls for touring predictor space, to focus on the part relevant to our discussion. Here, all section

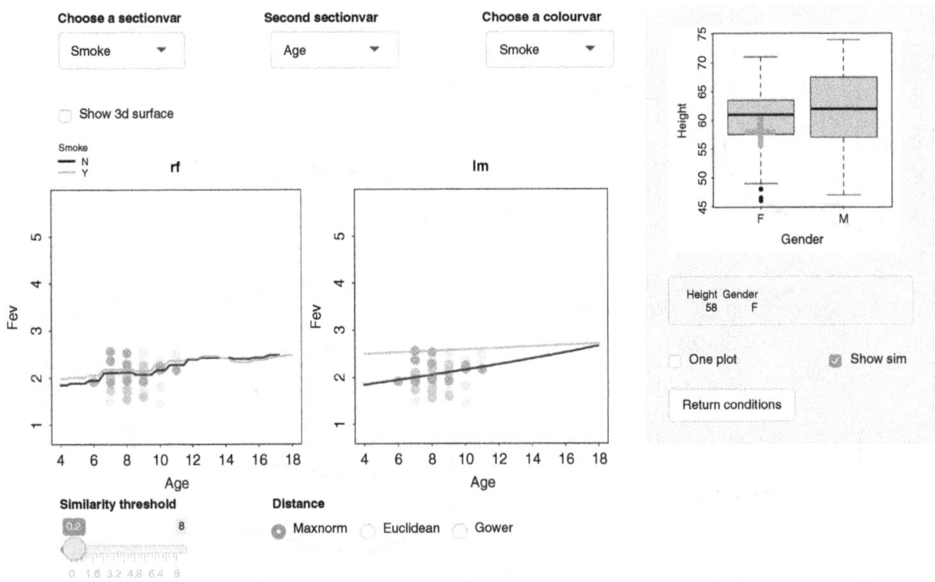

Figure 5 Condvis2 screenshot for a linear model and random forest fit to the FEV data, with age and smoke as section variables. The smoker fits are in gray, and the nonsmoker fits are in solid black.

plot observations shown are female, within $0.2 \times \text{sd}$ of height = 58, and at this location all are nonsmokers. The similarity threshold has been reduced to 0.2, so that the observations shown on the section plot vary in age but have little variation in height. The wild extrapolating behavior of the linear smoker fit is evident here. The apparent positive impact of smoking for young females evident in the linear fit of Figure 3a is not supported by the data as it is an artifact of extrapolation. The random forest produces similar fits for the smokers and nonsmokers, but both are overly rough and nonmonotonic.

4.2 Interactive Exploration of Pima Classification Models

Next, we return to the Pima dataset of Section 2. We build two classification models relating diabetes to all predictors, a random forest and a Bayesian additive regression tree (BART) [12]. Using a 60/40 split, we divide the data into a training and a test set. The random forest has a far lower misclassification rate on the training set (4% vs 17%), but the misclassification rates on the test sets are the same at 16%. We compare the two fits on the training data with our interactive software, focusing on glucose and age as section variables, as they were deemed strongly related to the presence of diabetes in our preliminary analysis of Section 2. (In fact, the random forest variable importance measure produces a predictor ranking almost identical to that of Figure 1.)

Comparing the two fits with condvis2, we first look at the predicted probabilities for the variable glucose; see Figure 6a,b. In Figure 6a, the section point is the medoid of the training observations. The random forest and BART curves both show that diabetes probability generally increases with glucose, but the random forest fit has a jump around glucose = 125.

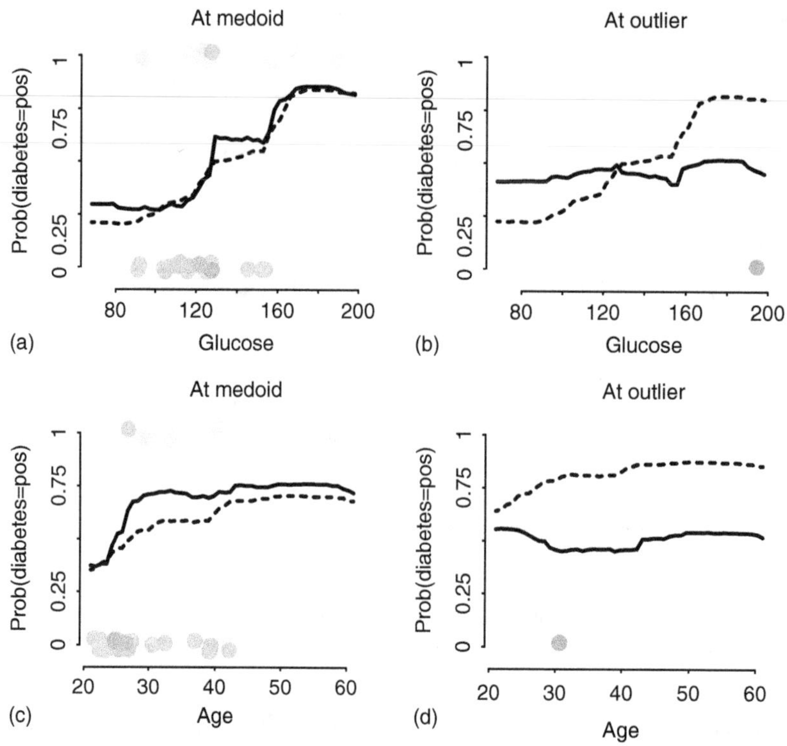

Figure 6 Condvis2 section plots for glucose and age from a BART (dashed line) and random forest fit (solid line) to the Pima training data. The observations are set to 1 for diabetes positive (light gray) and 0 for negative (dark gray) with some jitter. Visible observations are within 1 sd on the remaining predictors. Plots (a) and (c) show the fit at the medoid (b) and (d) at an extreme observation.

As we explore different section points, we see that the BART curves (in dashed line) are smoother throughout, and their shape does not change much. Some of the random forest curves are flat, or even have a slight downward trend in places. For example, the plot in Figure 6b shows the predicted probability curves at an extreme training observation, where there are no other nearby observations. Here, BART wrongly classifies the observation as diabetes positive, but the random forest is likely overfitting here.

In Figure 6c,d, we explore the effect of age, first for the medoid observation and then for an extreme observation. The random forest and BART fitted curves at the medoid are similar and indicate that the probability of diabetes increases with age and then flattens out at about age 45. But at the extremes of the data shown in Figure 6d, the random forest has fit a downward trend with age, again suggestive of overfitting. In Figure 7, we investigate the classification boundary as a function of age and glucose at the same two section points used in Figure 6. In Figure 7a, the two fits give similar predicted classes. In Figure 7b once again, we see the signs of overfitting in the random forest fit, with the dark gray island on the right ensuring that the outlying observation is correctly classified as diabetes negative.

Our model explorations show that the BART fit gives smoother curves than the random forest. The random forest is guilty of overfitting, a fact we might have suspected from the

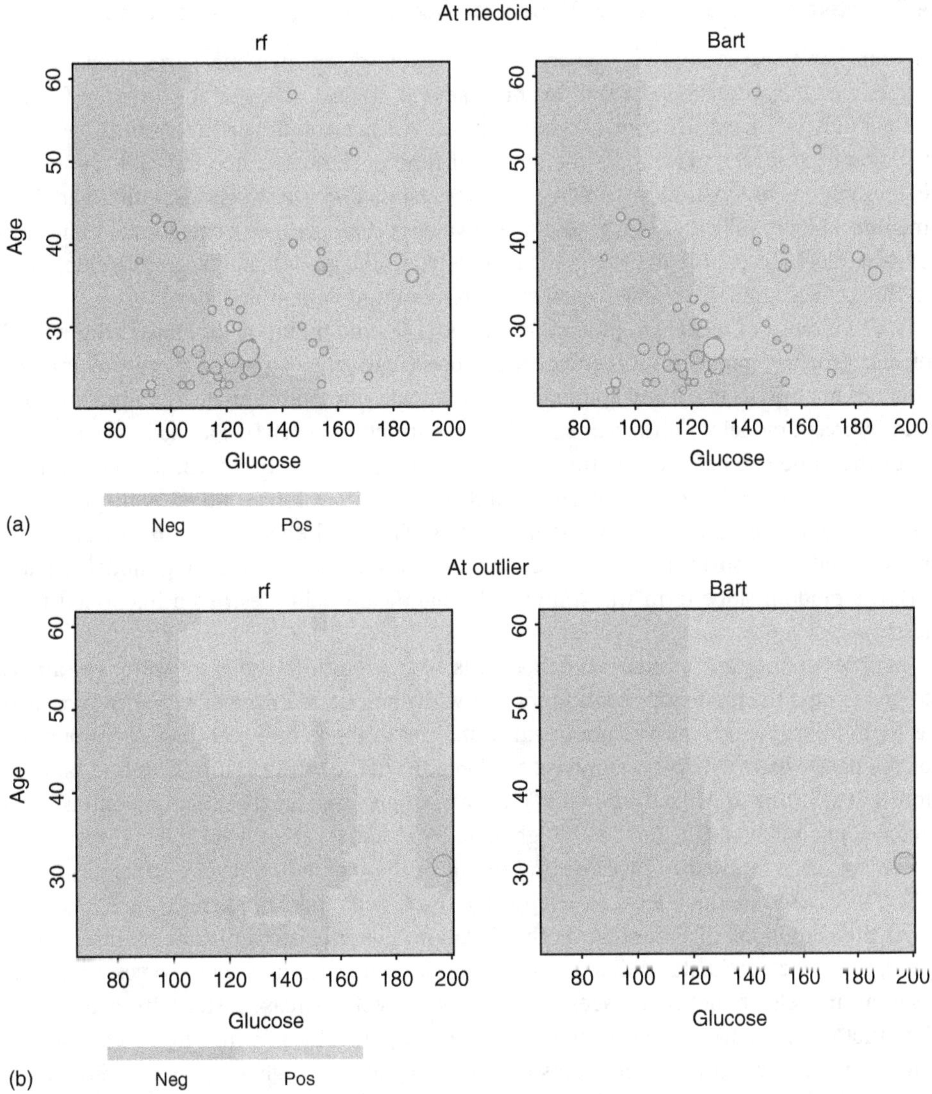

Figure 7 Condvis2 section plots for glucose and age showing classification boundaries from BART and random forest fits to the Pima training data. The observations are set to 1 for diabetes positive (light gray) and 0 for negative (dark gray) with some jitter. Visible observations are within 1 sd on the remaining predictors. Plot (a) shows the fit at the medoid, and (b) at an extreme observation.

big discrepancy between the training and test misclassification rates. The overfitting occurs predominantly at unusual observations in predictor space, which we can verify by touring through section points which are observations where the fits differ. Furthermore, the BART fitted curves for age and glucose have a similar shape as the section point changes, and further investigation shows that this holds for the other six predictors also. A series of ICE plots on a logit scale for all predictors confirms that the BART fit is in fact additive in all predictors, a property that leads to a simple model explanation.

4.3 Interactive Exploration of Models for Wages Repeated Measures Data

In our final case study, we use condvis2 to explore repeated measures models. The wages data [13] comprises natural log of wages (ln_wages), adjusted for inflation for 888 individuals (id) recorded over varying number of time points per individual. The time variable experience (xp) records the length of time in the workforce in years. Time zero is the first day at work, and the values of time points differ between ids. Other variables include a binary indicator of a graduate equivalency diploma (ged), categorical indicators of race (black, hispanic), highest grade completed (high_grade), and unemployment rates in the local geographic region at each measurement time (unemploy_rate).

We fit a range of models to predict ln_wages: (i) a random intercept model whose fixed effects structure includes a three-way interaction (high_grade:xp:black), two-way interaction (xp:unemploy_rate), and spline terms for xp and unemploy_rate; (ii) a random slope and intercept model with the same fixed effects structure as (i); (iii) a fixed effects model with the same fixed effect structure as (i) and with an interaction term for (id:xp), thus fitting separate regressions for each id; and (iv) a random forest model with predictors xp, high_grade, black, ged, and unemploy_rate. The fixed effects structure in the parametric models (i)–(iii) was chosen using likelihood ratio tests. Comparing models (i) and (ii), the random slope term was found to be significant using restricted likelihood ratio tests.

In order to study and compare the model fits, we use our interactive software. The aim is to understand the predictor effects in models with complicated fixed effects structure such as high-order interactions and nonlinear terms and to assess how well they are supported by the data. Figure 8 displays condvis2 section plots of the two mixed models and random forest fits, with xp and black (shown in gray) as section variables. The section point is set to ged = 1, unemploy_rate = 2.62, with high_grade = 12 and 6 in Figure 8a and b, respectively. Here, we are interested in investigating the fit of the population level coefficients (fixed effects) only, so random effects are dropped in the mixed model predictions.

At the conditions in Figure 8a (high education, living in areas of low unemployment rate), the models estimate similar starting wages for the two groups (black = yes, no); however, all models predict that wages of nonblack subjects increase faster with experience. The first two fits have a decrease in wages after six years of experience for black subjects, but there are few nearby observations beyond this point. At high_grade = 6 in Figure 8b, the fit of all models is shifted downward, and the fitted curves for black and nonblack subjects are almost parallel. The effect of the three-way interaction high_grade:xp:black in the mixed models is evident here: at low education level, the effect of experience on wages is the same for the two groups, but experience is far more valuable for the non-black group at higher education. The random forest fit follows the same pattern, indicating that the three-way interaction is supported by the data. The two mixed effects models give similar fits, and in fact, both have comparable prediction performance when tested on out-of-sample ids.

We can also use condvis2 to investigate how well the models fit the data at subject level. Figure 9a has condvis2 section plots of models (i)–(iii) and all observations for id = 1508, a nonblack person. The dark-gray curve shows the fit for this subject, and the light-gray curve shows the fit obtained if this subject were black. Similarly, Figure 9b shows the fit

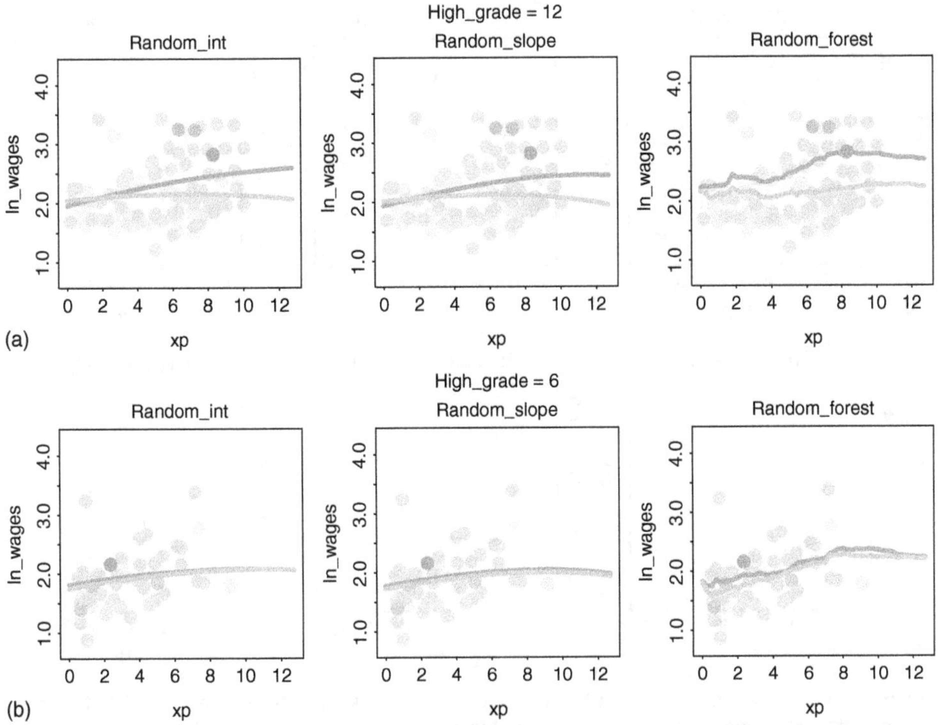

Figure 8 Condvis2 section plots for mixed effects models and random forest fit to the wages data, with xp and black (in gray) as section variables. Condition variables are high_grade = 12 (a) and 6 (b), ged = 1, and unemploy_rate = 2.62.

and observations for a black id. The fixed effects model does not benefit from partial pooling as can be seen from the displays. Double clicking to select id on the condition selector plots makes the section point the medoid of all observations for that id, calculated over all variables in the conditioning set. Selecting different ids shows that there is a lot of variability in how experience affects wages at individual level. Figure 9a is an example showing that the random slopes model captures this aspect better than the random intercepts model.

5 Discussion

As a precursor to the modeling stage, we demonstrated in Section 2 that seriated data plots are useful to highlight properties of the data relevant to the modeling problem. If the dimension of the feature space is beyond about 10, parallel coordinate plots and scatterplot matrices become impractical due to space limitations, but with seriated plots such as that in Figure 1 one could just focus on the first 10 variables, which are the most relevant for class discrimination.

Through a number of case studies, we illustrated interactive exploration of supervised statistical learning fits. Our interactive methods overcome the main limitation of PD/ICE plots, which is that the curves in Equations (1, 2) rely on broad extrapolations of possibly

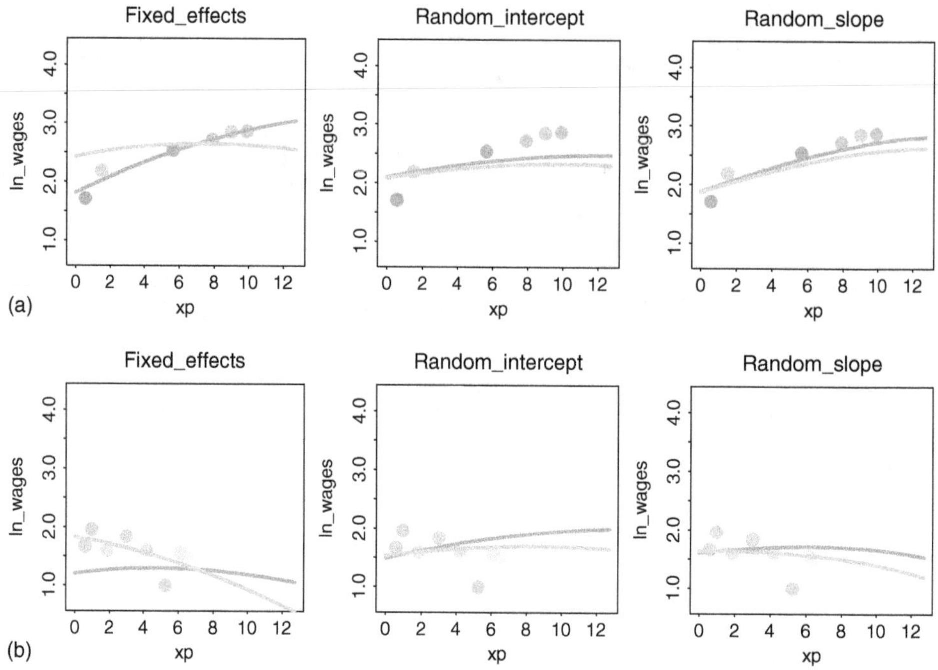

Figure 9 Condvis2 section plots of two mixed effects models and a fixed effects model fit to the wages data, with experience as a section variable and id included in the condition selector plots. The section point is the medoid of conditioning variables at each id. (a) id = 1508. (b) id = 4181.

wild model fits, a limitation that is even more severe when predictors are correlated. With condvis2, the analyst can investigate predictor effects and identify higher order interactions, explore goodness of fit to training or test datasets, and compare multiple fits. Interval estimates may also be added to the display, if offered by the model fit. In our experience, condvis2 is useful for datasets with up to 100 000 cases and 30 predictors. The package on CRAN offers vignettes which demonstrate the use of our software for a wide range of supervised and unsupervised machine learning fits.

Condvis2 is one of many efforts to use visualization to understand machine learning fits in a model-agnostic way. Many of these show how features locally explain a fit [14]. The books Molnar [15], Biecek and Burzykowski [16] are a good starting point for reading on explanatory model analysis. In our opinion, interactivity such as that offered by condvis2 adds super powers to model visualizations. See Baniecki and Biecek [17] for further compelling arguments in favor of model explanations derived from interactive model exploration.

References

1 O'Connell, M., Hurley, C., and Domijan, K. (2017) Conditional visualization for statistical models: an introduction to the condvis package in R. *J. Stat. Soft.*, 81 (5), 1–20.

2 Hurley, C.B., O'Connell, M., and Domijan, K. (2021) Interactive slice visualization for exploring machine learning models. *arXiv 2101.06986*.

3 Hurley, C., O'Connell, M., and Domijan, K. (2019) Condvis2: Conditional Visualization for Supervised and Unsupervised Models in Shiny. R Package Version 0.1.1.

4 Earle, D. and Hurley, C.B. (2015) Advances in dendrogram seriation for application to visualization. *J. Comput. Graph. Stat.*, 24 (1), 1–25.

5 Hurley, C.B. and Earle, D. (2013) DendSer: Dendrogram Seriation: Ordering for Visualisation. R Package Version 1.0.1.

6 Friedman, J.H. (2001) Greedy function approximation: a gradient boosting machine. *Ann. Stat.*, 29 (5), 1189–1232.

7 Leisch, F. and Dimitriadou, E. (2010) Mlbench: Machine Learning Benchmark Problems. R Package Version 2.1-1.

8 Hurley, C.B. (2004) Clustering visualizations of multidimensional data. *J. Comput. Graph. Stat.*, 13 (4), 788–806.

9 Hahsler, M., Hornik, K., and Buchta, C. (2008) Getting things in order: an introduction to the R package seriation. *J. Stat. Soft.*, 25 (3), 1–34.

10 Goldstein, A., Kapelner, A., Bleich, J., and Pitkin, E. (2015) Peeking inside the black box: visualizing statistical learning with plots of individual conditional expectation. *J. Comput. Graph. Stat.*, 24 (1), 44–65.

11 Rosner, B. (2010) *Fundamentals of Biostatistics*, Cengage Learning.

12 Chipman, H.A., George, E.I., and McCulloch, R.E. (2010) BART: Bayesian additive regression trees. *Ann. Appl. Stat.*, 4 (1), 266–298.

13 Singer, J.D. and Willett, J.B. (2003) *Applied Longitudinal Data Analysis*, Oxford University Press, Oxford, UK.

14 Ribeiro, M.T., Singh, S., and Guestrin, C. (2016) *"Why Should I Trust You?": Explaining the Predictions of Any Classifier*. Proceedings of the 22nd ACM SIGKDD International Conference on Knowledge Discovery and Data Mining, August 13–17, 2016, San Francisco, CA, pp. 1135–1144.

15 Molnar, C. (2019) Interpretable Machine Learning: A Guide for Making Black Box Models Explainable, https://christophm.github.io/interpretable-ml-book/ (accessed 07 June 2021).

16 Biecek, P. and Burzykowski, T. (2021) *Explanatory Model Analysis: Explore, Explain, and Examine Predictive Models*, CRC Press, Boca Raton, FL.

17 Baniecki, H. and Biecek, P. (2020) The grammar of interactive explanatory model analysis. *CoRR, abs/2005.00497*.

24

Functional Data Visualization

Marc G. Genton and Ying Sun

King Abdullah University of Science and Technology, Thuwal, Saudi Arabia

1 Introduction

Data in many fields are collected through a process naturally recorded as functional, such as curves, surfaces/images, and trajectories. Figure 1 shows one example of functional data introduced and discussed by Ramsay and Silverman [1]. The data consist of the angles formed by the hip and knee of 39 children over their gait cycles. The cycle begins and ends when the heel under observation strikes the ground. Each curve is one functional observation measured from each child's gait cycle, and the time is translated into values in $[0, 1]$, proportional to the gait cycle. For such datasets, functional data analysis (FDA) provides a considerable number of methodologies and techniques and has become a popular branch of statistics. A broad overview of FDA and data examples can be found in Ramsay and Silverman [1]. A more recent review of FDA and discussions can be found in Wang *et al.* [2] and references therein.

Visualizing functional data can be a challenge. Genton *et al.* [3] discussed and promoted dynamic visualization, coined visuanimation, including functional data visualization. Recently, Castruccio *et al.* [4] presented a suite of apps for spatiotemporal data visualization using stereoscopic view and virtual reality on smartphones. In statistics, exploratory data analysis, as the first step of data analysis, requires informative visualization tools. Classical visualization tools designed for scalar data may not be appropriate for functional data. For example, the well-known boxplot is a graphical tool to summarize the distribution of a scalar random variable. Its construction is based on the ranks of the observations. However, for functional data, there is no unique ordering. Data depth is a key idea to generalize ranks to the functional setting. The depth value for each observation measures how central an observation is with respect to the entire sample. The modified band depth proposed by López-Pintado and Romo [5] is one of the most popular depth notions for functional data due to its intuitive geometric interpretation. Other ways for ordering functional data exist, such as the tilting approach introduced by Genton and Hall [6]. With a given ranking, Sun and Genton [7] proposed the functional boxplot as a strong analog to the classical boxplot.

Computational Statistics in Data Science.
Edited by Walter W. Piegorsch, Richard A. Levine, Hao Helen Zhang and Thomas C. M. Lee.
 ISBN 978-1-11956107-1

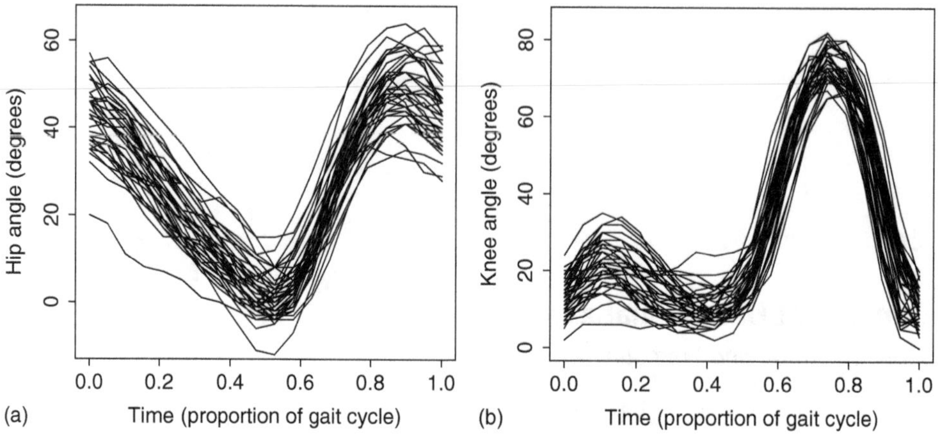

Figure 1 Functional data: the hip (a) and knee (b) angles of each of the 39 children as they go through their gait cycles. The time interval [0, 1] represents a single cycle.

Functional data are often multivariate. For example, Figure 1 shows the hip and knee angles separately, and it does not indicate how the hip angle is paired with the knee angle from the same child. To be able to generalize the boxplot to the multivariate functional setting, multivariate depth is needed. Ieva and Paganoni [8], Claeskens *et al.* [9], López-Pintado *et al.* [10], and Dai and Genton [11] have proposed various methodologies for multivariate functional data ranking.

Rank-based methods are robust against outliers and can be used for outlier-detection purposes. The classical boxplot detects potential outliers by the 1.5 times IQR (interquartile range) empirical rule. For functional data, this empirical rule was generalized by Sun and Genton [7]. However, outlier detection is much more complicated for functional data because the observations can be outlying either in magnitude or in shape. Huang and Sun [12] proposed the total variation depth and its decomposition to visualize both magnitude and shape outliers. They compared the outlier-detection performance with several existing methods, including the functional boxplot with different depth notions for ranking, such as the modified band depth [5] and the extremal depth [13], the outliergram [14], the functional outlier map (FOM) with the adjusted outlyingness proposed by Hubert *et al.* [15], and the extended FOM with the "directional outlyingness" proposed by Rousseeuw *et al.* [16]. Recently, Harris *et al.* [17] proposed a notion of elastic depth to improve shape outlier detection.

In Section 2, we describe tools for univariate functional data visualization, whereas in Section 3 we focus on tools for multivariate functional data visualization. We end with some discussions in Section 4.

2 Univariate Functional Data Visualization

2.1 Functional Boxplots

The functional boxplot first proposed by Sun and Genton [7] has been proven a valuable visualization tool for exploratory FDA. Inspired by the classical boxplot introduced as the

"Box-and-Whisker plot" by Tukey [18], the functional boxplot uses robust summary statistics to visualize the distribution of any given functional dataset. Instead of the original "Five Number Summary" in a classical boxplot, that is, the sample minimum, the first quartile, the median, the third quartile, and the sample maximum, the functional boxplot displays the functional median, the envelope of the 50% central region, and the maximum envelope.

To create a functional boxplot, the first step is to assign ranks to functional observations. Unlike for univariate scalar observations, the ranking for functional data is not unique. The functional boxplot by default uses the ranks induced by the modified band depth [5], while the depth values calculated using other depth notions can be provided instead. Denote the order statistics by $y_{[1]}(t), \dots, y_{[n]}(t)$ according to decreasing depth values. Since the ordering is from the center outward, the first-order statistics, $y_{[1]}(t)$ or the functional observation with the largest depth value, has the most central position and thus called the *functional median*. The sample 50% central region is then defined as $C_{0.5} = \{(t, y(t)) : \min_{r=1,\dots,\lfloor n/2 \rfloor} y_{[r]}(t) \leq y(t) \leq \max_{r=1,\dots,\lfloor n/2 \rfloor} y_{[r]}(t)\}$, where $\lfloor n/2 \rfloor$ is the smallest integer not less than $n/2$. The 50% central region indicates the variability of the central 50% of the data, and its border is called the *envelope* representing the box as in the classical boxplot. The functional median is displayed in the central region, which presents the centrality of the data. It is worth to point out that the functional median is one of the observations (or the average of the deepest curves if not unique). In contrast, the envelope of the 50% central region consists of pieces from different observations and is not an original observation anymore. The maximum envelope can be constructed in a similar way and is the analog to the whiskers in the classical boxplot.

In the functional boxplot, the maximum envelope is constructed after removing outliers by the 1.5 times the 50% central region empirical rule, where the fences are determined by inflating the envelope of the 50% central region by 1.5 times its range for each t. The constant factor F is set to be 1.5 by default as in the 1.5 times IQR rule in the classical boxplot. However, it can be modified by users. For instance, Sun and Genton [19] proposed an adjusted functional boxplot for outlier detection in spatiotemporal data, where the factor F is adjusted according to the dependence among the functional observations.

To illustrate the construction of the functional plot, we consider the gait data, as shown in Figure 1. Figure 2 shows the functional boxplots of hip and knee angles for two choices of F for the illustration of outlier detection. Compared to the plots of original curves, the functional median curves for the hip and knee angles show more clearly the different phases of the data. For example, we see that at the beginning, the hip angle decreases from its maximum angle to its minimum near zero, while the knee angle first increases from zero to a local maximum near zero and then decreases to near zero. When the hip angle switches to a phase of increase, the knee angle increases sharply to its maximum as the leg prepares to leave the floor. During the last phase, the knee angle decreases rapidly to zero as the leg extends, and the hip angle levels off. The gray area boxes the 50% central regions with the corresponding envelopes displayed in dark gray, which indicate the variability of these gait cycles among the 39 children. When $F = 1.5$, there were no outliers in both functional boxplots. For the hip angle data, one outlier is detected when decreasing F to 1.2, and for the knee angle data, when $F = 0.61$, one outlier is identified. These two outliers in short-dashed curve show slightly different features from the majority of the curves. Finally, the maximum envelopes in dark gray present the range of the nonoutlying observations. As pointed out by Sun and Genton [19], the adjustment of F is not so crucial for visualization purposes.

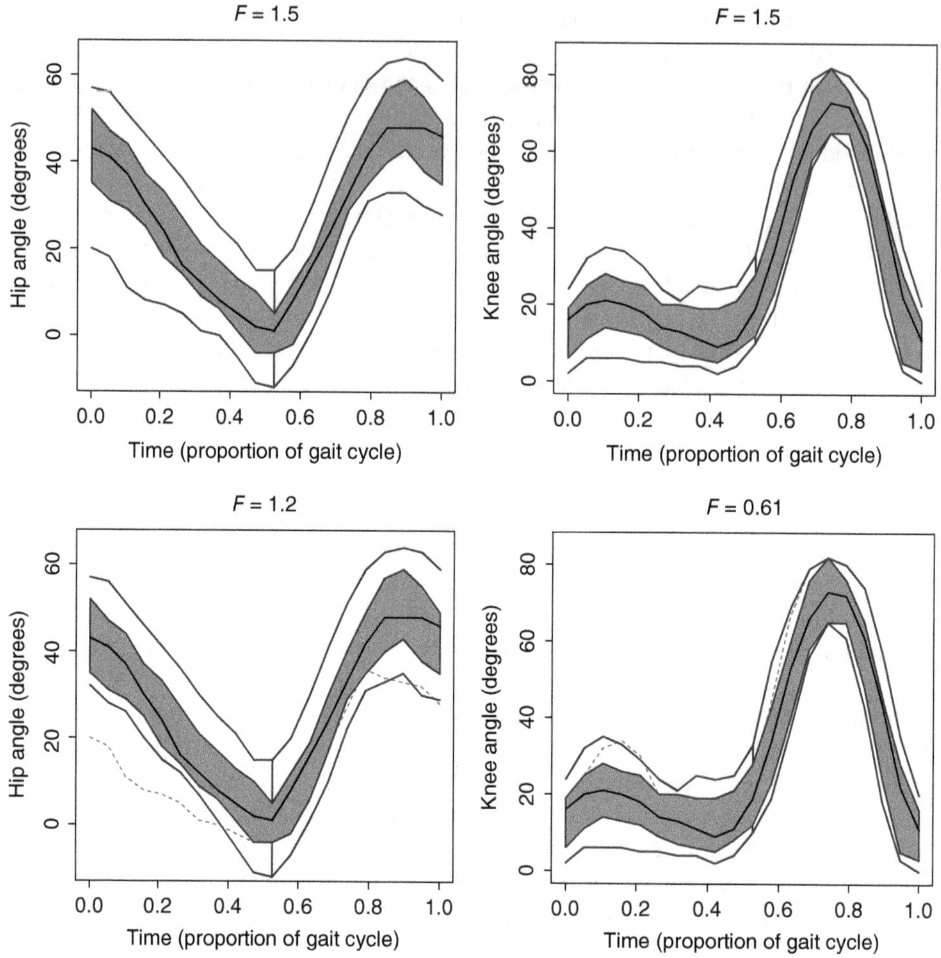

Figure 2 The functional boxplots for the hip and knee angles of each of the 39 children. The black curve in each functional boxplot is the median, the gray box is the 50% central region, and the short dashed curves are the potential outliers identified by the *F* times the 50% central region rule.

However, it needs to be carefully selected for outlier detection. More discussion on the selection of the adjustment factor *F* and various examples can be found in Sun and Genton [19].

Thanks to the fast algorithm developed in Sun *et al.* [20], the functional boxplot has been widely used and applied to various types of functional datasets, from visualization, outlier detection to simulation, and model diagnostics. For instance, the functional boxplot was used to detect outliers in functional autoregressive models by Martínez-Hernández *et al.* [21]; Huang and Sun [22] proposed to assess spatiotemporal covariance properties using functional boxplots; Sun and Genton [23] first used the functional boxplot to summarize simulation results of robust functional median polish for univariate ANOVA, and later it was extended to MANOVA by Qu *et al.* [24]. It has also been employed to analyze and visualize the log periodograms of EEG time series data in the spectral domain by Ngo *et al.* [25], and Euan and Sun [26] used functional boxplots to visualize clustering results for time series data and proposed the directional functional boxplot for visualizing the directional spectrum. Sun and Stein [27] utilized the functional boxplot to assess the statistical properties of their spatiotemporal rainfall model, and La Vecchia and Ronchetti [28] made use of the functional boxplot to examine the variability of the PP plot for comparing different methods when approximating the distribution of estimators and test statistics.

Other visualization tools exist for functional data, such as the functional bagplot and the functional highest density region boxplot [29] based on the first two robust principal component scores, and boxplots for amplitude, phase, and vertical translation [30] in the context of functional data registration.

2.2 Surface Boxplots

Sun and Genton [7] presented an illustration of the surface boxplot as a potential extension of the functional boxplot when observations are surfaces or images. The surface boxplot can be constructed similar to the functional boxplot once the ranking of the sample surfaces is obtained. However, the 3D visualization is not trivial. A user-friendly visualization tool was developed by Genton *et al.* [31] and illustrated on several real-world data examples. Yan *et al.* [32] utilized this tool to visualize estimated covariance matrices from simulations by treating the matrices as images.

3 Multivariate Functional Data Visualization

In the gait cycle data example, the knee and hip angles of each child must be dependent and can be visualized jointly as multivariate functional data, that is, a random vector indexed by time. For multivariate functional data, the outlier detection is even more complicated because multivariate functional outliers are not necessarily marginal outliers, and the outlyingness could also occur in magnitude, shape, or both, as in the univariate case. There is no clear definition for each type of outliers. Dai *et al.* [33] proposed a set of transformation operations on functional data for outlier-detection purposes, where potential outliers are either magnitude or shape outliers after suitable transformations. Next, we describe several tools for the visualization of multivariate functional data as well as their outlier detection.

3.1 Magnitude–Shape Plots

In order to rank multivariate functional data, Dai and Genton [11] introduced the notion of directional outlyingness, a vector $\mathbf{O}(t)$ which measures at time t the centrality of functional data by assessing the level and the direction of their deviation from the central region together. The mean vector **MO** and scalar variance VO of $\mathbf{O}(t)$ over time were defined to quantify the magnitude outlyingness and the shape outlyingness, respectively, of curves. Dai and Genton [34] proposed a new graphical tool, the magnitude–shape (MS) plot of (**MO**,VO), for visualizing both the magnitude and shape outlyingness of multivariate functional data. They also provided a dividing elliptical curve (or ellipsoidal surface) to separate nonoutlying data from the outliers. They demonstrated through Monte Carlo simulations and data applications that the MS plot is superior to the existing tools for visualizing centrality and detecting outliers for functional data, such as the FOM of Hubert *et al.* [15] and Rousseeuw *et al.* [16], both of which fail to detect shape outliers. Dai and Genton [34] also proposed the MS-plot array which displays marginal MS plots on the diagonal and pairwise bivariate MS plots on the off-diagonals.

Figure 3 depicts the bivariate and marginal data of hip and knee angles of each of the 39 children, as well as the associated bivariate and marginal MS plots. The light-gray ellipsoid and ellipses represent the thresholds to flag potential outliers. The dark-gray curves/points are the potential outliers identified. They are joint outliers but not marginal outliers. Hence, it is important to apply joint rather than marginal methods in the multivariate case.

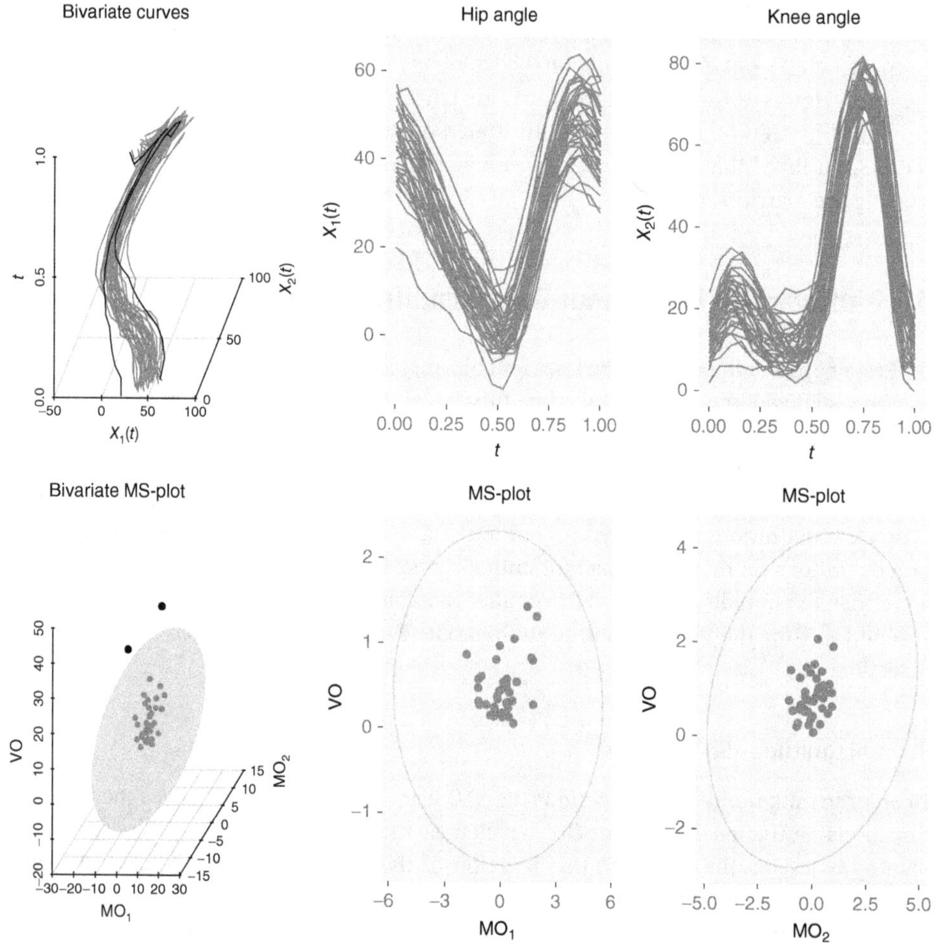

Figure 3 The bivariate and marginal MS plots for the hip and knee angles of each of the 39 children. The light-gray ellipsoid and ellipses represent the thresholds to flag potential outliers. The dark-gray curves/points are the potential outliers identified. They are joint outliers but not marginal outliers.

3.2 Two-Stage Functional Boxplots

Dai and Genton [35] proposed a two-stage functional boxplot for the visualization and exploratory FDA of multivariate curves. They showed that the shape of the original functional boxplot is sensitive to shape outliers and proposed to combine it with an outlier-detection procedure based on the aforementioned functional directional outlyingness in order to account for both the magnitude and shape outlyingness of functional data. In a first stage, shape outliers are identified, and in a second stage, the traditional functional boxplot is constructed on the remaining non-shape-outlying curves. This combination is robust to various types of outliers and allows to more precisely capture the data structures than does the functional boxplot alone. Moreover, it can provide both marginal and joint analyses of the multivariate curves.

We apply the two-stage functional boxplot to the hip and knee angles of each of the 39 children in Figure 4. The dashed curves are the potential shape outliers identified in the first stage. The shapes of the traditional functional boxplots in the second stage become slightly different from the shapes seen in the first row of Figure 2, in particular for the 50% central regions in gray and the whiskers.

3.3 Trajectory Functional Boxplots

An important subset of multivariate functional data consists of the so-called trajectory functional data, which are viewed without a dimension/axis for time. Examples of trajectory functional data include the paths of hurricanes or migrating birds in the 2D plane. One could also view the gait data in the hip/knee 2D plane, see Figure 5a. Note that such data could be open or closed curves.

Mirzargar *et al.* [36] introduced the curve boxplot as a generalization of the boxplot to an ensemble of curves. Yao *et al.* [37] proposed two exploratory tools for trajectory functional data: the trajectory functional boxplot and the modified simplicial band depth (MSBD)

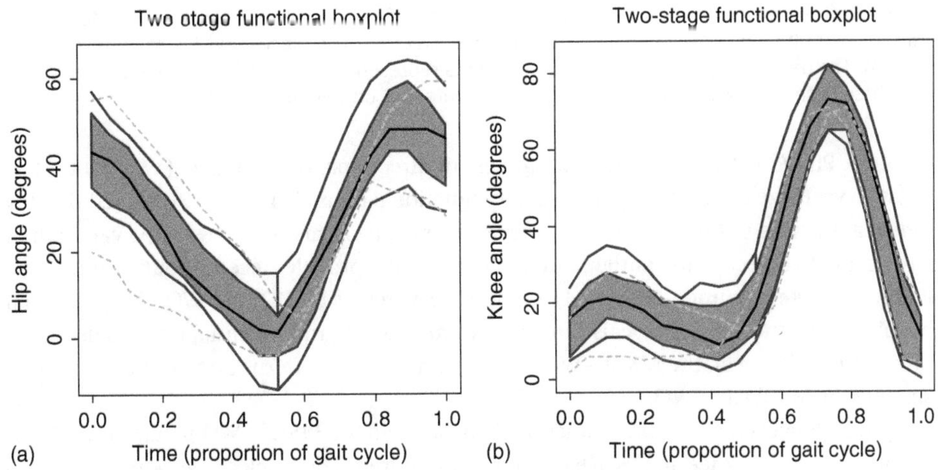

Figure 4 The two-stage functional boxplots for the hip (a) and knee (b) angles of each of the 39 children. The dashed curves are the potential bivariate shape outliers identified.

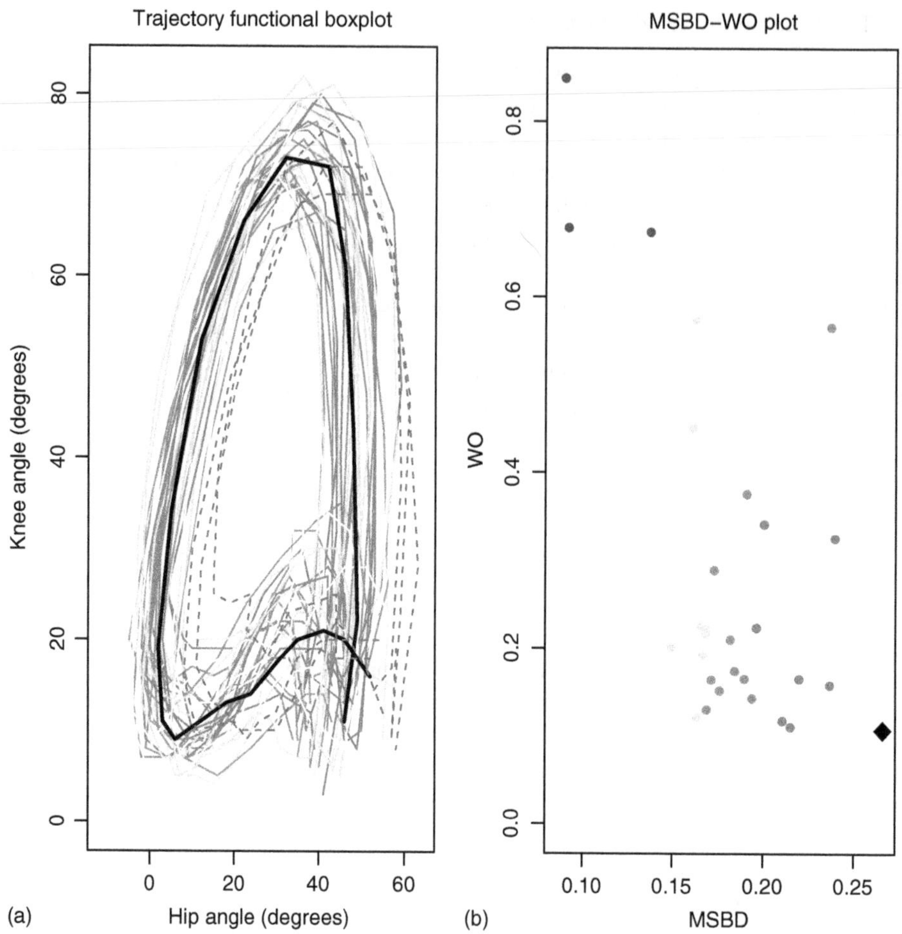

Figure 5 The trajectory functional boxplot (a) and the MSBD–WO plot (b) for the bivariate trajectories of hip and knee angles of each of the 39 children. The black curve/point represents the median. The dark-gray/gray/light-gray curves/points represent the 25%/50%/75% central regions. The three dashed curves/dark-gray points are the potential outliers identified.

of López-Pintado *et al.* [10] versus wiggliness of directional outlyingness (WO) plot. They allow to visualize the centrality of trajectory functional data. Yao *et al.* [37] defined the WO index, a measure of the roughness of curves, to effectively measure the shape variation of curves and to detect shape outliers. In addition, MSBD provides a center-outward ranking and identifies magnitude outliers. Using the measures of MSBD and WO, the functional boxplot of the trajectories reveals center-outward patterns and potential outliers using the raw curves. Moreover, the MSBD–WO plot illustrates such patterns and outliers in a space spanned by MSBD and WO.

Figure 5 depicts the trajectory functional boxplot and the MSBD–WO plot for the 2D trajectories of hip and knee angles of each of the 39 children. The black curve/point represents the median. The dark-gray/gray/light-gray curves/points represent the 25%/50%/75% central regions. The three dashed curves/dark-gray points are the potential outliers identified.

Note that the outliers detected with WO are not necessarily the same as those detected with directional outlyingness. Yao *et al.* [37] showed that WO is a more effective measure for detecting outliers in trajectory functional data.

4 Conclusions

The visualization of functional data is essential for many applications. This chapter focused on visualization tools and outlier-detection techniques based on functional data ranking, from univariate to multivariate cases. The reviewed methods have been illustrated on a dataset consisting of the angles formed by the hip and knee recorded from 39 children over their gait cycles.

These rank-based methods are robust and do not require any model or distribution assumptions. However, the computational cost might limit their capability in practice. The functional boxplot is implemented by the user-friendly `fbplot` command available in the *fda* packages [38] in R [39] and in MATLAB [40]. It is straightforward to extract important statistics from the `fbplot` command for future analysis and visualization purposes. Although it is possible to use depth values from other notions to rank functional data, the functional boxplot by default uses the fast algorithm of Sun *et al.* [20] for calculating the modified band depth values, which makes many applications feasible.

We have only illustrated univariate and bivariate curves, as well as trajectories, in this chapter. With modern data collection instruments and various applications from different research fields, the observations can be even more complex functional objects. To understand the features of complicated data from all the aspects, visualization is particularly useful for exploratory FDA, yet challenging. There are many open problems related to robust statistics and computations that require further research. Moreover, we have only focused on discussing the visualization of functional observations. Many core methods in FDA have common graphics to visualize the results from standard analyses, such as scree plots for functional principal component analysis and coefficient function plots in regression models. Wrobel *et al.* [41] illustrated how to create interactive graphics for functional data analyses with the *refund.shiny* package [42] in R.

Acknowledgment

This research was supported by the King Abdullah University of Science and Technology (KAUST).

References

1 Ramsay, J.O. and Silverman, B.W. (2005) *Functional Data Analysis*, Springer Series in Statistics, 2nd edn, Springer, New York.

2 Wang, J.L., Chiou, J.M., and Müller, H.G. (2016) Functional data analysis. *Ann. Rev. Stat. Appl.*, 3 (1), 257–295.

3 Genton, M.G., Castruccio, S., Crippa, P., *et al.* (2015) Visuanimation in statistics. *Stat*, 4, 81–96.

4 Castruccio, S., Genton, M.G., and Sun, Y. (2019) Visualizing spatiotemporal models with virtual reality: from fully immersive environments to applications in stereoscopic view. *J. R. Stat. Soc. A*, 182 (2), 379–387.

5 López-Pintado, S. and Romo, J. (2009) On the concept of depth for functional data. *J. Am. Stat. Assoc.*, 104 (486), 718–734.

6 Genton, M.G. and Hall, P. (2016) A tilting approach to ranking influence. *J. R. Stat. Soc. B. Stat. Methodol.*, 78 (1), 77–97.

7 Sun, Y. and Genton, M.G. (2011) Functional boxplots. *J. Comput. Graph. Stat.*, 20 (2), 316–334.

8 Ieva, F. and Paganoni, A.M. (2013) Depth measures for multivariate functional data. *Commun. Stat. Theory Methods*, 42 (7), 1265–1276.

9 Claeskens, G., Hubert, M., Slaets, L., and Vakili, K. (2014) Multivariate functional halfspace depth. *J. Am. Stat. Assoc.*, 109 (505), 411–423.

10 López-Pintado, S., Sun, Y., Lin, J.K., and Genton, M.G. (2014) Simplicial band depth for multivariate functional data. *Adv. Data Anal. Classif.*, 8 (3), 321–338.

11 Dai, W. and Genton, M.G. (2019) Directional outlyingness for multivariate functional data. *Comput. Stat. Data Anal.*, 131, 50–65.

12 Huang, H. and Sun, Y. (2019) A decomposition of total variation depth for understanding functional outliers. *Technometrics*, 61 (4), 445–458.

13 Narisetty, N.N. and Nair, V.N. (2016) Extremal depth for functional data and applications. *J. Am. Stat. Assoc.*, 111 (516), 1705–1714.

14 Arribas-Gil, A. and Romo, J. (2014) Shape outlier detection and visualization for functional data: the outliergram. *Biostatistics*, 15 (4), 603–619.

15 Hubert, M., Rousseeuw, P.J., and Segaert, P. (2015) Multivariate functional outlier detection. *Stat. Methods Appl.*, 24 (2), 177–202.

16 Rousseeuw, P.J., Raymaekers, J., and Hubert, M. (2018) A measure of directional outlyingness with applications to image data and video. *J. Comput. Graph. Stat.*, 27 (2), 345–359.

17 Harris, T., Tucker, J.D., Li, B. and Shand, L. (2019) Elastic depths for detecting shape anomalies in functional data. *arXiv:1907.06759*.

18 Tukey, J.W. (1977) *Exploratory Data Analysis*, Addison-Wesley, Reading, PA.

19 Sun, Y. and Genton, M.G. (2012) Adjusted functional boxplots for spatio-temporal data visualization and outlier detection. *Environmetrics*, 23 (1), 54–64.

20 Sun, Y., Genton, M.G., and Nychka, D.W. (2012) Exact fast computation of band depth for large functional datasets: how quickly can one million curves be ranked? *Stat*, 1, 68–74.

21 Martínez-Hernández, I., Genton, M.G., and González-Farías, G. (2019) Robust depth-based estimation of the functional autoregressive model. *Comput. Stat. Data Anal.*, 131, 66–79.

22 Huang, H. and Sun, Y. (2019) Visualization and assessment of spatio-temporal covariance properties. *Spat. Stat.*, 34, 100272.

23 Sun, Y. and Genton, M.G. (2012) Functional median polish. *J. Agric. Biol. Environ. Stat.*, 17 (3), 354–376.

24 Qu, Z., Dai, W., and Genton, M.G. (2020) Robust functional multivariate analysis of variance with environmental applications. *Environmetrics*, in press. doi: 10.1002/env.2641
25 Ngo, D., Sun, Y., Genton, M.G., *et al.* (2015) An exploratory data analysis of electroencephalograms using the functional boxplots approach. *Front. Neurosci.*, 9 (282), 1–18.
26 Euan, C. and Sun, Y. (2019) Directional spectra-based clustering for visualizing patterns of ocean waves and winds. *J. Comput. Graph. Stat.*, 28 (3), 659–670.
27 Sun, Y. and Stein, M.L. (2015) A stochastic space-time model for intermittent precipitation occurrences. *Ann. Appl. Stat.*, 9 (4), 2110–2132.
28 La Vecchia, D. and Ronchetti, E. (2019) Saddlepoint approximations for short and long memory time series: a frequency domain approach. *J. Econom.*, 213 (2), 578–592.
29 Hyndman, R.J. and Shang, H.L. (2010) Rainbow plots, bagplots, and boxplots for functional data. *J. Comput. Graph. Stat.*, 19 (1), 29–45.
30 Xie, W., Kurtek, S., Bharath, K., and Sun, Y. (2017) A geometric approach to visualization of variability in functional data. *J. Am. Stat. Assoc.*, 112 (519), 979–993.
31 Genton, M.G., Johnson, C., Potter, K., *et al.* (2014) Surface boxplots. *Stat*, 3, 1–11.
32 Yan, Y., Huang, H.C., and Genton, M.G. (2020) Vector autoregressive models with spatially structured coefficients for time series on a spatial grid. *arXiv:2001.02250*.
33 Dai, W., Mrkvička, T., Sun, Y., and Genton, M.G. (2020) Functional outlier detection and taxonomy by sequential transformations. *Comput. Stat. Data Anal.*, 149, 106960.
34 Dai, W. and Genton, M.G. (2018) Multivariate functional data visualization and outlier detection. *J. Comput. Graph. Stat.*, 27 (4), 923–934.
35 Dai, W. and Genton, M.G. (2018) Functional boxplots for multivariate curves. *Stat*, 7, e190.
36 Mirzargar, M., Whitaker, R.T., and Kirby, R.M. (2014) Curve boxplot: generalization of boxplot for ensembles of curves. *IEEE Trans. Vis. Comput. Graph.*, 20 (12), 2654–2663.
37 Yao, Z., Dai, W. and Genton, M.G. (2020) Trajectory functional boxplots. *Stat*, in press. doi: 10.1002/sta4.289
38 Ramsay, J.O., Wickham, H., Graves, S., and Hooker, G. (2018) fda: functional data analysis. R package version 2.4.8.
39 R Core Team (2019) *R: A Language and Environment for Statistical Computing*, R Foundation for Statistical Computing, Vienna, Austria.
40 MATLAB (2018) version 9.4.0 (R2018a). The MathWorks Inc., Natick, MA.
41 Wrobel, J., Park, S.Y., Staicu, A.M., and Goldsmith, J. (2016) Interactive graphics for functional data analyses. *Stat*, 5 (1), 108–118.
42 Wrobel, J. and Goldsmith, J. (2016) refund.shiny: Interactive Plotting for Functional Data Analyses, R package version 0.3.0.

Part VI

Numerical Approximation and Optimization

25

Gradient-Based Optimizers for Statistics and Machine Learning

Cho-Jui Hsieh

University of California, Los Angeles, CA, USA

1 Introduction

In many statistical and machine learning problems, the goal is to find the model parameters that best fit the training data. These problems can be naturally formulated as optimization: If we define a loss function to measure "how good do the parameters fit training data," then the parameter estimation problem is equivalent to finding the best parameter to minimize the prediction loss defined on the training set, which can be solved by optimization. Examples include but not limited to classification, regression, clustering, and dimensional reduction. Furthermore, neural network training is also based on these optimization methods. In this chapter, we introduce several commonly used methods for unconstrained optimization and use empirical risk minimization (ERM) problems, including classification and regression, as running examples to demonstrate how to apply these optimization algorithms in practice.

Formally, let $f : \mathbb{R}^d \to \mathbb{R}$ be the objective function to be minimized, the unconstrained optimization problem that will be discussed in this chapter can be written as

$$\min_x f(x) \tag{1}$$

We introduce two commonly used algorithms: gradient descent and stochastic gradient for solving unconstrained minimization problems. To make the discussions more concrete, we are going to focus on the following representative problems:

- Ridge regression: Given n training samples $\{x_i, y_i\}_{i=1}^n$, where each $x_i \in \mathbb{R}^d$ and $y_i \in \mathbb{R}$, ridge regression aims to build a linear model parameterized by w, such that $w^T x_i \approx y_i$. The model parameter w is computed by solving the following optimization problem:

$$\arg\min_w \sum_{i=1}^{n} \|w^T x_i - y_i\|_2^2 + \lambda \|w\|_2^2 \tag{2}$$

  where λ is a balancing parameter between the training error (the first term) and regularization (the second term). Clearly, this is a simple unconstrained quadratic programming problem, and in fact, it has a closed form solution. However, we show that numerical

Computational Statistics in Data Science.
Edited by Walter W. Piegorsch, Richard A. Levine, Hao Helen Zhang and Thomas C. M. Lee.
 ISBN 978-1-11956107-1

optimization can still be more efficient in practice, especially when dealing with large problems.

- We also discuss how to deal with the general loss function instead of the square loss. The objective function can be written as

$$\ell_2\text{-regularized logistic regression } \arg\min_w \sum_{i=1}^{n} \ell(w^T x_i, y_i) + \lambda \|w\|_2^2 \tag{3}$$

where $\ell(\hat{y}, y)$ is a loss function measuring the difference between the predicted label $\hat{y}$ and the observed label y. For binary classification problem when $y = \{+1, -1\}$, the loss function can be the logistic loss or the hinge loss:

$$\text{Logistic loss: } \ell_{\text{logistic}}(\hat{y}, y) = \log(1 + e^{-y\hat{y}})$$
$$\text{Hinge loss: } \ell_{\text{hinge}}(\hat{y}, y) = \max(0, 1 - \hat{y}y)$$

- *Lasso regression*: To make the learned model w sparse (with many zero elements), a typical way is to replace the ℓ_2 regularization term (the second term) in (2) by an ℓ_1 regularization. The resulting optimization problem can be written as

$$\arg\min_w \sum_{i=1}^{n} \|w^T x_i - y_i\|_2^2 + \lambda \|w\|_1 \tag{4}$$

In this Lasso formulation, a larger λ leads to more sparse solution. This objective function is nondifferentiable with respect to w_i when $w_i = 0$, which introduces difficulty to the optimizer. We thus discuss how to handle this kind of nonsmooth or nondifferentiable regularizations in this chapter.

- Finally, the methods introduced in this chapter can also be used for solving a general ERM problem, where the model, represented as $f_w(\cdot)$ that maps x to y, may not be linear. The objective function will then become

$$\text{General ERM problem: } \arg\min_w \sum_{i=1}^{n} \ell(f_w(x_i), y_i) + R(x) \tag{5}$$

where $R(x)$ is the regularization term. This includes many state-of-the-art machine learning models. For example, when $f_w(\cdot)$ is a neural network function, (5) becomes the neural network training problem, and we discuss which algorithm is suitable for this general form.

2 Convex Versus Nonconvex Optimization

Before introducing the optimization algorithms for solving (1), we first discuss some background in optimization. Based on the property of the objective function, we can classify the problems into convex and nonconvex optimization. For convex problems, the objective function is convex, and the formal definition of convexity is that

$$f(\alpha x_1 + (1-\alpha)x_2) \leq \alpha f(x_1) + (1-\alpha) f(x_2), \quad \forall x_1, x_2 \in \mathbb{R}^d \tag{6}$$

and for twice differentiable functions, it can be proved that this condition is equivalent to

$$\nabla^2 f(x) \succeq 0 \tag{7}$$

which means the Hessian matrix $\nabla^2 f(x)$ is positive semidefinite. On the other hand, we say the objective function is nonconvex if it does not satisfy (6), or equivalently, if the objective function is twice differentiable, there exist some points that the Hessian is not positive semidefinite.

A convex objective function enjoys several nice properties. The main benefit of having a convex objective function is that convexity implies

$$\nabla f(x^*) = 0 \text{ if and only if } x^* \text{ is a minimizer of } f(x)$$

Therefore, if the optimization algorithm guarantees finding an x with zero gradient, this is guaranteed to be a global minimizer of the problem. In contrary, for nonconvex problems, a point x^* with $\nabla f(x^*) = 0$ can be global minimum, local minimum, or saddle point. Unfortunately, it will become extremely difficult to find global minimizers in the nonconvex cases, and most algorithms (including gradient descent and stochastic gradient descent) are only able to find points with vanishing gradient, which could be local minimums or saddle points.

3 Gradient Descent

3.1 Basic Formulation

To optimize an objective function, a numerical optimizer usually starts from an *initial solution* x_0 and iterative updates the solution to generate a sequence $x_0, x_1, x_2, \ldots$ such that the sequence will converge to minimizers (for convex cases) or stationary points (for nonconvex cases). For both of these cases, we would like the sequence to converge to a point x^* such that $\nabla f(x^*) = 0$. This motivates gradient descent, one of the most simple and classical optimization algorithms. The gradient descent algorithm updates x_t to x_{t+1} by

$$x_{t+1} \leftarrow x_t - \eta \nabla f(x_t) \tag{8}$$

The intuition is to update x based on the negative gradient direction, which is supposed to be the direction that can maximally decrease the objective function value locally, multiplied with a learning rate $\eta > 0$.

To explain why gradient descent work, we introduce the following view of "successive function approximation." Assume x_t is the current solution; ideally, we want to find an update p such that $f(x_t + p)$ is minimized. However, this is equivalent to the original minimization problem which we do not know how to solve. Therefore, instead of minimizing $f(\cdot)$ directly, gradient descent builds a local function approximation based on Taylor expansion:

$$f(x_t + p) = f(x_t) + \nabla f(x_t)^T p + \frac{1}{2} p^T \nabla^2 f(z) p \tag{9}$$

$$\approx \underbrace{\left\{ f(x_t) + \nabla f(x_t)^T p + \frac{1}{2} p^T \left(\frac{1}{\eta} I \right) p \right\}}_{q_{x_t}(p)} \tag{10}$$

where z is some vector on line$(x, x+p)$, and in the approximation function, the second-order term $\nabla^2 f(z)$ is approximated by $(\frac{1}{\eta}I)$ with a constant η. The update rule of gradient descent is then equivalent to finding the minimizer of this local approximate function $q_w(p)$. To see this, let

$$p^* = \arg\min_p q_{x_t}(p) \tag{11}$$

then since the optimal solution of (11) satisfies $\nabla q_{x_t}(p^*) = 0$, we have $p^* = -\eta \nabla f(x_t)$, leading to the gradient descent update. Therefore, this is what gradient descent is doing when minimizing $f(x)$: at each iteration, builds a local (quadratic) approximate function $q_{x_t}(p) \approx f(x_t + p)$ for small p and then updates based on minimizing this approximate function. Clearly, this procedure does not work for any step size η, and thus, it is important to select a good step size. In fact, we have the following theorem:

Assumption 1. *The objective function $f(\cdot)$ is twice differentiable, and $\nabla^2 f(w) \preccurlyeq LI$.*

Theorem 1. *If unconstrained optimization with objective function satisfies Assumption 1, gradient descent converges to stationary points if $\eta < \frac{2}{L}$.*

We omit the proof for this theorem, but the main message here is that gradient descent will converge as long as the step size is small enough, which means the local approximation function $q_w(\cdot)$ is a conservative approximation of the original Hessian.

3.2 How to Find the Step Size?

We already know that gradient descent converges when the step size $\eta < \frac{2}{L}$, but the constant L is unknown in practice. It is thus tricky to choose the step size. When the step size is too small, despite the convergence of the gradient descent (based on Theorem 2), the convergence speed will be very slow. On the other hand, using a too large step size will make the gradient descent algorithm diverge. Therefore, choosing a good step size is important in practice. One way is to do a hyperparameter search, where we try a grid of η from large to small until finding a η that will not diverge. However, this is often time consuming and requires a lot of human efforts.

To better choose a step size and automate the whole procedure, several *line search* approaches have been proposed to automatically choose the step size at each gradient descent iteration. The main idea is also simple. Given the current solution x_t, we try a series of step sizes from large to small, such as $\{\bar{\eta}, \bar{\eta}/2, \bar{\eta}/4, \ldots\}$, and stop when we find a step size that can "sufficiently decrease" the objective function value. This is called the "backtracking" line search, and the following equation can be used to judge whether the current step size sufficiently decrease the objective function value:

$$\text{Sufficient decrease condition}: f(x - \eta g) > f(x) - c\eta\|g\|^2 \tag{12}$$

where g is the gradient, and $c \in (0, 1)$ is a small constant. The second term in the sufficient decrease condition is to guarantee that the step size not only decreases the objective function but also decreases it sufficiently to guarantee the convergence, and whether it is sufficient enough is judged by the norm of the gradient. The backtracking line search procedure can be summarized in Algorithm 1.

Algorithm 1. Gradient descent with backtracking line search

Initial $\boldsymbol{x}_0$
for $t = 0, 1, \ldots$ **do**
  Compute $\mathbf{g} = \nabla f(\boldsymbol{x}_t)$
  $\eta = \overline{\eta}$
  if $f(\boldsymbol{x}_t - \eta \mathbf{g}) \leq f(\boldsymbol{x}_t) - c\eta \|\mathbf{g}\|^2$ **then**
    $\eta \leftarrow \eta/2$
  end if
  $\boldsymbol{x}_{t+1} = \boldsymbol{x}_t - \eta \mathbf{g}$
end for

The convergence of Algorithm 1 is guaranteed by the following theorem:

Theorem 2. *If the objective function satisfies Assumption 1, then Algorithm 1 converges to stationary points of the objective function.*

3.3 Examples

Gradient descent can be easily applied to solve any objective function as long as the gradient is easy to compute. For example, in ridge regression (Equation 2), the gradient can be computed by

$$\text{Gradient of ridge regression} \quad \nabla f(w) = 2 \sum_{i=1}^{n} (w^T x_i - y_i) x_i + 2\lambda w \tag{13}$$

which can be done by a single pass of the dataset. Indeed, for any general ERM problem in (5), the gradient can be computed by

$$\nabla f(w) = \sum_{i=1}^{n} \left(\frac{\partial \ell(z, y_i)}{\partial z} |_{z=w^T x_i} \right) x_i + 2\lambda w \tag{14}$$

so gradient descent can be easily applied once the first derivative of loss function is computable.

4 Proximal Gradient Descent: Handling Nondifferentiable Regularization

In high-dimensional statistics, ℓ_1 regularization is widely used for obtaining a sparse solution, leading to the Lasso regression problem (4) or other ℓ_1-regularized ERM problems (5). Since ℓ_1 norm is nondifferentiable at 0, it imposes difficulties when applying the gradient descent. In this section, we discuss how to deal with ℓ_1 or other kinds of nonsmooth terms in the objective function in the gradient descent.

In particular, we discuss the following composite minimization problem:

$$\arg\min_{x} f(x) := \{g(x) + h(x)\} \tag{15}$$

where g(x) is a convex and differentiable function, and h(x) is also convex but may be nondifferentiable. Recall that we have shown the "successive function approximation" view of gradient descent in the previous section. Here, we show how to deal with composite minimization problems under the same framework. In (10), we are able to form the Taylor expansion of $f(x)$ since the whole objective function is differentiable. To extend this approach to the composite function, we conduct Taylor expansion to the differentiable part g(x) and keep the nondifferentiable part h(x) unchanged, leading to the following approximate function at iteration t:

$$f(x) \approx g(x_t) + \nabla g(x_t)^T(x - x_t) + \frac{1}{2\eta}||x - x_t||^2 + h(x) \tag{16}$$

$$= \frac{1}{2\eta}||(x - x_t) + \eta \nabla g(x_t)||^2 + h(x) + \text{constant} \tag{17}$$

$$= \frac{1}{2\eta}||x - (x_t - \eta \nabla g(x_t))||^2 + h(x) + \text{constant} \tag{18}$$

If we define $b = x_t - \eta \nabla g(x_t)$, then finding the optimal of (18) is equivalent to solving the following problem:

$$\arg\min_x \frac{1}{2}||x - b||^2 + \eta h(x) \tag{19}$$

This is the so-called proximal operator, formally defined as

$$\text{prox}_\eta(w) = \arg\min_x \frac{1}{2}||w - x||^2 + \eta h(x) \tag{20}$$

And therefore the proximal gradient descent update rule can be written as

$$x_{t+1} \leftarrow \text{prox}_\eta(x_t - \eta \nabla g(x_t)) \tag{21}$$

For example, for ℓ_1-regularized problem, where $h(x) = \lambda||x||_1$, the proximal operation can be computed by

$$\text{prox}_\eta(x)_i = \begin{cases} x_i - \eta\lambda & \text{if } x_i > \eta\lambda \\ x_i + \eta\lambda & \text{if } x_i < \eta\lambda \\ 0 & \text{if } |x_i| \leq \eta\lambda \end{cases} \tag{22}$$

Similar to gradient descent, the proximal gradient descent is guaranteed to converge to stationary point under certain conditions, and a line search method can be similarly applied.

5 Stochastic Gradient Descent

Despite having nice convergence properties, gradient descent needs to conduct an exact gradient evaluation at each iteration, which is time consuming for many large-scale problems. For example, many problems in machine learning have millions or billions of samples, and each gradient computation, which requires going through all the training samples, may need hours or days. Therefore, instead of applying gradient descent, most of the current machine learning algorithms use *stochastic gradient descent* (SGD) for training. In this section, we introduce the SGD algorithm and the current challenges when applying them in large-scale machine learning applications.

5.1 Basic Formulation

We introduce the SGD algorithm for minimizing the following finite-sum function:

$$\min_x f(x) := \left\{ \frac{1}{n} \sum_{i=1}^{n} f_i(x) \right\} \tag{23}$$

where in ERM problems each f_i is the loss defined on a training sample. The gradient of this function can be written as $\nabla f(x) = \sum_{i=1}^{n} \nabla f_i(x)$, so each iteration needs to go through the whole function.

The main idea of SGD is to use an unbiased estimator to approximate gradient for each update. For the finite-sum minimization problem, we can easily use subsampled gradient to estimate the full gradient. Therefore, the SGD update rule can be written as

$$x_{t+1} \leftarrow x_t - \eta_t \frac{1}{|B|} \sum_{i \in B} \nabla f_i(x_t) \tag{24}$$

where $B \subseteq \{1, \dots, n\}$ is a randomly sampled subset, where the average gradient of this subset is used to estimate the full gradient. $|B|$ denotes the size of the subset, also known as the *batch size* in machine learning. When $|B|$ is very small (in the extreme case, 1), each update of SGD is very efficient, but the gradient estimation is very noisy and tends to make less progress. On the contrary, if $|B|$ is very large, each update will be slower but more closer to the gradient descent update.

Intuitively, SGD works since the gradient estimation is unbiased. However, the noise introduced by SGD will break several nice convergence properties of the gradient descent. First, SGD does not converge to a stationary point when using a fixed step size. We can easily see this by assuming that x^* is a stationary point, and if we start at x^* and apply the SGD update rule (24) with a fixed learning rate η, it will actually move away from the stationary point. Therefore, to ensure the convergence of SGD, the step size η_t has to be a decrease sequence such that

$$\lim_t \eta_t = 0 \text{ and } \sum_{t=1}^{\infty} \eta_t = \infty$$

where the second property is to ensure that SGD can converge to the stationary point even when the initial point is far away. A common choice is to use polynomial decay: $\eta_t = \eta_0 t^{-a}$, where η_0 is the initial step size, and $a > 0$ is a decay factor. SGD has the following convergence property:

Theorem 3. *For convex differentiable functions with a bounded Lipschitz constant, when applying the SGD update, we have* $f(x_t) - f(x^*) = O(\frac{1}{\sqrt{t}}$, *and the rate will be* $O(\frac{1}{K})$ *in the strongly convex case.*

Note that these convergence speeds are strictly worse than the gradient descent – gradient descent achieves $O(\frac{1}{t})$ convergence rate for convex functions and linear convergence for strongly convex functions. However, SGD is still widely used in machine learning when facing large amount of training samples, since each update is much more efficient than the full gradient descent if we use a small batch size.

Although the vanilla SGD is still useful for many machine learning training tasks, people have found that in many cases adding momentum to SGD or using adaptive step size for each parameter can significantly improve the performance. This leads to several variations of SGD including RMSProp [1], Adagrad [2], and Adam [3].

5.2 Challenges

Despite being a dominating technique in large-scale machine learning, especially training deep neural networks, many challenges remain when applying SGD to real-world problems. We give two examples below.

- *Learning rate schedule*: The performance of SGD and its variants is very sensitive to the learning rate schedule. For some applications, not only the initial learning rate has to be carefully tuned, the decay rate is also important for achieving good performance. Furthermore, it has been observed that in many applications the learning rate needs to have a "warm-up" schedule, which increases the learning rate at first few iterations before entering the decay phase. Other fancier schedules such as cyclical schedule have also been proposed [4]. How to best tune the learning rate schedule is still an open problem for SGD.
- *Large-batch training*: The batch size of SGD is usually chosen according to the computing resource available. If we have more workers (e.g., CPUs and GPUs), we can usually increase the batch size linearly to the number of workers to fully utilize the computation resources. However, in deep learning training it has been observed that using a large batch size in SGD training will lead to degraded performance. This is a bit counterintuitive, since large batch should speed up convergence in the convex case. However, deep learning training objectives are highly nonconvex, and the algorithms often converge to local minimums, where different local minimums may have different generalization performance on test samples. Keskar *et al.* [5] observed that when increasing the batch size of SGD, it will often converge to "sharp local minimums," which leads to worse test accuracy. Many works since then have been trying to maintain the test performance when increasing the batch size; for instance, Goyal *et al.* [6] showed that heavy data augmentation and some batch size scaling techniques can help large batch training; You *et al.* [7–9] showed that a layer-wise learning rate scaling can scale up ImageNet and BERT training. However, all these current approaches still have limitations, and it is still an open problem how to further increase the batch size while maintaining the same test performance on large datasets.

References

1 Tieleman, T. and Hinton, G. (2012) *Lecture 6.5—RmsProp: Divide the Gradient by a Running Average of its Recent Magnitude*. COURSERA: Neural Networks for Machine Learning.

2 Duchi, J., Hazan, E., and Singer, Y. (2011) Adaptive subgradient methods for online learning and stochastic optimization. *J. Mach. Learn. Res.*, 12 (7), 2121–2159.

3 Kingma, D.P. and Ba, J. (2014) Adam: a method for stochastic optimization. *arXiv preprint arXiv:1412.6980*.
4 Smith, L.N. (2017) *Cyclical Learning Rates for Training Neural Networks*. 2017 IEEE Winter Conference on Applications of Computer Vision (WACV), pp. 464–472. IEEE.
5 Keskar, N.S., Mudigere, D., Nocedal, J. *et al.* (2016) On large-batch training for deep learning: generalization gap and sharp minima. *arXiv preprint arXiv:1609.04836*.
6 Goyal, P., Dollár, P., Girshick, R. *et al.* (2017) Accurate, large minibatch SGD: training ImageNet in 1 hour. *arXiv preprint arXiv:1706.02677*.
7 You, Y., Gitman, I., and Ginsburg, B. (2017) Scaling SGD Batch Size to 32k for ImageNet Training. *Tech. Report No. UCB/EECS-2017-156*.
8 You, Y., Zhang, Z., Hsieh, C-J. *et al.* (2018) *ImageNet Training in Minutes*. Proceedings of the 47th International Conference on Parallel Processing, pp. 1–10.
9 You, Y., Li, J., Reddi, S. *et al.* (2019) Large batch optimization for deep learning: training BERT in 76 minutes. *arXiv preprint arXiv:1904.00962*.

26

Alternating Minimization Algorithms

David R. Hunter

Penn State University, State College, PA, USA

1 Introduction

It must be stated from the outset that there seems to be no universal agreement in the literature on the precise meaning of the phrase "alternating minimization algorithms." That said, the basic idea is both very simple and very general: Suppose that we are given a real-valued function of two variables, $D(P, Q)$, and the goal is to minimize this function. An alternating minimization algorithm operates by switching back and forth between minimizing over Q while holding P fixed and then minimizing over P while holding Q fixed. Though there are myriad algorithms that might be classified as alternating minimization, they share a common rationale: Each of the separate minimizations is simpler – more mathematically tractable, more computationally efficient, and amenable to closed-form solutions – than the direct minimization of $D(P, Q)$. In many cases, the price paid for this simplicity is iteration, as alternating minimization switches back and forth repeatedly between the simpler subproblems.

Csiszár and Tusnády [1] introduced a convenient notation to express how an alternating minimization algorithm operates. We are given a starting value Q_0 of the variable Q (or P_0 of P), then we find $(P_r, Q_r)_{r\geq 1}$ according to

$$Q_{r-1} \stackrel{2}{\rightarrow} P_r \stackrel{1}{\rightarrow} Q_r \stackrel{2}{\rightarrow} \cdots \text{ or } P_{r-1} \stackrel{1}{\rightarrow} Q_r \stackrel{2}{\rightarrow} P_r \stackrel{1}{\rightarrow} \cdots$$

for $r = 1, 2, \ldots$, where

$$P \stackrel{1}{\rightarrow} Q \text{ means that } Q = \arg\min_q D(P, q) \tag{1}$$

$$Q \stackrel{2}{\rightarrow} P \text{ means that } P = \arg\min_p D(p, Q) \tag{2}$$

Naturally, one could extend the alternating minimization idea to a real-valued function of more than two variables, say, $D(P_1, P_2, \ldots, P_k)$, and here we consider this generalization to fall under the general category of alternating minimization. Yet we also find that there

Computational Statistics in Data Science.
Edited by Walter W. Piegorsch, Richard A. Levine, Hao Helen Zhang and Thomas C. M. Lee.
 ISBN 978-1-11956107-1

is a surprising range of problems in which the two-variable setup applies directly. In this chapter, we devote a great deal of attention to the particular sense of "alternating minimization" intended by Csiszár and Tusnády [1], who take P and Q to be not scalar, vector, or even matrix quantities but rather probability distributions, where $D(P, Q)$ is a measure of the statistical distance between P and Q. This chapter shows how to view the well-known class of expectation–maximization (EM) algorithms as alternating minimization in this sense before exhibiting several other instances of alternating minimization such as various matrix factorization, matrix completion, and clustering algorithms. First, however, we describe the most basic form that alternating minimization might take, namely, the methods known collectively as block coordinate descent (BCD) or block coordinate minimization (BCM).

2 Coordinate Descent

Perhaps the simplest example of alternating minimization is coordinate descent (CD), in which P and Q are taken to be the coordinates of a bivariate vector, say, $\beta_1 = P$ and $\beta_2 = Q$, and our goal is to minimize the objective function $D(\boldsymbol{\beta})$ for $\boldsymbol{\beta} \in \mathbb{R}^2$. CD is an iterative algorithm that starts at step $r = 0$ from some point $\boldsymbol{\beta}^0$, finding successive points $\boldsymbol{\beta}^1, \boldsymbol{\beta}^2, \ldots$ satisfying $D(\boldsymbol{\beta}^{r+1}) \leq D(\boldsymbol{\beta}^r)$ for each $r \geq 0$. CD operates not on the $\boldsymbol{\beta}$ vector all at once but rather on each of its coordinates individually, switching back and forth between β_1 and β_2, changing one's value while holding the other constant (relative to the current iterate) so as to decrease the objective function's value.

There is no reason that the number of coordinates in a CD algorithm should be limited to 2; if $\boldsymbol{\beta}$ is d-dimensional, then we may cycle through the d coordinates one at a time. Using a slight expansion of the usual meaning of "alternating," we still call this idea "alternating minimization." On the other hand, if $\boldsymbol{\beta}$ is d-dimensional, then we may partition its coordinates into some number B of blocks of coordinates, minimizing or decreasing the objective function over each block as a unit as we cycle through the blocks. Such an algorithm may be referred to as BCM or BCD, depending on whether we minimize the function's value or merely decrease it for each of the blocks. Regardless of the number of blocks, or of whether we consider BCM or BCD, we again stretch the meaning of "alternating minimization" to include all such algorithms because they all employ the same basic philosophy: Break the main problem into an iterative series of simpler ones. While within one iteration and cycling through the blocks of coordinates making up the $\boldsymbol{\beta}$ vector, one might hold blocks already visited during the same iteration constant at their newly updated values or at the values they held prior to the start of the iteration. The updates are sometimes called Gauss–Seidel updates in the former case and Jacobi updates in the latter. It is often the case that Gauss–Seidel updates result in fewer iterations before convergence, whereas Jacobi updates may be more easily parallelized.

The topics of BCD/BCM have spawned a huge amount of literature, which we will not attempt to summarize here. Shi *et al.* [2] provide a primer on these algorithms, while Lange *et al.* [3] describe many of them in contexts more specifically tailored to statistics. Here, we merely illustrate the main idea by considering a particularly common modern statistical

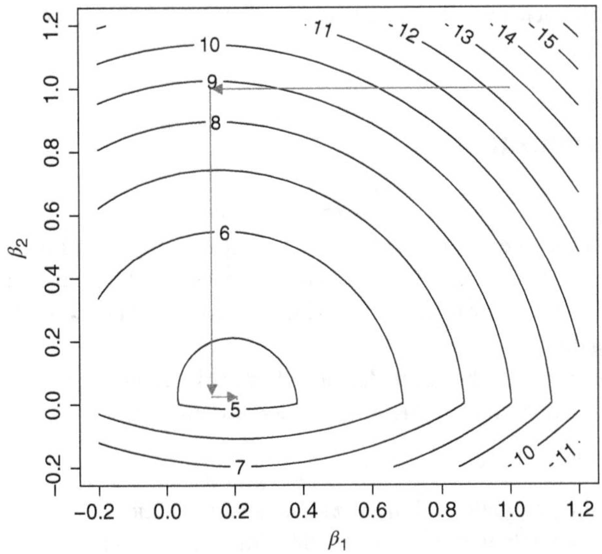


```
set.seed(123)
x <- matrix(rnorm(10))
y <- rnorm(10)

library(CDLasso)
l2.reg(t(x), y, lambda=5)
```

Figure 1 Minimizing $D(\beta_1, \beta_2)$ of Equation (3) via coordinate descent starting from the initial point $(\beta_1^0, \beta_2^0) = (1, 1)$. The level curves of the objective function with $\lambda = 5$ are shown along with the iteratively updated estimates that converge to the minimizer. As is characteristic of CD, each iterative update change occurs along only one coordinate direction at a time. At the right is the R code used to generate the data and find the minimizer.

problem, namely, penalized least-squares regression, in which the objective function takes the form

$$D(\boldsymbol{\beta}) = \sum_{i=1}^{n} (y_i - \mathbf{x}_i^\top \boldsymbol{\beta})^2 + \sum_{j=1}^{d} p_j(\beta_j)$$

for observed data $(\mathbf{x}_1, y_1), \ldots, (\mathbf{x}_n, y_n)$. As an example, we consider the case of linear regression with an intercept term and a lasso penalty on the slope parameters. For ease of depicting the objective function graphically, we assume that $\boldsymbol{\beta}$ has only two dimensions, so that

$$D(\beta_1, \beta_2) = \frac{1}{2} \sum_{i=1}^{n} (y_i - \beta_1 - x_i\beta_2)^2 + \lambda|\beta_2| \tag{3}$$

where λ is the tuning parameter that controls the amount of penalization, and the factor 1/2 is inserted in front of the sum-of-squares term in Equation (3) so that the objective function matches the one used by Wu and Lange [4]. We use R [5] to construct a simple simulated example, where we take $n = 10$, and every (x_i, y_i) pair consists of two independent standard normal random variables. The objective function (3) with tuning parameter $\lambda = 5$ is depicted in Figure 1.

The function $D(\beta_1, \beta_2)$ has no closed-form minimizer, and because it is not differentiable it is not amenable to many types of algorithms that require derivatives. Wu and Lange [4] describe a CD algorithm for minimizing it and claim that this method, despite its simplicity, exhibits state-of-the-art performance for lasso-penalized regression problems. We take

$\lambda = 5$ and employ these authors' R package called `CDLasso`, for "CD lasso," to implement the coordinate descent algorithm.

3 EM as Alternating Minimization

In the EM algorithm framework as formulated by Dempster *et al.* [6], we wish to maximize a likelihood function $L(\theta)$ from data x where we do not observe $\mathbf{x}$ but rather $\mathbf{y} = \mathbf{y}(\mathbf{x})$, a known function of $\mathbf{x}$. In this context, we refer to $\mathbf{x}$ as the complete data and $\mathbf{y}$ as the observed data. In many cases, there is a natural way to express $\mathbf{x}$ as $(\mathbf{y}, \mathbf{z})$, where $\mathbf{z}$ is totally unobserved; in these cases, we call $\mathbf{z}$ the missing data.

Typically, we assume that the form of the joint density of $(\mathbf{y}, \mathbf{z})$ is known up to an unknown parameter θ. Letting Θ denote the parameter space, let us define the set

$$\mathcal{Q} = \{Q_\theta(\mathbf{y}, \mathbf{z}) : \theta \in \Theta\}$$

of joint distributions indexed by the parameter θ. If $q_\theta(\mathbf{y}, \mathbf{z})$ is a density function, the likelihood function based on the observed data $\mathbf{y}$ may be obtained by integrating it with respect to the missing data, that is,

$$L(Q_\theta) = \int q_\theta(\mathbf{y}, \mathbf{z})\, \mathbf{dz}$$

Typically, the EM framework is most useful in cases where $q_\theta(\mathbf{y}, \mathbf{z})$ is mathematically tractable, but $L(Q)$ is either difficult to maximize directly or impossible to obtain in closed form. Instead, we view this problem in the alternating minimization context of Csiszár and Tusnády [1]. We first define a second set $\mathcal{P}$ that consists of all conditional distributions on the missing data $\mathbf{z}$, given the observed data $\mathbf{y}$. Recall that our overall goal is to minimize a function $D(P, Q)$ for all $P \in \mathcal{P}$ and $Q \in \mathcal{P}$. We now define the function $D(P, Q)$ as in Equation (3) of Neal and Hinton [7], who explain directly how to view EM as alternating minimization:

$$D(P, Q) = \mathrm{KL}[P \,||\, Q(\cdot \mid \mathbf{y})] - L(Q) \tag{4}$$

In the expression above, $Q(\cdot \mid \mathbf{y})$ is the conditional distribution of $\mathbf{z}$ given $\mathbf{y}$ arising from the joint distribution $Q(\mathbf{y}, \mathbf{z})$, and $\mathrm{KL}[\cdot \,||\, \cdot]$ is the Kullback–Leibler divergence:

$$\mathrm{KL}[P \,||\, Q(\cdot \mid \mathbf{y})] = E_P \log\left[\frac{p(\mathbf{z})}{q(\mathbf{z} \mid \mathbf{y})}\right]$$

Equation (4) makes clear that a minimizer $(\hat{P}, \hat{Q})$ of $D(P, Q)$ also gives a maximizer $\hat{Q}$ of $L(Q)$, as desired, since the class $\mathcal{P}$ contains all conditional distributions of $\mathbf{z}$ given $\mathbf{y}$, so the KL divergence may be made zero. The power of the alternating minimization framework is that there are many problems in which explicit expressions for P satisfying $P \overset{1}{\rightarrow} Q$ and for Q satisfying $Q \overset{2}{\rightarrow} P$, in the sense of (1) and (2), exist, even when no such expression exists for $\hat{P}$ or $\hat{Q}$.

In the case of an EM algorithm, one may show that the E-step, or expectation step, is equivalent to the alternating minimization step $Q \overset{2}{\rightarrow} P$; similarly, the M-step, or maximization step, is equivalent to $P \overset{1}{\rightarrow} Q$. To prove these facts is beyond the scope of this chapter, and we refer interested readers to the original papers of Csiszár and Tusnády [1] and Neal and Hinton [7]. Yet in the following subsection, we demonstrate how these ideas work in a particularly common example of EM.

3.1 Finite Mixture Models

To make these ideas more concrete, we show how they work in one of the best-known examples of EM, the finite mixture model case. This is precisely the case treated in the alternating minimization framework by Hathaway [8], and here we follow the development in that paper closely. Under the finite mixture model, $\mathbf{y}$ consists of a random sample of observations $y_1, \ldots, y_n$, each with density function

$$g[y \mid (\lambda, \boldsymbol{\phi})] = \sum_{j=1}^{m} \lambda_j f_j(y \mid \phi_j)$$

where the λ_j are positive and sum to one, and the functional form of the jth density function, given the jth parameter, $f_j(\cdot \mid \phi_j)$, is fully known. In an EM context, we imagine that the "complete data" consist of an independent and identically distributed sample $\mathbf{X}_1, \ldots, \mathbf{X}_n$, where each $\mathbf{X}_i = (Y_i, \mathbf{Z}_i)$. In this formulation, the Y_i are observed, whereas each Z_i is the m-vector of the component indicators,

$$Z_{ij} = I\{\text{observation } i \text{ comes from component } j\}, 1 \leq i \leq n, 1 \leq j \leq m$$

Here, the set $\mathcal{Q}$ consists of all possible joint distributions of $(Y, \mathbf{Z})$ in the model, one for every possible value of the parameters $(\lambda, \boldsymbol{\phi})$. In other words, we may use the notation $Q \in \mathcal{Q}$ interchangeably with $(\lambda, \boldsymbol{\phi}) \in \mathcal{Q}$. The set $\mathcal{P}$ consists of all distributions on the missing data, which means $\mathbf{Z}_1, \ldots, \mathbf{Z}_n$. Since every conditional distribution $Q(\mathbf{z} \mid \mathbf{y})$ implies conditionally independent $\mathbf{Z}_i$, we may assume without loss of generality that each $P \in \mathcal{P}$ imposes independence on the $\mathbf{Z}_i$. Therefore, we may identify every P with vectors $\mathbf{p}_1, \ldots, \mathbf{p}_n$, in which $\mathbf{p}_i$ is the multinomial parameter governing the distribution of $\mathbf{Z}_i$.

With this setup, $D[P, Q] = D[P, (\lambda, \boldsymbol{\phi})]$ in Equation (4) takes the form

$$D[P, (\lambda, \boldsymbol{\phi})] = \sum_{i=1}^{n} \sum_{j=1}^{m} p_{ij}[\log p_{ij} - \log \lambda_j f_j(y_i \mid \phi_j)] \tag{5}$$

Hathaway [8] shows explicitly how to minimize with respect to P when given $(\lambda^{r-1}, \boldsymbol{\phi}^{r-1})$:

$$p_{ij}^r = \frac{\lambda_j^{r-1} f_j(y_i \mid \phi_j^{r-1})}{g[y_i \mid (\lambda^{r-1}, \boldsymbol{\phi}^{r-1})]}$$

This is identical to the well-known E-step for finite mixtures, which demonstrates that the E-step at the rth iteration coincides with the alternating minimization step $Q^{r-1} \stackrel{2}{\rightarrow} P^r$. Similarly, the M-step at the rth iteration coincides with the step $P^r \stackrel{1}{\rightarrow} Q^r$, which according to Equation (5) entails maximizing

$$\sum_{i=1}^{n} p_{ij}^r \log \lambda_j f_j(y_i \mid \phi_j)$$

for $j = 1, \ldots, m$.

3.2 Variational EM

The objective function (4) may be rewritten as

$$D(P, Q) = E_P[\log p(\mathbf{z}) - \log q(\mathbf{z} \mid \mathbf{y}) - \log q(\mathbf{y})]$$

since the first two terms on the right side together equal $\mathrm{KL}[P \mid\mid Q(\cdot \mid \mathbf{y})]$, and the third term is the expectation of the constant (with respect to $\mathbf{z}$) log-likelihood $L(Q)$. Grouping the terms differently, we see that adding the log density of $\mathbf{y}$ to the log conditional density of $\mathbf{z}$ given $\mathbf{y}$ yields the log joint density of $(\mathbf{y}, \mathbf{z})$. In other words, we obtain

$$D(P, Q) = -H(P) - E_P \log q(\mathbf{y}, \mathbf{z}) \tag{6}$$

where $H(P) = -E_P \log p(\mathbf{z})$ is the entropy of the P distribution. In Equation (6), all EM algorithm practitioners will recognize the familiar expectation of the complete data log-likelihood that gives the E-step its name. In an EM framework, the expectation E_P is always taken with respect to the conditional distribution of $\mathbf{z}$ given $\mathbf{y}$. However, there are some problems in which this conditional distribution is intractable.

In such cases, an increasingly common technique is to implement the alternating minimization algorithm suggested by $D(P, Q)$ in which $\mathcal{P}$ is taken to be a tractable set of distributions. The nonnegativity of Kullback–Leibler divergence guarantees by Equation (4) that $-D(P, Q) \le L(Q)$; that is, $-D(P, Q)$ is a lower bound on the log-likelihood function. Furthermore, if, for a particular choice of Q, P may be chosen so as to minimize $\mathrm{KL}[P \mid\mid Q(\cdot \mid \mathbf{y})]$, we obtain

$$\tilde{D}(Q) \overset{\text{def}}{=} \max_{P \in \mathcal{P}} -D(P, Q)$$

The approximation of $L(Q)$ by the more easily maximized $\tilde{D}(Q)$ is called the *variational approach.*

Combining the variational approach with an alternating minimization algorithm is analogous to using an EM algorithm in which $Q^{r-1} \overset{2}{\rightarrow} P^r$ may be called the "Variational E-step" and $P^r \overset{1}{\rightarrow} Q^r$ may be called the "Variational M-step." As shown above, in the case where $\mathcal{P}$ always contains the exact conditional distribution of $\mathbf{z}$ given $\mathbf{y}$, the variational EM is the same as the actual EM.

Although an in-depth exploration of variational inference is beyond the scope of this chapter, there are multiple recent surveys in the literature, such as Blei *et al.* [9].

4 Matrix Approximation Algorithms

Many statistical problems may be formulated as approximating one matrix, partially or fully observed, by the product of two different matrices. This section introduces several problems that fit this description, citing a reference or two in each case and revealing just enough of the structure of the problem to show how its solution may be amenable to alternating minimization. Throughout the section, we denote by $S^{a \times b}$ the set of all $a \times b$ matrices whose entries are elements of S. In addition, $||M||_F^2$ will denote the square of the Frobenius norm $\sum_{i,j} M_{ij}^2$ of M.

4.1 k-Means Clustering

In the standard k-means clustering framework, we observe n points in $\mathbb{R}^m$, and the goal is to find a partition of those points into k categories, where each point is assigned to the category whose sample mean is closest to it in $\mathbb{R}^m$. This partition should ideally satisfy some sort of minimization criterion.

Let us suppose, therefore, that the matrix $X \in \mathbb{R}^{n \times m}$ contains the n points as rows, so the goal is to partition the rows of X. We may define the function to be minimized as

$$D(P, Q) = \|X - PQ\|_F^2$$

where $P \in \{0, 1\}^{n \times k}$ and $Q \in \mathbb{R}^{k \times m}$. We impose the additional constraint that each row in P contains exactly one 1, and all other entries are 0. Thus, P is the cluster assignment matrix in which $P_{ij} = I\{\text{observation } i \text{ is in category } j\}$. We conclude that $P^\top P$ is a $k \times k$ diagonal matrix whose jth diagonal entry is the number of observations in category j.

In this formulation, the step $P \overset{1}{\to} Q$ is straightforward: With P fixed and thus each point's category membership known, the point in $\mathbb{R}^m$ minimizing the sum of the squared distances to all points in a given cluster is the sample mean. Thus, for given P, Q satisfying $P \overset{1}{\to} Q$ is the matrix whose jth row is the m-dimensional sample mean of the points in the jth category according to P. Similarly, the step $Q \overset{2}{\to} P$ simply entails assigning each of the n points to the cluster whose mean, as coded as one of the rows of Q, is closest.

The standard k-means clustering algorithm therefore alternates between the $P \overset{1}{\to} Q$ and $Q \overset{2}{\to} P$ steps, following an initial starting value (say, a random set of rows of Q or some arbitrary partition of the points as expressed by P). This algorithm, while simple to define and clearly an instance of alternating minimization, is not guaranteed to arrive at a global minimizer of the $D(P, Q)$ function.

There are many generalizations of the basic k-means algorithm. Indeed, the finite mixture models seen in Section 3.1 of this very chapter are sometimes considered model-based clustering and may be viewed as a generalization. Another generalization is obtained by allowing a penalized version of $D(P, Q)$ in which the penalty is a weighted sum of the squared distances between all pairs of rows of the Q matrix. By choosing the weights properly, $D(Q, P)$ may be made convex, which in turn provides guarantees about converging to a minimizer. We do not explore this generalization here except to mention that Chi and Lange [10] discuss it in depth and even present an alternating minimization algorithm to solve it.

4.2 Low-Rank Matrix Factorization

Several high-dimensional problems may be expressed as some form of low-rank matrix factorization problems, that is, approximating data $X \in \mathbb{R}^{m \times n}$ by a product PQ, where $P \in \mathbb{R}^{m \times k}$ and $Q \in \mathbb{R}^{k \times n}$ with k considerably smaller than m or n.

$$D(P, Q) = \|X - PQ\|_F^2 = \sum_{i=1}^{m} \sum_{j=1}^{n} (X_{ij} - (PQ)_{ij})^2 \tag{7}$$

One way to view this problem is as one of matrix compression. That is, we want to retain an approximation of X using only P and Q, which together require much less storage than X. The smaller $D(P, Q)$ may be made, the better the approximation becomes.

Without modifications, the problem of minimizing $D(P, Q)$ is fairly straightforward, and we may employ a standard technique such as singular value decomposition. The problem of minimizing $D(P, Q)$ becomes more interesting in the presence of certain types of constraints. For instance, suppose that X contains only nonnegative entries, which we denote by $X \geq 0$. Then, we might intend that both P and Q contain only nonnegative entries. The corresponding problem of nonnegative matrix factorization (NMF) has a sizable literature. In some applications, the assumption of nonnegative or even positive entries allows for $D(P, Q)$ to measure a different statistical distance between X and PQ, such as Kullback–Leibler divergence.

Certain approaches to NMF that may seem natural have drawbacks. For instance, if we imagine defining $P^{r-1} \xrightarrow{1} Q^r$ as

$$Q^* = \arg \min_{Q \in \mathbb{R}^{k \times n}} \|X - P^{r-1}Q\|_F^2$$

followed by $Q^r = \max(Q^*, 0)$ simply shifting any negative entries to zero, we do not necessarily preserve descent at each iteration. This method of alternating least squares can sometimes be effective nonetheless. On the other hand, the problem of finding

$$Q^r = \arg \min_{Q \in \mathbb{R}^{k \times n}} \|X - P^{r-1}Q\|_F^2 \text{ subject to } Q \geq 0$$

which does in fact guarantee that the value of $D(P, Q)$ decreases at each iteration, is shown by Vavasis [11] to be NP-hard. Gillis [12] provides a survey of some of the many approaches to NMF in the literature.

A different sort of constraint takes the form of a data matrix X that contains many missing entries. Once again, the goal is to obtain a low-rank approximation to X, but the goal of this matrix completion problem is to fill in the missing entries of X. A particularly well-known instance of this type of problem is the Netflix Prize [13], in which, essentially, a subset of entries in a huge matrix was given, and the problem was to predict a known but hidden subset of the other entries as well as possible.

If the goal is to exactly reproduce the observed entries of X using a complete matrix with the smallest rank possible, then this problem is known to be NP-hard [14]. One way to circumvent this problem is to minimize the sum of the singular values of X among those matrices that match the observed entries [15], which is mathematically appealing because it makes the problem convex. However, this approach incurs a sizable computational cost since singular value decomposition can be time consuming.

An alternative approach is to define an objective function such as

$$D(P, Q) = \sum_{X_{ij} \text{ observed}} (X_{ij} - (PQ)_{ij})^2$$

where the sum is taken over all pairs (i, j) for which X_{ij} is known. As with many of the problems mentioned in this section, myriad approaches to minimizing $D(P, Q)$ exist; yet as Tanner and Wei [16] state, "algorithms for the solution … usually follow an alternating minimization scheme." The steps implied by $P \xrightarrow{1} Q$ and $Q \xrightarrow{2} P$ are both least-squares problems, though they are potentially computationally expensive. In Tanner and Wei [16], these steps are referred to as the "power factorization," or PF, problem; these authors propose two different approaches to its solution based on the idea of steepest descent.

4.3 Reduced Rank Regression

We close this section with an instance of low-rank matrix factorization that is more explicitly statistical than those presented earlier. In a regression setting where the response is multivariate, say $\mathbf{y} \in \mathbb{R}^p$, linear regression based on a given covariate vector $\mathbf{x} \in \mathbb{R}^q$ consists of predicting $\mathbf{y}$ based on $\boldsymbol{\beta}^\top \mathbf{x}$, where $\boldsymbol{\beta} \in \mathbb{R}^{q \times p}$. In the case where p and q are large, a low-rank representation of $\boldsymbol{\beta}$ might be desired.

To this end, given data $(y_1, x_1), \ldots, (y_n, x_n)$, we may define an objective function as

$$D(P, Q) = \frac{1}{2} \sum_{i=1}^{n} \|\mathbf{y}_i - PQ^\top \mathbf{x}_i\|_2^2$$

where $P \in \mathbb{R}^{p \times k}$ and $Q \in \mathbb{R}^{q \times k}$ as in Equation (3) of Zhao and Palomar [17]. Modern regression applications frequently involve some sort of regularization, or penalization, of the least-squares objective function for the purpose of selecting important predictors in the $\mathbf{x}$ vector. In an alternating minimization context, such penalization might be applied to either P or Q individually rather than the product PQ. This is precisely the approach considered by Zhao and Palomar [17], where the alternating minimization problem is formulated as follows: The $Q^{r-1} \overset{2}{\rightarrow} P^r$ step is the constrained least-squares problem

$$P^r = \arg\min_P D(P, Q^{r-1}) \text{ subject to } P^\top P = I$$

where the constraint is imposed to preserve identifiability of the parameter values in the product PQ. This $Q \overset{2}{\rightarrow} P$ step may be shown to have a closed-form solution. On the other hand, the penalization is applied to the Q matrix in the form

$$Q^r = \arg\min_Q \left[D(P^r, Q) + \sum_{i=1}^{q} \rho_i(\|Q_i\|) \right]$$

where Q_i refers to the ith row of the Q matrix. There is no closed-form solution to this problem, but it is amenable to numerical solution. As in the example of Section 2, the lasso penalty $\rho_i(a) = \lambda_i |a|$ leads to some of the rows of Q being set to exactly zero, and in this way sparsity in the regression solution may be obtained.

5 Conclusion

As there is no way to give a comprehensive account of all algorithms in the overbroad class of "alternating minimization," the particular choice of topics presented in any treatment of the subject necessarily reflects the personal biases and expertise of its author. The hope is that this introduction sparks some interest in the reader even if it cannot possibly serve as a comprehensive survey of the entire subject.

Among the many topics not covered here are the various convergence properties of alternating minimization algorithms. That said, certain types of algorithms often inherit desirable convergence properties, and the literature treating these types often provide details. For instance, the seminal work of Tseng [18] gives conditions under which any limit point of a CD algorithm may be guaranteed to be a minimizer. Similarly, Csiszár and Tusnády [1] present several results on convergence; yet, as alternating minimization often creates only the condition that a decrease in the value of $D(P, Q)$ is guaranteed at each iteration, it is frequently the case that convergence comes with some caveats. A useful lesson in what a guaranteed decrease at each iteration does and, equally importantly, does not imply is provided by the well-known paper about EM algorithms by Wu [19].

In the end, the surprisingly simple formulation of alternating between $P \overset{1}{\rightarrow} Q$ and $Q \overset{2}{\rightarrow} P$, even if each of those steps may be computationally tractable in isolation, leads to a dizzying array of versions of the generally challenging question of how to find a minimizer of the function $D(P, Q)$. This chapter has attempted to explain the central idea in the seminal paper by Csiszár and Tusnády [1] and how it applies to perhaps the best known of all statistical algorithms, while also providing a glimpse at the incredible breadth of problems amenable to algorithms satisfying the general criteria of alternating minimization.

References

1 Csiszár, I. and Tusnády, G. (1984) Information geometry and alternating minimization problems. *Stat. Decis., Supplement Issue*, 1, 205–237.

2 Shi, H.J.M., Tu, S., Xu, Y., and Yin, W. (2016) A primer on coordinate descent algorithms. *arXiv preprint arXiv:1610.00040*.

3 Lange, K., Chi, E.C., and Zhou, H. (2014) A brief survey of modern optimization for statisticians. *Int. Stat. Rev.*, 82 (1), 46–70.

4 Wu, T.T. and Lange, K. (2008) Coordinate descent algorithms for lasso penalized regression. *Ann. Appl. Stat.*, 2 (1), 224–244. doi: 10.1214/07-AOAS147.

5 R Core Team (2020) *R: A Language and Environment for Statistical Computing*, R Foundation for Statistical Computing, Vienna, Austria.

6 Dempster, A.P., Laird, N.M., and Rubin, D.B. (1977) Maximum likelihood from incomplete data via the EM algorithm. *J. R. Stat. Soc., Ser. B*, 39 (1), 1–22.

7 Neal, R.M. and Hinton, G.E. (1998) A view of the EM algorithm that justifies incremental, sparse, and other variants, in *Learning in Graphical Models* (ed. M.I. Jordan), Kluwer Academic Publications, Dordrecht, the Netherlands, pp. 355–368.

8 Hathaway, R.J. (1986) Another interpretation of the EM algorithm for mixture distributions. *Stat. Probab. Lett.*, 4 (2), 53–65.

9 Blei, D.M., Kucukelbir, A., and McAuliffe, J.D. (2017) Variational inference: a review for statisticians. *J. Am. Stat. Assoc.*, 112 (518), 859–877. doi: 10.1080/01621459.2017.1285773.

10 Chi, E.C. and Lange, K. (2015) Splitting methods for convex clustering. *J. Comput. Graph. Stat.*, 24 (4), 994–1013.

11 Vavasis, S.A. (2009) On the complexity of nonnegative matrix factorization. *SIAM J. Optim.*, 20(3), 1364–1377.

12 Gillis, N. (2014) The why and how of nonnegative matrix factorization, in *Regularization, Optimization, Kernels, and Support Vector Machines* (eds J.A.K., Suykens, M. Signoretto., and A Argyriou), Chapman & Hall/CRC, Boca Raton, FL, pp. 257–291.

13 Bennett, J. and Lanning, S. (2007) *The Netflix Prize*. Proceedings of KDD Cup and Workshop, vol. 2007, pp. 3–6.

14 Harvey, N.J.A., Karger, D.R., and Yekhanin, S. (2006) *The Complexity of Matrix Completion*. Proceedings of the Seventeenth Annual ACM-SIAM Symposium on Discrete

Algorithm, SODA 2006, pp. 1103–1111. Society for Industrial and Applied Mathematics, USA.

15 Candès, E.J. and Tao, T. (2010) The power of convex relaxation: near-optimal matrix completion. *IEEE Trans. Inf. Theory*, 56 (5), 2053–2080.

16 Tanner, J. and Wei, K. (2016) Low rank matrix completion by alternating steepest descent methods. *Appl. Comput. Harmon. Anal.*, 40 (2), 417–429.

17 Zhao, Z. and Palomar, D.P. (2018) *Sparse Reduced Rank Regression with Nonconvex Regularization*. 2018 IEEE Statistical Signal Processing Workshop (SSP), pp. 811–815. IEEE.

18 Tseng, P. (2001) Convergence of a block coordinate descent method for nondifferentiable minimization. *J. Optim. Theory Appl.*, 109 (3), 475–494.

19 Wu, C.F.J. (1983) On the convergence properties of the EM algorithm. *Ann. Stat.*, 11, 95–103.

27

A Gentle Introduction to Alternating Direction Method of Multipliers (ADMM) for Statistical Problems

Shiqian Ma[1] *and Mingyi Hong*[2]

[1] *University of California, Davis, CA, USA*
[2] *University of Minnesota, Minneapolis, MN, USA*

1 Introduction

The alternating direction method of multipliers (ADMM) has been widely used for solving problems in science and engineering in the past decade. The ADMM is closely related to the Douglas–Rachford operator splitting (DROS) method, which dates back to the 1950s [1] for solving variational problems arising from numerical PDEs. Later studies of DROS and ADMM include Refs 2–5, among others. The renaissance of ADMM that is widely used now happened in 2007–2009, when researchers found that it can efficiently solve image- and signal-processing problems [6–8]. The literature on ADMM is now vast, and we do not intend to exhaust it nor to cover the latest developments on this topic. Instead, this chapter gives a gentle introduction to ADMM for solving problems arising from statistics. For more comprehensive reviews on ADMM, we refer the readers to Eckstein [9], Boyd *et al.* [10], Eckstein and Yao [11].

We start our discussion on ADMM for solving the following two-block convex minimization problem:

$$\min_{x\in\mathbb{R}^{n_1}, y\in\mathbb{R}^{n_2}} f(x) + g(y), \text{ s.t., } Ax + By = b \tag{1}$$

where both f and g are proper and closed convex functions and might be nonsmooth, $A \in \mathbb{R}^{m\times n_1}$, $B \in \mathbb{R}^{m\times n_2}$, and $b \in \mathbb{R}^m$. A typical iteration of ADMM for solving (1) can be described as follows:

$$\begin{cases} x^{k+1} := \operatorname*{argmin}_x \mathcal{L}(x, y^k; \lambda^k) \\ y^{k+1} := \operatorname*{argmin}_y \mathcal{L}(x^{k+1}, y; \lambda^k) \\ \lambda^{k+1} := \lambda^k - \beta(Ax^{k+1} + By^{k+1} - b) \end{cases} \tag{2}$$

where the augmented Lagrangian function of (1) is defined as

$$\mathcal{L}(x, y; \lambda) := f(x) + g(y) - \langle \lambda, Ax + By - b\rangle + \frac{\beta}{2}\|Ax + By - b\|_2^2 \tag{3}$$

with λ being the Lagrange multiplier associated with the linear equality constraint, and $\beta > 0$ being a penalty parameter. That is, in each iteration, ADMM alternatingly minimizes

Computational Statistics in Data Science.
Edited by Walter W. Piegorsch, Richard A. Levine, Hao Helen Zhang and Thomas C. M. Lee.
 ISBN 978-1-11956107-1

the augmented Lagrangian function for one block variable with the other block being fixed and then updates the Lagrange multiplier. After discarding constant terms, the two subproblems in (2) reduce to

$$x^{k+1} := \underset{x}{\operatorname{argmin}}\ f(x) + \frac{\beta}{2}\|Ax + By^k - b - \lambda^k/\beta\|_2^2 \tag{4}$$

and

$$y^{k+1} := \underset{y}{\operatorname{argmin}}\ g(y) + \frac{\beta}{2}\|Ax^{k+1} + By - b - \lambda^k/\beta\|_2^2 \tag{5}$$

The complete description of ADMM (2) is summarized in Algorithm 1.

Algorithm 1. ADMM for Solving Two-Block Convex Minimization Equation (1)

Require: $y^0 \in \mathbb{R}^{n_2}$, $\lambda^0 \in \mathbb{R}^m$, parameter $\beta > 0$
  while Stopping criteria are not met **do**
    Update $(x^{k+1}, y^{k+1}, \lambda^{k+1})$ by Equation (2)
    $k \leftarrow k + 1$
  end while
  return (x^k, y^k)

The efficiency of ADMM (2) depends on whether (4) and (5) can be efficiently solved. In general, one may still need an iterative solver to solve (4) and (5). However, when A and B are identity matrices, (4) and (5) are equivalent to the proximal mappings of f and g, respectively. The proximal mapping of a function h is defined as

$$\operatorname{prox}_h(z) = \underset{x}{\operatorname{argmin}}\ h(x) + \frac{1}{2}\|x - z\|_2^2$$

For many functions that are commonly used in practice, their proximal mappings are easy to obtain, as we see in the examples in the following sections.

2 Two Perfect Examples of ADMM

In this section, we discuss two problems, robust PCA and graphical Lasso, that are perfectly suitable for ADMM (2).

Robust PCA [12, 13] seeks to decompose a given matrix $M \in \mathbb{R}^{m\times n}$ to the superposition of a low-rank matrix L and a sparse matrix S. Using the nuclear norm $\|\cdot\|_*$ to promote low rankness of L, and ℓ_1 norm $\|\cdot\|_1$ to promote the sparsity of S, robust PCA can be formulated as the following convex minimization problem:

$$\min_{L,S\in\mathbb{R}^{m\times n}} \|L\|_* + \mu\|S\|_1; \text{ s.t., } L + S = M \tag{6}$$

where $\mu > 0$ is a weighting parameter. When ADMM (2) is applied to solve (6), a typical iteration is (here, we use Λ to denote the Lagrange multiplier)

$$L^{k+1} := \underset{L}{\text{argmin}}\ \|L\|_* + \frac{\beta}{2}\|L + S^k - M - \Lambda^k/\beta\|_F^2 \tag{7a}$$

$$S^{k+1} := \underset{S}{\text{argmin}}\ \mu\|S\|_1 + \frac{\beta}{2}\|L^{k+1} + S - M - \Lambda^k/\beta\|_F^2 \tag{7b}$$

$$\Lambda^{k+1} := \Lambda^k - \beta(L^{k+1} + S^{k+1} - M) \tag{7c}$$

It is known that the solution of (7a) is the proximal mapping of the nuclear norm, which is given by the matrix shrinkage operation through a singular value decomposition (SVD) [14],

$$L^{k+1} := \text{MatShrink}(M - S^k + \Lambda^k/\beta, 1/\beta) \tag{8}$$

where the matrix shrinkage operator MatShrink(Z, ξ) is defined as

$$\text{MatShrink}(Z, \xi) := U\text{Diag}(\max\{\sigma - \xi, 0\})V^\top \tag{9}$$

and $U\text{Diag}(\sigma)V^\top$ is the SVD of matrix Z. The solution of (7b) is the proximal mapping of the ℓ_1 norm, which also admits an easy closed-form solution given by the ℓ_1 shrinkage operation:

$$S^{k+1} := \text{Shrink}(M - L^{k+1} + \Lambda^k/\beta, \mu/\beta) \tag{10}$$

where the ℓ_1 shrinkage operation Shrink(Z, ξ) is defined as

$$[\text{Shrink}(Z, \xi)]_{ij} := \text{sgn}(Z_{ij}) \cdot \max\{|Z_{ij}| - \xi, 0\} \tag{11}$$

Therefore, both subproblems in (7) can be easily solved.

Graphical Lasso considers to estimate a sparse inverse covariance matrix of a multivariate Gaussian distribution from sample data. Let $X = \{x^{(1)}, \dots, x^{(n)}\}$ be an n-dimensional random vector following an n-variate Gaussian distribution $\mathcal{N}(\mu, \Sigma)$, and let $G = (V, E)$ be a Markov network representing the conditional independence structure of $\mathcal{N}(\mu, \Sigma)$. Specifically, the set of vertices $V = \{1, \dots, n\}$ corresponds to the set of variables in X, and the edge set E contains an edge (i, j) if and only if $x^{(i)}$ is conditionally dependent on $x^{(j)}$, given all remaining variables; that is, the lack of an edge between i and j denotes the conditional independence of $x^{(i)}$ and $x^{(j)}$, which corresponds to a zero entry in the inverse covariance matrix Σ^{-1} [15]. Thus, learning the structure of this graphical model is equivalent to the problem of learning the zero pattern of Σ^{-1}. The following convex formulation for estimating this sparse inverse covariance matrix has been suggested by Yuan and Lin [16], Banerjee *et al.* [17], Friedman *et al.* [18]:

$$\min_{S \in S_+^n} \langle \hat{\Sigma}, S \rangle - \log\det(S) + \mu\|S\|_1 \tag{12}$$

where S_+^n denotes the set of $n \times n$ positive semidefinite matrices, $\hat{\Sigma}$ is the sample covariance matrix, and $\mu > 0$ is a weighting parameter. To apply ADMM (2), we introduce an auxiliary variable T and rewrite (12) as

$$\min_{S,T} \langle \hat{\Sigma}, S \rangle - \log\det(S) + \mu\|T\|_1, \quad \text{s.t.,}\ S - T = 0 \tag{13}$$

A typical iteration of ADMM for solving (13) is given by

$$S^{k+1} := \underset{S}{\operatorname{argmin}} \ \langle \hat{\Sigma}, S \rangle - \log\det(S) + \frac{\beta}{2}\|S - T^k - \Lambda^k/\beta\|_F^2 \tag{14a}$$

$$T^{k+1} := \underset{T}{\operatorname{argmin}} \ \mu\|T\|_1 + \frac{\beta}{2}\|S^{k+1} - T - \Lambda^k/\beta\|_F^2 \tag{14b}$$

$$\Lambda^{k+1} := \Lambda^k - \beta(S^{k+1} - T^{k+1}) \tag{14c}$$

The subproblem (14a) corresponds to the proximal mapping of $-\log\det(S)$, and its first-order optimality condition is given by

$$0 = \hat{\Sigma} - S^{-1} + \beta(S - T^k - \Lambda^k/\beta) \tag{15}$$

It is easy to verify that

$$S^{k+1} := U\mathrm{Diag}(\gamma)U^\top \tag{16}$$

satisfies (15) and thus is the optimal solution to (14a), where $U\mathrm{Diag}(\sigma)U^\top$ is the eigenvalue decomposition of $\hat{\Sigma}/\beta - T^k - \Lambda^k/\beta$ and $\gamma_i = (-\sigma_i + \sqrt{\sigma_i^2 + 4/\beta})/2, i = 1, \ldots, n$. The subproblem (14b) corresponds to the proximal mapping of $\|T\|_1$, whose solution is given by the ℓ_1 shrinkage operation

$$T^{k+1} := \mathrm{Shrink}(S^{k+1} - \Lambda^k/\beta, \mu/\beta) \tag{17}$$

where the ℓ_1 shrinkage operation is defined in (11).

Note that the ADMM (2) is very suitable to robust PCA (6) and graphical Lasso (12), because the subproblems are all easy to solve. In fact, they all admit closed-form solutions, and no iterative solver is needed to solve them. There are two reasons why these subproblems are easy: (i) in the equality constraints, the matrices A and B are identity matrices; (ii) the functions f and g have easily computable proximal mappings. If one of these two properties does not hold, then we need to perform some necessary manipulations to the problem or to the ADMM algorithm, so that the subproblems still admit easily computable closed-form solutions. This is our main task in the following section.

3 Variable Splitting and Linearized ADMM

In this section, we discuss the variable-splitting technique and linearized ADMM. We start our discussion with the Lasso problem [19]:

$$\min_{x \in \mathbb{R}^n} \ \|Ax - b\|_2^2, \ \text{s.t.,} \ \|x\|_1 \leq \tau \tag{18}$$

where $A \in \mathbb{R}^{m \times n}$, $b \in \mathbb{R}^m$, and $\tau > 0$ controls the sparsity of x. To apply ADMM (2), we first need to apply the variable-splitting technique which introduces an auxiliary variable y and rewrite (18) as

$$\min_{x,y \in \mathbb{R}^n} \ \|Ax - b\|_2^2 + \mathbf{1}(\|y\|_1 \leq \tau), \ \text{s.t.,} \ x - y = 0 \tag{19}$$

where the indicator function $\mathbf{1}(z \in S) = 0$, if $z \in S$, and $\mathbf{1}(z \in S) = +\infty$, otherwise. A typical iteration of ADMM (2) for solving (19) is given by

$$x^{k+1} := \underset{x}{\operatorname{argmin}} \ \|Ax - b\|_2^2 + \frac{\beta}{2}\|x - y^k - \lambda^k/\beta\|_2^2 \tag{20a}$$

$$y^{k+1} := \underset{y}{\operatorname{argmin}} \ \mathbf{1}(\|y\|_1 \leq \tau) + \frac{\beta}{2}\|x^{k+1} - y - \lambda^k/\beta\|_2^2 \tag{20b}$$

$$\lambda^{k+1} := \lambda^k - \beta(x^{k+1} - y^{k+1}) \tag{20c}$$

The two subproblems (20a and 20b) are both relatively easy to solve. Specifically, the solution of (20a) is given by solving the following linear system:

$$x^{k+1} := (2A^\top A + \beta I)^{-1}(\beta y^k + \lambda^k + 2A^\top b) \tag{21}$$

and the solution of (20b) corresponds to projecting $(x^{k+1} - \lambda^k/\beta)$ onto the ℓ_1 norm ball $(\|\cdot\| \leq \tau)$, which can be done efficiently [20, 21].

Another way to split the variables in (18) leads to the following reformulation of it:

$$\min_{x\in\mathbb{R}^n, z\in\mathbb{R}^m} \ \|z\|_2^2 + \mathbf{1}(\|x\|_1 \leq \tau), \ \text{s.t.,} \ Ax - z = b \tag{22}$$

ADMM for solving (22) is given by

$$x^{k+1} := \underset{x}{\operatorname{argmin}} \ \mathbf{1}(\|x\|_1 \leq \tau) + \frac{\beta}{2}\|Ax - z^k - b - \lambda^k/\beta\|_2^2 \tag{23a}$$

$$z^{k+1} := \underset{z}{\operatorname{argmin}} \ \|z\|_2^2 + \frac{\beta}{2}\|Ax^{k+1} - z - b - \lambda^k/\beta\|_2^2 \tag{23b}$$

$$\lambda^{k+1} := \lambda^k - \beta(Ax^{k+1} - z^{k+1} - b) \tag{23c}$$

The subproblem (23b) admits a closed-form solution given by

$$z^{k+1} := \frac{1}{2+\beta}(\beta A x^{k+1} - \beta b - \lambda^k)$$

Because of the existence of matrix A, the subproblem (23a) does not correspond to the proximal mapping of $\mathbf{1}(\|x\|_1 \leq \tau)$ and thus is not easy to solve. Fortunately, an iterative solver can still be avoided, if one changes (23a) to the following one:

$$x^{k+1} := \underset{x}{\operatorname{argmin}} \ \mathbf{1}(\|x\|_1 \leq \tau) + \frac{1}{2\tau_1}\|x - (x^k - \tau_1\beta A^\top(Ax^k - z^k - b - \lambda^k/\beta))\|_2^2 \tag{24}$$

The rationale of using (24) to replace (23a) is as follows. Note that $(x^k - \tau_1 \beta A^\top(Ax^k - z^k - b - \lambda^k/\beta))$ can be regarded a gradient step with step size τ_1 for $\frac{\beta}{2}\|Ax - z^k - b - \lambda^k/\beta\|_2^2$, the smooth part of the objective function of (23a). Therefore, (24) can be viewed as using one proximal gradient step to replace the exact minimization in (23a). This technique leads to the so-called linearized ADMM, and its global convergence is guaranteed when τ_1 is small (see Section 7 for more details).

The linearized ADMM for solving the general problem (1) is given below:

$$\begin{cases} x^{k+1} := \underset{x}{\operatorname{argmin}} f(x) + \frac{1}{2\tau_1}\|x - (x^k - \tau_1\beta A^\top(Ax^k + By^k - b - \lambda^k/\beta))\|_2^2 \\ y^{k+1} := \underset{y}{\operatorname{argmin}} g(y) + \frac{1}{2\tau_2}\|y - (y^k - \tau_2\beta B^\top(Ax^{k+1} + By^k - b - \lambda^k/\beta))\|_2^2 \\ \lambda^{k+1} := \lambda^k - \beta(Ax^{k+1} + By^{k+1} - b) \end{cases} \tag{25}$$

where $\tau_1 > 0$ and $\tau_2 > 0$ are the step sizes of the proximal gradient steps. Note that the two subproblems in (25) correspond to proximal mappings of f and g, respectively, and are thus easy to solve for many commonly seen functions such as ℓ_1 norm, ℓ_2 norm, and nuclear norm. For more applications of the linearized ADMM (25), we refer the readers to Ma [22].

Before closing this section, we use the fused Lasso problem to further illustrate how to utilize the variable-splitting and linearized ADMM techniques in practice. The fused Lasso problem can be formulated as follows [23]:

$$\min_{x\in\mathbb{R}^n} \|Ax-b\|_2^2, \text{ s.t., } \|x\|_1 \le s_1, \sum_{i=2}^{n} |x_i - x_{i-1}| \le s_2 \tag{26}$$

Note that this problem is not readily solved by ADMM, because the constraint $\sum_{i=2}^{n} |x_i - x_{i-1}| \le s_2$ does not admit an easy projection, which is the same as the proximal mapping of the indicator function $\mathbf{1}(\sum_{i=2}^{n} |x_i - x_{i-1}| \le s_2)$. However, (26) can be solved via the linearized ADMM. To see this, we first introduce a new variable $y \in \mathbb{R}^{n-1}$, and let $y_i = x_{i+1} - x_i$, $i = 1, \dots, n-1$. Hence, (26) can be rewritten as

$$\min_{x\in\mathbb{R}^n} \|Ax-b\|_2^2, \text{ s.t., } \|x\|_1 \le s_1, \ \|y\|_1 \le s_2, \ y = Lx \tag{27}$$

where $L \in \mathbb{R}^{(n-1)\times n}$ with $L_{ii} = -1$, $L_{i,i+1} = 1$, $i = 1, \dots, n-1$, and all other entries being zeros. By associating a Lagrange multiplier λ to the linear equality constraint $y = Lx$, the augmented Lagrangian function of (27) can be written as

$$\mathcal{L}(x,y;\lambda) := \|Ax-b\|_2^2 + \mathbf{1}(\|x\|_1 \le s_1) + \mathbf{1}(\|y\|_1 \le s_2) - \langle \lambda, Lx - y\rangle + \frac{\beta}{2}\|Lx - y\|_2^2$$

The ADMM for solving (27) is given by

$$x^{k+1} := \operatorname*{argmin}_x \ \|Ax-b\|_2^2 + \mathbf{1}(\|x\|_1 \le s_1) + \frac{\beta}{2}\|Lx - y^k - \lambda^k/\beta\|_2^2 \tag{28a}$$

$$y^{k+1} := \operatorname*{argmin}_y \ \mathbf{1}(\|y\|_1 \le s_2) + \frac{\beta}{2}\|Lx^{k+1} - y - \lambda^k/\beta\|_2^2 \tag{28b}$$

$$\lambda^{k+1} := \lambda^k - \beta(Lx^{k+1} - y^{k+1}) \tag{28c}$$

Note that (28b) corresponds to a projection onto the ℓ_1 norm ball, which is easy to compute. However, (28a) needs an iterative solver. Therefore, we apply the linearized ADMM and modify (28a–28c) to the following one:

$$\begin{cases} x^{k+1} := \operatorname*{argmin}_x \mathbf{1}(\|x\|_1 \le s_1) + \frac{1}{2\tau}\|x - (x^k - \tau(\beta L^\top(Lx^k - y^k - \lambda^k/\beta) + 2A^\top(Ax^k - b)))\|_2^2 \\ y^{k+1} := \operatorname*{argmin}_y \mathbf{1}(\|y\|_1 \le s_2) + \frac{\beta}{2}\|Lx^{k+1} - y - \lambda^k/\beta\|_2^2 \\ \lambda^{k+1} := \lambda^k - \beta(Lx^{k+1} - y^{k+1}) \end{cases} \tag{29}$$

Note that $\tau > 0$ is a step size, and $(\beta L^\top(Lx^k - y^k - \lambda^k/\beta) + 2A^\top(Ax^k - b))$ is the gradient of the smooth part of the objective function of (28a). Now the two subproblems in (29) both correspond to a projection onto the ℓ_1 norm ball, which can be easily done.

4 Multiblock ADMM

A natural extension of the two-block ADMM (2) for solving (1) is its multiblock version for solving the following convex minimization with separable objective:

$$\min \sum_{j=1}^{N} f_j(x_j), \text{ s.t., } \sum_{j=1}^{N} A_j x_j = b, \; x_j \in \mathbb{R}^{n_j}, j = 1, \dots, N \tag{30}$$

It is easy to see that (1) is a special case of (30) when $N = 2$. There are many statistical problems that have the multiblock structure as (30) with $N \geq 3$, for example, the stable principal component pursuit [24] and latent variable graphical Lasso [25].

By associating a Lagrange multiplier λ to the equality constraint, the augmented Lagrangian function of (30) is given by

$$\mathcal{L}(x_1, \dots, x_N; \lambda) := \sum_{j=1}^{N} f_j(x_j) - \left\langle \lambda, \sum_{j=1}^{N} A_j x_j - b \right\rangle + \frac{\beta}{2} \left\| \sum_{j=1}^{N} A_j x_j - b \right\|_2^2$$

A typical iteration of the multiblock ADMM for solving (30) can be described as

$$\begin{cases} x_1^{k+1} := \operatorname*{argmin}_{x_1} \mathcal{L}(x_1, x_2^k, \dots, x_N^k; \lambda^k) \\ \vdots \\ x_j^{k+1} := \operatorname*{argmin}_{x_j} \mathcal{L}(x_1^{k+1}, \dots, x_{j-1}^{k+1}, x_j, x_{j+1}^k, \dots, x_N^k; \lambda^k) \\ \vdots \\ x_N^{k+1} := \operatorname*{argmin}_{x_N} \mathcal{L}(x_1^{k+1}, \dots, x_{N-1}^{k+1}, x_N; \lambda^k) \\ \lambda^{k+1} := \lambda^k - \beta(\sum_{j=1}^N A_j x_j^{k+1} - b) \end{cases} \tag{31}$$

That is, in each iteration, ADMM (31) alternatingly minimizes the augmented Lagrangian function for one block variable with all other $N - 1$ block variables being fixed and then updates the Lagrange multiplier.

Since it is known that the two-block ADMM (2) globally converges for convex minimization (1), one would expect that the multiblock ADMM (31) converges as well. However, this is not the case. A counterexample was given in Ref. 26, which shows that the 3-block ADMM (i.e., (31) with $N = 3$) fails to converge for any $\beta > 0$ even when all f_js are linear functions. Specifically, it is shown in Ref. 26 that multiblock ADMM iteration (31) can indeed diverge for the following problem with three block variables (which has a unique solution $x_1 = x_2 = x_3 = 0$):

$$\begin{aligned} \text{Find} \quad & \{x_1, x_2, x_3\} \\ \text{s.t.,} \quad & A_1 x_1 + A_2 x_2 + A_3 x_3 = 0 \\ & \text{with} \quad [A_1 \; A_2 \; A_3] = \begin{bmatrix} 1 & 1 & 1 \\ 1 & 1 & 2 \\ 1 & 2 & 2 \end{bmatrix} \end{aligned}$$

Therefore, special care needs to be taken when using multiblock ADMM (31). There exists vast literature on modifying the multiblock ADMM (31) so that global convergence for convex problem can be guaranteed under certain conditions. We refer the readers to

Hong *et al.* [27], Hong and Luo [28], Lin *et al.* [29–32], Deng *et al.* [33], Sun *et al.* [34], He *et al.* [35], Chen *et al.* [36] for a partial list of results on this topic.

Here, we discuss some variable-splitting trick that transforms the multiblock problem (30) to a two-block problem. The two-block reformulation can then be solved by a linearized ADMM whose convergence is guaranteed under mild conditions. We refer the readers to Ma [22], Wang *et al.* [37], Ma *et al.* [38] for more details on using these ideas to solve stable principal component pursuit and latent variable graphical Lasso problems. Note that the general multiblock convex minimization (30) can be reformulated as

$$\begin{aligned} \min\ & \sum_{j=1}^{N} f_j(x_j) \\ \text{s.t.}\ & A_j x_j - y_j = b/N,\ j = 1, \dots, N \\ & \sum_{j=1}^{N} y_j = 0 \end{aligned} \tag{32}$$

By associating the Lagrange multiplier λ_j to the equality constraint $A_j x_j - y_j = b/N$, the augmented Lagrangian function of (32) can be written as

$$\begin{aligned} \mathcal{L}(x, y; \lambda) := & \sum_{j=1}^{N} f_j(x_j) + \mathbf{1}\left(\sum_{j=1}^{N} y_j = 0 \right) \\ & - \sum_{j=1}^{N} \langle \lambda_j, A_j x_j - y_j - b/N \rangle + \frac{\beta}{2} \sum_{j=1}^{N} \|A_j x_j - y_j - b/N\|_2^2 \end{aligned}$$

After discarding constant terms, the two-block ADMM for solving (32) can be described as

$$x_j^{k+1} := \underset{x_j}{\operatorname{argmin}}\ f_j(x_j) + \frac{\beta}{2}\|A_j x_j - y_j^k - b/N - \lambda_j^k/\beta\|_2^2,\ j = 1, \dots, N \tag{33a}$$

$$(y_1^{k+1}, \dots, y_N^{k+1}) := \underset{y_1, \dots, y_N}{\operatorname{argmin}}\ \mathbf{1}\left(\sum_{j=1}^{N} y_j = 0 \right) + \frac{\beta}{2} \sum_{j=1}^{N} \|A_j x_j^{k+1} - y_j - b/N - \lambda_j^k/\beta\|_2^2 \tag{33b}$$

$$\lambda_j^{k+1} := \lambda_j^k - \beta(A_j x_j^{k+1} - y_j^{k+1} - b/N),\ j = 1, \dots, N \tag{33c}$$

Note that (33a)–(33c) are two-block ADMM, because there are two block variables, $(x_1, \dots, x_N)$ and $(y_1, \dots, y_N)$. The $(x_1, \dots, x_N)$ in the first block can be updated in parallel, while the $(y_1, \dots, y_N)$ in the second block needs to be updated together. Moreover, the subproblem (33b) can be solved in closed-form [37]. Though the subproblem (33a) does not admit a closed-form solution due to the existence of matrices A_j, we can again apply the linearized ADMM to make it easier to solve. The linearized ADMM replaces (33a) by the following one:

$$x_j^{k+1} := \operatorname{argmin}_{x_j} f_j(x_j) + \frac{1}{2\tau}\|x_j - (x_j^k - \tau\beta A_j^\top (A_j x_j^k - y_j^k - b/N - \lambda_j^k/\beta))\|_2^2$$

which again corresponds to the proximal mapping of f_j.

5 Nonconvex Problems

Our discussion so far has been focused on convex problems. This section discusses applications of ADMM in the nonconvex setting.

To motivate our discussion, note that in many statistical problems, nonconvexity does arise. For example, when designing regularizers to find maximum-likelihood estimators, it is often desired that the resulting estimator is unbiased when the true parameter is large. In other words, the penalty function, denoted as $p(\cdot)$, should be (nearly) constant for large argument [39]. Typical forms of such regularizers are smoothly clipped absolute deviation (SCAD) [40] and minimax concave penalty (MCP) [41]. For example, for some scalar ϕ, and fixed parameters ν and b, let us define the scalar function below

$$p_\nu(\phi) = \begin{cases} \nu|\phi| - \frac{\phi^2}{2b} & \text{if } |\phi| \le b\nu \\ \frac{1}{2}b\nu^2 & \text{otherwise} \end{cases} \tag{34}$$

Then, for a given matrix variable $X \in \mathbb{R}^{m\times n}$, the MCP penalty is given as $p_{\text{MCP}}(X) := \sum_{i,j} p_\nu(x_{ij})$. One particular characterization for these nonconvex penalties is that they can be decomposed as a sum of an ℓ_1-norm function and a *concave* function $q_\nu(x)$ as $p_\nu(\phi) = \nu|\phi| + q_\nu(\phi)$ for some $\nu \ge 0$. When such kinds of penalties are used, many problems we discussed so far become *nonconvex* problems. As an example, problem (6) becomes

$$\min_{L,S\in\mathbb{R}^{m\times n}} \|L\|_* + \mu p(S), \text{ s.t., } L + S = M \tag{35}$$

Another related example is the sparse subspace estimation problem introduced in Ref. 42, where nonconvex regularizers are used to improve the subspace recovery probability.

Another popular application is distributed training of nonconvex learning models. Suppose that there are N computational nodes in the system, and the entire data set D is partitioned into data pieces $D(1), \ldots, D(N)$. The objective is to learn a model parameter $x \in \mathbb{R}^m$ from the distributed data. Assume that for each node $i \in [N]$, the local loss function is given by $f(x; D(i)) := f_i(x)$, which can be highly nonconvex if sophisticated models such as neural networks are used. Let $g(\cdot)$ be a convex and possibly nonsmooth regularizer. Then, we can formulate the empirical risk minimization problem as [10]:

$$\min_{y\in\mathbb{R}^m} \sum_{i=1}^{N} f_i(y) + g(y) \tag{36}$$

Because data is distributed, it is often useful to introduce N *local variables* $x_1, \ldots, x_N$ and rewrite the above problem as

$$\min_{x_i,y\in\mathbb{R}^m} \sum_{i=1}^{N} f_i(x_i) + g(y), \text{ s.t., } y = x_i, \forall\, i = 1, \cdots, N \tag{37}$$

Clearly, this problem is suitable for two-block ADMM (2) with $[x_1, \ldots, x_N]$ and y being the two block variables.

Next, we illustrate how to apply ADMM to problem (37) and discuss some potential algorithm design challenges. In this case, the augmented Lagrangian function becomes

$$\mathcal{L}(x, y, \lambda) = \sum_{i=1}^{N} f_i(x_i) + g(y) - \sum_{i=1}^{N} \langle \lambda_i, y - x_i \rangle + \sum_{i=1}^{N} \frac{\beta}{2}\|y - x_i\|^2 \tag{38}$$

A direct application of the two-block ADMM (2) yields the following iteration:

$$x_i^{k+1} = \arg\min_{x_i} f_i(x_i) - \langle \lambda_i^k, y^k - x_i \rangle + \frac{\beta}{2}\|x_i - y^k\|^2, \; i = 1, \dots, N \tag{39a}$$

$$y^{k+1} = \arg\min_{y} g(y) + \sum_{i=1}^{N} \left(-\langle \lambda_i^k, y - x_i^{k+1} \rangle + \frac{\beta}{2}\|y - x_i^{k+1}\|^2 \right) \tag{39b}$$

$$\lambda_i^{k+1} = \lambda_i^k - \beta(y^{k+1} - x_i^{k+1}) \; i = 1, \dots, N \tag{39c}$$

First, we note that despite the fact that f_is can be nonconvex, the subproblem (39a) can often be solved to global optimality. The reason is that, if we assume $\nabla f_i(x_i)$ is Lipschitz, then we can choose β to be large enough, so that the sum $f_i(x_i) + \frac{\beta}{2}\|x_i - y^k\|^2$ is *strongly* convex in x_i. Second, it is interesting to see that the x subproblem (39a) is completely decomposed into N local problems; therefore, it can be carried out in parallel by the local agents, by only utilizing local data. The above setting can also be extended to generic network topologies in which the nodes are not necessarily directly connected to a single server; see Hong *et al.* [43] and a recent survey [44]. We refer the readers to Hong *et al.* [45], Li and Pong [46], Wang *et al.* [47] for discussion of more applications of ADMM to nonconvex applications.

6 Stopping Criteria

Now we discuss the stopping criteria of ADMM. As a primal–dual algorithm, the termination of ADMM needs to take into account both primal and dual residuals. A widely used stopping criterion of ADMM (2) for solving the convex problem (1) is given in Ref. 10. The authors of Ref. 10 suggested to measure the primal residual using

$$r^{k+1} := Ax^{k+1} + By^{k+1} - b$$

and the dual residual using

$$s^{k+1} := \beta A^\top B(y^{k+1} - y^k)$$

The stopping criterion suggested in Boyd *et al.* [10] is

$$\|r^k\|_2 \le \epsilon^{\text{pri}} \quad \text{and} \quad \|s^k\|_2 \le \epsilon^{\text{dual}}$$

where

$$\begin{aligned} \epsilon^{\text{pri}} &:= \sqrt{m}\epsilon^{\text{abs}} + \epsilon^{\text{rel}}\max\{\|Ax^k\|_2, \|By^k\|_2, \|b\|_2\} \\ \epsilon^{\text{dual}} &:= \sqrt{n_1}\epsilon^{\text{abs}} + \epsilon^{\text{rel}}\|A^\top \lambda^k\|_2 \end{aligned}$$

Here, ϵ^{abs} and ϵ^{rel} are pregiven tolerance parameters whose values depend on specific problems.

7 Convergence Results of ADMM

In this section, we briefly discuss convergence conditions for different variants of ADMM. Note that there has been extensive recent literature on the convergence of this method, so it is not possible to provide an exhaustive discussion. Therefore, we choose to present some relatively basic results.

7.1 Convex Problems

7.1.1 Convex case

Generally speaking, the ADMM iteration (2) (Algorithm 1) for convex problems converges under quite mild conditions. Below, we provide a set of basic conditions [10]:

1. Problem (1) is feasible;
2. Both f and g are proper convex and lower semicontinuous functions;
3. A and B both have full column rank.

When these three conditions are satisfied, the sequence (x^k, y^k, λ^k) is bounded, and every limit point of (x^k, y^k) is an optimal solution for problem (1).

7.1.2 Strongly convex case

If the objective functions f and/or g are *strongly convex*, then the ADMM iteration (2) for solving (1) converges globally *linearly*. To be precise, define $w := [x; y; \lambda]$. Then, under any one of the four conditions given in Table 1, the following holds [48]:

$$||w^{k+1} - w^*||_G^2 \leq \frac{1}{\delta + 1}||w^k - w^*||_G^2 \tag{40}$$

for some $\delta > 0$, and for some w^* that belongs to the optimal solution set of (1), and some $G \succeq 0$.

7.1.3 Linearized ADMM

The global convergence of linearized ADMM (25) for solving convex problem (1) is guaranteed when $\tau_1 < 1/\lambda_{\max}(A^\top A)$ and $\tau_2 < 1/\lambda_{\max}(B^\top B)$ [22], where $\lambda_{\max}(X)$ represents the maximum eigenvalue of a matrix X.

7.2 Nonconvex Problems

Let us give a simple demonstration of how to show the convergence of ADMM in the nonconvex setting. Our presentation is based on the analysis in Ref. 45. Consider the simple

Table 1 The conditions for linear convergence.

Strong convexity	Lipschitz continuity	Full row rank
f	∇f	A, B^T
f, g	∇f	A
f	$\nabla f, \nabla g$	B^T
f, g	$\nabla f, \nabla g$	

Source: Modified from Wei Deng and Wotao Yin. On the global and linear convergence of the generalized alternating direction method of multipliers. Journal of Scientific Computing, 66(3): 889–916, 2016.

multiagent consensus problem given in (37). Let us assume that each local nonconvex cost function has Lipschitz gradient:

$$\|\nabla f_i(x_i) - \nabla f_i(z_i)\| \leq L_i \|x_i - z_i\|, \quad \forall\, i = 1, \ldots, N, \forall\, x_i,\, z_i \in \mathbb{R}^m$$

Further, assume that each function involved is lower bounded, that is, for some constant c, the following holds:

$$g(x) \geq c, \quad f_i(x) \geq c, \quad \forall\, i, \forall\, x \in \mathbb{R}^m$$

Then, the ADMM algorithm (39) can be analyzed by the following steps:

Step 1. First show that after one round of x, y, λ update, the augmented Lagrangian function is decreased in the following manner:

$$\begin{aligned} &\mathcal{L}(x^{k+1}, y^{k+1}; \lambda^{k+1}) - \mathcal{L}(x^k, y^k; \lambda^k) \\ &\quad \leq -c_1 \|y^{k+1} - y^k\|^2 - c_2 \sum_{i=1}^{N} \|x_i^{k+1} - x_i^k\|^2 + c_3 \sum_{i=1}^{N} \|x_i^{k+1} - y^{k+1}\|^2 \end{aligned}$$

where c_1, c_2, c_3 are positive constants. That is, the augmented Lagrangian function decreases proportional to the size of the distance traveled by the primal iterations, while it increases by the size of the constraint violation.

Step 2. By analyzing the optimality condition of (39b) as well as the dual update (39c), we can show the following:

$$\|x_i^{k+1} - y^{k+1}\|^2 \leq c_4 \|x_i^{k+1} - x_i^k\|^2, \quad \forall\, i = 1, \cdots, N$$

where $c_4 > 0$ is a constant. That is, the constraint violation can be upper bounded by the size of the successive differences of x_is.

Step 3. Combining the previous two steps, one can show that by properly adjusting the penalty parameter β, the augmented Lagrangian function is always decreasing.

Step 4. In the last step, one can show that $\mathcal{L}(x^{k+1}, y^{k+1}; \lambda^{k+1})$ is always lower bounded; therefore, by combing with the fact that it is decreasing, one can show that the algorithm will eventually converge.

With a few more simple steps, one can conclude that the nonconvex ADMM algorithm (39) converges to a *first-order* stationary solution of the original problem (36), as k goes to infinity. That is, the following holds:

$$\nabla\left(g(y^*) + \sum_{i=1}^{N} f_i(x_i^*)\right) = 0, \; x_i^* = y^*, \; i = 1, \cdots, N$$

where (x^*, y^*) is a limit point of the sequence $\{(x^k, y^k)\}_k$ generated by (39).

Acknowledgments

The research of Shiqian Ma is supported in part by NSF grants DMS-1953210 and CCF-2007797 and UC Davis CeDAR (Center for Data Science and Artificial Intelligence Research) Innovative Data Science Seed Funding Program. The research of Mingyi Hong is supported in part by NSF grant CMMI-1727757.

References

1 Douglas, J. and Rachford, H.H. (1956) On the numerical solution of the heat conduction problem in 2 and 3 space variables. *Trans. Am. Math. Soc.*, 82, 421–439.

2 Glowinski, R. and Marrocco, A. (1975) Sur l'approximation par èlèments finis et la rèsolution par pènalisation-dualitè d'une classe de problèmes de dirichlet non linèaires. *R.A.I.R.O.*, R2, 41–76.

3 Lions, P.L. and Mercier, B. (1979) Splitting algorithms for the sum of two nonlinear operators. *SIAM J. Numer. Anal.*, 16, 964–979.

4 Gabay, D. (1983) Applications of the method of multipliers to variational inequalities, in *Augmented Lagrangian Methods: Applications to the Solution of Boundary Value Problems* (eds M. Fortin and R. Glowinski), North-Holland, Amsterdam, 299–331.

5 Eckstein, J. and Bertsekas, D.P. (1992) On the Douglas–Rachford splitting method and the proximal point algorithm for maximal monotone operators. *Math. Program.*, 55, 293–318.

6 Combettes, P.L. and Pesquet, J-C. (2007) A Douglas-Rachford splitting approach to nonsmooth convex variational signal recovery. *IEEE J. Sel. Top. Signal Process.*, 1 (4), 564–574.

7 Goldstein, T. and Osher, S. (2009) The split Bregman method for L1-regularized problems. *SIAM J. Imaging Sci.*, 2, 323–343.

8 Yang, J. and Zhang, Y. (2011) Alternating direction algorithms for ℓ_1 problems in compressive sensing. *SIAM J. Sci. Comput.*, 33 (1), 250–278.

9 Eckstein, J. (1989) Splitting methods for monotone operators with applications to parallel optimization. PhD thesis. Massachusetts Institute of Technology.

10 Boyd, S., Parikh, N., Chu, E. *et al.* (2011) Distributed optimization and statistical learning via the alternating direction method of multipliers. *Found. Trends Mach. Learn.*, 3 (1), 1–122.

11 Eckstein, J. and Yao, W. (2012) Augmented Lagrangian and Alternating Direction Methods for Convex Optimization: A Tutorial and Some Illustrative Computational Results. Technical report. *RUTCOR Res. Rep.*, 2012.

12 Candès, E.J., Li, X., Ma, Y., and Wright, J. (2011) Robust principal component analysis? *J. ACM*, 58 (3), 1–37.

13 Chandrasekaran, V., Sanghavi, S., Parrilo, P., and Willsky, A. (2011) Rank-sparsity incoherence for matrix decomposition. *SIAM J. Optim.*, 21 (2), 572–596.

14 Ma, S., Goldfarb, D., and Chen, L. (2011) Fixed point and Bregman iterative methods for matrix rank minimization. *Math. Program. Ser. A*, 128, 321–353.

15 Lauritzen, S. (1996) *Graphical Models*, Oxford University Press.

16 Yuan, M. and Lin, Y. (2007) Model selection and estimation in the Gaussian graphical model. *Biometrika*, 94 (1), 19–35.

17 Banerjee, O., El Ghaoui, L., and d'Aspremont, A. (2008) Model selection through sparse maximum likelihood estimation for multivariate gaussian for binary data. *J. Mach. Learn. Res.*, 9, 485–516.

18 Friedman, J., Hastie, T., and Tibshirani, R. (2008) Sparse inverse covariance estimation with the graphical lasso. *Biostatistics*, 9 (3), 432–441.

19 Tibshirani, R. (1996) Regression shrinkage and selection via the Lasso. *J. R. Stat. Soc., Ser. B*, 58 (1), 267–288.

20 van den Berg, E. and Friedlander, M.P. (2008) Probing the Pareto frontier for basis pursuit solutions. *SIAM J. Sci. Comput.*, 31 (2), 890–912.

21 Duchi, J., Shalev-Shwartz, S., Singer, Y., and Chandra, T. (2008) *Efficient Projections onto the l1-Ball for Learning in High Dimensions*. ICML.

22 Ma, S. (2016) Alternating proximal gradient method for convex minimization. *J. Sci. Comput.*, 68 (2), 546–572.

23 Tibshirani, R., Saunders, M., Rosset, S. *et al.* (2005) Sparsity and smoothness via the fused lasso. *J. R. Stat. Soc. Ser. B Stat. Methodol.*, 67 (1), 91–108.

24 Zhou, Z., Li, X., Wright, J. *et al.* (2010) *Stable Principal Component Pursuit*. Proceedings of International Symposium on Information Theory.

25 Chandrasekaran, V., Parrilo, P.A., and Willsky, A.S. (2012) Latent variable graphical model selection via convex optimization. *Ann. Stat.*, 40 (4), 1935–1967.

26 Chen, C., He, B., Ye, Y., and Yuan, X. (2016) The direct extension of ADMM for multi-block convex minimization problems is not necessarily convergent. *Math. Program.*, 155, 57–79.

27 Hong, M., Chang, T.-H., Wang, X. *et al.* (2019) A block successive upper bound minimization method of multipliers for linearly constrained convex optimization. *Math. Oper. Res.*, 45, 797–1192.

28 Hong, M. and Luo, Z.-Q. (2017) On the linear convergence of the alternating direction method of multipliers. *Math. Program.*, 162 (1), 165–199.

29 Lin, T., Ma, S., and Zhang, S. (2016) Iteration complexity analysis of multi-block ADMM for a family of convex minimization without strong convexity. *J. Sci. Comput.*, 69, 52–81.

30 Lin, T., Ma, S., and Zhang, S. (2015) On the sublinear convergence rate of multi-block ADMM. *J. Oper. Res. Soc. China*, 3 (3), 251–274.

31 Lin, T., Ma, S., and Zhang, S. (2015) On the global linear convergence of the ADMM with multiblock variables. *SIAM J. Optim.*, 25 (3), 1478–1497.

32 Lin, T., Ma, S., and Zhang, S. (2018) Global convergence of unmodified 3-block ADMM for a class of convex minimization problems. *J. Sci. Comput.*, 76 (1), 69–88.

33 Deng, W., Lai, M., Peng, Z., and Yin, W. (2017) Parallel multi-block ADMM with $o(1/k)$ convergence. *J. Sci. Comput.*, 71 (2), 712–736.

34 Sun, R., Luo, Z.-Q., and Ye, Y. (2020) On the efficiency of random permutation for ADMM and coordinate descent. *Math. Oper. Res.*, 45 (1), 233–271.

35 He, B., Tao, M., and Yuan, X. (2012) Alternating direction method with Gaussian back substitution for separable convex programming. *SIAM J. Optim.*, 22, 313–340.

36 Chen, L., Sun, D., Toh, K.-C., and Zhang, N. (2019) A unified algorithmic framework of symmetric Gauss–Seidel decomposition based proximal ADMMs for convex composite programming. *J. Comput. Math.*, 37, 739–757.

37 Wang, X., Hong, M., Ma, S., and Luo, Z.-Q. (2015) Solving multiple-block separable convex minimization problems using two-block alternating direction method of multipliers. *Pacific J. Optim.*, 11 (4), 645–667.

38 Ma, S., Xue, L., and Zou, H. (2013) Alternating direction methods for latent variable Gaussian graphical model selection. *Neural Comput.*, 25 (8), 2172–2198.

39 Antoniadis, A., Gijbels, I., and Nikolova, M. (2009) Penalized likelihood regression for generalized linear models with non-quadratic penalties. *Ann. Inst. Stat. Math.*, 63 (3), 585–615.

40 Fan, J. and Li, R. (2001) Variable selection via nonconcave penalized likelihood and its oracle properties. *J. Am. Stat. Assoc.*, 96 (456), 1348–1360.

41 Zhang, C.-H. (2010) Nearly unbiased variable selection under minimax concave penalty. *Ann. Stat.*, 38(2), 894–942.

42 Gu, Q., Wang, Z., and Liu, H. (2014) *Sparse PCA with Oracle Property*. Proceedings of the 27th International Conference on Neural Information Processing Systems (NIPS), pp. 1529–1537.

43 Hong, M., Hajinezhad, D., and Zhao, M.-M. (2017) *Prox-PDA*: The Proximal Primal-Dual Algorithm for Fast Distributed Nonconvex Optimization and Learning Over Networks. ICML.

44 Chang, T.-H., Hong, M., Wai, H.-T. *et al.* (2020) Distributed learning in the non-convex world: from batch to streaming data, and beyond. *IEEE Signal Process. Mag.*, 37, 26–38.

45 Hong, M., Luo, Z.-Q., and Razaviyayn, M. (2016) Convergence analysis of alternating direction method of multipliers for a family of nonconvex problems. *SIAM J. Optim.*, 26 (1), 337–364.

46 Li, G. and Pong, T.K. (2015) Global convergence of splitting methods for nonconvex composite optimization. *SIAM J. Optim.*, 25 (4), 2434–2460.

47 Wang, Y., Yin, W., and Zeng, J. (2019) Global convergence of ADMM in nonconvex non-smooth optimization. *J. Sci. Comput.*, 78 (1), 29–63.

48 Deng, W. and Yin, W. (2016) On the global and linear convergence of the generalized alternating direction method of multipliers. *J. Sci. Comput.*, 66(3), 889–916.

28

Nonconvex Optimization via MM Algorithms: Convergence Theory

Kenneth Lange[1], Joong-Ho Won[2], Alfonso Landeros[1], and Hua Zhou[1]

[1] *University of California, Los Angeles, CA, USA*
[2] *Seoul National University, Seoul, South Korea*

1 Background

The majorization–minimization (MM) principle for constructing optimization algorithms [1–3] finds broad range of applications in

- statistics: multidimensional scaling [4], quantile regression [5], ranking sports teams [6], variable selection [7–10], multivariate distributions [11, 12], variance component models [13], robust covariance estimation [14], and survival models [15, 16];
- optimization: geometric and sigmoid programming [17] and proximal distance algorithm [18–20];
- imaging: transmission and positron tomography [21], wavelets [22], magnetic resonance imaging, and sparse deconvolution; and
- machine learning: nonnegative matrix factorization [23], matrix completion [24, 25], clustering [26, 27], discriminant analysis [28], and support vector machines [29].

The recent book [30] and survey papers [31, 32] give a comprehensive overview of MM algorithms.

The MM principle involves majorizing the objective function $f(\boldsymbol{x})$ by a surrogate function $g(\boldsymbol{x} \mid \boldsymbol{x}_n)$ around the current iterate $\boldsymbol{x}_n$ of a search. Majorization is defined by the two conditions

$$f(\boldsymbol{x}_n) = g(\boldsymbol{x}_n \mid \boldsymbol{x}_n) \tag{1}$$

$$f(\boldsymbol{x}) \leq g(\boldsymbol{x} \mid \boldsymbol{x}_n), \quad \boldsymbol{x} \neq \boldsymbol{x}_n \tag{2}$$

In other words, the surface $\boldsymbol{x} \mapsto g(\boldsymbol{x} \mid \boldsymbol{x}_n)$ lies above the surface $\boldsymbol{x} \mapsto f(\boldsymbol{x})$ and is tangent to it at the point $\boldsymbol{x} = \boldsymbol{x}_n$. Construction of the majorizing function $g(\boldsymbol{x} \mid \boldsymbol{x}_n)$ constitutes the first M of the MM algorithm. The second M of the algorithm minimizes the surrogate $g(\boldsymbol{x} \mid \boldsymbol{x}_n)$ rather than $f(\boldsymbol{x})$. If $\boldsymbol{x}_{n+1}$ denotes the minimizer of $g(\boldsymbol{x} \mid \boldsymbol{x}_n)$, then this action forces the

Computational Statistics in Data Science.
Edited by Walter W. Piegorsch, Richard A. Levine, Hao Helen Zhang and Thomas C. M. Lee.
 ISBN 978-1-11956107-1

descent property $f(\boldsymbol{x}_{n+1}) \leq f(\boldsymbol{x}_n)$. This fact follows from the inequalities

$$f(\boldsymbol{x}_{n+1}) \leq g(\boldsymbol{x}_{n+1} \mid \boldsymbol{x}_n) \leq g(\boldsymbol{x}_n \mid \boldsymbol{x}_n) = f(\boldsymbol{x}_n)$$

reflecting the definition of $\boldsymbol{x}_{n+1}$ and the tangency condition.

The same principle applied to the maximization problems leads to the minorization–maximization algorithms that monotonically increase the objective values. The celebrated expectation–maximization (EM) algorithm in statistics is a special case of the minorization–maximization algorithm as the E-step constructs a Q-function that satisfies the minorization properties. Derivation of EM algorithm hinges upon the notion of missing data and conditional expectation while that of MM algorithm upon clever use of inequalities. For most problems where an EM algorithm exists, the MM derivation often leads to the same algorithm. Notable exceptions include the maximum-likelihood estimation (MLE) of the Dirichlet-multinomial model [11, 33] and the variance components model [13]. However, the MM principle has much wider applications as it applies to both minimization and maximization problems and does not rely on the notion of missing data.

2 Convergence Theorems

Throughout, we denote by $\mathcal{X} \subset \mathbb{R}^d$ the subset underlying our problems. All of the functions we consider have domain $\mathcal{X}$ and are extended real valued with range $\mathbb{R} \cup \{\infty\}$. The interior of set S is denoted by **int**S, and its closure by **cl**S.

The following concepts are useful.

Definition 1 (Effective domain). *The effective domain of a function f is defined and denoted by*

$$\mathrm{dom} f = \{\boldsymbol{x} \in \mathcal{X} : f(\boldsymbol{x}) < \infty\}$$

Definition 2 (Properness). *Function $f(\boldsymbol{x})$ is called* proper *if* $\mathrm{dom} f \neq \emptyset$.

Definition 3 (Directional derivatives). *The directional derivative of function f at $\boldsymbol{x} \in \mathcal{X}$ is defined and denoted as*

$$d_{\boldsymbol{v}} f(\boldsymbol{x}) = \lim_{t \downarrow 0} \frac{f(\boldsymbol{x} + t\boldsymbol{v}) - f(\boldsymbol{x})}{t}$$

if the limit exists.

If f is differentiable at $\boldsymbol{x}$, then $d_{\boldsymbol{v}} f(\boldsymbol{x}) = \langle \nabla f(\boldsymbol{x}), \boldsymbol{v} \rangle$.

Definition 4 (L-smoothness). *Function f is said to be L-smooth with respect to a norm $\|\cdot\|$ if it is differentiable on* **int** $\mathrm{dom} f$ *and the gradient ∇f is Lipschitz continuous with a Lipschitz constant L:*

$$\| \nabla f(\boldsymbol{x}) - \nabla f(\boldsymbol{y}) \| \leq L \| \boldsymbol{x} - \boldsymbol{y} \|, \quad \forall \boldsymbol{x}, \boldsymbol{y} \in \mathbf{int}\, \mathrm{dom} f$$

It can be shown that $f(\boldsymbol{x})$ is L-smooth if and only if

$$f(\boldsymbol{x}) \leq f(\boldsymbol{y}) + \langle \nabla f(\boldsymbol{y}), \boldsymbol{x} - \boldsymbol{y} \rangle + \frac{L}{2} \| \boldsymbol{x} - \boldsymbol{y} \|^2, \quad \forall \boldsymbol{x}, \boldsymbol{y} \in \textbf{int}\, \mathrm{dom} f$$

Definition 5 (Strong convexity). *Function f is called μ-strongly convex with respect to a norm $\| \cdot \|$, $\mu \geq 0$, if $f(\boldsymbol{x}) - \frac{\mu}{2} \|\boldsymbol{x}\|^2$ is convex.*

It can be shown that if $f(\boldsymbol{x})$ is μ-strongly convex and has its minimum at $\boldsymbol{y}$, then

$$f(\boldsymbol{x}) - f(\boldsymbol{y}) \geq \frac{\mu}{2} \|\boldsymbol{x} - \boldsymbol{y}\|^2$$

Definition 6 (Tangent vector, tangent cone). *For a closed nonempty set $C \subset \mathcal{X}$, the* tangent cone *of C at $\boldsymbol{x}$ is*

$$T_C(\boldsymbol{x}) = \left\{ \boldsymbol{v} \in \mathcal{X} : \exists \{\boldsymbol{x}_n\} \subset C, \{t_n\} \subset \mathbb{R} \text{ such that } t_n \downarrow 0, \boldsymbol{x}_n \to \boldsymbol{x} \text{ and } \frac{\boldsymbol{x}_n - \boldsymbol{x}}{t_n} \to \boldsymbol{v} \right\}$$

where the notation $t_n \downarrow 0$ means that t_n approaches 0 from above. A vector $\boldsymbol{v} \in T_C(\boldsymbol{x})$ is said to be a tangent vector *of C at $\boldsymbol{x}$.*

2.1 Classical Convergence Theorem

Consider the problem of minimizing the objective function f over a closed nonempty set $C \subset \mathcal{X}$. The following is immediate from the decent property of the MM algorithms:

Proposition 1. *Let $\{\boldsymbol{x}_n\} \subset \mathcal{X}$ be the iterates generated by an MM algorithm. Assume (a) $\boldsymbol{x}_n \in C$ for each n. Then, the sequence of objective values $\{f(\boldsymbol{x}_n)\}$ monotonically decreases. Furthermore, if (b) $p^* = \inf_{x \in C} f(\boldsymbol{x}) > -\infty$, then $\{f(\boldsymbol{x}_n)\}$ converges.*

Whether the limit is the desired minimum and whether the iterate $\{\boldsymbol{x}_n\}$ will converge to a minimizer is more subtle. For the latter, a classical theory of convergence in nonlinear optimization algorithms is due to Zangwill and Mond [34]. We first recap Zangwill's theory following the modern treatment of Luenberger and Ye [35]. Note that most of the iterative optimization algorithms, including the MM algorithms, generate a sequence $\{\boldsymbol{x}_n\}$ by mapping $\boldsymbol{x}_n \in \mathcal{X}$ to another point $\boldsymbol{x}_{n+1} \in \mathcal{X}$. For example, in MM algorithms, $\boldsymbol{x}_{n+1}$ is a point that minimizes the surrogate function $g(\boldsymbol{x}|\boldsymbol{x}_n)$ in $\mathcal{X}$. However, such a minimizer may not be unique unless the $g(\boldsymbol{x}|\boldsymbol{x}_n)$ satisfies certain assumptions. Rather, $\boldsymbol{x}_{n+1}$ is *one of the minimizers* of $g(\boldsymbol{x}|\boldsymbol{x}_n)$ and can be written as $\boldsymbol{x}_{n+1} \in \mathrm{argmin}_{x \in C} g(\boldsymbol{x}|\boldsymbol{x}_n)$. Thus, we may in general define an algorithm map as a *set-valued map*:

Definition 7 (Algorithm map). *An* algorithm map *M is a mapping defined on $\mathcal{X}$ that assigns to every point $\boldsymbol{x} \in \mathcal{X}$ a subset of $\mathcal{X}$.*

Among which point $M(\boldsymbol{x}_n)$ to choose as $\boldsymbol{x}_{n+1}$ depends on the specific details of the actual optimization algorithm. If M is a single-valued map, that is, $M(\boldsymbol{x})$ is singleton for all $\boldsymbol{x} \in \mathcal{X}$, we write $\boldsymbol{x}_{n+1} = M(\boldsymbol{x}_n)$.

A desirable property of an algorithm map is closure, which extends continuity of single-valued maps to set-valued ones:

Definition 8 (Closure). *A set-valued map M from $\mathcal{X}$ to $\mathcal{Y}$ is said to be* closed *at $\boldsymbol{x} \in \mathcal{X}$ if $\boldsymbol{y} \in M(\boldsymbol{x})$ whenever $\{\boldsymbol{x}_n\} \subset \mathcal{X}$ converges to $\boldsymbol{x}$ and $\{\boldsymbol{y}_n : \boldsymbol{y}_n \in \boldsymbol{M}(\boldsymbol{x}_n)\}$ converges to $\boldsymbol{y}$. The map M is said to be closed on $\mathcal{X}$ if it is closed at each point of $\mathcal{X}$.*

The celebrated Zangwill's global convergence theorem is phrased in terms of an algorithm map M, a solution set Γ, and a *descent function* u:

Lemma 1 (Convergence Theorem A, Zangwill and Mond [34]). *Let the point-to-set map $M : \mathcal{X} \to \mathcal{X}$ determine an algorithm that given a point $\boldsymbol{x}_0 \in \mathcal{X}$ generates the sequence $\{\boldsymbol{x}_n\}$. Also, let a solution set $\Gamma \subset \mathcal{X}$ be given. Suppose that:*

1. *all points $\boldsymbol{x}_n$ are in a compact set $C \subset \mathcal{X}$;*
2. *there is a continuous function $u : \mathcal{X} \to \mathbb{R}$ such that (a) if $\boldsymbol{x} \notin \Gamma$, $u(\boldsymbol{y}) < u(\boldsymbol{x})$ for all $\boldsymbol{y} \in M(\boldsymbol{x})$, and (b) if $\boldsymbol{x} \in \Gamma$, then either the algorithm terminates or $u(\boldsymbol{y}) \leq u(\boldsymbol{x})$ for all $\boldsymbol{y} \in M(\boldsymbol{x})$;*
3. *the map M is closed at $\boldsymbol{x}$ if $\boldsymbol{x} \notin \Gamma$.*

Then, either the algorithm stops at a solution, or the limit of any convergent subsequence is a solution.

In applying Lemma 1 to specific algorithms, one usually needs to show the closure of the algorithm map M and carefully choose the solution set Γ and the descent function u. For example, in an MM algorithm, we can choose u as the objective function f and the solution set

$$\Gamma = \{\boldsymbol{x} \in \mathcal{X} : f(\boldsymbol{y}) \geq f(\boldsymbol{x}),\ \forall \boldsymbol{y} \in M(\boldsymbol{x})\}$$

for $M(\boldsymbol{x}) = \operatorname{argmin}_{\boldsymbol{z} \in \mathcal{X}} g(\boldsymbol{z}|\boldsymbol{x})$. Since $f(\boldsymbol{y}) \leq f(\boldsymbol{x})$ for all $\boldsymbol{y} \in M(\boldsymbol{x})$ by the descent property of MM, in fact

$$\Gamma = \{\boldsymbol{x} \in \mathcal{X} : f(\boldsymbol{y}) = f(\boldsymbol{x}),\ \forall \boldsymbol{y} \in M(\boldsymbol{x})\} =: \mathcal{P}$$

which we call a set of *no-progress points*. The final requirement that $\{\boldsymbol{x}_n\}$ is contained within a compact set is satisfied whenever f is lower semicontinuous and coercive.

We summarize the above discussion as the following proposition: see also Proposition 8 of Keys *et al.* [20].

Proposition 2 (Global convergence to no-progress points). *Suppose that the objective f is lower semicontinuous and coercive, and the algorithm map M defined by the MM algorithm is closed. Then, all the limit points of the iterates $\boldsymbol{x}_{n+1} \in M(\boldsymbol{x}_n)$ generated by the MM algorithm are no-progress points.*

This general result is slightly disappointing. Even though the objective values do not change within $\mathcal{P}$, the iterate $\{\boldsymbol{x}_n\}$ may not even converge – it may cycle through distinct no-progress points.

Example 1 (EM algorithm). As a classical example of cycling, Vaida [36] showed that in minimizing

$$f(\rho, \sigma^2) = 8\log\sigma^2 + \frac{18}{\sigma^2} + 2\log(1 - \log\rho^2) + \frac{4}{\sigma^2(1-\rho^2)}$$

over $\sigma^2 \leq 0$ and $-1 \leq \rho \leq 1$ (this objective function originates from the maximum-likelihood estimation of the variance and correlation coefficient of bivariate normal data with missing observations), the following particular surrogate function

$$g(\rho, \sigma^2 \mid \rho_n, \sigma_n^2) = f(\rho, \sigma^2) + 2\left(\log\frac{\sigma^2(1-\rho^2)}{\sigma_n^2(1-\rho_n^2)} + \frac{\sigma_n^2(1-\rho_n^2)}{\sigma^2(1-\rho^2)} - 1\right)$$

obtained by applying the EM algorithm, a special case of the MM algorithms, has two symmetric minima, $(\sigma_{n+1}^2, \rho_{n+1}) = (3, \pm\sqrt{2/3 - \sigma_n^2(1-\rho_n^2)/6})$. If we take $\sigma_0^2 = 3$ and

$$\rho_{n+1} = -\text{sgn}(\rho_n)\sqrt{2/3 - 3(1-\rho_n^2)/6)}$$

then the sequence $\{(\sigma_n^2, \rho_n)\}$ oscillates between two minima $(3, \pm 1/\sqrt{3})$ of f in the limit.

Although the above cycling can be considered desirable as it reveals multiple optima, the next example shows that this is not always the case:

Example 2 (Generalized CCA). The popular MAXDIFF criterion [37–39] for generalizing the CCA into $m > 2$ sets of (partially) orthogonal matrices solves

$$\text{maximize} \sum_{i<j} \text{tr}(\boldsymbol{O}_i^T \boldsymbol{A}_i^T \boldsymbol{A}_j \boldsymbol{O}_j) \text{ subject to } \boldsymbol{O}_i^T \boldsymbol{O}_i = \boldsymbol{I}_r,\ i = 1, \ldots, m \tag{3}$$

where $\boldsymbol{I}_r$ is an $r \times r$ identity matrix, and $\boldsymbol{O}_i \in \mathbb{R}^{d_i \times r}$; $\boldsymbol{A}_i \in \mathbb{R}^{n \times d_i}$ are n observations of variables of possibly different dimensions. A standard algorithm for solving the MAXDIFF problem is Ten Berge's block relaxation algorithm [38, 40], shown as Algorithm 1. This is an MM algorithm (here minorization–maximization), since at the update of the ith block in the kth sweep, the surrogate function

$$\begin{aligned} &g(\boldsymbol{O}_1, \ldots, \boldsymbol{O}_m \mid \boldsymbol{O}_1^{k+1}, \ldots, \boldsymbol{O}_{i-1}^{k+1}, \boldsymbol{O}_i^k, \ldots, \boldsymbol{O}_m^k) \\ &\quad = \frac{1}{2}\sum_{i=1}^{m} \text{tr}\left[\boldsymbol{O}_i^T\left(\sum_{j=1}^{i-1} \boldsymbol{A}_i^T \boldsymbol{A}_j \boldsymbol{O}_j^{k+1} + \sum_{j=i+1}^{m} \boldsymbol{A}_i^T \boldsymbol{A}_j \boldsymbol{O}_j^k\right)\right] \end{aligned}$$

minorizes the objective function of problem (3) at $(\boldsymbol{O}_1^{k+1}, \ldots, \boldsymbol{O}_{i-1}^{k+1}, \boldsymbol{O}_i^k, \ldots, \boldsymbol{O}_m^k)$ and is maximized based on the von Neuman–Fan inequality

$$\text{tr}(\boldsymbol{A}^T\boldsymbol{B}) \leq \sum_{l} \sigma_l(\boldsymbol{A})\sigma_l(\boldsymbol{B})$$

which holds for any two matrices $\boldsymbol{A}$ and $\boldsymbol{B}$ of the same dimensions with the lth largest singular values $\sigma_l(\boldsymbol{A})$ and $\sigma_l(\boldsymbol{B})$, respectively; equality is attained when $\boldsymbol{A}$ and $\boldsymbol{B}$ share a simultaneous ordered Singular Value Decomposition (SVD) [41].

While each iteration monotonically improves the objective function. Won *et al.* [42] show that Algorithm 1 may oscillate between suboptimal no-progress points. Set $m = 3, d_1 = d_2 = d_3 = d = r$ and $\boldsymbol{A}_1 = [\boldsymbol{I}_d, \boldsymbol{I}_d, \boldsymbol{0}]^T, \boldsymbol{A}_2 = [-\boldsymbol{I}_d, \boldsymbol{0}, \boldsymbol{I}_d]^T$, and $\boldsymbol{A}_3 = [\boldsymbol{0}, \boldsymbol{I}_d, \boldsymbol{I}_d]^T$ [43]. If Algorithm 1 is initialized with $(\boldsymbol{J}, \boldsymbol{K}, \boldsymbol{J})$ where

$$\boldsymbol{J} = \begin{bmatrix} 1 & 0 \\ 0 & 1 \\ 0 & 0 \end{bmatrix} \quad \text{and} \quad \boldsymbol{K} = \begin{bmatrix} 0 & 1 \\ 1 & 0 \\ 0 & 0 \end{bmatrix}$$

then both $\boldsymbol{J} - \boldsymbol{K}$ and $\boldsymbol{J} + \boldsymbol{K}$ have rank 1, and we see that $-\boldsymbol{K}$ is one of the maximizers of $\text{tr}[\boldsymbol{O}^T(\boldsymbol{J} - \boldsymbol{K})]$, and likewise, $\boldsymbol{J}$ maximizes $\text{tr}[\boldsymbol{O}^T(\boldsymbol{J} + \boldsymbol{K})]$. Taking these values as the outputs of Line 5 of Algorithm 1, we have the following cycling sequence at the end of each sweep:

$$(\boldsymbol{J}, \boldsymbol{K}, \boldsymbol{J}) \to (-\boldsymbol{K}, \boldsymbol{J}, -\boldsymbol{K}) \to (-\boldsymbol{J}, -\boldsymbol{K}, -\boldsymbol{J}) \to (\boldsymbol{K}, -\boldsymbol{J}, \boldsymbol{K}) \to (\boldsymbol{J}, \boldsymbol{K}, \boldsymbol{J}) \to \cdots$$

All four limit points yield the same objective values of 1. However, the global maximum of f can be shown to be 3.

The main reason for this oscillatory behavior is that the map $\boldsymbol{B} = \sum_{j \neq i} \boldsymbol{A}_i^T \boldsymbol{A}_j \boldsymbol{O}_j \mapsto \boldsymbol{P}_i \boldsymbol{Q}_i^T$ in Lines 5 and 6 is set valued. If $\boldsymbol{B}$ is rank deficient, any orthonormal basis of the null space of $\boldsymbol{B}^T$ (*resp.* $\boldsymbol{B}$) can be chosen as left (*resp.* right) singular vectors corresponding to the zero singular value. Furthermore, the product $\boldsymbol{P}_i \boldsymbol{Q}_i^T$ may not be unique [44, Proposition 7].

More satisfying "solution sets" are in order.

- Fixed points:

$$\mathcal{F} = \{\boldsymbol{x} \in \mathcal{X} : \boldsymbol{x} = M(\boldsymbol{x})\}$$

if M is single valued.

- Stationary points:

$$\mathcal{S} = \{\boldsymbol{x} \in \mathcal{X} : d_{\boldsymbol{v}} f(\boldsymbol{x}) \geq 0, \text{ for all tangent vectors } \boldsymbol{v} \text{ of } C \text{ at } \boldsymbol{x}\}$$

Algorithm 1. Ten Berge's algorithm for generalized CCA

1: Initialize $\boldsymbol{O}_1, \dots, \boldsymbol{O}_m$
2: For $k = 1, 2, \dots$
3:     For $i = 1, \dots, m$
4:         Set $\boldsymbol{B} = \sum_{j \neq i} \boldsymbol{A}_i^T \boldsymbol{A}_j \boldsymbol{O}_j$
5:         Compute SVD of $\boldsymbol{B}$ as $\boldsymbol{P}_i \boldsymbol{D}_i \boldsymbol{Q}_i^T$
6:         Set $\boldsymbol{O}_i = \boldsymbol{P}_i \boldsymbol{Q}_i^T$
7:     End For
8:     If there is no progress, then break
9: End For
10: Return $(\boldsymbol{O}_1, \dots, \boldsymbol{O}_m)$

All fixed points are no-progress points, that is, $\mathcal{F} \subset \mathcal{P}$, but not vice versa. Note that $\boldsymbol{y} \in \mathcal{S}$ is a *necessary* condition that $\boldsymbol{y}$ is a local minimizer of f in C. No-progress points and fixed points depend on the algorithm map M, whereas the stationary points depend on the problem itself. To make M single valued, note that any convex (and weakly convex) surrogate $g(\boldsymbol{x}|\boldsymbol{x}_n)$ can be made strongly convex, thus attains a unique minimum, by adding the viscosity penalty $\frac{\mu}{2}\| \boldsymbol{x} - \boldsymbol{x}_n\|^2$ majorizing 0; see Section 2.4. If M is closed and single valued, then it is continuous.

The classical global convergence results for MM algorithms [30, 45], which we summarize below, hinge on continuity of the map M:

Proposition 3. *If the MM algorithm map M is continuous, then $\mathcal{F} = \mathcal{P}$ and $\mathcal{F}$ is closed.*

Proposition 4. *If (i) f is continuous, (ii) f is coercive, or the set $\{\boldsymbol{x} : f(\boldsymbol{x}) \leq f(\boldsymbol{x}_0)\}$ is compact, and (iii) the algorithm map M is continuous, then every limit point of an MM sequence $\{\boldsymbol{x}_n\}$ is a fixed point of M. Furthermore, $\lim_{n\to\infty} \mathrm{dist}(\boldsymbol{x}_n, \mathcal{F}) = 0$.*

Proposition 5. *Under the same assumptions of Proposition 4, the MM sequence $\{\boldsymbol{x}_n\}$ satisfies*

$$\lim_{n\to\infty} \|\boldsymbol{x}_{n+1} - \boldsymbol{x}_n\| = 0$$

Furthermore, the set $\mathcal{T}$ of the limit points of $\{\boldsymbol{x}_n\}$ is compact and connected.

Note Proposition 4 states that $\mathcal{T} \subset \mathcal{F}$. Proposition 5 ensures there is no cycling.

Connecting the fixed points $\mathcal{F}$, which coincides with the no-progress points $\mathcal{P}$ for continuous M, with the stationary points $\mathcal{S}$ needs more assumptions. To equate stationary points of f to those of $g(\cdot|\boldsymbol{x})$, we require a stronger tangency condition than the usual tangency condition (1):

Definition 9 (Strong tangency). *An MM surrogate function $g(\cdot|\cdot)$ is said to be* strongly tangent *to f at $\boldsymbol{x} \in C$ if $d_{\boldsymbol{v}}g(\boldsymbol{x}|\boldsymbol{x}) = d_{\boldsymbol{v}}f(\boldsymbol{x})$ for all $\boldsymbol{x} \in C$ and tangent vector $\boldsymbol{v}$ in C at $\boldsymbol{x}$.*

Proposition 6. *Suppose that (i) the surrogate function $g(\boldsymbol{y}|\boldsymbol{x})$ is strongly tangent to the objective function f, (ii) the algorithm map M is closed and single valued, and (iii) stationary points and minimizers of $g(\boldsymbol{y}|\boldsymbol{x})$ are equivalent. Then, $\mathcal{F} = \mathcal{P} = \mathcal{S}$, that is, the sets of fixed points, no-progress points, and stationary points of f coincide.*

Proposition 7. *In addition to the assumptions of Proposition 6, if $\mathcal{S}$, the set of all stationary points of the objective function f, consists of isolated points, then the set $\mathcal{T}$ of the limit points of the MM sequence $\{\boldsymbol{x}_n\}$ is singleton, that is, an MM sequence $\{\boldsymbol{x}_n\}$ possesses a limit, and that limit is a stationary point of f as well as a fixed point of M.*

Strong tangency holds when $g(\boldsymbol{y}|\boldsymbol{x}) = f(\boldsymbol{x}) + h(\boldsymbol{y}|\boldsymbol{x})$ and $h(\boldsymbol{y}|\boldsymbol{x})$ is differentiable with $\nabla h(\boldsymbol{x}|\boldsymbol{x}) = \mathbf{0}$. See Vaida [36] and Yu *et al.* [46] for examples of these results in action.

We next present results that extend to nonasymptotic analysis and more general settings such as nonsmooth objectives.

2.2 Smooth Objective Functions

The following proposition gives a weak form of convergence for MM algorithms. The proposition features minimization and majorization by Lipschitz smooth functions.

Proposition 8. *Let $f(\boldsymbol{x})$ be a coercive differentiable function majorized by a uniformly L-Lipschitz surrogate $g(\boldsymbol{x} \mid \boldsymbol{x}_n)$ anchored at $\boldsymbol{x}_n$. If $\boldsymbol{y}$ denotes a minimum point of $f(\boldsymbol{x})$, then the iterates $\boldsymbol{x}_n$ delivered by the corresponding MM algorithm satisfy the sublinear bound*

$$\min_{0 \le k \le n} \| \nabla f(\boldsymbol{x}_k) \|^2 \le \frac{2L}{n+1}[f(\boldsymbol{x}_0) - f(\boldsymbol{y})] \tag{4}$$

When $f(\boldsymbol{x})$ is continuously differentiable, any limit point of the sequence $\boldsymbol{x}_n$ is a stationary point of $f(\boldsymbol{x})$.

Proof. Given that the surrogate $g(\boldsymbol{x} \mid \boldsymbol{x}_n)$ satisfies the tangency condition $\nabla g(\boldsymbol{x}_n \mid \boldsymbol{x}_n) = \nabla f(\boldsymbol{x}_n)$, the L-smoothness assumption entails the quadratic upper bound

$$\begin{aligned} f(\boldsymbol{x}_{n+1}) - f(\boldsymbol{x}_n) &\le g(\boldsymbol{x}_{n+1} \mid \boldsymbol{x}_n) - g(\boldsymbol{x}_n \mid \boldsymbol{x}_n) \\ &\le g(\boldsymbol{x} \mid \boldsymbol{x}_n) - g(\boldsymbol{x}_n \mid \boldsymbol{x}_n) \\ &\le \langle \nabla g(\boldsymbol{x}_n \mid \boldsymbol{x}_n), \boldsymbol{x} - \boldsymbol{x}_n \rangle + \frac{L}{2} \| \boldsymbol{x} - \boldsymbol{x}_n \|^2 \\ &= \langle \nabla f(\boldsymbol{x}_n), \boldsymbol{x} - \boldsymbol{x}_n \rangle + \frac{L}{2} \| \boldsymbol{x} - \boldsymbol{x}_n \|^2 \end{aligned}$$

for any $\boldsymbol{x}$. The choice $\boldsymbol{x} = \boldsymbol{x}_n - L^{-1} \nabla f(\boldsymbol{x}_n)$ yields the sufficient decrease condition

$$f(\boldsymbol{x}_n) - f(\boldsymbol{x}_{n+1}) \ge \frac{1}{2L} \| \nabla f(\boldsymbol{x}_n) \|^2 \tag{5}$$

A simple telescoping argument now gives

$$\begin{aligned} \frac{n+1}{2L} \min_{0 \le k \le n} \| \nabla f(\boldsymbol{x}_k) \|^2 &\le \frac{1}{2L} \sum_{k=0}^{n} \| \nabla f(\boldsymbol{x}_k) \|^2 \\ &\le f(\boldsymbol{x}_0) - f(\boldsymbol{x}_{n+1}) \\ &\le f(\boldsymbol{x}_0) - f(\boldsymbol{y}) \end{aligned}$$

which is equivalent to the bound (4). The second assertion follows directly from condition (5), the convergence of the sequence $f(\boldsymbol{x}_n)$, and the continuity of $\nabla f(\boldsymbol{x})$.

As a prelude to our next result, we state and prove a simple result of independent interest.

Proposition 9. *Suppose that $f(\boldsymbol{x})$ is convex with surrogate $g(\boldsymbol{z} \mid \boldsymbol{x})$ at the point $\boldsymbol{x}$. Then, $f(\boldsymbol{x})$ is differentiable at $\boldsymbol{x}$, and $\nabla f(\boldsymbol{x})$ equals $\nabla g(\boldsymbol{x} \mid \boldsymbol{x})$ wherever $\nabla g(\boldsymbol{x} \mid \boldsymbol{x})$ exists.*

Proof. Suppose that $\boldsymbol{x}$ is such a point. Let $\boldsymbol{v} \in \partial f(\boldsymbol{x})$ be a subgradient of $f(\boldsymbol{x})$. It suffices to show that $\boldsymbol{v}$ is uniquely determined as $\boldsymbol{v} = \nabla g(\boldsymbol{x} \mid \boldsymbol{x})$. For any direction $\boldsymbol{u}$, consider the forward difference quotient

$$\frac{g(\boldsymbol{x} + t\boldsymbol{u} \mid \boldsymbol{x}) - g(\boldsymbol{x} \mid \boldsymbol{x})}{t} \ge \frac{f(\boldsymbol{x} + t\boldsymbol{u}) - f(\boldsymbol{x})}{t} \ge \langle \boldsymbol{v}, \boldsymbol{u} \rangle$$

Taking limits produces $\langle \nabla g(\boldsymbol{x} \mid \boldsymbol{x}), \boldsymbol{u}\rangle \geq \langle \boldsymbol{v}, \boldsymbol{u}\rangle$. This cannot be true for all $\boldsymbol{u}$ unless the condition $\boldsymbol{v} = \nabla g(\boldsymbol{x} \mid \boldsymbol{x})$ holds.

Imposing strong convexity on $f(\boldsymbol{x})$ recovers linear convergence. The ratio $\frac{\mu}{L}$ makes an appearance, but, unlike convergence theorems for gradient descent, this ratio is not the condition number of either the objective or the surrogate.

Proposition 10. *Let $f(\boldsymbol{x})$ be a μ-strongly convex function majorized by a uniformly L-Lipschitz surrogate $g(\boldsymbol{x} \mid \boldsymbol{x}_n)$. If the global minimum occurs at $\boldsymbol{y}$, then the MM iterates $\boldsymbol{x}_n$ satisfy*

$$f(\boldsymbol{x}_n) - f(\boldsymbol{y}) \leq \left[1 - \left(\frac{\mu}{L}\right)^2\right]^n [f(\boldsymbol{x}_0) - f(\boldsymbol{y})]$$

thus establishing linear convergence of $\boldsymbol{x}_n$ to $\boldsymbol{y}$.

Proof. Existence and uniqueness of $\boldsymbol{y}$ follow from strong convexity. Because $\nabla g(\boldsymbol{y} \mid \boldsymbol{y}) = \boldsymbol{0}$, the smoothness of $g(\boldsymbol{x} \mid \boldsymbol{y})$ gives the quadratic upper bound

$$\begin{aligned} f(\boldsymbol{x}) - f(\boldsymbol{y}) &\leq g(\boldsymbol{x} \mid \boldsymbol{y}) - g(\boldsymbol{y} \mid \boldsymbol{y}) \\ &\leq \langle \nabla g(\boldsymbol{y} \mid \boldsymbol{y}), \boldsymbol{x} - \boldsymbol{y}\rangle + \frac{L}{2} \| \boldsymbol{x} - \boldsymbol{y}\|^2 \\ &= \frac{L}{2} \| \boldsymbol{x} - \boldsymbol{y}\|^2 \end{aligned} \tag{6}$$

which incidentally implies $\mu \leq L$. By the previous proposition, $f(\boldsymbol{x})$ is everywhere differentiable with $\nabla f(\boldsymbol{x}) = \nabla g(\boldsymbol{x} \mid \boldsymbol{x})$. In view of the strong convexity assumption, we have the lower bound

$$\begin{aligned} \| \nabla f(\boldsymbol{x}) \| \cdot \| \boldsymbol{y} - \boldsymbol{x} \| &\geq -\langle \nabla f(\boldsymbol{x}), \boldsymbol{y} - \boldsymbol{x}\rangle \\ &\geq f(\boldsymbol{y}) - f(\boldsymbol{x}) - \langle \nabla f(\boldsymbol{x}), \boldsymbol{y} - \boldsymbol{x}\rangle \\ &\geq \frac{\mu}{2} \| \boldsymbol{y} - \boldsymbol{x}\|^2 \end{aligned} \tag{7}$$

It follows that $\| \nabla f(\boldsymbol{x}) \| \geq \frac{\mu}{2} \| \boldsymbol{y} - \boldsymbol{x} \|$. Combining inequalities (6) and (7) furnishes the Polyak–Łojasiewicz (PL) bound

$$\| \nabla f(\boldsymbol{x})\|^2 \geq \frac{\mu^2}{2L}[f(\boldsymbol{x}) - f(\boldsymbol{y})]$$

We now turn to the MM iterates and take $\boldsymbol{x} = \boldsymbol{x}_n - \frac{1}{L}\nabla f(\boldsymbol{x}_n)$. The PL inequality implies

$$\begin{aligned} f(\boldsymbol{x}_{n+1}) - f(\boldsymbol{x}_n) &\leq g(\boldsymbol{x}_{n+1} \mid \boldsymbol{x}_n) - g(\boldsymbol{x}_n \mid \boldsymbol{x}_n) \\ &\leq g(\boldsymbol{x} \mid \boldsymbol{x}_n) - g(\boldsymbol{x}_n \mid \boldsymbol{x}_n) \\ &\leq \langle \nabla g(\boldsymbol{x}_n \mid \boldsymbol{x}_n), -\frac{1}{L}\nabla g(\boldsymbol{x}_n \mid \boldsymbol{x}_n)\rangle + \frac{L}{2}\left\|\frac{1}{L}\nabla g(\boldsymbol{x}_n \mid \boldsymbol{x}_n)\right\|^2 \\ &= -\frac{1}{2L} \| \nabla f(\boldsymbol{x}_n)\|^2 \\ &\leq -\frac{2\mu^2}{L^2}[f(\boldsymbol{x}_n) - f(\boldsymbol{y})] \end{aligned}$$

Subtracting $f(\boldsymbol{y})$ from both sides of the previous inequality and rearranging gives

$$f(\boldsymbol{x}_{n+1}) - f(\boldsymbol{y}) \leq \left[1 - \frac{\mu^2}{2L^2}\right] [f(\boldsymbol{x}_n) - f(\boldsymbol{y})]$$

Iteration of this inequality yields the claimed linear convergence.

2.3 Nonsmooth Objective Functions

Consider an MM minimization algorithm with objective $f(\boldsymbol{x})$ and surrogates $g(\boldsymbol{x} \mid \boldsymbol{x}_n)$. If $f(\boldsymbol{x})$ is coercive and continuous, and the $g(\boldsymbol{x} \mid \boldsymbol{x}_n)$ are μ-strongly convex, then we know that the MM iterates $\boldsymbol{x}_{n+1} = \text{argmin}_{\boldsymbol{x}}\, g(\boldsymbol{x} \mid \boldsymbol{x}_n)$ remain within the compact sublevel set $S = \{\boldsymbol{x} : f(\boldsymbol{x}) \leq f(\boldsymbol{x}_0)\}$ [30]. Furthermore, the strong convexity inequality

$$f(\boldsymbol{x}_n) - f(\boldsymbol{x}_{n+1}) \geq g(\boldsymbol{x}_n \mid \boldsymbol{x}_n) - g(\boldsymbol{x}_{n+1} \mid \boldsymbol{x}_n) \geq \frac{\mu}{2} \| \boldsymbol{x}_n - \boldsymbol{x}_{n+1} \|^2 \tag{8}$$

implies that

$$\sum_{n=0}^{\infty} \| \boldsymbol{x}_n - \boldsymbol{x}_{n+1} \|^2 \leq \frac{1}{\mu} [f(\boldsymbol{x}_0) - \bar{f}]$$

where $\bar{f} = \lim_{n \to \infty} f(\boldsymbol{x}_n)$. It follows from a well-known theorem of Ostrowski [30] that the set W of limit points of the $\boldsymbol{x}_n$ is compact and connected. It is also easy to show that $f(\boldsymbol{x})$ takes the constant value $\bar{f}$ on W and that $\lim_{n \to \infty} \text{dist}(\boldsymbol{x}_n, W) = 0$.

We will need the concept of a Fréchet subdifferential. If $f(\boldsymbol{x})$ is a function mapping $\mathbb{R}^p$ into $\mathbb{R} \cup \{+\infty\}$, then its Fréchet subdifferential at $\boldsymbol{x} \in \text{dom} f$ is the set

$$\partial^F f(\boldsymbol{x}) = \left\{ \boldsymbol{v} : \liminf_{\boldsymbol{y} \to \boldsymbol{x}} \frac{f(\boldsymbol{y}) - f(\boldsymbol{x}) - \boldsymbol{v}^t(\boldsymbol{y} - \boldsymbol{x})}{\| \boldsymbol{y} - \boldsymbol{x} \|} \geq 0 \right\}$$

The set $\partial^F f(\boldsymbol{x})$ is closed, convex, and possibly empty. If $f(\boldsymbol{x})$ is convex, then $\partial^F f(\boldsymbol{x})$ reduces to its convex subdifferential. If $f(\boldsymbol{x})$ is differentiable, then $\partial^F f(\boldsymbol{x})$ reduces to its ordinary differential. At a local minimum $\boldsymbol{x}$, Fermat's rule $\boldsymbol{0} \in \partial^F f(\boldsymbol{x})$ holds.

Proposition 11. *In an MM algorithm, suppose that $f(\boldsymbol{x})$ is coercive, $g(\boldsymbol{x} \mid \boldsymbol{x}_n)$ is differentiable, and the algorithm map $M(\boldsymbol{x})$ is closed. Then, all points $\boldsymbol{z}$ of the convergence set W are critical in the sense that $\boldsymbol{0} \in \partial^F(-f)(\boldsymbol{z})$.*

Proof. Let the subsequence x_{n_m} of the MM sequence $\boldsymbol{x}_{n+1} \in M(\boldsymbol{x}_n)$ converge to $\boldsymbol{z} \in W$. By passing a subsubsequence if necessary, we may suppose that $\boldsymbol{x}_{n_m+1}$ converges to $\boldsymbol{y}$. Owing to our closedness assumption, $\boldsymbol{y} \in M(\boldsymbol{z})$. Given that $f(\boldsymbol{y}) = f(\boldsymbol{z})$, it is obvious that $\boldsymbol{z}$ also minimizes $g(\boldsymbol{x} \mid \boldsymbol{z})$ and that $\boldsymbol{0} = \nabla g(\boldsymbol{z} \mid \boldsymbol{z})$. Since the difference $h(\boldsymbol{x} \mid \boldsymbol{z}) = g(\boldsymbol{x} \mid \boldsymbol{z}) - f(\boldsymbol{x})$ achieves it minimum at $\boldsymbol{x} = \boldsymbol{z}$, the Fréchet subdifferential $\partial^F h(\boldsymbol{x} \mid \boldsymbol{z})$ satisfies

$$\boldsymbol{0} \in \partial^F h(\boldsymbol{z} \mid \boldsymbol{z}) = \nabla g(\boldsymbol{z} \mid \boldsymbol{z}) + \partial^F(-f)(\boldsymbol{z})$$

It follows that $\boldsymbol{0} \in \partial^F(-f)(\boldsymbol{z})$.

We will also need to invoke Łojasiewicz's inequality. This deep result depends on some rather arcane algebraic geometry [47, 48]. It applies to semialgebraic functions and their more inclusive cousins semianalytic functions and subanalytic functions. For simplicity, we focus on semialgebraic functions. The class of semialgebraic subsets of $\mathbb{R}^p$ is the smallest class such that:

a) It contains all sets of the form $\{\boldsymbol{x} : q(\boldsymbol{x}) > 0\}$ for a polynomial $q(\boldsymbol{x})$ in p variables.
b) It is closed under the formation of finite unions, finite intersections, and set complementation.

A function $a : \mathbb{R}^p \mapsto \mathbb{R}^r$ is said to be semialgebraic if its graph is a semialgebraic set of $\mathbb{R}^{p+r}$. The class of real-valued semialgebraic contains all polynomials $p(\boldsymbol{x})$. It is closed under the formation of sums, products, absolute values, reciprocals when $a(\boldsymbol{x}) \neq 0$, roots when $a(\boldsymbol{x}) \geq 0$, and maxima $\max\{a(\boldsymbol{x}), b(\boldsymbol{x})\}$ and minima $\min\{a(\boldsymbol{x}), b(\boldsymbol{x})\}$. For our purposes, it is important to note that $\text{dist}(\boldsymbol{x}, S)$ is a semialgebraic function whenever S is a semialgebraic set.

Łojasiewicz's inequality in its modern form [49] requires that $f(\boldsymbol{x})$ be continuous and subanalytic with a closed domain. If $\boldsymbol{z}$ is a critical point of $f(\boldsymbol{x})$, then

$$|f(\boldsymbol{x}) - f(\boldsymbol{z})|^{\theta(\boldsymbol{z})} \leq c(\boldsymbol{z}) \parallel \boldsymbol{v} \parallel$$

for some constant $c(\boldsymbol{z})$, all $\boldsymbol{x}$ in some open ball $B_{r(\boldsymbol{z})}(\boldsymbol{z})$ around $\boldsymbol{z}$ of radius $r(\boldsymbol{z})$, and all $\boldsymbol{v}$ in $\partial^F f(\boldsymbol{x})$. This inequality applies to semialgebraic functions since they are automatically subanalytic. We apply Łojasiewicz's inequality to the points in the limit set W.

2.3.1 MM convergence for semialgebraic functions

Proposition 12. *Suppose that $f(\boldsymbol{x})$ is coercive, continuous, and subanalytic and all $g(\boldsymbol{x} \mid \boldsymbol{x}_n)$ are continuous, μ-strongly convex, and satisfy the Lipschitz condition*

$$\parallel \nabla g(\boldsymbol{u} \mid \boldsymbol{x}_n) - \nabla g(\boldsymbol{v} \mid \boldsymbol{x}_n) \parallel \leq L \parallel \boldsymbol{u} - \boldsymbol{v} \parallel$$

on the compact sublevel set $\{\boldsymbol{x} : f(\boldsymbol{x}) \leq f(\boldsymbol{x}_0)\}$. Then, the MM iterates $\boldsymbol{x}_{n+1} = \text{argmin}_{\boldsymbol{x}} g(\boldsymbol{x} \mid \boldsymbol{x}_n)$ converge to a critical point in W.

Proof. Because $h(\boldsymbol{x} \mid \boldsymbol{y}) = g(\boldsymbol{x} \mid \boldsymbol{y}) - f(\boldsymbol{x})$ achieves it minimum at $\boldsymbol{x} = \boldsymbol{y}$, the Fréchet subdifferential $\partial^F h(\boldsymbol{x} \mid \boldsymbol{y})$ satisfies

$$\boldsymbol{0} \in \partial^F h(\boldsymbol{y} \mid \boldsymbol{y}) = \nabla g(\boldsymbol{y} \mid \boldsymbol{y}) + \partial^F(-f)(\boldsymbol{y}).$$

It follows that $-\nabla g(\boldsymbol{y} \mid \boldsymbol{y}) \in \partial^F(-f)(\boldsymbol{y})$. By assumption

$$\parallel \nabla g(\boldsymbol{u} \mid \boldsymbol{x}_n) - \nabla g(\boldsymbol{v} \mid \boldsymbol{x}_n) \parallel \leq L \parallel \boldsymbol{u} - \boldsymbol{v} \parallel$$

for all $\boldsymbol{u}$ and $\boldsymbol{v}$ and $\boldsymbol{x}_n$. In particular, because $\nabla g(\boldsymbol{x}_{n+1} \mid \boldsymbol{x}_n) = \boldsymbol{0}$, we have

$$\parallel \nabla g(\boldsymbol{x}_n \mid \boldsymbol{x}_n) \parallel \leq L \parallel \boldsymbol{x}_{n+1} - \boldsymbol{x}_n \parallel \tag{9}$$

According to the Łojasiewicz inequality applied for the subanalytic function $\bar{f} - f(\boldsymbol{x})$, for each $\boldsymbol{z} \in W$ there exists a radius $r(\boldsymbol{z})$ and an exponent $\theta(\boldsymbol{z}) \in [0, 1)$ with

$$|f(\boldsymbol{u}) - f(\boldsymbol{z})|^{\theta(\boldsymbol{z})} = |\bar{f} - f(\boldsymbol{u}) - \bar{f} + \bar{f}|^{\theta(\boldsymbol{z})} \leq c(\boldsymbol{z}) \parallel \boldsymbol{v} \parallel$$

for all $\boldsymbol{u}$ in the open ball $B_{r(\boldsymbol{z})}(\boldsymbol{z})$ around $\boldsymbol{z}$ of radius $r(\boldsymbol{z})$ and all $\boldsymbol{v} \in \partial^F(\bar{f} - f)(\boldsymbol{u}) = \partial^F(-f)(\boldsymbol{u})$. We apply this inequality to $\boldsymbol{u} = \boldsymbol{x}_n$ and $\boldsymbol{v} = -\nabla g(\boldsymbol{x}_n \mid \boldsymbol{x}_n)$. In doing so, we would like to assume that the exponent $\theta(\boldsymbol{z})$ and constant $c(\boldsymbol{z})$ do not depend on $\boldsymbol{z}$. With this end in mind, cover W by a finite number of balls $B_{r(\boldsymbol{z}_i)}(\boldsymbol{z}_i)$ and take $\theta = \max_i \theta(\boldsymbol{z}_i) < 1$ and $c = \max_i c(\boldsymbol{z}_i)$. For a sufficiently large N, every $\boldsymbol{x}_n$ with $n \geq N$ falls within one of these balls and satisfies $|\bar{f} - f(\boldsymbol{x}_n)| < 1$. Without loss of generality assume $N = 0$. The Łojasiewicz inequality now entails

$$|\bar{f} - f(\boldsymbol{x}_n)|^{\theta} \leq c \parallel \nabla g(\boldsymbol{x}_n \mid \boldsymbol{x}_n) \parallel \tag{10}$$

In combination with the concavity of the function $t^{1-\theta}$ on $[0, \infty)$, inequalities (8), (9), and (10) imply

$$\begin{aligned}[f(\boldsymbol{x}_n) - \bar{f}]^{1-\theta} - [f(\boldsymbol{x}_{n+1}) - \bar{f}]^{1-\theta} &\geq \frac{1-\theta}{[f(\boldsymbol{x}_n) - \bar{f}]^{\theta}}[f(\boldsymbol{x}_n) - f(\boldsymbol{x}_{n+1})] \\ &\geq \frac{1-\theta}{c \parallel \nabla g(\boldsymbol{x}_n \mid \boldsymbol{x}_n) \parallel} \frac{\mu}{2} \parallel \boldsymbol{x}_{n+1} - \boldsymbol{x}_n \|^2 \\ &\geq \frac{(1-\theta)\mu}{2cL} \parallel \boldsymbol{x}_{n+1} - \boldsymbol{x}_n \parallel \end{aligned}$$

Rearranging this inequality and summing over n yield

$$\sum_{n=0}^{\infty} \|\boldsymbol{x}_{n+1} - \boldsymbol{x}_n\| \leq \frac{2cL}{(1-\theta)\mu}[f(\boldsymbol{x}_0) - \bar{f}]^{1-\theta}$$

Thus, the sequence $\boldsymbol{x}_n$ is a fast Cauchy sequence and converges to a unique limit in W.

2.4 A Proximal Trick to Prevent Cycling

Consider minimizing a function $f(\boldsymbol{x})$ bounded below and possibly subject to constraints. The MM principle involves constructing a surrogate function $g(\boldsymbol{x} \mid \boldsymbol{x}_n)$ that majorizes $f(\boldsymbol{x})$ around $\boldsymbol{x}_n$. For any $\rho > 0$, adding the penalty $(\rho/2) \parallel \boldsymbol{x} - \boldsymbol{x}_n\|^2$ to the surrogate produces a new surrogate

$$g(\boldsymbol{x} \mid \boldsymbol{x}_n) + \frac{\rho}{2} \parallel \boldsymbol{x} - \boldsymbol{x}_n\|^2$$

Rearranging the inequality

$$g(\boldsymbol{x}_{n+1} \mid \boldsymbol{x}_n) + \frac{\rho}{2}\|\boldsymbol{x}_{n+1} - \boldsymbol{x}_n\|^2 \leq g(\boldsymbol{x}_n \mid \boldsymbol{x}_n)$$

yields

$$\frac{\rho}{2}\|\boldsymbol{x}_{n+1} - \boldsymbol{x}_n\|^2 \leq g(\boldsymbol{x}_n \mid \boldsymbol{x}_n) - g(\boldsymbol{x}_{n+1} \mid \boldsymbol{x}_n) \leq f(\boldsymbol{x}_n) - f(\boldsymbol{x}_{n+1})$$

Thus, the MM iterates induced by the new surrogate satisfy

$$\lim_{n\to\infty} \|\boldsymbol{x}_{n+1} - \boldsymbol{x}_n \| = 0$$

This property is inconsistent with algorithm cycling between distant limit points.

3 Paracontraction

Another useful tool for proving iterate convergence of MM algorithms is paracontraction. Recall that a map $T : \mathcal{X} \to \mathbb{R}^d$ is contractive with respect to a norm $\|\boldsymbol{x}\|$ if $\| T(\boldsymbol{y}) - T(\boldsymbol{z}) \| < \| \boldsymbol{y} - \boldsymbol{z} \|$ for all $\boldsymbol{y} \neq \boldsymbol{z}$ in $\mathcal{X}$. It is strictly contractive if there exists a constant $c \in [0, 1)$ with $\| T(\boldsymbol{y}) - T(\boldsymbol{z}) \| \leq c \| \boldsymbol{y} - \boldsymbol{z} \|$ for all such pairs. If $c = 1$, then the map is nonexpansive.

Definition 10 (Paracontractive map). *A map $T : \mathcal{X} \to \mathbb{R}^d$ is said to be* paracontractive *if for every fixed point $\boldsymbol{y}$ of T (i.e., $\boldsymbol{y} = T(\boldsymbol{y})$), the inequality $\| T(\boldsymbol{x}) - \boldsymbol{y} \| < \| \boldsymbol{x} - \boldsymbol{y} \|$ holds unless $\boldsymbol{x}$ is itself a fixed point.*

A strictly contractive map is contractive, and a contractive map is paracontractive.

An important result regarding paracontractive maps is the theorem of Elsner, Koltract, and Neumann [50], which states that whenever a continuous paracontractive map T possesses one or more fixed points, then the sequence of iterates $\boldsymbol{x}_{n+1} = T(\boldsymbol{x}_n)$ converges to a fixed point regardless of the initial point $\boldsymbol{x}_0$. More formal statement is as follows:

Proposition 13. *Suppose that the continuous maps $T_0, \cdots, T_{r-1}$ of a set into itself are paracontractive under the norm $\| x \|$. Let F_i denote the set of fixed points of T_i. If the intersection $F = \cap_{i=0}^{r-1} F_i$ is nonempty, then the sequence*

$$\boldsymbol{x}_{n+1} = T_{n \bmod r}(\boldsymbol{x}_n)$$

converges to a limit in F. In particular, if $r = 1$ and $T = T_0$ has a nonempty set of fixed points F, then $\boldsymbol{x}_{n+1} = T(\boldsymbol{x}_n)$ converges to a point in F.

A simple proof is given in Lange [51].

Proposition 13 converts the task of proving convergence of MM iterates to that of showing (i) continuity, (ii) paracontractivity, and (iii) existence of a fixed point, of the MM algorithm map, and that (iv) any fixed point is a stationary point of the objective. A nice example is the recent work by Won *et al.* [52] on Euclidean projection onto the Minkowski sum of sets. The Minkowski sum of two sets A and B in $\mathbb{R}^d$ is

$$A + B = \{\boldsymbol{a} + \boldsymbol{b} : \boldsymbol{a} \in A,\ \boldsymbol{b} \in B\}$$

It is easy to show that $A + B$ is convex whenever A and B are both convex and is closed if at least one of the two sets is compact and the other is closed. When $A + B$ is closed with A and B convex, we may employ a block descent algorithm, an instance of MM algorithms, for finding the closest point to $\boldsymbol{x} \notin A + B$, which consists of alternating

$$\begin{aligned} \boldsymbol{b}_{n+1} &= P_B(\boldsymbol{x} - \boldsymbol{a}_n) \\ \boldsymbol{a}_{n+1} &= P_A(\boldsymbol{x} - \boldsymbol{b}_{n+1}) \end{aligned} \tag{11}$$

assuming that the projection operators P_A and P_B onto A and B are both known or easy to compute.

In order to show that the sequence $\{\boldsymbol{a}_n + \boldsymbol{b}_n\}$ converges to the closest point using Proposition 13, we first need to show the continuity of the map

$$T(\boldsymbol{a}) = P_A[\boldsymbol{x} - P_B(\boldsymbol{x} - \boldsymbol{a})]$$

The obtuse angle property of Euclidean projection [51, Example 6.5.3] yields

$$\langle \boldsymbol{a} - P_A(\boldsymbol{a}), P_A(\tilde{\boldsymbol{a}}) - P_A(\boldsymbol{a})\rangle \leq 0$$
$$\langle \tilde{\boldsymbol{a}} - P_A(\tilde{\boldsymbol{a}}), P_A(\boldsymbol{a}) - P_A(\tilde{\boldsymbol{a}})\rangle \leq 0$$

for any $\boldsymbol{a}, \tilde{\boldsymbol{a}} \in \mathbb{R}^d$. Adding these inequalities, rearranging, and applying the Cauchy–Schwarz inequality give

$$\begin{aligned} \| P_A(\boldsymbol{a}) - P_A(\tilde{\boldsymbol{a}})\|^2 &\leq \langle \boldsymbol{a} - \tilde{\boldsymbol{a}}, P_A(\boldsymbol{a}) - P_A(\tilde{\boldsymbol{a}})\rangle \\ &\leq \| \boldsymbol{a} - \tilde{\boldsymbol{a}} \| \| P_A(\boldsymbol{a}) - P_A(\tilde{\boldsymbol{a}}) \| \end{aligned} \tag{12}$$

Thus, $\| P_A(\boldsymbol{a}) - P_A(\tilde{\boldsymbol{a}}) \| \leq \| \boldsymbol{a} - \tilde{\boldsymbol{a}} \|$. That is, P_A is nonexpansive, and the inequality holds if and only if

$$P_A(\boldsymbol{a}) - P_A(\tilde{\boldsymbol{a}}) = c(\boldsymbol{a} - \tilde{\boldsymbol{a}}) \tag{13}$$

for some constant c. Likewise, P_B is nonexpansive. Therefore,

$$\begin{aligned} &\| P_A[\boldsymbol{x} - P_B(\boldsymbol{x} - \boldsymbol{a})] - P_A[\boldsymbol{x} - P_B(\boldsymbol{x} - \tilde{\boldsymbol{a}})] \| \\ &\leq \| P_B(\boldsymbol{x} - \boldsymbol{a}) - P_B(\boldsymbol{x} - \tilde{\boldsymbol{a}}) \| \leq \| \boldsymbol{a} - \tilde{\boldsymbol{a}} \| \end{aligned} \tag{14}$$

This proves that T is nonexpansive, hence continuous.

Next, we show that T is paracontractive. Suppose that $\tilde{\boldsymbol{a}}$ is a fixed point, $\boldsymbol{a} \neq \tilde{\boldsymbol{a}}$, and equality holds throughout inequalities (14). Inequalities (12) and Equation (13) indicate that equality is achieved in the previous two inequalities only if

$$\begin{aligned} &P_A[\boldsymbol{x} - P_B(\boldsymbol{x} - \boldsymbol{a})] - [\boldsymbol{x} - P_B(\boldsymbol{x} - \boldsymbol{a})] \\ &= P_A[\boldsymbol{x} - P_B(\boldsymbol{x} - \tilde{\boldsymbol{a}})] - [\boldsymbol{x} - P_B(\boldsymbol{x} - \tilde{\boldsymbol{a}})] \end{aligned}$$

and

$$P_B(\boldsymbol{x} - \boldsymbol{a}) - (\boldsymbol{x} - \boldsymbol{a}) = P_B(\boldsymbol{x} - \tilde{\boldsymbol{a}}) - (\boldsymbol{x} - \tilde{\boldsymbol{a}})$$

Subtracting the second of these equalities from the first gives

$$P_A[\boldsymbol{x} - P_B(\boldsymbol{x} - \boldsymbol{a})] - \boldsymbol{a} = P_A[\boldsymbol{x} - P_B(\boldsymbol{x} - \tilde{\boldsymbol{a}})] - \tilde{\boldsymbol{a}} = \boldsymbol{0}$$

It follows that equality in inequalities (14) is achieved only if $\boldsymbol{a}$ is also a fixed point.

To show that T possesses a fixed point, note that given the closedness of $A + B$, there exists a closest point $\tilde{\boldsymbol{a}} + \tilde{\boldsymbol{b}}$ to $\boldsymbol{x}$, where $\tilde{\boldsymbol{a}} \in A$ and $\tilde{\boldsymbol{b}} \in B$. Since block descent cannot improve the objective $f(\boldsymbol{a}, \boldsymbol{b}) = \frac{1}{2} \| \boldsymbol{x} - \boldsymbol{a} - \boldsymbol{b}\|^2$ on the set $A \times B$ starting from $(\tilde{\boldsymbol{a}}, \tilde{\boldsymbol{b}})$, it is clear that $\tilde{\boldsymbol{a}} = T(\tilde{\boldsymbol{a}})$.

Finally, suppose that $\tilde{\boldsymbol{a}}$ is any fixed point, and define $\tilde{\boldsymbol{b}} = P_B(\boldsymbol{x} - \tilde{\boldsymbol{a}})$. To prove that $\tilde{\boldsymbol{a}} + \tilde{\boldsymbol{b}}$ minimizes the distance to $\boldsymbol{x}$, it suffices to show that for every tangent vector $\boldsymbol{v} = \boldsymbol{a} + \boldsymbol{b} - \tilde{\boldsymbol{a}} - \tilde{\boldsymbol{b}}$ at $\tilde{\boldsymbol{a}} + \tilde{\boldsymbol{b}}$, the directional derivative

$$\begin{aligned} d_{\boldsymbol{v}} \frac{1}{2} \| \boldsymbol{x} - \tilde{\boldsymbol{a}} - \tilde{\boldsymbol{b}}\|^2 &= -\langle \boldsymbol{x} - \tilde{\boldsymbol{a}} - \tilde{\boldsymbol{b}}, \boldsymbol{v}\rangle \\ &= -\langle \boldsymbol{x} - \tilde{\boldsymbol{a}} - \tilde{\boldsymbol{b}}, \boldsymbol{a} - \tilde{\boldsymbol{a}}\rangle - \langle \boldsymbol{x} - \tilde{\boldsymbol{a}} - \tilde{\boldsymbol{b}}, \boldsymbol{b} - \tilde{\boldsymbol{b}}\rangle \end{aligned}$$

is nonnegative. However, the inequalities $-\langle \boldsymbol{x} - \tilde{\boldsymbol{a}} - \tilde{\boldsymbol{b}}, \boldsymbol{a} - \tilde{\boldsymbol{a}}\rangle \geq 0$ and $-\langle \boldsymbol{x} - \tilde{\boldsymbol{a}} - \tilde{\boldsymbol{b}}, \boldsymbol{b} - \tilde{\boldsymbol{b}}\rangle \geq 0$ hold because $\tilde{\boldsymbol{a}}$ minimizes $\boldsymbol{a} \mapsto \frac{1}{2} \| \boldsymbol{x} - \boldsymbol{a} - \tilde{\boldsymbol{b}}\|^2$, and $\tilde{\boldsymbol{b}}$ minimizes $\boldsymbol{b} \mapsto \frac{1}{2} \| \boldsymbol{x} - \tilde{\boldsymbol{a}} - \boldsymbol{b}\|^2$. Thus, any fixed point of T furnishes a minimum of the convex function $f(\boldsymbol{a}, \boldsymbol{b})$ on the set $A \times B$.

4 Bregman Majorization

Bregman majorization is a technique for constructing a sequence of surrogate functions pertinent to an MM algorithm. Let us first define the notion of Bregman divergence.

Definition 11 (Bregman divergence). *For a proper convex function $\phi(\boldsymbol{x})$ that is continuously differentiable on* **int** dom ϕ, *the Bregman divergence $B_\phi : \mathcal{X} \times \mathcal{X} \to \mathbb{R}$ is defined as*

$$B_\phi(\boldsymbol{x} \parallel \boldsymbol{y}) = \phi(\boldsymbol{x}) - \phi(\boldsymbol{y}) - \langle \nabla\phi(\boldsymbol{y}), \boldsymbol{x} - \boldsymbol{y} \rangle, \quad \boldsymbol{x}, \boldsymbol{y} \in \mathbf{int}\,\mathrm{dom}\,\phi$$

We are concerned with the following optimization problem:

$$\min_{\boldsymbol{x} \in C} f(\boldsymbol{x}), \quad C \subset \mathcal{X} \text{ is closed and convex} \tag{15}$$

where $f(\boldsymbol{x})$ is convex, proper, and lower semicontinuous. In order to solve this problem, the Bregman majorization method constructs the sequence of surrogate functions

$$g(\boldsymbol{x} \mid \boldsymbol{x}_n) = f(\boldsymbol{x}) + B_\phi(\boldsymbol{x} \parallel \boldsymbol{x}_n)$$

and successively minimizes these. This is a valid MM algorithm since the following properties of the Bregman divergence are immediate from definition:

1. $B_\phi(\boldsymbol{x} \parallel \boldsymbol{y}) \geq 0$;
2. $B_\phi(\boldsymbol{x} \parallel \boldsymbol{x}) = 0$;
3. If ϕ is strictly convex, then $B_\phi(\boldsymbol{x} \parallel \boldsymbol{y}) = 0$ if and only if $\boldsymbol{x} = \boldsymbol{y}$.

Thus, $g(\boldsymbol{x} \mid \boldsymbol{x}_n) \geq f(\boldsymbol{x})$ for all $\boldsymbol{x}$ and $g(\boldsymbol{x}_n \mid \boldsymbol{x}_n) = f(\boldsymbol{x}_n)$. We can choose $\phi(\boldsymbol{x})$ so that for $\mathbf{cl}\,\mathrm{dom}\,\phi = C$.

The subsequent section studies the convergence property of the Bregman majorization.

4.1 Convergence Analysis via SUMMA

The sequential unconstrained minimization method algorithm (SUMMA) [53] is a class of algorithms for solving optimization problems of the form

$$\min_{\boldsymbol{x} \in C} f(\boldsymbol{x}), \quad C \subset \mathcal{X} \text{ is closed} \tag{16}$$

by minimizing a sequence of auxiliary functions

$$G_n(\boldsymbol{x}) = f(\boldsymbol{x}) + g_{n+1}(\boldsymbol{x}), \quad n = 1, 2, \ldots$$

over $\mathcal{X}$. The minimizer of $G_n(\boldsymbol{x})$ is denoted by $\boldsymbol{x}_n$. The conditions imposed on the sequence of functions $g_n(\boldsymbol{x})$ are:

1. $g_n(\boldsymbol{x}) \geq 0$ for all $\boldsymbol{x} \in \mathcal{X}$;
2. $g_n(\boldsymbol{x}_{n-1}) = 0$;
3. $G_n(\boldsymbol{x}) - G_n(\boldsymbol{x}_n) \geq g_{n+1}(\boldsymbol{x})$ for all $\boldsymbol{x} \in C$.

If $g_n(\boldsymbol{x})$ depends on n only through the iterate x_n, then this method coincides with the MM algorithm by identifying $G_n(\boldsymbol{x}) = g(\boldsymbol{x} \mid \boldsymbol{x}_n)$ and $g_n(\boldsymbol{x}) = g(\boldsymbol{x} \mid \boldsymbol{x}_{n-1}) - f(\boldsymbol{x})$, with the additional requirement

$$g(\boldsymbol{x} \mid \boldsymbol{x}_n) - g(\boldsymbol{x}_{n+1} \mid \boldsymbol{x}_n) \geq g(\boldsymbol{x} \mid \boldsymbol{x}_{n+1}) - f(\boldsymbol{x}) \tag{17}$$

for all $\boldsymbol{x} \in C$.

Let us show that condition (17) is satisfied by the Bregman majorization $g(\boldsymbol{x} \mid \boldsymbol{x}_n) = f(\boldsymbol{x}) + \phi(\boldsymbol{x}) - \phi(\boldsymbol{x}_n) - \langle \nabla\phi(\boldsymbol{x}_n), \boldsymbol{x} - \boldsymbol{x}_n \rangle$. The optimality condition for minimizing $g(\boldsymbol{x} \mid \boldsymbol{x}_n)$ is

$$\mathbf{0} \in \partial f(\boldsymbol{x}_{n+1}) + \nabla\phi(\boldsymbol{x}_{n+1}) - \nabla\phi(\boldsymbol{x}_n)$$

For the appropriate choice of $\boldsymbol{s}_{n+1} \in \partial f(\boldsymbol{x}_{n+1})$, it follows that

$$\begin{aligned} g(\boldsymbol{x} \mid \boldsymbol{x}_n) - g(\boldsymbol{x}_{n+1} \mid \boldsymbol{x}_n) &= f(\boldsymbol{x}) - f(\boldsymbol{x}_{n+1}) + \phi(\boldsymbol{x}) - \phi(\boldsymbol{x}_{n+1}) \\ &\quad - \langle \nabla\phi(\boldsymbol{x}_n), \boldsymbol{x} - \boldsymbol{x}_{n+1} \rangle \\ &= f(\boldsymbol{x}) - f(\boldsymbol{x}_{n+1}) - \langle \boldsymbol{s}_{n+1}, \boldsymbol{x} - \boldsymbol{x}_{n+1} \rangle \\ &\quad + \phi(\boldsymbol{x}) - \phi(\boldsymbol{x}_{n+1}) - \langle \nabla\phi(\boldsymbol{x}_{n+1}), \boldsymbol{x} - \boldsymbol{x}_{n+1} \rangle \\ &\geq B_\phi(\boldsymbol{x} \parallel \boldsymbol{x}_{n+1}) = g(\boldsymbol{x} \mid \boldsymbol{x}_{n+1}) - f(\boldsymbol{x}) \end{aligned}$$

where the last inequality is a consequence of the convexity of $f(\boldsymbol{x})$.

The following propositions concern convergence of MM algorithms satisfying condition (17).

Proposition 14. *Assume (a)* $p^* = \inf_{x \in C} f(\boldsymbol{x}) > -\infty$ *and (b)* $\boldsymbol{x}_n \in C$ *for each n. If condition (17) holds, then any MM sequence generated by the map* $\boldsymbol{x}_{n+1} \in \operatorname{argmin}_{x \in \mathcal{X}} g(\boldsymbol{x} \mid \boldsymbol{x}_n)$ *satisfies* $\lim_{n \to \infty} f(\boldsymbol{x}_n) = p^*$.

Proof. By the descent property of MM and the bound $f(\boldsymbol{x}_n) \geq p^* > -\infty$ given $\boldsymbol{x}_n \in C$, the sequence $f(\boldsymbol{x}_n)$ converges to a limit $d \geq p^*$. Suppose for some $\boldsymbol{x} \in C$ that $f(\boldsymbol{x}) < d$. Then, by condition (17),

$$\begin{aligned} [g(\boldsymbol{x} \mid \boldsymbol{x}_n) - f(\boldsymbol{x})] - [g(\boldsymbol{x} \mid \boldsymbol{x}_{n+1}) - f(\boldsymbol{x})] &\geq g(\boldsymbol{x}_{n+1} \mid \boldsymbol{x}_n) - f(\boldsymbol{x}) \\ &\geq f(\boldsymbol{x}_{n+1}) - f(\boldsymbol{x}) \\ &\geq d - f(\boldsymbol{x}) \\ &> 0 \end{aligned}$$

Thus, the sequence $g(\boldsymbol{x} \mid \boldsymbol{x}_n) - f(\boldsymbol{x})$ decreases, and its successive differences are bounded away from zero. The latter property contradicts the requirement for the surrogate function that $g(\boldsymbol{x} \mid \boldsymbol{x}_n) \geq f(\boldsymbol{x})$, and therefore $d = p^*$.

Proposition 15. *In addition to the assumptions of Proposition 14, further assume that (c) the minimum* p^* *is attained and the set F of the minimizers of* $f(\boldsymbol{x})$ *in C is nonempty, (d)* $f(\boldsymbol{x})$ *is continuous on* $D \subset \mathcal{X}$ *such that* $\mathbf{cl}D = C$*, (e) for each n,* $g(\boldsymbol{x} \mid \boldsymbol{x}_n)$ *is* μ*-strongly convex with* *respect to the norm* $\|\cdot\|$ *and* $\operatorname{dom} g(\cdot \mid \boldsymbol{x}_n) = D$, *and (f)* $g(\boldsymbol{x} \mid \boldsymbol{x}_n) - f(\boldsymbol{x}) \le \frac{L}{2}\|\boldsymbol{x} - \boldsymbol{x}_n\|^2$ *for all* $\boldsymbol{x} \in D$ *and each n. If condition (17) holds, then the MM sequence* $\boldsymbol{x}_{n+1} = \operatorname{argmin}_{\boldsymbol{x}\in\mathcal{X}} g(\boldsymbol{x} \mid \boldsymbol{x}_n)$ *converges to a point in F.*

Proof. Because of strong convexity, the minimum of $g(\boldsymbol{x} \mid \boldsymbol{x}_n)$ is uniquely attained for each n. Furthermore, for any $\boldsymbol{x} \in D$,

$$g(\boldsymbol{x} \mid \boldsymbol{x}_n) - g(\boldsymbol{x}_{n+1} \mid \boldsymbol{x}_n) \ge \frac{\mu}{2}\|\boldsymbol{x} - \boldsymbol{x}_{n+1}\|^2 \tag{18}$$

Let $\boldsymbol{y} \in F$ be a minimizer of $f(\boldsymbol{x})$ in C. Since $f(\boldsymbol{x}_{n+1}) \le g(\boldsymbol{x}_{n+1} \mid \boldsymbol{x}_n)$,

$$g(\boldsymbol{y} \mid \boldsymbol{x}_n) - f(\boldsymbol{x}_{n+1}) \ge g(\boldsymbol{y} \mid \boldsymbol{x}_n) - g(\boldsymbol{x}_{n+1} \mid \boldsymbol{x}_n) \ge \frac{\mu}{2}\|\boldsymbol{y} - \boldsymbol{x}_{n+1}\|^2 \tag{19}$$

where the last inequality follows from the strong convexity of $g(\boldsymbol{x} \mid \boldsymbol{x}_n)$. Condition (17) also implies

$$\begin{aligned}[g(\boldsymbol{y} \mid \boldsymbol{x}_n) - f(\boldsymbol{y})] - [g(\boldsymbol{y} \mid \boldsymbol{x}_{n+1}) - f(\boldsymbol{y})] &\ge g(\boldsymbol{x}_{n+1} \mid \boldsymbol{x}_n) - f(\boldsymbol{y}) \\ &\ge f(\boldsymbol{x}_{n+1}) - p^* \ge 0\end{aligned}$$

Hence, the decreasing nonnegative sequence $g(\boldsymbol{y} \mid \boldsymbol{x}_n) - f(\boldsymbol{y})$ has a limit. In addition, $f(\boldsymbol{y}) - f(\boldsymbol{x}_{n+1})$ tends to zero by Proposition 14. It follows that the leftmost side of inequality (19) tends to a limit, and the sequence $\boldsymbol{x}_n$ is bounded.

Suppose that the convergent subsequence $\boldsymbol{x}_{n_m}$ of $\boldsymbol{x}_n$ has a limit $\boldsymbol{z}$. By continuity, $f(\boldsymbol{z}) = \lim_{m\to\infty} f(\boldsymbol{x}_{n_m}) = p^*$, so $\boldsymbol{z}$ is also optimal. Now,

$$\begin{aligned}0 &\le g(\boldsymbol{z} \mid \boldsymbol{x}_n) - g(\boldsymbol{x}_{n+1} \mid \boldsymbol{x}_n) \\ &= [g(\boldsymbol{z} \mid \boldsymbol{x}_n) - f(\boldsymbol{z})] + f(\boldsymbol{z}) - f(\boldsymbol{x}_{n+1}) - [g(\boldsymbol{x}_{n+1} \mid \boldsymbol{x}_n) - f(\boldsymbol{x}_{n+1})] \\ &\le g(\boldsymbol{z} \mid \boldsymbol{x}_n) - f(\boldsymbol{z}) \\ &\le \frac{L}{2}\|\boldsymbol{x}_n - z\|^2\end{aligned}$$

due to $f(\boldsymbol{z}) \le f(\boldsymbol{x}_{n+1})$, $g(\boldsymbol{x}_{n+1} \mid \boldsymbol{x}_n) - f(\boldsymbol{x}_{n+1}) \ge 0$, and assumption (f). Again by Condition (17), we further have

$$0 \le g(\boldsymbol{z} \mid \boldsymbol{x}_n) - g(\boldsymbol{x}_{n+1} \mid \boldsymbol{x}_n) \le g(\boldsymbol{z} \mid \boldsymbol{x}_n) - f(\boldsymbol{z}) \le g(\boldsymbol{z} \mid \boldsymbol{x}_{n-1}) - g(\boldsymbol{x}_n \mid \boldsymbol{x}_{n-1}) \tag{20}$$

Thus, the nonnegative sequence $g(\boldsymbol{z} \mid \boldsymbol{x}_n) - g(\boldsymbol{x}_{n+1} \mid \boldsymbol{x}_n)$ is monotonically decreasing and convergent. Its subsequence $g(\boldsymbol{z} \mid \boldsymbol{x}_{n_m}) - g(\boldsymbol{x}_{n_m+1} \mid \boldsymbol{x}_{n_m})$ is also bounded by $\frac{L}{2}\|\boldsymbol{x}_{n_m} - \boldsymbol{z}\|^2$, which converges to zero. Thus, the whole sequence tends to zero. By inequality (20), it follows that the sequence $g(\boldsymbol{z} \mid \boldsymbol{x}_n) - f(\boldsymbol{z})$ converges to zero.

The final inequality

$$\begin{aligned}g(\boldsymbol{z} \mid \boldsymbol{x}_n) - f(\boldsymbol{z}) &= g(\boldsymbol{z} \mid \boldsymbol{x}_n) - g(\boldsymbol{x}_{n+1} \mid \boldsymbol{x}_n) + g(\boldsymbol{x}_{n+1} \mid \boldsymbol{x}_n) - f(\boldsymbol{z}) \\ &\ge \frac{\mu}{2}\|\boldsymbol{z} - \boldsymbol{x}_{n+1}\|^2 + f(\boldsymbol{x}_{n+1}) - f(\boldsymbol{z})\end{aligned}$$

now proves that the entire sequence $\boldsymbol{x}_n$ converges to $\boldsymbol{z} \in F$.

Remark 1. Assumption (e) (uniform strong convexity of the surrogate functions) is much less restrictive than assuming strong convexity on the objective $f(\boldsymbol{x})$. For example, assumption (e) is satisfied when $f(\boldsymbol{x})$ is convex, and the convex function $\phi(\boldsymbol{x})$ defining the Bregman divergence is μ-strongly convex.

Remark 2. Assumption (f) is satisfied if $\phi(\boldsymbol{x})$ is L-smooth. Assumption (f) can be replaced by

(f') $g(\boldsymbol{x} \mid \boldsymbol{y})$ is continuous in $\boldsymbol{y}$ in D.

This is the condition implicitly imposed in the proof of Proposition 7.4.1 in Lange [41]. (This assumption is not made perfectly clear in the statement of the proposition.) Assumption (f') is satisfied, when $\phi(\boldsymbol{x})$ is a Bregman–Legendre function [53, 54].

4.2 Examples

4.2.1 Proximal gradient method

The proximal gradient method minimizes $f(\boldsymbol{x}) = f_0(\boldsymbol{x}) + h(\boldsymbol{x})$ over $C = \mathcal{X}$, where both $f_0(\boldsymbol{x})$ and $h(\boldsymbol{x})$ are convex, proper, and lower semicontinuous. It is further assumed that $f_0(\boldsymbol{x})$ is L-smooth. The algorithm iteratively solves

$$\boldsymbol{x}_{n+1} = \text{argmin}_{\boldsymbol{x}} \left\{ f_0(\boldsymbol{x}_n) + \langle \nabla f_0(\boldsymbol{x}_n), \boldsymbol{x} - \boldsymbol{x}_n \rangle + h(\boldsymbol{x}) + \frac{1}{2\alpha} \| \boldsymbol{x} - \boldsymbol{x}_n \|^2 \right\} \tag{21}$$

for a step size $0 < \alpha < L^{-1}$. To see that the proximal gradient algorithm is an instance of Bregman majorization, set $\phi(\boldsymbol{x}) = \frac{1}{2\alpha} \| \boldsymbol{x}\|^2 - f_0(\boldsymbol{x})$. Then,

$$\begin{aligned} f(\boldsymbol{x}) + B_{\phi}(\boldsymbol{x} \| \boldsymbol{x}_n) &= f_0(\boldsymbol{x}) + h(\boldsymbol{x}) + \frac{1}{2\alpha} \| \boldsymbol{x}\|^2 - f_0(\boldsymbol{x}) - \frac{1}{2\alpha} \| \boldsymbol{x}_n\|^2 + f_0(\boldsymbol{x}_n) \\ &\quad - \langle \frac{1}{\alpha}\boldsymbol{x}_n - \nabla f_0(\boldsymbol{x}_n), \boldsymbol{x} - \boldsymbol{x}_n \rangle \\ &= f_0(\boldsymbol{x}_n) + \langle \nabla f_0(\boldsymbol{x}_n), \boldsymbol{x} - \boldsymbol{x}_n \rangle + h(x) + \frac{1}{2\alpha} \| \boldsymbol{x} - \boldsymbol{x}_n \|^2 \end{aligned} \tag{22}$$

as desired. It remains to verify that $f(\boldsymbol{x})$ and $\phi(\boldsymbol{x})$ satisfy conditions (a) through (f) of Propositions 14 and 15. Conditions (a) and (c) are assumed; (b) and (d) are true. Condition (e) is satisfied since $\alpha \in (0, 1/L)$. The following fact is well known:

Lemma 2. *A differentiable convex function $f(\boldsymbol{x})$ is L-smooth $\nabla f(\boldsymbol{x})$ if and only if $\frac{L}{2} \|\boldsymbol{x}\|^2 - f(\boldsymbol{x})$ is convex.*

Then, since $\phi(\boldsymbol{x}) = \frac{1}{2}\left(\frac{1}{\alpha} - L\right) \|\boldsymbol{x}\|^2 + \frac{L}{2} \|\boldsymbol{x}\|^2 - f(\boldsymbol{x})$ and $\frac{1}{\alpha} > L$, ϕ is $(\frac{1}{\alpha} - L)$-strongly convex.

To check condition (f), we invoke the Baillon–Haddad theorem:

Lemma 3. *If function $f(\boldsymbol{x})$ is convex, differentiable, and L-smooth, then*

$$\langle \nabla f(\boldsymbol{x}) - \nabla f(\boldsymbol{y}), \boldsymbol{x} - \boldsymbol{y} \rangle \geq \frac{1}{L} \| \nabla f(\boldsymbol{x}) - \nabla f(\boldsymbol{y}) \|^2$$

Note $\nabla\phi(\boldsymbol{x}) = \frac{1}{\alpha}\boldsymbol{x} - \nabla f_0(\boldsymbol{x})$. Then,

$$\begin{aligned} \| \nabla\phi(\boldsymbol{x}) - \nabla\phi(\boldsymbol{y})\|^2 &= \|\alpha^{-1}(\boldsymbol{x}-\boldsymbol{y}) - [\nabla f_0(\boldsymbol{x}) - \nabla f_0(\boldsymbol{y})]\|^2 \\ &= \frac{1}{\alpha^2}\|\boldsymbol{x}-\boldsymbol{y}\|^2 + \|\nabla f_0(\boldsymbol{x}) - \nabla f_0(\boldsymbol{y})\|^2 - \frac{2}{\alpha}\langle \boldsymbol{x}-\boldsymbol{y}, \nabla f_0(\boldsymbol{x}) - \nabla f_0(\boldsymbol{y})\rangle \\ &\leq \frac{1}{\alpha^2}\|\boldsymbol{x}-\boldsymbol{y}\|^2 + \|\nabla f_0(\boldsymbol{x}) - \nabla f_0(\boldsymbol{y})\|^2 - \frac{2}{\alpha L}\|\nabla f_0(\boldsymbol{x}) - \nabla f_0(\boldsymbol{y})\|^2 \\ &\leq \frac{1}{\alpha^2}\|\boldsymbol{x}-\boldsymbol{y}\|^2 \end{aligned}$$

The first inequality is due to Lemma 3. The last inequality holds since $\alpha \in (0, 1/L)$ implies $1 - \frac{2}{\alpha L} \leq 0$. Therefore, $\nabla\phi(\boldsymbol{x})$ is $1/\alpha$-Lipschitz continuous and condition (f) is satisfied.

We summarize the discussion above as follows:

Proposition 16. *Suppose that $f_0(\boldsymbol{x})$ and $h(\boldsymbol{x})$ are convex, proper, and lower semicontinuous. If $f_0(\boldsymbol{x})$ is L-smooth, then for $0 < \alpha < 1/L$, the proximal gradient iteration (21) converges to a minimizer of $f(\boldsymbol{x}) = f_0(\boldsymbol{x}) + h(\boldsymbol{x})$ if it exists.*

Remark 3. Lemma 3 suggests that $\nabla\phi$ is $1/\alpha$-Lipschitz continuous if $0 < \alpha < 2/L$; in other words, the step size may be doubled. Indeed, employing monotone operator theory [55, 56] it can be shown that iteration (21) converges for $1/L \leq \alpha < 2/L$ as well. Even though the MM interpretation is lost for this range of step size, the descent property remains intact [57, 58].

Remark 4. The assumption that $h(\boldsymbol{x})$ is convex can be relaxed: if $h(\boldsymbol{x})$ is ρ-weakly convex, which means $h(\boldsymbol{x}) + \frac{\rho}{2}\|\boldsymbol{x}\|^2$ is convex, and $f_0(\boldsymbol{x})$ is ρ-strongly convex as well as L-smooth (this implies $\rho \leq L$), then the objective $f(\boldsymbol{x})$ remains convex. The inner optimization problem in iteration (21) is also strongly convex if $\rho\alpha < 1$ and $\boldsymbol{x}_{n+1}$ is uniquely determined. The latter condition is guaranteed if $\alpha \in (0, 1/L)$, and the conclusion of Proposition 16 holds. In fact, using monotone operator theory, a larger step size $\alpha \in (0, \frac{2}{L+\rho})$ is allowed [58]. Statistical applications include nonconvex sparsity-inducing penalties such as the MCP [59].

4.2.2 Mirror descent method

For the constrained problem (16) and the Euclidean norm $\|\cdot\|_2$, the proximal gradient method takes the form of *projected gradient*

$$\begin{aligned} \boldsymbol{x}_{n+1} &= \text{argmin}_{\boldsymbol{x}\in C}\left\{f(\boldsymbol{x}_n) + \langle\nabla f(\boldsymbol{x}_n), \boldsymbol{x}-\boldsymbol{x}_n\rangle + \frac{1}{2\alpha}\|\boldsymbol{x}-\boldsymbol{x}_n\|_2^2\right\} \\ &= P_C(\boldsymbol{x}_n - \alpha\nabla f(\boldsymbol{x}_n)) \end{aligned} \tag{23}$$

This method relies heavily on the Euclidean geometry of $\mathbb{R}^d$, not C: $\|\cdot\|_2 = \langle\cdot,\cdot\rangle$. If the distance measure $\frac{1}{2}\|\mathbf{x}-\mathbf{y}\|_2^2$ is replaced by something else (say $d(\mathbf{x}, \mathbf{y})$) that better reflects the geometry of C, then update such as

$$\boldsymbol{x}_{n+1} = P_C^d\left(\arg\min_{\boldsymbol{x}\in\mathbb{R}^d}\left\{f(\boldsymbol{x}_n) + \langle\nabla f(\boldsymbol{x}_n), \boldsymbol{x}-\boldsymbol{x}_n\rangle + \frac{1}{\alpha}d(\boldsymbol{x}, \boldsymbol{x}_n)\right\}\right) \tag{24}$$

may converge faster. Here,

$$P_C^d(\mathbf{y}) = \text{argmin}_{x \in C} d(\mathbf{x}, \mathbf{y})$$

is a new (non-Euclidean) projection operator that reflects the geometry of C.

To see that iteration (24) is a Bregman majorization for an appropriately chosen $d(\cdot,\cdot)$, let

$$d(\boldsymbol{x},\boldsymbol{y}) = B_\psi(\boldsymbol{x} \parallel \boldsymbol{y}) = \psi(\boldsymbol{x}) - \psi(\boldsymbol{y}) - \langle \nabla\psi(\boldsymbol{y}), \boldsymbol{x} - \boldsymbol{y}\rangle \geq \frac{1}{2} \parallel \boldsymbol{x} - \boldsymbol{y}\|^2$$

for a 1-strongly convex (with respect to some norm $\|\cdot\|$) and continuously differentiable function ψ in C, and set $\phi(\boldsymbol{x}) = \frac{1}{\alpha}\psi(\boldsymbol{x}) - f(\boldsymbol{x})$. Similar to Equation (22), we have

$$\begin{aligned} f(\boldsymbol{x}) + B_\phi(\boldsymbol{x} \parallel \boldsymbol{x}_n) &= f(\boldsymbol{x}_n) + \langle \nabla f(\boldsymbol{x}_n), \boldsymbol{x} - \boldsymbol{x}_n\rangle \\ &\quad + \frac{1}{\alpha}[\psi(\boldsymbol{x}) - \psi(\boldsymbol{x}_n) - \langle \nabla\psi(\boldsymbol{x}_n), \boldsymbol{x} - \boldsymbol{x}_n\rangle] \\ &= f(\boldsymbol{x}_n) + \langle \nabla f(\boldsymbol{x}_n), \boldsymbol{x} - \boldsymbol{x}_n\rangle + \frac{1}{\alpha} d(\boldsymbol{x}, \boldsymbol{x}_n) \end{aligned}$$

Let $\tilde{\boldsymbol{x}}_{n+1}$ be the *unconstrained* minimizer of $f(\boldsymbol{x}) + B_\phi(\boldsymbol{x} \parallel \boldsymbol{x}_n)$ (which is unique since $d(\boldsymbol{x}, \boldsymbol{x}_n)$ is strongly convex in $\boldsymbol{x}$). The associated optimality condition entails

$$\nabla\psi(\tilde{\boldsymbol{x}}_{n+1}) = \nabla\psi(\boldsymbol{x}_n) - \alpha\nabla f(\boldsymbol{x}_n) \tag{25}$$

Then,

$$\begin{aligned} \boldsymbol{x}_{n+1} &= \text{argmin}_{x\in C} d(\boldsymbol{x}, \tilde{\boldsymbol{x}}_{n+1}) \\ &= \text{argmin}_{x\in C}\{\psi(\boldsymbol{x}) - \psi(\tilde{\boldsymbol{x}}_{n+1}) - \langle \nabla\psi(\tilde{\boldsymbol{x}}_{n+1}), \boldsymbol{x} - \tilde{\boldsymbol{x}}_{n+1}\rangle\} \\ &= \text{argmin}_{x\in C}\{\psi(\boldsymbol{x}) - \langle \nabla\psi(\tilde{\boldsymbol{x}}_{n+1}), \boldsymbol{x}\rangle\} \\ &= \text{argmin}_{x\in C}\{\psi(\boldsymbol{x}) - \langle \nabla\psi(\boldsymbol{x}_n) - \alpha\nabla f(\boldsymbol{x}_n), \boldsymbol{x} - \boldsymbol{x}_n\rangle\} \\ &= \text{argmin}_{x\in C}\{f(\boldsymbol{x}) + \phi(\boldsymbol{x}) - \langle \nabla\phi(\boldsymbol{x}_n), \boldsymbol{x} - \boldsymbol{x}_n\rangle - \phi(\boldsymbol{x}_n)\} \\ &= \text{argmin}_{x\in C}\{f(\boldsymbol{x}) + B_\phi(\boldsymbol{x} \parallel \boldsymbol{x}_n)\} \end{aligned}$$

as sought. To establish iterate convergence via SUMMA, we see that just as the proximal gradient method, $f(\boldsymbol{x})$ and $\phi(\boldsymbol{x})$ satisfy conditions (a) through (e) of Propositions 14 and 15 if f is L-smooth and $\alpha \in (0, 1/L)$. In particular,

$$\phi(\boldsymbol{x}) = \frac{1}{\alpha}\psi(\boldsymbol{x}) - f(\boldsymbol{x}) \geq \frac{1}{2\alpha} \parallel \boldsymbol{x}\|^2 - f(\boldsymbol{x})$$

to check condition (e). Condition (f') is fulfilled since $B_\phi(\boldsymbol{x} \parallel \boldsymbol{y}) = \phi(\boldsymbol{x}) - \phi(\boldsymbol{y}) - \langle \nabla\phi(\boldsymbol{y}), \boldsymbol{x} - \boldsymbol{y}\rangle$ is continuous in $\boldsymbol{y}$ by construction.

Computation of $\boldsymbol{x}_{n+1}$ can be further analyzed. It is well known that if ψ is μ-strongly convex, then ψ^* is $1/\mu$-smooth, where ψ^* is the Fenchel conjugate function of ψ [55]:

$$\psi^*(\boldsymbol{y}) = \sup_{\boldsymbol{x}\in \text{dom}\psi} \langle \boldsymbol{x}, \boldsymbol{y}\rangle - \phi(\boldsymbol{x})$$

Hence, $\nabla\psi^*$ is well defined. Furthermore, $\nabla\psi^*(\nabla\psi(\boldsymbol{x})) = \boldsymbol{x}$. Therefore, the unconstrained optimality condition (25) is equivalent to

$$\tilde{\boldsymbol{x}}_{n+1} = \nabla\psi^*(\nabla\psi(\boldsymbol{x}_n) - \alpha\nabla f(\boldsymbol{x}_n))$$

and we decompose the update (24) into three steps:

$$\begin{aligned}
\boldsymbol{y}_{n+1} &= \nabla\psi(\boldsymbol{x}_n) - \alpha\nabla f(\boldsymbol{x}_n) \quad \text{(gradient step)}\\
\tilde{\boldsymbol{x}}_{n+1} &= \nabla\psi^*(\boldsymbol{y}_{n+1}) \quad \text{(mirroring step)}\\
\boldsymbol{x}_{n+1} &= P_C^d(\tilde{\boldsymbol{x}}_{n+1}) \quad \text{(projection step)}
\end{aligned}$$

Hence, Bregman majorization with $\phi(\boldsymbol{x}) = \frac{1}{\alpha}\psi(\boldsymbol{x}) - f(\boldsymbol{x})$ coincides with the mirror descent method under B_ψ [60]. The first step performs the gradient descent step in the dual space $\mathcal{X}^*$ of $\mathcal{X}$, and the second step maps the dual vector back to the primal space by the inverse mapping $\nabla\psi^* = (\nabla\psi)^{-1}$. The final step projects (in a non-Euclidean fashion) the mapped primal vector onto the constraint set C.

Example 3 (Exponentiated gradient). As a concrete instance of mirror descent, consider optimization over probability simplex $C = \Delta^{d-1} = \{\mathbf{x} \in \mathcal{X} = \mathbb{R}^d : \sum_{i=1}^d x_i = 1,\ x_i \geq 0,\ i = 1, \dots, d\}$. An appropriate Bregman divergence is the Kullback–Leibler divergence, that is, we use negative entropy $\psi(\boldsymbol{x}) = \sum_{i=1}^d x_i \log x_i - \sum_{i=1}^d x_i$. It is easy to check, using the Taylor expansion and the Cauchy–Schwarz inequality, that ψ is 1-strongly convex with respect to the ℓ_1 norm $\|\boldsymbol{x}\|_1 = \sum_{i=1}^d |x_i|$ within C. Furthermore, we have $\nabla\psi(\boldsymbol{x}) = (\log x_1, \dots, \log x_d)^T =: \log\boldsymbol{x}$ and $\nabla\psi^*(\boldsymbol{y}) = (\nabla\phi)^{-1}(\boldsymbol{y}) = (e^{y_1}, \dots, e^{y_d})^T =: \exp(\boldsymbol{y})$. The mirror descent or Bregman MM update is then

$$\begin{aligned}
\boldsymbol{y}_{n+1} &= \log\boldsymbol{x}_n - \alpha\nabla f(\boldsymbol{x}_n)\\
\tilde{\boldsymbol{x}}_{n+1} &= \exp(\boldsymbol{y}_n) = \boldsymbol{x}_n \odot \exp(-\alpha\nabla f(\boldsymbol{x}_n))\\
\boldsymbol{x}_{n+1} &= \tilde{\boldsymbol{x}}_{n+1}/Z_t
\end{aligned}$$

where $\odot$ denotes an elementwise product, and

$$Z_t = \sum_{i=1}^d x_{n,i}\exp(-\alpha\nabla f(\boldsymbol{x}_n)_i)$$

is the normalization constant. The last step is because

$$\begin{aligned}
P_C^d(\boldsymbol{y}) &= \operatorname{argmin}_{\boldsymbol{x}\in\Delta^{d-1}} B_\psi(\boldsymbol{x} \,\|\, \boldsymbol{y})\\
&= \operatorname*{argmin}_{x_i\geq 0, \sum_{i=1}^d x_i = 1} \sum_{i=1}^d \left(x_i \log\frac{x_i}{y_i} - x_i + y_i\right)\\
&= \operatorname*{argmin}_{x_i\geq 0, \sum_{i=1}^d x_i = 1} \sum_{i=1}^d \left(x_i \log\frac{x_i}{y_i}\right)
\end{aligned}$$

and the associated Lagrangian

$$\mathcal{L}(\boldsymbol{x}, \mu) = \sum_{i=1}^d \left(x_i \log\frac{x_i}{y_i}\right) + \mu\left(\sum_{i=1}^d x_i - 1\right)$$

yields

$$x_i = y_i \exp(-\mu - 1) = c y_i, \quad i = 1, \dots, d$$

for some $c > 0$. Summing these over all i yields $c = 1/(\sum_{i=1}^{d} y_i)$ to have

$$x_i = \frac{y_i}{\sum_{j=1}^{d} y_j}, \quad i = 1, \dots, d$$

This special case is called the *exponentiated gradient method* [61, 62].

References

1 Becker, M.P., Yang, I., and Lange, K. (1997) EM algorithms without missing data. *Stat. Methods Med. Res.*, 6 (1), 38–54.

2 Lange, K., Hunter, D.R., and Yang, I. (2000) Optimization transfer using surrogate objective functions. *J. Comput. Graph. Statist.*, 9 (1), 1–59. [With discussion, and a rejoinder by Hunter and Lange].

3 Hunter, D.R. and Lange, K. (2004) A tutorial on MM algorithms. *Am. Stat.*, 58, 30–37.

4 Borg, I. and Groenen, P.J.F. (2005) *Modern Multidimensional Scaling: Theory and Applications*, Springer Series in Statistics, 2nd edn, Springer, New York.

5 Hunter, D.R. and Lange, K. (2000) Quantile regression via an MM algorithm. *J. Comput. Graph. Statist.*, 9 (1), 60–77.

6 Hunter, D.R. (2004) MM algorithms for generalized Bradley-Terry models. *Ann. Statist.*, 32 (1), 384–406.

7 Hunter, D.R. and Li, R. (2005) Variable selection using MM algorithms. *Ann. Statist.*, 33 (4), 1617–1642.

8 Yen, T.-J. (2011) A majorization-minimization approach to variable selection using spike and slab priors. *Ann. Statist.*, 39 (3), 1748–1775.

9 Bien, J. and Tibshirani, R.J. (2011) Sparse estimation of a covariance matrix. *Biometrika*, 98 (4), 807–820.

10 Lee, S. and Huang, J.Z. (2013) A coordinate descent MM algorithm for fast computation of sparse logistic PCA. *Comput. Statist. Data Anal.*, 62, 26–38.

11 Zhou, H. and Lange, K. (2010) MM algorithms for some discrete multivariate distributions. *J. Comput. Graph. Stat.*, 19, 645–665.

12 Zhang, Y., Zhou, H., Zhou, J., and Sun, W. (2017) Regression models for multivariate count data. *J. Comput. Graph. Stat.*, 26 (1), 1–13.

13 Zhou, H., Hu, L., Zhou, J., and Lange, K. (2019) MM algorithms for variance components models. *J. Comput. Graph. Statist.*, 28 (2), 350–361.

14 Sun, Y., Babu, P., and Palomar, D.P. (2015) Regularized robust estimation of mean and covariance matrix under heavy-tailed distributions. *IEEE Trans. Signal Process.*, 63 (12), 3096–3109.

15 Hunter, D.R. and Lange, K. (2002) Computing estimates in the proportional odds model. *Ann. Inst. Statist. Math.*, 54 (1), 155–168.

16 Ding, J., Tian, G.-L., and Yuen, K.C. (2015) A new MM algorithm for constrained estimation in the proportional hazards model. *Comput. Statist. Data Anal.*, 84, 135–151.

17 Lange, K. and Zhou, H. (2014) MM algorithms for geometric and signomial programming. *Math. Program. Series A*, 143, 339–356.

18 Chi, E.C., Zhou, H., and Lange, K. (2014) Distance majorization and its applications. *Math. Program.*, 146 (1–2), 409–436.

19 Xu, J., Chi, E., and Lange, K. (2017) Generalized linear model regression under distance-to-set penalties, in *Advances in Neural Information Processing Systems*, Curran Associates, Inc., Red Hook, NY, pp. 1385–1395.

20 Keys, K.L., Zhou, H., and Lange, K. (2019) Proximal distance algorithms: theory and examples. *J. Mach. Learn. Res.*, 20 (66), 1–38.

21 Lange, K. and Carson, R. (1984) EM reconstruction algorithms for emission and transmission tomography. *J. Comput. Assist. Tomogr.*, 8 (2), 306–316.

22 Figueiredo, M.A.T., Bioucas-Dias, J.M., and Nowak, R.D. (2007) Majorization–minimization algorithms for wavelet-based image restoration. *IEEE Trans. Image Process.*, 16 (12), 2980–2991.

23 Lee, D.D. and Seung, H.S. (1999) Learning the parts of objects by non-negative matrix factorization. *Nature*, 401 (6755), 788–791.

24 Mazumder, R., Hastie, T., and Tibshirani, R. (2010) Spectral regularization algorithms for learning large incomplete matrices. *J. Mach. Learn. Res.*, 11, 2287–2322.

25 Chi, E.C., Zhou, H., Chen, G.K. *et al.* (2013) Genotype imputation via matrix completion. *Genome Res.*, 23 (3), 509–518.

26 Chi, E. and Lange, K. (2015) Splitting methods for convex clustering. *J. Comput. Graph. Stat.*, 24 (4), 994–1013.

27 Xu, J. and Lange, K. (2019) By all means, k-means, *under review*.

28 Wu, T.T. and Lange, K. (2010) Multicategory vertex discriminant analysis for high-dimensional data. *Ann. Appl. Stat.*, 4 (4), 1698–1721.

29 Nguyen, H.D. (2017) An introduction to majorization-minimization algorithms for machine learning and statistical estimation. *WIREs Data Min. Knowl. Discov.*, 7 (2), e1198.

30 Lange, K. (2016) *MM Optimization Algorithms*, Society for Industrial and Applied Mathematics, Philadelphia, PA.

31 Sun, Y., Babu, P., and Palomar, D.P. (2017) Majorization-minimization algorithms in signal processing, communications, and machine learning. *IEEE Trans. Signal Process.*, 65 (3), 794–816.

32 Nguyen, H.D. (2017) An introduction to Majorization-Minimization algorithms for machine learning and statistical estimation. *WIREs Data Min. Knowl. Discov.*, 7 (2), e1198.

33 Zhou, H. and Zhang, Y. (2012) EM vs MM: a case study. *Comput. Stat. Data Anal.*, 3909–3920. 56:

34 Zangwill, W.I. and Mond, B. (1969) *Nonlinear Programming: A Unified Approach*, Prentice-Hall International Series in Management, Prentice-Hall Inc., Englewood Cliffs, N.J.

35 Luenberger, D.G. and Ye, Y. (2008) Linear and Nonlinear Programming, in *International Series in Operations Research & Management Science*, vol. 116, 3rd edn, Springer, New York.

36 Vaida, F. (2005) Parameter convergence for EM and MM algorithms. *Stat. Sin.*, 15, 831–840.

37 van de Geer, J.P. (1984) Linear relations among *k* sets of variables. *Psychometrika*, 49 (1), 79–94.

38 Ten Berge, J.M.F. (1988) Generalized approaches to the maxbet problem and the maxdiff problem, with applications to canonical correlations. *Psychometrika*, 53 (4), 487–494.

39 Hanafi, M. and Kiers, H.A. (2006) Analysis of k sets of data, with differential emphasis on agreement between and within sets. *Comput. Stat. Data Anal.*, 51 (3), 1491–1508.

40 Ten Berge, J.M.F. and Knol, D.L. (1984) Orthogonal rotations to maximal agreement for two or more matrices of different column orders. *Psychometrika*, 49 (1), 49–55.

41 Lange, K. (2016) MM Optimization Algorithms, SIAM.

42 Won, J.-H., Zhou, H., and Lange, K. (2018) Orthogonal trace-sum maximization: applications, local algorithms, and global optimality, *arXiv preprint arXiv:1811.03521*.

43 Ten Berge, J.M.F. (1977) Orthogonal procrustes rotation for two or more matrices. *Psychometrika*, 42 (2), 267–276.

44 Absil, P.-A. and Malick, J. (2012) Projection-like retractions on matrix manifolds. *SIAM J. Optim.*, 22 (1), 135–158.

45 Lange, K. (2010) *Statistics and Computing: Numerical Analysis for Statisticians*, 2nd edn, Springer, New York.

46 Yu, D., Won, J.-H., Lee, T. *et al.* (2015) High-dimensional fused lasso regression using majorization–minimization and parallel processing. *J. Comput. Graph. Stat.*, 24 (1), 121–153.

47 Bierstone, E. and Milman, P.D. (1988) Semianalytic and subanalytic sets. *Inst. Hautes Études Sci. Publ. Math.*, 67, 5–42.

48 Bochnak, J., Coste, M., and Roy, M.-F. (1998) *Real Algebraic Geometry*, vol. 36, Ergebnisse der Mathematik und ihrer Grenzgebiete (3), Springer-Verlag, Berlin. [Translated from the 1987 French original, Revised by the authors].

49 Attouch, H. and Bolte, J. (2009) On the convergence of the proximal algorithm for nonsmooth functions involving analytic features. *Math. Program.*, 116 (1-2, Ser. B), 5–16.

50 Elsner, L., Koltracht, I., and Neumann, M. (1992) Convergence of sequential and asynchronous nonlinear paracontractions. *Numerische Mathematik*, 62 (1), 305–319.

51 Lange, K. (2013) *Optimization*, 2nd edn, Springer, New York, NY.

52 Won, J.-H., Xu, J., and Lange, K. (2019) *Projection Onto Minkowski Sums with Application to Constrained Learning*. International Conference on Machine Learning, pages 3642–3651.

53 Byrne, C.L. (2008) Sequential unconstrained minimization algorithms for constrained optimization. *Inverse Prob.*, 24 (1), 015013.

54 Byrne, C.L. (2014) Lecture Notes on Iterative Optimization Algorithms. http://faculty.uml.edu/cbyrne/IOIPNotesOct2014.pdf.

55 Bauschke, H.H. and Combettes, P.L. (2011) *Convex Analysis and Monotone Operator Theory in Hilbert Spaces*, vol. 408, Springer, New York, NY.

56 Ryu, E.K. and Boyd, S. (2016) Primer on monotone operator methods. *Appl. Comput. Math.*, 15 (1), 3–43.

57 She, Y. (2009) Thresholding-based iterative selection procedures for model selection and shrinkage. *Electron. J. Stat.*, 3, 384–415.

58 Bayram, I. (2015) On the convergence of the iterative shrinkage/thresholding algorithm with a weakly convex penalty. *IEEE Trans. Signal Process.*, 64 (6), 1597–1608.

59 Zhang, C.-H. (2010) Nearly unbiased variable selection under minimax concave penalty. *Ann. Statist.*, 38 (2), 894–942.

60 Juditsky, A. and Nemirovski, A. (2011) First order methods for nonsmooth convex large-scale optimization I: general purpose methods. *Optim. Mach. Learn.*, 121–148.

61 Helmbold, D.P., Schapire, R.E., Singer, Y., and Warmuth, M.K. (1997) A comparison of new and old algorithms for a mixture estimation problem. *Mach. Learn.*, 27 (1), 97–119.

62 Azoury, K.S. and Warmuth, M.K. (2001) Relative loss bounds for on-line density estimation with the exponential family of distributions. *Mach. Learn.*, 43 (3), 211–246.

Part VII

High-Performance Computing

29

Massive Parallelization

Robert B. Gramacy

Virginia Polytechnic Institute and State University, Blacksburg, VA, USA

1 Introduction

Computing advances in the late twentieth century were primarily about clock speed and size of random access memory (RAM). CPU clock speed roughly doubled, and RAM capacity increased 10-fold year over year, among other advances. RAM and other memory capacity continue to grow, but clock speed asymptoted. The number of instructions that could be carried out in a serial, vertical manner peaked. Numbers of transistors have continued to grow exponentially in the twenty-first century, however, seeming to thwart Moore's law[1] by adopting a more horizontal architecture – allowing for multiple instructions to be carried out in parallel. First this meant clusters of nodes with single computing cores, then came multicore workstations (even laptops), and networks thereof, followed by the adoption of specialized architectures such as graphical processing units (GPUs).

Codes taking advantage of these new computing regimes have lagged, and this is true in almost every corner of computing. Some areas, such as gaming, were quick to adopt GPUs for advances in graphics but have been slower to adopt *symmetric multicore/shared memory parallelization* (SMP).[2] In SMP, processor cores reside on the same motherboard, often on the same chip, and therefore share the much of the same high-speed memory (i.e., RAM). Hyperthreading, a virtualized extension of SMP effectively doubling the number of "cores," remains an underutilized resource.[3] Such features are differentiated from nodes in a cluster which, while likely also being multicore/hyperthreaded, are physically distinct and have separate hardware and memory. Some areas of scientific computing, such as in finite element analysis and solving coupled systems of differential equations, have successfully exploited SMP (and cluster) parallelization triggering exponential advances in the size of problems and fidelity of analysis compared to decades ago.

Statistical computing has not enjoyed as much of a renaissance. Most of our software packages are stuck in the serial paradigm of 30 years ago. Facilities for parallel calculation are in abundance, for example, https://cran.r-project.org/web/views/HighPerformanceComputing.html but very few packages for R on CRAN [1] implement methodology in a natively parallel fashion as a means of expanding data sizes and enhancing fidelity. Deep neural networks (DNNs), which are arguably more of a machine

Computational Statistics in Data Science.
Edited by Walter W. Piegorsch, Richard A. Levine, Hao Helen Zhang and Thomas C. M. Lee.
 ISBN 978-1-11956107-1

learning than mainstream stats tool, are an important exception. DNNs famously tax some of the world's largest supercomputers [2, 3] leading to high-powered predictors – for climate science and cosmology and numerous AI tasks from speech to image recognition and reinforcement learning for autonomous vehicles – by exploiting data subsets through stochastic gradient descent [4, Chapter 8]. A downside to DNNs is that why they work so well is not well understood, and uncertainty quantification (UQ; i.e., error bars with good coverage properties) is notably absent. The main trick, of inducing independence in training data through data subsetting, appears to allow massive distribution of parallel computational instances without any deleterious effect on inferential or predictive accuracy. In fact, breaking things up seems to help: leading to more accurate and stable predictors. This serendipitous state of affairs is well documented empirically, but why and how to port this to other methodology is less clear.

Pace of development for DNNs, both theoretical and practical, is feverish, and its review is best left to another chapter. However, the idea of data subsetting represents an attractive "scale-up" tactic in the more general setting of nonparametric nonlinear regression of which neural networks are a special case. Divide and conquer may port to models whose fits offer a better-understood inferential framework and UQ properties. Gaussian process (GP) regression [5] leads to powerful predictors with excellent out-of-sample coverage properties but is famously limited to training data sizes N in the small thousands owing to cubic runtime and quadratic storage requirements for $N \times N$ covariance matrices. Fitting GPs to data subsets represents a seductive alternative; however it is easy to do that with lukewarm results or in a way that is not readily parallelizable to a degree amenable to modern-scale distributed computing. Leveraging fancy hardware such as GPUs remains limited to bespoke settings. What is missing are general paradigms that set up methodology for success in parallel implementation.

This chapter summarizes a spatial divide-and-conquer scheme that is massively parallelizable and, similar to DNNs, often works better than full-scale (nonapproximate) alternatives (when those are tractable). Most importantly, the scheme is defensible technically and does not compromise UQ. All it gives up is a little smoothness, which can be spun from bug into feature. The underlying methodology is an example of *transductive learning* [6], where fitting is tailored to prediction goals. This is as opposed to the more familiar inductive sort where fitting and prediction transpire in two distinct stages. Although perhaps not conceived with modern computing architectures in mind, transductive learners offer a natural framework for establishing a weak statistical independence which is just strong enough to accommodate computational considerations, especially as regards parallelization. A primary aim in this exposition is to allow the reader to imagine similar ideas being deployed more widely, for a large range of inference, prediction, supervised, and unsupervised learning tasks.

Our motivating class of regression problems are ones that entail a careful decoupling of strong (spatial) dependencies and emphasize distribution of tasks from a single source rather than the other way around, that is, not applications where data collection is itself distributed (say geographically) and large and complex enough to necessitate reduction before communication to a centralized repository or inferential apparatus. A great example of this would be geographically dispersed data centers recording customer transactions in an e-commerce setting (e.g., Amazon), with information or summaries queried for learning and experimental purposes. For more on that paradigm and data subsetting/divide and conquer therein, see., for example, Kleiner *et al.* [7]. Our setting will be simpler: we have a big data set. Not so big that it does not fit into memory but too big to work with computationally. Big memory computing is common, even in the desktop/workstation setting. Big computation is only accessible on supercomputers that are massively parallelized.

As a final disclaimer, many advances in distributed linear algebra have been made over the years. Methods such as GPs, which involve big matrix manipulation and decomposition, benefit from customized and parallelized linear algebra libraries such as those available from Intel's Math Kernel Library[4] for SMP and NVIDIA's Math Libraries[5] for GPUs. These will not be discussed in much detail. The former can help expand full-scale GP regression capabilities by a factor of 10 [8], Appendix A] and the latter perhaps by a factor of 4 alone [9] or 50 when combined with distributed (multinode SMP) computation [10]. This chapter targets a scaling that is much more ambitious than that, although some relevant subroutines will be mentioned in due course.

The development is organized as follows. Section 2 reviews GP regression with emphasis on the challenges inherent in that methodological paradigm, drawing connections to other mainstream statistical tasks desperate for a fidelity boost. Section 3 introduces the local approximate GP as a transductive, divide-and-conquer alternative that offers a quick approximation with appropriate UQ and that also is amenable to massive SMP and cluster parallelization. The discussion focuses on how expensive subroutines can be off-loaded for GPU calculation and provides details on a cascade of distributed computing common in contemporary supercomputing environments. Empirical work in that setting is summarized in Section 4. Section 5 finishes with thoughts on the scope for divide and conquer as a tool for bringing statistical implementation into the twenty-first century.

2 Gaussian Process Regression and Surrogate Modeling

The Gaussian process (GP) regression model, sometimes called *Gaussian spatial processes* (GaSP), has been popular for decades in spatial data contexts such as geostatistics [11] where they are known as *kriging* [12] and in computer experiments where they are deployed as *surrogate models* or *emulators* [8, 13, 14]. More recently, they have become a popular prediction engine in the machine learning literature [5]. The reasons are many, but the most important are probably: Gaussian structure imparts a degree of analytic capability not enjoyed by other general-purpose approaches to nonparametric nonlinear modeling; they perform well in out-of-sample tests; and offer sensible UQ. They are not, however, without drawbacks. Two important ones are computational tractability and nonstationary flexibility, which we will return to shortly.

In all three settings – geo/spatial statistics, computer experiments, and machine learning – training data sets are getting big. Our examples will come primarily from computer experiments where GPs are the canonical meta model for simulation campaigns. Computer simulation of a system under varying conditions represents a comparatively inexpensive alternative to actual physical experimentation and/or monitoring. Examples include aeronautics (designing a new aircraft wing) and climate science (collecting atmospheric ozone data). In some cases, simulations/surrogates are the only (ethical) alternative, for example, in epidemiology. One reason GPs have been popular in such contexts is that they can both interpolate – connect the dots between runs when appropriate – and at the same time offer sensible UQ in the form of sausage-shaped (or football) error bars with predictive variance growing organically away from training data runs. A downside of working with (multivariate) Gaussian distributions is the cubic cost of matrix decomposition, which limits training data sizes. Yet, however vastly expanded computing capabilities may be in modern times, enabling orders of magnitude larger computer experiments than decades ago, the ability to collect data is still limited, and so fitted surrogate models are essential to applications such as Bayesian optimization [15], calibration to field data [16], and input sensitivity analysis [17]. A key requirement of GP surrogates in such applications is to be able to provide predictive summaries at much lower computational expense than running new simulations. Thus, cubic fitting and prediction costs limit GP application on both fidelity and scale.

2.1 GP Basics

To provide more detail, let $D_N = (X_N, Y_N) = (x_1, y_1), \dots, (x_N, y_N)$ denote a data set encapsulating a corpus of computer model simulations, recording input conditions x_i, and producing outputs y_i. Given the data D_N, a surrogate/emulator provides a distribution over possible responses $Y(x) \mid D_N$ for new inputs x. By jointly modeling data outputs Y_N at X_N with predictive outputs $Y(x)$ at new x under a unified multivariate normal (MVN) structure, a predictive distribution for $Y(x) \mid D_N$ can be derived by a simple application of MVN conditioning identities, for example, from Wikipedia.[6] If that MVN has a mean of zero and a covariance structure defined as $\tau^2 K_\theta(\cdot,\cdot)$ where, for example, the covariance kernel K is defined in terms of "hyperparameterized" scale-inverse Euclidean distance

$$K_\theta(x, x') = \exp\left\{ -\sum_{k=1}^{p} (x_k - x'_k)^2/\theta_k \right\} \tag{1}$$

then predictive equations $p(y(x) \mid D_N, K_\theta)$ may be derived in closed form as Gaussian with

$$\text{mean} \qquad \mu(x|D_N, \theta) = k^\top(x) K^{-1} Y \tag{2}$$

$$\text{and scale} \qquad \sigma^2(x|D_N, \theta) = \frac{\psi[K(x, x) - k^\top(x) K^{-1} k(x)]}{N} \tag{3}$$

where $k^\top(x)$ is the N-vector whose ith component is $K_\theta(x, x_i)$, K is an $N \times N$ matrix whose entries are $K_\theta(x_i, x_j)$, and $\psi = Y^\top K^{-1} Y$. Figure 1 shows "sausage/football-shaped" predictive intervals, which are wide away from data locations x_i, for a small synthetic computer experiment based on a sinusoid sampled uniformly in the span of one period.

The maximum-likelihood inference for *lengthscale* hyperparameters such as θ, controlling the rate of decay of spatial correlation in terms of coordinatewise distance, is straightforward via the log likelihood arising from an MVN density, emitting closed-form derivatives for Newton-like optimization. Together, analytic prediction and straightforward optimization for inference make for a relatively easy implementation of a nonparametric regression. Open-source libraries abound. For R [1], these include `mlegp` [18], `GPfit` [19], `spatial` [20], `fields` [21], `RobustGaSP` [22], `kernlab` [23], and `hetGP` [24] – all performing maximum-likelihood (or maximum *a posteriori*/Bayesian regularized) point inference; or `tgp` [25], `emulator` [26], `plgp` [27], and `spBayes` [28] – performing fully

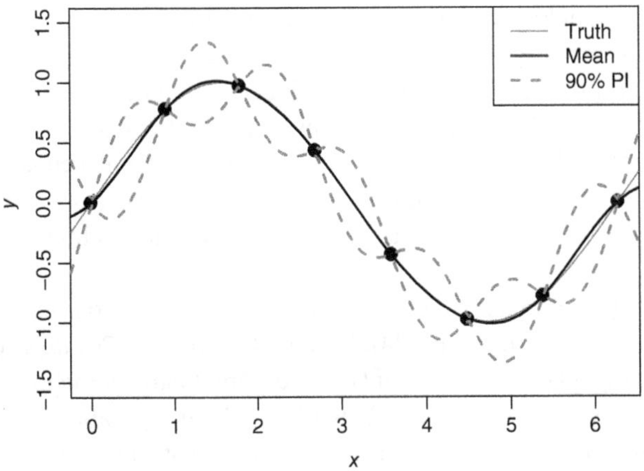

Figure 1 Simple computer surrogate model example where the response, $y = \sin(x)$, is measured at $N = 8$ equally spaced inputs x_i. Predictive mean is solid black curve, 90% intervals in short dashes, and truth in thin gray.

Bayesian inference. For Python, see GPy[7], and for MATLAB/Octave, see gpstuff [29].[8] Erickson *et al.* [30] provide a nice review and comparison of several libraries. The choice of correlation structure, $K_\theta(\cdot, \cdot)$, can have a substantial impact on the nature of inference and prediction, restricting the smoothness of the functions and controlling a myriad of other aspects. The version above is built for interpolating deterministic computer model simulations. Introducing a so-called nugget parameter can facilitate smoothing instead. There are several simple default choices/embellishments that are popular in the literature. The general methodology we present is independent of this choice. For some review and more detail, see, for example, Chapter 5 of Gramacy [8].

2.2 Pushing the Envelope

Unfortunately, Equations (2) and (3) reveal a computational expense that depends on the size of the correlation matrix, K. In spite of many attractive features, the inverse and determinant[9] calculations are $O(N^3)$ which, even for modest N, can mean that GPs may not satisfy the key requirement of being fast relative to the computer simulation being emulated. Advances in hardware design, for example, multicore machines and GPUs, may offer some salvation. Recently, several authors [9, 31, 32] have described custom GP prediction and inference schemes which show a potential to handle much larger problems than ever before but generally not more than an order of magnitude or so.

Other authors have sought approximations, for example, by inducing sparsity in the covariance structure either explicitly or implicitly [33–41]. A downside to these approaches is that, while similarly expanding on capability computationally, none emit readily parallelizable implementations and sacrifice on fidelity in order to facilitate approximation. Although enabling larger data sets, their nonparametric flexibility does not expand substantially as data get big. For example, all are limited to the so-called stationary covariance structures like in Equation (1) where dynamics are determined only by relative distance and can therefore not evolve dynamically in space. In fact, many of the approaches cited above gain computational tractability by compromising on long-range dependence without appreciably enhancing short-range reactivity. An important exception may be the recent work by Gardner *et al.* [42] combining linear conjugate gradients for matrix solves ($K^{-1}Y$) with stochastic Lanczos quadrature [43] to approximate log determinant evaluations ($\log |K|$). Although not yet known to be distributable across nodes of a cluster, a key feature of the method is that it avoids storage of large K, requiring only access kernel evaluations $k(\cdot, \cdot)$. This dramatically reduces communication overheads in off-loading data and calculations to GPUs.

One class of models which has the potential to tackle both drawbacks in one go is based on partition modeling. Examples include treed GPs [44], Voronoi tessellated GPs [45, 46], and related methods [47–49]. Fitting quasi-independent GPs in different parts of the input space offers potential for both reactive nonstationary dynamics and parallelizable inference in one fell swoop. A downside to these, however, is that the degree of computational independence is not deliberate. Strangely, the parallelization problem gets harder when the regression problem is easier – more stationarity means less divide and conquer. Other, more deliberate ways of data subsetting, such as the block bootstrap Latin hypercube (BLHS) [50, 51], allow hyperparameter inference to remain tractable, but prediction (e.g., by bootstrap aggregating/bagging) remains dulled by lack of local dynamics [52, 53].

3 Divide-and-Conquer GP Regression

Contemporary supercomputing architectures offer the potential to subdivide calculations in a hierarchical fashion: clusters of nodes with multiple cores and hyperthreaded pipelines which can off-load specialized, labor-intensive tasks to custom hardware (e.g., multiple onboard GPUs). It makes sense to develop statistical methodology which can exploit these resources, with a degree of scaling resilience, that is, which work commensurately well when more or less of them is available, or when configurations change. Local approximate Gaussian process (LAGP) regression is one such framework [54]. Even when parallel computing capability is modest, implicit sparsity in the divvying mechanism leads to vastly greater fidelity and capability by avoiding large-matrix decompositions.

3.1 Local Approximate Gaussian Processes

The core idea is to focus expressly on deriving predictor(s) for particular location(s), x. Gramacy and Apley [54] (G&A below) recognized, as many others have before, that training data whose inputs are far from x have negligible influence on GP prediction at x when inverse exponentiated (Euclidean) distance-based correlation functions are used. Nearly identical GP-based predictions could instead be obtained from data subsets $D_n(x) \equiv D_n(X_n(x))$ obtained on a subdesign of nearby $X_n(x) \subset X \equiv X_N$, with $n \ll N$, pretending no other data exist. One option is a so-called nearest neighbor (NN) subdesign, where D_n is composed of the inputs in X which are closest to x, measured relative to the chosen correlation function. The best reference for this idea is Emery [55]. Computational costs are in $\mathcal{O}(n^3)$ and $\mathcal{O}(n^2 + N)$ for decomposition(s) and storage, respectively, and NNs can be found in $\mathcal{O}(n \log N)$ time with k-d trees[10] after an up-front $\mathcal{O}(N \log N)$ build cost.

In practice, one can choose local n as large as computational constraints allow, although there may be reasons to prefer smaller n on reactivity grounds. Predictors may potentially be more accurate at x if they are not burdened by information from training data far from x. Inducing sparsity in this way deliberately imposes statistical independence in the resulting predictive process, which means that computations at distinct x can occur in parallel. But let us table that discussion for a moment. There are a few more details to attend to first.

Note that this is different, and much simpler, than the so-called nearest neighbor GPs (NNGP) [56], developed around the same time. NNGP does not utilize NNs in the canonical, machine learning sense,[11] that is, with reference to predictive locations x. Rather, neighbors are used to anchor an approximate Cholesky decomposition leading to a joint distribution similar to what could be obtained at greater computational expense under a full conditioning set. This trick is known as *Vecchia approximation* [57, 58], inducing sparsity in the inverse covariance structure. Also see Katzfuss and Guinness [59] for a more general treatment of conditioning sets toward that end. Which (LAGP or NNGP) offers better inference/prediction is the topic of another study. For example, see Heaton *et al.* [60] for a geostatistical comparison. Relevant here is that the nature of sparsity in NNGP is not as amenable to exploiting a hierarchical cascade of parallel capability. Packages `spBayes` [28] and `GpGp` [61] on CRAN support a degree of threading, but a modest one by comparison to what is showcased momentarily.

NN selection for $D_n(X)$ in the LAGP context is known to be suboptimal. It is better to take at least a few design points farther away in order to obtain good estimates of the lengthscale hyperparameter θ [62]. However, searching for the optimal design $\hat{D}_n(x)$, according to almost any criteria, is a combinatorially huge undertaking. The interesting pragmatic research question that remains is: is it possible to do better than the NN scheme without much extra computational effort? G&A showed that it is indeed possible, with the following greedy scheme. Suppose that a local design $X_j(x)$, $j < n$, has been built up already, and that a GP predictor has been inferred from data $D_j(x)$. Then, choose x_{j+1} by searching among the remaining unchosen design candidates $X_N \setminus X_j(x)$ according to a criterion, discussed momentarily. Augment the data set $D_{j+1}(x) = D_j \cup (x_{j+1}, y(x_{j+1}))$ to include the chosen design point and its corresponding response and update the GP predictor. Updating a GP predictor is possible in $O(j^2)$ time [35] with judicious application of partitioned inverse equations [63]. So as long as each search for x_{j+1} is fast, and involves no new operations larger than $O(j^2)$, then the final scheme, repeating for $j = n_0, \ldots, n$, will require $O(n^3)$ time, just like the NN analog.

G&A considered two criteria in addition to NN, one being a special case of the other. The first is to minimize empirical Bayes mean-square prediction error (MSPE):

$$J(x_{j+1}, x) = \mathbb{E}\{[Y(x) - \mu_{j+1}(x|D_{j+1}, \hat{\theta}_{j+1})]^2 | D_j(x)\}$$

where $\hat{\theta}_{j+1}$ is the estimate for θ based on D_{j+1}. The predictive mean $\mu_{j+1}(x|D_{j+1}, \hat{\theta}_{j+1})$ follows equation (2), except that a $j+1$ subscript has been added in order to indicate dependence on x_{j+1} and the future, unknown y_{j+1}. They then derive the approximation

$$J(x_{j+1}, x) \approx V_j(x|x_{j+1}; \hat{\theta}_j) + \left(\left. \frac{\partial \mu_j(x; \theta)}{\partial \theta} \right|_{\theta = \hat{\theta}_j} \right)^2 \Big/ \mathcal{G}_{j+1}(\hat{\theta}_j) \qquad (4)$$

The first term in (4) estimates predictive variance at x after x_{j+1} is added into the design,

$$V_j(x|x_{j+1};\theta) = \frac{(j+1)\psi_j}{j(j-1)} v_{j+1}(x;\theta)$$
$$\text{where} \quad v_{j+1}(x;\theta) = [K_{j+1}(x,x) - k_{j+1}^\top(x)K_{j+1}^{-1}k_{j+1}(x)] \tag{5}$$

Minimizing predictive variance at x is a sensible goal. The second term in (4) estimates the rate of change of the predictive mean at x, weighted by the expected *future* inverse information, $\mathcal{G}_{j+1}(\hat{\theta}_j)$, after x_{j+1}, and the corresponding y_{j+1} are added into the design. Note that this weight does not depend on x, but in weighting the rate of change (derivative) of the predictive mean at x, it is "commenting" on the value of x_{j+1} for estimating the parameter of the correlation function, θ. So, this MSPE criterion balances reducing predictive variance with learning local wigglyness of the surface.

It turns out that the contribution of the second term, beyond the new reduced variance, is small. G&A show that the full MSPE criterion leads to qualitatively similar local designs $X_n(x)$ as ones obtained using just $V_j(x|x_{j+1};\hat{\theta}_j)$, which provides indistinguishable out-of-sample predictive performance at a fraction of the computational cost (since no derivative calculations are necessary). This simplified criterion is equivalent to choosing x_{j+1} to maximize *reduction* in variance:

$$\begin{aligned} v_j(x;\theta) - v_{j+1}(x;\theta) &= k_j^\top(x)G_j(x_{j+1})m_j^{-1}(x_{j+1})k_j(x) + 2k_j^\top(x)g_j(x_{j+1})K(x_{j+1},x) \\ &\quad + K(x_{j+1},x)^2 m_j(x_{j+1}) \end{aligned} \tag{6}$$

where $G_j(x') \equiv g_j(x')g_j^\top(x')$,

$$g_j(x') = -m_j(x')K_j^{-1}k_j(x') \quad \text{and} \quad m_j^{-1}(x') = K_j(x',x') - k_j^\top(x')K_j^{-1}k_j(x') \tag{7}$$

Seo *et al.* [64] first used a similar criterion toward porting an active learning criterion over from neural networks [65], dubbing it ALC for *active learning Cohn.*

Algorithm 1 provides the details involved in building up a subdesign $X_n(x)$, completing with responses to obtain local data $D_n(x)$, and maximizing the likelihood in order to obtain predictions based on hyperparameters $\hat{\theta}_n(x)$ ideal for inference local to x. That algorithm is implemented by the `laGP` function in the `laGP` package for R on CRAN [66]. For more details, see the package vignette [67]. It is worth remarking that the scheme is completely deterministic, calculating the same local designs for prediction at x, given identical inputs (n, initial θ_0, and data D_N) in repeated executions. It also provides local uncertainty estimates – a hallmark of any approximation – via Eq. (3) with $D_n(x)$, which are organically inflated relative to their full data (D_N) counterparts. Empirically, those uncertainty estimates overcover, as they are perhaps overly conservative.

Figure 2 shows two local designs $X_n(x)$ coming out of Algorithm 1 using $n_0 = 6$ initializing NNs followed by 44 iterations of selection until $n = 50$ total data elements have been chosen. Available candidates X_N number about 40 000 in a regular 2d grid filling out the input space $[-2, 2]^2$, which is only partly visualized within the plotting window. The reference location x is indicated as a gray dot. Numbers plotted indicate location and order in which each site was chosen. Although the two criteria do not select the same local design, differences are subtle. Both contain a clump of nearby points with "satellite" sites emanating loosely along rays from x.

Algorithm 1. Local Approximate GP Regression

Assume criterion $J(x_{j+1}, x; \theta)$, e.g., MSPE (4) or ALC, on distance-based covariance through hyperparameters θ which are vectorized below.

Require large-N training data $D_N = (X_N, Y_N)$ and predictive/testing location x; local design size $n \ll N$ with NN init size $n_0 < n$ and NN search window size $n \le N' \ll N$.

Then

1. Initialize $X_{n_0}(x)$ with n_0 nearest $X_N(x)$ to x and establish a candidate set of remaining design elements $X^{\text{cand}}_{N'-n_0}(x) = X_N \setminus X_{n_0}$.
2. **For** $j = n_0, \ldots, n-1$, acquire the next local design element.
   (a) Optimize criterion J to select

   $$x_{j+1} = \text{argmin}_{x' \in X^{\text{cand}}_{N'-j}} J(x', x; \theta)$$

   (b) Update $X_{j+1}(x) \leftarrow X_n \cup \{x_{j+1}\}$ and $X^{\text{cand}}_{N'-j-1}(x) \leftarrow X^{\text{cand}}_{N'-j}(x) \setminus x_j$.

   End For
3. Pair $X_n(x)$ with Y_n-values to form local data $D_n(x)$.
4. Optionally update hyperparameters $\theta \leftarrow \hat{\theta}_n(x)$ where

   $$\hat{\theta}_n(x) = \text{argmin}_\theta - \ell(\theta; D_n(x))$$

Return predictions, e.g., pointwise mean and variance (2) given θ and $D_n(x)$.

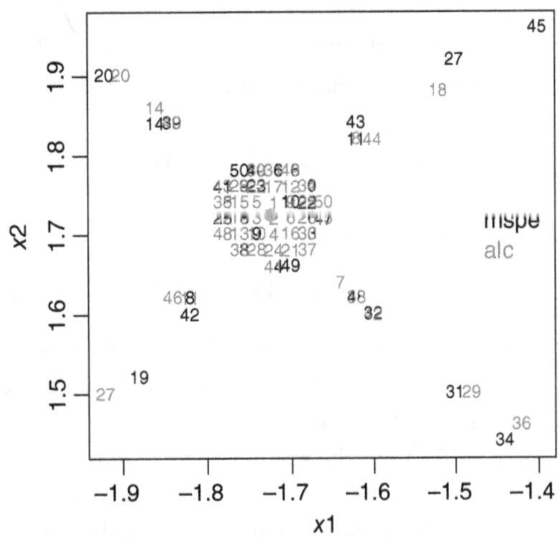

Figure 2 Example local designs $X_n(x)$ under MSPE and ALC criteria. Numbers plotted indicate the iteration, from the "for" loop in Algorithm 1, that the candidate was added into the design.

It is perhaps intriguing that the greedy local designs differ from full NN ones, which are easy to imagine in the negative space. An exponentially decaying $K_\theta(\cdot, \cdot)$ should substantially devalue locations far from x. Gramacy and Haaland [68] offer an explanation, which surprisingly has little to do with the particular choice of K_θ. The explanation lies in the form of Equation (6). Although quadratic in $K_\theta(x_{j+1}, x)$, the "distance" between the x and the potential new local design location x_{j+1}, it is also quadratic in $g_j(x_{j+1})$, a vector measuring "inverse distance," via K_j^{-1}, between x_{j+1} and the current local design $X_j(x)$. So, the criterion makes a trade-off: minimize "distance" to x while maximizing "distance" (or minimizing "inverse distance") to the existing design. Or, in other words, the potential value of new design element (x_{j+1}, y_{j+1}) depends not just on its proximity to x but also on how potentially different that information is to where we already have (lots of) it, at $X_j(x)$.

Several alternative Step 2a implementations in the `laGP` package offer potential for further speedups from shortcuts under the ALC criterion. Providing `method="alcray"` replaces iteration over candidates with a 1d line search over rays emanating from x [68], with solutions snapped back to candidates. Derivatives offer another way to replace discrete with continuous search. Sun *et al.* [69] provide details, with implementation as `method="alcopt"`. Both options, as well as the original, can be sped up by short-circuiting an exhaustive search of the remaining, unselected candidates in X_N. Searching over as few as $N' = 1000$ nearest neighbors to x can lead to identical selections $X_n(x)$ when n is small, like $n = 50$. More care is needed when n is larger. As illustrated momentarily, search over large candidate sets is ripe for off-loading to specialized hardware.

3.2 Massively Parallelized Global GP Approximation

Global emulation, which is predicting over a dense grid of x-values, can be done in serial by looping over the x's, or in parallel since each calculation of local $X_n(x)$'s is independent of the others. This kind of embarrassingly parallel scheme is most easily implemented on SMP machines via OpenMP pragmas,[12] allowing elements of a `for` loop run on unique threads. In `laGP`'s C implementation, that is as simple as a `parallel for` pragma.

```
#ifdef _OPENMP
  #pragma omp parallel for private(i)
#endif
  for(i = 0; i < npred; i++) { ...
```

To illustrate global emulation, consider "Herbie's tooth" [70]. Let

$$g(z) = \exp(-(z-1)2) + \exp(-0.8(z+1)2) - 0.05\sin(8(z+0.1)) \tag{8}$$

be defined for scalar inputs z. Then, for inputs x with m coordinates $x_1, \dots, x_m$, the response is $f(x) = -\prod_{j=1}^{m} g(x_j)$. Create a training data set based on X_N above in $[-2, 2]^2$ and gather $f(x_i)$ calculated thereupon.

Figure 3 shows the predictive mean surface derived, via eight parallel threads, by applying ALC-based `laGP` as in Algorithm 1 separately to each element of a regular 10,000-element grid in the input space. Parallel evaluation with OpenMP is automated by the `aGP` function in the `laGP` package. Observe that the predictive surface is smooth despite the independent (computationally and statistically) calculations. In fact, the surface is pathologically discontinuous, with "jumps" on an extremely small scale. Execution time on my 2016 Intel 8-core i7-6900K CPU at 3.20 GHz (hyperthreaded) machine is about 70 s. Each individual `laGP` requires about 0.05 s, leading to an almost linear scaling of OpenMP distribution. Overheads are trivial.

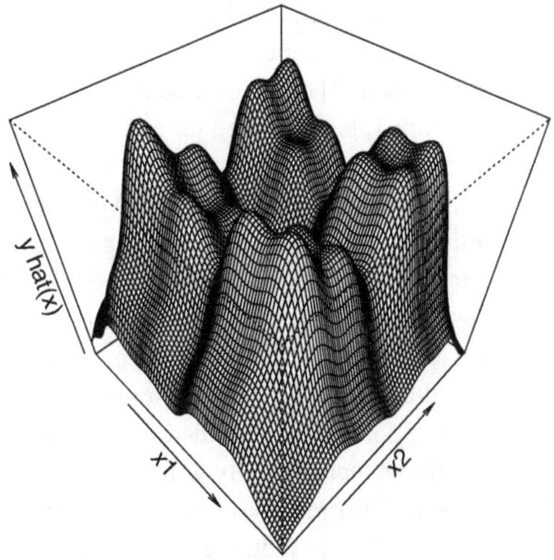

Figure 3 LAGP-calculated predictive mean on "Herbie's Tooth" data. Actually, a negated surface is plotted to ease visuals.

One can additionally divvy up predictions across nodes with the *simple network of workstations* (SNOW) cluster computing model, implementing a two-level hierarchical cascade of parallelism: across nodes (SNOW) and across codes within nodes (OpenMP). The function `aGP.parallel` in the `laGP` package can take a cluster instance from `makeCluster` in the `parallel` package (formerly `snow`) and act as a wrapper which partitions predictive locations $\mathcal{X}$ into chunks. Each chunk may be processed on a separate node via `clusterApply`. Subsequently, chunk outputs may be combined into a single object for passing back to the user. Socket (PSOCK)[13] and MPI (Message Passing Interface),[14] via the `Rmpi` package [71], have successfully been used with `aGP.parallel`. A demonstration, with data scaled up in both length (bigger N and n) and breadth (higher input dimension), is differed until Section 4.2 after introducing a third level into the hierarchy.

3.3 Off Loading Subroutines to GPUs

Under NVIDIA's `CUDA` programming model, work is off-loaded to a general purpose GPU device by calling a *kernel* function: specially written code that targets execution on many hundreds of GPU cores. Efficient kernel development for particular tasks requires a rather intimate knowledge of (rapidly evolving) GPU architecture, best reviewed elsewhere [72], and identifying tasks amenable distribution over that apparatus.

`CUDA` has gained widespread adoption since its introduction in 2007, and many "drop-in" libraries for GPU acceleration have been published, for example, the `CUBLAS` library, which contains a `cublasDgemm` function that is the GPU equivalent of the `DGEMM` matrix–matrix multiplication function from the C `BLAS` library. Such GPU-aware libraries allow for significant speedups at minimal coding investment, and most use of GPUs for acceleration in statistical applications has been accomplished by replacing calls to CPU-based library functions with the corresponding GPU kernel call from a GPU-aware library [9, 31, 32]. This can be an effective approach to GPU acceleration when very large matrices are involved, for example, of dimension $\geq$1000.

LAGP, as described in Algorithm 1, manipulates relatively small matrices by design and therefore does not benefit from this drop-in style approach to GPU acceleration. Instead, Gramacy *et al.* [73] describe a custom kernel that, in implementing the entirety of Step 2a in Algorithm 1, is optimized for a multitude of relatively small matrix situation(s) and also carries out many processing steps in a single invocation.

The most computationally intensive subroutine in Algorithm 1 is Step 2, looping over all remaining candidates and evaluating the reduction in variance (6) to find the next training data site to add into the design. Each reduction in variance calculation is $O(j^2)$, and in a design with N points, there are $N' = N - j$ candidates. Usually $N \gg j$, so the overall scheme for a single x is $O(Nn^3)$, a potentially huge undertaking. Fortunately, the structure of the evaluations of (6), independent for each of the N' candidates, is ideal for GPU computing. The details are provided in Gramacy *et al.* [73]. Also, fortunately, GPUs are set up to interface asynchronously with their (CPU) calling environment, allowing independent (e.g., OpenMP) threads to queue up kernels for evaluation on the GPU when resources become available. In fact, it can even be advantageous to have more threads queuing GPU kernels than processor cores in order to make sure GPUs remain engaged, minimizing idle time.

Identifying an appropriate subroutine for off-loading to GPUs, or other highly customized and vastly parallelized hardware components, can be crucial to effective statistical and scientific computing at scale. Entertaining candidates via ALC for LAGP is one such example. Finding nuggets such as these is likely to remain a comparative advantage for certain methodological frameworks for some time, at least until general-purpose hardware, machine instructions, and their compilers/preprocessors catch up. See, for example, OpenACC.[15]

4 Empirical Results

Here, the ultimate goal is to showcase the three-level cascade – cluster, SMP/OpenMP, and GPU/CUDA – of parallelism toward approximate GP emulation on very big computer experiment data. As an illustrative warm-up, Section 4.1 considers a classic, big-N/modest-p (number of input dimensions), example from the machine learning literature, focusing on SMP parallelization with `laGP`. Section 4.2 then revisits a scale-up exercise from Gramacy *et al.* [73], pushing the boundaries of problem sizes and accuracies that can be obtained in a matter of hours on a classic computer surrogate modeling example.

4.1 SARCOS

The SARCOS data[16] features as a prime example in Rasmussen and Williams [5] book on *Gaussian processes for Machine Learning* (GPML). The data comprise a prepartitioned set of about $N = 44\,000$ training runs and 4.4 thousand testing runs, with $p = 21$ inputs and seven outputs. Here, we consider only the first of the seven outputs. Matlab files storing these data have been converted to plain text and included with the supplementary material. Figure 4

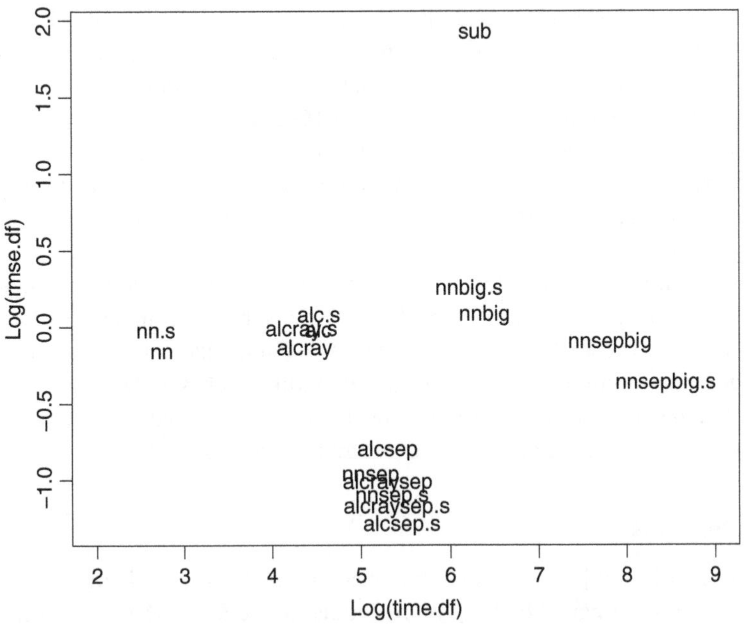

Figure 4 Time versus accuracy comparison on SARCOS data.

considers 17 variations of `laGP`-based comparators plotting accuracy in terms of (log) root mean-squared error (RMSE) on the testing set, versus (log) compute time, as measured on an 8-core i7 workstation. Combinations of comparators are enumerated below, and the code fitting these models is provided in the supplementary material.

i) NN-based LAGP with isotropic (all components $\theta_j = \theta_k$ and separable (6) alternatives (`aGP/aGPsep` with `method="nn"`), local design size $n \equiv$ `end=50` (default), and $n \equiv$ `end=200`);
ii) ALC-based LAGP, both isotropic and separable (`aGP/aGPsep` with default `method="alc"`);
iii) ALC-based LAGP with approximate ray search [68], both isotropic and separable (`aGP/aGPsep` with `method="alcray"`);
iv) Ordinary separable GP trained on a random subsample of 1000 input–output pairs;
v) Combining #i–iii with inputs prescaled by the MLE lengthscales from #iv for a multiresolution effect [69].

The experiment used coded inputs for #i–iii into $[0, 1]^{21}$ and priors on the lengthscale built via `darg` with `samp.size=10000`. There are several noteworthy results evident in the view provided by the figure. The best methods are based on ALC local design ("alc" prefix) with "sep"arable local lengthscales, primed by prescaling with global lengthscales (".s" suffix). NN analogues are competitive but slightly worse despite taking about the same amount of time. NN with isotropic local lengthscales is much faster but also much less accurate. NN with a larger training set (big) is not competitive based on time or accuracy grounds. Apparently, a smaller local design, with the right mix of neighbors and satellite points, offers the right trade-off between reactivity and computational effort. Ray-based search is not that much faster than exhaustive search, especially when "sep"arable local lengthscales are involved. An explanation here is that MLE calculations, requiring n^3 decompositions, dominate the flops required. Solving for MLE lengthscales in 21 dimensions requires a lot of work. The best RMSEs, which are near −1.2 in log space, translate into about 0.3 after exponentiating. RMSE of 0.3 on the scale of the y-values in the testing set, which ranges in $[-84, 121]$, is remarkably accurate. That 0.3 is about 0.1% on the scale of y-values.

Perhaps most importantly, observe how the best cohort of methods compare to a full GP fit to a random subset of the data (sub). Using a GP subset is roughly an order of magnitude slower – because it cannot leverage SMP parallelization – and three orders of magnitude less accurate because its global nature makes it far less reactive to shifting nonlinear dynamics in the data. Needless to say, full GP modeling on $N = 4.4 \times 10^5$ is a nonstarter. Local modeling by divide and conquer is a win.

4.2 Supercomputer Cascade

The borehole experiment [74, 75] involves an 8-dimensional input space, and our use of it here follows the setup of Kaufman *et al.* [37]; more details can be found therein. Results from two similar experiments are reported, involving out-of-sample prediction based on designs of increasing size N. The designs and predictive sets (also of size N) are from a joint random space-filling Latin hypercube sample [76]. As N is increased, so are the local design size n and candidate set N', so that there is steady reduction of out-of-sample MSE down the rows of Table 1. The numbers in the middle section of the table, between the vertical double bars, are from Gramacy *et al.* [73]. These rows/columns show results from a 96-node CPU cluster, where each node has 16 cores, alongside results from a 5-node GPU/CPU cluster, where each node has 2 GPUS and 16 cores. These nodes were part of the University of Chicago's `midway` supercomputer:[17] dual-socket 8-core 2.6 GHz Intel Sandy Bridge Xeons with 32 GB of main memory; GPU nodes are similar with NVIDIA Tesla M2090 GPU devices with 5 GB of global memory, and the L1 cache option is set to prefer shared memory (giving 48 KB per block). Both involved calls to `aGP.parallel` using a script like the one provided with the supplementary material.

To the right of the double bars in the table is the outcome of a similar experiment performed by Gramacy and Haaland [68] with the ray-based approximation. These combine a single-node experiment on a 4-core 2.93 GHz Intel i7 iMac with a distributed analog on the 96-core supercomputer. Ray-based search is not amenable to GPU parallelization. The goal of both experiments was to see how large of a design, and accurate a predictor, could be developed in about an hour of super/computing time.

Besides reproducing these tables here to demonstrate capability in our target hierarchical cascade of massive parallelization capability, these results – which are by now almost a decade old, taking into account publication lag times – can be updated using the rapidly expanding capabilities of modern kit. Table 2 reports on results obtained on 10 nodes of the `cascades` supercomputer at Virginia Tech's Advanced Research Computing Center.[18] These nodes are dual-socket 12-core 3.0 GHz Intel Xeon (Skylake) processors with 756 GB

Table 1 Timings and out-of-sample accuracy measures for increasing problem sizes on the borehole data.

			Exhaustive						**Via rays**	
			Intel Sandy Bridge		Nvidia Tesla		iMac		Intel SB	
			96× CPU		5× 2 GPUs		1×(4-core) CPU		96× CPU	
N	n	N'	Seconds	MSE	Seconds	MSE	Seconds	MSE	Seconds	MSE
1 000	40	100	0.48	4.88	1.95	4.63	8.00	6.30	0.39	6.38
2 000	42	150	0.66	3.67	2.96	3.93	17.83	4.47	0.46	4.10
4 000	44	225	0.87	2.35	5.99	2.31	40.60	3.49	0.62	2.72
8 000	46	338	1.82	1.73	13.09	1.74	96.86	2.24	1.31	1.94
16 000	48	507	4.01	1.25	29.48	1.28	222.41	1.58	2.30	1.38
32 000	50	760	10.02	1.01	67.08	1.00	490.94	1.14	4.65	1.01
64 000	52	1 140	28.17	0.78	164.27	0.76	1 076.22	0.85	9.91	0.73
128 000	54	1 710	84.00	0.60	443.70	0.60	3 017.76	0.62	17.99	0.55
256 000	56	2 565	261.90	0.46	1 254.63	0.46	5 430.66	0.47	40.16	0.43
512 000	58	3 848	836.00	0.35	4 015.12	0.36	12 931.86	0.35	80.93	0.33
1 024 000	60	5 772	2 789.81	0.26	13 694.48	0.27	32 866.95	0.27	188.88	0.26
2 048 000	62	-	-	-	-	-	-	-	466.40	0.21
4 096 000	64	-	-	-	-	-	-	-	1 215.31	0.19
8 192 000	66	-	-	-	-	-	-	-	4 397.26	0.17

The "MSE" columns are mean-squared predictive error to the true outputs on the $|\mathcal{X}| = N$ locations from separate runs (hence, the small discrepancies between the two columns). Both CPU and GPU nodes have 16 CPU cores. So, the "96× CPU" shorthand in the table indicates 1536 CPU cores.

of memory and dual NVIDIA V100 GPUs, a substantial upgrade compared to the older UChicago kit. In fact, runs similar to those from Table 1 were so fast that the experimental setup was adjusted somewhat to compensate/expand by allowing larger neighborhood sizes n and candidate windows N' in for Table 2. As you can see, this resulted in far more accurate calculations – as well as faster – for comparable computational effort.

Results in that table demonstrate that the `laGP` framework, with parallel distribution automated by `aGP` and `aGP.parallel`, adequately and organically expands to leverage expanded breath in hardware. Notice that although the training data sets are not as big as the ones in Table 1, the results are more accurate owing to the expansion of neighborhood sizes (n and N'), demanding heavier and more highly distributed computation. It is clear that doubling the training data size would result in marginal improvement on MSE at the expense of substantially higher compute times, unless more compute nodes can be brought to bear. The take-home message here is that LAGP offers a simple framework for utilizing vast computing resources on large problems. This is the direction that computing is going, with pipelines growing in breadth faster than in depth, which has all but plateaued.

Table 2 Updated GPU/CPU results based on a more modern cascade of supercomputing resources.

			10× Intel Sky Lake 2× Nvidia V100	
N	*n*	*N'*	**Seconds**	**MSE**
1 000	90	1 000	< 1	0.56
2 000	100	1 500	< 1	0.44
4 000	110	2 250	2.37	0.22
8 000	120	3 375	7.43	0.15
16 000	130	5 062	22.57	0.11
32 000	140	7 593	70.67	0.07
64 000	150	11 390	231.03	0.05
1 28 000	160	17 085	815.37	0.04
2 56 000	170	25 628	2 542.82	0.03
5 12 000	180	55 100	12 607.12	0.02

The 10 Sky Lake nodes each have 24 cores.

5 Conclusion

LAGP swaps a large problem for many small independent ones. This is perfect for modern distributed computing, by divvying up and off-loading smaller chunks of evaluations to a hierarchical cascade of processing units. Modern workstations have multiple cores and (sometimes) multiple GPUs. Supercomputers these days are not much more than enormous clusters of high-end multicore desktops and GPUs. With exascale computing on the frontier,[19] there is a dire need for parallelizable algorithms which scale with architecture. In the case of statistical modeling, and in particular nonlinear and nonparametric regression (as exemplified by GPs), leveraging a massive degree of parallel capability requires approximation. Transductive learning can offer a template into one way in which such enterprises can be implemented in practice.

Others have had similar success parallelizing non-GP models for computer emulation. For example, Pratola *et al.* [77] parallelized the Bayesian additive regression trees (BART) method using MPI and report-handling designs as large as $N = 7$M using hundreds of computing cores. Such efforts will likely remain in vogue as long as computing resources continue to grow "out" (with more nodes/cores, etc.) faster than they grow "up," which will be for quite some time to come.

Acknowledgments

RBG gratefully acknowledges funding from DOE LAB 17-1697 via subaward from Argonne National Laboratory for SciDAC/DOE Office of Science ASCR and High Energy Physics, and from National Science Foundation (NSF) award DMS-1821258.

Notes

1 https://en.wikipedia.org/wiki/Moore\stquotes_law
2 https://en.wikipedia.org/wiki/Symmetric_multiprocessing
3 https://en.wikipedia.org/wiki/Hyper-threading
4 https://software.intel.com/en-us/mkl
5 https://developer.nvidia.com/gpu-accelerated-libraries#linear-algebra
6 https://en.wikipedia.org/wiki/Multivariate_normal_distribution#Conditional_distributions
7 https://sheffieldml.github.io/GPy
8 https://research.cs.aalto.fi/pml/software/gpstuff
9 For MKL likelihood as well as for prediction.
10 https://en.wikipedia.org/wiki/K-d_tree
11 https://en.wikipedia.org/wiki/Nearest_neighbor_search
12 https://www.openmp.org
13 https://en.wikipedia.org/wiki/Berkeley_sockets#BSD_and_POSIX_sockets
14 https://en.wikipedia.org/wiki/Message_Passing_Interface
15 https://www.openacc.org
16 http://www.gaussianprocess.org/gpml/data
17 https://rcc.uchicago.edu
18 https://www.arc.vt.edu
19 https://en.wikipedia.org/wiki/Exascale_computing

References

1 R Core Team (2019) `R`: A Language and Environment for Statistical Computing, R Foundation for Statistical Computing, Vienna, Austria.

2 Kurth, T., Treichler, S., Romero, J., *et al.* (2018) Exascale Deep Learning for Climate Analytics. Proceedings of the International Conference for High Performance Computing, Networking, Storage, and Analysis, 51. IEEE Press.

3 Yang, L., Treichler, S., Kurth, T., *et al.* (2019) Highly-scalable, physics-informed GANs for learning solutions of stochastic PDEs. *arXiv preprint arXiv:1910.13444.*

4 Goodfellow, I., Bengio, Y., and Courville, A. (2016) *Deep Learning*, MIT Press. http://www.deeplearningbook.org.

5 Rasmussen, C.E. and Williams, C.K.I. (2006) *Gaussian Processes for Machine Learning*, The MIT Press, Cambridge, MA.

6 Vapnik, V. (1995) *The Nature of Statistical Learning Theory*, Springer Verlag, New York.

7 Kleiner, A., Talwalkar, A., Sarkar, P., and Jordan, M.I. (2014) A scalable bootstrap for massive data. *J. R. Stat. Soc. Series B*, 76 (4), 795–816.

8 Gramacy, R.B. (2020) *Surrogates: Gaussian Process Modeling, Design and Optimization for the Applied Sciences*, Chapman Hall/CRC, Boca Raton, FL. http://bobby.gramacy.com/surrogates.

9 Franey, M., Ranjan, P., and Chipman, H. (2012) A Short Note on Gaussian Process Modeling for Large Datasets using Graphics Processing Units. Tech. rep., Acadia University.

10 Paciorek, C., Lipshitz, B., Kaufman, C., *et al.* (2014) `bigGP`: Distributed Gaussian Process Calculations. R package version 0.1-3.

11 Cressie, N. (1991) *Statistics for Spatial Data*, Revised edn, John Wiley and Sons, Inc., Hoboken, NJ.

12 Matheron, G. (1963) Principles of Geostatistics. *Econ. Geol.*, 58, 1246–1266.

13 Sacks, J., Welch, W.J., Mitchell, T.J., and Wynn, H.P. (1989) Design and analysis of computer experiments. *Stat. Sci.*, 4, 409–435.

14 Santner, T., Williams, B., and Notz, W. (2018) *The Design and Analysis of Computer Experiments*, 2nd edn, Springer–Verlag, New York, NY.

15 Jones, D., Schonlau, M., and Welch, W.J. (1998) Efficient global optimization of expensive black box functions. *J. Global Optim.*, 13, 455–492.

16 Kennedy, M. and O'Hagan, A. (2001) Bayesian calibration of computer models (with discussion). *J. R. Stat. Soc. Series B*, 63, 425–464.

17 Oakley, J. and O'Hagan, A. (2004) Probabilistic sensitivity analysis of complex models: a Bayesian approach. *J. R. Stat. Soc. Series B*, 66 (3), 751–769.

18 Dancik, G. (2018) `mlegp`: Maximum Likelihood Estimates of Gaussian Processes. R package version 3.1.7.

19 MacDonald, B., Chipman, H., and Ranjan, P. (2019) `GPfit`: Gaussian Processes Modeling. R package version 1.0-8.

20 Ripley, B. (2015) `spatial`: Functions for Kriging and Point Pattern Analysis. R package version 7.3-11.

21 Nychka, D., Furrer, R., Paige, J., and Sain, S. (2019) `fields`: Tools for Spatial Data. R package version 9.7.

22 Gu, M., Palomo, J., and Berger, J. (2018) `RobustGaSP`: Robust Gaussian Stochastic Process Emulation. R package version 0.5.6.

23 Karatzoglou, A., Smola, A. and Hornik, K. (2018) `kernlab`: Kernel-Based Machine Learning Lab. R package version 0.9-27.

24 Binois, M. and Gramacy, R. (2019) `hetGP`: Heteroskedastic Gaussian process modeling and design under replication. `R` package version 1.1.1.

25 Gramacy, R. and Taddy, M. (2016) `tgp`: Bayesian Treed Gaussian Process Models. R package version 2.4-14.

26 Hankin, R. (2019) `emulator`: Bayesian Emulation of Computer Programs. R package version 1.2-20.

27 Gramacy, R. (2014) `plgp`: Particle Learning of Gaussian Processes. R package version 1.1-7.

28 Finley, A. and Banerjee, S. (2019) `spBayes`: Univariate and Multivariate Spatial-Temporal Modeling. R package version 0.4-2.

29 Vanhatalo, J., Riihimäki, J., Hartikainen, J., *et al.* (2012) Bayesian modeling with Gaussian processes using the GPstuff toolbox. *Preprint on arXiv:1206.5754.*

30 Erickson, C.B., Ankenman, B.E., and Sanchez, S.M. (2018) Comparison of Gaussian process modeling software. *Eur. J. Oper. Res.*, 266 (1), 179–192.

31 Eidsvik, J., Shaby, B.A., Reich, B.J., *et al.* (2014) Estimation and prediction in spatial models with block composite likelihoods. *J. Comput. Graph. Stat.*, 23, 295–315.

32 Paciorek, C.J., Lipshitz, B., Zhuo, W., *et al.* (2015) Parallelizing Gaussian process calculations in R. *J. Stat. Softw.*, 63 (10), 1–23.

33 Snelson, E. and Ghahramani, Z. (2006) Sparse Gaussian processes using pseudo-inputs, in *Advances in Neural Information Processing Systems*. MIT press, Cambridge, MA, pp. 1257–1264.

34 Haaland, B. and Qian, P. (2011) Accurate emulators for large-scale computer experiments. *Ann. Stat.*, 39 (6), 2974–3002.

35 Gramacy, R. and Polson, N. (2011) Particle learning of Gaussian process models for sequential design and optimization. *J. Comput. Graph. Stat.*, 20 (1), 102–118.

36 Cressie, N. and Johannesson, G. (2008) Fixed rank kriging for very large data sets. *J. R. Stat. Soc. Series B*, 70 (1), 209–226.

37 Kaufman, C., Bingham, D., Habib, S., *et al.* (2012) Efficient emulators of computer experiments using compactly supported correlation functions, with an application to cosmology. *Ann. Appl. Stat.*, 5 (4), 2470–2492.

38 Sang, H. and Huang, J.Z. (2012) A full scale approximation of covariance functions for large spatial data sets. *J. R. Stat. Soc. Series B*, 74 (1), 111–132.

39 Nychka, D., Wikle, C., and Royle, J. (2002) Multiresolution models for nonstationary spatial covariance functions. *Stat. Model.*, 2, 315–331.

40 Quiñonero–Candela, J. and Rasmussen, C. (2005) A unifying view of sparse approximate gaussian process regression. *J. Mach. Learn. Res.*, 6, 1939–1959.

41 Furrer, R., Genton, M., and Nychka, D. (2006) Covariance tapering for interpolation of large spatial datasets. *J. Comput. Graph. Stat.*, 15, 502–523.

42 Gardner, J., Pleiss, G., Weinberger, K., *et al.* (2018) Gpytorch: Blackbox matrix-matrix Gaussian process inference with GPU acceleration, in *Advances in Neural Information Processing Systems*, pp. 7576–7586.

43 Ubaru, S., Chen, J., and Saad, Y. (2017) Fast estimation of $\text{tr}(f(A))$ via stochastic Lanczos quadrature. *SIAM J. Matrix Anal. Appl.*, 38 (4), 1075–1099.

44 Gramacy, R. and Lee, H. (2008) Bayesian treed Gaussian process models with an application to computer modeling. *J. Am. Stat. Assoc.*, 103 (483), 1119–1130.

45 Kim, H., Mallick, B., and Holmes, C. (2005) Analyzing nonstationary spatial data using piecewise Gaussian processes. *J. Am. Stat. Assoc.*, 100 (470), 653–668.

46 Rushdi, A. P., Swiler, L. T., Phipps, E., *et al.* (2016) VPS: Voronoi piecewise surrogate models for high-dimensional data fitting. *Int. J. Uncertainty Quantification*, 7, 1–21.

47 Park, C., Huang, J.Z., and Ding, Y. (2011) Domain decomposition approach for fast Gaussian process regression of large spatial datasets. *J. Mach. Learn. Res.*, 12, 1697–1728.

48 Park, C. and Huang, J.Z. (2016) Efficient computation of Gaussian process regression for large spatial data sets by patching local Gaussian processes. *J. Mach. Learn. Res.*, 17 (174), 1–29.

49 Park, C. and Apley, D. (2017) Patchwork kriging for large-scale Gaussian process regression. *arXiv:1701.06655*.

50 Liu, Y. (2014) Recent advances in computer experiment modeling. Ph.D. thesis. Rutgers University.

51 Zhao, Y., Amemiya, Y. and Hung, Y. (2018) Efficient Gaussian process modeling using experimental design-based subagging. *Stat. Sinica*, 28 (3), 1459–1479.

52 Chen, T. and Ren, J. (2009) Bagging for Gaussian process regression. *Neurocomputing*, 72 (7), 1605–1610.

53 Cochran, W.G. (1954) The Combination of Estimates from Different Experiments. *Biometrics*, 10 (1), 101–129.

54 Gramacy, R. and Apley, D. (2015) Local gaussian process approximation for large computer experiments. *J. Comput. Graph. Stat.*, 24 (2), 561–578. See arXiv:1303.0383.

55 Emery, X. (2009) The kriging update equations and their application to the selection of neighboring data. *Comput. Geosci.*, 13 (3), 269–280.

56 Datta, A., Banerjee, S., Finley, A., and Gelfand, A. (2016) Hierarchical nearest-neighbor Gaussian process models for large geostatistical datasets. *J. Am. Stat. Assoc.*, 111 (514), 800–812.

57 Vecchia, A. (1988) Estimation and model identification for continuous spatial processes. *J. R. Stat. Soc. Series B*, 50, 297–312.

58 Stroud, J., Stein, M., and Lysen, S. (2017) Bayesian and maximum likelihood estimation for Gaussian processes on an incomplete lattice. *J. Comput. Graph. Stat.*, 26 (1), 108–120.

59 Katzfuss, M. and Guinness, J. (2018) A general framework for Vecchia approximations of Gaussian processes. *Preprint on arXiv:1708.06302.*

60 Heaton, M., Datta, A., Finley, A., *et al.* (2018) A case study competition among methods for analyzing large spatial data. *J. Agric. Biol. Environ. Stat.*, 1–28.

61 Guinness, J. and Katzfuss, M. (2019) `GpGp`: fast Gaussian process computation using Vecchia's approximation. R package version 0.1.1.

62 Stein, M.L., Chi, Z., and Welty, L.J. (2004) Approximating likelihoods for large spatial data sets. *J. R. Stat. Soc. Series B*, 66 (2), 275–296.

63 Barnett, S. (1979) *Matrix Methods for Engineers and Scientists*, McGraw-Hill.

64 Seo, S., Wallat, M., Graepel, T., and Obermayer, K. (2000) Gaussian Process Regression: Active Data Selection and Test Point Rejection. Proceedings of the International Joint Conference on Neural Networks, vol. III, IEEE, 241–246.

65 Cohn, D.A. (1996) Neural network exploration using optimal experimental design, in *Advances in Neural Information Processing Systems*, vol. 6.9, Morgan Kaufmann Publishers, 679–686.

66 Gramacy, R.B. and Sun, F. (2018) `laGP`: Local Approximate Gaussian Process Regression. R package version 1.5-2.

67 Gramacy, R. (2016) `laGP`: large-scale spatial modeling via local approximate Gaussian processes in R. *J. Stat. Softw.*, 72 (1), 1–46.

68 Gramacy, R. and Haaland, B. (2016) Speeding up neighborhood search in local Gaussian process prediction. *Technometrics*, 58 (3), 294–303.

69 Sun, F., Gramacy, R., Haaland, B., Lawrence, E. and Walker, A. (2019) Emulating satellite drag from large simulation experiments. *IAM/ASA J. Uncertainty Quantification*, 7 (2), 720–759. Preprint arXiv:1712.00182.

70 Lee, H., Gramacy, R., Linkletter, C. and Gray, G. (2011) Optimization subject to hidden constraints via statistical emulation. *Pacific J. Optim.*, 7 (3), 467–478.

71 Yu, H. (2002) `Rmpi`: Parallel Statistical Computing in R. *R News*, 2 (2), 10–14.

72 Kirk, D.B. and Wen-mei, W.H. (2010) *Programming Massively Parallel Processors: A Hands-on Approach*, Morgan Kaufmann.

73 Gramacy, R., Niemi, J., and Weiss, R. (2014) Massively Parallel Approximate Gaussian Process Regression. *SIAM/ASA J. Uncertainty Quantification*, 2 (1), 564–584.

74 Worley, B. (1987) Deterministic Uncertainty Analysis. Tech. Rep. ORN-0628, National Technical Information Service, 5285 Port Royal Road, Springfield, VA 22161, USA.

75 Morris, D., Mitchell, T. and Ylvisaker, D. (1993) Bayesian design and analysis of computer experiments: use of derivatives in surface prediction. *Technometrics*, 35, 243–255.

76 McKay, M., Conover, W. and Beckman, R. (1979) A comparison of three methods for selecting values of input variables in the analysis of output from a computer code. *Technometrics*, 21 (2), 239–245.

77 Pratola, M.T., Chipman, H., Gattiker, J., Higdon, D., McCulloch, R. and Rust, W. (2014) Parallel Bayesian Additive Regression Trees. *J. Comput. Graph. Stat.*, 23 (3), 830–852.

30

Divide-and-Conquer Methods for Big Data Analysis

Xueying Chen[1], *Jerry Q. Cheng*[2], *and Min-ge Xie*[3]

[1]*Novartis Pharmaceuticals Corp., East Hanover, NJ, USA*
[2]*New York Institute of Technology, New York, NY, USA*
[3]*Rutgers University, Piscataway, NJ, USA*

1 Introduction

With ever advancing computing and storage technologies, we frequently have access to large data sets gathered from a variety of sources with information sensing or collecting capabilities, such as the Internet of things devices (e.g., mobile devices), social media, and consumer activities. As a result, we are facing information explosions in the era of big data, which has imposed both opportunities and challenges. Extremely large data sets usually not only have high-dimensional exploratory variables but also large sample sizes. Often, it is impossible to analyze entirely or even store such data in a single computer. To solve this problem, tremendous efforts have been invested to reduce computational difficulties with various models. Among them, the divide-and-conquer approach was proposed and has been widely adopted as a general framework to handle and analyze extremely large data sets. Under this framework, we divide an original problem into several subproblems, analyze them separately, and then combine the results to provide an inference.

The divide-and-conquer approach is easy to implement. In the *divide* step, one could randomly split an entire data set into several subsets. Sometimes big data sets are available in separate chunks by nature. For example, they might be stored in multiple machines due to the storage limit of one machine, or generated continuously as streaming data. Next, each subset is analyzed separately to provide statistical results with no extra or less model fitting or programming efforts. In the *combine* step, a simple and straightforward method is to average (with or without weights) the results from the subset analyses to obtain estimators. This combining strategy has been applied in various models such as regularized estimators in generalized linear models, M-estimators, and functional estimators in kernel regression models and nonparametric additive models.

The divide-and-conquer methodology can be described in a general term as follows [1]. When an entire data set is split into K subsets, $\{\mathcal{D}_1, \dots, \mathcal{D}_K\}$, a low dimension statistic $T_k = g_k(\mathcal{D}_k)$ is constructed for each subset with some function $g_k(\cdot)$. Then, the final result is obtained by aggregating $\{T_k, k = 1, \dots, K\}$ through an aggregation function $G(\cdot)$. In many studies, $g_k(\cdot)$ is selected as the same estimation function or test statistics for the entire data,

Computational Statistics in Data Science.
Edited by Walter W. Piegorsch, Richard A. Levine, Hao Helen Zhang and Thomas C. M. Lee.
 ISBN 978-1-11956107-1

which saves efforts of additional model fitting. It may still be computationally intensive to analyze smaller data sets K times with some complex models.

Note that, if the simple average is used as the aggregation function, biases may not be reduced because each subset now has a much smaller sample size. Some recent works, for example, Chen *et al.* [1], Wang *et al.* [2], Jordan *et al.* [3], Wang *et al.* [4], Fan *et al.* [5], have proposed to apply the divide-and-conquer algorithm iteratively with linearization of the original estimation function. It is worthwhile noting that only one subset would have $g_k(\cdot)$ the same as the estimation function of the entire data set in each iteration. This multiround divide-and-conquer algorithm could further reduce computational burden but require additional model fitting to obtain $g_k(\cdot)$ and $G(\cdot)$ functions.

The remaining chapter is organized as follows. Section 2 reviews the divide-and-conquer algorithms in linear regression models, a simple case that can provide a key insight into the developments of the divide-and-conquer methodology. Section 3 presents divide-and-conquer algorithms and their properties in parametric models, including sparse high-dimensional models, M-estimators, Cox regression, and quantile regression models. Performance on nonstandard problems is also reviewed. Section 4 explains the algorithm in nonparametric and semiparametric models. Section 5 describes divide-and-conquer applications in the setting of online sequential updating. Section 6 discusses a method in small n but large p situation, where the division is vertically among p number of covariates. Section 7 explains Bayesian divide-and-conquer and median-based combining methods. Section 8 provides some real data analyses using divide-and-conquer methods in various application areas. Section 9 concludes this chapter with discussions.

2 Linear Regression Model

We first consider a regular linear regression model

$$y_i = \boldsymbol{x}_i^T \boldsymbol{\beta} + \epsilon_i, \quad i = 1, \dots, n$$

where y_i is a response variable, $\boldsymbol{x}_i$ is a vector of p explanatory variables, $\boldsymbol{\beta}$ is a vector of p unknown parameters, and ϵ_i is an error noise of mean 0. The ordinary least-square (OLS) estimator using the entire data is $\hat{\boldsymbol{\beta}} = (\boldsymbol{X}^T\boldsymbol{X})^{-1}\boldsymbol{X}^T\boldsymbol{y}$, where $\boldsymbol{y} = (y_1, \dots, y_n)^T$ is the response vector, and $\boldsymbol{X} = (\boldsymbol{x}_1, \dots, \boldsymbol{x}_n)^T$ is the design matrix with the assumption that $\boldsymbol{X}^T\boldsymbol{X}$ is invertible.

Suppose that the entire data set is split into K subsets, the kth subset has n_k observations: $(\boldsymbol{X}_k, \boldsymbol{y}_k)$, and the OLS estimator using the kth subset is $\hat{\boldsymbol{\beta}}_k = (\boldsymbol{X}_k^T\boldsymbol{X}_k)^{-1}\boldsymbol{X}_k^T\boldsymbol{y}_k$. A combined estimator [6, 7] is proposed as

$$\hat{\boldsymbol{\beta}}^{(c)} = \left(\sum_{k=1}^{K} \boldsymbol{X}_k^T\boldsymbol{X}_k\right)^{-1} \sum_{k=1}^{K} \boldsymbol{X}_k^T\boldsymbol{X}_k\hat{\boldsymbol{\beta}}_k$$

By simple algebra, we can show that the combined estimator is exactly identical to the whole-data solution:

$$\hat{\boldsymbol{\beta}}^{(c)} = \left(\sum_{k=1}^{K} \boldsymbol{X}_k^T\boldsymbol{X}_k\right)^{-1} \sum_{k=1}^{K} \boldsymbol{X}_k^T\boldsymbol{y}_k = (\boldsymbol{X}^T\boldsymbol{X})^{-1}\boldsymbol{X}^T\boldsymbol{y} = \hat{\boldsymbol{\beta}} \tag{1}$$

Lin and Xi [6] extend the results of regular Gaussian linear regression models to the setting of general estimating equation (EE) estimation, where the EE estimator $\hat{\boldsymbol{\beta}}$ is the solution to the EE: $\boldsymbol{M}_n(\boldsymbol{\beta}) = \sum_{i=1}^{n} \psi(\boldsymbol{x}_i, y_i; \boldsymbol{\beta}) = 0$. Since ψ is often nonlinear, the explicit analytical form of $\hat{\boldsymbol{\beta}}$ is not available, and thus, the combined estimator cannot be obtained straightforwardly. To solve this problem, Lin and Xi [6] approximate the nonlinear EE by its first-order Taylor expansion at the EE estimator $\hat{\boldsymbol{\beta}}_k$ based on the data of the kth subset: $(\boldsymbol{x}_{ki}, y_{ki} : i = 1, \dots, n_i)$:

$$\boldsymbol{M}_{n_k}(\boldsymbol{\beta}) = \{\boldsymbol{A}_k(\hat{\boldsymbol{\beta}}_k)\}(\boldsymbol{\beta} - \hat{\boldsymbol{\beta}}_k) + \boldsymbol{R}_k \tag{2}$$

where $\boldsymbol{A}_k(\boldsymbol{\beta}) = -\sum_{i=1}^{n_k} \partial\psi(\boldsymbol{x}_{ki}, y_{ki}; \boldsymbol{\beta})/\boldsymbol{\beta}$, and $\boldsymbol{R}_k$ is the remainder. Now the combined EE estimator $\hat{\boldsymbol{\beta}}_k$ is the solution to $\sum_{k=1}^{K}\{\boldsymbol{A}_k(\hat{\boldsymbol{\beta}}_k)\}(\boldsymbol{\beta} - \hat{\boldsymbol{\beta}}_k) = 0$ when $\boldsymbol{R}_k$ is negligible. Lin and Xi [6] propose the combined estimator as

$$\hat{\boldsymbol{\beta}}^{(c)} = \left(\sum_{k=1}^{K} \boldsymbol{A}_k(\hat{\boldsymbol{\beta}}_k)\right)^{-1} \sum_{k=1}^{K} \{\boldsymbol{A}_k(\hat{\boldsymbol{\beta}}_k)\}\hat{\boldsymbol{\beta}}_k \tag{3}$$

3 Parametric Models

3.1 Sparse High-Dimensional Models

In supervised learning problems, both explanatory variables and response variables are observed. To handle extremely large-scale data sets with enormous sample sizes and often possibly a huge number of the explanatory variables at the same time, regularized regression models have been studied in depth with various approaches proposed, such as least absolute shrinkage and selection operator (LASSO) estimator [8, 9], least angle regression (LARS) algorithm [10], smoothly clipped absolute deviation penalty (SCAD) estimator [11], and minimax concave penalty (MCP) estimators [12]. However, these approaches are computationally intensive and have cubic or higher algorithmic complexity in the sample size dimension [7]. Divide-and-conquer strategy can substantially reduce computing time and computer memory requirement for such models.

Consider a generalized linear model. Given explanatory variables $\boldsymbol{X} = (\boldsymbol{x}_1, \dots, \boldsymbol{x}_n)^T$, the conditional distribution of response variable $\boldsymbol{y} = (y_1, \dots, y_n)^T$ is assumed to follow the canonical exponential distribution:

$$f(\boldsymbol{y}; \boldsymbol{X}, \boldsymbol{\beta}) = \prod_{i=1}^{n} f_0(y_i; \theta_i) = \prod_{i=1}^{n} \left\{ c(y_i) \exp\left[\frac{y_i\theta_i - b(\theta_i)}{\phi}\right]\right\} \tag{4}$$

where $\theta_i = \boldsymbol{x}_i^T\boldsymbol{\beta}, i = 1, \dots, n$, and ϕ is a nuisance dispersion parameter. The log-likelihood function $\log f(\boldsymbol{y}; \boldsymbol{X}, \boldsymbol{\beta})$ is then given by

$$\ell(\boldsymbol{\beta}; \boldsymbol{y}, \boldsymbol{X}) = [\boldsymbol{y}^T\boldsymbol{X}\boldsymbol{\beta} - \boldsymbol{1}^T\boldsymbol{b}(\boldsymbol{X}\boldsymbol{\beta})]/n$$

where $\boldsymbol{b}(\boldsymbol{\theta}) = (b(\theta_1), \dots, b(\theta_n))^T$ for $\boldsymbol{\theta} = (\theta_1, \dots, \theta_n)^T$, and the function $b(\cdot)$ is a smooth function with second derivatives. When p is large (or grows with n) and $\boldsymbol{\beta}$ is sparse (i.e., many elements of $\boldsymbol{\beta}$ are zero) with s nonzero entries, a regularized likelihood estimator is often used, which is defined as, in a general form,

$$\hat{\boldsymbol{\beta}}_\lambda = \text{argmax}_{\boldsymbol{\beta}}\{\ell(\boldsymbol{\beta}; \boldsymbol{y}, \boldsymbol{X})/n - \rho(\boldsymbol{\beta}; \lambda)\} \tag{5}$$

where ρ is the penalty function with tuning parameter λ.

Under the divide-and-conquer framework, after the entire data set is split into K subsets, the regularized estimator for the kth subset is

$$\hat{\boldsymbol{\beta}}_k = \text{argmax}_{\boldsymbol{\beta}}\{\ell(\boldsymbol{\beta};\boldsymbol{y}_k,\boldsymbol{X}_k)/n_k - \rho(\boldsymbol{\beta};\lambda_k)\} \tag{6}$$

where $\ell(\boldsymbol{\beta};\boldsymbol{y}_k,\boldsymbol{X}_k)/n_k$ is the log-likelihood function for the kth subset with sample size n_k, and $\rho(\boldsymbol{\beta};\lambda_k)$ is the penalty function with tuning parameter λ_k. Note that, since each $\hat{\boldsymbol{\beta}}_k$ is estimated from a different subset of data, $\{j : \hat{\beta}_{kj} \neq 0\}$ the set of selected variables (nonzero elements) of $\hat{\boldsymbol{\beta}}_k$ can be different from one to another.

In order to obtain a combined estimator, Chen and Xie [7] used a majority voting method to obtain the set of selected variables of the combined estimator, that is, $\hat{\mathcal{A}}^{(c)} \overset{\text{d}}{=} \left\{j : \sum_{k=1}^{K} \text{I}(\hat{\beta}_{kj} \neq 0) > w\right\}$, where $w \in [0, K)$ is a prespecified threshold, and I is the indicator function. Then, the following weighted average of $\hat{\boldsymbol{\beta}}_{k,\hat{\mathcal{A}}^{(c)}}$, $k = 1, \dots, K$, is proposed as the combined estimator:

$$\hat{\boldsymbol{\beta}}^{(c)} \overset{\text{d}}{=} \boldsymbol{A}\left(\sum_{k=1}^{K} \boldsymbol{A}^T\{\boldsymbol{X}_k^T\boldsymbol{\Sigma}(\hat{\boldsymbol{\theta}}_k)\boldsymbol{X}_k\}\boldsymbol{A}\right)^{-1} \sum_{k=1}^{K} \boldsymbol{A}^T\{\boldsymbol{X}_k^T\boldsymbol{\Sigma}(\hat{\boldsymbol{\theta}}_k)\boldsymbol{X}_k\}\boldsymbol{A}\hat{\boldsymbol{\beta}}_{k,\hat{\mathcal{A}}^{(c)}} \tag{7}$$

where $\hat{\boldsymbol{\theta}}_k = \boldsymbol{X}_k\hat{\boldsymbol{\beta}}_k$, $\hat{\boldsymbol{\beta}}_{k,\hat{\mathcal{A}}^{(c)}}$ is the subvector of $\hat{\boldsymbol{\beta}}_k$ confined by the majority voting set $\hat{\mathcal{A}}^{(c)}$, and $\boldsymbol{\Sigma}(\boldsymbol{\theta}) = \text{diag}(\sigma(\theta_1), \dots, \sigma(\theta_n))$ with $\sigma(\theta) = \partial^2 b(\theta)/\partial^2\theta$. Also, $\boldsymbol{E} = \text{diag}(v_1, \dots, v_p)$ is the $p \times p$ voting matrix with $v_j = 1$ if $\sum_{k=1}^{K} \text{I}(\hat{\beta}_{k,j} \neq 0) > w$ and 0 otherwise, and $\boldsymbol{A} = \boldsymbol{E}_{\hat{\mathcal{A}}^{(c)}}$ is the $p \times |\hat{\mathcal{A}}^{(c)}|$ selection matrix. Here, for any index subset S of $\{1, \dots, p\}$, $\boldsymbol{E}_S$ stands for a $p \times |S|$ submatrix of $\boldsymbol{E}$ formed by columns whose indices are in S.

Chen and Xie [7] show that the combined estimator $\hat{\boldsymbol{\beta}}^{(c)}$ is sign consistent under some regularity conditions and converges at the regular order of $O(\sqrt{s/n})$ under the L_2 norm. The combined estimator also obtains asymptotic normality with the same variance as the penalized estimator using the entire data. For the selection of the number of splits K, Chen and Xie [7] find that a stronger constraint on the growth rate of p would be imposed in order to detect the same signal strength as the corresponding complete data set analysis under the infinity norm.

Another strategy for combination is to debias or desparsify regularized estimators obtained from subsets, which has been adopted by Lee *et al.* [13], Battey *et al.* [14], and Tang *et al.* [15]. Using LASSO estimators in linear regression for illustration, the debiased LASSO estimator by Javanmard and Montanari [16] is

$$\hat{\boldsymbol{\beta}}^{d} \overset{\text{d}}{=} \hat{\boldsymbol{\beta}}_\lambda + n^{-1}(\boldsymbol{X}^T\boldsymbol{X}/n)^{-}\boldsymbol{X}^T(\boldsymbol{y} - \boldsymbol{X}\hat{\boldsymbol{\beta}}_\lambda)$$

where $\hat{\boldsymbol{\beta}}_\lambda$ is the regularized estimator defined in Equation (5) with L_1 norm penalty, and $(\boldsymbol{X}^T\boldsymbol{X}/n)^-$ is an approximate inverse of $\boldsymbol{X}^T\boldsymbol{X}/n$.

Both Lee *et al.* [13] and Battey *et al.* [14] propose the simple aggregated debiased LASSO estimator as the combined estimator:

$$\hat{\boldsymbol{\beta}}^{(c)} \overset{\text{d}}{=} \sum_{k=1}^{K} \hat{\boldsymbol{\beta}}_k^d/K = K^{-1}\sum_{k=1}^{K} \left[\hat{\boldsymbol{\beta}}_k + \{(\boldsymbol{X}_k^T\boldsymbol{X}_k)/n_k\}^{-}\boldsymbol{X}_k^T(\boldsymbol{y}_k - \boldsymbol{X}_k\hat{\boldsymbol{\beta}})\right] \tag{8}$$

where $\hat{\boldsymbol{\beta}}_k^d$ is the debiased LASSO estimator for the kth subset with sample size n_k. Lee *et al.* [13] show that with high probability when the rows of $\boldsymbol{X}$ are independent sub-Gaussian random vectors, the error of the aggregated debiased LASSO estimator in L_∞ norm is $O(\sqrt{\log(p/n)}) + O(sK\log(p)/n)$. When n is large enough, the latter term is negligible compared with the former term. The same results are obtained in Battey *et al.* [14]. To further reduce the computing cost, Lee *et al.* [13] use a single matrix $\hat{\boldsymbol{\Theta}}$ to replace all the terms $(\boldsymbol{X}_k^T\boldsymbol{X}_k/n_k)^-$, for $k = 1, \ldots, K$, which used to be solved for each subset k and thus made it the most computational expensive step. Following Van de Geer *et al.* [17], a common $\hat{\boldsymbol{\Theta}}$ is constructed by a nodewise regression on the explanatory variables.

Battey *et al.* [14] also tackle hypothesis testing problems using divide-and-conquer in the framework of the Wald and Rao's score tests. Consider a test of size α of the null hypothesis for any coefficient, $H_0 : \beta_j = \beta_j^H$ against the alternative, $H_1 : \beta_j \neq \beta_j^H$, $j = 1, \ldots, p$. A divide-and-conquer Wald statistic is proposed:

$$\overline{S_n} = \sqrt{n}\sum_{k=1}^{K}\left(\hat{\beta}_{k,j}^d - \beta_j^H\right) / \left(\overline{\sigma}\sqrt{\boldsymbol{b}_{k,j}^T\boldsymbol{b}_{k,j}}\right) \tag{9}$$

where $\overline{\sigma}$ is an estimator for the standard deviation of error based on K subsets, and $\boldsymbol{b}_{k,j}$ is the jth column of $(\boldsymbol{X}_k^T\boldsymbol{X}_k/n_k)^-\boldsymbol{X}_k^T$ which can be obtained from the following optimization algorithm:

$$\boldsymbol{b}_{k,j} = \text{argmin}_{\boldsymbol{b}}\boldsymbol{b}^T\boldsymbol{b}/n_k, s.t. \|\boldsymbol{X}_k^T\boldsymbol{b}/n_k - \boldsymbol{e}_j\|_\infty \le \vartheta_1, \|\boldsymbol{b}\|_\infty \le \vartheta_2$$

where $\boldsymbol{e}_j$ is a $p \times 1$ vector, with the jth entry being 1 and the others being 0; ϑ_1 and ϑ_2 are tuning parameters. A simple proposal for $\overline{\sigma}$ is given by

$$\overline{\sigma}^2 = K^{-1}\sum_{k=1}^{K} n_k^{-1}\|\boldsymbol{y}_k - \boldsymbol{X}_k^T\hat{\boldsymbol{\beta}}_k^d\|_2^2$$

Similarly, a simple average of the score estimators from K subsets is proposed as the divide-and-conquer score statistic.

Battey *et al.* [14] show that the limiting distribution of the divide-and-conquer estimator is asymptotically as efficient as the full sample estimator, that is,

$$\lim_{n\to\infty} \text{Var}\left(\hat{\beta}_j^{(c)}\right) / \text{Var}\left(\hat{\beta}_j^d\right) - 1 = 0$$

$j = 1, \ldots, p$. Note that the hypothesis testing method is only developed for low-dimensional parameters.

Tang *et al.* [15] utilize confidence distributions [18] to combine bias-corrected regularized estimators from subsets with the advantage that it provides a distribution estimator for various statistical inference, for example, estimation or hypothesis testing, that can be established straightforwardly. Particularly, in the setting of generalized linear model (4) with LASSO penalty, asymptotic confidence density for each subset is constructed as

$$\hat{h}_{n_k}(\boldsymbol{\beta}) \propto \exp[-(2\phi)^{-1}(\boldsymbol{\beta} - \hat{\boldsymbol{\beta}}_k^d)\{\boldsymbol{X}_k^T\boldsymbol{\Sigma}(\boldsymbol{X}_k\hat{\boldsymbol{\beta}}_k^d)\boldsymbol{X}_k\}(\boldsymbol{\beta} - \hat{\boldsymbol{\beta}}_k^d))] \tag{10}$$

where $\boldsymbol{\Sigma}(\boldsymbol{X}_k\hat{\boldsymbol{\beta}}_k)$ is the diagonal weight matrix based on the variance function of a generalized linear model as defined in Equation (7). Following Liu *et al.* [19], K confidence densities are combined to derive a combined estimator as the solution of $\hat{\boldsymbol{\beta}}^{(c)}$ as

$$\hat{\boldsymbol{\beta}}^{(c)} \stackrel{\text{d}}{=} \text{argmax}_{\boldsymbol{\beta}} \log \Pi_{k=1}^K \hat{h}_{n_k}(\boldsymbol{\beta}) \tag{11}$$

$$= \left\{\sum_{k=1}^{K}\boldsymbol{X}_k^T\boldsymbol{\Sigma}\left(\boldsymbol{X}_k\hat{\boldsymbol{\beta}}_k^d\right)\boldsymbol{X}_k\right\}^{-1}\left\{\sum_{k=1}^{K}\boldsymbol{X}_k^T\boldsymbol{\Sigma}\left(\boldsymbol{X}_k\hat{\boldsymbol{\beta}}_k^d\right)\boldsymbol{X}_k\hat{\boldsymbol{\beta}}_k^d\right\} \tag{12}$$

Tang *et al.* [15] show that the combined estimator Equation (11) is asymptotically equally efficient as the estimator using the entire data. Note that both Chen and Xie [7] and Tang *et al.* [15] have the combined estimator in the weighted average form, while Lee *et al.* [13] and Battey *et al.* [14] have simple average estimator as the combined estimator.

3.2 Marginal Proportional Hazards Model

In the setting of multivariate survival analysis, Wang *et al.* [20] apply a divide-and-combine approach in the marginal proportional hazards model [21] and the shared frailty model [22]. They use the similar combination estimator as Equation (3) with three different weight structures for $\mathbf{A}_k$: (i) the minus second derivative of the log likelihood; (ii) the inverse of variance–covariance matrix of the subset estimator; and (iii) the sample size. They prove that under mild regularity conditions, the divide-and-combine estimator is asymptotically equivalent to the full-data estimator.

Wang *et al.* [20] also proposed a confidence distribution-based [23] regularization approach for the regularized estimator by minimizing the following objective function:

$$Q(\boldsymbol{\beta}) = n\left(\boldsymbol{\beta} - \hat{\boldsymbol{\beta}}^{(c)}\right)^T \hat{\boldsymbol{\Sigma}}_c^{-1} \left(\boldsymbol{\beta} - \hat{\boldsymbol{\beta}}^{(c)}\right) + n \sum_{j=1}^{d} \lambda_j |\beta_j| \tag{13}$$

where $\lambda_1, \lambda_2, \dots, \lambda_d$ denote the tuning parameters, and $|\cdot|$ is the absolute value of a scalar. With a proper choice of $\lambda = (\lambda_1, \lambda_2, \dots, \lambda_d)^T$, the regularized estimator $\hat{\boldsymbol{\beta}}_\lambda^{(c)}$ has the selection consistency, estimation consistency, and an oracle property.

3.3 One-Step Estimator and Multiround Divide-and-Conquer

Consider the M-estimator for a parameter of interest θ obtained by maximizing empirical criterion function $m(x_i; \theta)$ of sample size n and data x_i, $i = 1, \dots, n$:

$$\hat{\theta} = \mathrm{argmax}_\theta \sum_{i=1}^{n} m(x_i; \theta)$$

When data is split into K subsets, each subset is analyzed separately to provide an estimator $\hat{\theta}_k = \mathrm{argmax}_\theta M_k(\theta)$, where $M_k(\theta) = \sum_{i=1}^{n_k} m(x_{k,i}; \theta)$ is the empirical criterion function of the kth subset with sample size n_k, for $k = 1, \dots, K$.

Shi *et al.* [24] consider the weighted average of estimators from subsets with weight depending on the subset sample size:

$$\hat{\theta}^{(c)} = \sum_{k=1}^{K} \omega_k \hat{\theta}_k = \sum_{k=1}^{K} n_k^{2/3} \hat{\theta}_k \Big/ \left(\sum_{k=1}^{K} n_k^{2/3}\right)$$

They establish the asymptotic distribution of the combined estimator and show that the combined estimator converges at a faster rate and has asymptotic normal distribution if the number of subsets diverges at a proper rate as the sample size of each subset grows.

The aforementioned divide-and-conquer approaches all have the combined estimator in the form of either simple average or weighted average. To further enhance the performance of the combined estimator or reduce the computational burden of solving K problems for complex models, one-step update approach has been developed. It basically utilizes Newton–Raphson update once or in iterations to obtain a final estimator.

For M-estimators, a simple average combined estimator is defined as

$$\hat{\theta}^{(0)} = \sum_{k=1}^{K} \hat{\theta}_k / K$$

On top of the simple average estimator for pooling, Huang and Huo [25] propose a one-step estimator $\hat{\theta}^{(1)}$ by performing a single Newton–Raphson update:

$$\hat{\theta}^{(1)} = \hat{\theta}^{(0)} - [\ddot{M}(\hat{\theta}^{(0)})]^{-1}[\dot{M}(\hat{\theta}^{(0)})] \tag{14}$$

where $M(\theta) = \sum_{k=1}^{K} M_k(\theta)$, and $\dot{M}(\theta)$ and $\ddot{M}(\theta)$ are the gradient and Hessian of $M(\theta)$, respectively. They show that the proposed one-step estimator has oracle asymptotic properties and mean-squared error of $O(n^{-1})$ under mild conditions. It is worth noting that the proposed method and results are only developed for low-dimensional cases. Numerical examples show that the one-step estimator has better performance than simple average estimators in terms of mean-square errors.

The strategy of one-step update is also used in sparse Cox regression models by Wang *et al.* [2] and quantile regression models by Chen *et al.* [1] in addition to linearization of the original optimization problem. Due to the complexity of these models, it takes a long time to solve the original problem for all subsets as well. Therefore, multiround divide-and-conquer is proposed to further reduce computational burden. The idea is that the original problem is only solved once for *one* subset, and its result is used to construct a statistic for every other subsets. Statistics from all subsets are aggregated. This divide-and-conquer process is then repeated iteratively.

Wang *et al.* [2] propose to start with a standard estimator that maximizes the partial likelihood for Cox proportional hazards model of *one* subset as an initial estimator. Then, the initial estimator is updated iteratively using all subsets linearly, in the same form of Equation (14) with corresponding matrices, to approximate the maximum partial likelihood estimator without penalty. Lastly, the final penalized estimator is obtained by applying least-square approximation to the partial likelihood function [26], given the estimator obtained in the second step. Since the maximization of partial likelihood function is only solved once on a subset in the first step, and the penalized estimator is based on linear approximation in the last step, computational time is reduced tremendously.

Chen *et al.* [1] propose a divide-and-conquer linear estimator for quantile regression (LEQR) which has a similar scheme of Wang *et al.* [2]. Using the idea of smoothing, a LEQR is developed, given a consistent initial estimator. To apply the divide-and-conquer approach, an initial estimator is calculated based on *one* subset using standard quantile regression method. Then, the corresponding weight matrices of all subsets are calculated and aggregated to update the estimator by solving a linear system. The second step is then repeated iteratively to provide a final estimator. Chen *et al.* [1] show that the divide-and-conquer LEQR achieves nearly the optimal rate of the Bahadur remainder term and achieves the same asymptotic efficiency as the estimator obtained based on the entire data set.

Jordan *et al.* [3] develop a general framework called Communication-efficient surrogate likelihood (CSL) which starts with an initial value, and gradients of the loss function are calculated for each subset at the initial value. Similarly, the loss function is simplified and linearized using Taylor expansion and gets updated from aggregated gradients from subsets. This process is repeated iteratively to provide a final result. Jordan *et al.* [3] illustrate this multiround divide-and-conquer approach in regular parametric models, high-dimensional penalized regression, and Bayesian analysis. A similar approach for penalized regression models is developed by Wang *et al.* [4] separately as well. For the multiround divide-and-conquer approach, the requirement for the number of splits or machines K is much relaxed to $K \preccurlyeq \text{poly}(n)$ in contrast to $K \ll n$ in one round divide-and-conquer approach.

The multiround divide-and-conquer by Wang *et al.* [2] and Chen *et al.* [1] relies heavily on good initiation that is already consistent due to the nature of the Newton-type methods. The framework by Jordan *et al.* [3] and Wang *et al.* [4] has no restriction on the initial value but still requires a moderate sample size of each subset. Fan *et al.* [5] improve CSL by adding a strict convex quadratic regularization to the updating step, and the regularization is adjusted according to the current solution during the iteration. This approach is called Communication-Efficient Accurate Statistical Estimators (CEASE) and can converge fast.

3.4 Performance in Nonstandard Problems

In the setting of noisy matrix recovery, Mackey *et al.* [27] propose an algorithmatic divide-factor-combine framework for large-scale matrix factorization. A matrix is partitioned into submatrices according to its rows or columns, and each submatrix can be factored using any standard factorization algorithm. Submatrix factorizations are combined to obtain a final estimate by matrix projection or spectral reconstruction approximation. In the setting of noisy matrix factorization, consider matrix $\mathbf{M} = \mathbf{L}_0 + \mathbf{S}_0 + \mathbf{Z}_0 \in R^{m*n}$, where a subset of $\mathbf{M}$ is available, $\mathbf{L}_0$ has rank $r \ll m, n$, $\mathbf{S}_0$ represents a sparse matrix of outliers of arbitrary magnitude, and $\mathbf{Z}_0$ is a dense noise matrix. Mackey *et al.* [27] show that if $\mathbf{L}_0$'s singular vector is not too sparse or too correlated ($(\mu, r) -$ *coherent* condition) and entries of $\mathbf{M}$ are observed at locations sampled uniformly without replacement, divide-factor-combine algorithms can recover $\mathbf{L}_0$ with high probability.

Banerjee *et al.* [28] study the performance of the divide-and-conquer approach in nonstandard problems where the rates of convergence are usually slower than $\sqrt{n}$ and the limit distribution is non-Gaussian, specifically in the monotone regression setting. Consider n i.i.d. observations $(y_i, x_i), i = 1, \dots, n$ from the model

$$y_i = \mu(x_i) + \epsilon_i$$

where μ is a continuous monotone (nonincreasing) function on $[0, 1]$ that is continuously differentiable with $0 < c < |\mu'(t)| < d < \infty$ for all $t \in [0, 1]$; $x_i \sim$ uniform(0, 1) and independent of ϵ_i with mean 0 and variance v^2. Let $\hat{\theta}$ denote the isotonic estimate of $\theta = \mu^{-1}(a)$ for any $a \in \mathbb{R}$. It is known that $n^{1/3}(\hat{\theta} - \theta) \to_d \tilde{\kappa} Z$, where Z is the Chernoff random variable, and $\tilde{\kappa} > 0$ is a constant.

If the entire data set is split into K subsets, and each provides an estimator $\hat{\theta}_k, k = 1, \dots, K$, Banerjee *et al.* [28] shows that the simple average combined estimator $\hat{\theta}^{(c)}$ outperforms the isotonic regression estimator using the entire data when K is a fixed integer:

$$\mathrm{E}[n^{2/3}(\hat{\theta}^{(c)} - \theta)^2] \to K^{-1/3}\, \mathrm{Var}(\tilde{\kappa} Z)$$

However, for a suitably chosen (large enough) class of models, that is, a neighborhood of μ, called $\mathcal{M}$, as the class of all continuous nonincreasing functions that coincide with μ outside of $(x_0 - \varepsilon_0, x_0 + \varepsilon_0)$ for some small $\varepsilon_0 > 0$, when $K \to \infty$,

$$\lim_{n\to\infty} \inf_{\mathcal{M}} \sup[n^{2/3}(\hat{\theta}^{(c)} - \theta)^2] = \infty$$

whereas for the estimator using the entire data set,

$$\lim_{n\to\infty} \sup_{\mathcal{M}} \sup[n^{2/3}(\hat{\theta}^{(c)} - \theta)^2] < \infty$$

It indicates that the combined estimator, that is, the simple average of estimators obtained from subsets, outperforms the estimator using the entire data set in the sense of pointwise inference under any fixed model. The combined estimator converges faster than the estimator using the entire data set and is asymptotically normal. However, for appropriately chosen classes of models, the performance of the combined estimators worsens when the number of splits increases.

4 Nonparametric and Semiparametric Models

Given a data set $\{(x_i, y_i)\}_{i=1}^n$ consisting of n i.i.d. samples drawn from an unknown distribution, the goal is to estimate the function that minimizes the mean-square error $E[(f(X) - Y)^2]$, where the expectation is taken jointly over (X, Y) pairs, and X is a univariate random variable. Consider the kernel ridge regression estimator of the optimal function $f^*(x) \stackrel{\text{d}}{=} E[Y|X = x]$:

$$\hat{f} \stackrel{\text{d}}{=} \text{argmin}_{f\in\mathcal{H}} \left\{ n^{-1} \sum_{i=1}^{n} (f(x_i) - y_i)^2 + \lambda \|f\|_{\mathcal{H}}^2 \right\} \tag{15}$$

where λ is a tuning parameter, $\mathcal{H}$ is a reproducing kernel Hilbert space which is endowed with an inner product $<\cdot,\cdot>_{\mathcal{H}}$, and $\|f\|_{\mathcal{H}} = \sqrt{<f, f>_{\mathcal{H}}}$ is the norm in $\mathcal{H}$.

Zhang *et al.* [29] propose to split the entire data set into K subsets and for each subset calculate the local kernel ridge regression estimate $\hat{f}_k$, $k = 1, \dots, K$ from Equation (15) using only data from the corresponding subsets. The combined estimate is the average of local estimates:

$$\hat{f}^{(c)} = \sum_{k=1}^{K} \hat{f}_k / K \tag{16}$$

Zhang *et al.* [29] establish the mean-squared error bounds for the combined estimate in the setting of $f^* \in \mathcal{H}$ as well as $f^* \not\subset \mathcal{H}$. They show that the combined estimate achieves the minimax rate of convergence over the underlying Hilbert space.

All approaches discussed so far in this chapter are developed in the context that homogeneous data are observed, either stored in different machines or split into subsets. In the case that the entire data is already split into subsets and heterogeneity exits in different subsets, Zhao *et al.* [30] and Wang *et al.* [31] consider partially linear models. Suppose that we have data with n observations $\{(y_i, \boldsymbol{x}_i, \boldsymbol{z}_i)\}_{i=1}^n$, there are K subpopulations, and the kth subpopulation has n_k observations: $(y_{k,i}, \boldsymbol{x}_{k,i}, \boldsymbol{z}_{k,i})$, $i = 1, \ldots, n_k$.

$$\boldsymbol{y}_k = \boldsymbol{X}_k^T \boldsymbol{\beta}_k + f(\boldsymbol{Z}_k) \tag{17}$$

where $\boldsymbol{y}_k = (y_{k,1}, \ldots, y_{k,n_k})^T$, $\boldsymbol{X}_k = (\boldsymbol{x}_{k,1}, \ldots, \boldsymbol{x}_{k,n_k})^T$, and $\boldsymbol{Z}_k = (\boldsymbol{z}_{k,1}, \ldots, \boldsymbol{z}_{k,n_k})^T$. Here, $f(\cdot)$ is common to all subpopulations. In this model, $\boldsymbol{y}_k$ depends on $\boldsymbol{X}_k$ through a linear function that may vary across subsets and depends on $\boldsymbol{Z}_k$ through a nonlinear function $f(\cdot)$ that is common to all subsets.

Wang *et al.* [31] choose $f(\boldsymbol{Z}_k) = \sum_{l=1}^L g_l(\boldsymbol{Z}_k)$, $k = 1, \ldots, K$, to be additive nonlinear functions, with $g_l(\cdot)$ as unknown smooth functions estimated by the regression spline method. Zhao *et al.* [30] use the kernel ridge regression method to estimate function f. In both approaches, $\boldsymbol{\beta}_k$ and f are estimated based on each subset providing $\hat{\boldsymbol{\beta}}_k$ and $\hat{f}_k$, $k = 1, \ldots, K$. Since $\boldsymbol{\beta}_k$ presents the heterogeneity among different subsets, no additional action is needed. Further combination is done for the commonality part by averaging to provide the final nonparametric estimate $\hat{f} = \sum_{k=1}^K \hat{f}_k / K$. Both approaches can be applied to homogeneous data as well, which can be handled with a divide-and-conquer approach.

5 Online Sequential Updating

For many divide-and-conquer approaches, it is assumed that all data are available at the same time although data may be stored in different machines or cannot be analyzed at once. However, in some applications, data may arrive in batches or in streams and exceed the capacity of a single machine for storage or analysis. The divide-and-conquer approach, generally referred as online sequential updating, can be extended to such cases.

In the case of OLS estimator Equation (1), suppose that we have the weight matrix $\boldsymbol{V}_{k-1} = \sum_{l=1}^{k-1} \boldsymbol{X}_l^T \boldsymbol{X}$ and the combined estimator $\hat{\boldsymbol{\beta}}_{k-1}^{(c)}$ available using data from subsets $l = 1, \ldots, k-1$. Once data in the kth subset come in, the online estimator can be updated [32] to

$$\hat{\boldsymbol{\beta}}_k^{(c)} = (\boldsymbol{X}_k^T \boldsymbol{X}_k + \boldsymbol{V}_{k-1})^{-1} (\boldsymbol{X}_k^T \boldsymbol{X}_k \hat{\boldsymbol{\beta}}_k + \boldsymbol{V}_{k-1} \hat{\boldsymbol{\beta}}_{k-1}^{(c)}) \tag{18}$$

where the initial values of $\hat{\boldsymbol{\beta}}_0^{(c)}$ and $\boldsymbol{V}_0$ are set to 0, and $\boldsymbol{V}_k$ is updated to $\boldsymbol{V}_k = \boldsymbol{V}_{k-1} + \boldsymbol{X}_k^T \boldsymbol{X}_k$.

Schifano *et al.* [32] also propose an online updating estimator for general EE estimators. Instead of performing Taylor expansion at the EE estimator of the kth subset $\hat{\boldsymbol{\beta}}_k$ [6], Schifano *et al.* [32] consider an intermediary estimator:

$$\tilde{\boldsymbol{\beta}}_k = \left\{\tilde{\boldsymbol{A}}_{k-1} + \boldsymbol{A}_k(\hat{\boldsymbol{\beta}}_k)\right\}^{-1} \left[\sum_{l=1}^{k-1} \boldsymbol{A}_k(\tilde{\boldsymbol{\beta}}_l)\tilde{\boldsymbol{\beta}}_k + \left\{\boldsymbol{A}_k(\hat{\boldsymbol{\beta}}_k)\right\} \hat{\boldsymbol{\beta}}_k\right]$$

where $\tilde{\boldsymbol{A}}_{k-1} = \sum_{l=1}^{k-1} \boldsymbol{A}_k(\tilde{\boldsymbol{\beta}}_l)$, with $\boldsymbol{A}_k(\boldsymbol{\beta})$ defined in Equation (2), and the initial values of $\tilde{\boldsymbol{\alpha}}_0$, $\tilde{\boldsymbol{\beta}}_0$ are set to 0. Plugging in $\tilde{\boldsymbol{\beta}}_k$ to the first-order Taylor expansion and by some algebra, one can obtain the online updating estimator as

$$\hat{\boldsymbol{\beta}}_k^{(c)} = \left\{\tilde{\boldsymbol{A}}_{k-1} + \boldsymbol{A}(\tilde{\boldsymbol{\beta}}_k)\right\}^{-1} \left\{\boldsymbol{a}_{k-1} + \boldsymbol{A}_k(\tilde{\boldsymbol{\beta}}_k)\tilde{\boldsymbol{\beta}}_k + \boldsymbol{b}_{k-1} + \boldsymbol{M}_{n_k}(\tilde{\boldsymbol{\beta}}_k)\right\}$$

where $\boldsymbol{a}_k = \sum_{l=1}^k \{\boldsymbol{A}_k(\tilde{\boldsymbol{\beta}}_k)\}\tilde{\boldsymbol{\beta}}_k = \boldsymbol{A}_k(\tilde{\boldsymbol{\beta}}_k)\tilde{\boldsymbol{\beta}}_k + \boldsymbol{a}_{k-1}$ and $\boldsymbol{b}_k = \sum_{l=1}^k M_{n_l}(\tilde{\boldsymbol{\beta}}_l) = M_{n_k}(\tilde{\boldsymbol{\beta}}_k) + \boldsymbol{b}_{k-1}$ with initial values of $\boldsymbol{a}_0 = 0$ and $\boldsymbol{b}_0 = 0$.

Wang *et al.* [33] address the online updating problem with the emergence of new variables, that is, new predictors become available midway through the data stream. Under the assumption that the true model contains these new variables, not only estimation of coefficients for newly available variables is needed, the bias for previously existing variables should be corrected as well. The bias of the existing variables for the online updating estimator $\hat{\boldsymbol{\beta}}_{k-1}^{(c)}$ up to block $k-1$ can be corrected using data in block k alone as the difference between OLS estimators with and without new variables. Then, a weighted average similar to Equation (18) is applied to update the cumulative estimator of the existing variables, with extra care of the variance of a bias term. Estimate of new variables is based on data in block k to start with. After that, updating for future blocks is a weighted average of full models.

Kong and Xia [34] consider online updating for various kernel-based nonparametric estimators. They propose weighted sum updating:

$$\hat{f}_k(\boldsymbol{x}) = (1-\alpha_k)\hat{f}_{k-1}(\boldsymbol{x}) + \alpha_k K_{h_k}(\boldsymbol{x}; \boldsymbol{X}_k)$$

where $\alpha_k \in (0,1)$ is a prespecified series of constants, and K_{h_k} is the kernel function with bandwidth h_k. Note that the bandwidth h_k is independent of the previous observed data and only depends on new data $\boldsymbol{X}_k$. They investigate the optimal choices of bandwidths and optimal choices of weights. The relative efficiencies of online estimation with regard to dimension p are also examined.

6 Splitting the Number of Covariates

Under the sparse high-dimensional setting in Section 3.1, the divide-and-conquer approach would split a data set (of size n) into subsets of smaller sample size (n_k) where each data point has all the information available, that is, response variable(s) and all explanatory variables. From a different perspective, Song and Liang [35] propose to split a high-dimensional data set into several lower dimensional subsets, each of which has the same sample size as the entire data set but only a portion of the explanatory variables. Furthermore, the explanatory variables in subsets are mutually exclusive. Once data is split, Bayesian variable selection is performed for each subset based on the marginal inclusion probability iteratively. Finally, variables selected from subsets are merged into a single set, and another Bayesian variable selection is performed on the merged data set. This procedure is named as split-and-merge (SAM).

The proposed SAM method can reduce computational cost tremendously in ultrahigh-dimensional settings where the number of explanatory variables is much larger than the sample size. This is because in the second step where the Bayesian variable selection is performed on the subsets, a great number of variables have been screened out. With extreme splitting where each subset only has one variable, SAM is similar to sure independence screening (SIS) [36]. However, unlike SIS which screens out uncorrelated explanatory variables individually, SAM utilizes joint information of all explanatory variables in a subset to filter explanatory variables, which leads to more accurate selection. Song and Liang [35] show that SAM can select true variables with nonzero coefficients correctly as the sample size becomes large.

7 Bayesian Divide-and-Conquer and Median-Based Combining

Minsker *et al.* [37, 38] propose a robust posterior distribution in Bayesian analysis which also utilizes the divide-and-conquer scheme. Let π be a prior distribution over the parameter space Θ and $\theta \in \Theta$. The entire sample is divided into K disjoint subsets $\{\boldsymbol{X}_k = (\boldsymbol{x}_{k,1}, \dots, \boldsymbol{x}_{k,n_k}), k = 1, \dots, K\}$. Suppose that $f_k(\theta|\boldsymbol{X}_k, \pi)$ is the posterior distribution depending on subset k. Minsker *et al.* [38] defines the M-posterior as

$$f^{(c)}(\theta|\boldsymbol{X}_1, \dots, \boldsymbol{X}_K, \pi) = \text{med}(f_1(\theta|\boldsymbol{X}_1, \pi), \dots, f_K(\theta|\boldsymbol{X}_k, \pi))$$

where the median is the *geometric median* defined for a probability measure μ:

$$x_* = \text{argmin}_{y \in \mathbb{Y}} \int_{\mathbb{Y}} (\|y - x\| - \|x\|)\mu(dx)$$

with $\mathbb{Y}$ be a normed space with norm $\|\cdot\|$ and μ be a probability measure on $(\mathbb{Y}, \|\cdot\|)$ equipped with Borel σ-algebra.

Due to the property of a geometric median, there exists $\alpha_1 \geq 0, \dots, \alpha_K \geq 0$ and $\sum_{k=1}^{K} \alpha_k = 1$ such that $f^{(c)}(\theta|\boldsymbol{X}_1, \dots, \boldsymbol{X}_K, \pi) = \sum_{k=1}^{K} \alpha_k f_k(\theta|\boldsymbol{X}_1, \pi)$, which leads to a weighted average of posterior distribution from subsets, and the weights depend on the norm used on probability measure space. Note that it is possible to have $\alpha_k = 1$ for one subset, and the rest of the weights are zero in which case the "median" is being selected as the combined posterior.

Minsker *et al.* [37, 38] further improve the robust posterior by replacing posterior distribution from subsets with *stochastic approximations*. The stochastic approximation can be obtained as a posterior distribution, given each data point in a subset is observed K times. Minsker *et al.* [38] show that the modified posterior yields credible sets with better coverage, but $f^{(c)}(\theta|\boldsymbol{X}_1, \dots, \boldsymbol{X}_K, \pi)$ often overestimates the uncertainty about θ. Numerical algorithms to calculate the geometric mean of probability distributions are also provided.

The "median"-based combing approach can be generalized to many other models, including non-Bayesian estimators. Minsker *et al.* [39] discuss that the averaging-based combing approach attains the optimal converging rate if the bias of each subset estimator is small enough. However, if one or more subset estimators are deviating from the norm, the combined estimator from averaging would be affected as well. Therefore, Minsker *et al.* [39] propose to use a more robust combining approach such as median or a robust M-estimator and investigate the performance of median combined estimators. They demonstrate that the median combined estimator has a much slower converging rate if subset estimators remain the standard converging rate at regular conditions unless the number of subsets K is limited and small. However, the converging rate can be improved with additional constraints when K is as large as $O(\sqrt{n})$. Detailed investigations and discussions are illustrated for the median-of-mean estimators and maximum-likelihood estimation.

Getting back to Bayesian divide-and-conquer, if the M-posterior by Minsker *et al.* [37, 38] combines posteriors from subsets through their median in the Wasserstein space of order one, Srivastava *et al.* [40, 41] combine the posteriors of subsets through the mean in the Wasserstein space of order two, which is called Wasserstein posterior. They demonstrate that the proposed posterior converges in expectation and provide numerical algorithms for computation.

Bayesian divide-and-conquer approaches include the prior distribution in each subset's inference. In many approaches, the prior is multiply counted when the inference or posterior distribution is combined. But if the prior distribution is divided into pieces as well, for example, fractional of prior $\pi(\theta)^{1/K}$ is used, it may be too weak to effectively regularize [42]. To solve this issue, Vehtari *et al.* [42] propose to use Expectation Propagation (EP) as a framework for Bayesian analysis in a distributed setting. EP is an iterative algorithm in which a target density $f(\theta)$ is approximated by a density $g(\theta)$ from some specified parametric family. The algorithm takes advantage of the natural factorization of likelihood function and the fact that the posterior distribution is proportional to the product of prior distribution and likelihood function:

$$f(\theta) \propto \prod_{k=0}^{K} f_k(\theta)$$

where $f_k(\theta)$ is the likelihood function for subset $k, k = 1, \dots, K$, and $f_0(\theta)$ is the prior distribution. Then, the iterative algorithm is applied treating the prior distribution and likelihood functions equally. Vehtari *et al.* [42] review the general EP algorithm and provide its implementation for various Bayesian models as well.

8 Real-World Applications

With the emerging of big data in different fields, the divide-and-conquer approach has a wide range of applications as demonstrated in many articles.

Advances in genetics and molecular biology have dramatically increased our ability to collect massive data such as gene expressions and structures of chemical compounds. Questions such as relationships between phenotypes and candidate genes and screening of chemical compounds often arise. Milanzi *et al.* [43] quantified expert opinions to assess 22 015 clusters of chemical compounds to identify those for further screening and development. Meng *et al.* [44] analyzed an Illumina HiSeq data set downloaded from the Cancer Genome Atlas (TCGA) Program (http://cancergenome.nih.gov) for 59 cancer patients with 20 529 genes using linear regression models. Song and Liang [35] illustrated the Bayesian SAM methods in a metabolic quantitative trait loci experiment, which links SNPs data to metabolomics data as well as a polymerase chain reaction data set which contains 60 mouse samples of 22 575 genes' expression levels.

Divide-and-conquer approach has also been applied in social sciences and civil applications such as the General Society Survey (GSS) (http//gss.norc.org), which has collected responses about evolution and the growing complexity of American society since 1972 with approximately 28 000 respondents [38]; the airline on-time performance data from the 2009 ASA Data Expo that includes flight arrival and departure details for all commercial flights within the United States from October 1987 to April 2008 [32, 45]; and manifest data, which is compiled from custom forms submitted by merchants or shipping companies from the US custom offices and the Department of Homeland Security (DHS) [7].

Online recommendation services of advertisements or news articles have received extensive attentions, and massive data can be easily collected via internet. Different large-scale advertisement data sets have been studied using the divided-and-conquer approach, for example, a public advertisement data set released by Criteo, which has 15 million instances with a binary outcome [46] and a Yahoo! Today Module user click log data sets with 45 811 883 user visits to the Today Module during the first 10 days in May 2009 [24].

Geographical and climate problems always involve big data as well. Guhaniyogi *et al.* [47] considered the problem of capturing the spatial trends and characterizing the uncertainties in the sea surface temperature data in the west coast of mainland United States, Canada, and Alaska from NODC World Ocean Database (http://www.nodc.noaa.gov/OC5/WOD/pr_wod.html). Liang *et al.* [48] analyzed more than 100 year data from the National Climatic Data Center from 1895 to 1997 (http://www.image.ucar.edu/GSP/Data/US.monthly.met).

Several publicly available movie scoring and music prediction data sets have been analyzed with divide-and-conquer approaches. Tang *et al.* [46] examined the MovieLens Data, which is a popular public movie rating data set containing 20 000 263 movie ratings by 138 493 users of 27 278 movies from 1995 to 2015. Meng *et al.* [44] and Zhang *et al.* [29] applied the divide-and-conquer approach to the Million Song Dataset (http://labrosa.ee.columbia.edu/millionsong/), which contains 515 345 songs with their years of release as the response.

9 Discussion

The divide-and-conquer approach is a general framework, and it has been implemented in various models. Theoretical and numerical results demonstrate that the divide-and-conquer approach works well for big data sets. In many models where a simple average or weighted average is used, the combined results show the same efficiency as the results obtained by analyzing the entire data set altogether. In more complex models such as Cox regression models, even the divide-and-conquer approach may not reduce the computational burden and time enough for practical use. An enhanced divide-and-conquer approach which includes linearization of original problem and one-step update strategy is utilized and demonstrates excellent performance. This is further extended to a multiround divide-and-conquer framework. In addition, the combining step can be viewed as an optimization problem for certain loss function with regard to inferences from subsets. When a nondifferentiable loss function is used, it can lead to median-based combining approaches.

One big challenge for the divide-and-conquer approach is how to choose K, the number of subsets. The choice of K has been discussed in different models, and the requirement of K depends on the model as well as the rate of the number of parameters. Several authors, for example, Tang *et al.* [15], provide practical suggestions on the selection of K. However, a universal investigation and guidance would further improve the understanding and implementation of the divide-and-conquer approach. The multiround divide-and-conquer framework relaxes the requirement on the number of subsets K, which can be at the same order of the total sample size n. Though the computational time can increase with the number of iterations, Jordan *et al.* [3] show that $O(\log n/\log(n/K))$ iterations would be sufficient.

Acknowledgment

The authors wish to thank the editor and reviewer for their constructive comments and suggestions. The work is supported in part by US NSF grants DMS1737857, DMS1812048, DMS2015373, and DMS2027855.

References

1 Chen, X., Liu, W., and Zhang, Y. (2019) Quantile regression under memory constraint. *Ann. Statist.*, 47 (6), 3244–3273.

2 Wang, Y., Hong, C., Palmer, N. *et al.* (2021) A fast divide-and-conquer sparse cox regression. *Biostatistics*, 22 (2), 381–401.

3 Jordan, M.I., Lee, J.D., and Yang, Y. (2019) Communication-efficient distributed statistical inference. *J. Am. Stat. Assoc.*, 114 (526), 668–681.

4 Wang, J., Kolar, M., Srebro, N., and Zhang, T. (2017) *Efficient Distributed Learning with Sparsity*. Proceedings of the 34th International Conference on Machine Learning-Volume 70. JMLR. org, pp.3636–3645.

5 Fan, J., Guo, Y., and Wang, K. (2019) Communication-efficient accurate statistical estimation. *arXiv preprint arXiv:1906.04870*.

6 Lin, N. and Xi, R. (2011) Aggregated estimating equation estimation. *Stat. Interface*, 4 (1), 73–83.

7 Chen, X. and Xie, M.-g. (2014) A split-and-conquer approach for analysis of extraordinarily large data. *Stat. Sin.*, 24 (4), 1655–1684.

8 Tibshirani, R. (1996) Regression shrinkage and selection via the lasso. *J. R. Stat. Soc. Ser. B (Methodol.)*, 58 (1), 267–288.

9 Chen, S., Donoho, D., and Saunders, M. (2001) Atomic decomposition by basis pursuit. *SIAM Rev.*, 43, 129–159.

10 Efron, B., Hastie, T., Johnstone, I., and Tibshirani, R. (2004) Least angle regression. *Ann. Stat.*, 32 (2), 407–451.

11 Fan, J. and Li, R. (2001) Variable selection via nonconcave penalized likelihood and its oracle properties. *J. Am. Stat. Assoc.*, 96 (456), 1348–1360.

12 Zhang, C. (2010) Nearly unbiased variable selection under minimax concave penalty. *Ann. Stat.*, 38 (2), 894–942.

13 Lee, J.D., Liu, Q., Sun, Y., and Taylor, J.E. (2017) Communication-efficient sparse regression. *J. Mach. Learn. Res.*, 18 (1), 115–144.

14 Battey, H., Fan, J., Liu, H. *et al.* (2015) Distributed estimation and inference with statistical guarantees. *arXiv preprint arXiv:1509.05457*.

15 Tang, L., Zhou, L., and Song, P.X.-K. (2016) Method of divide-and-combine in regularised generalised linear models for big data. *arXiv preprint arXiv:1611.06208*.

16 Javanmard, A. and Montanari, A. (2014) Confidence intervals and hypothesis testing for high-dimensional regression. *J. Mach. Learn. Res.*, 15 (1), 2869–2909.

17 Van de Geer, S., Bühlmann, P., Ritov, Y., and Dezeure, R. (2014) On asymptotically optimal confidence regions and tests for high-dimensional models. *Ann. Stat.*, 42 (3), 1166–1202.

18 Xie, M., Singh, K., and Strawderman, W.E. (2011) Confidence distributions and a unifying framework for meta-analysis. *J. Am. Stat. Assoc.*, 106 (493), 320–333.

19 Liu, D., Liu, R.Y., and Xie, M. (2015) Multivariate meta-analysis of heterogeneous studies using only summary statistics: efficiency and robustness. *J. Am. Stat. Assoc.*, 110 (509), 326–340.

20 Wang, W., Lu, S.-E., Cheng, J. *et al.* (2020) Multivariate survival analysis in big data: a divide-and-combine approach. *Biometrics*. doi: 10.1111/biom.13469.

21 Spiekerman, C.F. and Lin, D. (1998) Marginal regression models for multivariate failure time data. *J. Am. Stat. Assoc.*, 93 (443), 1164–1175.

22 Gorfine, M., Zucker, D.M., and Hsu, L. (2006) Prospective survival analysis with a general semiparametric shared frailty model: a pseudo full likelihood approach. *Biometrika*, 93 (3), 735–741.

23 Xie, M.-G. and Singh, K. (2013) Confidence distribution, the frequentist distribution estimator of a parameter: a review. *Int. Stat. Rev.*, 81 (1), 3–39.

24 Shi, C., Lu, W., and Song, R. (2018) A massive data framework for m-estimators with cubic-rate. *J. Am. Stat. Assoc.*, 113 (524), 1698–1709.

25 Huang, C. and Huo, X. (2019) A distributed one-step estimator. *Math. Program.*, 174 (1), 41–76.

26 Wang, H. and Leng, C. (2007) Unified lasso estimation by least squares approximation. *J. Am. Stat. Assoc.*, 102 (479), 1039–1048.

27 Mackey, L., Talwalkar, A., and Jordan, M.I. (2011) *Divide-and-Conquer Matrix Factorization*. Advances in neural information processing systems, vol. 24.

28 Banerjee, M., Durot, C., Sen, B. *et al.* (2019) Divide and conquer in nonstandard problems and the super-efficiency phenomenon. *Ann. Stat.*, 47 (2), 720–757.

29 Zhang, Y., Duchi, J., and Wainwright, M. (2015) Divide and conquer kernel ridge regression: a distributed algorithm with minimax optimal rates. *J. Mach. Learn. Res.*, 16 (1), 3299–3340.

30 Zhao, T., Cheng, G., and Liu, H. (2016) A partially linear framework for massive heterogeneous data. *Ann. Stat.*, 44 (4), 1400.

31 Wang, B., Fang, Y., Lian, H., and Liang, H. (2019) Additive partially linear models for massive heterogeneous data. *Electron. J. Stat.*, 13 (1), 391–431.

32 Schifano, E.D., Wu, J., Wang, C. *et al.* (2016) Online updating of statistical inference in the big data setting. *Technometrics*, 58 (3), 393–403.

33 Wang, C., Chen, M.-H., Wu, J. *et al.* (2018) Online updating method with new variables for big data streams. *Can. J. Stat.*, 46 (1), 123–146.

34 Kong, E. and Xia, Y. (2019) On the efficiency of online approach to nonparametric smoothing of big data. *Stat. Sin.*, 29 (1), 185–201.

35 Song, Q. and Liang, F. (2015) A split-and-merge Bayesian variable selection approach for ultrahigh dimensional regression. *J. R. Stat. Soc.: Ser. B (Stat. Methodol.)*, 77 (5), 947–972.

36 Fan, J. and Lv, J. (2008) Sure independence screening for ultrahigh dimensional feature space. *J. R. Stat. Soc.: Ser. B (Stat. Methodol.)*, 70 (5), 849–911.

37 Minsker, S., Srivastava, S., Lin, L., and Dunson, D. (2014) *Scalable and Robust Bayesian Inference Via the Median Posterior*. International Conference on Machine Learning, pp.1656–1664.

38 Minsker, S., Srivastava, S., Lin, L., and Dunson, D.B. (2017) Robust and scalable bayes via a median of subset posterior measures. *J. Mach. Learn. Res.*, 18 (1), 4488–4527.

39 Minsker, S. and Strawn, N. (2019) Distributed statistical estimation and rates of convergence in normal approximation. *Electron. J. Stat.*, 13 (2), 5213–5252.

40 Srivastava, S., Cevher, V., Dinh, Q., and Dunson, D. (2015) *Wasp: Scalable Bayes Via Barycenters of Subset Posteriors*. Artificial Intelligence and Statistics, pp. 912–920.

41 Srivastava, S., Li, C., and Dunson, D.B. (2018) Scalable bayes via barycenter in Wasserstein space. *J. Mach. Learn. Res.*, 19 (1), 312–346.

42 Vehtari, A., Gelman, A., Sivula, T. *et al.* (2020) Expectation propagation as a way of life: a framework for Bayesian inference on partitioned data. *J. Mach. Learn. Res.*, 21 (17), 1–53.

43 Milanzi, E., Alonso, A., Buyck, C. *et al.* (2014) A permutational-splitting sample procedure to quantify expert opinion on clusters of chemical compounds using high-dimensional data. *Ann. Appl. Stat.*, 8 (4), 2319–2335.

44 Meng, C., Wang, Y., Zhang, X. *et al.* (2017) Effective statistical methods for big data analytics, in *Handbook of Research on Applied Cybernetics and Systems Science* (eds S. Saha, A. Mandal, A. Narasimhamurthy, V. Sarasvathi, and S. Sangam), IGI Global, Hershey, PA, pp. 280–299.

45 Wang, C., Chen, M.-H., Schifano, E. *et al.* (2016) Statistical methods and computing for big data. *Stat. Interface*, 9 (4), 399.

46 Tang, L., Chaudhuri, S., Bagherjeiran, A., and Zhou, L. (2018) *Learning Large Scale Ordinal Ranking Model Via Divide-and-Conquer Technique*. Companion Proceedings of the Web Conference 2018, pp.1901–1909.

47 Guhaniyogi, R., Li, C., Savitsky, T.D., and Srivastava, S.(2017) A divide-and-conquer Bayesian approach to large-scale kriging. *arXiv preprint arXiv:1712.09767*.

48 Liang, F., Cheng, Y., Song, Q. *et al.* (2013) A resampling-based stochastic approximation method for analysis of large geostatistical data. *J. Am. Stat. Assoc.*, 108 (501), 325–339.

31

Bayesian Aggregation

Yuling Yao[1,2]

[1]*Columbia University, New York, NY, USA*
[2]*Center for Computational Mathematics, Flatiron Institute, New York, NY, USA*

1 From Model Selection to Model Combination

Bayesian inference provides a coherent workflow for data analysis, parameter estimation, outcome prediction, and uncertainty quantification. However, the model uncertainty is not automatically calibrated; the posterior distribution is always conditioning on the model we use in which the true data-generating mechanism is almost never included. No matter if viewed from the perspective of a group of modelers holding different subjective beliefs, or a single modeler revising belief models through the routine of model check and criticism, or the need of expanding plausible models for flexibility and expressiveness, it is common in practice to obtain a range of possible belief models.

In Section 1.1, we review the Bayesian decision theory through which the model comparison, model selection, and model combination are viewed in a unified framework. The estimation of the expected utility depends crucially on how the true data-generating process is modeled and is described by different $\mathcal{M}$-views in Section 1.2. We compare Bayesian model averaging (BMA) and leave-one-out (LOO)-based Bayesian stacking in Section 2, which corresponds to the $\mathcal{M}$-closed and $\mathcal{M}$-open views, respectively. To explain why these methods work, we discuss related asymptotic theories in Section 3. In Section 4, we investigate the computation efficiency and demonstrate an importance-sampling-based implementation in `Stan` and `R` package `loo`. We also consider several generalizations in non-iid data. The outline of the concepts is illustrated in Figure 1.

1.1 The Bayesian Decision Framework for Model Assessment

We denote $D = \{(y_1, x_1), \ldots, (y_n, x_n)\}$ a sequence of observed outcomes $y \in \mathcal{Y}$ and covariates $x \in \mathcal{X}$. The unobserved future observations are $(\tilde{x}, \tilde{y})$. In a predictive paradigm [1, 2], the statistical inference should be inference on observable quantities such as the future observation $\tilde{y}$, where Bayesian decision theory gives a natural framework for the prediction evaluation. Therefore, we can view model comparison, model selection, and model combination as formal Bayesian decision problems. At a higher level, whether to make a single model selection or model combination is part of the decision.

Computational Statistics in Data Science.
Edited by Walter W. Piegorsch, Richard A. Levine, Hao Helen Zhang and Thomas C. M. Lee.
 ISBN 978-1-11956107-1

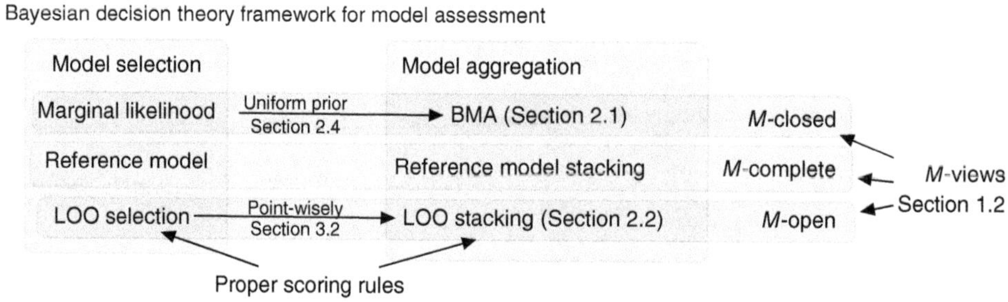

Figure 1 The organization and connections of concepts in this chapter.

Given a model M with its parameter vector θ, we compute the posterior predictive density $p(\tilde{y}|y, M) = \int p(\tilde{y}|\theta, M)p(\theta|y, M)d\theta$, where we have suppressed the dependence on x for brevity. To evaluate how close the prediction is to the *truth*, we construct the utility function of the predictive performance through *scoring rules*. In general, conditioning on $\tilde{x}$, the unobserved future outcome $\tilde{y}$ is the random variable in sample space $(\mathcal{Y}, \mathcal{A})$. $\mathcal{P}$ is a convex class of probability measure on $\mathcal{Y}$. Any member of $\mathcal{P}$ is called a *probabilistic forecast*. A scoring rule [3] is a function $S : \mathcal{P} \times \mathcal{Y} \to [\infty, \infty]$ such that $S(P, \cdot)$ is $\mathcal{P}$-quasi-integrable for all $P \in \mathcal{P}$. In the continuous case, every distribution $P \in \mathcal{P}$ is identified with its density function p.

For two probability measures P and Q, we write $S(P, Q) = \int_{\mathcal{Y}} S(P, \omega)dQ(\omega)$. A scoring rule S is called *proper* if $S(Q, Q) \geq S(P, Q)$ and *strictly proper* if equality holds only when $P = Q$ almost surely. A proper scoring rule defines the divergence $d : \mathcal{P} \times \mathcal{P} \to [0, \infty)$ as $d(P, Q) = S(Q, Q) - S(P, Q)$. For continuous variables, some popularly used scoring rules include:

- *Quadratic score*: $\text{QS}(p, \tilde{y}) = 2p(\tilde{y}) - ||p||_2^2$ with the divergence $d(p, q) = ||p - q||_2^2$.
- *Logarithmic score*: $\text{LogS}(p, \tilde{y}) = \log p(\tilde{y})$ with $d(p, q) = \text{KL}(q, p)$. The logarithmic score is the only proper local score assuming regularity conditions.
- *Continuous-ranked probability score*: $\text{CRPS}(F, \tilde{y}) = -\int_{\text{IR}} (F(\tilde{y}') - 1(y' \geq \tilde{y}))^2 dy'$ with $d(F, G) = \int_{\text{IR}} (F(\tilde{y}) - G(\tilde{y}))^2 d\tilde{y}$, where F and G are the corresponding distribution functions.
- *Energy score*: $\text{ES}(P, y) = \frac{1}{2}\mathbb{E}_P||Y - Y'||_2^\beta - \mathbb{E}_P||Y - y||_2^\beta$, where Y and Y' are two independent random variables from distribution P. When $\beta = 2$, this becomes $\text{ES}(P, \tilde{y}) = -||\mathbb{E}_P(\tilde{y}) - \tilde{y}||^2$. The energy score is strictly proper when $\beta \in (0, 2)$ but not when $\beta = 2$.
- *Scoring rules depending on first and second moments*: Examples include $S(P, \tilde{y}) = -\log\det(\Sigma_P) - (\tilde{y} - \mu_P)^T\Sigma_P^{-1}(y - \mu_P)$, where μ_P and Σ_P are the mean vector and covariance matrix of distribution P.

In such framework, the expected utility for any posterior predictive distribution $p(\cdot)$ is

$$\mathbb{E}_{\tilde{y}}S(p(\cdot), \tilde{y}) = \int_{\mathcal{Y}} S(p, \tilde{y})p_t(\tilde{y}|y)d\tilde{y} \tag{1}$$

where $p_t(\tilde{y}|y)$ is the unknown true data generating density of outcomes $\tilde{y}$, given current observations.

With the widely used logarithm score, the expected log predictive density (elpd) of model M is

$$\text{elpd} = \int_y \log p(\tilde{y}|y, M) p_t(\tilde{y}|y) d\tilde{y} \tag{2}$$

The general decision problem is an optimization problem that maximizes the expected utility within some decision space $\mathcal{P}$: $p^{\text{opt}} = \arg\max_{p \in \mathcal{P}} \int S(p, \tilde{y}) dp_t(\tilde{y})$. *Model selection* can be viewed as a subdecision space of *model combination* by restricting model weights to have only one nonzero entry. In such sense, model selection may be unstable and wasteful of information.

The expected scoring rule (Equation 1) depends on the generating process of $\tilde{y}$, which is unknown in the first place. How we estimate such expectation depends on how we view the relation between belief models and the true generating process, that is, three $\mathcal{M}$-views.

1.2 Remodeling: $\mathcal{M}$-Closed, $\mathcal{M}$-Complete, and $\mathcal{M}$-Open Views

Bernardo and Smith [1] classified model comparison problems into three categories: $\mathcal{M}$-closed, $\mathcal{M}$-complete, and $\mathcal{M}$-open.

- In $\mathcal{M}$-closed problems, the true data-generating process can be expressed by one of $M_k \in \mathcal{M}$, although it is unknown to researchers.
- $\mathcal{M}$-complete refers to the situation where the true model exists and is out of model list $\mathcal{M}$. But we still wish to use a model M^* because of tractability of computations or communication of results, compared with the actual belief model.
- The $\mathcal{M}$-open perspective acknowledges that the true model is not in $\mathcal{M}$, and we cannot specify the explicit form $p(\tilde{y}|y)$ because it is too difficult conceptually or computationally, we lack time to do so, or do not have the expertise, and so on.

Computing the integral (Equation 1) requires a model for $\tilde{y}$. The inference and model assessment can have different model assumptions, akin to the distinction between estimation and hypothesis testing in frequentist statistics. For $\mathcal{M}$-closed and $\mathcal{M}$-complete problems, we specify a belief model M^* that we believe to be or well approximate the data-generating process, and we describe all uncertainties related to future data in the belief model M^* through $p(\tilde{y}|y, M^*)$. The expected utility of any prediction Q is estimated by

$$\mathbb{E}_{\tilde{y}} S(Q, \tilde{y}) \approx \int_y S(Q, \tilde{y}) p(\tilde{y}|y, M^*) d\tilde{y} \tag{3}$$

$\mathcal{M}$-closed and $\mathcal{M}$-complete are a simplification of reality. No matter how flexible the belief model M^* is, there is little reason to believe it reflects the *truth*, unless in rare situations such as computer simulations. Although such simplification is sometimes useful, the stronger assumption may also result in an unverifiable and irretrievably bias in Equation (1), which will further lead to an undesired performance in model aggregation.

In $\mathcal{M}$-open problems, we still rely on models in $\mathcal{M}$ in inference and prediction. But we make minimal assumptions in the model assessment phase. Cross-validation is a widely used strategy to this end, where we reuse samples $y_1, \ldots, y_n$ as pseudo-Monte Carlo draws from the true data-generating process without having to model it explicitly.

For example, the LOO predictive density of a model M is a consistent estimation of Equation (2).

$$\text{elpd}_{\text{loo}} = \frac{1}{n}\sum_{i=1}^{n} \log p(y_i|y_{-i}, M) = \frac{1}{n}\sum_{i=1}^{n} \int \log p(y_i|\theta, M)p(\theta|M, y_1, \dots, y_{i-1}, y_{i+1}, \dots, y_n)d\theta$$

2 From Bayesian Model Averaging to Bayesian Stacking

We have a series of models $\mathcal{M} = \{M_1, \dots, M_K\}$, each having parameter vectors $\theta_k \in \Theta_k$. In general, θ_k can have different dimensions and interpretations, and some may be infinite dimensional too. We denote the likelihood and prior in the kth model by $p(y|\theta_k)$ and $p(\theta_k|M_k)$. The goal is to aggregate all component predictive distributions $\{p(\tilde{y}|y, M), M \in \mathcal{M}\}$. Adopting different $\mathcal{M}$-views, we solve the problem by various methods as follows.

2.1 $\mathcal{M}$-Closed: Bayesian Model Averaging

BMA assigns a prior both to the model space $p(M_k)$ and parameters $p(\theta_k|M_k)$. Through Bayes rule, the posterior probability of model k is proportional to the product of its prior and marginal likelihood,

$$p(M_k|y) = \frac{p(y|M_k)p(M_k)}{\sum_{k'=1}^{K} p(y|M_{k'})p(M_{k'})}$$

In particular, the aggregated posterior predictive distribution of new data $\tilde{y}$ is estimated by

$$p_{\text{BMA}}(\tilde{y}|y) = \sum_{k=1}^{K} p(\tilde{y}|M_k, y)p(M_k|y)$$

In $\mathcal{M}$-closed cases, BMA is optimal if the method is evaluated based on its frequency properties assessed over the joint prior distribution of the models and their internal parameters [4, 5]. In $\mathcal{M}$-open and $\mathcal{M}$-complete cases, BMA almost always asymptotically *selects* the one single model on the list that is closest in Kullback–Leibler (KL) divergence, compromising the extra expressiveness of model aggregation.

Furthermore, BMA is contingent on the marginal likelihood $p(y|M_k) = \int p(y|\theta_k) p(\theta_k|M_k)d\theta_k$, which will be sensitive to the prior $p(\theta_k|M_k)$. A correct specification of the model (an $\mathcal{M}$-closed view) is stronger than the asymptotic convergence to truth in some model as it also requires the prior to be correctly chosen in terms of reflecting the actual population distribution of the underlying parameter. For example, consider observations $y_1, \dots, y_n$ generated from $y \sim N(0, 0.1^2)$ and a normal–normal model: $y \sim N(\mu, 1)$ with a prior $\mu \sim N(0, 10^2)$. Such prior is effectively flat on the range of observed y. However, a change of prior to $\mu \sim N(0, 100^2)$ or $N(0,1000^2)$ would divide the marginal likelihood, and thereby the posterior probability, by roughly a factor of 10 or 100.

2.2 $\mathcal{M}$-Open: Stacking

Stacking is originated from machine learning for the purpose of pooling point estimates from multiple regression models [6–8]. Clyde and Iversen [9], Le and Clarke [10], and Yao *et al.* [11] develop and extend its Bayesian interpretation.

The ultimate goal of stacking a set of K predictive distributions built from the model list $\mathcal{M} = (M_1, \ldots, M_K)$ is to find the predictive distribution with the form of a linear pooling $C = \{\sum_{k=1}^{K} w_k p(\cdot|M_k) : \sum_k w_k = 1, w_k \geq 0\}$ that is optimal according to a specified utility. The decision to make is the model weights w, which has to be a length-K simplex $w \in \mathbb{S}_1^K = \{w \in [0,1]^K : \sum_{k=1}^{K} w_k = 1\}$. Given a scoring rule S, or equivalently the divergence d, the optimal stacking weight should solve

$$\max_{w \in \mathbb{S}_1^K} S\left(\sum_{k=1}^{K} w_k p(\cdot|y, M_k), p_t(\cdot|y)\right) \quad \text{or equivalently} \quad \min_{w \in \mathbb{S}_1^K} d\left(\sum_{k=1}^{K} w_k p(\cdot|y, M_k), p_t(\cdot|y)\right) \tag{4}$$

where $p(\tilde{y}|y, M_k)$ is the predictive density of new data $\tilde{y}$ in model M_k that has been trained on observed data y, and $p_t(\tilde{y}|y)$ refers to the true distribution.

With an $\mathcal{M}$-open view, we empirically estimate the optimal stacking weight in Equation (4) by replacing the full predictive distribution $p(\tilde{y}|y, M_k)$ evaluated at a new data point $\tilde{y}$ with the corresponding LOO predictive distribution $\hat{p}_{k,-i}(y_i) = \int p(y_i|\theta_k, M_k) p(\theta_k|y_{-i}, M_k) d\theta_k$.

Therefore, it suffices to solve the following optimization problem:

$$\hat{w}^{\text{stacking}} = \max_{w \in \mathbb{S}_1^K} \frac{1}{n} \sum_{i=1}^{n} S\left(\sum_{k=1}^{K} w_k \hat{p}_{k,-i}, y_i\right) \tag{5}$$

The aggregated predictive distribution on new data $\tilde{y}$ is $p_{\text{stacking}}(\tilde{y}|y) = \sum_{k=1}^{K} \hat{w}_k^{\text{stacking}} p(\tilde{y}|y, M_k)$.

In terms of Vehtari and Ojanen [2], (Section 3.3), stacking predictive distributions (Equation 5) is the M^*-optimal projection of the information in the actual belief model M^* to $\hat{w}$, where explicit specification of M^* is avoided by reusing data as a proxy for the predictive distribution of the actual belief model, and the weights w_k are the free parameters.

2.2.1 Choice of utility

The choice of the scoring rule should depend on the underlying application and researchers' interest. Generally, we recommend logarithmic score because (i) log score is the only proper local scoring rule and (ii) the easy interpretation of the underlying KL divergence. When using logarithmic score, we name Equation (5) as *stacking of predictive distributions*:

$$\max_{w \in \mathbb{S}_1^K} \frac{1}{n} \sum_{i=1}^{n} \log \sum_{k=1}^{K} w_k p(y_i|y_{-i}, M_k) \tag{6}$$

2.3 $\mathcal{M}$-Complete: Reference-Model Stacking

It is possible to replace cross-validation with a nonparametric reference model M^*. Plugging it into Equation (3), we compute the expected utility and further optimize over stacking weights, which we call reference-model stacking. We can stack either the component models $p(\tilde{y}|M_k)$ or the projected component models using a projection predictive approach which projects the information from the reference model to the restricted models [12]. However, in general, it is challenging to construct a useful reference model, as then there is probably no need for model averaging.

2.4 The Connection between BMA and Stacking

BMA, and more generally marginal likelihood-based model evaluation, can also be viewed as a special case of the utility-based model assessment.

First, under an $\mathcal{M}$-closed view, we believe the data is generated from one of the model $M^* \in \mathcal{M}$ in the candidate model list. We consider a zero-one utility by an indicator function of whether the model has been specified correctly:

$$u(M^*, M_k) = \mathbb{1}(M^* = M_k) \tag{7}$$

Then, the expected utility M_k is $\int \mathbb{1}(M^* = M_k)p(M^*|y)\mathrm{d}M^* = p(M_k|y)$, which is exactly the posterior model probability $p(M_k|y)$ in BMA. Hence, the decision-maker will pick the model with the largest posterior probability, which is equivalent to the approach of the Bayes factor. Interestingly, the model with the largest BMA weight is also the model to be selected under the zero-one utility, whereas in general, the model with the largest stacking weight is not necessarily single-model-selection optimal (see discussion in Section 3.3).

Second, under the $\mathcal{M}$-closed view, the information about unknownness is contained in the posterior distribution $p(M_k, \theta_k|y)$, and the actual beliefs about the future observations are described by the BMA predictive distribution. Using Equations (3) and (4), stacking over the logarithmic score reads

$$\max_{w \in \mathbb{S}_1^K} \int_{\mathcal{Y}} \log\left(\sum_{k'=1}^{K} w_{k'} p(\tilde{y}|M_{k'}, y)\right) \sum_{k=1}^{K} p(M_k|y)p(\tilde{y}|M_k, y)\mathrm{d}\tilde{y}$$

whose optimal solution is always the same as the BMA weight $w_k^{\text{opt}} = p(M_k|y)$, as the logarithmic score is strictly proper.

In practice, it is nearly impossible to either come up with an exhaustive list of possible candidate models that encompasses the true data-generating process or to formulate the true prior that reflects the population. It is not surprising that stacking typically outperforms BMA in various prediction tasks (see extensive simulations in Yao *et al.* [11], Clarke [13]). Notably, in the large sample limit, BMA assigns weight 1 to the closest model to the true data-generating process measured in KL divergence, regardless of how close other slightly more wrong models are. It effectively becomes model selection and yields practically spurious and overconfident results [14] in $\mathcal{M}$-open problems.

2.5 Hierarchical Stacking

Model averaging is more likely to be useful when candidate models are more dissimilar – different models perform better or worse in different subsets of data. This suggests that we can further improve the aggregated prediction by identifying which model can apply to which part of data, so that model averaging is a step toward model improvement rather than an end to itself.

Hierarchical stacking [15] allows the model weight w to vary by input covariate x, such that at any input location $\tilde{x} \in \mathcal{X}$, the "local" model weight $w(\tilde{x})$ is a length-K simplex vector. The aggregated conditional prediction becomes $p(\tilde{y}|\tilde{x}, w) = \sum_{k=1}^{K} \hat{w}_k(\tilde{x})p(\tilde{y}|\tilde{x}, M_k)$.

For example, if x is discrete and takes on J different values in the data, we need to construct a $J \times K$ matrix of weights such that $w(x = j) = w_{jk}$, which can be mapped to an unconstrained weight space $\alpha_{jk} \in \mathbb{R}^{J(K-1)}$ via softmax:

$$w_{jk} = \frac{\exp(\alpha_{jk})}{\sum_{k=1}^{K} \exp(\alpha_{jk})}, \; 1 \le k \le K-1, \; 1 \le j \le J; \quad \alpha_{jK} = 0, \; 1 \le j \le J$$

Because of the larger decision space, separately solving stacking (Equation 5) for all j leads to large variance. To partially pool the local weights across x, we can use a hierarchical prior conditional on hyperparameters $\mu \in \mathbb{R}^{K-1}$ and $\sigma \in \mathbb{R}_+^{K-1}$,

$$\begin{aligned} \text{prior}: \quad & \alpha_{jk} \mid \mu_k, \sigma_k \sim \text{normal}(\mu_k, \sigma_k), \; k = 1, \dots, K-1, \; j = 1, \dots, J \\ \text{hyperprior}: \quad & \mu_k \sim \text{normal}(\mu_0, \tau_\mu), \quad \sigma_k \sim \text{normal}^+(0, \tau_\sigma), \quad k = 1, \dots, K-1 \end{aligned}$$

Hierarchical stacking then folds the model averaging task into a hierarchical Bayesian inference problem. Up to a normalization constant, the log joint posterior density of all free parameters $\alpha \in \mathbb{R}^{J \times K}$, $\mu \in \mathbb{R}^{K-1}$, $\sigma \in \mathbb{R}_+^{K-1}$ is defined by

$$\begin{aligned} \log p(\alpha, \mu, \sigma | \mathcal{D}) = & \sum_{i=1}^{n} \log \left(\sum_{k=1}^{K} w_k(x_i) \hat{p}_{k,-i}(y_i) \right) + \sum_{k=1}^{K-1} \sum_{j=1}^{J} \log p^{\text{prior}}(\alpha_{jk} | \mu_k, \sigma_k) \\ & + \sum_{k=1}^{K-1} \log p^{\text{hyper}}_{\text{prior}}(\mu_k, \sigma_k) \end{aligned}$$

This formulation generalizes log-score-stacking (Equation 5), as the latter method equals the maximum-a-posteriori (MAP) solution of hierarchical stacking when all $\sigma_k = 0$. Yao *et al.* [15] discuss other extensions of hierarchical stacking, including regression for continuous predictors, nonexchangeable models for nested or crossed grouping factors, and nonparametric priors.

2.6 Other Related Methods and Generalizations

The aforementioned methods have multiple variants.

When the marginal likelihood in BMA is hard to evaluate, it can be approximated by information criterion. In pseudo-Bayes factors [16, 17], we replace the marginal likelihoods $p(y|M_k)$ by a product of Bayesian LOO cross-validation predictive densities $\prod_{i=1}^{n} p(y_i|y_{-i}, M_k)$. Yao *et al.* [11] propose another information criterion-based weighting scheme named pseudo-BMA weighting. The weight for model k is proportional to the exponential of the model's estimated elpd: $w_k \propto \exp(\widehat{\text{elpd}}^{k}_{\text{loo}})$. Alternatively, such quantity can be estimated using a nonparametric reference model in $\mathcal{M}$-complete views [18]. We may further take into account the sampling variance in cross-validation and average over weights in multiple Bayesian bootstrap resamples [11]. The information criterion weighting is computationally easier but should only be viewed as an approximation to the more desired stacking weights.

We may combine the cross-validation and BMA. Intrinsic Bayesian model averaging (iBMA) [19] enables improper prior, which is not allowed in BMA. It first partitions samples into a small training set $y(l)$ and remaining $y(-l)$ and replaces the marginal likelihood by partial likelihood $\int p(y(-l)|M_k, \theta_k) p(\theta_k|y(l), M_k) d\theta$. The final weight is the average across some or all possible training samples. An alternative is to avoid averaging over all subsets and use the fractional Bayes factor [20]. iBMA is more robust for models with vague priors but is reported to underperform stacking.

All model aggregation techniques introduced so far are two-step procedures, where we first fit individual models and then combine all predictive distributions. It is also possible to conduct both steps jointly, which can be viewed as a decision problem on both the model weights and component predictive distributions. Ideally, we may avoid the model combination problem by extending the model to include the separate models M_k as special cases. A finite-component mixture model is the easiest model expansion but is generally quite expensive to make inference. Further, if the sample size is small or several components in the mixture could do the same thing, the mixture model can face nonidentification or instability. In fact, the immunity to duplicate models is a unique feature of stacking, while many methods including BMA, information criterion weighting, and mixture models often have a disastrous performance in the face of many similar weak models.

Apart from combining *models*, when we fit one single model but unstable computation, model averaging techniques are also useful to combine *inference* results from multiple nonmixing runs. This is related to the idea of bagging [21]. In particular, when the posterior density $p(\theta|y)$ from a model contains multiple isolated modes, Markov chain Monte Carlo (MCMC) algorithms can have difficulty moving between modes. Yao *et al.* [22] propose to use parallel runs of randomly initialized MCMC, variational, or mode-based inference to hit as many modes or separated regions as possible and then reweigh and combine the posterior Monte Carlo draws using stacking (Equation 5). The result from multirun stacking is not necessarily equivalent, even asymptotically, to full-Bayesian inference, but it serves many of the same goals. With a misspecified model and multimodal posterior density, multirun stacking could lead to better predictive performance than the full-Bayesian inference.

3 Asymptotic Theories of Stacking

To better understand how stacking works, we outline three theory properties in the following subsections.

3.1 Model Aggregation Is No Worse than Model Selection

The stacking estimate (Equation 4) finds the optimal predictive distribution within the linear combination that is the closest to the data-generating process with respect to the chosen scoring rule. Solving for the stacking weights in Equation (6) is an M-estimation problem. To what extent shall we worry about the finite sample error in LOO cross-validation? Roughly speaking, as long as there is consistency for single model cross-validation, then asymptotically model averaging never does worse than model selection in terms of prediction [23]. Le and Clarke [10] further prove that under some mild conditions, for either the logarithmic scoring rule or the energy score (negative squared error) and a given set of weights $w_1 \dots w_K$, the weighted LOO-score is a consistent estimate as sample size $n \to \infty$,

$$\frac{1}{n}\sum_{i=1}^{n} S\left(\sum_{k=1}^{K} w_k \hat{p}_{k,-i}, y_i\right) - \mathbb{E}_{\tilde{y}|y} S\left(\sum_{k=1}^{K} w_k p(\tilde{y}|y, M_k), \tilde{y}\right) \to 0$$

In this sense, stacking gives optimal combination weights asymptotically and is an approximation to the Bayes action.

3.2 Stacking Viewed as Pointwise Model Selection

Besides justified by the decision theory, stacking weights also have a probabilistic interpretation. To see this, we divide the input–output product space $\mathcal{X} \times \mathcal{Y}$ into K disjoints subsets based on which model performs the best,

$$\mathcal{J}_k := \{(\tilde{x}, \tilde{y}) \in \mathcal{X} \times \mathcal{Y} : p(\tilde{y}|M_k, \tilde{x}) > p(\tilde{y}|M_{k'}, \tilde{x}), \forall k' \neq k\}.\ k = 1, \ldots, K$$

We call a family of predictive densities $\{p(\tilde{y}|M_k, \tilde{x})\}_{k=1}^{K}$ to be locally separable with a constant pair $L > 0$ and $0 \leq \epsilon < 1$, with respect to the true data-generating process $p_t(\tilde{y}, \tilde{x})$, if

$$\sum_{k=1}^{K} \int_{(\tilde{x},\tilde{y}) \in \mathcal{J}_k} \mathbb{1}(\log p(\tilde{y}|M_k, \tilde{x}) < \log p(\tilde{y}|M_{k'}, \tilde{x}) + L,\ \forall k' \neq k) p_t(\tilde{y}, \tilde{x}) \mathrm{d}\tilde{y} \mathrm{d}\tilde{x} \leq \epsilon \tag{8}$$

Yao *et al.* [15] show that under the separation condition (Equation 8), the log score stacking weight (Equation 5) is approximately the probability of the model being the locally best fit: $w_k^{\text{stacking}} \approx \Pr(\mathcal{J}_k)$, where the probability is taken with respect to the joint true data-generating process.

3.3 Selection or Averaging?

The advantage of model averaging comes from the fact that model can behave differently in different regions in (x, y) space. Let $\rho = \sup_k \Pr(\mathcal{J}_k)$, then $1 - \rho$ is a rough description of the diversity of models. In terms of the elpd, Yao *et al.* [15] show that under the separation condition (Equation 8), the gain from the optimally weighted models (against model selection) is lower bounded by

$$\text{elpd}_{\text{stacking}} - \sup_k \text{elpd}_k \geq L(1-\rho)(1-\epsilon) - \log K$$

One practical difficulty in model comparison is to determine how large the difference between model performance is “significant” and whether to discard bad models [24]. The probabilistic approximation in the previous subsection suggests that an overall weak model can still be useful in the aggregation. As long as a model is better than all the remaining models in some subset of data, this model possesses a nonzero stacking weight, no matter how poorly it fits everywhere else.

Lastly, a model with the largest BMA weight (assuming equal prior) is optimal under marginal likelihood model selection. In contrast, a model with the largest stacking weight is not necessarily optimal in terms of single model selection: it may outperform other models most of the time but also have arbitrarily low elpd in the remaining areas – stacking is not designed for model selection. Hence, we do not recommend to discard models with small weights from the average.

4 Stacking in Practice

4.1 Practical Implementation Using Pareto Smoothed Importance Sampling

Stacking (Equation 5) requires LOO predictive density $p(y_i|y_{-i}, M_k)$ whose exact evaluation needs to refit each model n times. k-fold cross-validation is computationally cheaper but may introduce higher bias. Vehtari *et al.* [25] proposed an approximate method for Bayesian LOO. It is based on the importance sampling identity:

$$p(\theta|y_{-i}) \propto \frac{1}{p(\theta|y_i)} p(\theta|y_1, \dots, y_n)$$

In the kth model, we fit to all the data, obtaining S simulation draws $\theta_k^s (s = 1, \dots S)$ from the full posterior $p(\theta_k|y, M_k)$ and calculate

$$r_{i,k}^s = \frac{1}{p(y_i|\theta_k^s, M_k)} \propto \frac{p(\theta_k^s|y_{-i}, M_k)}{p(\theta_k^s|y, M_k)} \quad (9)$$

A direct importance sampling often has high or infinite variance and we remedy it by Pareto smoothed importance sampling (PSIS) [26]. For each fixed model k and data y_i, we fit the generalized Pareto distribution to a set of largest importance ratios $r_{i,k}^s$ and calculate the expected values of the order statistics of the fitted generalized Pareto distribution. These values are used to obtain the smoothed importance weight $w_{i,k}^s$, which is used to replace $r_{i,k}^s$. PSIS–LOO importance sampling computes the LOO predictive density as

$$p(y_i|y_{-i}, M_k) = \int p(y_i|\theta_k, M_k) \frac{p(\theta_k|y_{-i}, M_k)}{p(\theta_k|y, M_k)} p(\theta_k|y, M_k) d\theta_k \approx \frac{\sum_{s=1}^{S} w_{i,k}^s p(y_i|\theta_k^s, M_k)}{\sum_{s=1}^{S} w_{i,k}^s}$$

An R package loo [27] provides model weights from the PSIS–LOO-based stacking and pseudo-BMA. Suppose that fit1, fit2, and fit3 are three model fit objects from the Bayesian inference package Stan [28], then we can compute their stacking weights as follows:

```
model_list <- list(fit1, fit2, fit3)
log_lik_list <- lapply(model_list, extract_log_lik)
# stacking:
wts <- loo_model_weights( log_lik_list, method = "stack-
ing",
optim_control = list(reltol=1e-10))
```

4.2 Stacking for Multilevel Data

Although the illustration in this chapter is focused on iid data, the LOO consistency only requires the conditional *exchangeability* of outcomes y given x (Bernardo and Smith [1], chapter 6). Roberts *et al.* [29] review cross-validation strategies for data with temporal, spatial, hierarchical, and phylogenetic structures. In general, the PSIS–LOO approximation applies to factorizable models $p(y|\theta,x) = \prod_{i=1}^{N} p(y_i|\theta,x_i)$ such that the pointwise log-likelihood can be obtained easily by computing $\log p(y_i|\theta,x_i)$.

Nonfactorizable models can sometimes be factorized by reparameterization. In a multilevel model with J groups, we denote the group-level and global parameters as θ_m and ψ. The joint likelihood is

$$p(y|x,\theta,\psi) = \prod_{j=1}^{J}\left[\prod_{n=1}^{N_j} p(y_{jn}|x_{jn},\theta_j)p(\theta_j|\psi)\right]p(\psi) \tag{10}$$

where y are partially exchangeable, that is, y_{mn} are exchangeable in group j, and θ_m are exchangeable. Rearrange the data and denote the group label of (x_i, y_i) by z_i, then Equation (10) can be reorganized into the long format $\prod_{i=1}^{N'} p(y_i|x_i,z_i,\theta,\psi)$, so the previous results follow. Depending on whether the prediction task is to predict a new observation within a group, or a new group, we should consider leave-one-point-out or leave-one-group-out cross-validation.

When the future data are known to come from a group j, there are two stacking strategies: (i) apply generic stacking only to observations from the jth group, which is asymptotically optimal with enough data but has large variance if the group size is small and (ii) apply stacking to all observations regardless of their group structure, which has smaller variance at the cost of less flexibility. The more preferred hierarchical stacking (Section 2.5) trades off between these two extremes. Its Bayesian hierarchical formulation shares information across groups, stabilizing model weights in small groups, while still allowing the flexibility of group-specific weighting.

4.3 Stacking for Time Series Data

When observations y_t come in sequence, and the main purpose is to make prediction for the next not-yet-observed data, we can use the prequential principle [30] to factorize the likelihood: $p(y_{1:N}|\theta) = \prod_{t=1}^{N} p(y_t|y_{1:t-1},\theta)$. In model averaging, we can replace the LOO density $p(y_i|y_{-i})$ in Equation (5) by the sequential predictive density leaving out all future data: $p(y_t|y_{<t}) = \int p(y_t|y_{1:t-1},\theta)p(\theta|y_{1:t-1})d\theta$ in each model and then stacking follows. The ergodicity of y will yield

$$\lim_{N\to\infty}\frac{1}{N}\sum_{t=1}^{N} S(p(\cdot|y_{<t}),y_t) - \lim_{N\to\infty}\frac{1}{N}\mathbb{E}_{Y_{1:N}}\sum_{t=1}^{N} S(p(\cdot|Y_{<t}),Y_t) \to 0$$

which implies a similar stacking optimality as discussed in Section 3.1. Geweke and Amisano [31] investigate this stacking approach in time series data.

When there is a particular horizon of interest for prediction, a model that is good at short-term forecast is not necessarily good for long-term forecast. We can extend the one-step-ahead $p(y_t|y_{<t})$ to m-step-ahead predictive density $p(y_{t<m}|y_{<t}) = p(y_t,\dots,y_{t+m-1}|y_1,\dots,y_{t-1}) = \int p(y_{t<m}|y_{<t},\theta)p(\theta|y_{<t})d\theta$ in the objective function [32].

In terms of computation, the exact prequential evaluation requires refitting each model for each t, which can be approximated by PSIS as $p(y_t|y_{<t}) = \int p(y_t|\theta,y_{<t})\frac{p(\theta|y_{<t})}{p(\theta|y)}p(\theta|y)d\theta$. We then start from the full data inference $p(\theta|y)$ and dynamically update $p(\theta|y_{<t})$ using PSIS approximation. When $p(\theta|y_{<t})$ reveals large discrepancy from $p(\theta|y)$ for some small

t, we refit the model $p(\theta|y_{<t})$ and update the proposal. Burkner *et al.* [33] verify that such approximation gives stable and accurate results with minimal number of refits in time series.

We can further extend the static stacking scheme to a dynamic model weighting, allowing the explanation power of models to change over time. Yao *et al.* [15] present an election forecast example that applies hierarchical stacking to longitudinal polling data. Another flexible model weighting strategy in time series forecasting is Bayesian predictive synthesis (BPS) [34, 35]: The predictive density has the form $\int \alpha(y|z)\prod_{k=1:K} h_k(z_k)dz$, where $z = z_{1:K}$ is the latent vector generated from predictive densities $h_k(\cdot)$ in each model and $\alpha(y|z)$ is the distribution for y given z that is designed to calibrate the model-specific biases and correlations.

4.4 The Choice of Model List

As we have discussed earlier, BMA and information criterion weighting are undesired against many similar weak models. We may remedy this by a careful construction of priors. For example, George [36] establishes dilution priors to compensate for model space redundancy in linear models, putting smaller weights on those models that are close to each other. Fokoue and Clarke [37] introduce prequential model list selection to obtain an optimal model space.

Stacking is prior invariant and immune to model duplication. Nevertheless, all methods discussed in the present chapter fit models separately and are thereby limited in that they do not pool information between the different model fits. The benefit of stacking depends only on the span of the model list [10], and models to be stacked should be as different as possible [7]. In light of discussion in Section 3.2, the ideal situation of stacking is when models can offer different predictive density pointwisely.

In general, we do not recommend constructing a large list of weak models (e.g., subset regression) and aggregate them in a black box way, as in that setting we would recommend moving to a continuous model space that encompasses all separate models. We prefer to carefully construct component models that would have individually fit the data as much as possible, and all admissible estimators for parameters should be considered before the optimization procedures.

5 Discussion

Along with an increasing number of statistical models and learning algorithms, ensemble methods have been appealing tools to expand the existing models and inferential procedures and to improve predictive performance. In addition, the popularity of ensemble methods in Bayesian statistics can be viewed as representing a modern shift in Bayesian data analysis: from a static model-based inference to a Bayesian workflow in which we are fitting many models while working on a single problem.

This chapter is mostly about BMA, stacking, and their variants. For these methods, the model weights are trained after model-specific inferences, and the cost of the former is typically much smaller than the latter. Another popular approach to construct ensembles is to train each model and the model weight simultaneously or iteratively, such as in boosting [38], gradient boosting [39], and mixture of experts [40]. These methods are computationally intensive for full-Bayesian inference but more useful to combine weak learners. On the other hand, different ensemble methods can be further aggregated, for example, to stack fits from BMA and mixture of experts.

Many of these ensemble methods had limited usage until enough computational resources and efficient approximation became available. Conversely, many model averaging strategies also help solve difficulties in statistical computing. For example, bagging stabilizes otherwise nonrobust point estimates, and stacking can be used in multimodal posterior sampling.

Looking forward, there are many open questions. To name a few, both BMA and stacking are restricted to a linear mixture form, would it be beneficial to consider other aggregation forms such as convolution of predictions and a geometric bridge of predictive densities? Stacking often relies on some cross-validation, how can we better account for the finite sample variance therein? While staking can be equipped with many other scoring rules, what is the impact of the scoring rule choice on the convergence rate and robustness? Beyond current model aggregation tools, can we develop an automated ensemble learner that could fully explore and expand the space of model classes – for example, using an autoregressive (AR) model and a moving-average (MA) model to learn an ARMA model? We leave these directions for future investigation.

References

1 Bernardo, J.M. and Smith, A.F.M. (1994) *Bayesian Theory*, John Wiley & Sons.

2 Vehtari, A. and Ojanen, J. (2012) A survey of Bayesian predictive methods for model assessment, selection and comparison. *Stat. Surv.*, 6, 142–228.

3 Gneiting, T. and Raftery, A.E. (2007) Strictly proper scoring rules, prediction, and estimation. *J. Am. Stat. Assoc.*, 102, 359–378.

4 Madigan, D., Raftery, A.E., Volinsky, C., and Hoeting, J. (1996) *Bayesian Model Averaging*, Proceedings of the AAAI Workshop on Integrating Multiple Learned Models, pp. 77–83.

5 Hoeting, J.A., Madigan, D., Raftery, A.E., and Volinsky, C.T. (1999) Bayesian model averaging: a tutorial. *Stat. Sci.*, 14, 382–401.

6 Wolpert, D.H. (1992) Stacked generalization.*Neural Netw.*, 5, 241–259.

7 Breiman, L. (1996) Stacked regressions. *Mach. Learn.*, 24, 49–64.

8 LeBlanc, M. and Tibshirani, R. (1996) Combining estimates in regression and classification. *J. Am. Stat. Assoc.*, 91, 1641–1650.

9 Clyde, M. and Iversen, E.S. (2013) Bayesian model averaging in the M-open framework, in *Bayesian Theory and Applications* (eds P. Damien, P. Dellaportas, N.G. Polson, and D.A. Stephens), Oxford University Press, pp. 483–498.

10 Le, T. and Clarke, B. (2017) A Bayes interpretation of stacking for M-complete and M-open settings. *Bayesian Anal.*, 12, 807–829.

11 Yao, Y., Vehtari, A., Simpson, D., and Gelman, A. (2018) Using stacking to average Bayesian predictive distributions (with discussion). *Bayesian Anal.*, 13, 917–1003.

12 Piironen, J. and Vehtari, A. (2017) Comparison of Bayesian predictive methods for model selection. *Stat. Comput.*, 27, 711–735.

13 Clarke, B. (2003) Comparing Bayes model averaging and stacking when model approximation error cannot be ignored. *J. Mach. Lear. Res.*, 4, 683–712.

14 Yang, Z. and Zhu, T. (2018) Bayesian selection of misspecified models is overconfident and may cause spurious posterior probabilities for phylogenetic trees. *Proc. Natl. Acad. Sci.*, 115, 1854–1859.

15 Yao, Y., Piršš, G., Vehtari, A., and Gelman, A. (2021) Bayesian hierarchical stacking: some models are (somewhere) useful. *arXiv:2101.08954.*

16 Geisser, S. and Eddy, W.F. (1979) A predictive approach to model selection. *J. Am. Stat. Assoc.*, 74, 153–160.

17 Gelfand, A.E. (1996) Model determination using sampling-based methods, in *Markov Chain Monte Carlo in Practice* (eds W.R. Gilks, S. Richardson, and D.J. Spiegelhalter), Chapman & Hall, pp. 145–162.

18 Li, M. and Dunson, D.B. (2020) Comparing and weighting imperfect models using D-probabilities. *J. Am. Stat. Assoc.*, 115 (e531), 1349–1360.

19 Berger, J.O. and Pericchi, L.R. (1996) The intrinsic Bayes factor for model selection and prediction. *J. Am. Stat. Assoc.*, 91, 109–122.

20 O'Hagan, A. (1995) Fractional Bayes factors for model comparison. *J. R. Stat. Soc. B*, 57, 99–118.

21 Breiman, L. (1996) Bagging predictors. *Mach. Learn.*, 24, 123–140.

22 Yao, Y., Vehtari, A., and Gelman, A. (2020) Stacking for non-mixing Bayesian computations: the curse and blessing of multimodal posteriors. *arXiv:2006.12335.*

23 Clarke, B. (2001) Combining model selection procedures for online prediction. *Sankhyā: Indian J. Stat., Ser. A*, 63 (2), 229–249.

24 Sivula, T., Magnusson, M., and Vehtari, A. (2020) Uncertainty in Bayesian leave-one-out cross-validation based model comparison. *arXiv:2008.10296.*

25 Vehtari, A., Gelman, A., and Gabry, J. (2017) Practical Bayesian model evaluation using leave-one-out cross-validation and WAIC. *Stat. Comput.*, 27, 1413–1432.

26 Vehtari, A., Simpson, D., Gelman, A. *et al.* (2019) Pareto smoothed importance sampling. *arXiv:1507.02646.*

27 Vehtari, A., Gabry, J., Yao, Y., and Gelman, A. (2019) LOO: efficient leave-one-out cross-validation and WAIC for Bayesian models. R package version 2.2.0.

28 Stan Development Team (2019) Stan modeling language. Version 2.25.0. http://mc-stan.org/ (accessed 28 February 2021).

29 Roberts, D.R., Bahn, V., Ciuti, S. *et al.* (2017) Cross-validation strategies for data with temporal, spatial, hierarchical, or phylogenetic structure. *Ecography*, 40, 913–929.

30 Dawid, A.P. (1984) Present position and potential developments: some personal views statistical theory the prequential approach. *J. R. Stat. Soc. A*, 147, 278–290.

31 Geweke, J. and Amisano, G. (2012) Prediction with misspecified models. *Am. Econ. Rev.*, 102 (3), 482–486.

32 Lavine, I., Lindon, M., and West, M. (2021) Adaptive variable selection for sequential prediction in multivariate dynamic models. *Bayesian Anal.*, 1–25. doi: 10.1214/20-BA1245.

33 Bürkner, P.-C., Gabry, J., and Vehtari, A. (2020) Approximate leave-future-out cross-validation for Bayesian time series models. *J. Stat. Comput. Simul.*, 90, 2499–2523.

34 McAlinn, K. and West, M. (2019) Dynamic Bayesian predictive synthesis in time series forecasting. *J. Econom.*, 210, 155–169.

35 McAlinn, K., Aastveit, K.A., Nakajima, J., and West, M. (2020) Multivariate Bayesian predictive synthesis in macroeconomic forecasting. *J. Am. Stat. Assoc.*, 115, 1092–1110.

36 George, E.I. (2010) Dilution priors: compensating for model space redundancy, in *Borrowing Strength: Theory Powering Applications – A Festschrift for Lawrence D. Brown* (eds J.O. Berger, T.T. Cai, and I.M. Johnstone), Institute of Mathematical Statistics, Beachwood, OH, pp. 158–165.

37 Fokoue, E. and Clarke, B. (2011) Bias-variance trade-off for prequential model list selection. *Stat. Pap.*, 52, 813–833.

38 Freund, Y. and Schapire, R.E. (1997) A decision-theoretic generalization of on-line learning and an application to boosting. *J. Comput. Syst. Sci.*, 55, 119–139.

39 Friedman, J.H. (2001) Greedy function approximation: a gradient boosting machine. *Ann. Stat.*, 29 (5), 1189–1232.

40 Jacobs, R.A., Jordan, M.I., Nowlan, S.J., and Hinton, G.E. (1991) Adaptive mixtures of local experts. *Neural Comput.*, 3, 79–87.

32

Asynchronous Parallel Computing

Ming Yan

Michigan State University, East Lansing, MI, USA

1 Introduction

"The free lunch is over" [1] around 2005 when the computer industry had to increase the number of cores in a central processing unit (CPU) to increase computing power. Also, the rapid proliferation of big data brings new challenges to the existing sequential statistical algorithms. Many existing algorithms are computationally expensive for large-scale problems. The much faster increase in data than that of single-thread performance makes parallel computing inevitable [2]. These algorithms need to be rewritten for parallel implementation, and additional efforts are required to take advantage of the parallel architecture.

There are many new challenges in the transition to parallel computing. One of the major challenges is the synchronization bottleneck, which requires the completion of the computation at all cores to start the next iteration (see Section 1.1 for the explanation). With synchronization, the performance using multiple cores could be worse than using only one core. Fortunately, many asynchronous parallel implementations work well on large-scale problems. Asynchronous parallel computing has achieved great success in many other applications, such as power systems [3] and reinforcement learning [4]. Asynchronous training is even supported in Tensorflow and PyTorch. In this chapter, we explain the benefits of asynchronous parallel computing specifically for machine learning problems and describe several existing asynchronous parallel implementations.

Many statistical and data science problems can be formulated as optimization problems in the following form:

$$\min_{w \in \mathbb{R}^p} \frac{1}{N} \sum_{i=1}^{N} f_i(w) + g(w) \tag{1}$$

In this problem, N is usually the number of data samples, and we solve for the coefficients (or weights) w. Functions $\{f_i\}_{i=1}^N$ are the data-fitting terms, and the function g is a regularization on w. Statistical learning examples that are in the form of (1) include:

- *Ordinary least squares (OLS)*: $f_i(w) = \|x_{(i)}^\top w - y_{(i)}\|^2/2$; $g(w) = 0$. It minimizes the residual sum of squares between the actual targets $y_{(i)}$ and the predicted targets $x_{(i)}^\top w$ from the linear approximation. Here, $\{x_{(i)}, y_{(i)}\}_{i=1}^N$ are the training data samples.

Computational Statistics in Data Science.
Edited by Walter W. Piegorsch, Richard A. Levine, Hao Helen Zhang and Thomas C. M. Lee.
 ISBN 978-1-11956107-1

- *Elastic net* [5]: $f_i(w,b) = \|x_{(i)}^\top w + b - y_{(i)}\|^2/2$; $g(w) = \lambda(\rho\|w\|_1 + (1-\rho)\|w\|^2/2)$ for $\rho \in [0,1]$. It trains the linear approximation with ℓ_1- and ℓ_2-norm regularizations. When $\rho = 1$, it is the least absolute shrinkage and selection operator (LASSO) [6], and it is the ridge regression if $\rho = 0$. Here, (w, b) are the coefficients.
- *Logistic regression*: $f_i(w,b) = \log(\exp(-y_{(i)}(x_{(i)}^\top w + b)) + 1)$; $g(w) = \lambda(\rho\|w\|_1 + (1-\rho)\|w\|^2/2)$. It trains the linear model to predict the classification of data samples. Logistic regression handles the binary response $y_{(i)}$ taking values $\{0, 1\}$ and follows the Bernoulli distribution with the log-odds of $Y = 1$ being $x_{(i)}^\top w$.
- *Soft-margin support vector machine (SVM)* [7]: $f_i(w,b) = \max(0, 1 - y_{(i)}(x_{(i)}^\top w - b))$; $g(w) = \lambda\|w\|^2$. SVM tries to find a hyperplane described by $x^\top w - b = 0$ to separate two groups of data samples.
- *Neural network models*: $f_i(w) = \|y_{(i)} - \hat{y}_{(i)}\|^2$, where $\hat{y}_{(i)}$ is the output of the neural network with the ith data sample, and $y_{(i)}$ is the actual target for the ith data sample. This model is usually nonlinear.
- *Nonnegative matrix factorization* (NMF): $f_i(w) = \|(WH)_{ij} - V_{ij}\|^2/2$; $g(w)$ returns 0 if both W and H are nonnegative and $+\infty$ otherwise. Here, V is a given nonnegative matrix, and (W, H) are the coefficients.

These problems usually have N data samples and P coefficients. In large-scale machine learning problems, at least one number is big. When N is big, that is, there are many data samples, we let each core process a small number of samples each time and update the coefficients. This is related to stochastic gradient descent (SGD). If P is big, we can let each core update a small portion of the coefficients each time. This is related to block coordinate update. We introduce asynchronous parallel implementations under both scenarios.

1.1 Synchronous and Asynchronous Parallel Computing

When applying iterative algorithms to solve these large-scale problems, we formulate one iteration as

$$w^{k+1} = T(w^k) \tag{2}$$

Here, w^k are the coefficients w at the kth iteration. During each iteration, we update the coefficients based on the algorithm we select (the operator T in this formulation). Not all iterative algorithms are in this form. For example, the update of the coefficients may depend on more than one previous iteration [8]. In this chapter, we focus on this form to illustrate the idea of asynchronous parallel computing.

Using multiple cores to update the coefficients, we assign the computation to multiple cores and let these cores work on the iteration. Ideally, we want these cores to work independently of each other to perform the computation in parallel. However, not all algorithms have this property (see Section 1.2 for an example). In this section, we use a shared-memory architecture to show the difference between synchronous and asynchronous parallel computing.

In synchronous parallel computing, we start the next iteration after all cores finish their computation (see Figure 1a). Here, we assume that the communication time is shorter

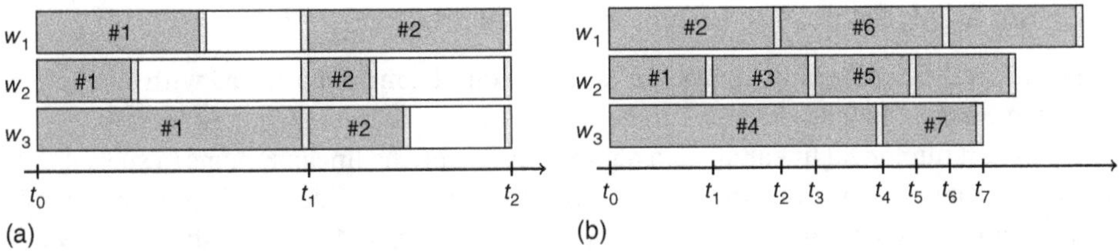

Figure 1 Synchronous versus asynchronous parallel computing with shared memory. (a) (Synchronous): fast cores have to wait for slow ones; (b) (asynchronous): cores do not wait for others. Dark-gray and light-gray bars stand for the computation and communication, respectively. White bars are the idle periods. In synchronous parallel computing, cores send the data to the shared memory immediately after finishing the computation and receive the data after all cores send the data to the shared memory. While in asynchronous parallel computing, the two transfers (both send and receive) between one core and the shared memory happen together. Except that the idle periods are removed, asynchronous parallel computing spreads the communication.

than the computation time. In fact, in many distributed computing scenarios with limited bandwidth, for example, federated learning with small smart devices, the communication time could be significantly longer than the computation time. In synchronous parallel computing, cores that finish early have to wait for slow ones, and there are many idle periods. It can be even worse when the number of cores increases. One major benefit of using synchronous parallel computing is the convergence guarantee of the iterative algorithms. Because the iterations are the same as their nonparallel versions, their convergence follows immediately. Indeed, synchronous parallel computing is very useful when all the cores finish their computational tasks simultaneously, and the bandwidth is not an issue. In this case, the idle time is minimal compared to the total time. However, these are not satisfied in many applications, and asynchronous parallel computing has to be considered.

In contrast, there is no idle time in asynchronous parallel computing (see Figure 1b). In asynchronous parallel computing, all cores perform computation and communication without knowing other cores' existence. Except that the idle time is removed, the communication also happens at different times, so we do not need large bandwidth as in synchronous parallel computing. However, the main challenge is its convergence. Because cores do not finish the computation simultaneously, we increase the iteration number whenever one core updates. Assume that we have three coefficients ($w = (w_1, w_2, w_3)$) in Figure 1 and let one core update one coefficient. The iterations w^1, w^2, w^4 are all based on w^0, while the iteration w^3 depends on w^1. Therefore, the simple iteration form in Equation (2) does not hold, and the convergence of these asynchronous algorithms require additional assumptions. When more than one core try to read/write the same number in the shared memory, the asynchronous iteration could be more complicated [9]. The iteration becomes

$$w^{k+1} = T^{k+1}(\overline{w}^k)$$

where the operator T^{k+1} depends on the $(k+1)$th update, and $\overline{w}^k$ may not equal w^j for any $j \leq k$ if inconsistent read happens when multiple cores are updating while one core is reading; see Peng *et al.* [9] for an example.

1.2 Not All Algorithms Can Benefit from Parallelization

Though parallel computing allows the acceleration of some algorithms with the help of multiple cores, not all algorithms can benefit from parallelization. In this section, we use two classical numerical linear algebra algorithms for solving linear systems (a special ordinary linear regression problem), that is, Jacobi and Gauss–Seidel, to demonstrate it. We want to find a vector $w \in \mathbb{R}^P$ such that $Xw = y$. Here, $X \in \mathbb{R}^{P\times P}$ and satisfies some conditions, for example, diagonal dominance. We decompose the matrix X as $X = L + D + U$, with D, L, and U being the diagonal, lower triangular, and upper triangular components of X, respectively. The two iterative algorithms to solve this linear system are:

$$\text{Jacobian:} \quad w^{k+1} = D^{-1}(y - (L+U)w^k) \tag{3}$$

$$\text{Gauss–Seidel:} \quad w^{k+1} = (L+D)^{-1}(y - Uw^k) \tag{4}$$

Assume that we have P cores and let one core update one coordinate of w. The Jacobi iteration for the ith coordinate is

$$w_i^{k+1} = D_{ii}^{-1}\left(y_i - \sum_j (L_{ij} + U_{ij})w_j^k\right)$$

Since it depends on the values from the previous iteration, we can apply parallel computing to speed up the algorithm. However, for the Gauss–Seidel iteration, we have

$$w_i^{k+1} = D_{ii}^{-1}\left(y_i - \sum_{j:j<i} L_{ij}w_j^{k+1} + \sum_{j:j>i} U_{ij}w_j^k\right)$$

The update of the ith coordinate relies on the updates of the previous $i - 1$ coordinates. Though Gauss–Seidel converges much faster than Jacobi because the new updates are used during the iteration, it is difficult to parallelize Gauss–Seidel for general linear systems. The P cores cannot be updated in parallel, and Figure 1 does not apply. If the matrix X has special structures, the coordinates can be divided into blocks, and each block can be updated in parallel [10]. Though Gauss–Seidel cannot be parallelized, asynchronous parallel Jacobi algorithms have superperformance. When P cores are used, it may have a $P\times$ speedup compared to Gauss–Seidel [11]. Also, they converge under weaker conditions than the standard Jacobi [12].

1.3 Outline

As mentioned previously, it is not easy to show the convergence of asynchronous parallel algorithms, and they rely on particularly assumed conditions. In this chapter, we describe the transition of some statistical algorithms to asynchronous parallel architecture with convergence guarantee, while referring the convergence to a recent review [13] and corresponding references.

We introduce two types of asynchronous parallel algorithms for solving large-scale problems: asynchronous parallel coordinate update (Section 2) and asynchronous parallel stochastic approaches (Section 3). For each type, we present how to perform asynchronous parallel computing and benefit from it with some examples. Then, we describe another asynchronous parallel algorithm that combines both approaches in Section 4. There are so many asynchronous parallel works that we can only cover a few of them in this short chapter.

1.4 Notation

We use uppercase characters (except scalars $E, M, N,$ and P) to denote matrices or an operator. For a matrix X, we use X_{ij}, $X_{i\cdot}$, and $X_{\cdot j}$ to denote its (i,j) component, ith row, and jth column, respectively. Lowercase characters are used to denote vectors including scalars. For a vector w, we use w_i to denote its ith component. Vectors are shown in columns, and we use $[x; y; z]$ and $[x^\top; y^\top; z^\top]$ to stack vectors together into one long vector and a matrix, respectively. We use superscript to denote iterations, for example, w^k is the value of w at the kth iteration.

2 Asynchronous Parallel Coordinate Update

Many iterative algorithms for solving Equation (1) can be formulated as fixed-point iterations. Let us consider the following fixed-point iteration:

$$v^{k+1} = T(v^k) \tag{5}$$

Here, T is an operator that maps from one space to itself, and we want to find a fixed-point v^* of T such that $v^* = T(v^*)$. Here, we choose v^k, instead of w^k, because the fixed-point operators in some algorithms may not be directly applied to the coefficient w. This fixed-point iteration covers many science and engineering problems, though we only focus on statistical and machine learning examples. Finding a fixed point of T is equivalent to finding a zero of $S := I - T$, with I being the identity operator that returns itself. That is, we want to find v^* such that $0 = S(v^*)$.

Coordinate update algorithms update one coordinate (or a block of coordinates) from v at a time [14–16]. We use one coordinate for simplicity, and the algorithms can be easily extended to blocks. Consider the following update:

$$v^{k+1} = v^k - \alpha_k S_{i_k}(v^k)$$

where α_k is a parameter, and $S_{i_k}(v) = (0, \ldots, 0, (S(v))_{i_k}, 0, \ldots, 0)$ keeps the update of the i_kth coordinate only. In this iteration, we updated the coordinate v_{i_k} only while keeping other coordinates unchanged. The many ways to choose i_k can be divided into three groups: (i) cyclic; (ii) randomized; and (iii) greedy. To apply the coordinate update, one needs to make sure that the update of one coordinate v_{i_k} is much faster than the update of the whole variable v. Operators that satisfy this property are called "*coordinate friendly*" by Peng *et al.* [17], and many statistical algorithms have this property. When it goes to parallel computing, multiple cores update the coordinates in parallel. We first consider the shared-memory architecture, where all cores are connected to the shared memory. Synchronous parallel methods just let the cores compute the coordinate updates for all coordinates and update the variable in the shared memory at the end of each iteration. In asynchronous parallel methods, each core reads the variable from the shared memory, computes the update for one coordinate, and updates the coordinate in the shared memory.

The convergence for deterministic asynchronous parallel fixed-point iterations was considered as early as the 1970s [18, 19]. In this work, the coordinates are assigned to multiple cores, and each node updates the assigned coordinate only. To show convergence, it requires a contracting assumption. The first asynchronous parallel coordinate update method for solving optimization problems was due to Bertsekas and Tsitsiklis [2]. Recently, Liu *et al.* [20], Liu and Wright [21] proposed an asynchronous parallel proximal coordinate descent algorithm for minimizing convex composite objective functions. Hsieh *et al.* [22] proposed an asynchronous parallel dual coordinate descent method for solving ℓ_2 regularized empirical risk minimization problems. The generalization to fixed-point iterations is also considered by Peng *et al.* [9, 23], Hannah and Yin [24]. Combettes and Eckstein [25] proposed asynchronous block-iterative primal–dual decomposition methods based on choosing a subset of monotone operators.

Next, we present four asynchronous parallel coordinate update algorithms: proximal gradient for a convex problem (Section 2.1); projected gradient for a nonconvex problem (Section 2.2); a primal–dual algorithm (Section 2.3); and a primal–dual algorithm after reformulation (Section 2.4). For each algorithm, we reformulate the algorithm into an iteration such that the coordinates are updated independently, and the update of one coordinate is significantly faster than the update of the whole variable.

2.1 Least Absolute Shrinkage and Selection Operator (LASSO)

We consider the unconstrained LASSO problem in its Lagrangian form:

$$\min_{w\in\mathbb{R}^P} \frac{1}{N}\sum_{i=1}^{N} \|x_{(i)}^{\top}w - y_{(i)}\|_2^2 + \lambda\|w\|_1 \tag{6}$$

It minimizes the sum of a smooth function and a simple nonsmooth function. So, we can apply the following iterative shrinkage-thresholding algorithm [8] to solve it:

$$w^{k+1} = \mathrm{Sh}\left(w^k - \frac{2\eta}{N}X^{\top}(Xw^k - y), \lambda\eta\right) \tag{7}$$

Here, $X = [x_{(1)}^{\top};\cdots;x_{(N)}^{\top}]$, $y = [y_{(1)};\cdots;y_{(N)}]$, η is the stepsize, and $\mathrm{Sh}(\cdot,\cdot)$ is the soft thresholding, which is defined as

$$(\mathrm{Sh}(w,t))_j = \begin{cases} w_j - t, & \text{if } w_j > t \\ 0, & \text{if } |w_j| \le t \\ w_j + t, & \text{if } w_j < -t \end{cases}$$

The update of the jth coordinate of w is

$$\widetilde{w}_j = \mathrm{Sh}\left(w_j - \frac{2\eta}{N}(X_{:j})^{\top}(Xw - y), \lambda\eta\right) \tag{8}$$

The update of one coordinate in Equation (8) requires $O(NP + N)$ operations, and the update of the whole variable w in Equation (7) also requires $O(NP)$ operations. At first glance, this operator is not coordinate friendly because we need to compute Xw even if we only update one coordinate. To make it coordinate friendly, we store $Xw - y$ in the shared memory and update it after updating one coordinate. In this way, both the computation of the coordinate update and the update of $Xw - y$ require $O(N)$ operations. Thus, it becomes coordinate friendly because the time to update one coordinate is about $1/P$ of the time to update the whole variable. In addition, we do not need additional storage because we do not store y, which has the same size as $Xw - y$.

Then we are ready to implement the asynchronous parallel algorithm based on ARock [9]. We store w, X, and $Xw - y$ in the shared memory. Each machine performs the following steps:

1. pick an index j, read w_j, $Xw - y$, and the jth columns of X from the shared memory;
2. perform the update for the jth coordinate using Equation (8), which takes $O(N)$ operations;
3. add $\Delta w_j := \alpha(\widetilde{w}_j - w_j)$ and $\Delta(Xw - y) := X(\Delta w_j)$ to the shared memory, which takes $O(N)$ operations.

Note that, the updates of w_j and $Xw - y$ are performed in the shared memory to avoid inconsistency. Between the time a machine reads from and writes to the shared memory, other machines may have updated w_j and $Xw - y$. This machine only adds the changes to the shared memory, so other machines' changes are kept. Interested readers are referred to Peng *et al.* [23] for the numerical performance of this asynchronous parallel algorithm. Another method to solve this LASSO problem asynchronously is given in Section 4.

2.2 Nonnegative Matrix Factorization

NMF is a dimensionality reduction method and has applications in computer vision, document clustering, recommendation systems, signal processing, and bioinformatics. Given a large nonnegative matrix $V \in \mathbb{R}_+^{M\times N}$, NMF tries to find two small nonnegative matrices W and H such that WH is close to V. Using the Frobenius norm distance, we have the following optimization problem:

$$\min_{W,H} \|WH - V\|_F^2 \qquad \text{s.t. } W \in \mathbb{R}_+^{M\times P},\ H \in \mathbb{R}_+^{P\times N} \tag{9}$$

Here, P is the rank of the target matrix WH. We apply the coordinate projected gradient descent to solve this problem. In the asynchronous parallel architecture, each machine picks one column from W or one row from H and performs the update. For example, the update of the jth column of W is just

$$\widetilde{W}_{\cdot j} = \text{proj}_{\mathbb{R}_+^M}(W_{\cdot j} - \eta(WH - V)H_{j\cdot}^{\top})$$

where η is the stepsize, and $\text{proj}_{\mathbb{R}_+^M}$ is the projection to the set of nonnegative vectors in $\mathbb{R}^M$. Similarly to the LASSO problem, we store $WH - V$ in the shared memory to save the computation of WH in this iteration. Therefore, the update of one column of W requires $O(MN)$ operations because of the matrix–vector multiplication. The update of $WH - V$ requires $O(MN)$ operations as well. On the other hand, the update of all columns of X requires $O(MNP)$ operations because of the matrix–matrix multiplication WH. The update of one row of H has similar performance. Also, we can update the jth column of W and the jth row of H together. Interested readers are referred to Peng *et al.* [23] for the numerical performance of this asynchronous parallel algorithm.

2.3 Kernel Support Vector Machine

Given the training data $\{(x_{(i)}, y_{(i)})\}_{i=1}^N$ with $y_{(i)} \in \{+1, -1\}$, the kernel SVM [26] finds a hyperplane, denoted as $w^\top \varphi(x) = b$, to separate two groups of data in their transformed space. Here, the function $\varphi(\cdot)$ is provided, and the optimization problem is

$$\min_{w,s,b} \lambda \|w\|_2^2 + \frac{1}{N} \sum_{i=1}^N s_i \quad \text{s.t. } y_{(i)}(w^\top \varphi(x_{(i)}) - b) \geq 1 - s_i, s_i \geq 0, \forall i = 1, \cdots, N \tag{10}$$

It is difficult to apply coordinate update on this formulation because w and $\{s_i\}_{i=1}^N$ are coupled in the constraints. So, we look at its dual problem

$$\min_c f(c) := \frac{1}{2} c^\top Qc - \sum_{i=1}^N c_i \quad \text{s.t. } \sum_{i=1}^N c_i y_{(i)} = 0, \ 0 \leq c_i \leq \frac{1}{2N\lambda}, \ \forall i = 1, \cdots, N$$

where $Q_{ij} = y_{(i)} y_{(j)} k(x_{(i)}, x_{(j)}) = y_{(i)} y_{(j)} \varphi(x_{(i)})^\top \varphi(x_{(j)})$. After the dual problem is solved, we let $w = \sum_{i=1}^N c_i y_{(i)} \varphi(x_{(i)})$. To obtain b, we find some index i such that $0 < c_i < 1/(2N\lambda)$ and let $b = w^\top \varphi(x_{(i)}) - y_{(i)}$. Then $\varphi(x_{(i)})$ lies on the boundary of the margin in the transformed space. Still, we cannot apply the coordinate projected gradient descent as in the previous subsections because of the constraint $\sum_{i=1}^N c_i y_{(i)} = 0$. This constraint couples all the coordinates, and the projection to this constraint requires the gradient descent of all coordinates. Therefore, we apply a primal–dual splitting scheme [27, 28]. The update of all coordinates is, with $y = [y_{(1)}; \cdots; y_{(N)}]$,

$$d^{k+1} = d^k + \gamma \sum_{i=1}^N y_{(i)} c_i^k$$

$$c^{k+1} = \text{proj}_{[0,1/(2N\lambda)]}(c^k - \eta(\nabla f(c^k) + (2d^{k+1} - d^k)y))$$

In this iteration, d is the dual variable, $\text{proj}_{[0,1/(2N\lambda)]}$ is the projection to the box constraint, and η and γ are primal and dual stepsizes, respectively. We consider the combination of the primal and dual variables together to update coordinates. When the dual variable is chosen, we can update d easily. However, when a coordinate c_i is chosen, the update of this coordinate also requires the update of d. To ensure convergence, we can just compute the update in the dual variable without updating it in the shared memory. It takes $O(N)$ computation. Alternatively, we store an additional variable $s = \gamma \sum_{i=1}^N y_{(i)} c_i$ in the shared memory, and the computation of the dual update needs $O(1)$ operations. In asynchronous parallel computing, each machine picks one coordinate from c or the dual variable. If the dual variable is chosen, read s and add $\Delta d := \alpha s$ to the shared memory. Otherwise, one coordinate c_i from c is chosen, and then the machine performs the following steps:

1. read $c, y_{(i)}, Q_{i\cdot}, s$, and d;
2. perform the update of c_i using

$$\widetilde{c}_i = \text{proj}_{[0,1/(2N\lambda)]}(c_i - \eta(Q_{i\cdot}c - 1 + y_{(i)}(2s + d)))$$

3. add $\Delta c_i := \alpha(\widetilde{c}_i - c_i)$ and $\Delta s = \gamma y_{(i)} \Delta c_i$ to the shared memory.

2.4 Decentralized Algorithms

The previously described asynchronous algorithms require shared memory or master nodes that are connected to all worker nodes. Another type of distributed architecture does not have center nodes, and all worker nodes communicate with their immediate neighbors. Because of the absence of master nodes, these algorithms rely on the worker nodes' local computation and communication and are called decentralized. Decentralized optimization has applications in wireless sensor networks, resource allocation, and distributed learning. Interested readers are referred to Nedic [29] for a review of decentralized optimization. In this architecture, these worker nodes form a connected network, and they cooperatively solve the following optimization problem:

$$\min_{w \in \mathbb{R}^P} \sum_{i=1}^{N} f_i(w)$$

For simplicity, we assume that the function f_i is differentiable and kept privately at node i, though nonsmooth functions are also considered in the literature.

Many asynchronous decentralized algorithms are proposed [29–31]. Here, we describe the asynchronous algorithm proposed by Wu *et al.* [32] because it can be formulated as a primal–dual fixed-point iteration before ARock can be applied. The optimization problem is reformulated as

$$\min_{W} f(W) := \sum_{i=1}^{N} f_i(w_{(i)}), \quad \text{s.t. } WV = 0 \tag{11}$$

Here, $W = [w_{(1)}, \cdots, w_{(N)}] \in \mathbb{R}^{P\times N}$, with the ith column storing the estimate of w at node i. The constraint $WV = 0$ ensures that $w_{(1)} = w_{(2)} = \cdots = w_{(N)}$. There are many possible choices for the matrix V, and it was chosen to incorporate the communication between the nodes. A primal–dual algorithm [27, 28] for solving Equation (11) is

$$\begin{aligned} W^{k+1} &= W^k(I - 2(VV^\top)) - \eta\nabla f(W^k) - Y^k V^\top \\ Y^{k+1} &= Y^k + W^k V \end{aligned}$$

Let E be the number of edges in the network formed by the worker nodes. Wu *et al.* [32] chose the matrix $V \in \mathbb{R}^{N\times E}$ such that there are only two nonzero elements in each column, and $V_{ie} \neq 0$ if and only if node i belongs to edge e. Because of this special construction, $VV^\top \in \mathbb{R}^{N\times N}$ is sparse, and $(VV^\top)_{ij} \neq 0$ if and only if nodes i and j are connected through an edge. In this way, columns of W and Y can be updated independently, and ARock can be applied. Node i stores w_i and $Y_{\cdot e}$ for all incident edges e of nodes i. Let $\mathcal{N}(i)$ and $\mathcal{E}(i)$ be the set of neighbor nodes and the set of incident edges of node i. Then, node i performs the following steps:

1. collect $w_{(j)}$ and $Y_{\cdot e}$ received from its neighbors;
2. update $w_{(i)}$ and $Y_{\cdot e}$ for all incident edges of node i using

$$\begin{aligned} \widetilde{w}_{(i)} &= \sum_{j\in\mathcal{N}(i)} (I - 2VV^\top)_{ji} w_{(j)} - \eta\nabla f_i(w_{(i)}) - \sum_{e\in\mathcal{E}(i)} V_{ie} Y_{\cdot e} \\ \widetilde{Y}_{\cdot e} &= Y_{\cdot e} + V_{ie} w_{(i)} + V_{je} w_{(j)}, \qquad \forall e = (i,j) \in \mathcal{E}(i) \end{aligned}$$

3. send $w_{(i)}$ and $Y_{\cdot e}$ to its neighbors. Only $w_{(i)}$ and one column of Y are transferred through one edge.

Because there is no shared memory, there will be no conflict in the updates. Instead of sending $w_{(i)}$ and $Y_{\cdot e}$ to its neighbors, the node can choose to send the changes in the variables to save the computation and communication. Interested readers are referred to Wu *et al.* [32] for the numerical performance.

3 Asynchronous Parallel Stochastic Approaches

In the previous section, we considered the big-P case with a large number of coefficients. We divide the coefficients into multiple blocks and let one core update one block each time. In this section, we consider the big-N case with a large number of data samples. Because of the large number of data samples, it is impossible or inefficient to process all the data samples each time. Instead, we pick one data sample or a small number of data samples each time. Again, we use one sample in this chapter, and all the results can be extended to a small number of data samples. In addition, we consider the case without the function g for simplicity.

SGD [33] uses a random gradient based on one data sample to replace the true gradient $\frac{1}{N}\sum_{i=1}^{N}\nabla f_i(w^k)$ in gradient descent. Then, the update becomes

$$w^{k+1} = w^k - \eta_k \nabla f_{i_k}(w^k) \tag{12}$$

where η_k is the stepsize, and i_k is the index of the randomly selected data sample. Because of the variance in the stochastic gradient, SGD with a constant step only converges to a neighborhood of the optimum whose radius is proportional to the stepsize and the variance in the stochastic gradients. Therefore, variance reduction techniques and diminishing stepsizes are applied to obtain the optimum.

SGD can be easily parallelized with a master node and many worker nodes. The data samples are partitioned and stored at worker nodes. Each node randomly picks one data sample and computes the stochastic gradient. Then, the gradients are aggregated at the master node to obtain the update of the coefficient w. With synchronization, the update becomes $w^{k+1} = w^k - \frac{1}{|E_k|}\sum_{i\in E_k}\nabla f_i(w^k)$, where E_k is the set of indices of selected data samples by all worker nodes. In this case, the variance of the averaged stochastic gradient is smaller than that in SGD, and the number of iterations to find the optimum decreases. It is exactly synchronous parallel minibatch SGD if the data samples at all worker nodes follow the same distribution.

As shown in Figure 1(b), the master node in asynchronous computing updates the coefficients w whenever it receives the update from one worker node. The worker nodes perform the following three steps independently: (i) receive the coefficients from the master node; (ii) compute the stochastic gradient from its own data samples; (iii) send the computed gradient to the master node. On the other hand, the master node updates the coefficients based on the gradients received from the worker nodes using a round-robin scheduler (i.e., the master node applies the updates from worker nodes one by one) and sends the updated coefficients to the worker nodes. Between the time the master node sends the coefficients to a worker node and the time it uses the gradient from this worker node to update the coefficients, the master node has already updated the coefficients multiple times based on the information from other worker nodes [34, 35]. Therefore, the iteration becomes

$$w^{k+1} = w^k - \eta_k \nabla f_{i_k}(w^{k-\tau_k})$$

where τ_k is the number of updates between the two times. There are many ways to modify the updates in each iteration because of the delay τ_k. They include delay compensation [36], variance reduction [37], and stepsizes based on maximal delay [38], expected delay [39, 40], or actual delay [41]. Asynchronous SGD is also implemented in the Spark-based cloud computing framework ASYNC [42].

3.1 Hogwild!

When the number of coefficients P is also big, updating the coefficients w takes time, and it is ineffective to update them with the round-robin scheduler. When there are a large number of worker nodes, the queue of gradients to be added could be very long. This situation could be even worse in the shared-memory architecture. While one node is updating the coefficients, it requires the lock of the coefficients, and other worker nodes cannot read/write. The lock-free algorithm Hogwild! [43] lets the worker nodes update the coefficients in the shared memory using the atomic write operations on individual coefficients. It is especially effective when the gradients are sparse such that the chance for two worker nodes to update the same coefficient is very low. It solves the following optimization problem:

$$\min_{w\in\mathbb{R}^P} f(w) = \sum_{e\in\mathcal{E}} f_e(w_e) \tag{13}$$

where e is a small subset of $\{1, \ldots, P\}$, and w_e is the subvector of w with only the coordinates indexed by e. It has applications in sparse SVM, matrix completion, and graph cuts. The steps for each machine are

1. pick e from $\mathcal{E}$, read w_e and f_e, and evaluate $g_e(w_e) = \nabla f_e(w_e)$.
2. update $x_e \leftarrow x_e - \eta g_e(w_e)$ in the shared memory using the atomic operations on individual coefficients.

Even if the gradient is not sparse, sparsification can be applied to send some components of the gradient only to reduce the communication cost [44]. Variance reduction techniques are applied as well [45].

3.2 Federated Learning

Most asynchronous parallel algorithms require the same distribution of data samples at the worker nodes. This can be fulfilled with a shared memory storing all the data samples. Without a shared memory, we can distribute the data samples to all worker nodes such that their distributions are close. However, there are scenarios where the data samples at each worker node are private and cannot be transferred to other nodes.

Federated learning [46] enables the learning of the massive private data at many edge devices, including smartphones, wearable devices, and sensors without exchanging data samples. In this case, we need to overcome several challenges including limited communication, heterogeneous data samples, and infrequent task scheduling. There are two major differences between the standard federated learning algorithm FedAvg [47] and the previously described parallel SGD. First, not all devices perform the computation; second, there are multiple local update steps before the updates are sent to the master node (server).

This synchronous implementation has the same limitations as parallel computing shown in Figure 1(a). Because of the heterogeneous data at worker nodes, its convergence relies on the selection of updating devices in each iteration. This heterogeneity has to be taken into account during the development of asynchronous federated learning algorithms [48, 49].

4 Doubly Stochastic Coordinate Optimization with Variance Reduction

We describe another asynchronous algorithm that does not fit in the previous frameworks. In fact, it is a combination of both approaches. Doubly stochastic coordinate optimization with variance reduction (DSCOVR) [50] solves a group of optimization problems in the form of

$$\min_{w} \frac{1}{N}\sum_{i=1}^{N} f_i(x_{(i)}^\top w) + \sum_{k=1}^{P} g_k(w_k) \tag{14}$$

where $w = [w_1; \cdots ; w_P]$. This problem has an equivalent saddle-point formulation:

$$\min_{w} \max_{v} \frac{1}{N}\sum_{i=1}^{N} v_i x_{(i)}^\top w - \frac{1}{N}\sum_{i=1}^{N} f_i^*(v_i) + \sum_{k=1}^{P} g_k(w_k) \tag{15}$$

which can be solved by the following primal–dual algorithm:

$$\widetilde{v}_i = \mathrm{prox}_{\gamma f_i^*}(v_i + \gamma x_{(i)}^\top w), \quad \forall i = 1, \cdots, N$$

$$\widetilde{w}_k = \mathrm{prox}_{\eta g_k}\left(w_k - \frac{\eta}{N}\sum_{i=1}^{N} X_{ik} v_i \right), \quad \forall k = 1, \cdots, P$$

Here, $X = [x_1^\top; \cdots ; x_N^\top]$, and η and γ are primal and dual stepsizes, respectively. The primal and dual variables are updated independently, which is different from standard primal–dual algorithms [27, 51], where the primal and dual variables are updated alternatively. However, the convergence of this algorithm requires the additional strong convexity of f_i^* and g_k. Though this algorithm fits into the framework of ARock, the implementation requires too much communication. We store Xw and $X^\top v$ in the shared memory. Assume that v_i is chosen to be updated. We need to

1. read v_i and $x_{(i)}^\top w$ from the shared memory, which requires $O(1)$ data transfer;
2. perform the computation; add Δv_i and $(X^\top)_i \Delta v_i$ for all k, which requires $O(P)$ data transfer.

In addition, the updates of Xw and $X^\top v$ may require a lock on the vectors to avoid conflicts. $O(P + N)$ data transfer is required if one primal coordinate and one dual coordinate are updated together. To reduce the communication in each iteration, DSCOVR computes variance-reduced stochastic gradients to replace $x_{(i)}^{\top}w$ and $(X^{\top})_{k\cdot}v$. These stochastic updates require w_k, v_i, and $X_{i,k}$ only, along with some additional variables that can be easily computed [50]. DSCOVR is implemented with master and worker nodes. The updates of primal–dual coordinate pairs are computed at worker nodes, and the primal variable w is updated at the master node. Also, the vector $x_{(i)}$ is kept at worker node i, so the updates of α_i and w_k require w_k from the master node only. Depending on the choice of the variance reduction technique, additional simple steps are required. A scheduler is included to maintain the indices of primal coordinates not being updated to avoid the conflict that two workers update the same primal coordinate at the same time. This algorithm requires $O(1)$ data transfer in each iteration, and more iterations are needed to converge. However, the total time can be shorter, especially when the communication is the bottleneck. Accelerated variants of DSCOVR are also introduced in Xiao *et al.* [50].

5 Concluding Remarks

Asynchronous parallel algorithms have shown promising results in solving large-scale data science problems. This short chapter describes two types of parallelization techniques with several examples. The coordinate update approach reformulates the algorithm such that the update of one coordinate is much faster than the update of all coordinates and their updates are independent of each other. The stochastic approach lets each core update the coefficients based on one data sample. The theoretical understanding of the performance of these algorithms, including their convergence, is far from complete. Current results usually require restrictive assumptions, while these algorithms can work under very mild conditions [52]. Also, many practical empirical algorithms lack theoretical justification. We would like to mention that synchronous parallel algorithms could perform better than asynchronous parallel algorithms [53] because their performance depends on the environment.

References

1 Sutter, H. (2005) The free lunch is over: a fundamental turn toward concurrency in software. *Dr. Dobb's J.*, 30 (3), 202–210.

2 Bertsekas, D.P. and Tsitsiklis, J.N. (1989) *Parallel and Distributed Computation: Numerical Methods*, vol. 23, Prentice Hall, Englewood Cliffs, NJ.

3 Ramanan, P., Yildirim, M., Chow, E., and Gebraeel, N. (2019) An asynchronous, decentralized solution framework for the large scale unit commitment problem. *IEEE Trans. Power Syst.*, 34 (5), 3677–3686.

4 Mnih, V., Badia, A.P., Mirza, M. *et al.* (2016) *Asynchronous Methods for Deep Reinforcement Learning*. International Conference on Machine Learning. PMLR, pp. 1928–1937.

5 Zou, H. and Hastie, T. (2005) Regularization and variable selection via the elastic net. *J. R. Stat. Soc.: Ser. B (Stat. Methodol.)*, 67 (2), 301–320.

6 Tibshirani, R. (1996) Regression shrinkage and selection via the lasso. *J. R. Stat. Soc.: Ser. B (Stat. Methodol.)*, 58 (1), 267–288.

7 Cortes, C. and Vapnik, V. (1995) Support-vector networks. *Mach. Learn.*, 20 (3), 273–297.
8 Beck, A. and Teboulle, M. (2009) A fast iterative shrinkage-thresholding algorithm for linear inverse problems. *SIAM J. Imag. Sci.*, 2 (1), 183–202.
9 Peng, Z., Xu, Y., Yan, M., and Yin, W. (2016) ARock: an algorithmic framework for asynchronous parallel coordinate updates. *SIAM J. Sci. Comput.*, 38 (5), A2851–A2879.
10 Trottenberg, U., Oosterlee, C.W., and Schuller, A. (2000) *Multigrid*, Elsevier.
11 Anzt, H., Tomov, S., Dongarra, J., and Heuveline, V. (2013) A block-asynchronous relaxation method for graphics processing units. *J. Parallel Distrib. Comput.*, 73 (12), 1613–1626.
12 Wolfson-Pou, J. and Chow, E. (2019) Modeling the asynchronous Jacobi method without communication delays. *J. Parallel Distrib. Comput.*, 128, 84–98.
13 Assran, M., Aytekin, A., Feyzmahdavian, H. R., Johansson, M., and Rabbat, M. G. (2020). Advances in asynchronous parallel and distributed optimization. *Proceedings of the IEEE*, 108 (11), 2013–2031.
14 Lange, K., Chi, E.C., and Zhou, H. (2014) A brief survey of modern optimization for statisticians. *Int. Stat. Rev.*, 82 (1), 46–70.
15 Wright, S.J.(2015) Coordinate descent algorithms. *Math. Program.*, 151 (1), 3–34.
16 Shi, H.J.M., Tu, S., Xu, Y., and Yin, W. (2016) A primer on coordinate descent algorithms,. *arXiv preprint arXiv:1610.00040.*
17 Peng, Z., Wu, T., Xu, Y. *et al.* (2016) Coordinate friendly structures, algorithms and applications. *Ann. Math. Sci. Appl.*, 1(1), 57–119.
18 Baudet, G.M. (1978) Asynchronous iterative methods for multiprocessors. *J. ACM*, 25 (2), 226–244.
19 Frommer, A. and Szyld, D.B. (2000) On asynchronous iterations. *J. Comput. Appl. Math.*, 123 (1–2), 201–216.
20 Liu, J., Wright, S., Ré, C. *et al.* (2014) *An Asynchronous Parallel Stochastic Coordinate Descent Algorithm.* International Conference on Machine Learning, pp. 469–477.
21 Liu, J. and Wright, S.J. (2015) Asynchronous stochastic coordinate descent: parallelism and convergence properties. *SIAM J. Optim.*, 25 (1), 351–376.
22 Hsieh, C.J., Yu, H.F., and Dhillon, I. (2015) *Passcode: Parallel Asynchronous Stochastic Dual Co-Ordinate Descent.* International Conference on Machine Learning, pp. 2370–2379.
23 Peng, Z., Xu, Y., Yan, M., and Yin, W. (2019) On the convergence of asynchronous parallel iteration with unbounded delays. *J. Oper. Res. Soc. China*, 7 (1), 5–42.
24 Hannah, R. and Yin, W. (2018) On unbounded delays in asynchronous parallel fixed-point algorithms. *J. Sci. Comput.*, 76 (1), 299–326.
25 Combettes, P.L. and Eckstein, J. (2018) Asynchronous block-iterative primal-dual decomposition methods for monotone inclusions. *Math. Program.*, 168 (1–2), 645–672.
26 Boser, B.E., Guyon, I.M., and Vapnik, V.N. (1992) *A Training Algorithm for Optimal Margin Classifiers.* Proceedings of the Fifth Annual Workshop on Computational Learning Theory, pp. 144–152.
27 Yan, M. (2018) A new primal–dual algorithm for minimizing the sum of three functions with a linear operator. *J. Sci. Comput.*, 76 (3), 1698–1717.
28 Condat, L., Kitahara, D., Contreras, A., and Hirabayashi, A. (2019) Proximal splitting algorithms: overrelax them all! *arXiv preprint arXiv:1912.00137.*

29 Nedić, A. (2018) *Distributed Optimization Over Networks*. Multi-Agent Optimization. Springer, pp. 1–84.

30 Wei, E. and Ozdaglar, A. (2013) *On the o(1/k) Convergence of Asynchronous Distributed Alternating Direction Method of Multipliers*. 2013 IEEE Global Conference on Signal and Information Processing. IEEE, pp. 551–554.

31 Eisen, M., Mokhtari, A., and Ribeiro, A. (2017) Decentralized quasi-Newton methods. *IEEE Trans. Signal Process.*, 65 (10), 2613–2628.

32 Wu, T., Yuan, K., Ling, Q. *et al.* (2017) Decentralized consensus optimization with asynchrony and delays. *IEEE Trans. Signal Inf. Process. Netw.*, 4 (2), 293–307.

33 Bottou, L., Curtis, F.E., and Nocedal, J. (2018) Optimization methods for large-scale machine learning. *Siam Rev.*, 60 (2), 223–311.

34 Agarwal, A. and Duchi, J.C. (2011) *Distributed Delayed Stochastic Optimization*. Advances in Neural Information Processing Systems, pp. 873–881.

35 Zinkevich, M., Langford, J., and Smola, A.J. (2009) *Slow Learners are Fast*. Advances in Neural Information Processing Systems, pp. 2331–2339.

36 Zheng, S., Meng, Q., Wang, T. *et al.* (2017) *Asynchronous Stochastic Gradient Descent with Delay Compensation*. International Conference on Machine Learning, pp. 4120–4129.

37 Reddi, S.J., Hefny, A., Sra, S. *et al.* (2015) *On Variance Reduction in Stochastic Gradient Descent and Its Asynchronous Variants*. Advances in Neural Information Processing Systems, pp. 2647–2655.

38 Feyzmahdavian, H.R., Aytekin, A., and Johansson, M. (2016) An asynchronous mini-batch algorithm for regularized stochastic optimization. *IEEE Trans. Autom. Control*, 61 (12), 3740–3754.

39 Alistarh, D., De Sa, C., and Konstantinov, N. (2018) *The Convergence of Stochastic Gradient Descent in Asynchronous Shared Memory*. Proceedings of the 2018 ACM Symposium on Principles of Distributed Computing, pp. 169–178.

40 De Sa, C.M., Zhang, C., Olukotun, K., and Ré, C. (2015) *Taming the Wild: A Unified Analysis of Hogwild-Style Algorithms*. Advances in Neural Information Processing Systems, pp. 2674–2682.

41 Sra, S., Yu, A.W., Li, M., and Smola, A. (2016) *Adadelay: Delay Adaptive Distributed Stochastic Optimization*. Artificial Intelligence and Statistics, pp. 957–965.

42 Soori, S., Can, B., Gurbuzbalaban, M., and Dehnavi, M.M. (2020, May). ASYNC: A Cloud Engine with Asynchrony and History for Distributed Machine Learning. In 2020 IEEE International Parallel and Distributed Processing Symposium (IPDPS) (pp. 429–439). IEEE.

43 Recht, B., Re, C., Wright, S., and Niu, F. (2011) *Hogwild: A Lock-Free Approach to Parallelizing Stochastic Gradient Descent*. Advances in Neural Information Processing Systems, pp. 693–701.

44 Grishchenko, D., Iutzeler, F., Malick, J., and Amini, M.R. (2018) Asynchronous distributed learning with sparse communications and identification. *arXiv preprint arXiv:1812.03871*.

45 Leblond, R., Pedregosa, F., and Lacoste-Julien, S. (2017) *ASAGA: Asynchronous Parallel SAGA*. Artificial Intelligence and Statistics, pp. 46–54.

46 Konečný, J., McMahan, H.B., Yu, F.X. *et al.* (2016) Federated learning: strategies for improving communication efficiency. *arXiv preprint arXiv:1610.05492.*

47 McMahan, B., Moore, E., Ramage, D. *et al.* (2017) *Communication-Efficient Learning of Deep Networks from Decentralized Data*. Artificial Intelligence and Statistics, pp. 1273–1282.

48 Wu, W., He, L., Lin, W. *et al.* (2021) SAFA: a semi-asynchronous protocol for fast federated learning with low overhead. *IEEE Trans. Comp.*, 70 (5), 655–668.

49 Xie, C., Koyejo, S., and Gupta, I. (2019) Asynchronous federated optimization. *arXiv preprint arXiv:1903.03934.*

50 Xiao, L., Yu, A.W., Lin, Q., and Chen, W. (2019) DSCOVR: randomized primal-dual block coordinate algorithms for asynchronous distributed optimization. *J. Mach. Learn. Res.*, 20, 1–58.

51 Chambolle, A. and Pock, T. (2011) A first-order primal-dual algorithm for convex problems with applications to imaging. *J. Math. Imaging Vision*, 40 (1), 120–145.

52 Sun, T., Hannah, R., and Yin, W. (2017) *Asynchronous Coordinate Descent Under More Realistic Assumptions*. Advances in Neural Information Processing Systems, pp. 6182–6190.

53 Chen, J., Pan, X., Monga, R. *et al.* (2016) Revisiting distributed synchronous SGD. *arXiv preprint arXiv:1604.00981.*

Index

Computational Statistics in Data Science.
Edited by Walter W. Piegorsch, Richard A. Levine, Hao Helen Zhang and Thomas C. M. Lee.
 ISBN 978-1-11956107-1

i

j

k

l

Abbreviations and Acronyms

ABC	approximate Bayesian computation
ACOSSO	adaptive COSSO
ACS	American Community Survey
ADMM	alternating direction method of multipliers
AIC	Akaike's Information Criterion
AIS	adaptive importance sampling
AM	adaptive Metropolis
APT	adaptive parallel tempering
AR	autoregressive
ART	acceptance-rejection trees
ASM	adaptive scaling Metropolis
ATLAS	automatically tuned linear algebra software
AUC	area under curve
AVX	advance vector extensions
BART	Bayesian additive regression trees
BCD	block coordinate descent
BCM	block coordinate minimization
BEAST	Bayesian evolutionary analysis by sampling trees
BIC	Bayesian Information Criterion
BIRCH	balanced iterative reducing and clustering using hierarchies
BLAS	basic linear algebra subprograms
BLHS	block bootstrap Latin hypercube
BLUE	best linear unbiased estimator
BMA	Bayesian model averaging
BOD	biomedical oxygen demand
BPS	Bayesian predictive synthesis
BUGS	Bayesian inference using Gibbs sampling
CAP	composite absolute penalty
CD	coordinate descent
CDA	coordinate descent algorithm
CEASE	Communication-Efficient Accurate Statistical Estimators

Computational Statistics in Data Science.
Edited by Walter W. Piegorsch, Richard A. Levine, Hao Helen Zhang and Thomas C. M. Lee.
 ISBN 978-1-11956107-1

CEDAS	clustering of evolving data-streams into arbitrary shapes
CERFIT	causal effect random forest of interaction trees
CEVOT	Concept-adapting Evolutionary Algorithm for Decision Tree
CG	conjugate gradient
CLR	centered log-ratio
CLT	central limit theorem
CNN	convolutional neural network
CP	CANDECOMP/PARAFAC
CPS	covering point set
CPU	central processing units
CR	contribution ratio
CRAN	Comprehensive R Archive Network
CSL	communication-efficient surrogate likelihood
cSMC	conditional SMC
CURE	clustering using representatives
CV	cross-validation
DBSCAN	Density-Based Spatial Clustering with Noise
DC	difference-of-convex
DCA	DC algorithm
DHS	Department of Homeland Security
DIC	deviance information criterion
DID	difference in differences
DNN	deep neural network
DROS	Douglas–Rachford operator splitting
DSCLU	data stream clustring
DSCOVR	doubly stochastic coordinate optimization with variance reduction
EBIC	extended BIC
EDF	effective degrees of freedom
EE	estimating equation
EEG	electroencephalography
EJML	efficient Java matrix library
elpd	expected log predictive density
EM	expectation–maximization
EMA	exponential moving average
EP	expectation propagation
ERM	empirical risk minimization
ERT	extremely randomized trees
ESS	Emacs Speaks Statistics
FDA	functional data analysis
FDR	false discovery rate
FEV	forced expiratory volume
FFBSi	forward-filtering backward-simulation
fMRI	functional magnetic resonance imaging
FOM	functional outlier map
GAM	generalized additive model

GaSP	Gaussian spatial processes
GC	generalized CV
GDA	gradient descent algorithm
GED	graduate equivalency diploma
GI	generalized information
GIC	generalized information criterion
GLM	generalized linear model
GMM	Gaussian mixture model
GP	Gaussian process
GPML	Gaussian processes for Machine Learning
GPU	graphical processing unit
GSS	General Society Survey
HMC	Hamiltonian Monte Carlo
HMM	hidden Markov model
HMM-VB	hidden Markov model on variable blocks
HOP	hypothetical outcome plot
HPX	high-performance parallex
iBMA	intrinsic Bayesian model averaging
IC	irrepresentable condition
ICE	individual conditional expectation
IDC	International Data Cooperation
IDE	integrated development environment
IID	independent and identically distributed
IJ	infinitesimal jackknife
INIS	iterative NIS
IoT	internet of things
IQR	interquartile range
IS	importance sampling
ISIS	iterate SIS
ISTA	iterative shrinkage-thresholding algorithm
IT	Interaction tree
ITE	individualized treatment effect
JAGS	just another Gibbs sampler
JAMA	Java matrix
JIT	just-in-time
JSC	Java Statistical Classes
KKT	Karush–Kuhn–Tucker
KL	Kullback–Leibler
KRR	kernel ridge regression
LAGP	Local approximate Gaussian process
LAND	linear and nonlinear discover method
LAPACK	linear algebra package
LARS	least angle regression
LASSO	least absolute shrinkage and selection operator
LDA	linear discriminant analysis

LEQR	linear estimator for quantile regression
LIFT	Least Impact First Targeted removal
LLA	local linear approximation
LLN	law of large numbers
LOO	leave-one-out
LOO-CV	leave-one-out CV
LP	linear programming
LQA	local quadratic approximation
LSane	Linked Stream Annotation Engine
LSE	least-squared estimator
LSF	load sharing facility
LSTM	long short-term memory
LUM	large-margin unified machine
MA	moving-average
MAP	maximum-a-posteriori
MARS	multivariate adaptive regression splines
MC	minimax concave
MCMC	Markov chain Monte Carlo
MCP	minimax concave penalty
MCP	minimax concave plus
MDP	Markov decision process
MEM	modal EM
MERF	mixed-effects random forest
M–H	Metropolis–Hastings
MIO	mixed-integer optimization
MIS	multiple importance sampling
MKL	Math Kernel Library
ML	machine learning
MLE	maximum-likelihood estimation
MLP	multilayer perceptron
MM	majorization–minimization
MR	multivariate random forest
MRI	magnetic resonance imaging
MS	magnitude–shape
MSBD	modified simplicial band depth
MSE	mean squared error
MSPE	mean-square prediction error
MVN	multivariate normal
NICTA	National ICT Australia
NIS	nonparametric independence screening
NLP	natural language processing
NMF	nonnegative matrix factorization
NNGP	nearest neighbor GPs
NUTS	no-U-turn sampler
ODAC	open distributed application construction

OLS	ordinary least square
OOB	out-of-bag
OpenMP	open multiprocessing
OPTICS	ordering points to identify clustering structure
ORF	oblique random forest
OT	optimal transport
OWL	Web Ontology Language
PD	partial dependence
PDMP	partly deterministic Markov process
PE	prediction error
PF	particle filter
PF	power factorization
PG	particle Gibbs
PL	Polyak-Łojasiewicz
PLEASE	Pediatric Longitudinal Study of Elemental Diet and Stool Microbiome Composition
PLR	penalized logistic regression
PMC	population Monte Carlo
PMMH	particle marginal Metropolis–Hastings
PMSE	prediction mean-squared error
PO	proportional-odds
PPL	probabilistic programming language
PSD	positive semidefinite
PSIS	Pareto smoothed importance sampling
PT	parallel tempering
QDA	quadratic discriminant analysis
QMC	Quasi-Monte Carlo
QP	quadratic programming
QUINT	QUalitative INteraction Trees
RAM	random access memory
RAM	robust adaptive Metropolis
RAMP	Regularization Algorithm under Marginality Principle
RCDM	random coordinate descent method
RDF	Resource Description Framework
ReLU	rectified linear unit
RFIT	random forest of interaction trees
RKHS	reproducing kernel Hilbert space
RL	reinforcement learning
RMRF	repeated measures random forest
RMSE	root mean-squared error
RNN	recurrent neural network
RSVM	robust SVM
RUF	random uniform forests
RWM	random-walk Metropolis
SAM	split-and-merge

SAREF	smart appliances reference
SCAD	smoothly clipped absolute deviation
SDCA	Stochastic Dual Coordinate Ascent
SE	standard error
SGD	stochastic gradient descent
SGDA	stochastic-gradient descent algorithm
SIDE	subgroup identification based on differential effect search
SimC	similarity-based data stream classification
SIMD	single instruction, multiple data
SIR	sequential importance resampling
SIS	sequential importance sampling
SIS	sure independence screening
SMC	sequential Monte Carlo
SMP	symmetric multicore/shared memory parallelization
SNOW	simple network of workstations
SOA	Stream Annotation Ontology
SpAM	sparse additive model
SPLOMs	scatterplot matrices
SQL	Structured Query Language
SRIG	sparse regression incorporating graphical structure among predictors
SS-ANOVA	smoothing spline ANOVA
SSE	streaming SIMD extensions
SSM	state-space model
SSN	Semantic Sensor Network
SSS	smooth sigmoid surrogate
SUMMA	sequential unconstrained minimization method algorithm
SV	support vector
SVD	singular value decomposition
SVM	support vector machine
SWEM	sliding window with expectation maximization
TCGA	The Cancer Genome Atlas
TE	training error
UFFT	Ultra-Fast Forest Tree system
UQ	uncertainty quantification
VAE	variational autoencoder
VAR	vector autoregressive
VAT	Virtual Adversarial Training
VFDT	Very Fast Decision Tree learner
VI	variational inference
VPU	vector processing unit